Computer-Aided Analysis and Design of Analog Electronic Circuits

Current and future electronic designs rely on circuit design and modeling software. This book guides engineering students and industry practitioners through simulation-based analog circuit design, from fundamental op-amp and diode applications to multi-transistor-multi-stage amplifier circuits. A basic design that integrates simulations may start with manual calculations that utilize simplified formulas and assumed parameters. It is followed by a simulation that allows the designer to contrast the approximations with an exact solution. It then ends with circuit tweaking and precise parameter tuning. In more advanced designs, simulations play a more central role in the design process itself.

Key Features:

- Makes extensive use of the Advanced Design System (ADS) software by Keysight Technologies, Inc., with a focus on simulation, yield, and optimization features of the tool
- The text is supported by over 750 illustrations and tables
- Includes an ADS process design kit and ADS workspaces that contain all the schematics used in this book

Computer-Aided Analysis and Design of Analog Electronic Circuits

Wilfredo Rivas-Torres and Zvi S Roth

CRC Press
Taylor & Francis Group
Boca Raton London New York

CRC Press is an imprint of the
Taylor & Francis Group, an **informa** business

Designed cover image: Wilfredo Rivas-Torres and Zvi S Roth

First edition published 2026
by CRC Press
2385 NW Executive Center Drive, Suite 320, Boca Raton, FL 33431

and by CRC Press
4 Park Square, Milton Park, Abingdon, Oxon, OX14 4RN

CRC Press is an imprint of Taylor & Francis Group, LLC

ISBN: 9781032999326 (hbk)
ISBN: 9781032999333 (pbk)
ISBN: 9781003606772 (ebk)

DOI: 10.1201/9781003606772

Typeset in Times
by codeMantra

Access the Instructor and Student Resources: www.routledge.com/9781032999326

Contents

Preface

The Advanced Design System (ADS) by Keysight Technologies is an electronic design automation (EDA) platform intended primarily for RF (Radio Freuency) and communication circuits and high-speed digital physical layer components. ADS has comprehensive design and simulation capabilities. It offers a wide range of simulation tools, including circuit, electromagnetic, electro-thermal, and statistical simulations for PCB, MMIC, and IC design. ADS provides superior simulation technology, ensuring accurate results prior to hardware prototyping.

This book focuses on the applications of the ADS platform for electronic circuits. Although ADS is a complex software, one of the book's aims is to lower the learning curve required of the users. ADS is an industry-grade design platform that provides multiple modules as enterprise commercial software. Generous educational donations and discounts have made ADS very affordable to all higher-education institutions and accessible for free to all their students. As of 2024, no less than 680 universities worldwide have subscribed to ADS.

This book is not intended as a primary textbook on electronic circuits. Readers are referred to monumental texts such as [Gray1], [Razavi1], and [Sedra1]. The latter, for instance, has been adopted as a required text at no less than three undergraduate courses, namely Electronics 1, Electronics 2, and Digital Electronics, at Florida Atlantic University (FAU). This book certainly aims to be a secondary textbook on analog electronic circuits, teaching students how to use simulation tools, specifically ADS, as effective analysis and design tools.

Many books have been written about SPICE simulation of linear and electronic circuits; the most comprehensive among them is probably the book by [Roberts1] which served as a companion book to [Sedra1]. Nor is the book the first of its kind about ADS. The book [Behagi1] focuses on RF and microwave circuits design using ADS.

This book is not intended to be a mere user manual of electronic circuit features of ADS, even though it is very graphical, and it takes the reader on a step-by-step textbook-style approach to electronic circuit analysis and design using ADS. The authors wish to make a case for integrating theory with simulations for a better understanding of the operation of any electronic circuit.

The basic concepts of electronic circuit simulation are universal, shared by almost all SPICE-based simulators. The differences among individual simulators may lie in the user-friendliness of simulation data manipulation, the availability of advanced analyses and design tools, and the richness of the components' library.

The basic simulation controllers, relevant for electronic circuits, that ADS provides are **transient simulation** that allows an oscilloscope-like monitoring of time signals; **AC simulation** that allows the user to observe frequency response plots of linear or linearized circuits; **DC simulation** that enables basic voltage, current and power solution of DC circuits, and data collection from multiple runs, allowing users to plot components' input-output curves and devices' parametric studies; **parametric**

sweep that allows multiple parameter sweeps to run simultaneously; and **harmonic balance simulation** that performs spectral analysis of circuits.

More advanced tools that ADS offers include **yield analysis** that allows the user to introduce tolerances into each component as well as internal parameters; **sensitivity analysis** that ranks circuit components in terms of the sensitivity of the circuit's desired output to variations in each component; and **optimizer** that allows the circuit designer to specify design goals, design parameters, and constraints for autonomous (yet closely guided and supervised) circuit design solutions.

All the above, and more, is covered in this book in much detail. Every circuit design example is first solved analytically, often using approximations of variables and simplified formulas. The analytic design is then contrasted against the exact results generated by the circuit simulator. The reasons for the performance differences are discussed. Finally, design tweaking techniques are introduced to allow the actual designs to meet the design specifications. This is why this book is not an ordinary user guide for the design software. A brief theory is attached to each of the book chapters to facilitate the analytic solution part of every design example.

In order to better explain the history of this book, let us first introduce the authors.

Dr. Wilfredo Rivas-Torres is an EDA engineer at Keysight Technologies. Over the years, he has been involved in customer consulting of ADS and has provided ADS training and support to industry designers all over the world. He currently works as a Senior R&D Solution Architect & Development Scientist in the Keysight EDA division, developing ESL (Electronic System Level) software products. His current interests are RF; phased array; communications systems design; scenario modeling, with an emphasis on digital twinning; and model based design. He pioneered the concept of extracted behavioral modeling (EBM), which pertains to creating electronic component models used in system designs from measurements and simulations data.

Dr. Zvi Roth has been teaching many undergraduate and graduate Electrical Engineering courses, having been a faculty member inthe Electrical Engineering Department at Florida Atlantic University (FAU) since 1982. He won several teaching awards and co-authored two robotics books (1992 and 1996) and a control book (2019). He served as the Director of the FAU Robotics Center (1985–1993), Chair of the FAU EE Department (1993–1997), and Co-Director of the Florida-Israel Institute (2005–2008).

His interest in teaching undergraduate electronics courses started when he served as department chair. He was trying to get to the bottom of what made electronics a stumbling block to so many students working on their senior design projects.

In the late 1990s, Dr. Roth began to integrate SPICE-based circuit simulations into all undergraduate electronics circuits' lectures. The in-class demos and the simulation-based homework projects were well received by the FAU students.

Inspired by Prof. Razavi's 2001 first edition textbook *Design of Analog CMOS Integrated Circuits* [Razavi1], a new graduate-level course, EEE 5321 CMOS Amplifiers, started to be offered in 2003. That is where Wilfredo Rivas-Torres, a very experienced engineer at Agilent Technologies (which later became Keysight Technologies), and Zvi Roth met. At that time, Wilfredo was completing his MSEE degree at FAU [Rivas-Torres2] and later on went on to complete a PhD degree in EE at FAU [Rivas-Torres1]. Wilfredo, who attended the CMOS Amplifiers class, introduced Dr. Roth and many of the classmates to ADS. The CMOS Amplifiers course was originally designed to utilize OrCAD Lite PSPICE simulation activities, supported by basic optimizers and realized using the Microsoft EXCEL Solver tool. Several of the other class students, working in local industry, routinely used ADS as an industry-grade platform for RF and electronics circuit design, as well as circuit board layout design. Wilfredo opted to submit every one of his eight homework projects as a detailed ADS tutorial. These very instructive tutorials were posted on the course's web page and were very well received. Wilfredo and Zvi became friends.

In 2009, Wilfredo and Zvi decided to team-teach the CMOS Amplifiers course, this time around with ADS as the main design platform. Every Tuesday, Dr. Roth delivered a theory session based primarily on Razavi's book, and every Thursday, Wilfredo delivered ADS demonstrations of the theory covered that week. The course was very fast-paced, and the ADS demos reached topics such as Feedback, Noise in Electronic Amplifiers, Switched Capacitor Amplifiers, and Design of OTA Applications. In subsequent years, as Wilfredo was pursuing his doctoral studies at FAU, he was often invited to fully teach some of the FAU undergraduate electronics courses. He created ADS Process Design Kits for commercial electronic components that students used in the FAU electronics labs.

During the Summer semester of 2021, Dr. Roth was scheduled to teach Electronics 1, after several years of a layover. The course, for many years, used OrCAD Lite PSPICE. He decided to use ADS instead, but to facilitate the instruction, for himself and for successor instructors, he decided to create detailed PowerPoint ADS tutorials. He included not just collections of schematics and sample results but also design considerations and contrasting of manual design (which usually includes necessary approximations) with the exact simulation results. Such comparisons often lead to design tweaking. He wanted the students to fully appreciate that simulation is not just a tool for verifying the manual design. It should be an active and hands-on analysis and design tool. No less than 21 tutorials, averaging each 15–20 slides, were created during the semester. In the Fall 2022 semester, Dr. Roth happened to teach Electronics 2, and that is how nine more detailed ADS tutorials were added, and the total number of PowerPoint slides exceeded 500. That was how the idea for the book titled *ADS for Analog Electronic Circuits* germinated.

Dr. Rivas-Torres and Dr. Roth teamed up again. Dr. Rivas-Torres first took on himself the revision of all 30 tutorials, adding, to all the examples, industry perspective stemming from his many years of engineering experience. He also added many more needed ADS features and then set out to create new tutorials aimed at the graduate-level courses such as CMOS Amplifiers, drawing from his doctoral dissertation ideas on the use of ADS for submicron analog electronics design. He demonstrated some of these design ideas on simple CMOS amplifiers. Dr. Rivas-Torres and Roth then drew the line, deciding against going deeper into Razavi's book, as was once done in 2009. This book, whose main purpose was to teach students how to utilize the ADS platform intelligently, was already large enough.

This book consists of seven chapters, listed below in terms of the electronic circuits theory that is covered. Of course, the ADS concepts advance gradually. Each example, presented in the book, is intended to further advance the reader's simulation proficiency.

Chapter 1: Basic Op-Amp Circuits
Chapter 2: Diode Circuits
Chapter 3: Basic BJT Circuits
Chapter 4: Basic MOSFET Circuits
Chapter 5: Single- and Two-Stage BJT Amplifiers
Chapter 6: Single- and Multistage Differential Amplifiers
Chapter 7: Introductory Concepts in ADS Simulation-Based Submicron Design of MOSFET Amplifiers

Each chapter features brief theory sections, followed by ADS concepts and ADS-based analysis or design examples, and end-of-chapter problems. The ADS part is very graphic, illustrating in successive pictures the progressing ADS syntax and usage.

For a first course in electronics, such as the junior-level FAU EEE 3300 Electronics 1 course, Chapters 1–4 cover all the needed basic analog electronics concepts, coupled with ADS material such as transient, AC, DC and harmonic balance simulations, sweep studies, measurement equations, and data display equations.

Chapters 5–7 can support courses such as the senior-level FAU EEE 4361 Electronics 2. A second course in analog electronics typically involves multi-transistor multistage amplifiers. Such courses cover concepts such as differential amplification, amplifier's bandwidth and Miller effect, and feedback amplifiers. Electronic design activities, in such a second course in analog electronics, often require more advanced ADS topics, including Yield Analysis, Sensitivity Analysis, Optimization, Tuning, and Data Manipulation. This is included in Chapters 5 and 6. Chapter 7 supports the addition of basic CMOS Amplifiers subjects to the senior-level electronics course. Advanced ADS subjects that are needed for Chapter 7 include Detailed Device Operating Point and Small-Signal Parameters and Swept Optimization.

Much online material was made available to readers via the (www.routledge.com/9781032999326):

1. Process Design Kits needed for all the book activities (available to all users).
2. Detailed Solution Manual. Almost all end-of-chapter problems were used as homework assignments in recent years' versions of the FAU electronics courses. The solution manual is based on the best student's solutions: detailed analytic solutions and detailed ADS design and analysis (available fully only to instructors who adopt the book as a textbook. Selected solutions, though, are made available to all users).
3. 34 PowerPoint tutorials with a total of over 700 detailed slides (available only to instructors who adopt the book as a textbook).
4. There are seven ADS workspaces, one for each of the book chapters. The workspaces contain the schematics of all the book examples, and they can be downloaded (available to all users).

REFERENCES

[Behagi1] Ali Behagi, *RF and Microwave Circuit Design: with 100 Keysight (ADS) Workspaces*, 2nd Edition, Techno Search, 2017.

[Gray1] Paul R. Gray, Paul J. Hurst, Stephen H. Lewis and Robert G. Meyer, *Analysis and Design of Analog Integrated Circuits*, 4th Edition, John Wiley & Sons, 2001.

[Razavi1] Behzad Razavi, *Design of Analog CMOS Integrated Circuits*, 2nd Edition, McGraw-Hill, 2017.

[Rivas-Torres1] Wilfredo Rivas-Torres, "An Empirical Methodology for Foundry Specific Submicron CMOS Analog Circuit Design", PhD Dissertation, Florida Atlantic University, December 2013.

[Rivas-Torres2] Wilfredo Rivas-Torres, "Submicron CAD Design and Analysis of MOS Current Mirrors", MS Thesis, Florida Atlantic University, May 2004.

[Roberts1] Gordon W. Roberts and Adel S. Sedra, *SPICE for Microelectronic Circuits*, 3rd Edition, Oxford University Press, 1995.

[Sedra1] Adel S. Sedra and Kenneth C. Smith, *Microelectronic Circuits*, 7th Edition, Oxford University Press, 2015.

Ackowledgments

The authors wish to thank Keysight Technologies for their encouragement and support.

We further wish to acknowledge the outstanding Electronics homework solutions, combining analytical approach with ADS usage, that were selected for the book's solution manual. We especially thank the contributions of the FAU students, Chase Gallo and Jancarlos Gomez. Special thanks also go to the following FAU students: Morgan Brady, Jose Chavez, Benjamin Coleman, Alex Daniello, Paulina Devito, Endian Escott, Nikita Ostro, Andrei Pielea, John Serrano, and Kirk Shields.

Finally, we thank our families for their patience and support. Special thanks go to our wives Haydee Torres and Eva Roth.

Dr. Wilfredo Rivas-Torres (Keysight Technologies) and Dr. Zvi S Roth (Florida Atlantic University), May 2025.

Authors

Wilfredo Rivas-Torres earned his PhD in Electrical Engineering from Florida Atlantic University, Boca Raton, Florida, in 2013. He has been with Agilent and now with Keysight Technologies for over 20 years. Presently, he is a senior R&D solution architect & development scientist in the Design Engineering Software (DES) Division at Keysight Technologies, Inc. His research interests are advanced behavioral modeling of circuits and phased array systems modeling and design. His research interests also include scenario modeling of aerospace defense and communication systems with an emphasis on design exploration and optimization.

Zvi S Roth earned his BSc and MSc in Electrical Engineering from the Technion, Israel Institute of Technology, in 1975 and 1979 and his PhD in Systems Engineering from Case Western Reserve University in 1983. He is presently a professor of Electrical Engineering at the Department of Electrical Engineering and Computer Science at Florida Atlantic University in Boca Raton, Florida. His teaching and research interests are in control systems, robotics and automation, electronic circuit design, and bioengineering. He co-authored two books in robotics (1991 and 1996) and one in control systems (2019), and he has over 70 refereed publications. He served as the director of the FAU Robotics Center (1985–1993), chair of the FAU EE Department (1993–1997), and co-director of the Florida-Israel Institute (2005–2008). He was the winner of multiple FAU teaching awards.

1 Basic Op-Amp Circuits

1.1 INVERTING OP-AMP AMPLIFIERS

An operational amplifier (op-amp in short) is a multi-transistor circuit designed to have the following properties: (i) It is a **differential amplifier**. That is, its output $v_o(t)$ is proportional to the difference between the non-inverting "+ input" and the inverting "–input" (shown in Figure 1.1), (ii) The op-amp's **differential voltage gain is large** (in the order of magnitude of 10^6), (iii) The op-amp's **differential input resistance is large** (in the order of magnitude of 1 MΩ) and (iv) The op-amp's **output resistance is small** (in the order of magnitude of tens of Ω).

The circuit of the op-amp typically requires dual and symmetric DC voltage supplies. These supply levels define the possible maximum and minimum levels of $v_o(t)$.

Whenever a resistive (or capacitive) path is created between the op-amp's output node and its inverting input node, we say that the op-amp operates in **negative feedback**. The combined effects of negative feedback, high op-amp differential gain, and large op-amp input resistance create the "**virtual ground effect**": *The voltage signals at the op-amp's inverting and non-inverting input nodes become very close to each other.*

In the inverting op-amp amplifier (shown in Figure 1.1), the "+ input" is connected to ground. Therefore, by the virtual ground effect, the voltage at the "– input" is forced to become close to zero (hence the name "virtual ground"). The main result of an inverting amplifier is equation (1.1) below:

$$v_o(t) = -\frac{R_f}{R_s} v_s(t) \tag{1.1}$$

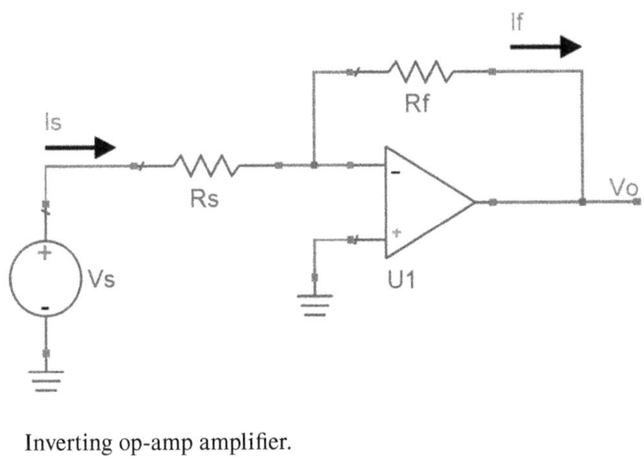

FIGURE 1.1 Inverting op-amp amplifier.

DOI: 10.1201/9781003606772-1

In an inverting op-amp amplifier, the input signal $v_s(t)$ is amplified (or attenuated) by a ratio of two resistors R_f and R_s. The output signal $v_o(t)$ retains the shape of $v_s(t)$, but it is inverted (due to the negative sign in equation (1.1)).

Formula (1.1) is explained as follows: Because of the virtual ground at the "– input", the input current signal is $v_s(t)/R_s$. Almost the entire input current flows into R_f, because R_{in} of the op-amp is much larger than R_f. The current flowing through R_f creates a voltage drop of $(v_s(t)/R_s) \cdot R_f$. Finally, $v_o(t)$ equals the difference between the voltage of the virtual ground node and the voltage drop on R_f. Formula (1.1) was derived assuming an "**ideal op-amp**" (that is, **infinite differential voltage gain, infinite input resistance, and zero output resistance**). For a more detailed derivation of equation (1.1) and about the crucial role of the negative feedback, refer to [Sedra1] and [Gray1]. The theory of op-amp inverting amplifiers is revisited and expanded in Chapter 6.

1.2 UNARCHIVING THE FAU EE ADS DESIGN KIT

Let us explore the high-level requirements for any circuit simulator. The first and foremost requirement is an environment to capture the design schematic. This implies that the environment has parts that can be connected to each other. Part types include sources, devices (resistors, capacitors, transistors, etc.), and measurement devices such as current probes. The parts will have models that are internally a set of mathematical relationships that determine the input and output currents and voltages of the devices and parts that they represent.

Practical circuit simulation tools such as advanced design system (ADS) or OrCAD PSPICE have facilities to include vendor parts. Access to vendor parts is typically found in vendor libraries, which are containers inside the simulation software that contain the parts and corresponding models. Advanced circuit simulators implement design kits (more generally known as process design kits or PDKs). Design kits include more information than a regular library. Things like process limits, design rules, connectivity rules, layout artwork, etc., can be included in the design kit. Many of these advanced design kit features are beyond the scope of this chapter and are mentioned here for the reader's awareness.

This chapter is accompanied by the FAU EE Design Kit. The design kit was prepared for FAU students by Dr. Wilfredo Rivas-Torres of Keysight Technologies back in 2011. This design kit is used in many of our examples and end-of-chapter problems.

The FAU EE Design Kit contains the 741 op-amp, several more op-amps, several BJT transistors like the 2N3904 (NPN) and 2N3906 (PNP), the 1N4148 diode, a variety of Zener diodes (such as the 1N750), the 555 timer and a few more components. These components are needed for some of the circuits that we shall discuss.

How do we get the design kit, and where do we place it?

The design kit comes as a zipped file named **FAU_EE_Kit_v1.0.zip**. It is available for downloading from the book's web page and from the web pages of the authors. The file must be placed in a folder of your choice (say, C:\Users\name\ads_libraries), and it has to be unzipped ("unarchived" is another word for the process).

An ADS user should also designate another folder to which the ADS designs are placed. Let us call it the working directory and name it **ads_work_files**. It is highly recommended that "ads_work_files" and "ads_libraries" be inside the same folder.

Let us demonstrate the entire step-by-step PDK unarchiving process.

Step 1: Creating the two folders.

For this demo: **ads_work_files** is the folder for the ADS work files. The folder for the design kit is **ads_libraries**. Both folders are located inside the **C:\Users\name** folder. See Figure 1.2. Later, one may want to use one's own names and locations. Choosing the C: drive and the **C:\Users\name** folder is just for illustration. One may use the Z: drive (or any other drive where all folders and files are kept) and place the two folders (above) inside another folder there.

So, we save the zipped design kit file into **ads_libraries**. The unzipping of the design kit zipped file is a one-time action.

Step 2: Going to the ADS main page in order to unarchive the zip file. Start ADS. See Figure 1.3.

On the Start menu, we find Advanced Design System 2023 and click it. One may have another ADS year version – all are okay.

On the Get Started Welcome window in Figure 1.3, we select the "Don't show …" to have this window dismissed permanently for future sessions. We then close the Getting Started window.

We are now at the Advanced Design System 2023 Main Window. See Figure 1.4.

Step 3: Finding the unzipping (unarchive) tool and the zipped file.

In the ADS main page, we expand the File menu.

We click on the Unarchive … option in the Commands section (of the expanded File menu).

ADS is now looking for the zipped file.

We go to the **ads_libraries** folder and open it. See Figure 1.5. Inside the folder, we should have the **FAU_EE_v1.0.zip** file. We click on the zipped file. Then, we click on the Open button.

Step 4: Unarchive Wizard starts.

This step may not always be needed. It depends on whether one is inside an open workspace or not.

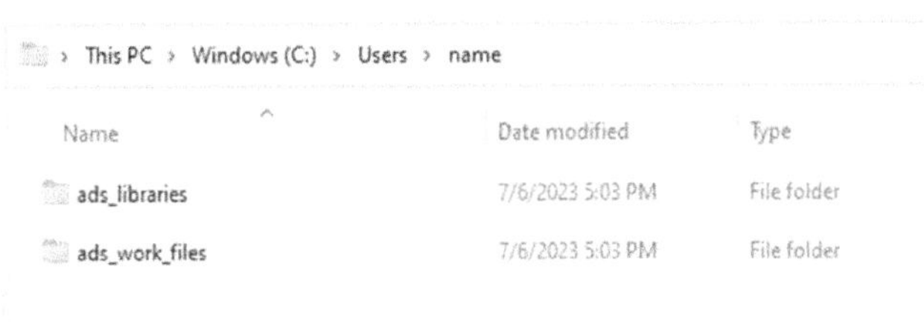

FIGURE 1.2 Folders placement.

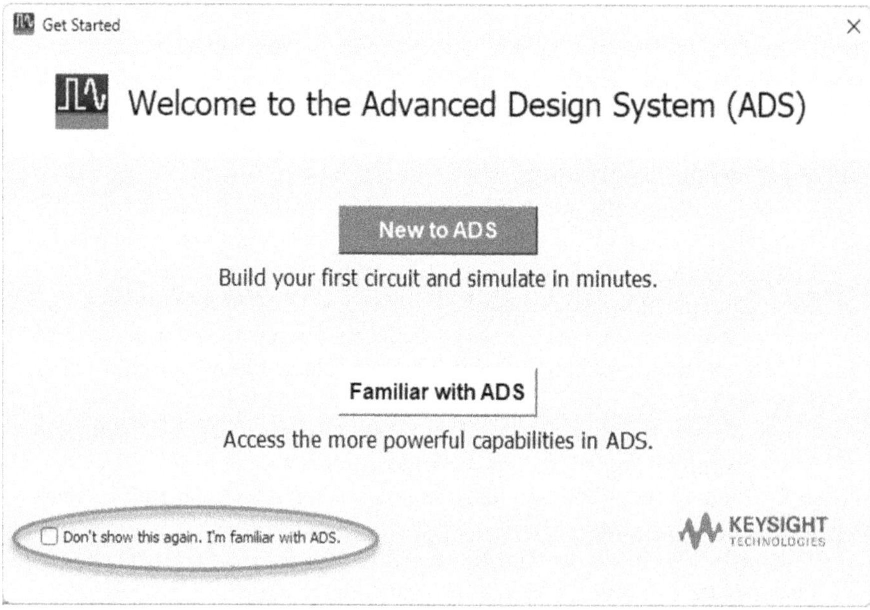

FIGURE 1.3 ADS welcome page.

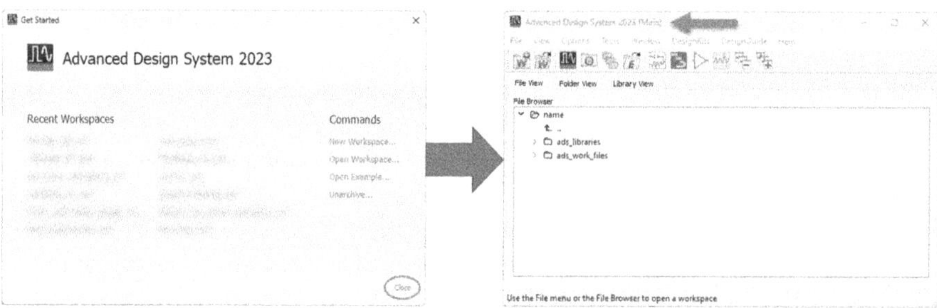

FIGURE 1.4 From ADS getting started to the ADS main window.

As the unarchive Wizard starts, you need to select "Continue with a normal unarchive". See Figure 1.6.

We click the "Next >" button.

Step 5: Setting a destination path.

Need to enter the destination path. In our demo, it is C:\Users\name\ads_libraries. We click the "Next >" button. See Figure 1.7.

Step 6: Selecting Items to Unarchive.

Need to make sure that under Design Kits the FAU_EE_Kit_v1.0 design kit is checked. See Figure 1.8. We click the "Next >" button.

Step 7: One may skip the rest of the Unarchive Wizard steps.

The next several steps in the Unarchive Wizard are not important for our book.

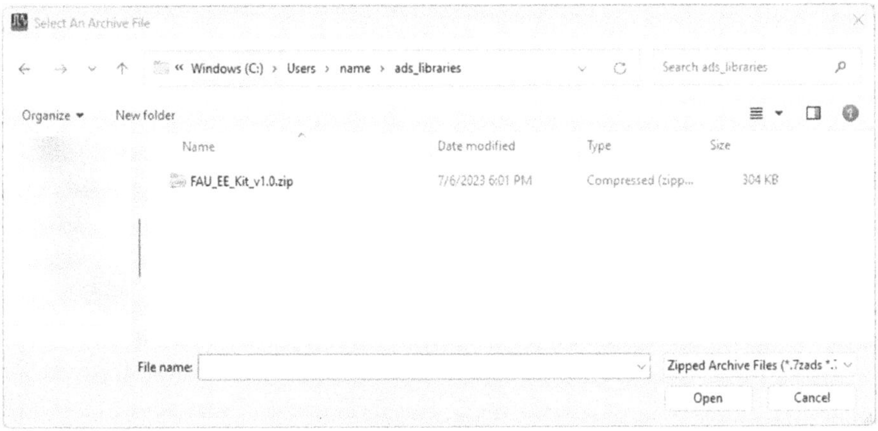

FIGURE 1.5 Finding the zipped design kit file.

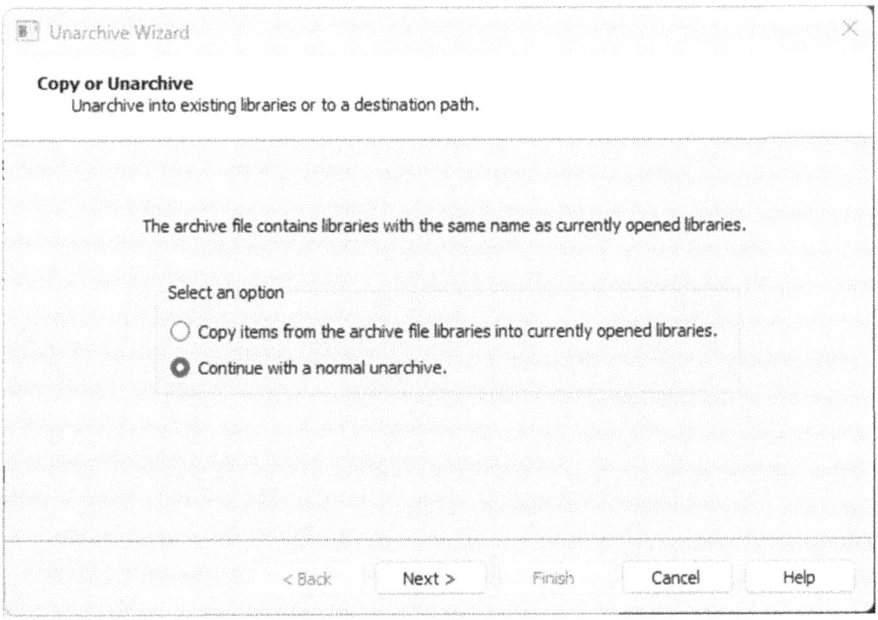

FIGURE 1.6 Unarchive wizard.

It can be skipped by pressing the Finish button. Note: if we wish to see the rest of the steps, we need to click the "Next >" button on all subsequent pages until we reach the Finish page.

Final Step: A message announcing a successful unarchiving of the library is shown in Figure 1.9.

Now that the design kit is unarchived, we shall see in the next section how to use it to bring into a circuit one or more of the kit's components.

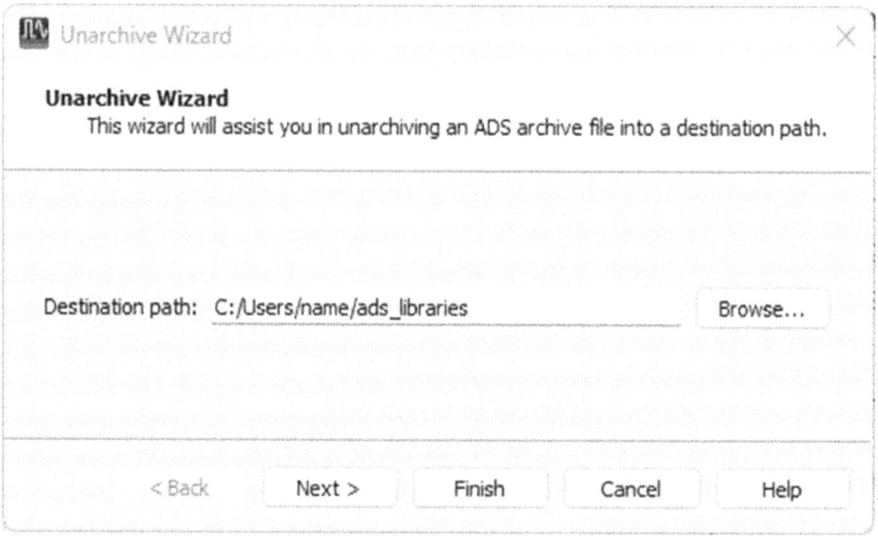

FIGURE 1.7 Setting the destination path via the unarchive wizard.

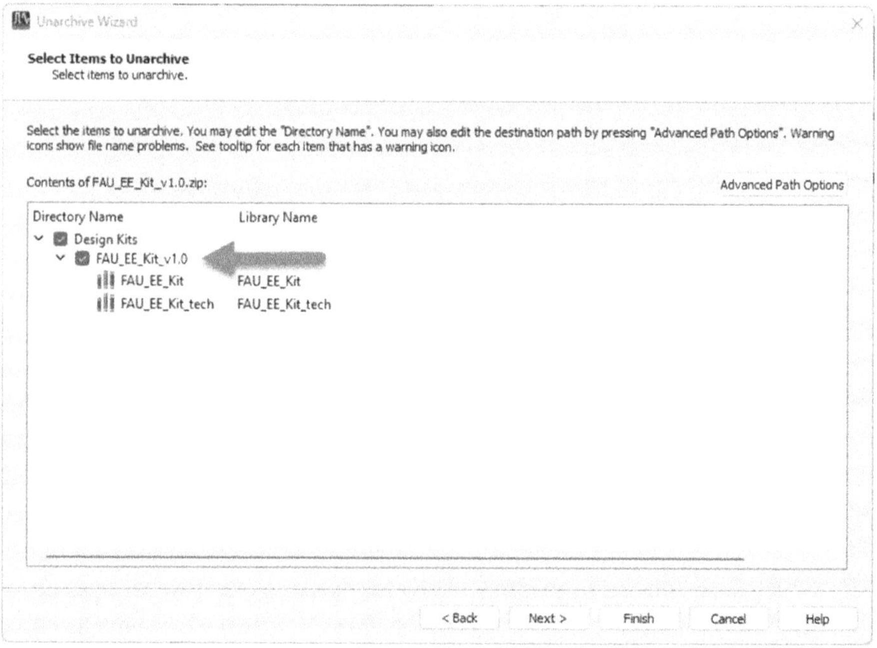

FIGURE 1.8 Selection of items to be unarchived via the unarchive wizard.

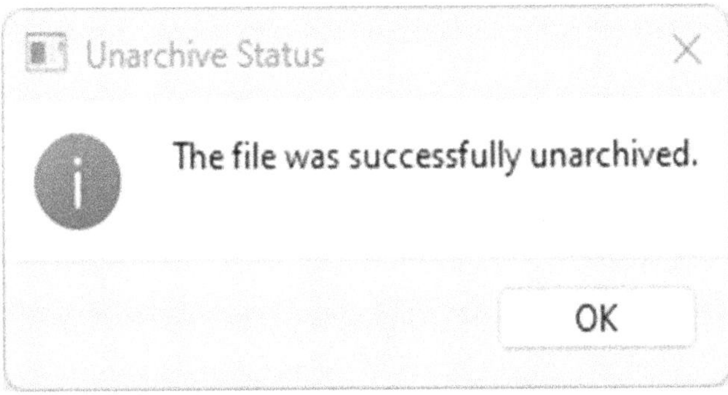

FIGURE 1.9 Successful unarchive status.

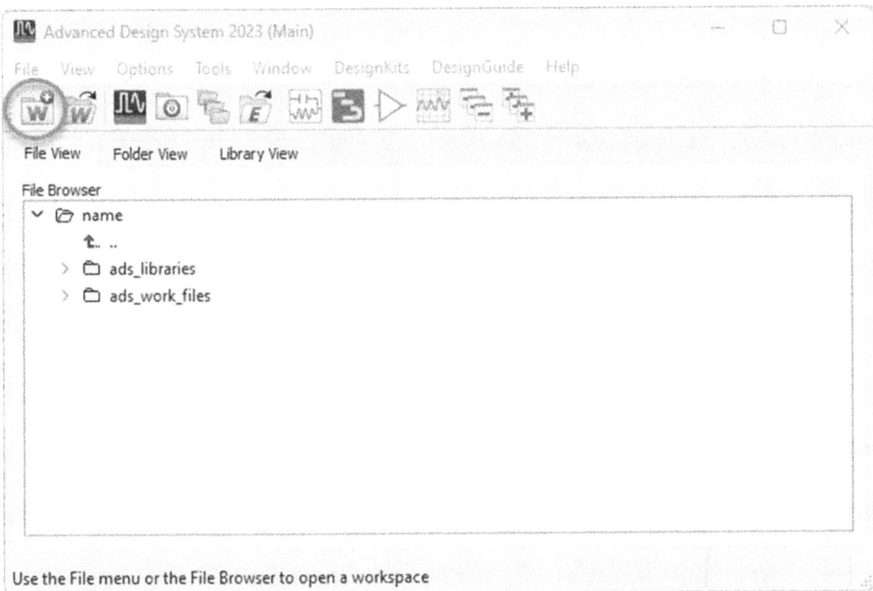

FIGURE 1.10 Main ADS window – circled "Create a New Workspace" icon.

1.3 CREATING AN ADS WORKSPACE

ADS uses the concept of a workspace as the main container for all users' design work.

Creating and setting up a workspace is the first step of any project. How many circuits one place into a workspace is not limited by the software, but a reasonable approach is to place those circuits that are related or will be connected to each other.

To create a new workspace, we start ADS. We close the "Get Started" window when it pops up. From the ADS main window, we click-select the "Create a New Workspace" icon on the upper left corner. See Figure 1.10.

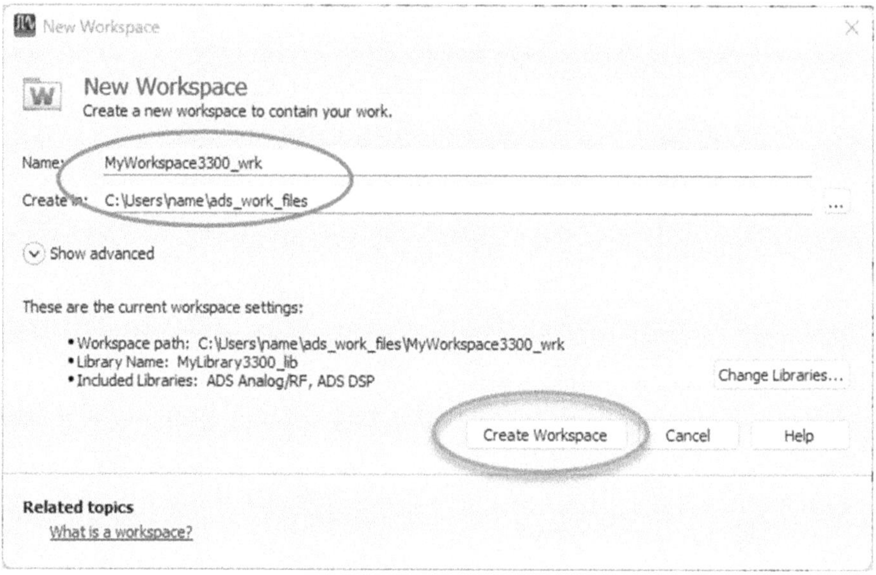

FIGURE 1.11 Create a new ADS workspace window.

Once one clicks on the icon, a "New Workspace Wizard", a window pops up as seen in Figure 1.11. If we are strictly following the steps in this chapter, we name the workspace "MyWorkspace3300_wrk" and create it in the same folder as shown in Figure 1.11. One can create the workspace elsewhere, if it is more convenient, or needed in another location, but we keep track of this location since the chapter always assumes that we are following the default location established in this chapter.

If you have followed our instructions, you have created the MyWorkspace3300_wrk workspace container. See Figure 1.12. Congratulations, you have created your first ADS workspace!

1.4 ADDING THE DESIGN KIT TO THE WORKSPACE

So far, we have learned how to unarchive a design kit and how to create a new workspace. Say that the workspace's name is MyWorkspace3300_wrk and it is created in say C:\Users\name\ads_work_files. A library is automatically created by ADS, and it is named MyLibrary3300_lib. Initially, this library was empty.

As we create designs, these will be placed in this library by default.

The unarchived design kit (PDK) contains several common lab parts such as µA741 op-amp, D1N4148 diode, and Q2N3904 NPN transistor. The PDK was placed in say C:\Users\name\ads_libraries folder. The next thing that we will learn is how to bring the PDK into the workspace, so that we can use the parts contained in it in our designs.

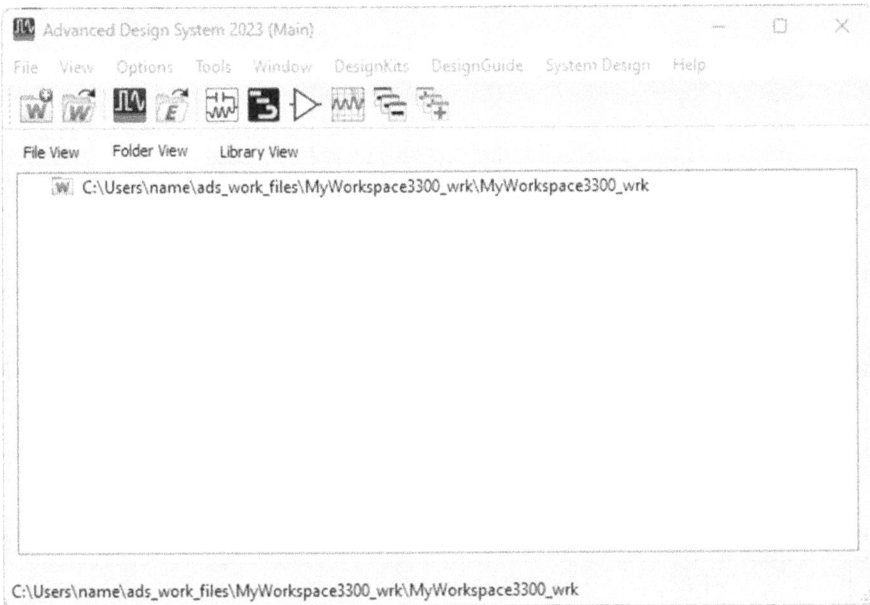

FIGURE 1.12 Successful creation of the MyWorkspace3300_wrk workspace.

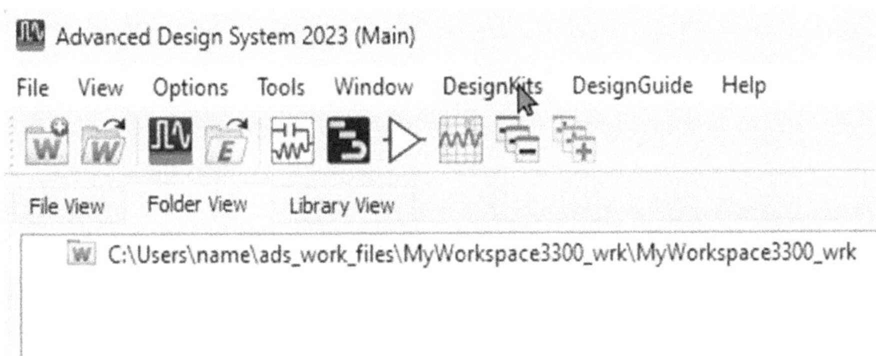

FIGURE 1.13 ADS main window design kits menu pick.

Step 1: We start at the Main ADS window with the workspace open. Need to click on the DesignKits menu pick. See Figure 1.13.

Step 2: We choose the "Manage Libraries" option. In the Manage Libraries window, we click on "Add Library Definition File …" See Figure 1.14.

Step 3: We searched for the folder where the FAU_EE_Kit_v1.0 design kit was unarchived. If we follow the chapter folders naming, it will be in the C:\User\name\ads_libraries\FAU_EE_Kit_v1.0 folder. We select the lib.defs file in this folder and click Open. See Figure 1.15.

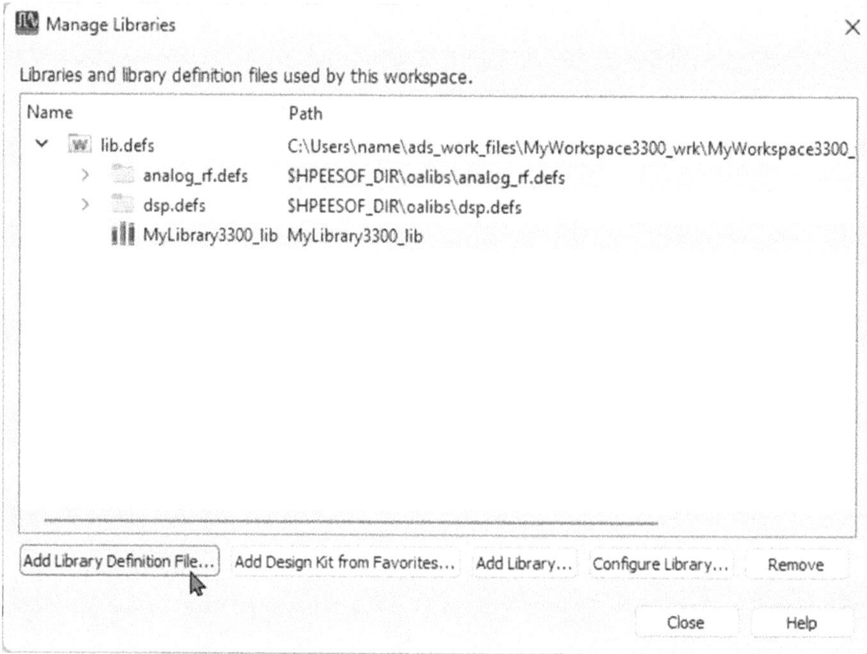

FIGURE 1.14 ADS manage libraries window.

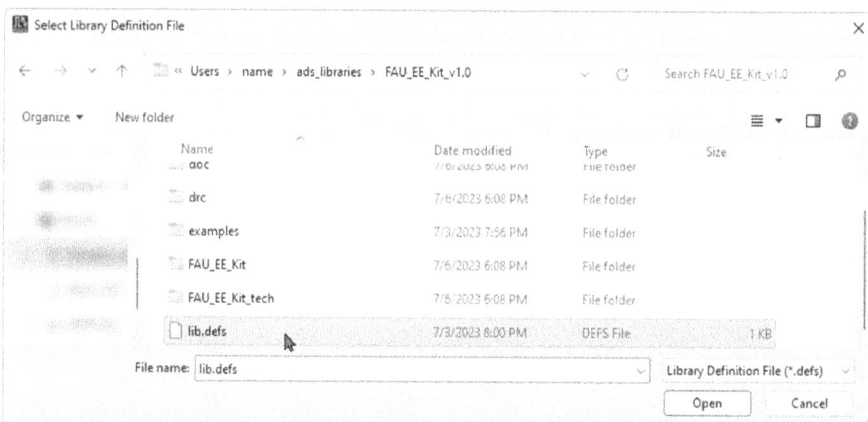

FIGURE 1.15 ADS select library definition file window.

Step 4: Done adding the design kit to our workspace. We close the Manage Libraries window. The FAU EE kit is a read-only library. For its contents, we follow steps A–C (Figure 1.16):

A: We select the "Library View" tab below the toolbar in the Main ADS window.

B: We click on the arrow to the left of "Read-Only Libraries" to make its content visible.

C: We click the arrow next to the FAU_EE_Kit.

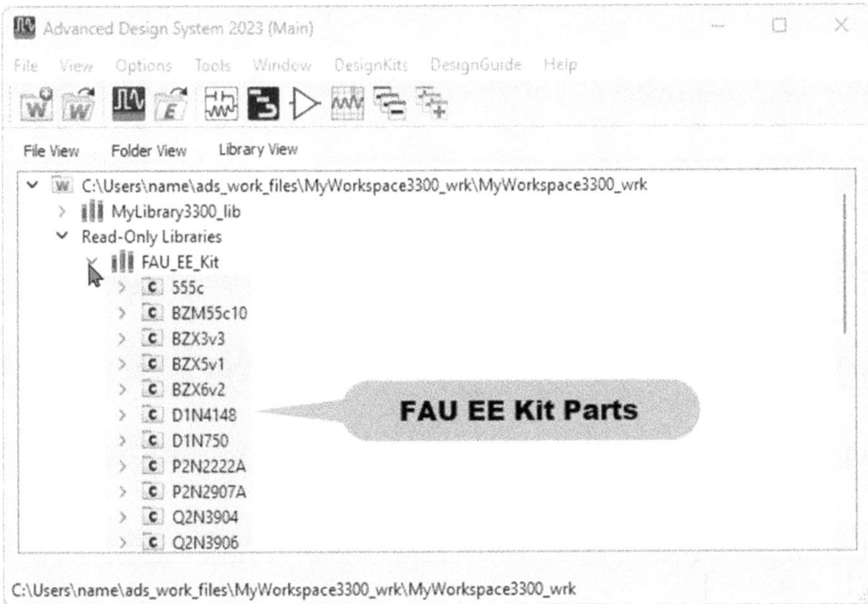

FIGURE 1.16 Library view showing FAU EE kit parts.

1.5 ADS TECHNOLOGY SETUP

ADS, being a large enterprise-like tool, needs to have a *Technology Setup*. Technology Setup is where one specifies units, resolution, layers, etc., related to the technology (board, MMIC, IC, etc.) that one is involved with. In our library, which is simple in nature, the units are "mils", and we have a few custom layers. As such, when one starts a workspace, one must set up the technology.

Step 1: We select the following from the Main ADS window. See Figure 1.17.
Step 2: The Technology Setup, see Figure 1.18, will pop up. Notice that this window (as an example) indicates issues with a Red X box in the bottom left.
Step 3: To correct the problem, we add the Technology Setup from the FAU_EE Kit. We click the "Add Reference Library …" button in the Technology Setup window, noting the green arrow in Figure 1.18.
Step 4: The ADS Add Reference Library window pops up. We select the FAU_EE_Kit_tech and select OK. See Figure 1.19.
Step 5: The red boxes will become green checkmarks. Click OK in the Technology Setup window. We are done setting up the technology (Figure 1.20).

1.6 BUILDING, SIMULATING, AND RESULTS OF THE ADS DESIGN

Let us now set up the schematic of the inverting amplifier and then simulate it. In the Main ADS window, we click on the "new schematic" icon, shaped as a resistor in parallel to a capacitor, which is the fifth icon from the left. See Figure 1.21. A dialog

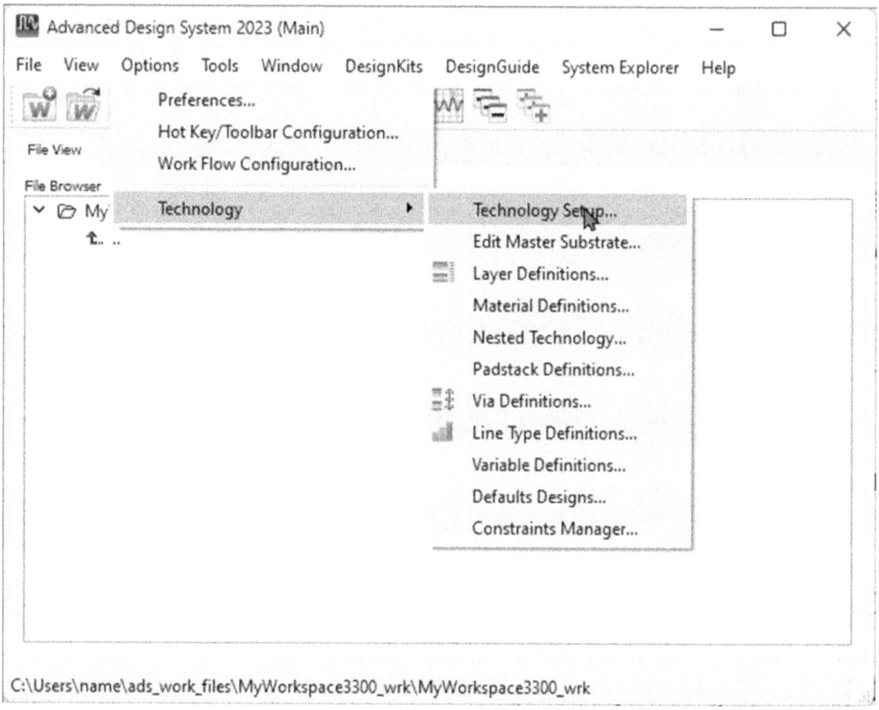

FIGURE 1.17 ADS main window technology setup selection.

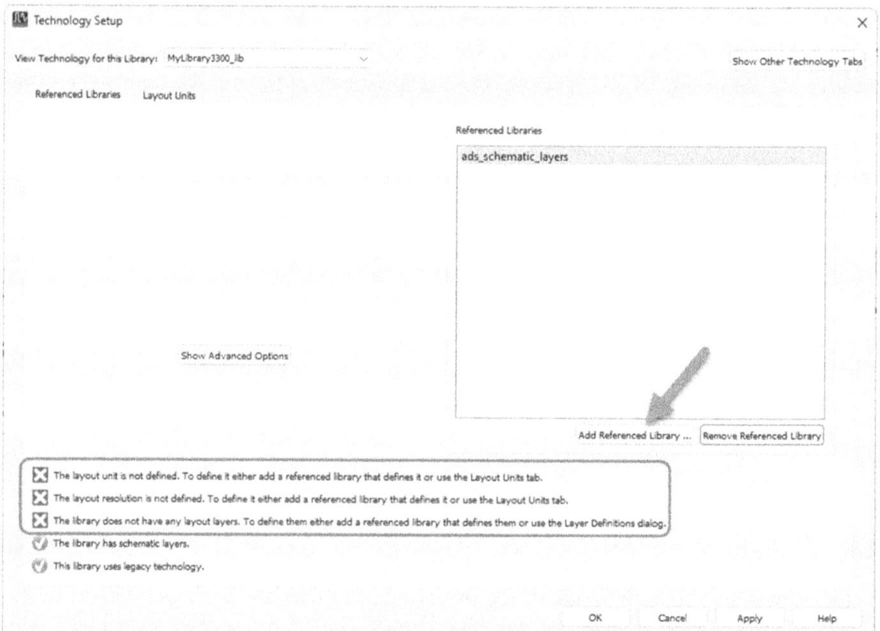

FIGURE 1.18 ADS technology setup indicating the problem with the red X boxes.

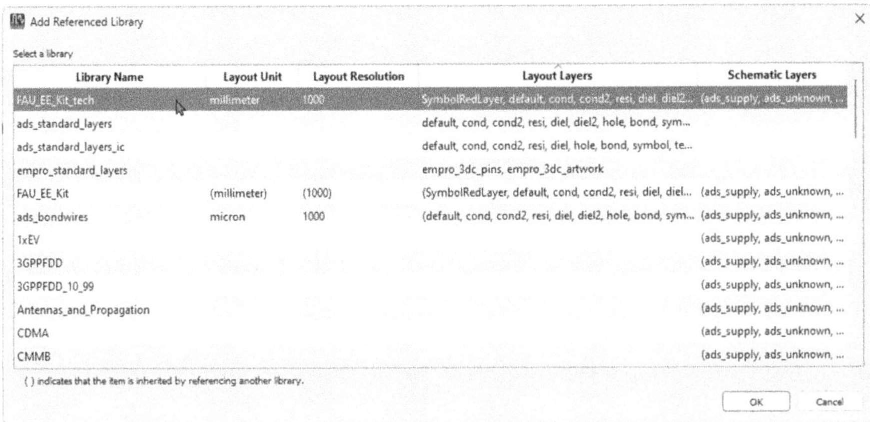

FIGURE 1.19 ADS add reference library window.

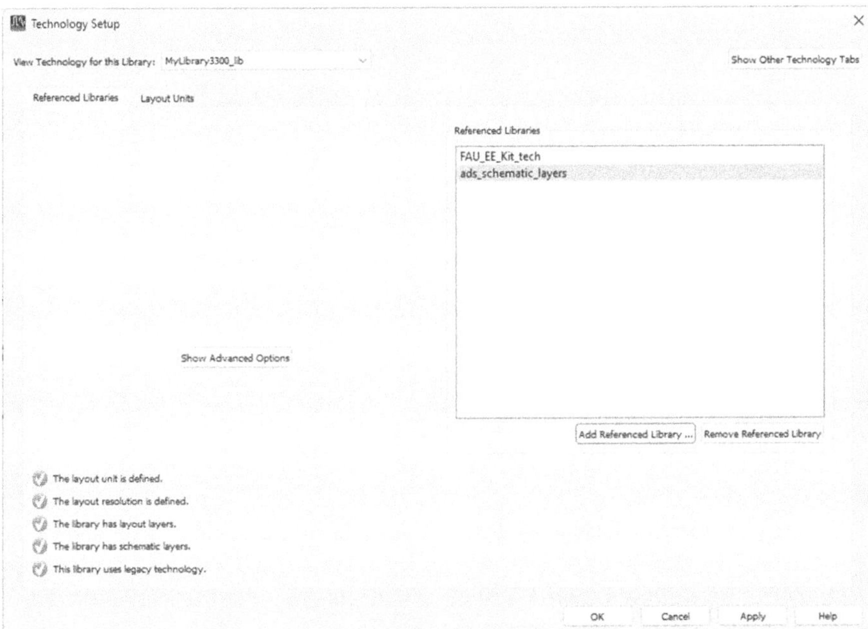

FIGURE 1.20 Successful ADS technology setup.

box "new schematic" opens up (Figure 1.22), allowing us to replace the generic schematic cell name "cell_1" by a new name, say, "OpAmp_Inverting_Amp_ µA741". As we OK the new schematic's name, a small blank schematic diagram opens up, which we promptly expand (Figure 1.23).

Let us first bring into the schematic diagram the components that are needed for the construction of the inverting amplifier circuit. There are two ways of bringing ADS Libraries components, such as resistors and capacitors, into the schematic.

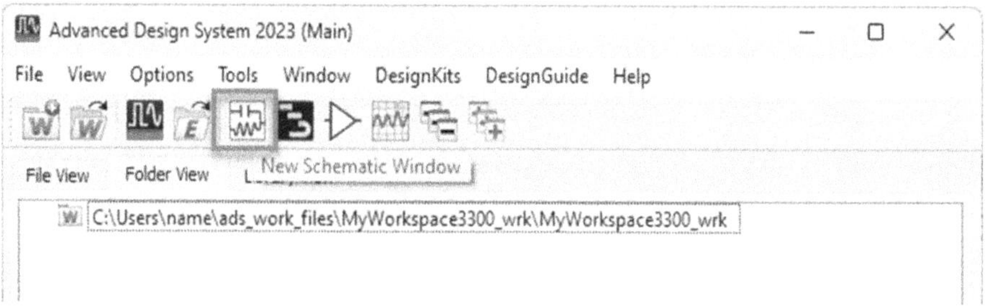

FIGURE 1.21 ADS main is ready to open a new schematic.

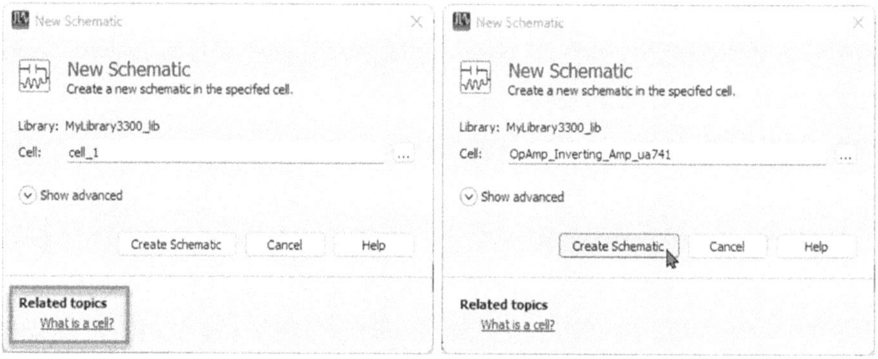

FIGURE 1.22 New schematic dialog box: (Left) default, (right) with new name.

The first, fastest and most common, is to utilize the palette located on the left-hand side of the schematic. See Figure 1.23.

The other method is to use the "Display Component Library List" (with an icon shaped as a file of books), which is the leftmost tab just above the schematic diagram. Again, see Figure 1.23.

The "Display Component Library List" can be used to bring in PDK components such as op-amps.

Clicking on the Open Library Browser icon brings us to the page shown in Figure 1.24. The ADS Analog/RF Libraries list is very long, but let us start with the much shorter Read-Only Libraries list of components in order to bring to the schematic an op-amp component shown in Figure 1.24.

Right-click on the "μA741" component's name and then left-click on "Place Component" (from the list). Minimizing the view of the Components Library takes us back to the schematic diagram. Over the schematic page, one sees a "ghost image" of the μA741. As one moves the ghost image of the 741 op-amp, a left-click places the component at a desired location on the diagram (where other left-clicks will place more components). A right-click, followed by "End Command", stops the production of components, see Figure 1.25.

As seen in Figure 1.1, which shows the full implementation of the inverting amplifier, we need, in addition to the 741 op-amp, two resistors, two DC supplies, a signal source, and ground nodes.

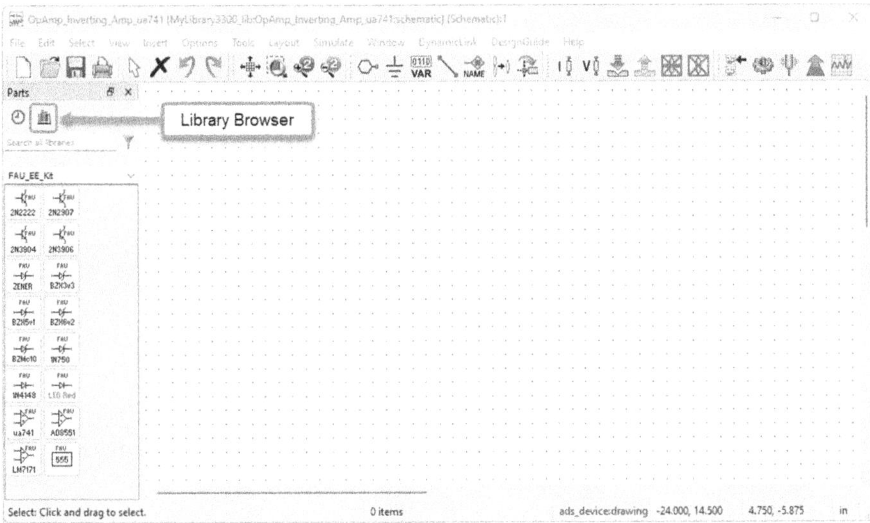

FIGURE 1.23 The blank schematic diagram that opens up after creating the new schematic name.

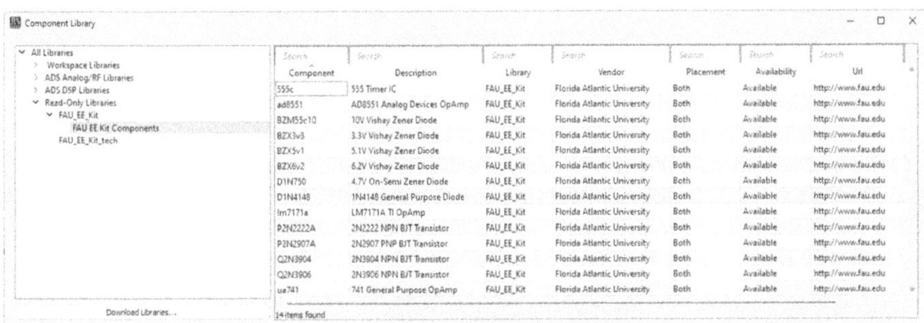

FIGURE 1.24 The default view of the component library.

Let us demonstrate how to pick up components from the palette. From the "Lumped-Components", palette we click on R and drag it to the schematic. A sequence of left-clicks places resistors at desired locations.

As seen in Figure 1.26, three resistors were placed before right-clicking End Command stopped the production of resistors. We see that each resistor receives a default name (R_1, R_2, etc.) and a 50 Ω default value. We shall modify component names, values, and orientation later on. The third resistor (R_3) is intended as an optional load (between the v_o node and ground).

An input sinusoidal source is needed. We will use the ADS VtSine source for this input. The VtSine source is picked up from the Sources-Time Domain palette > Sine. From the same component palette, we picked up two V_DCs, which will provide the DC power for our circuit. See Figure 1.27.

We complete placing the needed parts by adding five ground nodes, as seen in Figure 1.28.

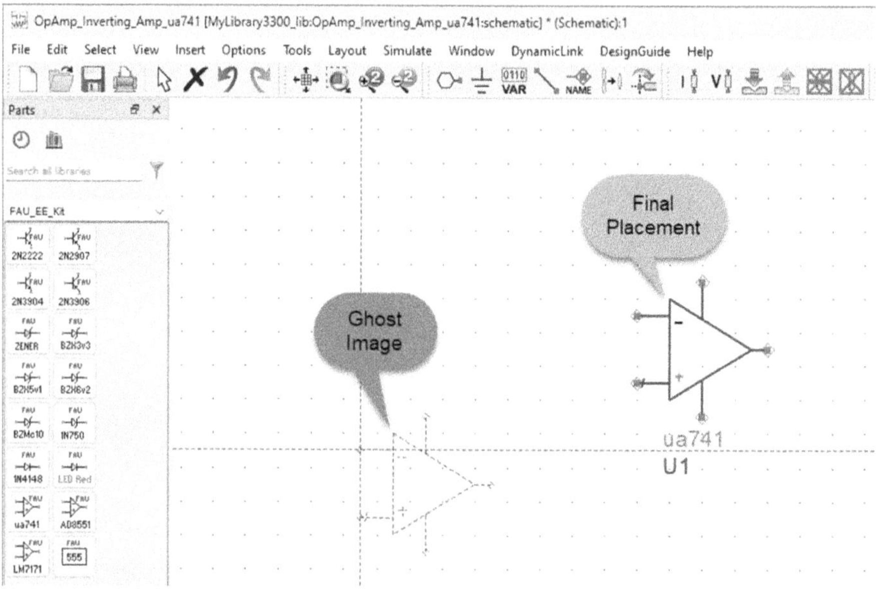

FIGURE 1.25 A 741 op-amp placed in the schematic diagram.

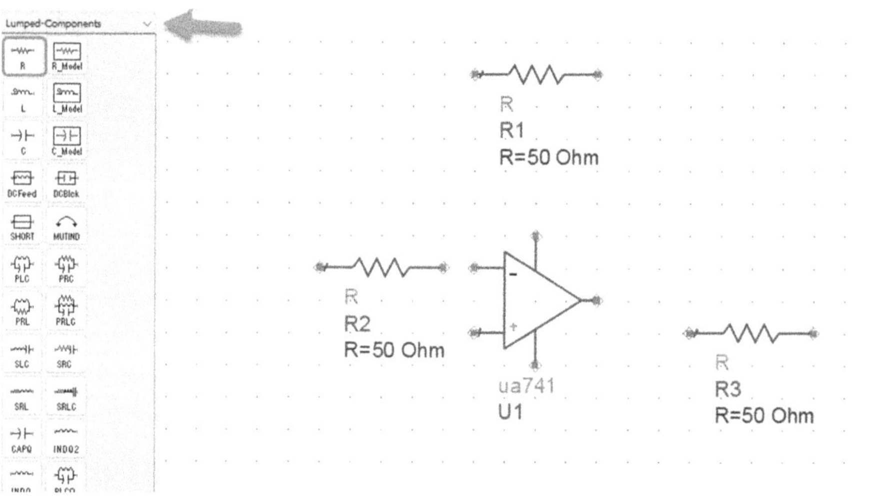

FIGURE 1.26 Placing resistors around the op-amp.

Note that arranging the components in positions and orientations that are similar to those of the original circuit diagram (Figure 1.1) is recommended, as it may help with the debugging of the schematic. When one selects a component, we note that ADS makes available a Rotate and a Move Text handle, see Figure 1.29. Using these handles, we move the text of the μA741 to a more convenient location.

To allow wiring the negative supply, we move the text for R_1 to the top side and rotate R_3. See Figure 1.30 for the final positions.

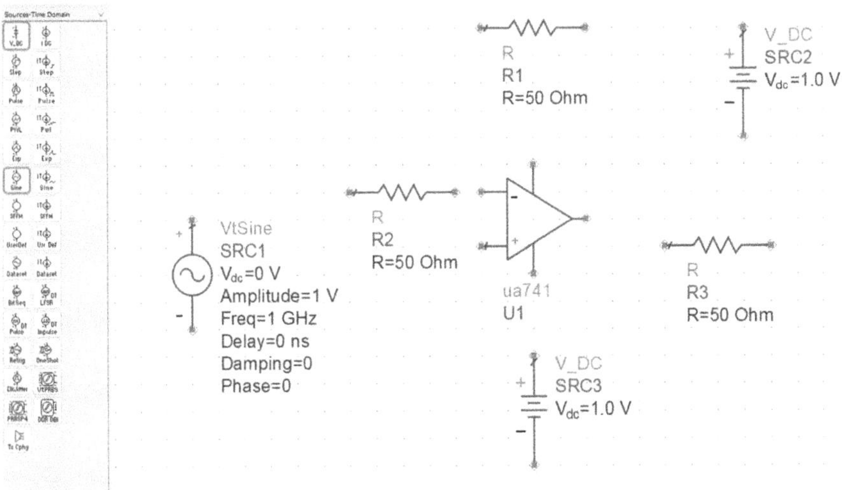

FIGURE 1.27 Adding an AC signal source and DC bias voltage sources.

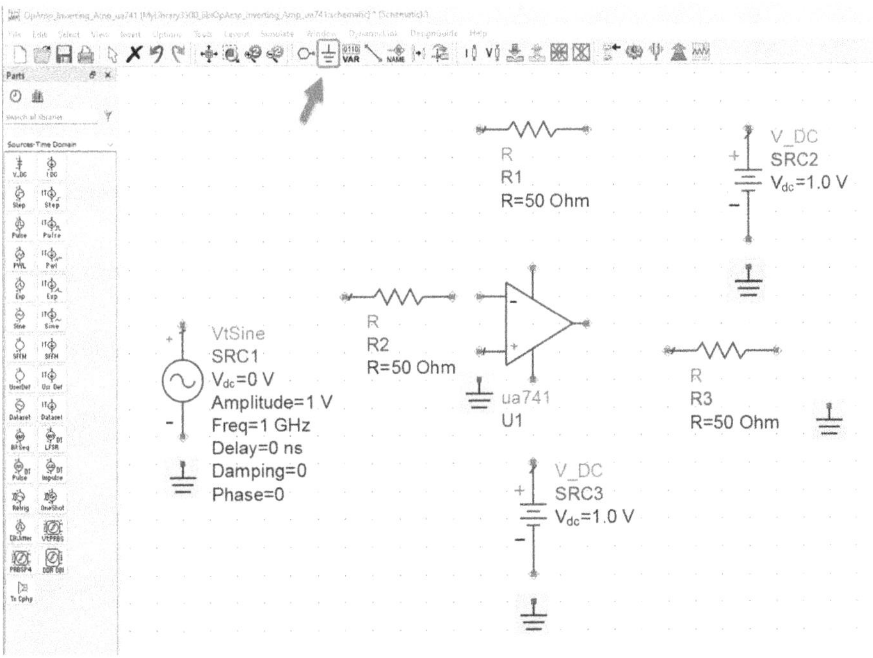

FIGURE 1.28 Completed parts placement with multiple ground nodes.

The ADS simulator does not always show all the component parameters on the schematic. This is done typically to not burden the schematic display with too much information. This means it is up to the designer to display those parameter values that are most significant.

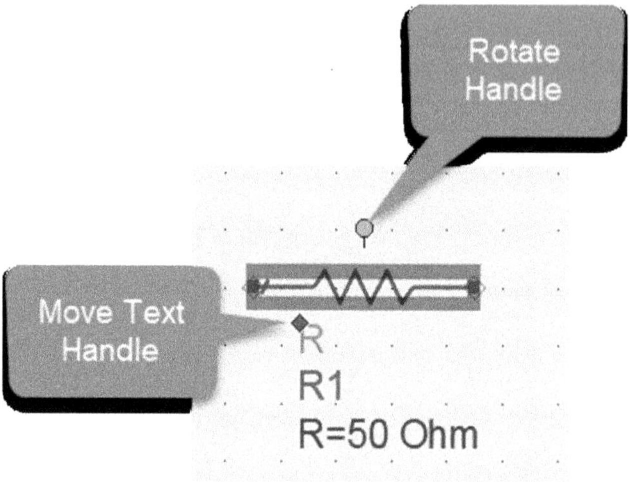

FIGURE 1.29 ADS move text and rotate part handles.

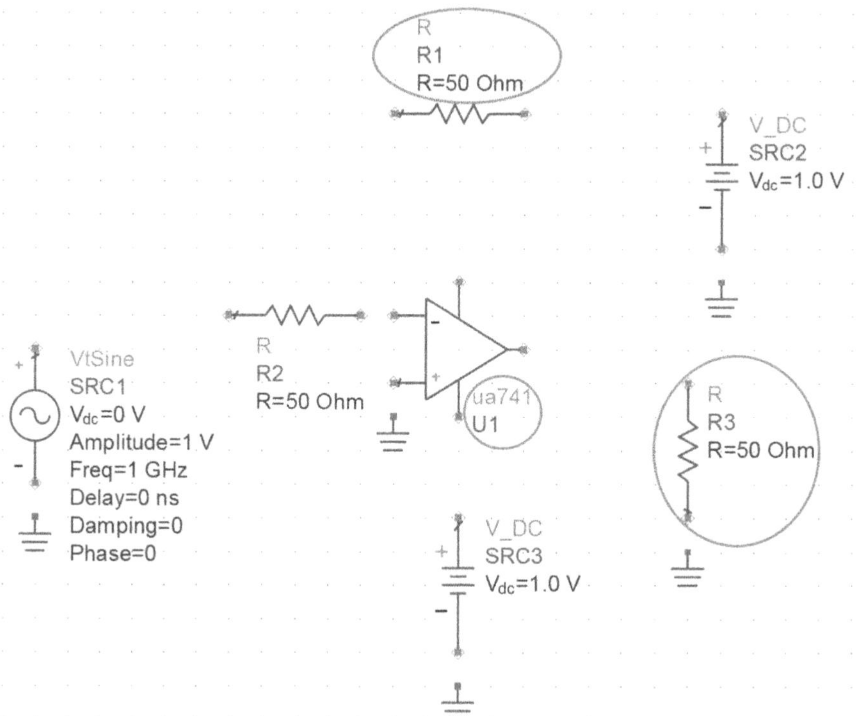

FIGURE 1.30 Final components position, orientation, and text.

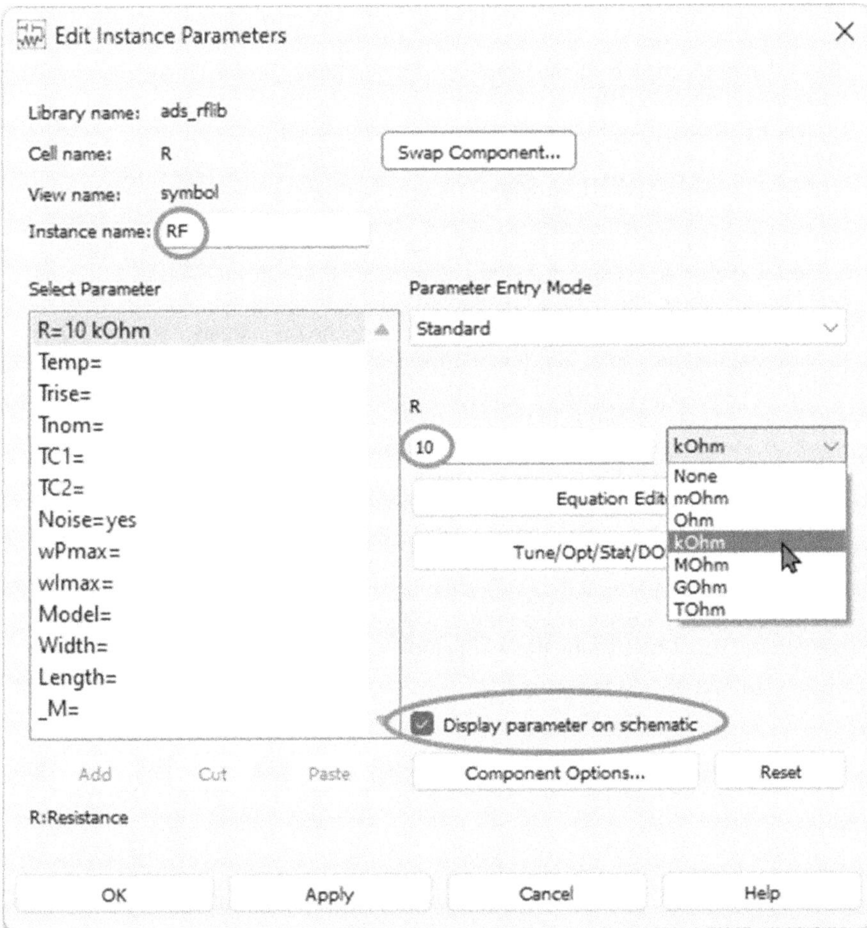

FIGURE 1.31 Parts instance parameters, changes made to instance name, resistor value, resistor units, and schematic display turned on.

Double-clicking on a component brings a dialog box that allows us to change the component's name, parameter values, and choose units for the values. See Figure 1.31. One can also select which component parameters and values will be visible on the schematic.

At this point, we need to decide on some specifics of the simulation.

Example 1.1

Say, we want to amplify an AC signal, with an amplitude of 1 V and frequency of 1 kHz, by a factor of –10. The desired voltage gain of –10 is obtained by letting the ratio of R_f:R_s = 10:1. Double-clicking on a resistor brings up a dialog box that allows the user to assign a new name and new value to the resistor. Let us initially set the DC supply values to ±10 V. The performance of the inverting amplifier is, in theory, independent of the R_L load value. In practice, whenever R_L is too small (say 100 Ω, thus it may require too much peak current from the op-amp), it does affect the performance of the amplifier. Voltage and current limitations of the op-amp will be demonstrated in the end-of-chapter problems. At this point in our design, we update the component parameter values and finalize the design as seen in Figure 1.32 using the component parameters dialogue box.

In the unmarked μA741 op-amp component, the top vertical terminal (located next to the "– input") is for the op-amp's positive DC supply. The bottom vertical terminal (located next to the "+ input") is for the op-amp's negative DC supply.

Let us elaborate on the negative battery named V^-. In real life, as one may build an electronic circuit on a breadboard, short-circuiting the minus terminal of battery 1 to the plus terminal of battery 2, defines the 0 V reference for the circuit (that we name "ground"). The plus terminal of battery 1 has a positive voltage (with respect to ground), and the minus terminal of battery 2 has a negative voltage (with respect to ground). In simulations though, as a battery may be assigned a negative voltage. It may be safe practice (especially for beginner students) to always connect the ground to the minus terminal of DC supplies and let the plus terminals be connected to the appropriate op-amp terminals (as shown in Figure 1.33).

The schematic wires are added next. In ADS the wiring process begins by clicking on the Insert Wire icon to obtain a wiring tool. In order to wire the components, one to another, we use the wiring tool located 11th from the left on the toolbar. As we see in Figure 1.32, all yet unwired components' terminals are colored red. Left-clicking the wiring tool allows the user to visit and left-click

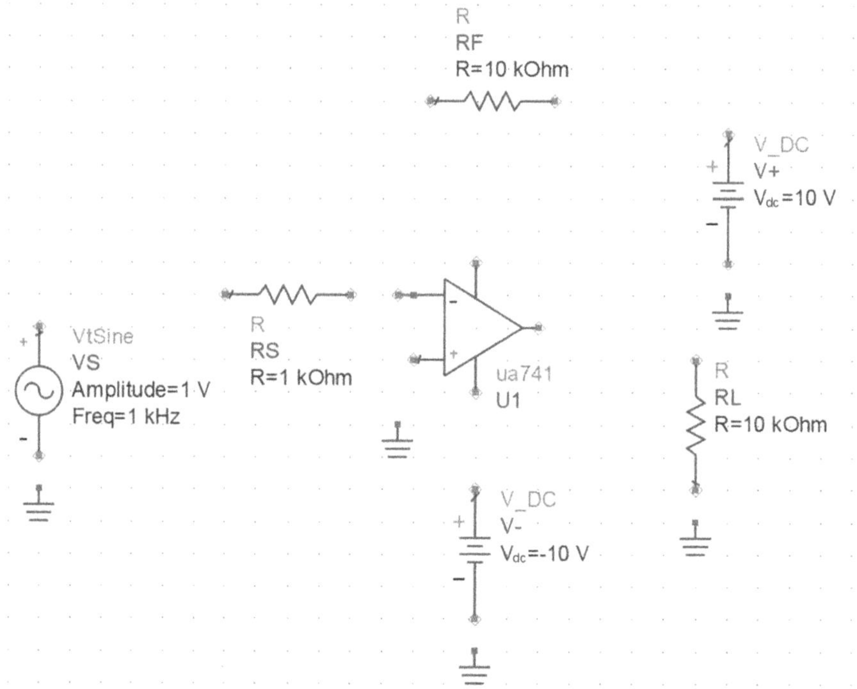

FIGURE 1.32 Finalizing component names and values.

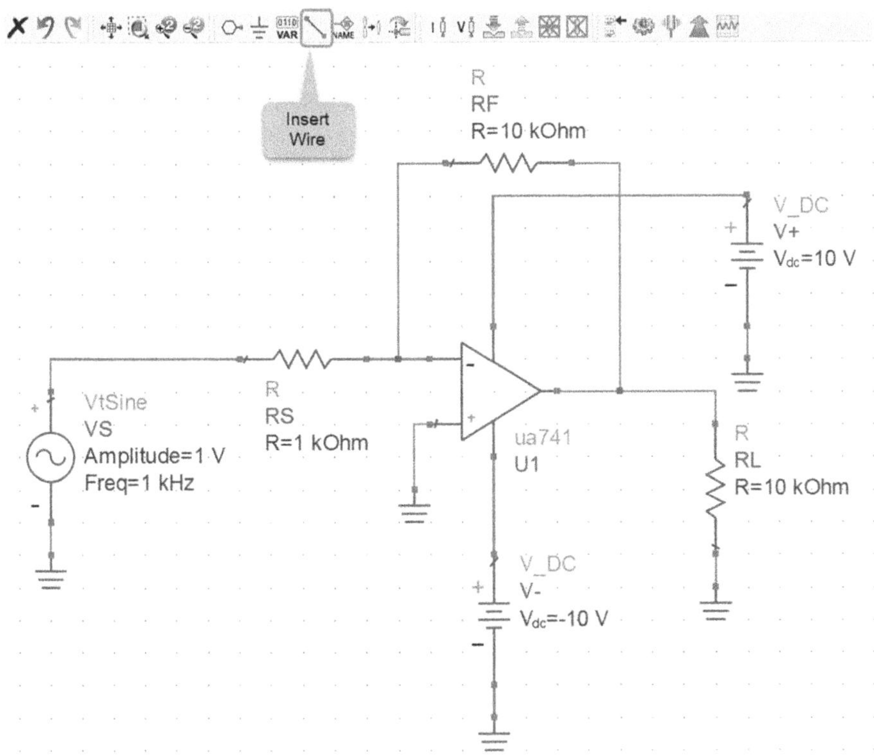

FIGURE 1.33 The wired circuit diagram.

on an unwired terminal, then on another unwired terminal, which establishes a connection between the (now colored blue) terminals, as seen in Figure 1.33. The visitation of all unwired terminals proceeds continuously using repeated left-clicking.

A simple wiring move creates a straight-line wire (as shown in Figure 1.33) between the minus terminal node of V_S source and ground. The next simple wiring move involves a single 90° turn, as shown in the connection between the plus terminal of V_S and the left terminal of R_S. Left-clicking in middleway allows the creation of a wire that has more than one vertical turn, as shown in the wire between the plus terminal of V^+ and the positive supply op-amp terminal.

Say (in Figure 1.33) that the output of the op-amp is already wired to R_L. Now, in order to connect R_F to the same node, we wire from the right terminal of R_F and left-click on the wire that connects the op-amp to R_L. The resulting blue dot indicates that indeed R_F, R_L, and the op-amp's output are connected electrically. Note that the same wire (between R_F and R_L) crosses another wire that connects V^+ to the op-amp, but there is no blue dot. Hence, the two wires are not electrically shorted.

Beginners using ADS are advised to carefully examine their circuit wiring for unconnected terminals, which remain red, incorrectly connected wires that have a unnecessary blue dot, and wires that should be electrically connected but are not, as shown by the lack of blue dot.

We are almost done. The inverting amplifier shown in Figure 1.33 seems ready for simulation; however, two crucial details are still missing, as seen from Figure 1.34.

These are (i) the simulation type and (ii) the selected nodes' names. What is "simulation type"? Any ADS simulation requires a "simulation engine", such as transient simulation, DC simulation, AC simulation, etc., to be included in the schematic. Why name some nodes? Some of the circuit nodes must be named (or "aliased"), otherwise, their voltage signals may not be plotted or presented in any other way. Schematic nodes must be named wherever we anticipate a voltage signal, at that node, that should be plotted.

Let us imagine that we build an inverting amplifier circuit in the lab. How do we intend to test the circuit? There is a signal $v_s(t)$ that comes from a function generator and needs to be amplified by the tested circuit, and there is the output signal $v_o(t)$. Both voltage time signals are observed on an oscilloscope. Transient simulation is the equivalent of watching time voltage signals on an oscilloscope. As the ADS simulation concludes, a list of variables ready to be plotted is presented to the user. Named nodes are included in the list. Unnamed nodes are not included. We are obviously interested in watching the input signal, so we name its node "V_{in}", and the output signal "V_{out}". The "eee" node may enable us to have a closer look at the virtual ground signal. See Figure 1.34.

The node-naming tool is marked "NAME" on the toolbar. Clicking on this tool brings up a window titled "Wire/Pin Label". We click on a desired node and enter the node name in the "Net Name" field. We click enter when done, and the node name will appear on the schematic.

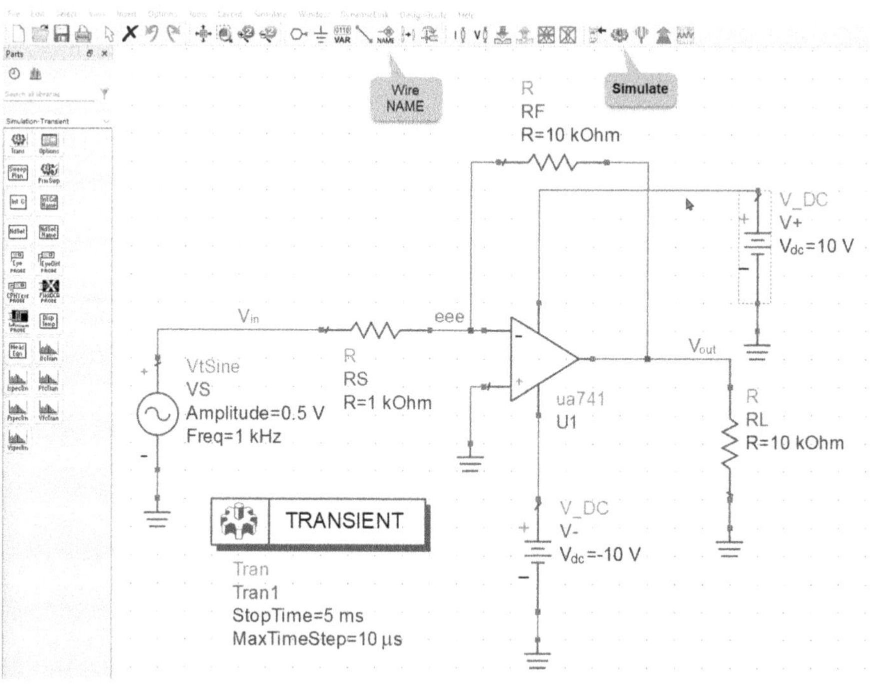

FIGURE 1.34 A schematic that is ready to be simulated.

Simulation-type blocks, such as transient (shown in Figure 1.34), may be obtained like obtaining any other electronic part: From the palette or from the Component Library. Transient simulation is for observing time signals. There are two simulation parameters that must be specified: StopTime and MaxTimeStep. Let us explain the choices shown in Figure 1.34: StopTime=5 ms and MaxTimeStep=10 μs (written as 10 μsec). Recalling that the input signal was chosen to have f=1 kHz (or a period T=1/f =1 ms). Most simulations assume that the simulation start time is t=0. The choice of StopTime=5 ms will enable the user to view the first five periods of the input's AC signal and the output's AC signal. The maximum time step parameter is related to the numerical solution of the circuit. It presents a compromise between accuracy and speed of the simulation. A numerical solution involves time sampling and updating of all the variables from any sampling instance to the next. The sampling period is not necessarily constant. The solution algorithm can detect periods at which the solution is fast-changing. In such a case, the sampling rate increases, and the solution is recalculated. At the other extreme, as the algorithm detects a slowly varying solution, the sampling rate decreases for the sake of reducing the number of unnecessary computations. The MaxTimeStep parameter sets a limit on how large the sampling period is allowed to be. A too large value may result in an inaccurate simulation (often seen as a discontinuous and grainy plot). Too small a value causes the simulation to be very slow and accumulate too much data. **A good rule of thumb for selecting the MaxTimeStep parameter is to take it as 1/100 of the fastest period or time constant in the circuit.** In our example (above), there is only one input signal that has a period of T=1 ms. Hence, MaxTimeStep=0.01 ms (or 10 μsec, as written).

One last change prior to simulation. The amplifier is expected to have a voltage gain of −10. The DC supplies are ±10 V. We initially set the input signal's amplitude to 1 V, thus expecting the output signal's amplitude to be 10 V. Remembering that the supply voltage levels define the dynamic range of the amplifier, a question arises whether or not it is okay to bring the output voltage signal all the way up to the allowed limit. We shall revisit op-amp voltage and current limits in some of the end-of-chapter problems, but meanwhile, let us simulate with an input signal's amplitude of 0.5 V (as shown in Figure 1.34).

In Figure 1.34, we note the gear-shaped icon on the toolbar. Clicking on this icon starts the simulation. If the schematics contain no errors, the simulation will end with what is shown in Figure 1.35. This is the first page of any post-processing, called ADS **data display**, that may be conducted based on the collected simulation data and results. Post-processing typically includes plots, equations for the calculation of circuit features based on the collected measurements, and tables of calculated results. In transient simulation, the most common output is plots of circuit variables as functions of time. The plotting process is initiated by clicking "Insert" followed by selection of "Plot". The next step is to click near the upper left corner of the blank workspace area and then move the mouse toward the bottom-right corner of the workspace in order to define a rectangular frame for the first plot. This results in the Plot Traces & Attributes dialog box shown in Figure 1.36. We notice that the three node names (defined earlier on the schematic) V_{in}, V_{out}, and eee are indeed included in the menu of to-be-plotted variables. We click-select a variable from the left menu (say V_{in}), and then click ">>Add >>" to place this variable's name in the right field.At this point, if we click OK, a plot of $V_{in}(t)$ appears inside the rectangular frame defined earlier. See Figure 1.37.

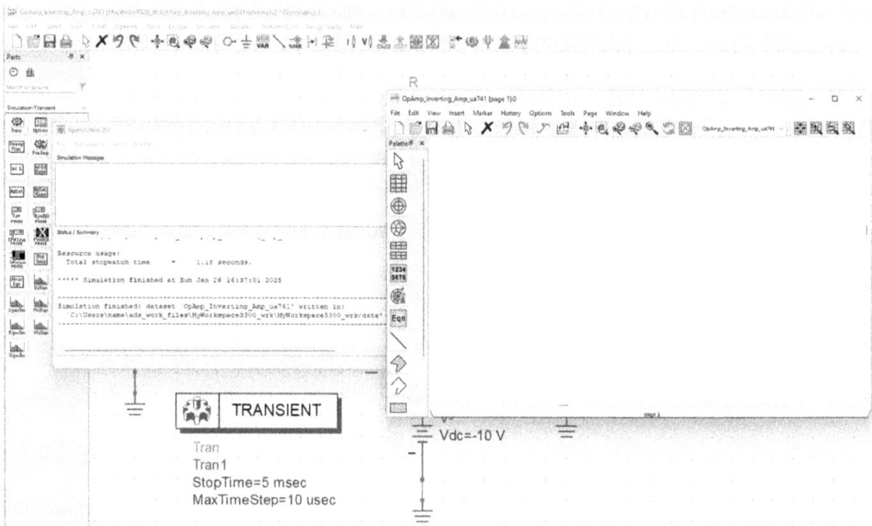

FIGURE 1.35 Simulation is done. ADS is ready for post-processing of the collected data.

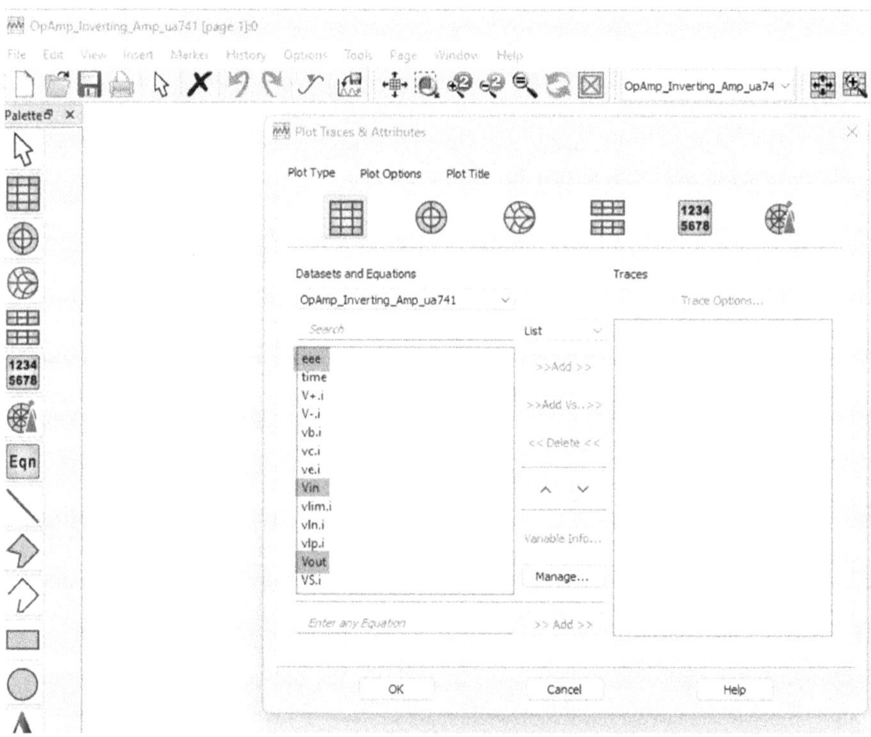

FIGURE 1.36 Plot traces and attributes: Selection of variables from the left menu and moving the selection to the trace options space.

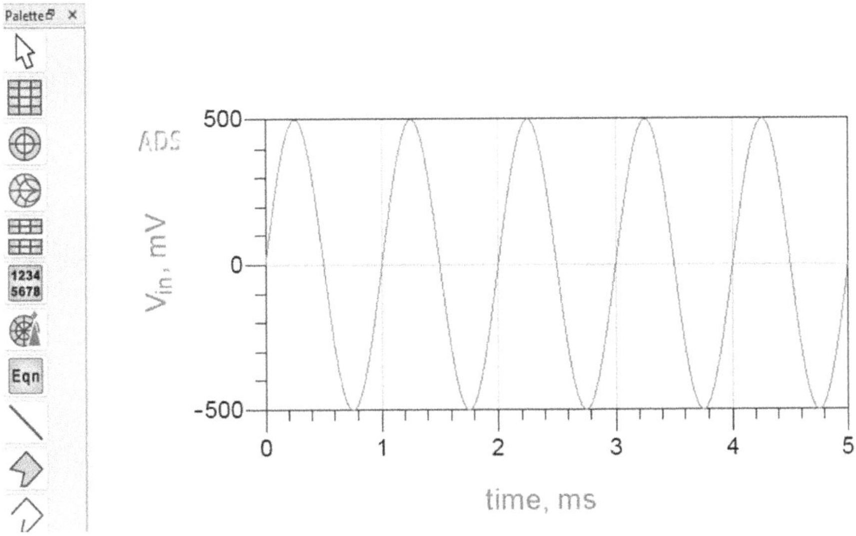

FIGURE 1.37 First plot: V_{in} vs. t plotted inside the rectangle frame. Axes titles are created automatically.

Another option is to select from the menu a second variable, say V_{out}, and add it to the Trace Options. As a result, both V_{in} and V_{out} will be listed in the Trace Options. This will cause $V_{in}(t)$ and $V_{out}(t)$ to be plotted on the same scale.

Whenever we wish each variable to have its own scale, it is a good idea to separate the plots, as done in Figure 1.38.

From the plots of $V_{in}(t)$ and $V_{out}(t)$ seen in Figure 1.38, we can measure/calculate the voltage gain of the amplifier as the ratio between the output signal's peak-to-peak (ptp) amplitude and the input signal's ptp amplitude. The ptp amplitude of $V_{in}(t)$ is known (1 V, as generated by the signal source). The ptp amplitude of $V_{out}(t)$, by crude inspection, is 10 V. Notice that $V_{out}(t)$ is inverted from $V_{in}(t)$. So, yes, the voltage gain is approximately −10 as expected. Is it exactly equal to −10? After all, the 741 op-amp is a "good" op-amp but not an "ideal" op-amp. We can measure the accurate ptp amplitude of $V_{out}(t)$ with the help of measurement markers.

On the left side of Figure 1.39, there is a collection of available markers. The simplest one is the topmost – it is a marker that is selected and placed manually on a desired point that belongs to the plot. For an accurate measurement of ptp amplitude, there is a need to place markers exactly at the peaks. The second marker on the left is a Peak Marker. It is inserted near an upper peak of the signal, and it can seek a local maximum and correct its own position. See marker m_1 on the plot of $V_{out}(t)$. Marker m_2 is chosen from the third left marker, and it is a Valley Marker. It seeks a local minimum near its initial placement. An insertion of any marker creates a label that contains the measurement information. By default, the labels show up one above the other above the relevant plot, but they can be moved, as done in Figure 1.39.

Subtracting the V_{out} values of markers m_1 and m_2 results in a ptp of $V_{out}(t)$ of 9.994. Is that the exact voltage gain of the amplifier? Note that the subtraction of

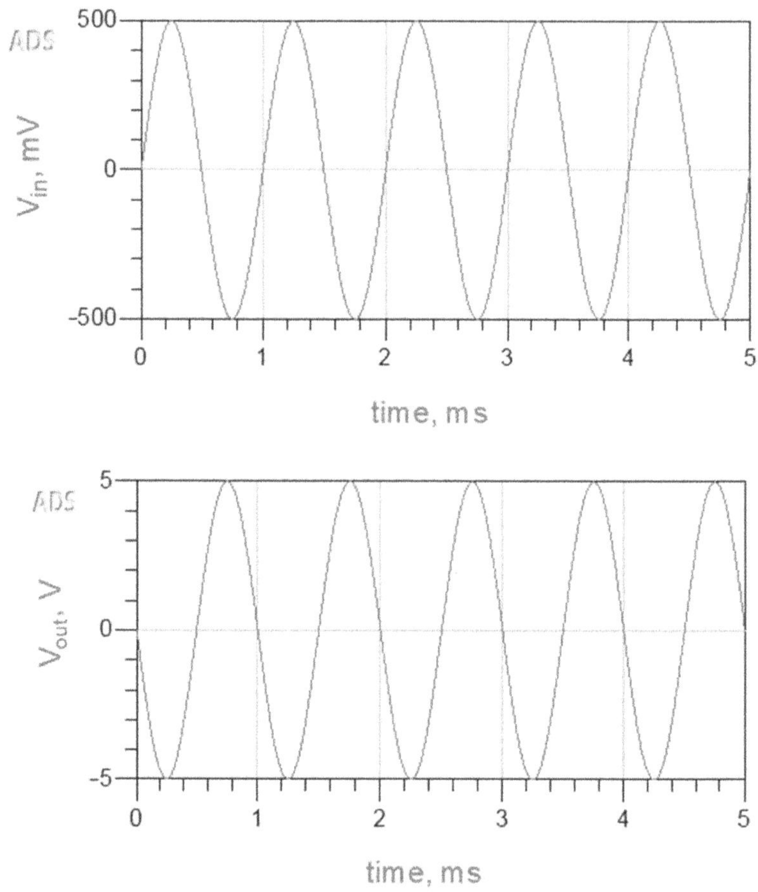

FIGURE 1.38 Second plot added below the first plot. It is easy to keep the rectangles to have the same width (not done here).

the time information of m_1 and m_2 provides 0.5001 ms. The deviation from the exact half-period value of 0.5 ms is due to maximum and minimum search errors of the markers. For all practical purposes, the voltage gain of the μA741 op-amp inverting amplifier is remarkably close to the ideal theoretical value.

Finally, Figure 1.40 allows us to observe the contents of the "virtual ground" node. In this example, we see that eee(t) is a small AC signal of frequency 1 kHz and an amplitude that is close to 5 mV.

Recall the explanation of the gain formula (1.1): We assumed an input current of $V_s(t)/R_s$, but the current is actually $[V_s(t) - eee(t)]/R_s$. Can eee(t) be considered negligible compared to $V_s(t)$? The ratio of amplitudes is 500–5 mV. So, yes, practically speaking, the voltage at the "– input" of the op-amp is approximately a "ground" voltage.

For more information on workspaces, schematics, and design kits, the reader is referred to the ADS User Guide [Keysight1].

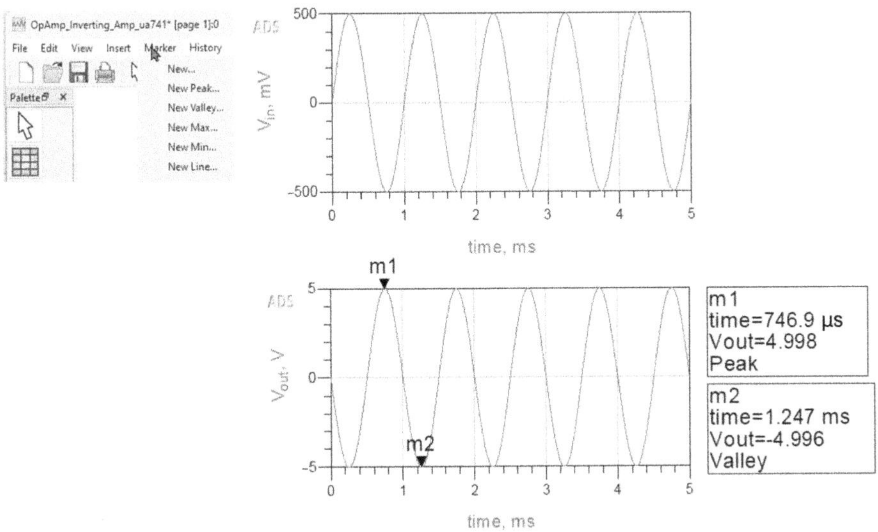

FIGURE 1.39 Accurate peak-to-peak amplitude measurement using markers.

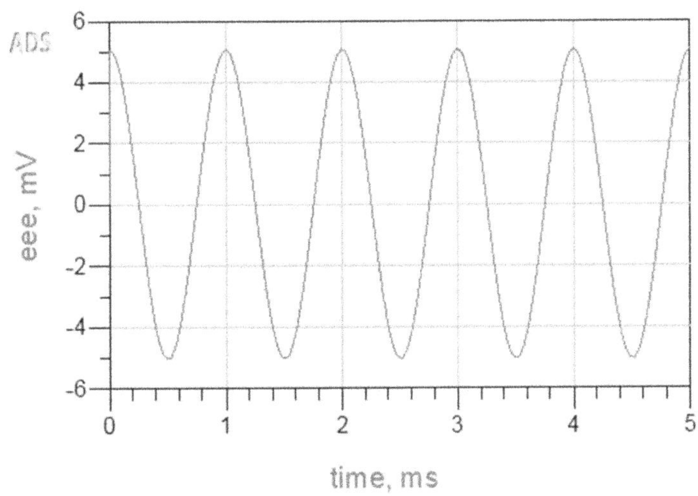

FIGURE 1.40 A closer look at the voltage of the virtual ground node.

1.7 NON-INVERTING OP-AMP AMPLIFIERS

An op-amp non-inverting amplifier can be built as shown in Figure 1.41: The input signal $v_s(t)$ is connected to the op-amp's "+ input". There is negative feedback as $v_o(t)$ is connected by a resistor R_2 to the "– input" of the op-amp. Again, the op-amp's large differential voltage gain, large differential input resistance and the negative feedback all team up to create a "virtual ground effect": The voltage signals at the

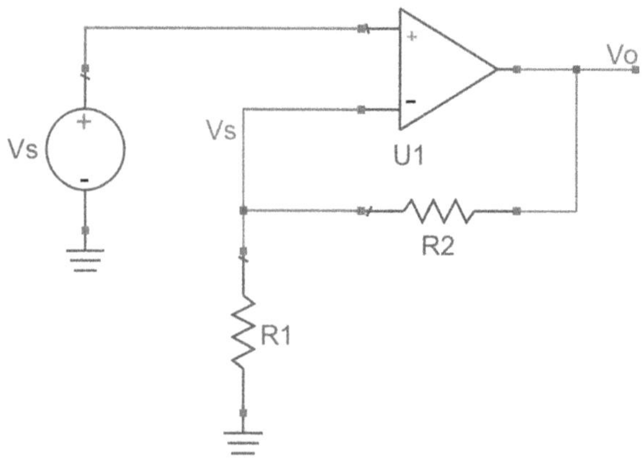

FIGURE 1.41 Non-inverting op-amp amplifier.

op-amp's inverting and non-inverting input nodes become very close to each other. The input signal $v_s(t)$, connected to the "+ input" "jumps over" to the "– input", as shown in Figure 1.41. We refer readers to [Sedra1] and [Franco1] for more theoretical details.

A current $v_s(t)/R_1$ is created. This current cannot flow into the input of the op-amp because of the large op-amp's input resistance. It does flow through the feedback resistor R_2. A closed electrical circuit to the ground is established through the op-amp's output terminal and internal circuitry inside the op-amp. Formula (1.2a) that shows the equal currents that flow through R_1 and R_2 establishes the non-inverting amplifier's voltage gain, Formula (1.2b):

$$\frac{v_s(t)}{R_1} = \frac{v_O(t) - v_s(t)}{R_2} \tag{1.2a}$$

$$v_O(t) = \left(1 + \frac{R_2}{R_1}\right) v_s(t) \tag{1.2b}$$

The non-inverting amplifier has a positive voltage gain of $1+R_2/R_1$. Let us end this brief summary with two comments: (i) If we want the positive gain to be less than 1, we can feed $v_s(t)$ to the op-amp via a resistive voltage divider (as shown in the example of Figure 1.42), (ii) A non-inverting amplifier with a gain of 1 is called a **buffer**. It is obtained by eliminating both R_1 and R_2, and short-circuiting $v_o(t)$, to the op-amp's "– input", as shown in Figure 1.43.

1.8 ADS SCHEMATIC REUSE, MEASUREMENT EQUATIONS, AND TRANSIENT SIMULATION OF A NON-INVERTING AMPLIFIER

In Section 1.6, we learned how to construct a schematic from scratch, bringing all the needed components from the palette of components. We placed the components in the schematic and wired them. We then assigned name and value to each component,

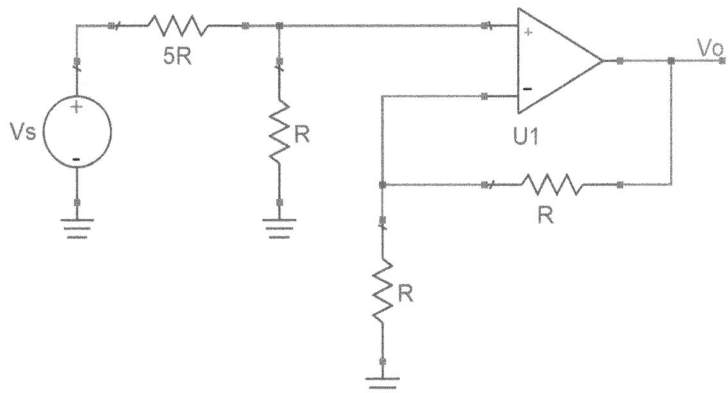

FIGURE 1.42 An example of a voltage divider and non-inverting amplifier combo with an overall voltage gain of 1/3: The voltage divider creates a gain of 1/6 whereas the op-amp non-inverting amplifier has a gain of 2.

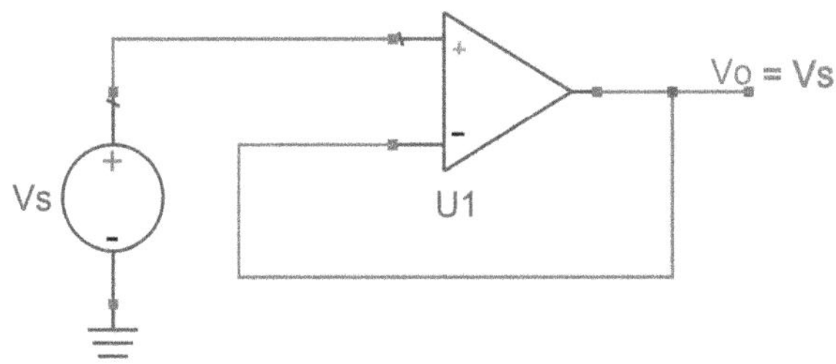

FIGURE 1.43 An op-amp voltage buffer.

and we assigned names to certain important circuit nodes. We also learned how to use transient simulation: How to set up the simulation timing parameters and how to plot time function simulation results. In order to analyze the results, some precise measurements are sometimes needed. We learned how to use markers on the obtained plots.

Very often, one can save some time by building upon an existing schematic that resides in the same workspace.

Example 1.2

Let the next assignment be to design a non-inverting amplifier with a voltage gain of 14. The design will be tested in transient simulation. Again, we need one μA741 op-amp, two DC supplies, three resistors, and an input signal source.

We look for our workspace in the ads_work_files folder. We open our MyWork3300_wrk workspace. See Figure 1.44.

Let us use the inverting amplifier schematic as the starting point. First, let us File> "Save as" the inverting amplifier schematic under a new name "OpAmp_Noninverting_Amp_μA741". The process is shown in Figure 1.45.

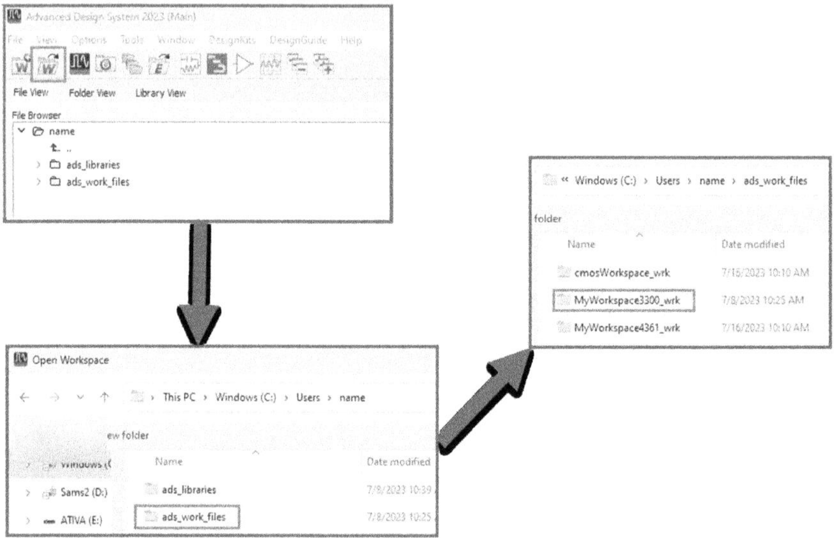

FIGURE 1.44 Finding the desired workspace.

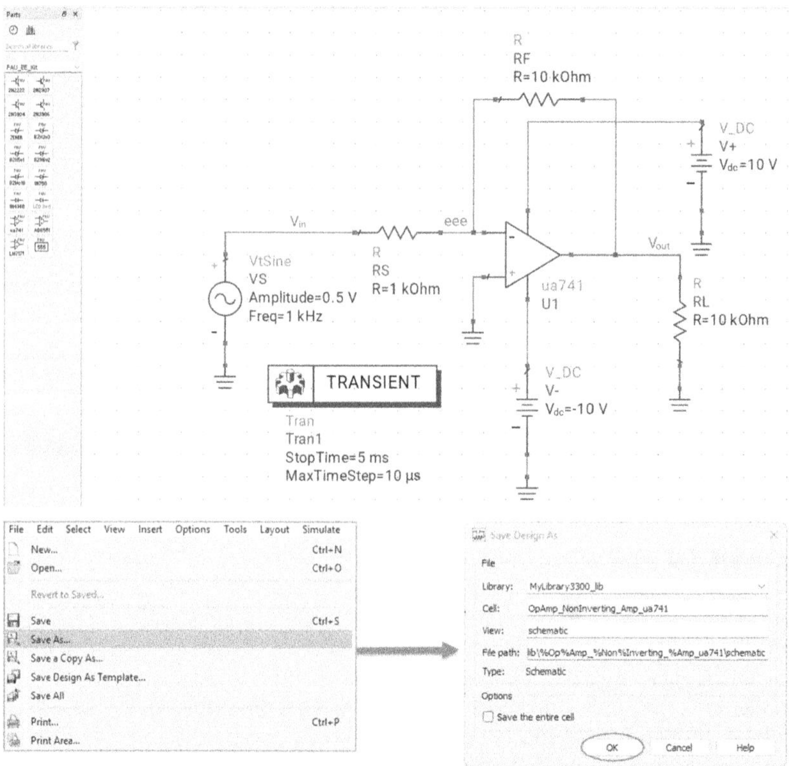

FIGURE 1.45 Saving the inverting amplifier schematics as a new non-inverting amplifier schematic.

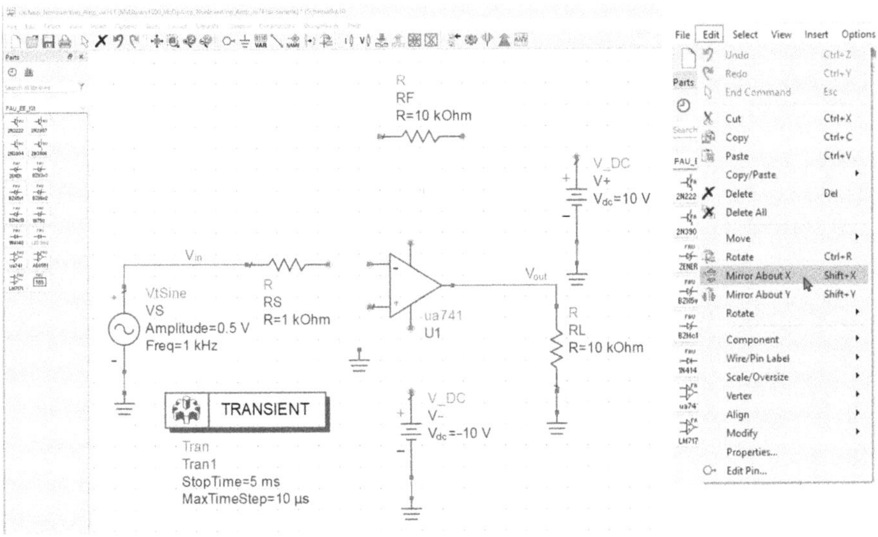

FIGURE 1.46 Op-amp's wires removed in preparation for the op-amp being flipped horizontally.

Next, we shall remove some wires and reorient the op-amp. First, we disconnect all the wires attached to the op-amp. Then we use the menu selection "Edit>Mirror About X Axis" to flip the μA741, such that its+input goes on top, and – input goes to the bottom. The supply ports flip too: Positive to the bottom and negative to the top. As we click on the mirror editing, a ghost cross hair is created. We move it to the 741 op-amp, and we click the 741 to flip it. The op-amp, stripped of its wires, is shown in Figure 1.46. The horizontally flipped op-amp is shown in Figure 1.47.

The changes, from the inverting amplifier's schematic, are displayed in Figure 1.47. The reader is invited to try to figure out all the other changes that were made to the inverting amplifier schematic: Moving and rotating components, moving component text, value changes, etc.

Remember to save your work frequently.

The reason for horizontally flipping the op-amp is to make the schematic mimic the full circuit diagram (Figure 1.41) for easier debugging. In Figure 1.46, contrasted with Figure 1.47, the flipping of the op-amp is just one of several modifications (changing respectively the signs of each DC supply voltage, moving the resistors to tie with the op-amp's output and the "– input" of the op-amp, changing the feedback resistor value such that the ratio between R_2 and R_1 will be 13:1, in order to provide a voltage gain of 14, and changing one of the nodes' names, for reasons that will become clear soon). The schematic diagram of Figure 1.47 is ready for simulation.

Transient settings stay the same as the input signal stays the same. With an input amplitude of 0.5 V and a voltage gain of 14, the amplitude of V_{out} is expected to be 7 V.

To observe the "virtual ground effect", let us also watch the signal Vminus (which is expected to be identical to V_{in}).

Note that with Insert>Plot, we can actually select the placement and size of each plot on the data display page, viewed in Figure 1.48.

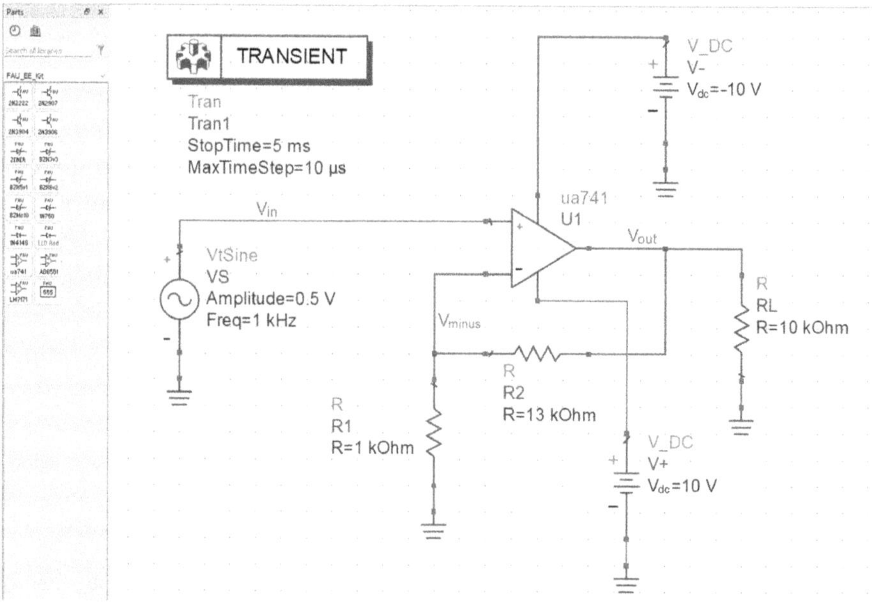

FIGURE 1.47 The completed non-inverting amplifier schematic.

Let us comment on the results. As expected, $V_{out}(t)$ is non-inverted and has a ptp amplitude of 14 V. The ratio of $ptp(V_{out})/ptp(V_{in})$ yields a gain of 14.

Note that $V^-(t)$ and $V_{in}(t)$ look the same. Will it be a good idea to observe the error signal $e(t) = V_{in}(t) - V^-(t)$?

Hint: Peak Marker. See Figure 1.49.

There is a better way though, involving the addition of measurement equations (MeasEqn) to the diagram. See Figure 1.50.

We are doing it to create a new variable, named eee (and note, which is already used as an ADS internal variable). The new variable will appear in the simulation results dataset. The variable eee is created by using the existing named nodes variables V_{in} and V^-.

Let us bring a "Measurement Equation" component from a simulation palette. Select any simulation palette, for example, simulation-transient. Select the MeasEqn block and drag and drop it on the schematic.

Note: See Section 1.9 for another method to instantiate a MeasEqn block.

By double-clicking on the MeasEqn block, we shall open a dialog box for defining eee, and also for getting rid of the dummy equation Meas1 = 1. See Figures 1.51 and 1.52.

We write the eee variable expression over the default expression. Click Apply to update the expression.

The Add button allows us to add more measurement expressions whenever needed.

OK button applies the changes and closes the window.

Now, when we simulate again, the variable eee will be added to the dataset, and we can use it with the Insert> Plot menu (Figures 1.53 and 1.54).

See ADS Tips and Tricks 1B (Section 1.10) on how to Add a Plot Title and Label Axes.

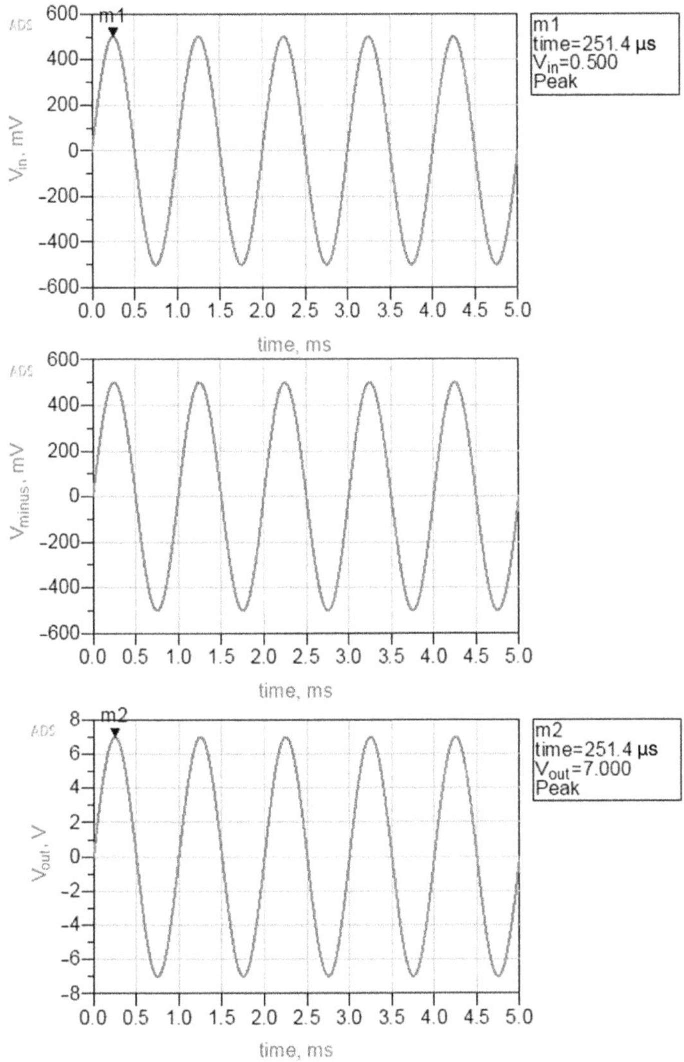

FIGURE 1.48 Simulation plots for the non-inverting amplifier circuit.

If no units are shown on the plot of a voltage signal, then it must mean "Volt". We see that eee(t) amplitude is around 7 mV. So, V^- and V_{in} are not identical, but the error is practically small (as we compare 7–500 mV) (Figure 1.54).

1.9 ADS TIPS AND TRICKS 1A: MEASEQN INSTANTIATION USING LIBRARY BROWSER

Let us bring a "Measurement Equation" component, using the Component Library.

The list of components shown in Figure 1.56 is very long. Therefore, above the Component column, we begin to search by keying in the first few letters.

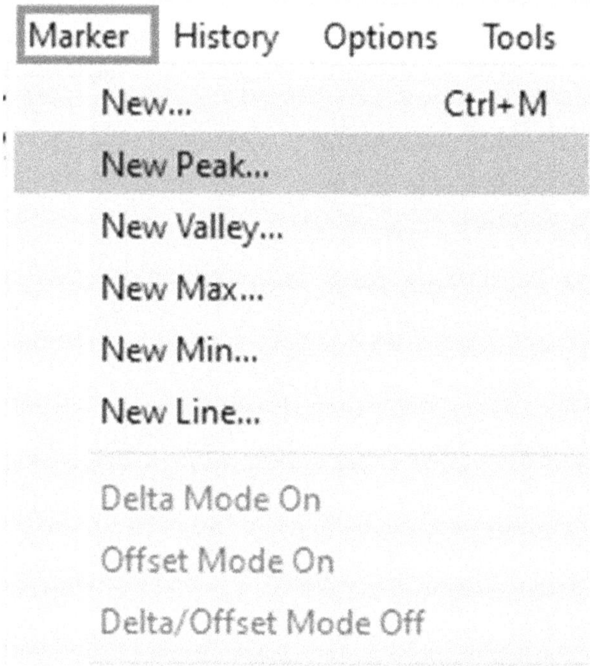

FIGURE 1.49 Invoking new peak markers for precision measurements on the plots.

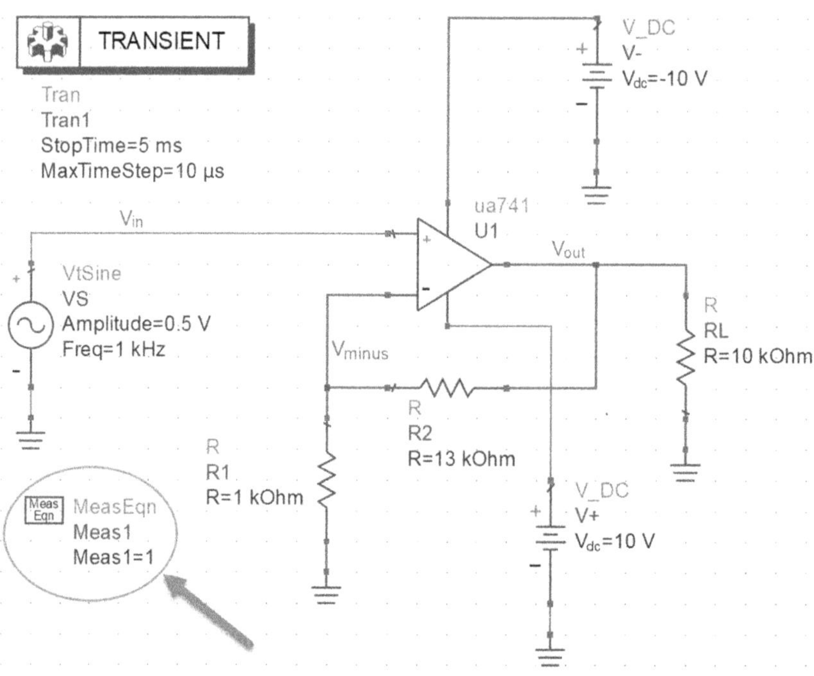

FIGURE 1.50 A measurement equation block added to the schematic.

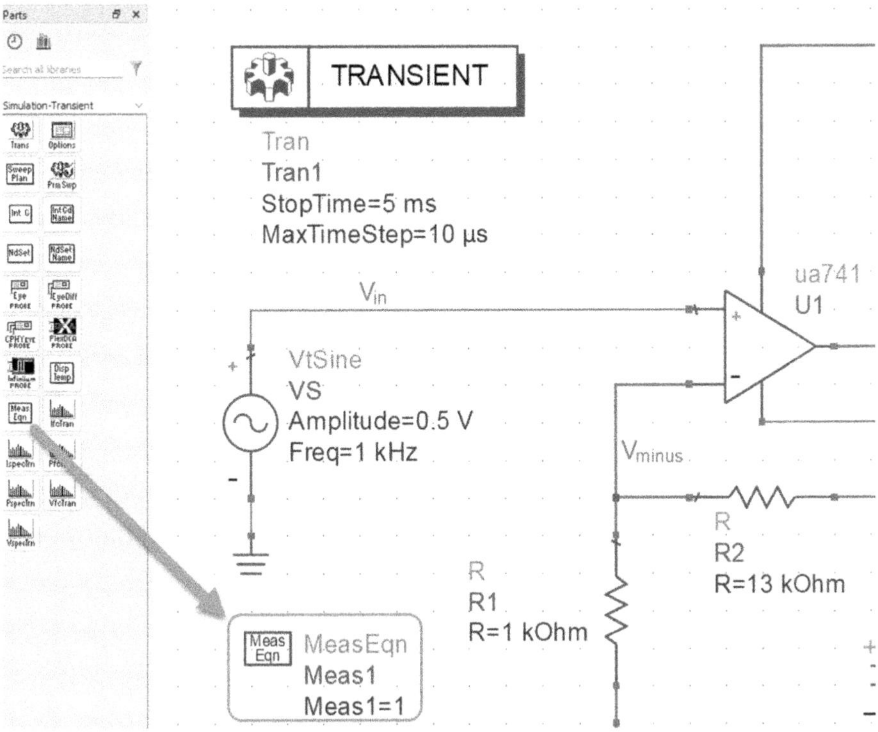

FIGURE 1.51 Adding the MeasEqn “component” to the diagram.

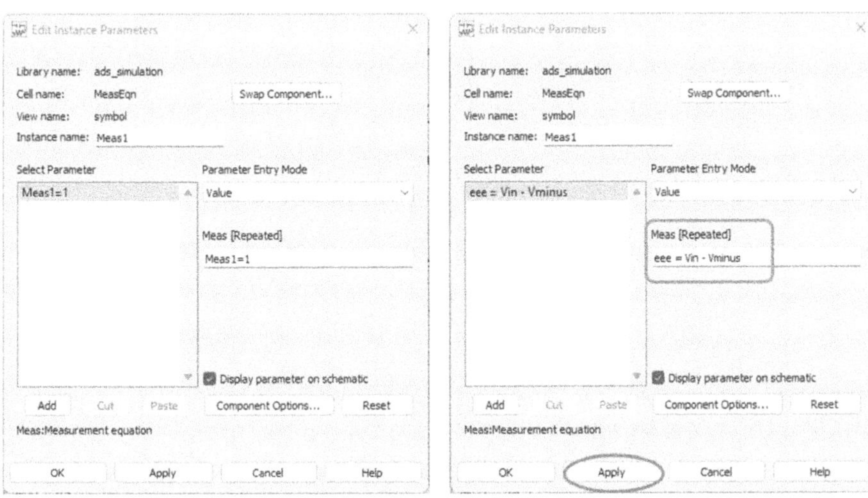

FIGURE 1.52 (Left) Initial view of the MeasEqn dialog box window. (Right) Final view of the MeasEqn dialog box window.

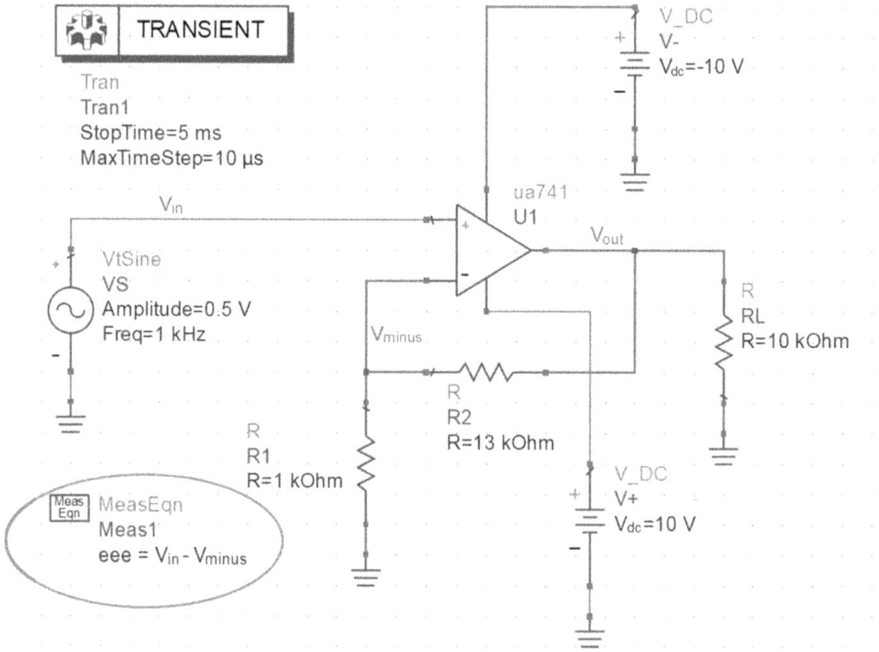

FIGURE 1.53 The updated schematic (with the eee variable) of the non-inverting amplifier.

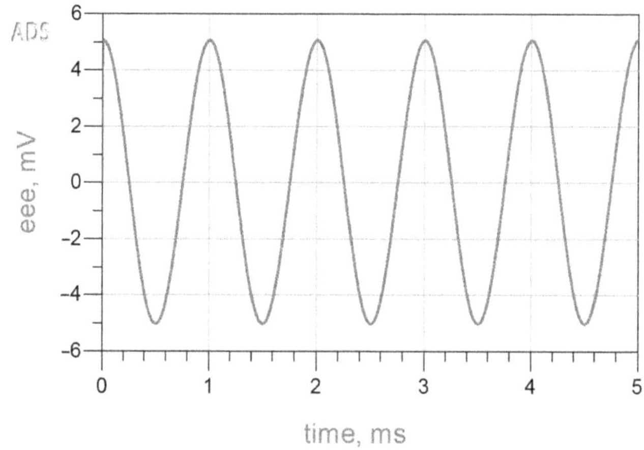

FIGURE 1.54 Plot of the added variable eee(t).

As we type the information, a segment of the list of components shows up.

We left-click on the top MeasEqn and "Place component". Finally, we minimize the component library.

Working in ADS sometimes we open many windows.

To close all these windows with one simple action, go to the Main ADS window and select "File>Close All".

A similar simple action exist to save all open windows, very handy to save all your work at once. Again, from the Main ADS window now select "File>Save All".

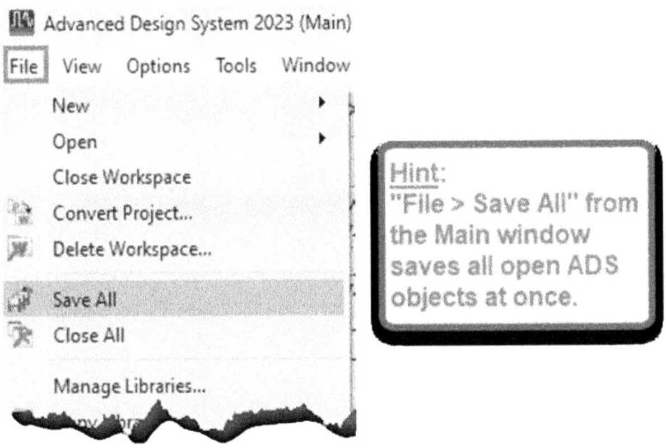

FIGURE 1.55 Saving ADS objects.

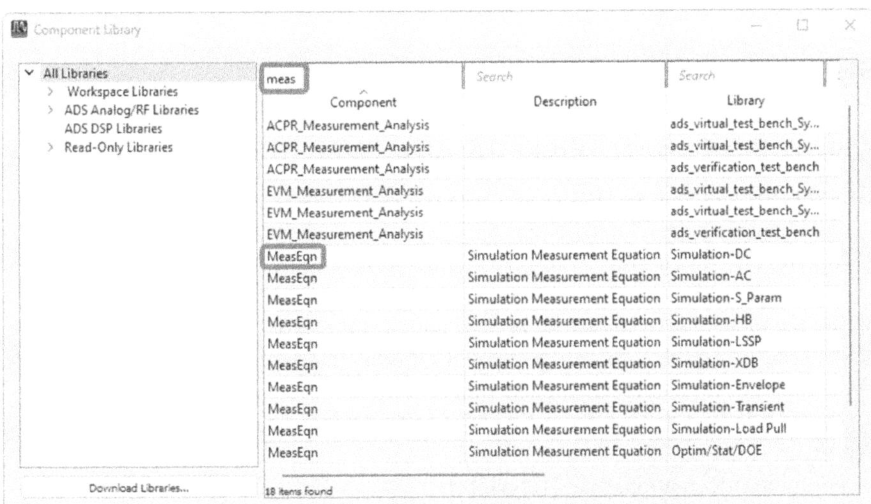

FIGURE 1.56 View of the component library after typing "meas".

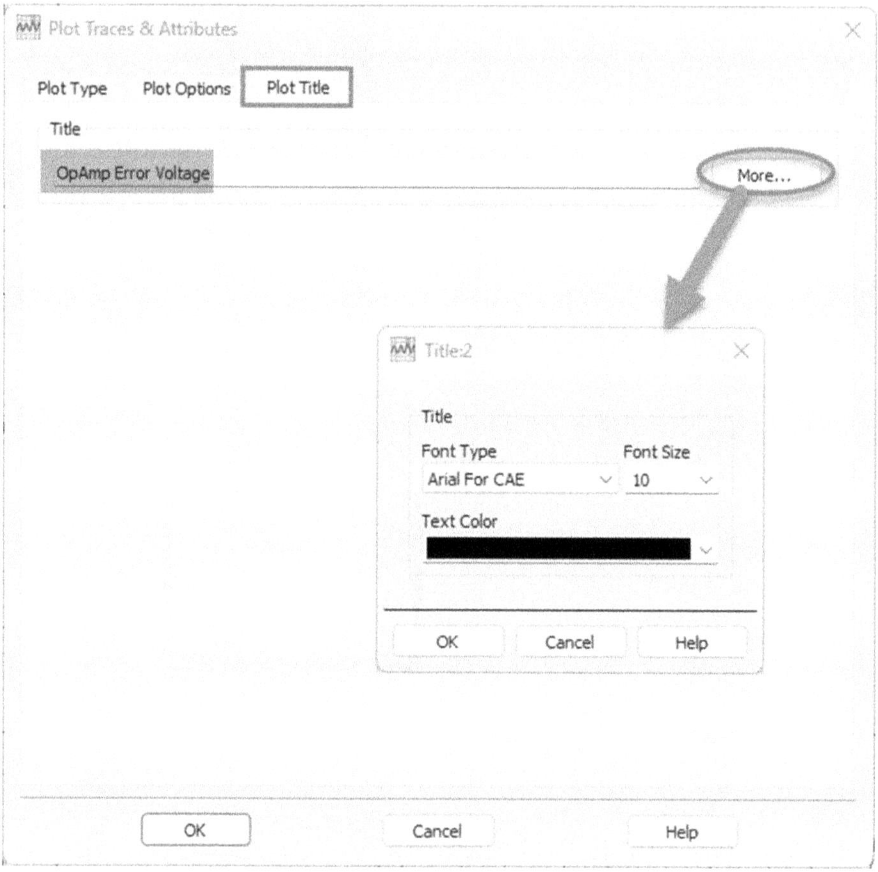

FIGURE 1.57 Adding a title to a plot.

1.10 ADS TIPS AND TRICKS 1B: ADD PLOT TITLE AND LABEL AXES

To add a title to a plot, we double-click on the plot to bring up the properties window.

We then select the Plot Title tab.

We enter the title in the corresponding field. See Figure 1.57.

We noted earlier that the eee expression did not have units; a more common type of plot uses legends for the variables and units for the axes.

As seen in Figure 1.58, we go to Plot Options and select (say) the Y axis. Add "Volts" in the Axis label field. Then we click OK.

To add a legend to our plot, we right-click on the plot and select "Insert Legend". See Figure 1.59.

For more information, read [Keysight1].

1.11 LINEAR COMBINERS

A simple extension of the inverting amplifier circuit is the addition of two or more input signals. The goal is to create a weighted summation of the amplified signals.

FIGURE 1.58 Adding units to the axes of a plot.

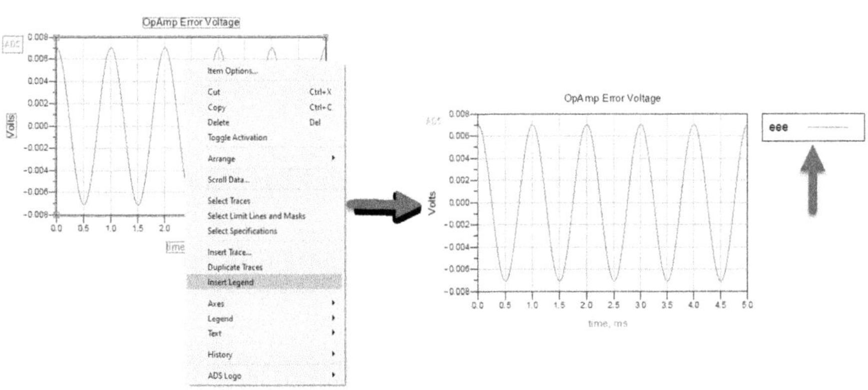

FIGURE 1.59 Adding a legend to the plot.

The example shown in Figure 1.60 involves two independent signal sources $v_{s1}(t)$ and $v_{s2}(t)$. Each input signal is connected to the "virtual ground" node via a resistor (R_{s1} and R_{s2}, respectively).

Two input currents are created: $v_{s1}(t)/R_{s1}$ and $v_{s2}(t)/R_{s2}$. The two currents combine into a single current that flows into resistor R_f and then into the output node of the op-amp, as reflected by equation (1.3a). The voltage drop of R_f sets the linear combiner's output signal $v_o(t)$ to be a weighted sum of the two inputs, with weights that are determined by the three resistors, as shown in equation (1.3b).

$$v_o(t) = -R_f\left(\frac{v_{s1}(t)}{R_{s1}} + \frac{v_{s2}(t)}{R_{s2}}\right) \tag{1.3a}$$

$$v_o(t) = -\frac{R_f}{R_{s1}}v_{s1}(t) - \frac{R_f}{R_{s2}}v_{s2}(t) \tag{1.3b}$$

The output voltage of the linear combiner consists of the superposed inverted amplified outputs of each individual input signal. Readers are referred to [Sedra1] for more information.

1.12 ADS TRANSIENT SIMULATION OF LINEAR COMBINERS

Example 1.3

Let us create a simulation of the two-input linear combiner shown in Figure 1.60, assuming that $v_{s1}(t)$ is a 1 KHz AC signal with an amplitude of 0.5 V, and that $v_{s2}(t)$ is a 2 KHz AC signal with an amplitude of 1 V and a phase shift of −75°. Input signal $v_{s1}(t)$ is to be amplified by a factor of −5. Input signal $v_{s2}(t)$ is to be amplified by a factor of −2. In other words, the two signals are to be inverted and combined by a ratio of 5:2. It is assumed that the two op-amp DC supplies are ±10 V.

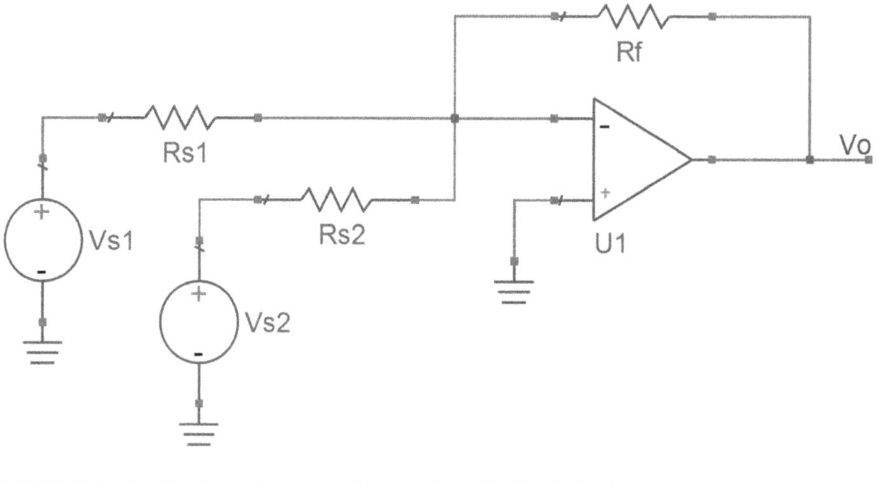

FIGURE 1.60 Two-input linear combiner, also called weighted summer.

Inverting and non-inverting amplifiers operate linearly as long as the output signal does not attempt to rise above any of the DC supply levels. That is, if the input signal's amplitude and the voltage gain are such that their product exceeds one of the DC supply levels, then $v_o(t)$ "clips" at a level that is slightly below the supply voltage level. Voltage clipping destroys the linear operation of the amplifier. The same may happen in a linear combiner. Let us do a quick check that there is no clipping in this example. The first input signal, of amplitude 0.5 V and voltage gain of −5, creates a 1 KHz component of $v_o(t)$ of amplitude 2.5 V. The second input signal, of amplitude of 1 V and voltage gain of −2, creates a 2 KHz component of $v_o(t)$ of amplitude of 2 V. In the worst case, when the two output components become synchronized, the combined effect will be an output signal of an amplitude of 4.5 V, well below the supply levels. Actually, the two output components are not synchronized, so each attains its peak at a different time, most of the time.

Again, the starting point for the schematic construction will be the schematic for an inverting amplifier. See Figure 1.34. All we do is to add one more signal source and one more resistor (Figure 1.61).

Need to find the workspace containing the inverting amplifier schematic, open it, and "Save as" for another name related to the linear combiner. See Figure 1.62.

Recall that it has been suggested to use the "C:\Users\name\ads_works_files" folder to save our workspaces. If one used a different folder, needs to go to that other folder.

As seen in Figure 1.63, the design is tested in transient simulation.
The only new ADS feature is the determination of the correct transient simulation parameters (Stop Time, Max Step Size), now that there are multiple signals with different frequencies.

As will be seen:

- The slowest signal determines the Stop Time.
- The fastest signal determines the Max Step Size.

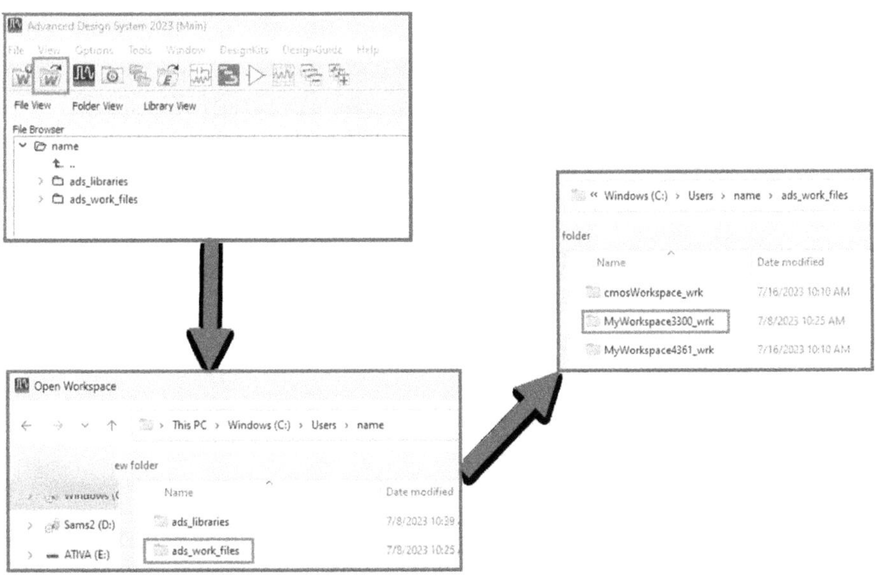

FIGURE 1.61 Opening a desired workspace (a repetition of Figure 1.44).

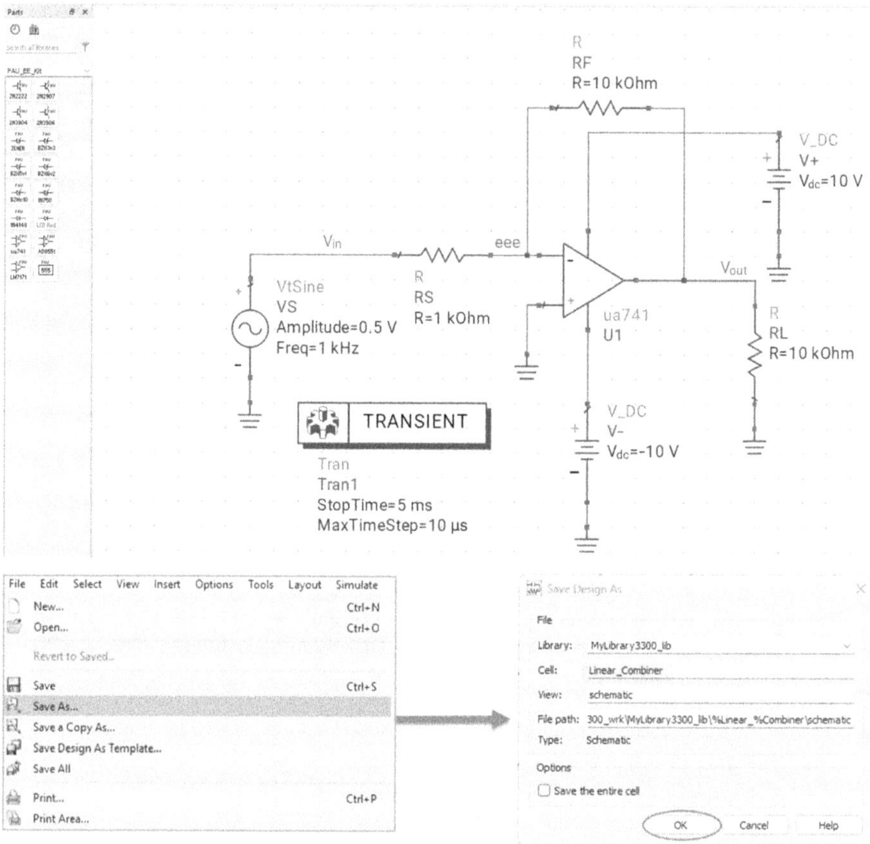

FIGURE 1.62 Saving the inverting amplifier's schematic as a new schematic named "Linear_Combiner", in the same workspace.

There are now two input signals. V_{S1} is AC with 0.5 V amplitude and 1 kHz frequency. V_{S2} is AC with 1 V amplitude, 2 kHz frequency, and –75° phase angle. See Figure 1.63.

The slower signal (1 kHz) determines the Stop Time = 3 periods (3 ms). The faster signal (2 kHz) determines the MaxTimeStep, using 1/100 of the 0.5 ms period, which is 5 μs. See Figure 1.63.

Note that to amplify V_{S1} by a factor of –5 we chose R_{S1} = 1 kΩ and R_F = 5 kΩ. To amplify V_{S2} by a factor of –2, as R_F = 5 kΩ, we have to choose R_{S2} = 2.5 kΩ. Again, see Figure 1.63. The resulting plots are in Figure 1.64.

1.13 VOLTAGE-LEVEL SHIFTERS

An important particular case of a two-input linear combiner is that of an arbitrary input signal $v_s(t)$ that is linearly combined with a DC input signal V_R. Such a circuit is called a voltage-level shifter.

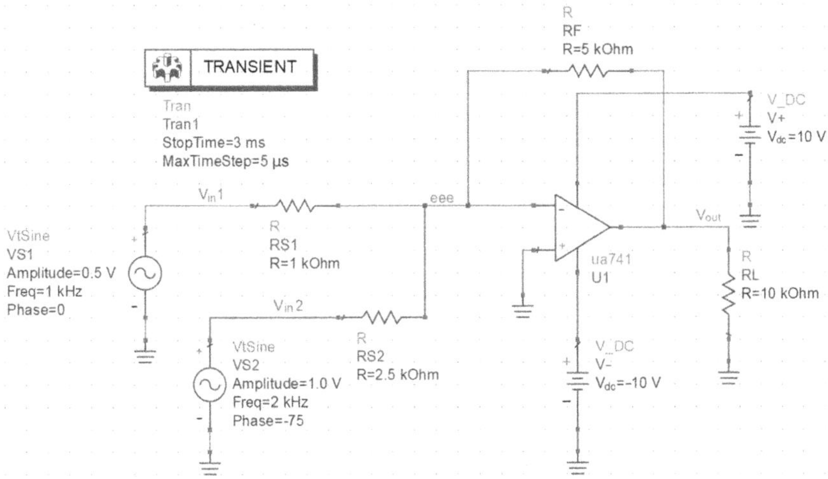

FIGURE 1.63 The complete schematic of the linear combiner circuit.

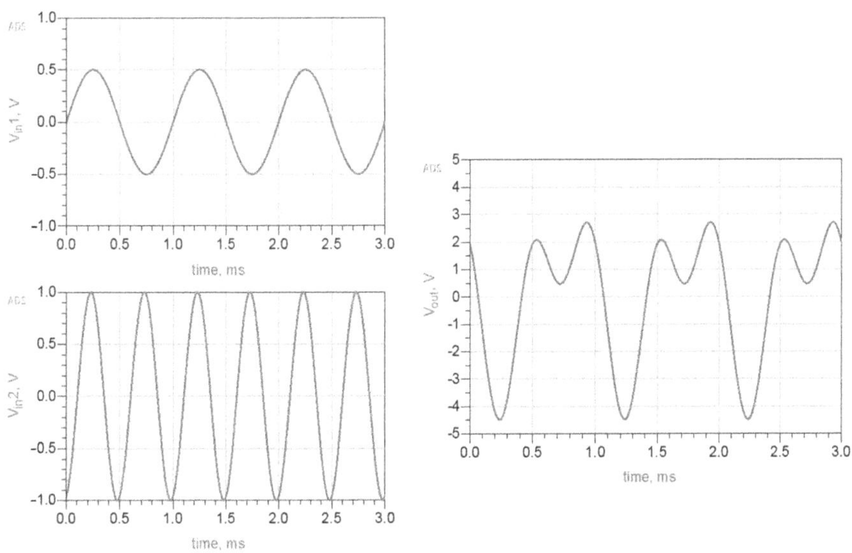

FIGURE 1.64 Linear combiner simulation results.

Applying the linear combining equation (1.3b) to the circuit shown in Figure 1.65 yields:

$$v_o(t) = -\frac{R_f}{R_s} v_s(t) - \frac{R_f}{R_1} V_R \tag{1.4}$$

The first term of $v_o(t)$ is the result of the amplification of $v_s(t)$ by a factor of $-R_f/R_s$. The second term $-(R_f/R_1)\cdot V_R$ is a DC signal that may cause the amplified term to move up or down, as shown in Figure 1.66.

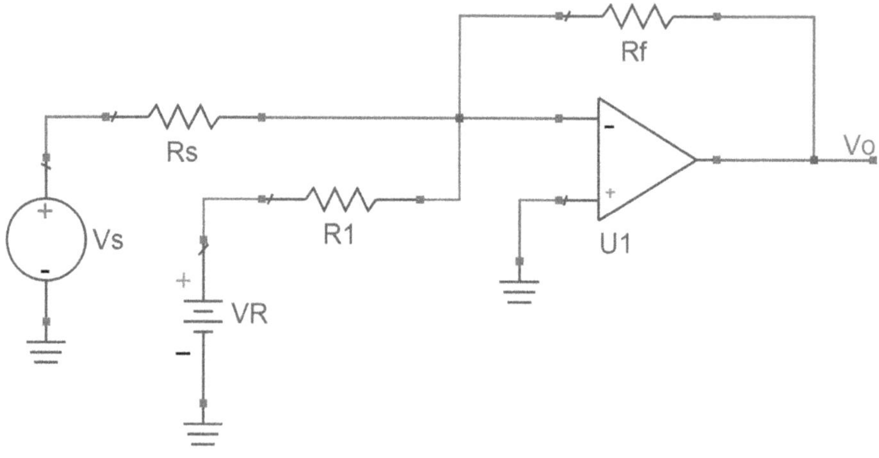

FIGURE 1.65 Voltage-level shifter.

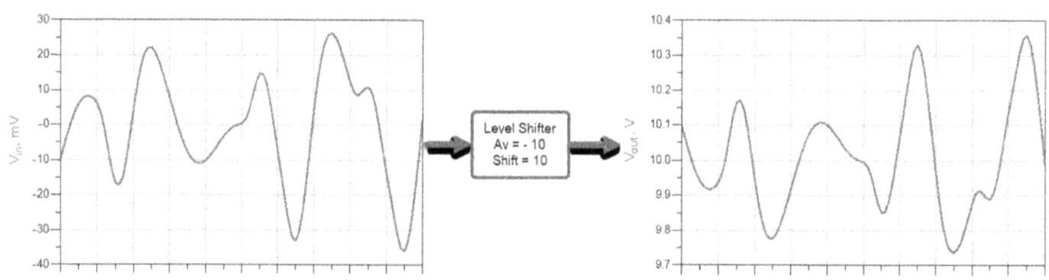

FIGURE 1.66 Voltage-level shift consists of voltage amplification (A_v) combined with DC shifting.

Mathematically speaking, a voltage-level shifting circuit is characterized by four input–output voltage values: v_{smin}, v_{smax}, v_{omin}, and v_{omax}. The output $v_o(t)$ can be an inverted, or non-inverted, replica of the input $v_s(t)$. For an implementation that is based on the linear combiner circuit (of Figure 1.63), the output signal is an inverted version of the input, and it preserves the input's shape.

The following numerical example will add more insight into the engineering significance of voltage-level shifting.

Example 1.4

Analog-to-Digital Conversion of Sensory Information

A temperature sensor creates a voltage that runs between $v_{smin} = 30\,mV$ and $v_{smax} = 50\,mV$ over the range of measured temperatures. The sensor voltage is sent to an unsigned 8-bit analog-to-digital converter (ADC). The ADC accepts input signals in the range of 0–5 V. Need to implement a voltage-level shifting circuit, shown in Figure 1.65, taking supplies of ±10 V.

Let us explain first why a voltage-level shifter is needed. There are $2^8 = 256$ voltage level "strips" (located between 0 and 5 V). Say that the lowest strip is assigned an 8-bit 00…0 binary number, and the highest strip is assigned 11…1. We see that every 19.53 mV, the binary representation of the digitized signal must be incremented. An input signal range of 30–50 mV can be represented by two binary numbers only. In order to utilize the full dynamic range of the ADC, a voltage-level shift must be done.

The first step in the design of a voltage-level shifter is the determination of the needed voltage gain. We recall that a voltage gain can be measured from the ratio of the output voltage's ptp amplitude and the input's ptp amplitude. In our example, the absolute value of the voltage gain A_V is therefore computed as $|A_V| = |v_{omax} - v_{omin}|/|v_{smax} - v_{smin}| = |5 - 0|/|0.05 - 0.03| = 250$. In other words, $v_s(t)$ that has a range of [0.03, 0.05] V must be stretched, maintaining its shape, to a new range of width of 5 V. Referring to the implementation diagram shown in Figure 1.65, the first step of the design allows the designer to select the resistors R_f and R_s. Say $R_s = 1$ kΩ and $R_f = 250$ kΩ.

The second step in the design of the voltage-level shifter is the determination of the required amount of DC level shifting of the output signal $v_o(t)$. By assuming that $V_R = 0$, the voltage-level shifting circuit is reduced to an inverting amplifier circuit. Let us observe how the extreme input signal levels, v_{smin} and v_{smax}, are mapped to corresponding levels of the output signal. A level of $v_s = v_{smin} = 0.03$ V, under a voltage gain of −250, is mapped to a value of −7.5 V. Likewise, $v_s = v_{smax} = 0.05$ V is mapped to a level of −12.5 V. We see that, because the voltage-level shifter contains an inverting amplifier, it is v_{smin} that is mapped to v_{omax}, whereas v_{smax} is mapped to v_{omin}. We further see that the output levels obtained when $V_R = 0$ need to be DC level shifted upward: The level of −7.5 V is shifted to a level of $v_{omax} = 5$ V, whereas the level of −12.5 V is DC shifted to $v_{omin} = 0$. Both cases indicate that the amount needed of DC shifting is +12.5 V.

The final step in the design is the implementation of the DC level shifting. We have not yet selected the values of V_R and R_1. By equation (1.4) the DC shift component of $v_o(t)$, that is created by V_R and R_1, is equal to $-V_R \cdot (R_f/R_1)$. The DC source V_R can take any value, subject to a single constraint: If the needed DC shift is positive (that is, upward), then V_R must be negative. If the DC shift is downward, then V_R must be positive. Since we already have two DC supplies, V^+ and V^-, that power the op-amp, it is efficient to have V_R created by one of the existing supplies. The design is completed by selecting $V_R = -10$ V and $R_1 = 200$ kΩ, a value that creates a positive DC shift of 12.5 V (Figure 1.67).

No new ADS skills arise from studying voltage-level shifting as a particular case of linear combining. Readers will have an opportunity to perform ADS design of voltage-level shifting, as part of addressing the end-of-chapter problems.

1.14 OP-AMPS WITH IMPEDANCES AND LOW-PASS FILTERS

Another extension of the inverting amplifier circuit is to replace the input and feedback resistors with impedances: R_s is replaced by Z_s, and R_f is replaced by Z_f.

An impedance consists of any simple or complicated linear network. For example, a mix of resistors, capacitors, and inductors. We keep in mind that in a circuit that consists of impedances, all the voltages and currents can be represented as the Laplace transform functions of a complex variable s of the counterpart time signals.

At this early stage of picking up some electronics basics, there is no need to delve deep into the theory of Laplace transform. Let us just present a list of facts to allow us to move on. See [Gable1].

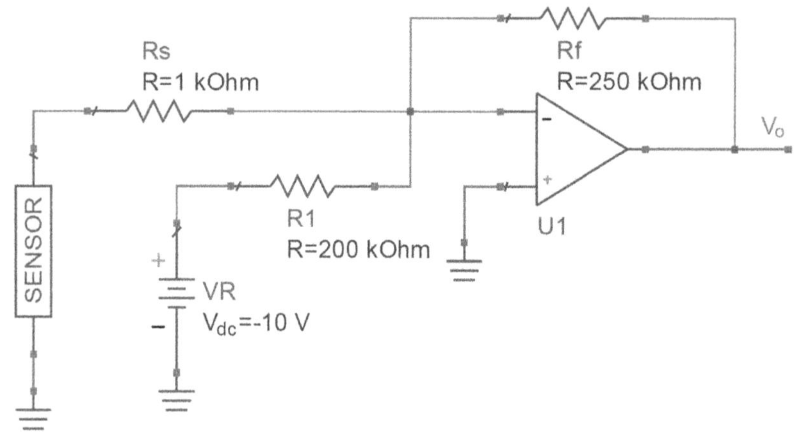

FIGURE 1.67 The completed design of the voltage-level shifting circuit.

a. Let i(t) and v(t) be, respectively, the time function current going into a component and the voltage drop of the component. Let I(s) and V(s) be the Laplace transforms of i(t) and v(t), respectively. Then the component's impedance is **Z(s) = V(s)/I(s)**. This is "Ohm's Law" for impedances. The unit of any impedance is Ω (Ohm)
b. The variable **s is a differentiation operator**. That is, if {v(t), V(s)} form a Laplace transform pair, then the time function that corresponds to s·V(s) (that is, the inverse Laplace transform of s·V(s)) is equal to dv(t)/dt.
c. The impedance Z_R of a resistor of resistance R is $Z_R = R$.
d. **The impedance Z_C of a capacitor C is $Z_C = 1/sC$**. This comes directly from the capacitor's time domain current–voltage relationship $i_c(t) = C \cdot dv_c(t)/dt$.
e. Like (d), the impedance Z_L of an inductor is $Z_L = sL$.
f. Let X(s) (a voltage or a current) be an external input of an impedance network. Let Y(s) be one of the outputs of the network. Then the transfer function T(s) between X(s) and Y(s) is defined as T(s) = Y(s)/X(s).
g. In (f) above, if we set s = jω then we call T(jω) the Frequency Response of T(s). T(jω) is a complex function of the frequency ω. T(jω) has a magnitude |T(jω)| and a phase arg[T(jω)].

The reader is referred to [Sedral]. Let us refer again to Figure 1.68. The extension (1.5) of the inverting amplifier voltage gain is that the transfer function $V_O(s)/V_S(s)$ is given in terms of the input impedance $Z_S(s)$ and the feedback impedance $Z_f(s)$, as follows:

$$T(s) = \frac{V_O(s)}{V_S(s)} = -\frac{Z_f(s)}{Z_S(s)} \tag{1.5}$$

The derivation of equation (1.5) is done following exactly the same ideas as those used for deriving the voltage gain of an inverting amplifier: A Laplace transform of the input current is $V_S(s)/Z_S(s)$. The current flows into the virtual ground node. Because of the infinite input impedance of the op-amp, the input current flows into

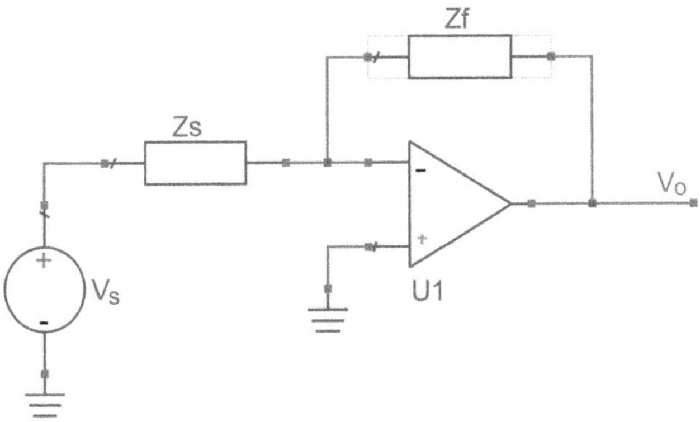

FIGURE 1.68 An inverting amplifier with input and feedback impedances.

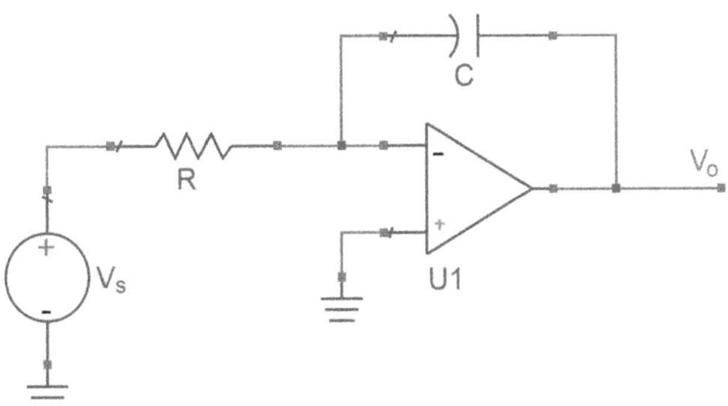

FIGURE 1.69 An op-amp integrator.

the feedback impedance $Z_f(s)$. The Laplace transform $V_O(s)$ is obtained from the voltage drop of the feedback impedance.

It is straightforward to perform the calculation of the transfer function T(s) for some given input and feedback impedance networks. It is more difficult though to understand what the resulting T(s) means in the time domain. Let us observe several simple examples of transfer functions T(s).

Example 1.5

An Op-Amp Integrator (Figure 1.69).

With $Z_S = R$ and $Z_f = 1/sC$, the transfer function $T(s) = 1/sRC$. The output $v_o(t)$ is the integral of $v_s(t)$ (with a gain of 1/RC) added on top of the initial capacitor's voltage $v_o(0)$. Any practical implementation of the integrator circuit must include an arrangement for an initial charging of the capacitor and a resetting mechanism. We shall not discuss the subject of integrators any further.

Can a choice of $Z_S = 1/sC$ and $Z_f = R$, yielding $T(s) = sRC$, be a basis for creating an op-amp differentiator? Well, no. The attempted differentiation interferes with the op-amp's internal stabilization, a topic that will be studied much later on in Chapter 6.

Example 1.6

A simple RC low-pass filter.

In the RC circuit, shown in Figure 1.70 and analyzed using impedances, the input voltage is $V_S(s)$ and the output, chosen to be $V_o(s)$, is the capacitor's voltage. Let us initially replace the load resistance R_L with an open circuit.

$V_o(s)$ is obtained from $V_S(s)$ by voltage division between R and 1/sC. The resulting transfer function is $T(s) = V_o(s)/V_S(s) = 1/(1 + sRC)$. Many questions come to mind: What does this s-function mean in the frequency domain, if we set $s = j\omega$? Why does the circuit work as a low-pass filter? What does this s-function mean in the time domain?

The frequency response is $T(j\omega) = 1/(1+j\omega RC)$. The magnitude frequency response is $|T(j\omega)| = 1/(1 + \omega^2R^2C^2)^{1/2}$. The radian frequency ω (in [rad/s]) and the frequency f (in [Hz]) are related through $\omega = 2\pi f$. An equivalent magnitude frequency response expression (in terms of f) can be written as in equation (1.6):

$$|T(f)| = \frac{1}{\sqrt{1+\left(\frac{f}{f_c}\right)^2}} \quad \text{where } f_c = \frac{1}{2\pi RC} \tag{1.6}$$

The frequency f_c is called the **cutoff frequency**. At an input frequency f equal to $f = f_c$, we obtain a magnitude of $|T(f_c)| = 2^{-1/2} \approx 0.707$.

Why is the frequency response equation (1.6) referred to as a "low-pass filter"? The best way to observe it is to plot |T(f)| against log(f) as shown in Figure 1.71:

We see that |T(f)| has two distinct regions: The low-frequency range in which the magnitude is close to 1, and the higher than $f = f_c$ region in which the magnitude is rapidly dropping as f increases. The filter "passes" AC signals of low frequency, with very little attenuation, and it attenuates AC signals of high frequency.

Equation (1.6) displays a "design" formula, the expression for f_c. That is, the setting of $f = f_c$ is user-defined via selection of R and C values. The formula for |T(f)| is a "performance" expression. For a given cutoff frequency f_c, the expression for |T(f)| displays how the filter attenuates input signals of any frequency f.

So, what is the time domain behavior of the RC network low-pass filter? If we test the filter with an input signal $v_s(t)$ that is an AC signal of frequency f, then if f is lower than f_c, the filter passes the input signal. That is, $v_o(t) \approx v_s(t)$. If f is larger than f_c then

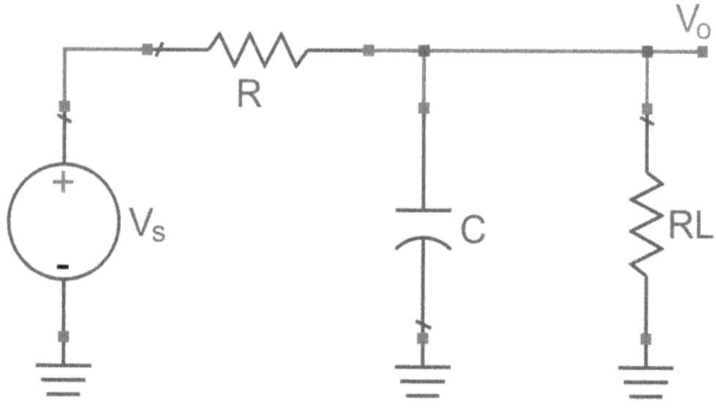

FIGURE 1.70 A simple RC network acting as a primitive low-pass filter.

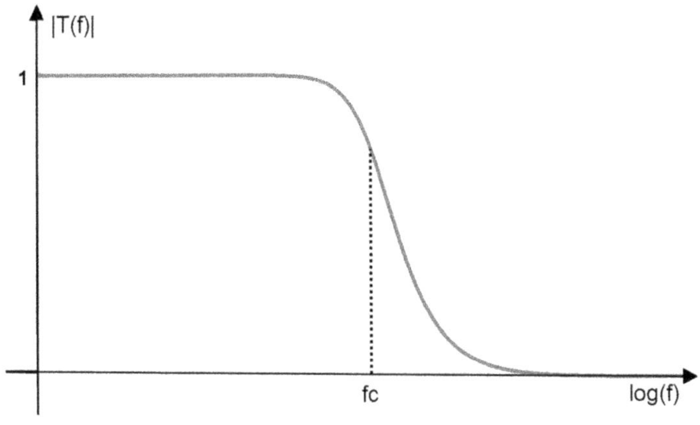

FIGURE 1.71 Magnitude frequency response, on a log scale, of an RC circuit.

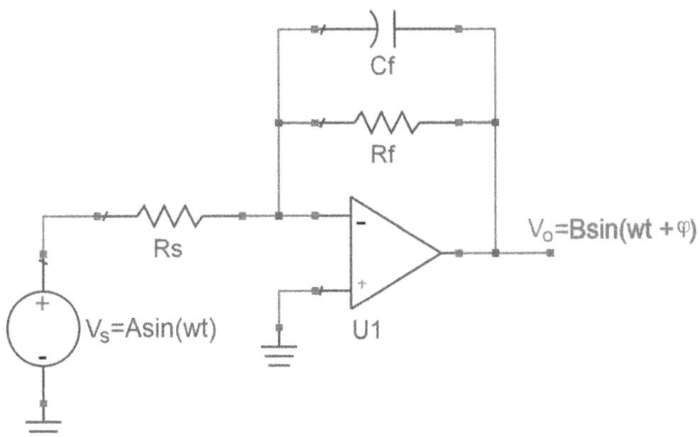

FIGURE 1.72 Op-amp low-pass filter.

the amplitude of the AC signal $v_o(t)$, which is of course of the same input frequency f, becomes small. If we test the filter with a signal that consists of many simultaneous AC components, then all the low-frequency components pass and all the high-frequency components become small. This is the essence of low-pass filtering.

In Example 1.7, we shall look at an op-amp low-pass filter. The reason why an op-amp low-pass filter is considered superior to the RC low-pass filter is because of the role played by the load resistor R_L. In the RC network, if we attach a load resistor R_L parallel to C (as shown in Figure 1.70), the value of R_L affects the filter's design. It can be shown that the filter's gain changes from 1 to $R_L/(R+R_L)$ and the filter's cutoff frequency f_c becomes $f_c=1/2\pi(R||R_L)C$. The op-amp filter, on the other hand, performs almost independently of any load resistance R_L.

Example 1.7

An Op-Amp First-Order Low-Pass Filter

Much, if not all, of the relevant theory for Example 1.7 has already been covered in Example 1.6 (Figure 1.72).

We apply $Z_S = R_S$ and $Z_f = R_f||1/sC_f = R_f/(1 + sR_fC_f)$ to equation (1.5). The resulting transfer function is $T(s) = V_O(s)/V_S(s) = -(R_f/R_S)/(1 + sR_fC_f)$. In other words, the transfer function's denominator represents "low-pass filtering" action with a cutoff frequency of $f_c = 1/2\pi R_fC_f$. The transfer function's numerator represents an inverting amplification action, with a voltage gain of $-R_f/R_S$. Combining a filter with an amplifier makes little sense, because a typical filtering situation always involves signals that are considered "unimportant" (or "noise"). Noise signals should not be amplified if we intend to attenuate them later on.

If we let $R_S = R_f = R$, we get a transfer function that, aside from the inverting sign, is identical to the transfer function of the RC circuit of Example 1.6. The design and performance formulas of equation (1.6) apply to the op-amp filter as well.

Example 1.8

Let us discuss a typical example of the low-pass filter design process, using specific numerical values. We take the input signal $v_s(t)$ to be the sum of two AC signals: A "desired" AC signal $v_{s1}(t)$ of amplitude 1 V and frequency 100 Hz, and a "noise" AC signal $v_{s2}(t)$ of amplitude 1 V and frequency of 7 kHz. Need to attenuate the noise signal.

The key step in the design is to select the filter's cutoff frequency f_c. The selection of the cutoff frequency presents a tradeoff between the passing performance of desired signals and the attenuation performance of noise signals. The larger f_c is with respect to the desired signals' frequencies f, the closer to 1 are the gains for these desired signals, but the larger the gains for the noise signals. See Table 1.1.

The gain values were computed using the performance part of equation (1.6). A good compromise value for f_c seems to be $f_c = 300$ Hz. Once we select f_c, we use the design part of equation (1.6) in order to select the component values. Let $R_s = R_f = 1$ kΩ, then $C_f = 0.53$ μF, and we are ready for ADS simulation of Example 1.8.

1.15 ADS AC AND TRANSIENT SIMULATION OF A LOW-PASS FILTER

Again, the starting point for the schematic construction will be the schematic for the inverting amplifier.

TABLE 1.1
Candidate f_c Values for a Low-Pass Filter That Needs to Pass a 100-Hz Signal and Attenuate a 7-kHz Signal

Cutoff f_c [Hz]	Gain at f = 100 Hz	Gain at f = 7 kHz
100	0.707	0.014
200	0.894	0.029
300	0.949	0.043
400	0.970	0.057
500	0.981	0.071
600	0.986	0.085

The input resistor RS and the feedback resistor RF should be equal, as it makes no sense to build a filter that has a gain different than 1. See Figure 1.73.

We have to add a capacitor CF in parallel to RF. The component values were suggested in Example 1.8.

The design will be tested in AC simulation by observing the frequency response of the filter. We need to check that $f_c = 300\,\text{Hz}$ comes out as chosen. Then we shall test the design in the time domain, with the desired and the noise signal coming in simultaneously.

In AC simulation, we use a special input signal V_AC (see Figure 1.73).

Signal $v_{s1}(t)$ is an AC with amplitude 1 V and frequency 100 Hz. It is considered "desired". We wish that it passes the filter unattenuated. Signal $v_{s2}(t)$ is AC with

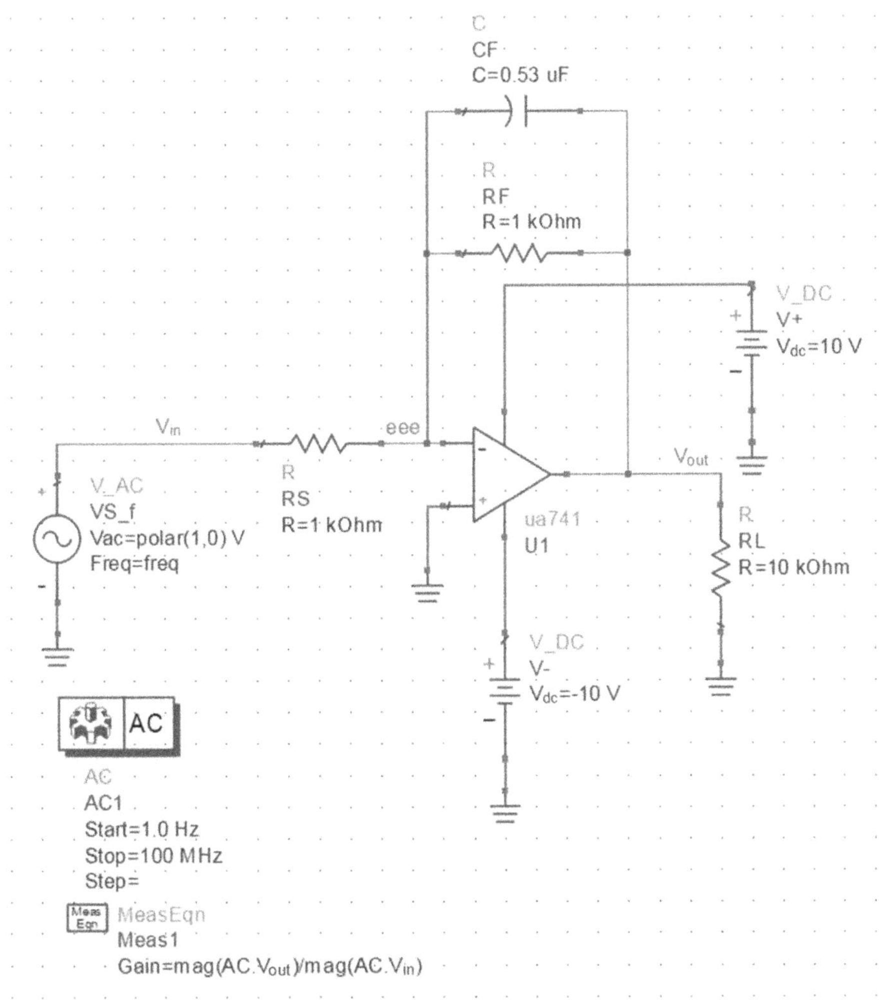

FIGURE 1.73 AC simulation of an op-amp low-pass filter.

amplitude 1 V and frequency 7 kHz. It is considered "noise". The filter should attenuate the noise signal.

The design of the filter consists of selection of the cutoff frequency f_c. The cutoff frequency should be larger than the desired signal's frequency and smaller than the noise's frequency. The selection of $f_c = 300\,Hz$ was explained at the end of the previous section. The larger is f_c from the desired f the better is the pass gain but the worse is the attenuation of the noise. The design tradeoff formulas are derived from equation (1.6):

$$f_{desired} < f_c = \frac{1}{2\pi \cdot R_F C_F} < f_{noise}$$

$$1 \approx \frac{\left|V_{out}\left(f_{desired}\right)\right|}{\left|V_{in}\left(f_{desired}\right)\right|} = \frac{1}{\sqrt{1+\frac{{f_{desired}}^2}{fc^2}}} >> \frac{\left|V_{out}\left(f_{noise}\right)\right|}{\left|V_{in}\left(f_{noise}\right)\right|} = \frac{1}{\sqrt{1+\frac{{f_{noise}}^2}{fc^2}}}$$

AC simulation is for studying the frequency response of a circuit. See Figure 1.73. The variable gain (defined as MeasEqn) is the ratio of the magnitude of the complex vector phasor of V_{out} and V_{in}.

The gain will be calculated at all the frequencies. Note that the MeasEqn uses the same V_{out} and V_{in} we used on the transient simulation, because these are the node names given in the schematic. For AC simulation, though these variables will become complex voltage phasors versus frequency. Adding "AC …" in MeasEqn, with the variables written as $AC.V_{out}$ and $AC.V_{in}$, is not strictly needed in stand-alone AC simulation. See Section 1.16 for mixed-mode simulation.

The setting of the AC simulation is shown in Figure 1.73. The setting of the plots is shown in Figure 1.74.

We shall always choose the Sweep Type to be "log" (and not linear). See Figure 1.74. Whenever one observes the frequency response of an amplifier, there will be many frequency decades: 1–10, 10–100, 100–1,000 Hz, etc.

Because of the selection $f_c = 300\,Hz$, we let the frequency response start at 1 Hz, so that we can watch part of the filter's flat gain region. Going to 100 MHz is probably not needed, as the 741 does not work right at such high frequencies.

An accurate frequency response plot can be obtained with Pts./decade of as low as 10. We chose 100 to increase the marker search resolution.

Again, as the simulation ends, we use Insert > Plot, but in AC simulation, there are additional steps, involving Plot Option (seen in Figure 1.75). We set the X axis (which is frequency) to a log scale. The Y axis (voltage gain) is kept linear. This is shown in Figure 1.75.

The resulting voltage gain frequency response plot is shown in Figure 1.76.

Note that the filter's pass band (which is flat with values near 1), and the Reject Band, starting a little prior to $f = f_c$.

How do we know that $f_c = 300\,Hz$ has been implemented correctly?

Need to measure with a marker.

After obtaining the frequency response plot, insert a marker from the toolbar. Place on the graph close to 300 Hz. Thereafter, you can move it along the graph.

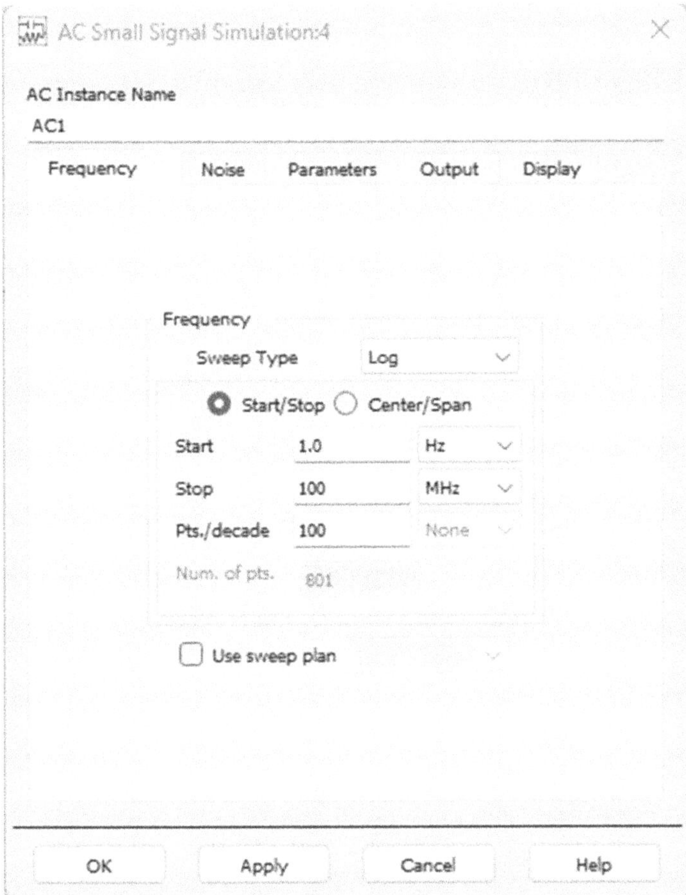

FIGURE 1.74 Dialog box obtained by double-clicking on the AC block. It is used to select the AC simulation settings.

It is convenient to move the marker from the keyboard by selecting it and using the ← and → arrow keys. With 10 points per decade the marker search resolution is somewhat poor (Figure 1.76). With 100 points per decade the 302 Hz result is the best that we could approach $f_c = 300\,\text{Hz}$ (see Figure 1.77). We see that $f_c = 300\,\text{Hz}$ has been implemented correctly.

Referring again to Figure 1.73, let us now add transient simulation to the schematic. We shall also deactivate the AC signals and the MeasEqn. There are two types of deactivation: Open circuit (as we choose to do soon) and short circuit. See Figure 1.78.

The deactivated objects and the added transient simulation objects are shown in Figure 1.79.

There is a need to set transient parameters according to the signals of 100 Hz (slow) and 7 kHz (fast). The period of the 100 Hz signal is 10 ms. Let us take two periods as the stop time. The period of the 7 kHz signal is 0.14 ms. One hundredth of 0.14 ms is 1.4 μsec. See the dialog box for transient simulation in Figure 1.80.

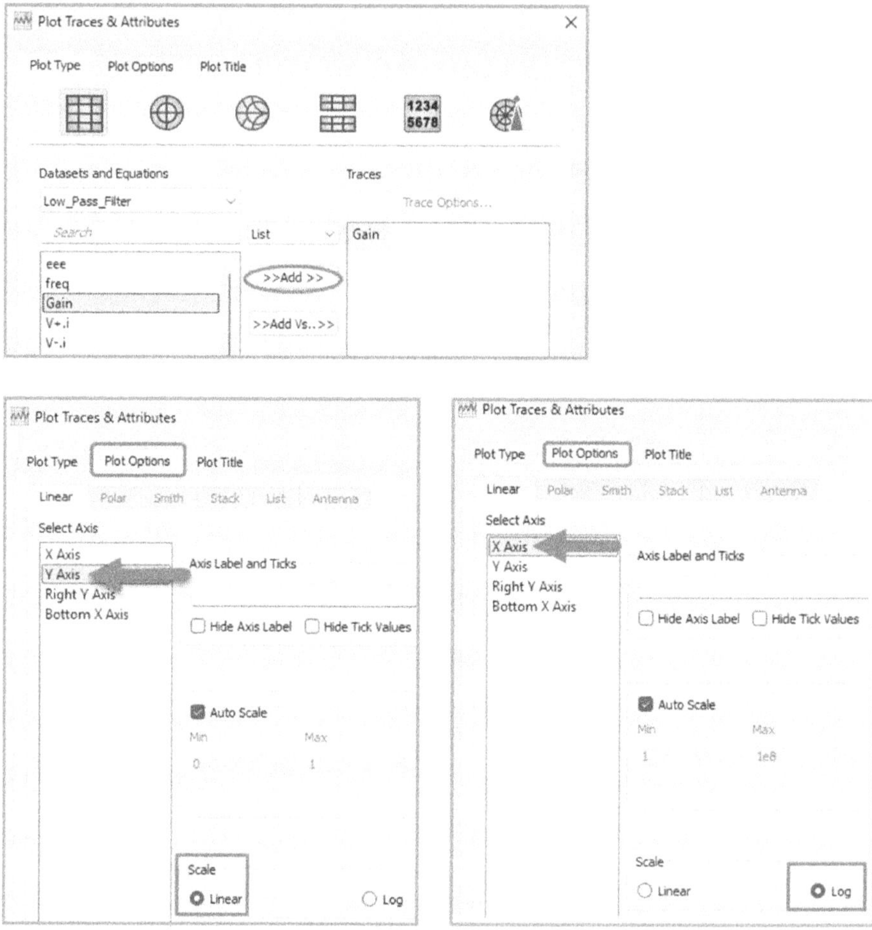

FIGURE 1.75 Plot options dialog box and selections.

The wider the gap between the two frequencies, the slower the simulation (because much more data is collected).

The transient simulation results are shown in Figure 1.81.

Initially, both signals were 1 V amplitude signals. After the filtering action, the amplitude of the 7 kHz signal became around 23 times smaller, whereas the amplitude of the 100 Hz signal hardly changed.

1.16 ADS TIPS AND TRICKS 1C: RUNNING MULTIPLE SIMULTANEOUS SIMULATIONS

One can set up the schematic such that it is possible to run multiple simulation types at once.

In ADS, the VtSine source (see Figure 1.82) acts as a short for AC simulations, and a V_AC source acts as a short during a transient simulation. This property of

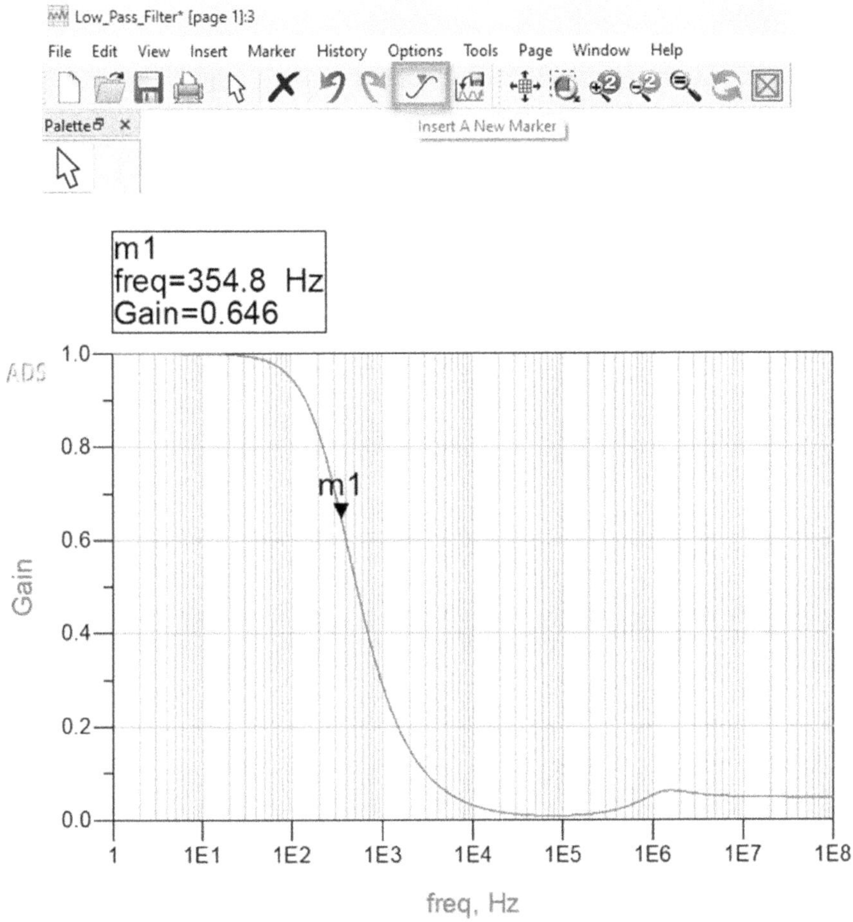

FIGURE 1.76 The voltage gain frequency response plot and the "Insert a New Marker" tab.

these two kinds of sources allows us to place them in series without them interfering with each other.

Notice that we reconnected the V_AC source differently, such that it is in series with the two VtSine sources.

In addition, the MeasEqn was modified and the "AC". Qualifier in front of V_{in} and V_{out} has to be used since the equation is only valid during AC analysis.

The effect of the simultaneous simulations on the output dataset is shown in Figure 1.83.

1.17 OP-AMP DIFFERENTIAL AMPLIFIERS

A differential amplifier is a two-input single-output device whose output depends only on the difference between the two inputs. If any of the input signals have components that exist in the other input signal, these common components have no effect on the amplifier's output.

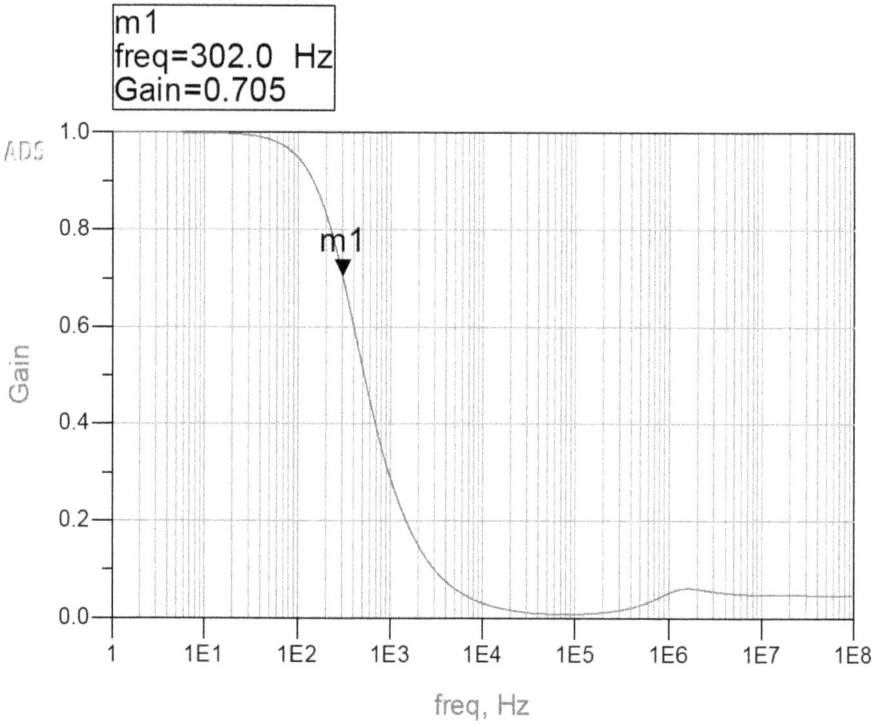

FIGURE 1.77 Searching with a marker along the frequency response plot for the $f = f_c$ point.

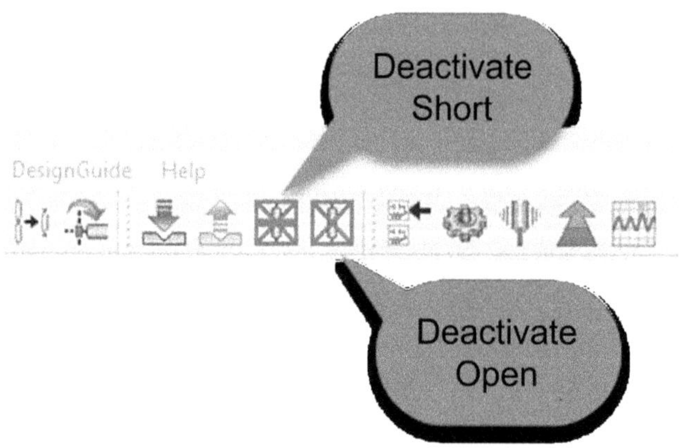

FIGURE 1.78 Deactivation tabs.

Figure 1.84 shows a candidate circuit for the possible implementation of an op-amp differential amplifier.

The two input signals, one that connects via a resistor R_3 to the op-amp's + input, and the other that connects via R_1 to the − input of the op-amp, have a common component $V_C(t)$. The input signals have components $V_A(t)$ and $V_B(t)$, respectively, that are

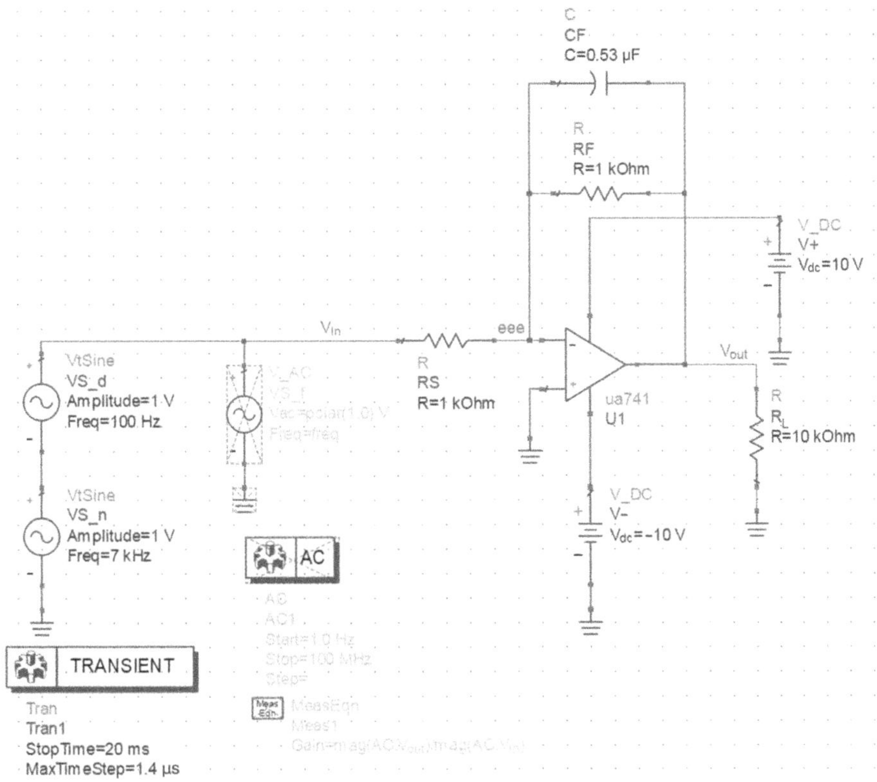

FIGURE 1.79 The low-pass filter schematic ready for transient simulation and having open deactivation of the AC simulation engine block, the V_AC signal that works only in AC simulation, and the MeasEqn formula that makes sense only in AC simulation.

in general, different from each other. The signal **$V_C(t)$ is called the common-mode input signal. The signal $V_B(t) - V_A(t)$ is called the differential input signal.**

We note that, due to the "virtual ground effect", the two op-amp input voltages become equal to one another, and we denote them as $V_1(t)$. It is the virtual ground effect that causes the two input signals $V_A(t)+V_C(t)$ and $V_B(t)+V_C(t)$ to mix with each other: By voltage division at the op-amp's + input, the signal $V_1(t)$ depends only on $V_B(t)$ and $V_C(t)$. Then the current through R_1 and R_2, due to the very large input resistance of the op-amp, depends on all three signals $V_A(t)$, $V_B(t)$, and $V_C(t)$. For the op-amp circuit to function as a differential amplifier, we need the output $v_o(t)$ to be independent of $V_C(t)$ and to be dependent only on the difference between $V_B(t)$ and $V_A(t)$. In general, for some arbitrary choice of the four resistors R_1, R_2, R_3, and R_4, the amplifier does not perform as a differential amplifier. The output $v_o(t)$ is in general, a function of the mixed three signals. The math shown in equations (1.7) below follows [Sedra1]:

$$V_1(t) = (V_B(t) + V_C(t))\frac{R_4}{R_3 + R_4} \tag{1.7a}$$

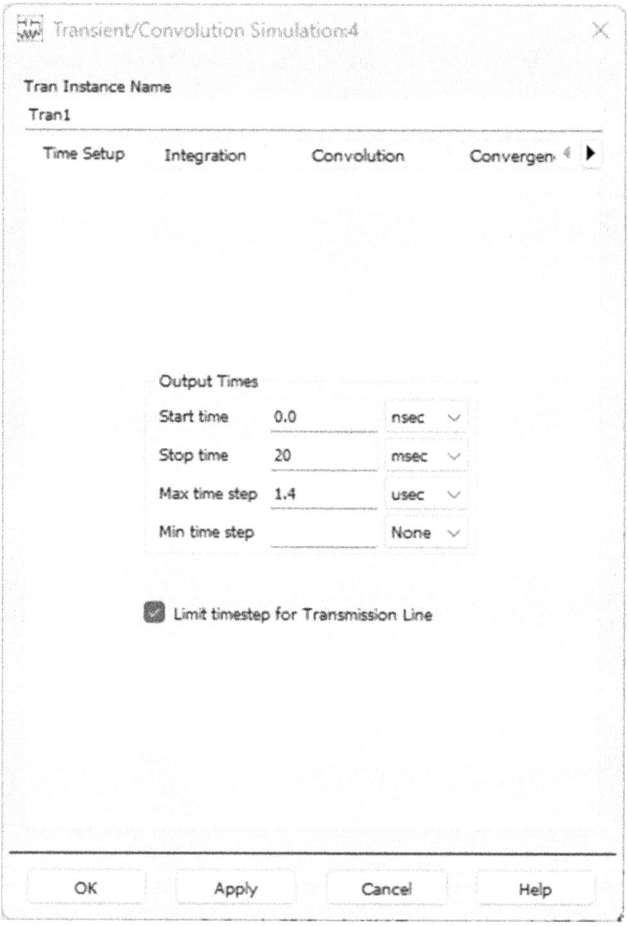

FIGURE 1.80 Transient simulation parameters setting dialog box.

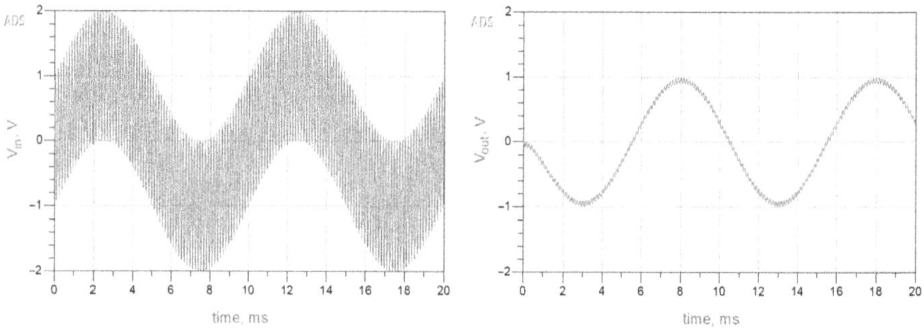

FIGURE 1.81 Transient simulation results illustrating the low-pass filter's performance in the time domain.

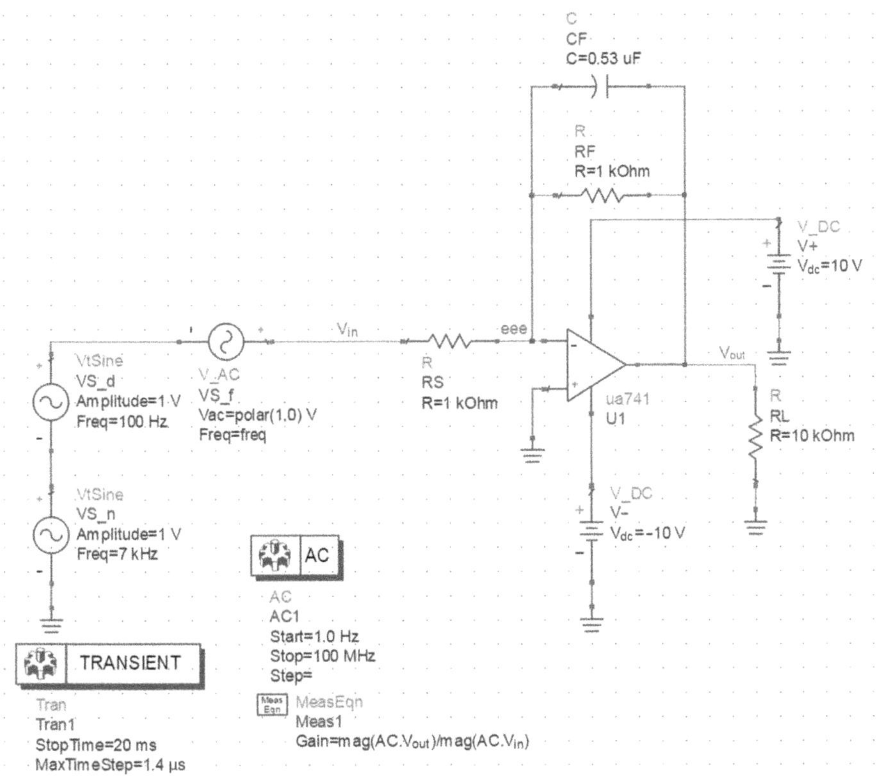

FIGURE 1.82 The low-pass filter schematic with AC and transient simulations coexistence.

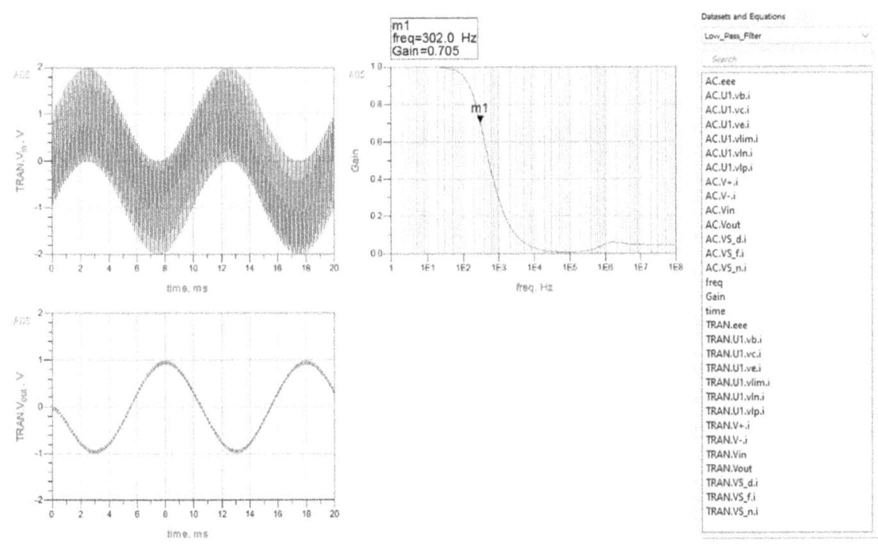

FIGURE 1.83 A longer dataset and equations menu (right) and altered Y-axis labels (left).

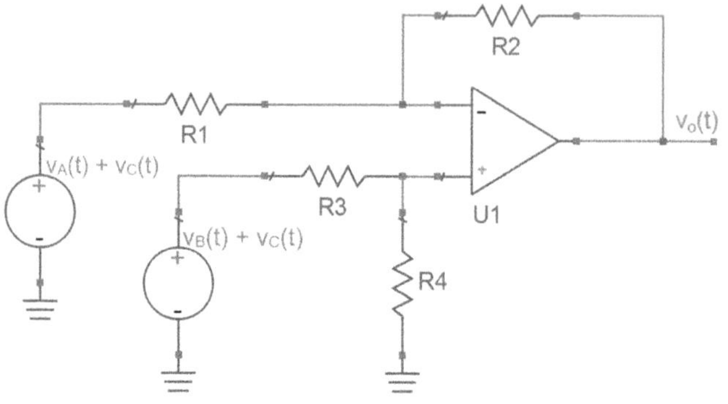

FIGURE 1.84 An op-amp circuit that has the potential of becoming a differential amplifier.

$$I_{in}(t) = \frac{V_A(t) + V_C(t) - V_1(t)}{R_1} \tag{1.7b}$$

$$I_{in}(t) = \frac{V_1(t) - v_o(t)}{R_2} \tag{1.7c}$$

By collecting terms in equation (1.7), we can express $v_o(t)$ as a linear combination of $V_A(t)$, $V_B(t)$, and $V_C(t)$, with coefficients that depend on the four resistor values, as shown in equation (1.8):

$$\frac{V_A(t)}{R_1} + V_C(t)\left[\frac{1}{R_1} - \frac{\frac{1}{R_1}+\frac{1}{R_2}}{1+\frac{R_3}{R_4}}\right] - V_B(t)\frac{\frac{1}{R_1}+\frac{1}{R_2}}{1+\frac{R_3}{R_4}} = -\frac{v_o(t)}{R_2} \tag{1.8}$$

The coefficient of $V_C(t)$ becomes zero if $\mathbf{R_3/R_4 = R_1/R_2}$. This is the condition on the four resistors that makes the op-amp circuit of Figure 1.5.1 a true differential amplifier. The output $v_o(t)$ becomes independent of the common-mode input signal $V_C(t)$. Furthermore, $v_o(t)$ becomes dependent on the differential input signal $V_B(t) - V_A(t)$, as shown in equation (1.9).

$$\text{If } \frac{R_1}{R_2} = \frac{R_3}{R_4} = K \quad \text{then} \quad v_o(t) = K(V_B(t) - V_A(t)) \tag{1.9}$$

The ratio K is the **differential voltage gain** of the amplifier. The **common-mode voltage gain** is ideally 0. Nonideally, there may be resistor tolerances and other small asymmetries, located inside the op-amp, that may cause the common-mode gain to be nonzero, but it is always expected to be very small. Engineers often characterize a differential amplifier's performance by a single number called the **Common-Mode Rejection Ratio** (CMRR): CMRR = |Differential Gain|/|Common-Mode gain|. The larger the CMRR, the better the amplifier. The differential gain may be large, as in an op-amp operating in open loop. Very often, the differential gain is kept modest because the main purpose of the differential amplifier is to remove background noises that present themselves as a common-mode input signals.

1.18 ADS DIFFERENTIAL-MODE AND COMMON-MODE DIFFERENTIAL AMPLIFIER TESTING; CREATING A SIMULATION VARIABLE USING THE VAR BLOCK

Example 1.9

Need to construct an op-amp differential amplifier with a differential voltage gain of 100 and a common-mode voltage gain of 0.

There are two input signals: One is a mix of two AC signals: (3 V amplitude, 900 Hz frequency) and (0.1 V amplitude, 600 Hz frequency) that goes to the (–) input via R_1. The other input signal is AC (3 V amplitude, 900 Hz frequency) that goes to the (+) input via R_3.

Obviously (3 V, 900 Hz) is a common-mode signal and (0.1 V, 600 Hz) is a differential signal. With a nonideal op-amp, will the common-mode gain be exactly zero?

We used the inverting amplifier schematic as a starting point. We added two signal sources and two resistors. The ratio of resistors in each input channel is 100 to meet the differential gain specification.

Three periods of the 600 Hz signal determine the 5 ms Stop Time of the transient simulation. One hundredth of the period of 900 Hz determined the 11 µsec MaxTimeStep. See Figure 1.85.

With 10 V supplies, will 0.1 V times a gain of 100 cause clipping? Let us observe the results, displayed in Figure 1.86.

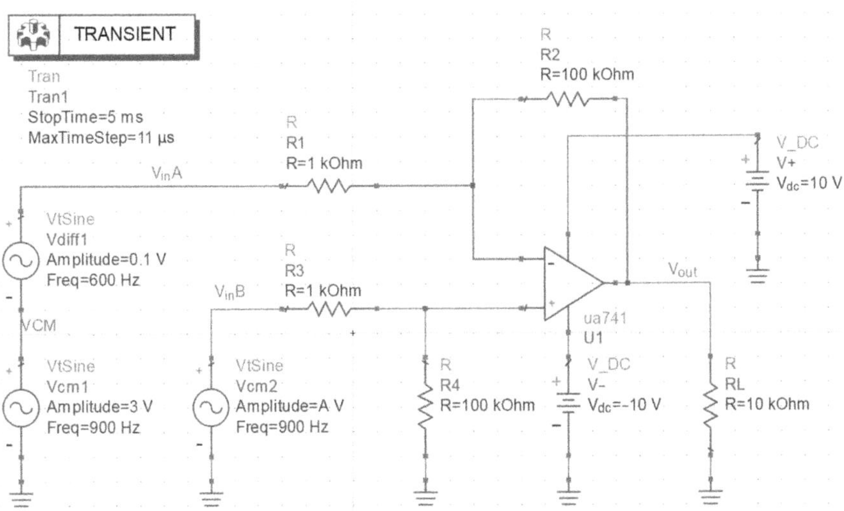

FIGURE 1.85 Schematic of a differential amplifier with a differential gain of 100.

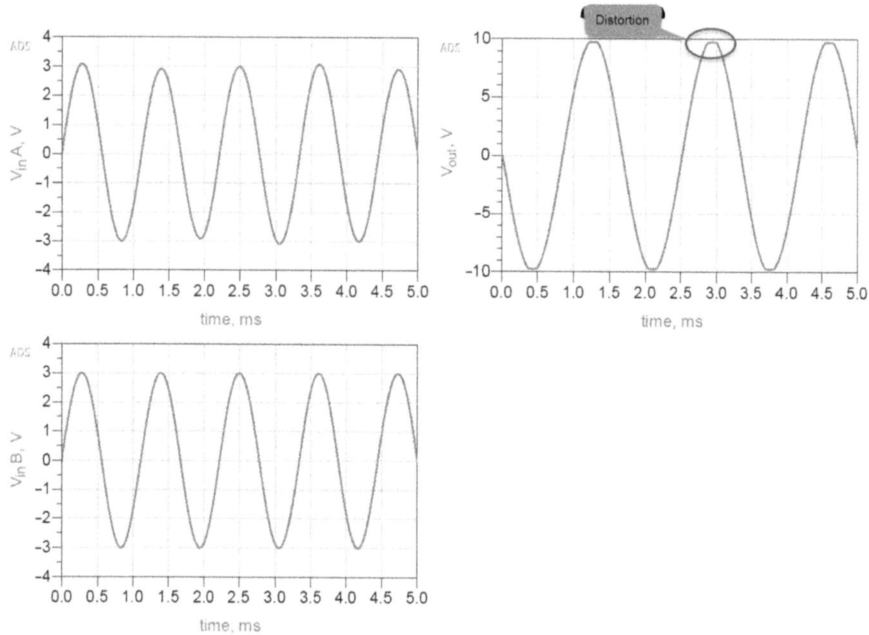

FIGURE 1.86 Input signals and output signal of the differential amplifier.

Both input channels show the same signal (3 V, 900 Hz). Where is the differential input signal (0.1 V, 600 Hz)? We can see it if we look hard at V_{inA} "Hard" because 0.1 V is much smaller than 3 V (the amplitude of the common-mode signal).

Yes, V_{out} is slightly distorted due to the clipping effect. The frequency of V_{out} is 600 Hz (as that of the differential input), and we observe no trace of the 900 Hz common-mode signal's frequency at the amplifier output.

Let us increase the DC supplies to 11 V and make some more changes. We shall insert variable equations and measurement equations (see Figures 1.87 and 1.88).

We raised the DC supplies to 11 V to avoid clipping distortion.

With the help of VAR, we created algebraic values for the input amplitudes A and B. In the VAR block, we must provide default values for A and B. See Figure 1.88.

In a differential amplifier, we run a differential-mode measurement, letting A = 0 (Figures 1.89–1.90). Letting B = 0 Figures 1.91–1.92 create a common-mode measurement. That is why we use VAR for convenience.

By means of MeasEqn we created the differential input signal that can now be plotted (see Figure 1.89).

Note that A = 0 nulls the common-mode input signal. The differential amplifier is fed by a pure differential input signal. Here: $VCM = V_{inB} = 0$. V_{inA} (or V_{diff}) represents the differential input signal. Let us leave the transient setup parameters unchanged.

The simulation results for the differential-mode measurement are shown in Figure 1.90. The 600 Hz differential input signal, with an amplitude of 0.1 V is amplified to 10 V amplitude (same frequency). V_{out} came inverted because V_{diff} is applied to the channel that leads to the (−) op-amp input.

Note that in Figure 1.91, A = 3 and B = 0 eliminates the differential input signal.

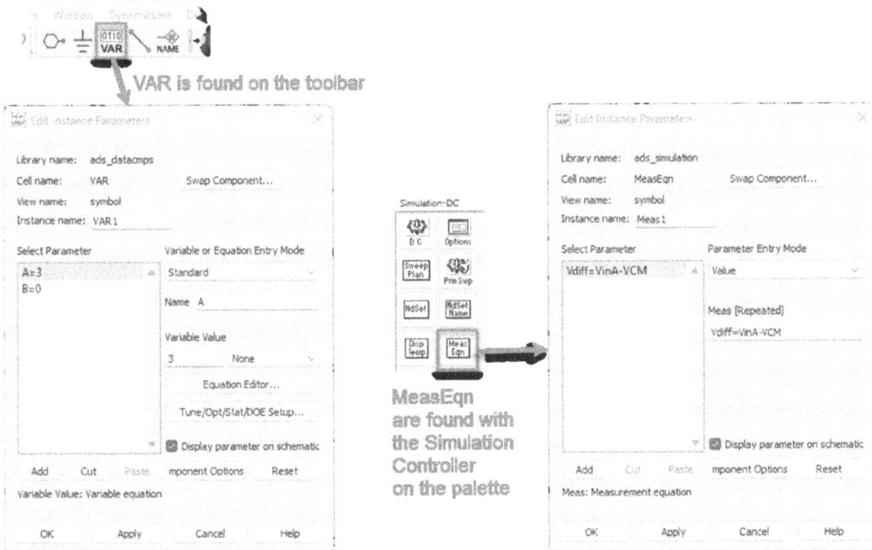

FIGURE 1.87 Adding VAR (left) and MeasEqn (right).

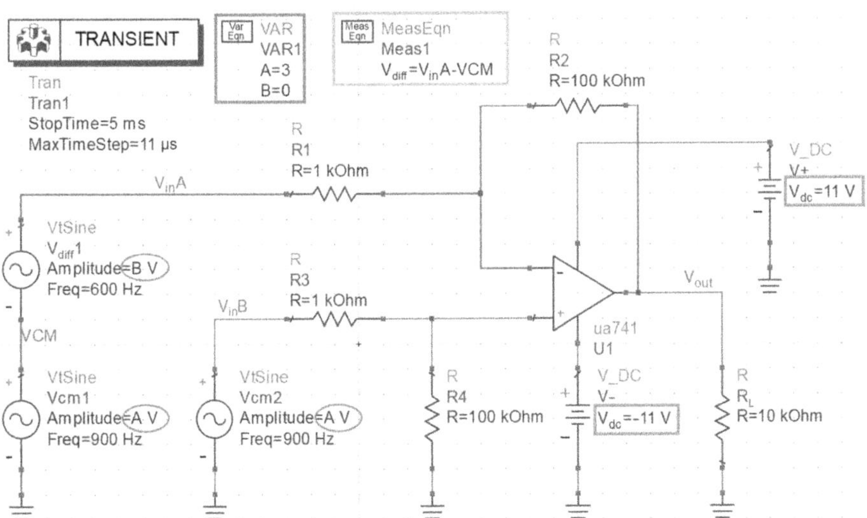

FIGURE 1.88 Differential amplifier schematics with VAR and MeasEqn, and new supply voltages.

We now apply the common-mode input to both amplifiers' input channels. Were the op-amp be ideal, the common-mode input, no matter how large, would be rejected at the output (V_{out}).

The results of the common-mode measurement are displayed in Figure 1.92. V_{out} is not zero! There is some small common-mode voltage gain.

Approximately (from the markers' measurements in Figure 1.92): CM_Gain = ptp(V_{out})/ptp(VCM) $\approx (11.5 - (-7.41)) \cdot 10^{-3}/6 = 0.0032$.

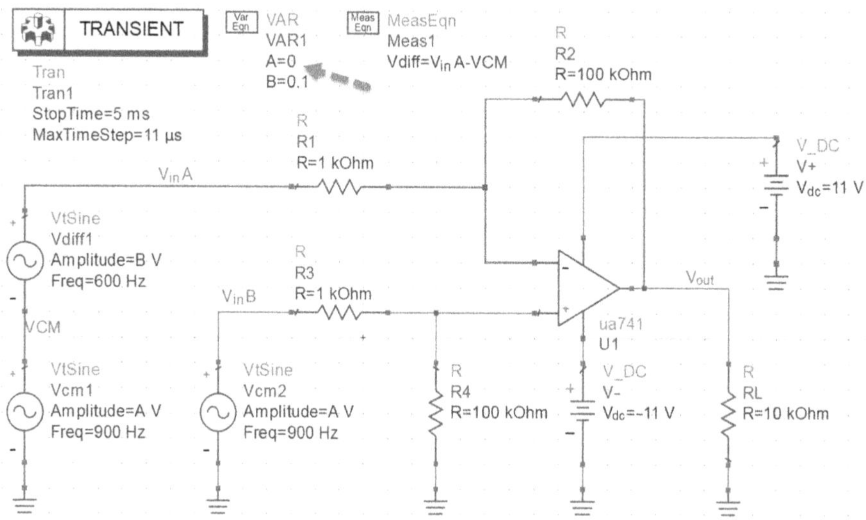

FIGURE 1.89 Differential-mode measurement.

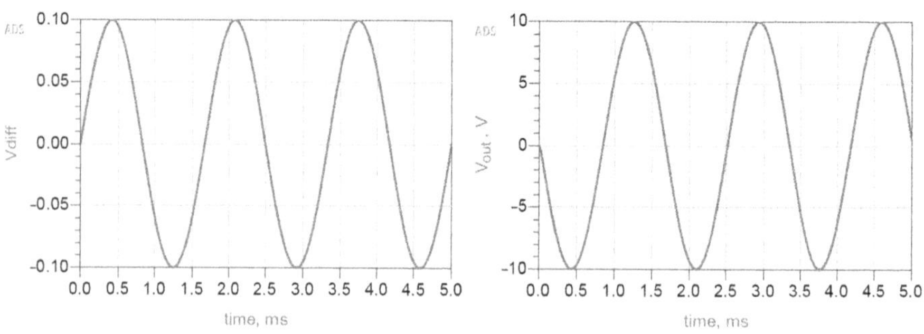

FIGURE 1.90 Differential-mode measurement results.

Note that this time the frequency of V_{out} is 900 Hz, as is that of the common-mode input signal.

Whenever both A and B are nonzero, it is called total measurement. "Total" stands for the differential and common-mode signals being active simultaneously.

In a total operation of the differential amplifier, the common-mode input signal very often dominates the differential input signal. However, the differential output signal convincingly dominates the output common-mode signal.

1.19 ADS TIPS AND TRICKS 1D: ALTERNATE DIFFERENTIAL/COMMON-MODE SOURCE SETUP

In Figure 1.93, we show a different way to set up the sources that results in the same common-mode and differential voltages. This setup can be found in textbooks and other literature, so we present it for completeness.

Note that the two V_{diff} sources now have voltage amplitude that is B/2.

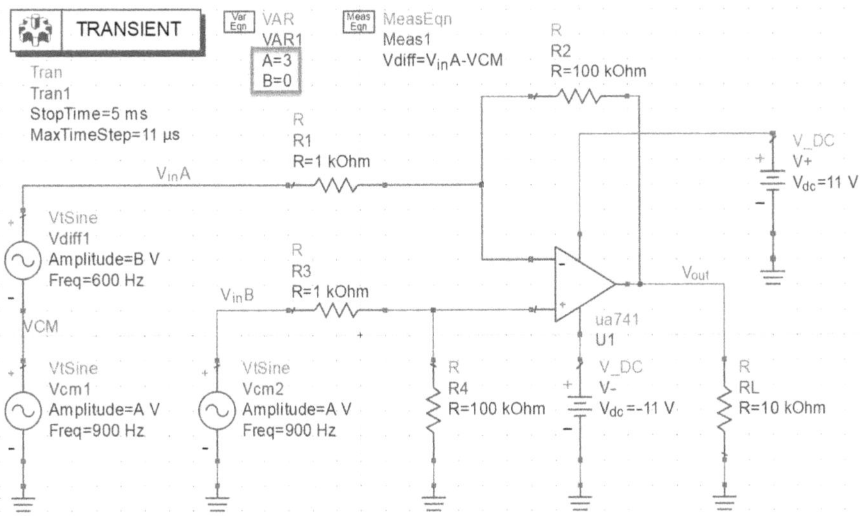

FIGURE 1.91 Common-mode measurement.

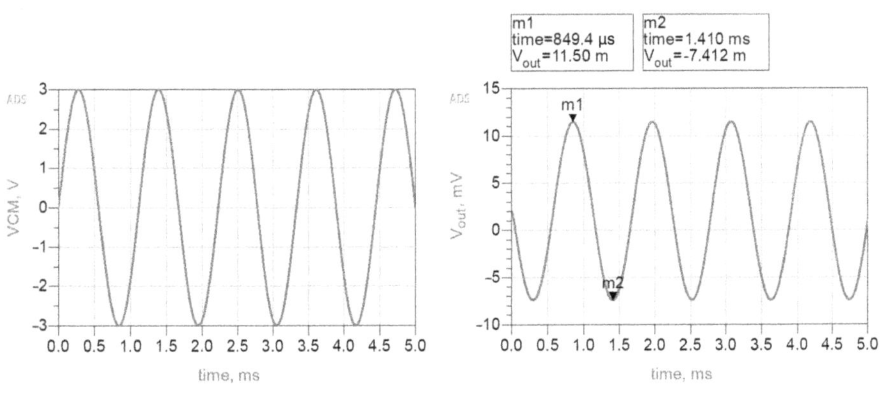

FIGURE 1.92 Common-mode measurement result.

1.20 SIMPLE COMPARATORS

So far, in all the op-amp applications that we have studied, there was a negative feedback, implemented by connecting a resistor between the op-amp's output terminal and the op-amp's inverting input terminal. This negative feedback, combined with specific properties of the op-amp, created the "virtual ground effect", which was key to understanding the operation of each application circuit. Are there any meaningful applications in which the op-amp operates in open loop (that is, with no negative feedback)? Are there applications in which the op-amp works with positive feedback? Yes, and yes. Readers are referred to [Franco1].

Let us discuss first an open-loop operation of an op-amp. Say that, at a certain arbitrary time t, v_2 is the input voltage value to the + input of the op-amp, and v_1 is the

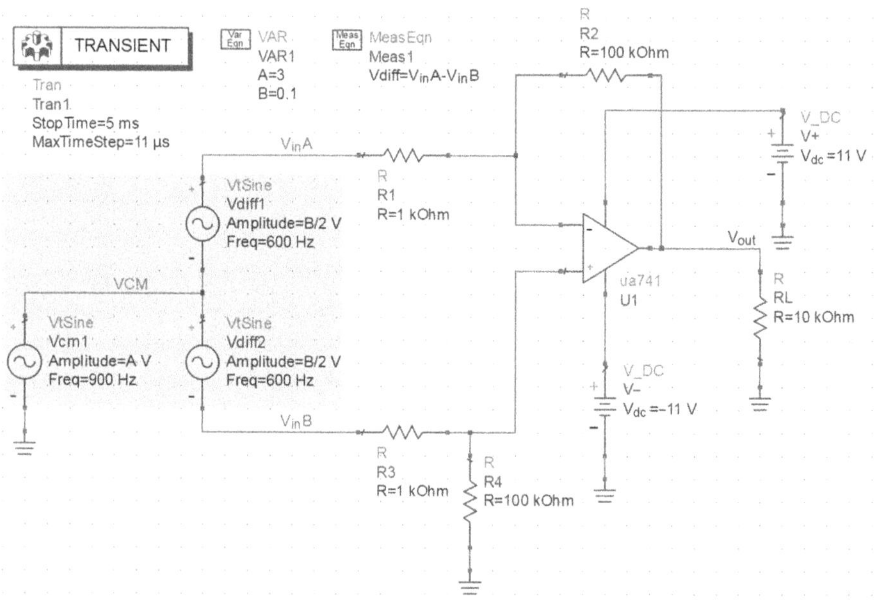

FIGURE 1.93 Alternate differential and common-mode sources setup.

input value to the − input. The op-amp is a differential amplifier with a large differential voltage gain A. Let the op-amp's DC supplies be V^+ and V^-, and let us assume that $V^- = -V^+$. Figure 1.94 shows how the op-amp's output voltage v_o is related to the differential input voltage $v_2 - v_1$.

With $v_2 - v_1$ being small enough, the op-amp operates linearly. That is $v_o = A(v_2 - v_1)$. As $v_2 - v_1$ becomes larger, almost reaching the value of V^+/A, the op-amp's output v_o saturates at a voltage level that is slightly smaller than the supply level of V^+. With A being large, the linear range of the open-loop op-amp is very narrow. For instance, if $V^+ = 10\,V$ and $A = 10^6$, then at a differential input of slightly less than 10 µV, the output clips at a voltage that is slightly below 10 V. If we consider typical input signals' noise levels of say 0.1–1 mV, then the linear zone of the op-amp is missed by a wide margin. Practically speaking, an op-amp in open loop has only two output voltage levels: Maximum positive and maximum negative, created by the op-amp's DC supplies.

Consider the simple circuit shown in Figure 1.95.

Whenever $v_s(t) > V_{ref}$, $v_o(t)$ equals the op-amp's maximum negative voltage. Whenever $v_s(t) < V_{ref}$, $v_o(t)$ equals the maximum positive voltage. So, at any given time, the voltage values of the two op-amp inputs are compared to each other. Hence, the name "comparator". The resulting output voltage is an indicator as to which op-amp input voltage is larger (that is, "more positive").

Here are several easy examples, all related to Figure 1.95, assuming that the op-amp's supplies are ±10 V, and assuming that the op-amp's maximum and minimum output voltage levels are ±9.5 V.

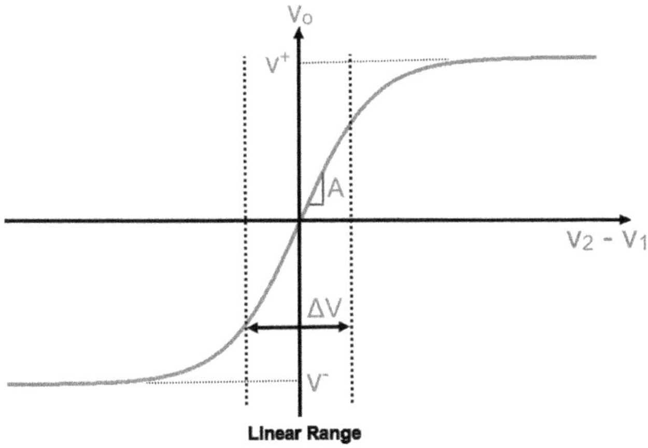

FIGURE 1.94 Input–output characteristic of an op-amp.

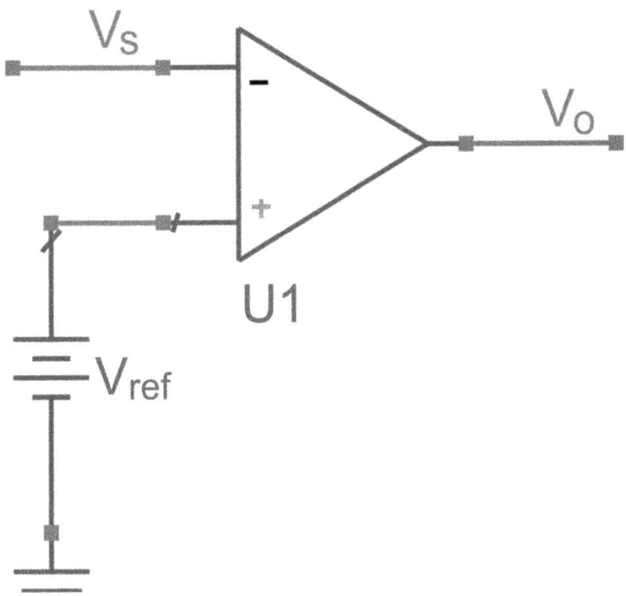

FIGURE 1.95 An open-loop op-amp driven by two input voltage signals: $v_s(t)$ and a DC voltage of V_{ref}.

Example 1.10

$V_s = 5$ VDC, $V_{ref} = 1$ VDC. Then $v_o = -9.5$ VDC.

Example 1.11

$V_s = 0.5$ VDC, $V_{ref} = 1$ VDC. Then $v_o = 9.5$ V.

Example 1.12

$V_s = -5$ VDC, $V_{ref} = -4$ VDC. Then $v_o = 9.5$ V. [Explanation: V_{ref} is "more positive" than V_s, and it connects to the op-amp's + input. Hence, the positive output.]

Example 1.13

$V_s(t)$ is a sine voltage signal with an amplitude of 5 V and a frequency of 50 Hz. Let $V_{ref} = 0$. Then $v_o(t)$ is a square wave of frequency of 50 Hz, and an amplitude of 9.5 V, starting at a level of −9.5 V, and having a duty cycle of 50%. [Duty cycle of a square wave is defined as the percent ratio of the positive part duration in one period divided by the period.]

Many op-amps (such as the μA741), due to input capacitance effects, exhibit a noticeable "slew rate" phenomenon. That is, the op-amp cannot jump instantly from its maximum positive voltage to its maximum negative voltage, or vice versa. It takes some finite time. The larger the input frequency (in Example 1.13), say 1 and 5 kHz, the more noticeable is the slew rate phenomenon, causing $v_o(t)$ to assume a trapezoidal wave shape rather than a square wave shape.

A design of a simple comparator involves decisions such as: If an input signal is compared against a DC threshold, what should be the threshold level? To which one of the op-amp's input terminals should the input signal be connected?

The performance of a simple comparator becomes problematic if the input signal is corrupted with noise, because the threshold may be tripped multiple times. Let us demonstrate the phenomenon before discussing the theory of more advanced comparators.

Example 1.14

A comparator's output should go HIGH whenever its input signal $v_s(t)$ falls below a threshold voltage of 3.3 V. Need to construct the comparator and test it first with a noise-free AC input of frequency 1 kHz. Then need to repeat the testing with a high-frequency noise added to the input signal. Need to assess the noisy-input comparator's performance. Let the DC supplies be ±10 V.

The design proceeds as follows: If $v_o(t)$ goes HIGH for $v_s(t) < V_{th} = 3.3$ V, then the DC threshold voltage must be connected to the op-amp's + input, and therefore the input signal is connected to the op-amp's − input.

1.21 ADS SIMULATION OF SIMPLE COMPARATORS; OVERLAYING A THRESHOLD WITH A SIGNAL

Let us continue with the design task started in Example 1.14. We shall test the comparator with an AC signal of amplitude 5 V and frequency of 1 kHz. The noise signal will be emulated by a 1-V amplitude, and a 10-KHz frequency AC signal connected in series to the input signal V_{in}.

The starting point for constructing the ADS schematic is again the inverting amplifier's schematic. Figure 1.96 shows the comparator circuit's schematic, with all the changes made.

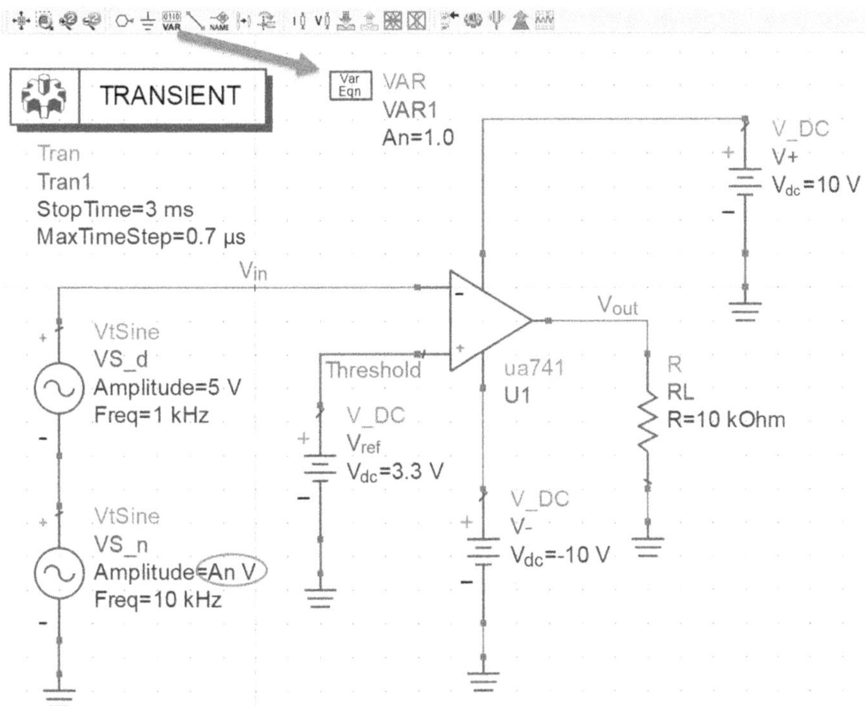

FIGURE 1.96 Simple comparator circuit involving a noisy-input signal.

The use of VAR for the noise amplitude makes it convenient to simulate the design at two conditions: (i) noise-free, setting $A_n = 0$, and (ii) noisy operation, setting $A_n = 1$.

Notice that the volts units of A_n are provided at the VS_n signal source. Never double-count the units.

Transient parameters were set as in other multiple-signal simulations. Figure 1.97 shows the schematic for a noise-free operation of the comparator, setting $A_n = 0$.

The only new feature of ADS, in this section, is the overlaying of two plots: V_{in} and threshold. V_{out} is plotted separately. See Figure 1.98, which includes the simulation results.

In Insert > Plot, just do >> Add >> twice to bring two variables (threshold and V_{in}) to the right menu. Both are then plotted on the same scale.

Note the slight "slew rate" of V_{out}. It takes a finite time for V_{out} to transition from one saturation level to the other, due to op-amp input capacitance effects.

Figure 1.99 (a repetition of Figure 1.96) displays the comparator's schematic for noise operation.

Imagine that V_{out} is a command to a large power unit (such as a home A/C compressor).

Can a noise switch the output HIGH and LOW several times?

If so, such an operation is deemed unacceptable. Let us study the simulation results shown in Figure 1.100. The noise does cause V_{out} to start heading in the direction of

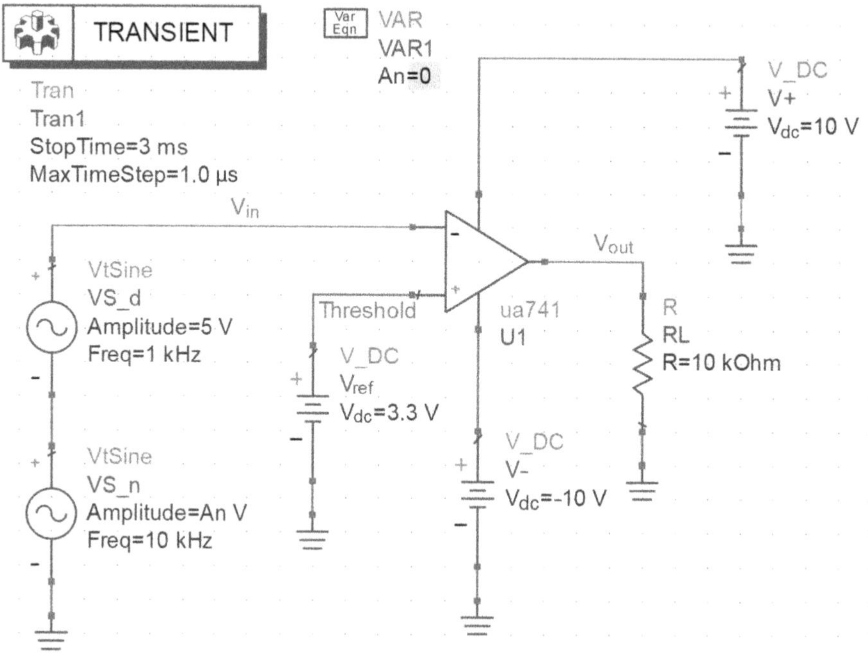

FIGURE 1.97 Noise-free operation of the comparator.

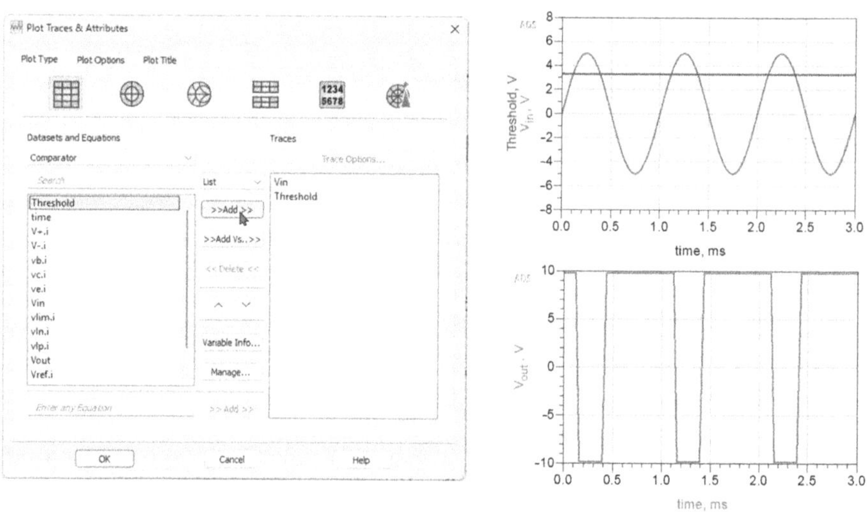

FIGURE 1.98 Setting of two plots sharing the same scale (left), simulation results (right).

the other extreme, however, due to the finite slew rate, another switching occurred before V_{out} went down all the way.

Let us now increase the noise amplitude from 1 to 1.5 V, and increase its frequency from 10 to 15 kHz. See the new schematic in Figure 1.101.

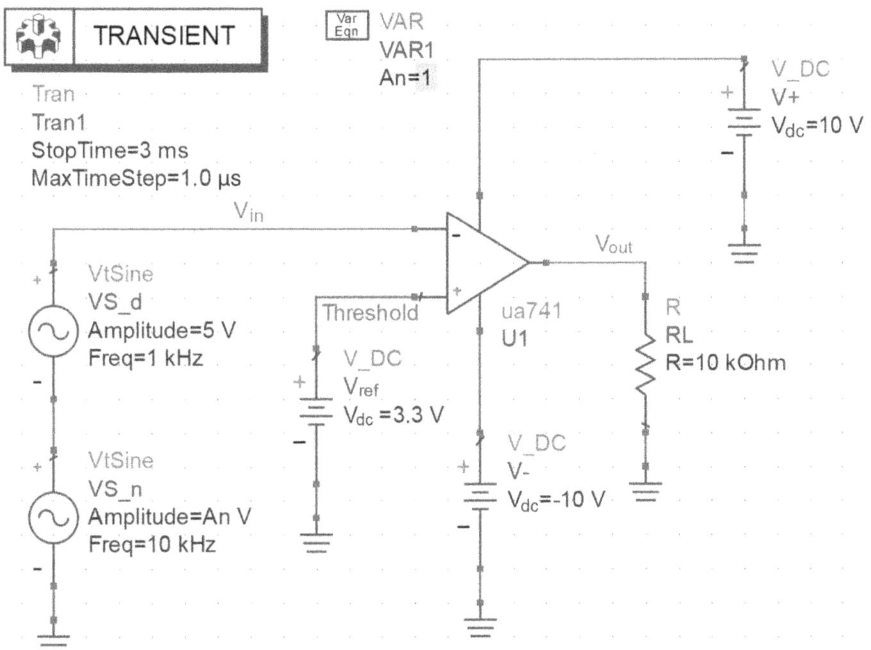

FIGURE 1.99 Comparator with noisy input.

As we study Figure 1.102, we see why A/C systems come with two thresholds:

As the temperature goes above the upper threshold, the A/C turns ON. For the A/C to turn OFF, the temperature needs to go below a lower threshold. A comparator device that has two thresholds is called a Schmitt trigger.

1.22 SCHMITT TRIGGERS

The Schmitt trigger circuit diagram is shown in Figure 1.103. See [Francol]. Let us make a few observations about the circuit and discuss its operation principles.

We see that the op-amp exhibits positive feedback (as R_1 is connected between the op-amp's output terminal and the op-amp's non-inverting input terminal). Let us stress that without negative feedback, there can be no "virtual ground effect". The input voltage $V_s(t)$, which enters the op-amp's inverting input and the voltage $V_2(t)$, of the op-amp's + input, are not equal one to another. In other words, op-amps that operate in open loop (like in simple comparators) or in a positive feedback (like in the Schmitt trigger circuit) are still basically "comparators". That is, the device has only two possible output levels, and the op-amp's input voltages are compared to each other at all times.

The output voltage levels are V_{max} and $V_{min} = -V_{max}$, where V_{max} is slightly smaller than the op-amp's supply V^+. The input signal $V_S(t)$ is compared against $V_2(t)$, acting like a threshold. There are two threshold levels because $V_2(t)$ is created by a superposition of two effects: the influence of V_R, R_1, and R_2, and the influence of $V_0(t)$, R_1 and R_2. $V_0(t)$ has two possible levels, hence $V_2(t)$ has two voltage levels too.

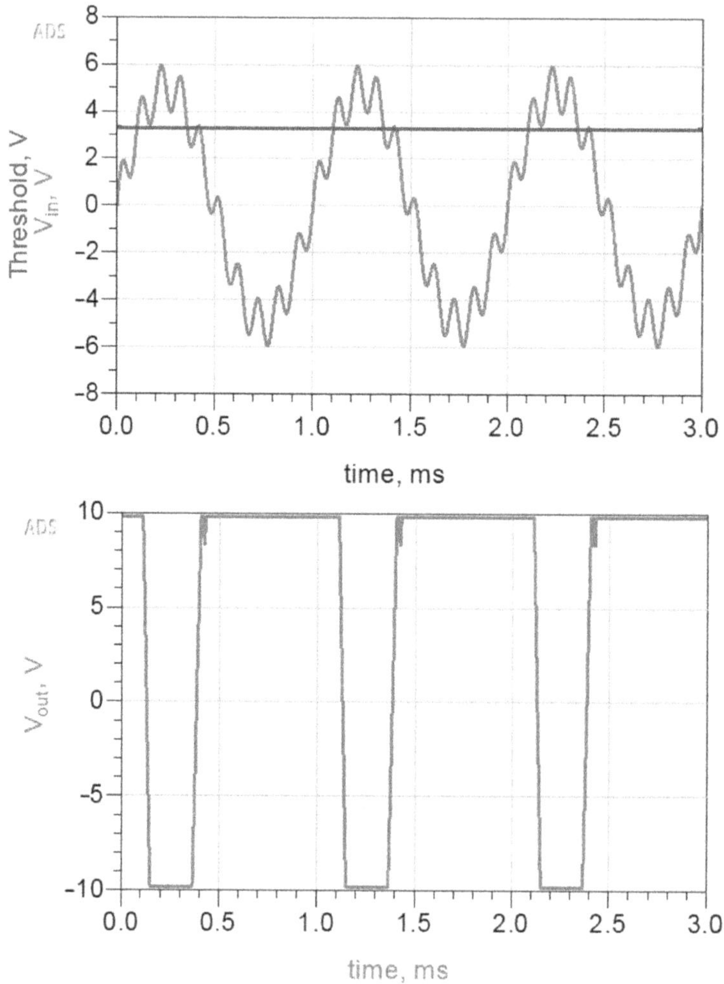

FIGURE 1.100 Simulation results of a comparator driven by a noisy input.

Let us assume first that $\mathbf{V_R = 0}$. Then V_2 is obtained from V_o via voltage division. $\mathbf{V_2 = \pm\, V_{max} \cdot R_2/(R_1 + R_2)}$. The two levels, V_{2L} and $V_{2H} = -V_{2L}$ of V_2 (where the subscript L stands for "low" and the subscript H stands for "high"), are symmetric with respect to $V_2 = 0$. By choosing a certain ratio R_2/R_1, we can arrange for any V_{2H} value that is larger than 0 and smaller than V_{max}. What does it practically mean to have two such threshold levels?

Let us envision a sweeping experiment of a DC input V_S. Say that we start at $V_S = 0$, and then we slowly increase the value of V_S. Let us assume that for $V_S = 0$, the op-amp's output voltage is V_{max}. It causes $V_2 = V_{2H} > 0$, justifying the assumption that V_o is indeed positive. As long as V_S is smaller than $V_2 = V_{2H}$, the output (and V_2) both remain unchanged. As V_S reaches the value of V_{2H}, the comparator changes state: V_o jumps from V_{max} to the negative symmetric value V_{min}. As that happens, V_2 changes instantly from $V_2 = V_{2H}$ to $V_2 = V_{2L}$. The comparator stays at this new state as V_S

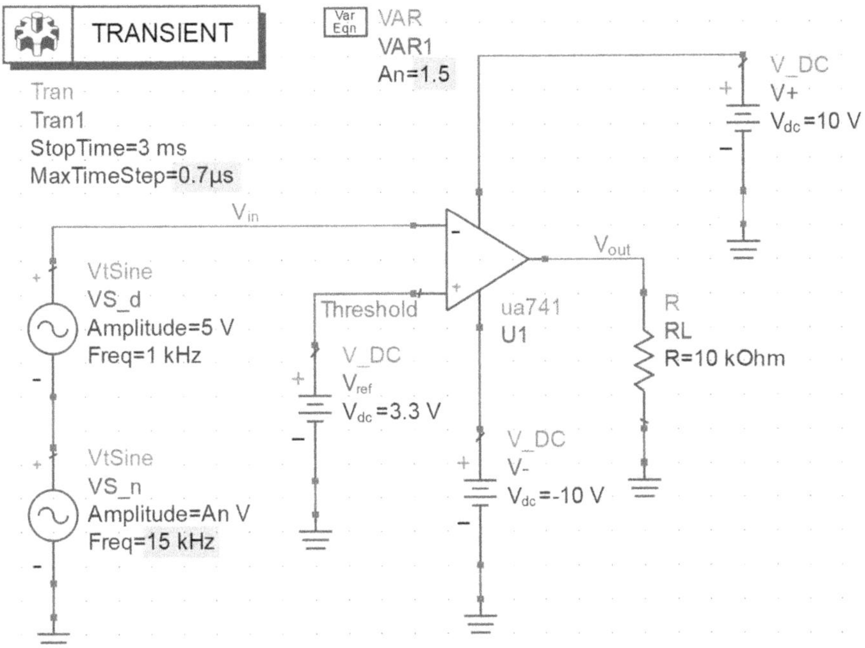

FIGURE 1.101 Comparator with a stronger and faster noise corrupting the input.

continue to climb to values larger than V_{2H}. Continuing the sweeping experiment, let us reverse the direction of changes in V_S. As V_S begins to go down, the comparator holds on to its negative output state. Nothing happens as V_S crosses $V_2=V_{2H}$ because V_2 at such an instance equals $V_2=V_{2L}$. Only when V_S reaches the second threshold, a comparator's change of state occurs. In summary, the upper threshold $V_2=V_{2H}$ is valid only as V_S increases, the lower threshold $V_2=V_{2L}$ is valid as V_S decreases.

Recall the simple comparator and the effect of high-frequency noise, causing potentially multiple quick changes of the comparator's states. If we let the upper threshold of the Schmitt trigger be equal to the original threshold of the simple comparator, we can design the double threshold device, such that the difference between the two thresholds is larger than the expected ptp input noise amplitude.

The two actual thresholds V_{2L} and V_{2H} are in general nonsymmetric, due to the effect of the DC input V_R, which creates a voltage-level shifting of the symmetric thresholds (for $V_R=0$). By superposition, if we now assume $\mathbf{V_o=0}$, then $\mathbf{V_2=V_R \cdot R_1/(R_1+R_2)}$.

By superposition, the two threshold levels of a Schmitt trigger are therefore:

$$V_{2H} = V_{max}\frac{R_2}{R_1+R_2} + V_R\frac{R_1}{R_1+R_2} \tag{1.10a}$$

$$V_{2L} = -V_{max}\frac{R_2}{R_1+R_2} + V_R\frac{R_1}{R_1+R_2} \tag{1.10b}$$

We see from equations (1.10) that if we subtract one equation from the other, the spacing between V_{2H} and V_{2L} depends only on V_{max} and the two resistor

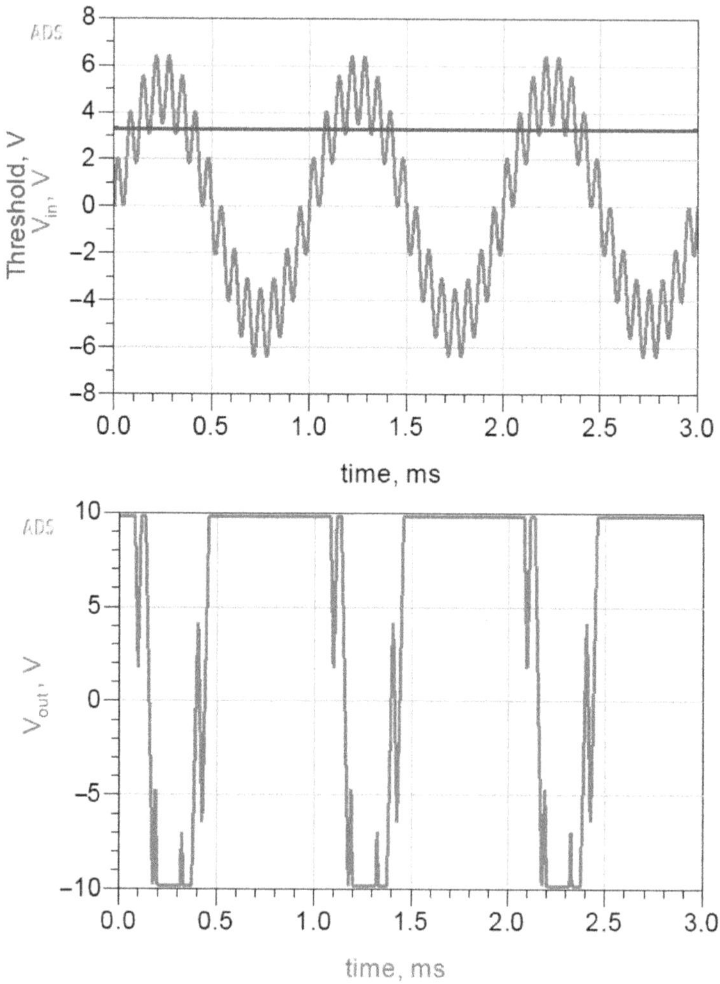

FIGURE 1.102 Simulation results for a larger amplitude and frequency input noise signal.

values: $V_{2H} - V_{2L} = 2V_{max} \cdot R_2/(R_1 + R_2)$. Such a spacing is accomplished by taking $\mathbf{R_1/R_2 = [2V_{max}/(V_{2H} - V_{2L})] - 1}$.

If we add up the two (equation 1.10) equations, and assume that R_1 and R_2 have already been selected to meet a thresholds spacing specification, then V_R is taken based on the mean value of the two threshold levels: $V_{2H} + V_{2L} = 2V_R \cdot R_1/(R_1 + R_2)$, yielding $\mathbf{V_R = (1 + R_2/R_1) \cdot (V_{2H} + V_{2L})/2}$.

Example 1.15

Let us go back to the simple comparator's Example 1.14. This time around, we shall replace the simple comparator with a Schmitt trigger. We let V_{2H} be equal to V_{ref} of

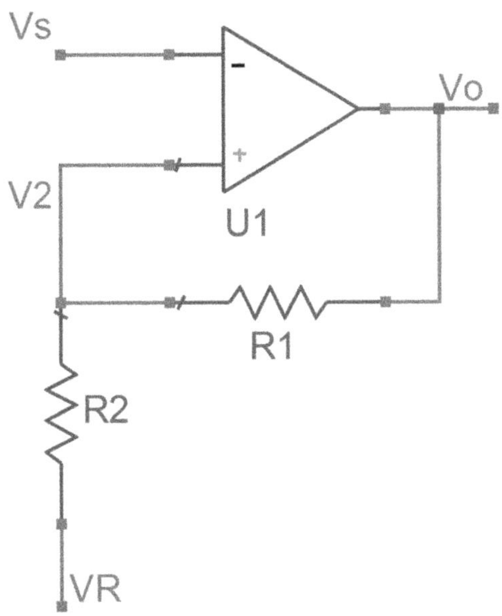

FIGURE 1.103 A Schmitt trigger circuit diagram. $V_S(t)$ is the input signal that needs to be compared against a threshold. V_R is a fixed DC input voltage. The op-amp's supplies are $\pm V^+$.

the simple comparator: $V_{2H} = V_{ref} = 3.3$ V. Let us assume that we design the Schmitt trigger to overcome an input noise of frequency 10 kHz and amplitude of 1 V.

We want the spacing between V_{2H} and V_{2L} to be slightly larger (say by an amount of $\delta = 0.1$ V) than the anticipated noise ptp amplitude. That is, we choose $V_{2L} = 3.3 - 2 - \delta = 1.2$ V.

We first find the resistors' ratio R_1/R_2, based on the desired spacing between V_{2H} and V_{2L}, which is 2.1 V. Let us assume approximately that $V_{max} \approx V^+ = 10$ V, which, by the way is not a needed approximation. One could find the resistors ratio using, say $V_{max} \approx 9.5$ V.

Thus, $R_1/R_2 = (20/2.1 - 1) = 8.52$. In the simulation, let us take $R_1 = 8.5$ kΩ and $R_2 = 1$ kΩ.

Now we find V_R, as follows: $V_R = (1 + 8.52^{-1})(3.3 + 1.2)/2 = 2.51$ V. In the simulation let us choose $V_R = 2.5$ V.

1.23 ADS SIMULATION OF A SCHMITT TRIGGER

The full schematic of the Schmitt trigger is shown in Figure 1.104.

The noisy-input simulation results are shown in Figure 1.105. We see that the noisy V_{in} overlayed with threshold, which is no longer constant, but a signal that jumps back and forth between $V_{2H} = 3.3$ V and $V_{2L} = 1.2$ V. The special selection of $V_{2H} - V_{2L}$ prevents any spurious triggering of V_{out}. Changing the input noise amplitude from, say, 1–1.5 V, necessitates a redesign of R_1, R_2, and V_R.

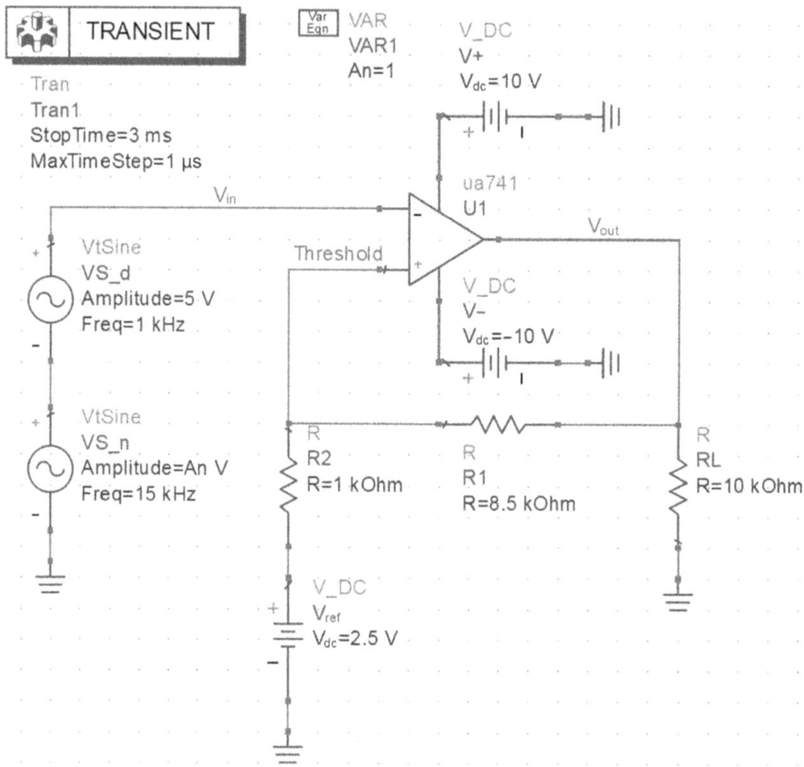

FIGURE 1.104 Schmitt trigger schematic.

REFERENCES

[Franco1] Sergio Franco, *Design with Operational Amplifiers and Analog Integrated Circuits*, 4th Edition, McGraw-Hill, 2014.

[Gable1] Robert A. Gable and Richard A. Roberts, *Signals and Linear Systems*, 3rd Edition, Wiley, 1986.

[Gray1] Paul R. Gray, Paul J. Hurst, Stephen H. Lewis and Robert G. Meyer, *Analysis and Design of Analog Integrated Circuits*, 4th Edition, John Wiley & Sons, 2001.

[Keysight1] *Keysight Advanced System Design (ADS) Release 2025*, Keysight Technologies, Inc. https://www.keysight.com/us/en/products/software/pathwave-design-software/pathwave-advanced-design-system.html

[Sedra1] Adel S. Sedra and Kenneth C. Smith, "Microelectronic Circuits", 7th Edition, Oxford University Press, 2015.

CHAPTER 1 PROBLEMS – OP-AMPS

Problem 1.1: Amplifier Design

Need to design and construct a two-stage amplifier with a total voltage gain of $A = A_1 \cdot A_2 = 35$. You should do it in two different ways, both leading to the same performance: (i) Using two inverting amplifiers in cascade, and (ii) Using two

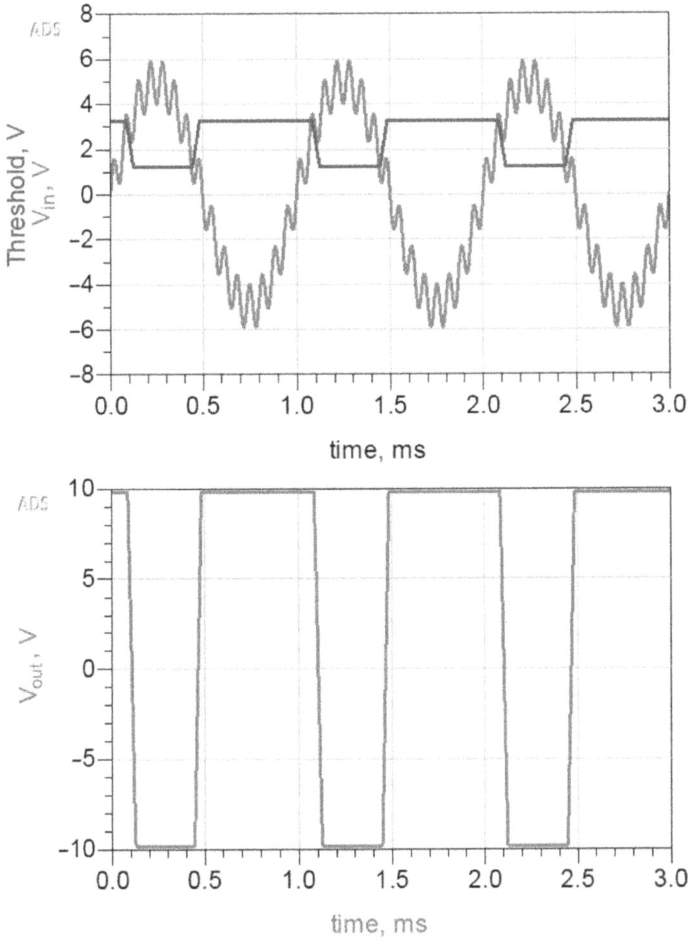

FIGURE 1.105 Simulation results where the noisy-input comparator is replaced by a Schmitt trigger.

non-inverting amplifiers in cascade. Let A_1 be the gain of the first amplifier and A_2 be the gain of the second amplifier. The output of Stage 1 becomes the input of Stage 2. Choose any suitable values $|A|_1 > 1$ and $|A_2| > 1$ that you wish. Note that each of the gains must be larger than 1. Assume that the available batteries are 9V and −9V.

Use ADS markers to check whether the overall gain that you get, and the gain in each stage, are exactly as planned. Observe the output for an AC voltage input of amplitude 0.1V and frequency of 1kHz. Use standard resistors that are in the kΩ range.

Problem 1.2: Amplifiers

As you know, op-amp amplifiers may be connected in cascade (that is, the output of one amplifier becomes the input of the next amplifier), such that the overall voltage gain is the product of the individual gains of each amplifying stage. The circuit given

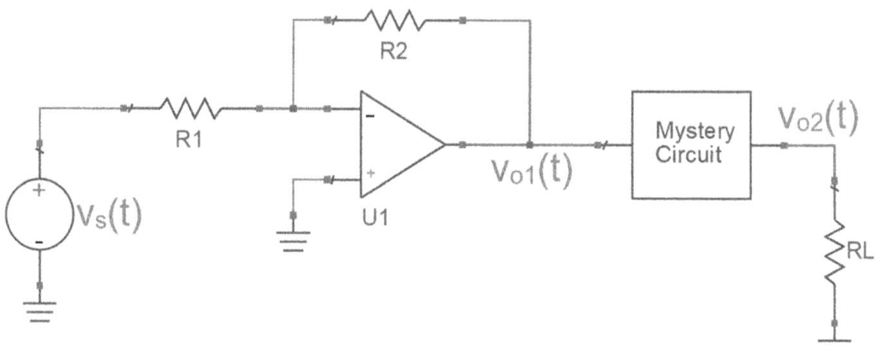

FIGURE 1.106 Problem 1.2 circuit.

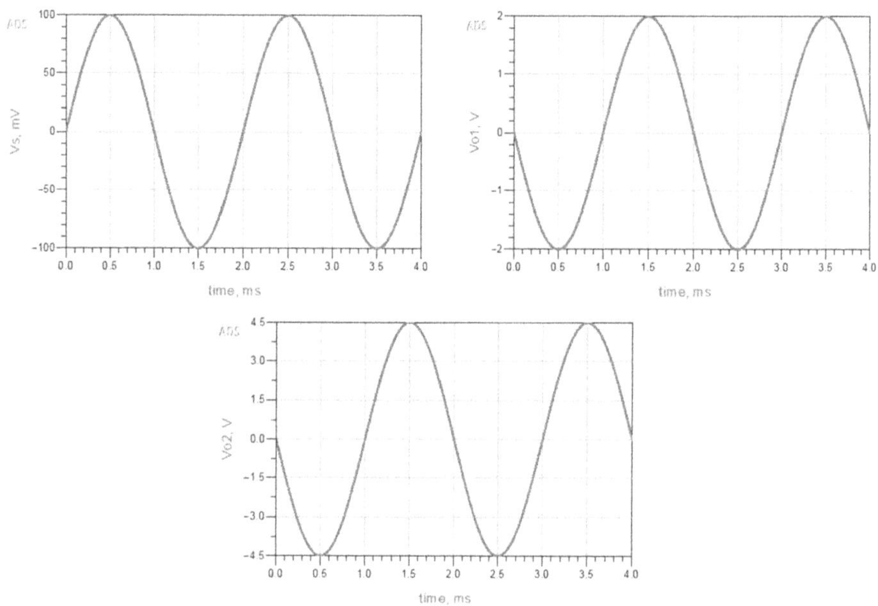

FIGURE 1.107 Problem 1.2 input and output waveforms.

below is that of a two-stage amplifier. It is given that $R_1 = 1.5$ kΩ, $R_L = 10$ kΩ and $V^+ = -V^- = 9$ V (Figure 1.106).

You need to find R_2 and the full "mystery circuit", based on the measurement data presented in Figure 1.107. Test your design in ADS. Are there any notable discrepancies between your design, based on ideal op-amps, and the actual circuit, using μA741 nonideal op-amps?

Problem 1.3: Amplifiers and Linear Combiners

3a: You have two DC voltage supplies ±9 V, an unlimited supply of standard resistors, two μA741 op-amps, and a single voltage signal of $v_1(t) = 2\sin(2\pi \cdot 600t)$ V.

Create two new signals $v_2(t) = 4.5\sin(2\pi \cdot 600t)$ V and $v_3(t) = -0.5\sin(2\pi \cdot 600t)$ V. Each of the new signals, showing up at each op-amp's output, is applied to a load resistor: $v_2(t)$ is connected to a resistor R_{L2} whose other side is at ground, and $v_3(t)$ is similarly connected to R_{L3}. How small is each of the loads, R_{L2} and R_{L3}, allowed to be before the op-amp's current limit problems arise? Demonstrate in ADS. What is the current limit of a μA741 op-amp?

3b: As a follow-up to (Problem 1.3a), assume that the circuits built in (Problem 1.3a) stay built. Consider the three voltages $v_1(t)$, $v_2(t)$, and $v_3(t)$ as available signals. Three new items are available as well: Two more DC voltage supplies ±1.5 V, a third μA741 op-amp, and one more voltage signal $v_4(t) = \cos(2\pi \cdot 800t)$ V. It is up to you to decide whether or not to remove any of the load resistors R_{L2} and/or R_{L3}, that each by now have a permitted resistance value. A reason to remove a load resistor may be to avoid the load resistor from interfering with the connection of one amplifier to the next amplifier stage. Can you create a new voltage signal $v_5(t) = -4\cos(2\pi \cdot 800t) + 6\sin(2\pi \cdot 600t)$ [V]?

Assume first that each op-amp has 9 and −9 V supplies. Demonstrate your design in ADS.

Hint: If some voltage clipping effect is created at the output of the third op-amp, see if adding the two new DC supplies to the third op-amp helps to remove the voltage limit effects. Stacking a 1.5 VDC battery on top of a 9-VDC battery creates a supply of 10.5 VDC.

PROBLEM 1.4: VOLTAGE-LEVEL SHIFTING

In ADS, create a VtSine AC sinusoidal signal V(in) with the following parameters: DC level of = −1 V, AMPLITUDE = 3.5 V, FREQUENCY = 2 kHz. Design a level shifting circuit such that the lower peak of V(in) is shifted to the voltage-level shifter's output of 6 V, and the higher peak of V(in) is shifted to −7 V. Show by means of ADS transient simulation that your circuit really works. Use two 9 V batteries.

PROBLEM 1.5: VOLTAGE-LEVEL SHIFTING AND OP-AMP VOLTAGE AND CURRENT LIMITATIONS

The following op-amp circuit performs voltage-level shifting (Figure 1.108).

Let $R_1 = 1$ kΩ, $R_2 = 5$ kΩ, $R_f = 5$ kΩ, $R_L = 20$ kΩ, $V^+ = -V^- = 10$ V and $V = V^+$ or V^- as needed.

5a: Let $v_s(t)$ be an AC signal of amplitude 1 V and frequency f = 3 kHz. How is $v_s(t)$ voltage-level shifted to create $v_o(t)$? That is, is $v_s(t)$ amplified, and if so, by how much? Is $v_s(t)$ DC level shifted up or down? Which one (up or down?) and by how much? Consider both options for V. Do both options work correctly? Use ADS for simulations.

5b: For the circuit of (Problem 1.5a), for one choice of V, as you see fit, assuming that the op-amp is μA741, what happens to $v_o(t)$ if R_L is reduced to $R_L = 50$ Ω? Why? Vary R_L and explore for what load value "trouble begins".

5c: For the circuit of (Problem 1.5b), using μA741, what happens to $v_o(t)$ if the amplitude of $v_s(t)$ is increased to 2 V? Why? Vary the amplitude of $v_s(t)$ and explore for what values of the amplitude "trouble begins".

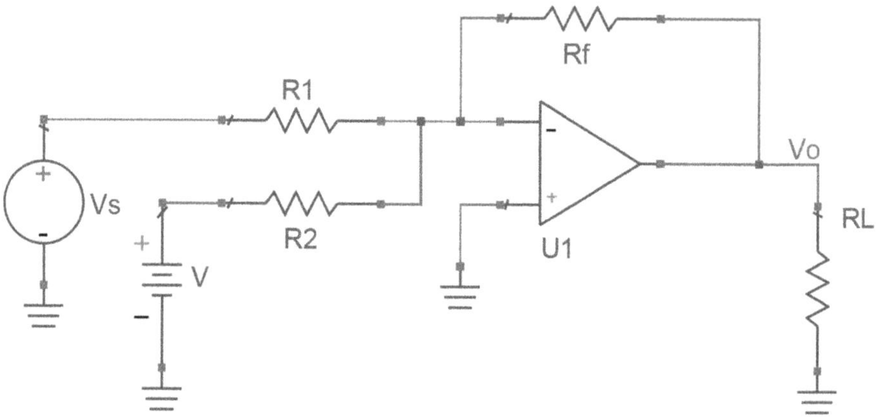

FIGURE 1.108 Problem 1.5 circuit.

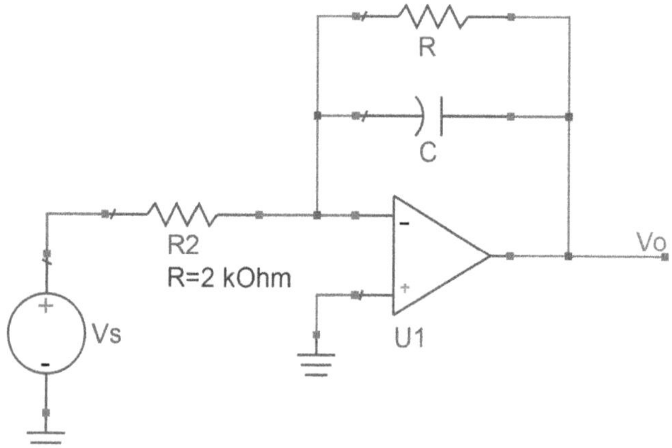

FIGURE 1.109 Problem 1.6 circuit.

Problem 1.6: Low-Pass Filter

An input signal $v_s(t)$ that consists of 6 AC components of frequencies 150, 200, 320, 5,250, 5,700, and 6,400 Hz needs to be processed by an op-amp low-pass filter, as shown in the circuit of Figure 1.109.

The objective of the processing is to enhance the low-frequency signals of {150, 200, 320 Hz} relative to the higher frequency signals of {5,250, 5,700, 6,400 Hz}.

The DC gain of the low-pass filter is −1. The absolute gain of the low-pass filter for each of the low-frequency components should be at least 0.92, whereas the absolute gain of the filter for each of the high-frequency components should be no larger than 0.33.

Your job is to design R and C and test the performance of the filter in ADS simulations, both in the time domain and in the frequency domain.

Assume that the op-amp is μA741 with supplies of ±9 V. Assume that each of the AC components of the input has an amplitude of 1 V.

Here is what you have to do: Once you decide what f_c should be, find R and C that implement such a cutoff frequency. Check that it works in AC simulation: Use a log scale for the frequency axis. Check that at $f = f_c$ the gain drops to 0.707. Then do transient simulation verification: Create $v_s(t)$ that consists of all the voltage signals, putting VtSine sources in series. Watch $v_s(t)$ and $v_o(t)$ and draw conclusions.

Problem 1.7: Differential Amplifier and Low-Pass Filter

Consider the following two voltage signals:

$$v_1(t) = 2.5\cos(2\pi \cdot 5000t) + 2\sin(2\pi \cdot 100t)\ [V]$$

$$v_2(t) = 2.5\cos(2\pi \cdot 5000t)\ [V]$$

You need to create an output signal $v_{out}(t) \approx 2\sin(2\pi \cdot 100t)$ [V]. That is, the output voltage should be as close as possible to an AC voltage with an amplitude of 2 V and a frequency of 100 Hz. The output signal should have no trace of a 5-kHz AC signal, or a 5-kHz trace that is as small as possible.

You have a single μA741 op-amp, DC supplies of ±9 V, and an unlimited supply of standard resistors.

7a: Accomplish the task by constructing an op-amp differential amplifier.

7b: You are also given no more than two standard capacitors of your choice. Accomplish the task by constructing an op-amp low-pass filter.

Both designs should be tested in ADS. Which of the designs works better in terms of getting rid of the higher frequency signals?

Problem 1.8: Differential Amplifier

You have to find the mystery circuit (shown below), based on the given input and output signals:

$$v_{s1}(t) = 4 + 0.05\cos(2\pi \cdot 750t) \quad v_{s2}(t) = 4.1 - 0.01\cos(2\pi \cdot 750t)$$

$$v_o(t) = -7 + 4.2\cos(2\pi \cdot 750t)$$

Is the mystery circuit a differential amplifier? Why? If so, is $v_{s1}(t)$ connected to the channel that leads to the op-amp's "+" input, or is it to the "–" input terminal? Likewise, for the other input signal. What are the common-mode and differential signals? What is the differential voltage gain of the amplifier? Finally, if the op-amp is μA741, what is the range of permitted DC supply values for this amplifier to assure that the output signal does not get clipped? (Figure 1.110).

Hint: You will find out that both the common-mode signal and the differential signal consist of DC and AC. Be sure to check that the differential gain for DC is the same as that for AC.

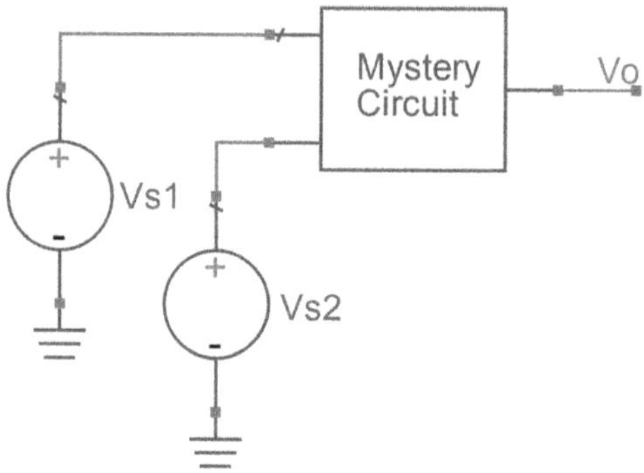

FIGURE 1.110 Problem 1.8 circuit.

Problem 1.9: Differential Amplifier

A light sensor can generate voltages in the range of 0 to +42 mV, depending on the light intensity that is sensed from the light source. This voltage must be amplified to fit the range of an A/D device (ADC), which is 0–3 V. There is a lot of background light that acts like noise. What can be done?

It was decided to use two identical sensors: One will face the light source and the other will be bent to face away from the light source. Both sensors are equally exposed to the background noise. For the sake of simulation, assume that the sensor's sensed voltage (from the desired source) is AC with a lower peak of 0 and a higher peak of 42 mV and a frequency of 250 Hz (Use an AC Time domain source and set its DC offset correctly). Assume that the noise signal is an AC of amplitude 2 V and frequency of 1 kHz. Design the differential amplifier and verify its performance. Measure the amplifier's differential gain and then run an experiment to measure the amplifier's common-mode gain. Use 9 V batteries and a μA741 op-amp.

Problem 1.10: Comparator

Design a simple μA741 comparator that compares an incoming AC signal, of amplitude 7 V and frequency of 100 Hz, to a DC threshold of 4 V. The supplies are 10 and −10 V. The comparator goes HIGH whenever parts of the AC signal become larger than the threshold.

10a: Show in ADS transient simulation that the comparator works as planned.

10b: The "slew rate" of a comparator is defined as the time that it takes the comparator to reach from LOW output to HIGH output (or from HIGH to LOW). Measure the slew rate.

10c: If we change the AC input frequency (say, from 100 Hz to 1 kHz, and then to 2 kHz) does the slew rate change? Is the slew rate a property of the op-amp, regardless of its inputs? Describe in a few words how the slew rate affects the comparator's performance.

Problem 1.11: Comparators

A sensor's output voltage can range between 1 V and 5 V. Any sensor output voltage that is between 3 V and 4 V implies "dangerous conditions" requiring activation of a red alarm light. Use two μA741 op-amps and two 9 V batteries to implement such a circuit.

The ADS Design Kit has a model of an LED. Alternately, one may emulate a GaAs LED by two regular diodes (use D1N4148 from the kit) connected in series. You will understand diodes better after studying Chapter 2, but for the time being, all you have to know is the following: The symbol of a diode is a triangle topped by a bar. The (+) side of the diode is called the *anode* and it is located at the triangle side. The (−) side of the diode is called the *cathode*, and it is located at the bar side. Current can flow only from the anode to the cathode, but not the other way around. Connect the output of op-amp 1 to a 1-kΩ resistor and the resistor to the anode of the diode. Connect the cathode of the diode to the output of op-amp 2. The only way that a current can flow through the diode is if op-amp 1 has a HIGH output and simultaneously op-amp 2 has a LOW output. Emulate the sensor by an AC source with a frequency of 200 Hz and appropriate amplitude and DC offset values. The design of the circuit is now nothing more than deciding for each op-amp what input of the op-amps to connect to the sensor and to what input to connect the DC threshold. The voltage references of 3 V and 4 V (needed for the alarm device that you are designing) are obtained by voltage division from the 9 V battery. Must we use op-amp buffers in order to connect each of the voltage division threshold voltages to the comparator?

Problem 1.12: Schmitt Trigger

An A/C system turns on whenever the room temperature rises above 80°F. Temperature control is done using a temperature sensor and some sort of comparator. The signal coming out of the comparator activates the A/C compressor. Assume that when the temperature is 80° the sensor creates a voltage of 4.7 V. When the temperature is 79°F the sensor's voltage becomes 4.3 V. As the room warms up, we want the A/C to turn ON at 80°, but as the room cools down, we wish that the A/C system turns OFF only when the temperature reaches 79°. The two-threshold comparator is implemented using a Schmitt trigger. Design the Schmitt trigger using a μA741 op-amp and 9 V batteries. Then demonstrate that the A/C system works correctly: On the same scale plot the threshold voltages (measured at the + input of the op-amp), the input voltage, and the op-amp's output voltage. Measure with cursors to verify that the thresholds are implemented accurately (at 4.3 V and at 4.7 V). Emulate the sensor's output as a 75-Hz AC signal that varies between the limits of 3.5 V and 6 V.

Problem 1.13: Amplifier, Buffer, Linear Combiner, and Differential Amplifier

In Figure 1.111 consider the linear combiner circuit of the top left op-amp. Let the two AC input signals be: $v_{s1}(t) = 1.1\cos(2\pi \cdot 200t)$ $v_{s2}(t) = 1.2\cos(2\pi \cdot 10{,}000t)$. The units are [V].

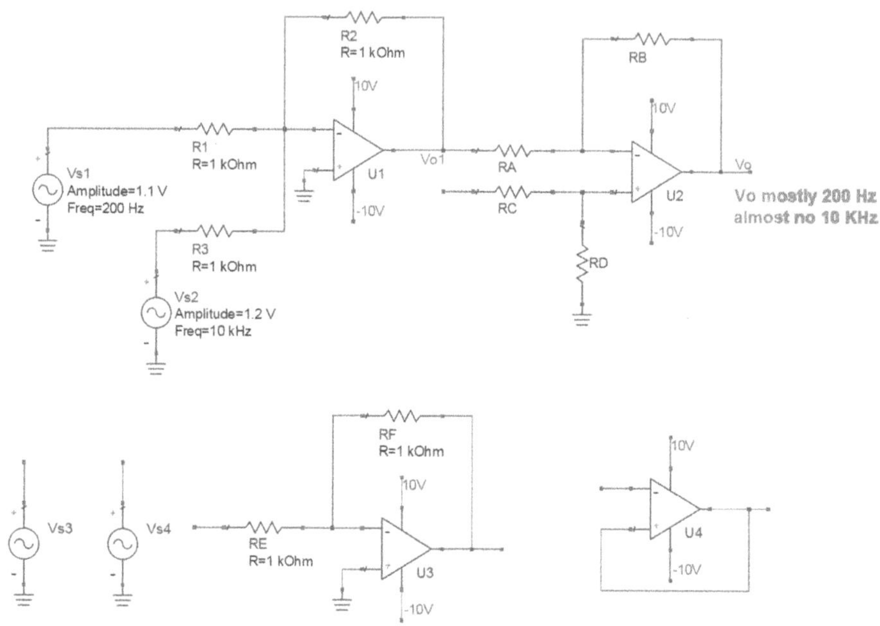

FIGURE 1.111 Problem 1.13 circuit.

Figure 1.111 shows the building blocks that are available for the full circuit. The output $v_o(t)$ has mostly a 200-Hz AC signal, and almost no 10kHz signal. The circuit has no capacitors. In this problem, you, the designer, have separate access to each of the input signals $v_{s1}(t)$ and $v_{s2}(t)$ (as shown by $v_{s3}(t)$ and $v_{s4}(t)$).

An unconnected op-amp inverter and op-amp buffer are also given, in case you need one or both. The signal $v_{o1}(t)$ is already connected to R_A. *Complete the circuit connections*, showing them on the diagram above. You don't have to use every available signal or sub-circuit. Some available signals or circuits may be irrelevant.

Also choose resistor values for R_A, R_B, R_C, and R_D, explaining your choices, and your circuit connections.

Demonstrate your design in ADS.

Problem 1.14: Comparators and Buffers

14a: Using a single μA741 op-amp, two ±12 V DC supplies, and as many resistors as you choose, design a **simple comparator** whose output goes from HIGH to LOW whenever the input $v_s(t)$ falls below $V_{threshold} = 5$ V. Draw the circuit and find the resistors that create the threshold voltage.

14b: Regarding the above design, one of the engineers on the team claims that two μA741 op-amps are needed – one to act as a comparator and another to act as a buffer, when creating the DC threshold voltage. Is the engineer right or is it okay to use a single 741 op-amp only? Explain your answer.

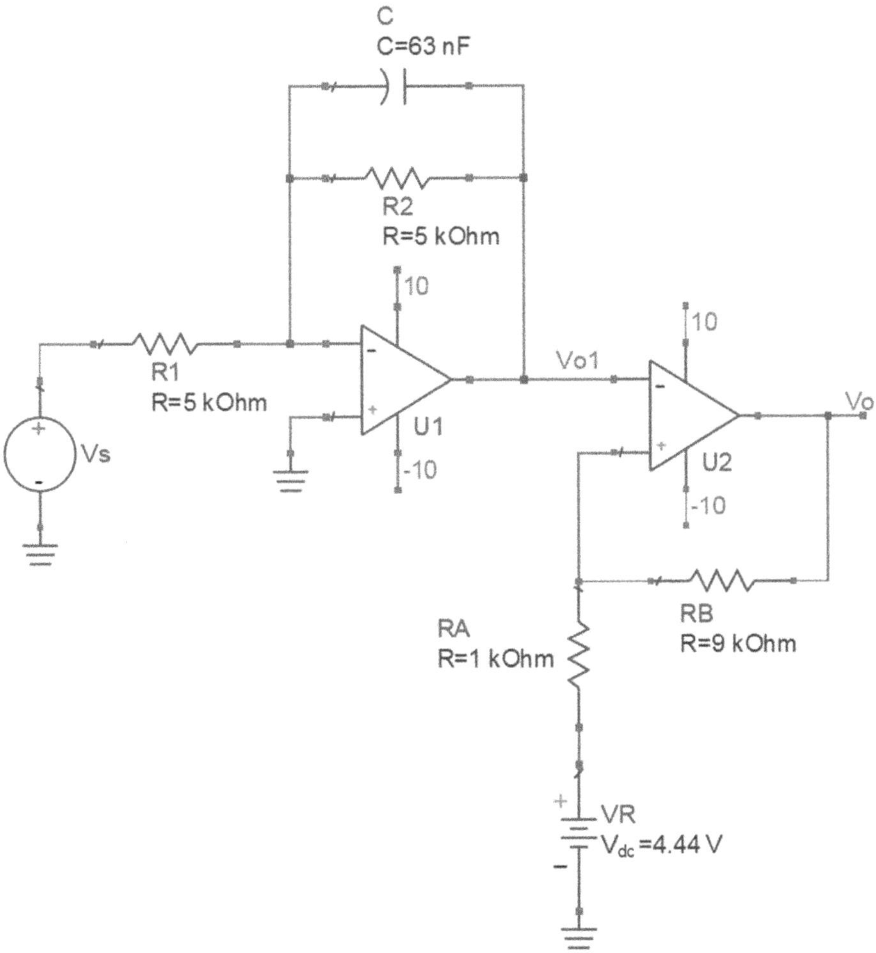

FIGURE 1.112 Problem 1.15 circuit.

14c: What is the output voltage of the comparator if $v_s(t)$ is the AC of amplitude 1.5 V and frequency 100 Hz? What is the output of the comparator if $v_s(t)$ is AC with a 5.2-V amplitude and 100 Hz frequency?

Demonstrate all the above in ADS.

Problem 1.15: Low-Pass Filter and Schmitt Trigger

Consider the following op-amp circuits connected in cascade. The input signal is undetermined yet (Figure 1.112).

15a: For $v_s(t) = 6\sin(2\pi \cdot 100t)$ [V] find $v_{o1}(t)$ and $v_o(t)$.

15b: For $v_s(t) = 6\sin(2\pi \cdot 50{,}000t)$ [V] find $v_{o1}(t)$ and $v_o(t)$.

2 Diode Circuits

2.1 INTRODUCTION TO DIODES: WHAT IS A DIODE?

A diode is a semiconductor device, often made of silicon, that allows electrical current to flow only in one direction, from the anode terminal to the cathode terminal (see Figure 2.1). This happens whenever the anode voltage is sufficiently more positive than the cathode voltage. If the cathode voltage is larger than the anode's voltage, the diode is OFF – its current is very small, practically zero.

Typical current–voltage characteristics of germanium and silicon diodes are shown in Figure 2.2. In Figure 2.2, the diode voltage V is the difference between the anode and cathode voltages.

I is the diode current, often defined as positive if it flows from the anode to the cathode. For a silicon diode the reverse bias region, in which the diode is OFF (that is $I \approx 0$), extends from some large negative value of V (say around –70 V) to a small "cut-in" positive value of about $V = V_{cut\text{-}in} \approx 0.5\,V$, where the diode turns ON and enters the forward bias region. The I vs. V curve in the forward bias region is exponential. Small changes in V cause large changes in I.

The third region, called the breakdown region happens when V becomes negative enough, allowing current to flow from the cathode to the anode. In a regular diode (contrary to a Zener diode, which will be explained later), the breakdown region is avoided.

Let us explain briefly, using semiconductor physics concepts and terminology, what makes a diode [Gray1], [Sedra1], [Sze1].

A silicon atom has four valence electrons. Each atom forms four covalent bonds with its neighboring atoms, thus creating a silicon crystalline structure. Due to thermal energy, a small number of such bonds may break up, releasing an equal number of free electrons and free "holes" into the crystal space. What are holes? Free electrons that move about may occupy free holes, causing the holes to wander about the crystal structure from place to place. This equal concentration of **free electrons** and **free holes**, in pure silicon, is called the **intrinsic concentration**. It depends strongly on the temperature. Pure silicon material is a poor conductor though. What makes silicon a "semiconductor" is the act of **doping**: Adding five-valent-electron **donor** atoms (such as phosphor) to the silicon material creates an **n-type silicon**, where

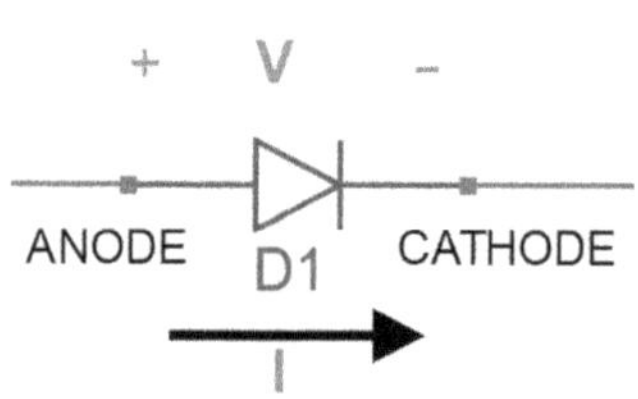

FIGURE 2.1 Diode electrical symbol and terminal names.

DOI: 10.1201/9781003606772-2

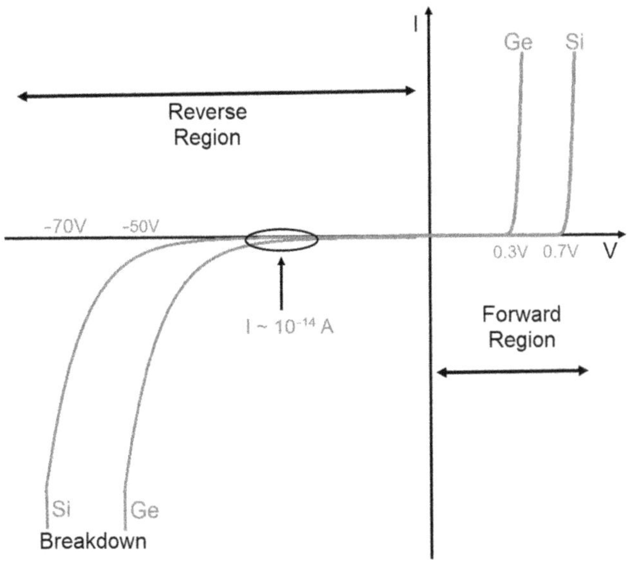

FIGURE 2.2 I vs. V curves, showing the three possible operation regions of a diode.

the concentration of free electrons is way larger than the concentration of holes. Likewise, adding three-valent-electron **acceptor** atoms (such as boron), to the pure silicon material, creates a **p-type silicon**, where the concentration of holes is much larger than the concentration of free electrons. The electrons in n-type silicon are called **majority carriers**, whereas the holes are referred to as **minority carriers**. Vice versa in p-type silicon, where holes are the majority carriers and electrons are minority carriers. What makes a diode is the surface contact created between n-type and p-type materials, called a **p–n junction**. For instance, a p–n junction is created if n-type silicon is grown into a p-type silicon wafer.

The p-type side of a diode is called an **anode**, and the n-type side of a diode is called a **cathode**. Let us explain why the phenomenon, of a current that is allowed to flow in one direction only, happens in a p–n junction.

The first fact to recognize is that there is, in general, a variety of currents in a p–n junction: Electron's current, holes current, which are both further classified into diffusion current and drift current. So, for instance, an electron **diffusion current** happens whenever there is an electron concentration gradient. For example, electrons on the n-type side of the junction have a tendency to diffuse toward the p-side, where the electron concentration is low. An example of another current, that may exist in a p–n junction, is a hole **drift current**: Say, that there is a positive applied voltage difference between the anode and the cathode, causing holes from the p-type side to drift toward the n-type side under the influence of the positive electrical field. That same positive field also creates an electron drift current from the cathode to the anode. The four types of currents combine to create the total current of the device.

The second fact to mention is the **depletion region**, which is created on both sides of the surface that separates the p-type and n-type regions. Holes that diffuse from the p-type side to the n-type side meet electrons that diffuse from the n-type side to the p-type side. Many free electrons and free holes recombine. The depletion region has almost no free carriers. What is left in that region are immobile positively ionized donors (on the n-type side of the junction) and immobile negatively ionized acceptors (on the p-type side of the junction). These charged regions create a built-in potential between the p-type and n-type sides of the p–n junction. So, for example, in an unconnected diode, the initial diffusion currents create the depletion region's electrical field, which creates counter drift currents. Hence, in an unconnected diode, the total current is, of course, zero, as the diffusion and drift currents balance one another.

In a diode that is **reverse biased**, the electrical field opposes the diffusion current of the p–n junction majority carriers, forcing the carriers away from the junction and further into their original side of the junction. This causes an increase in the width of the depletion region. The reverse field helps to create a small drift current of minority carriers. The total reverse current is very small. For all practical purposes, a reverse-biased diode is considered to be an OFF state.

In a diode that is **forward biased**, the diffussion currents and the drift currents all combine to create the diode's forward current. It can be proved [Sze1] that at the border of the depletion region with the n-type region, the concentration of minority carriers (holes) increases exponentially with the positive bias voltage, and likewise the concentration of minority carriers (electrons) on the other side of the depletion region.

A current–voltage equation for a diode (shown in equation 2.1a) is valid in both reverse and forward bias:

$$i_d = I_S\left(e^{\frac{v_d}{V_T}} - 1\right) \tag{2.1a}$$

$$V_T = \frac{KT}{q} \tag{2.1b}$$

The current I_S is called the **saturation current**. The voltage V_T is called the **thermal voltage**. K is Boltzmann's constant, q is the absolute value of an electron charge, and T is the temperature in Kelvin. At room temperature (T = 293 K), the thermal voltage equals approximately $V_T = 25.2\,\text{mV}$. The voltage v_d (in equation 2.1a) is the difference between the anode and the cathode voltages. In forward bias, v_d is positive and therefore $i_d \approx I_S \cdot \exp(v_d/V_T)$. In reverse bias, as v_d is negative, $i_d \approx -I_S$. As seen in Figure 2.2, the value of I_S is very small.

2.2 DC DIODE CIRCUITS

As seen by the curve of the diode current vs. diode voltage (shown in Figure 2.2), a diode is a nonlinear device. Whenever a diode is placed in a linear circuit, the whole circuit becomes nonlinear, defying many linear circuit analysis solution methods (such as superposition). Kirchhoff's voltage and current laws still work, of course, but as the diodes' i_d vs. v_d mathematical characteristics equation (2.1a) are substituted into the network equations, an analytic solution is usually not possible.

A practical solution (where a solution means finding all the component currents and all the node voltages) of any circuit, that contains diodes, necessitates an initial **assumption regarding the state of each diode**. Is a diode ON or OFF? Following up on the assumption, the circuit is solved, and then the initial assumptions must be either justified or disputed. For each diode, we need to check whether the anode and cathode voltages agree with the diode's state assumption, and whether the diode's current (if the diode is assumed to be ON) indeed flows from the anode to the cathode. If any of the assumptions result in a contradiction, we have to go back to the start and make a different set of assumptions.

Let us comment on the diodes' state assumption process. If a circuit contains N diodes, do we sometimes go through all 2^N combinatorial possibilities? Of course not. For example, diodes in series, all have the same state, either all are ON or all are OFF. Another helpful hint comes from the "water flow analogy": As water tends to flow from a higher place to a lower place, an electrical current tends to flow from a higher voltage node toward a lower voltage node. Few assumptions tend to be plausible, whereas most other combinatorial possibilities make no sense.

As we follow up on the diodes' states assumption, we often replace every diode's nonlinear i_d vs. v_d relationship with a simpler model. We neglect the current I_S (of equation 2.1a) and assume that the diode becomes an open circuit whenever it is in reverse bias. For forward bias, there are three possible mathematical models of the diode.

We use either the left model or the middle model in analytic solutions, which both yield approximated currents and voltages results, and we compare against the exact solution created by simulators, such as ADS. Almost always, solutions that use the ideal model and solutions that use the practical model, tend to agree on the "big picture" – which diodes are ON and which ones are OFF. The two approaches generate different sets of approximated currents and voltages, and almost always, the solutions created with the aid of practical diode models come closer to the exact simulation results.

Let us add several comments about the diode models.

In the ideal model, we assume that, for any negative diode voltage, the diode is an open circuit (that is, having a zero current). For any attempted positive diode voltage, the diode becomes a short circuit.

Figure 2.3 (middle) current vs. voltage graph is actually an ideal diode in series with a 0.7 V battery. In other words, the diode is OFF for negative voltages, as well as for small positive voltages (up to 0.7 V). Then, when the diode turns ON, it has a voltage drop of 0.7 V, no matter what its current size is. A slight refinement of the practical model is to assume that the diode is OFF up to some "cut-in" voltage, say 0.5 V. When the diode turns ON, we assume a voltage drop of 0.7 V.

The exponential curve of the exact model is steep – small changes in v_d result in large changes in the diode's current i_d. Vice versa, large changes in the diode's current result in slight changes in the diode's voltage, which indeed hovers around 0.7 V.

In the exact model, every current value corresponds to a unique diode's voltage. In the rare occasions when we need to use a diode's exact model in an analytic

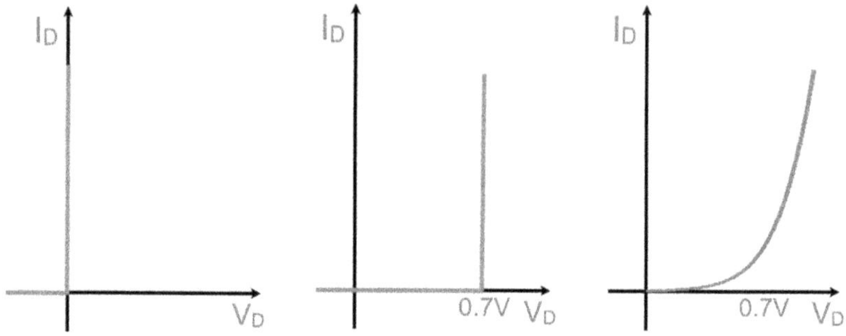

FIGURE 2.3 (Left) Ideal model (diode is either an open circuit or a short circuit). (Middle) Practical model (diode is OFF for negative voltages and for positive voltages up to some cut-in voltage), and when it turns ON, its voltage drop is constant and independent of the diode's current. (Right) Exact model (an exponential current–voltage relationship whenever the diode is ON, or an open circuit when the diode is OFF).

solution, it is helpful to recognize that a known point (i_{d1}, v_{d1}) on the exponential plot can help us find any other pair (i_{d2}, v_{d2}) in which one of the entries is unknown. From equation (2.1a):

$$\frac{i_{d1}}{i_{d2}} = e^{\frac{v_{d1}-v_{d2}}{V_T}} \quad \text{or} \quad v_{d1} - v_{d2} = V_T \ln\left(\frac{i_{d1}}{i_{d2}}\right) \tag{2.2}$$

We use the left formula whenever both voltages are known and one of the currents is unknown. We use the right formula whenever two currents are known and we seek to find the unknown voltage.

Let us end this theory part by solving a few illustrative examples:

Example 2.1

See Figure 2.4. By the "water flow analogy", we assume that the diode is ON. According to the "ideal model", the diode is replaced by a short circuit.

Therefore, $v_d = 0\,V$ and $i_d = 10\,mA$. As i_d flows from the anode to the cathode, there seem to be no contradictions in the solution. According to the "practical model", we replace the diode with a 0.7-V battery. Then $v_d = 0.7\,V$ and $i_d = 9.3\,mA$. Again, the solution is consistent with the assumption. We shall soon simulate this example and observe which of the above two solutions comes closer to the exact solution.

Example 2.2

The example is based on [Sedra1].

Each diode's anode is connected to a DC source (as shown in Figure 2.5). By the "water flow analogy", one may initially assume that every one of the diodes is ON; however, the shared cathode immediately implies a contradiction: Whether one

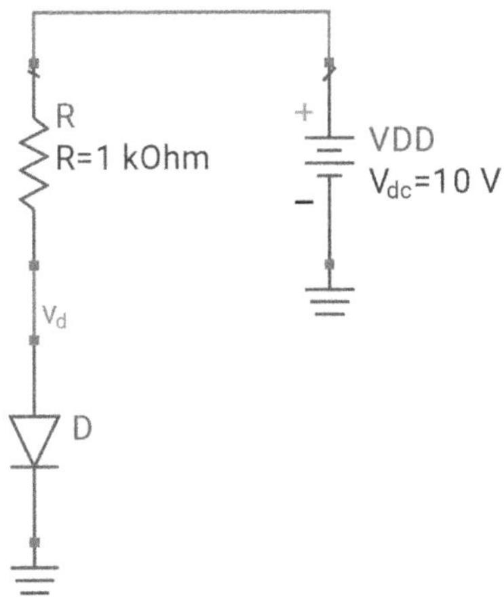

FIGURE 2.4 Single diode DC circuit.

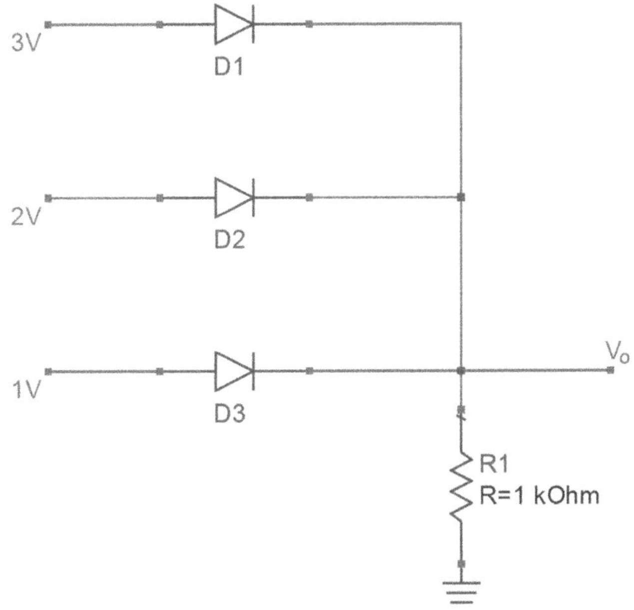

FIGURE 2.5 Multiple diodes in a DC circuit.

uses ideal models or practical models for the ON diodes, each cathode voltage equals the anode voltage minus the diode's voltage drop (either 0 or 0.7 V). Since the three cathodes are short-circuited, there can be only one cathode voltage.

Consequently, only one of the diodes can be ON. The others must be OFF. Which one? If we assume that D_3 is ON then $v_o = 1$ V (if we use the ideal model) or $v_o = 0.3$ V (if we use the practical model). Now that we know the anode and cathode voltages of each diode, we may justify or dispute the initial assumption: The voltage drop of D_1 is 2 V (for ideal models) or 2.7 V (for practical models) but with such a voltage drop, the diode should have been ON and not OFF (as assumed). Likewise for D_2.

Conclusion: Need to assume that diode D_1 is ON and the other diodes are OFF. Following up on the assumption: $v_o = 3$ V (ideal model) or $v_o = 2.3$ V (practical model).

Comparing such cathode voltage to the anode voltages of D_2 and D_3 justifies the assumption that both are OFF.

Once we find a working solution, there is no need to check other combinations. Here, for example, we have not checked the assumption that all diodes are OFF, even though it could have easily been disputed: If all diodes are OFF, then $v_o = 0$ V, but such a cathode voltage creates a large positive voltage drop in each diode, which contradicts the OFF assumption.

Finally, going back to the correct solution (above). The resistor's current is either 3 or 2.3 mA (depending on the adopted diode model), flowing from the 3 V battery through diode D_1.

2.3 ADS DC SIMULATION OF DIODE CIRCUITS; INTRODUCING CURRENT PROBES, RESISTOR, AND TEMPERATURE SWEEP; PLOTTING OF A VARIABLE AGAINST ANOTHER VARIABLE

Let us build a simple resistor and diode DC circuit. The diode D1N4148 is taken from the FAU EE design kit (see Figure 2.6 (left)).

Let us use DC simulation (see Figure 2.7).

We need to measure the diode's current. Let us bring in a current probe (see Figure 2.8 (right)).

Alternatively, one can go to the ADS component library, start typing "pro…" and the I_Probe component becomes visible (Figure 2.8).

We connect the I_Probe as shown in Figure 2.9. For now, let us make sure the I_Probe arrow points toward the D1N4148 diode as shown. More information about this is forthcoming.

No point in plotting time domain graphs of DC entities. A simple table will do. See Figures 2.10 and 2.11.

According to the practical model approximation, we expect the diode's voltage to be $v_d = 0.7$ V (compared to the actual value of 0.7195 V shown in Figure 2.11) and the diode's current to be $i_d = (V_1 - v_d)/R_1 = (10 - 0.7)/1 = 9.3$ mA (compared to the actual value of 9.28 mA).

A second method of watching DC voltages and currents is to annotate the circuit diagram itself, which we will discuss after discussing current polarity first.

The placement of component I_Probe needs to be done with care.

The ADS current polarity convention is:

1. A current leaving a component node is negative, such as the current leaving a voltage source.
2. A current entering a component node is positive, such as that entering the diode's anode.

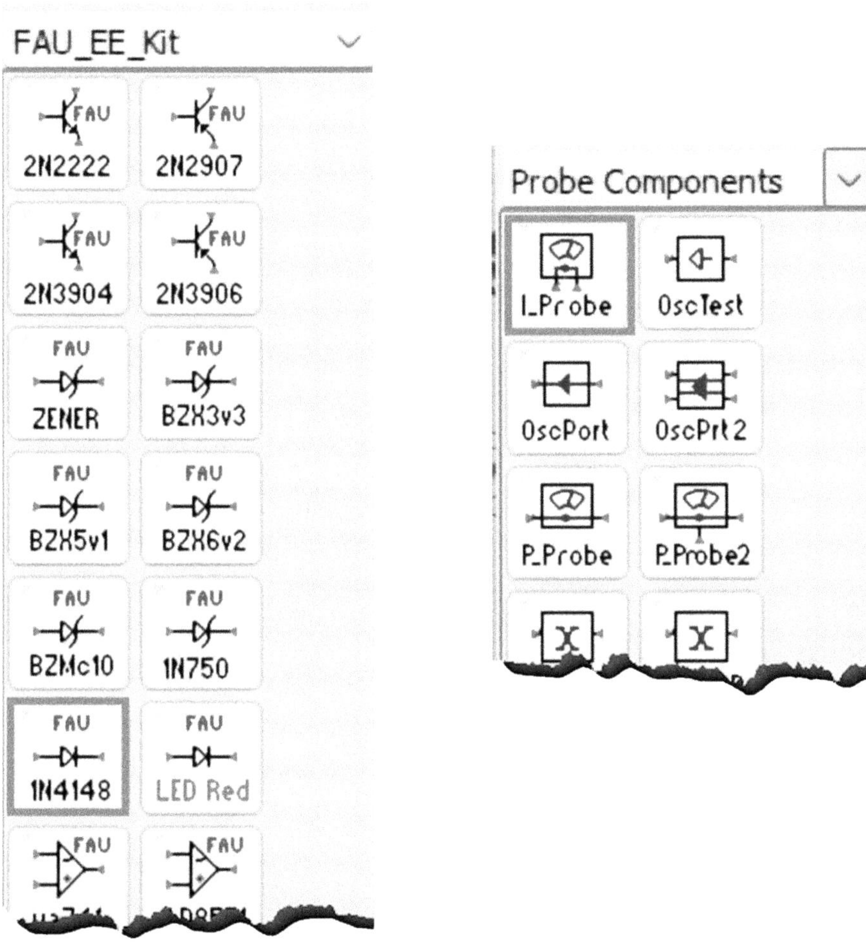

FIGURE 2.6 (Left) Segment of the FAU EE design kit integrated into the component's palette. (Right) Segment of the probe components palette.

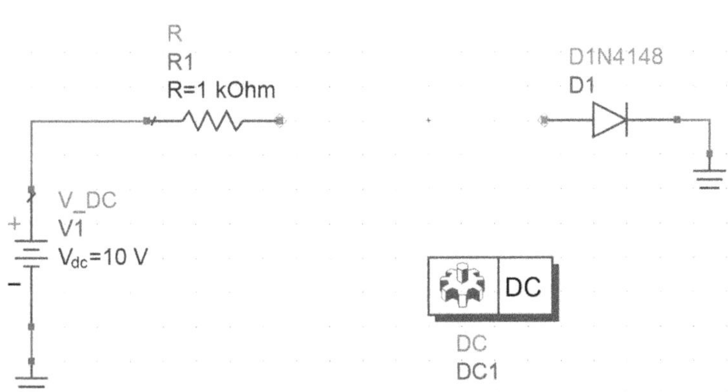

FIGURE 2.7 Construction of a resistor and diode circuit in DC simulation.

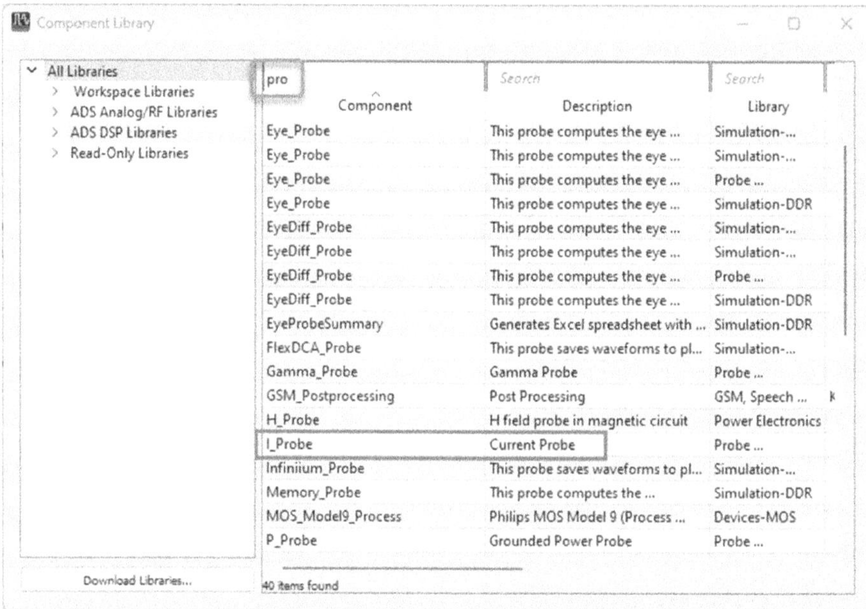

FIGURE 2.8 View of ADS component library.

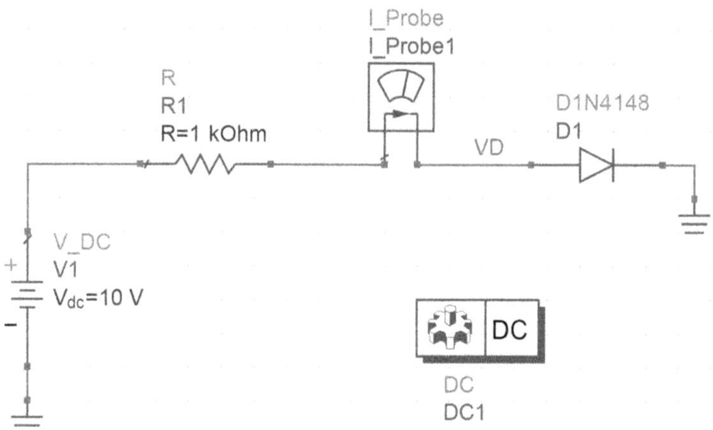

FIGURE 2.9 The full resistor and diode circuit.

For example, as shown in Figure 2.12, the current probe arrow now points away from the diode. The DC current result is now a negative polarity current because from the current probe perspective, the current is flowing in the opposite direction.

After the circuit is simulated, even if the data display has been closed, we select "Simulate" from the toolbar. From the Simulate menu, select "DC Annotation", as shown in Figure 2.13 (right). Then select "Annotate Voltage", for each node voltage to show up in the diagram. Select "Annotate Pin Current" and the currents show up. If we make a change in the circuit and simulate again, the I and V values get updated.

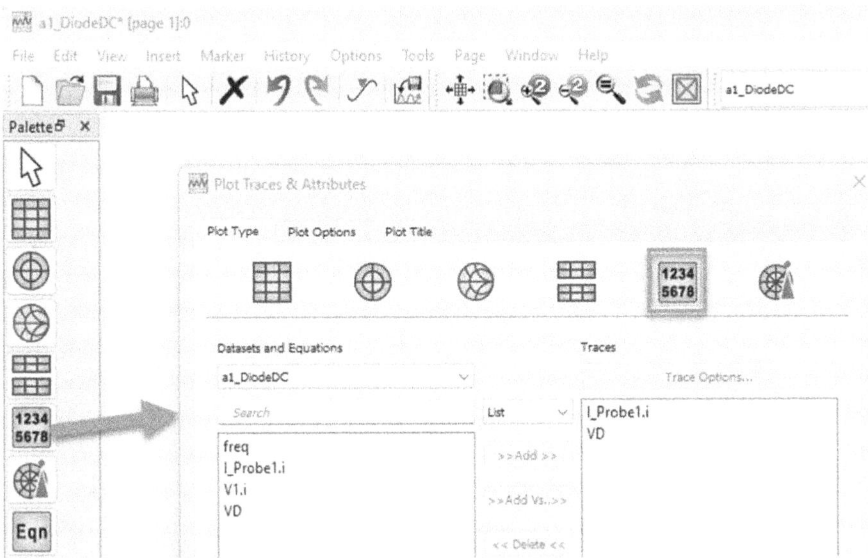

FIGURE 2.10 View of plot traces, showing the choice of a table for the results. In ADS simulation, currents are denoted with an ".i" ending.

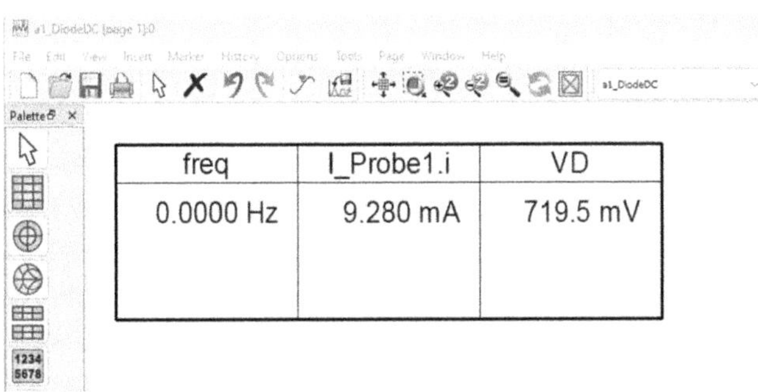

FIGURE 2.11 The resulting diode voltage and current.

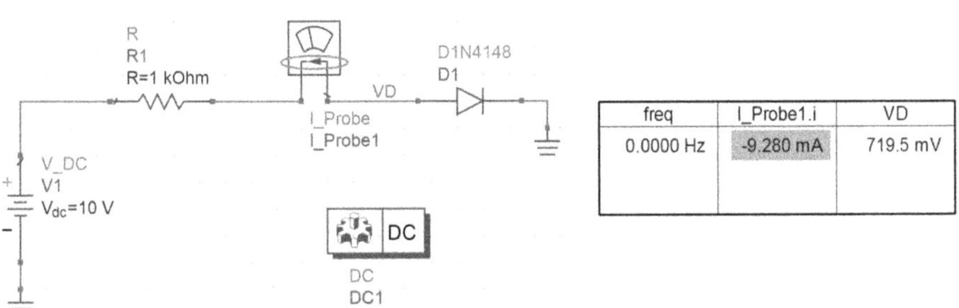

FIGURE 2.12 A look at a current probe having the wrong current polarity.

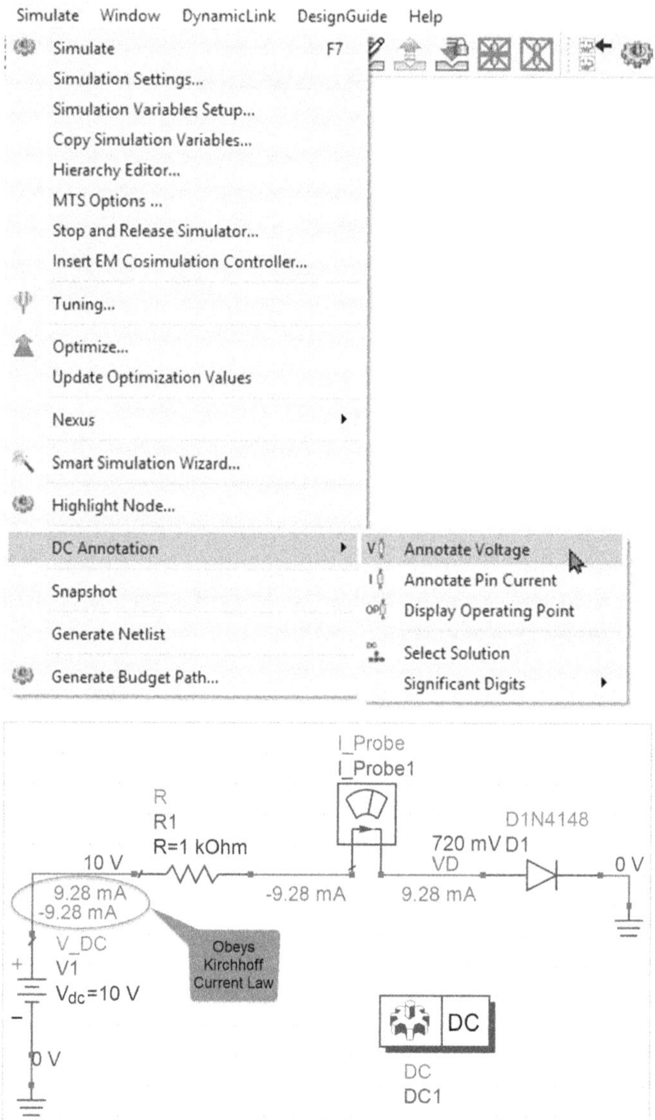

FIGURE 2.13 Currents and node voltages DC annotation.

To turn off DC voltage and current annotation, just select them from the Simulate >> DC Annotation menu again (these are toggle switches for this feature).

Say that we change R_1 from 1 to 2 kΩ. Need to simulate again. We can close the output file. The updated values appear in the circuit diagram (as shown in Figure 2.14):

I_D changed from 9.38 to 4.66 mA. V_D changed from 720 to 682 mV.

We see that for each diode's current value, there is a different value of the diode's voltage. How do we plot the full I_D vs. V_D curve?

We "save as" the previous schematic as a first step in creating a new one.

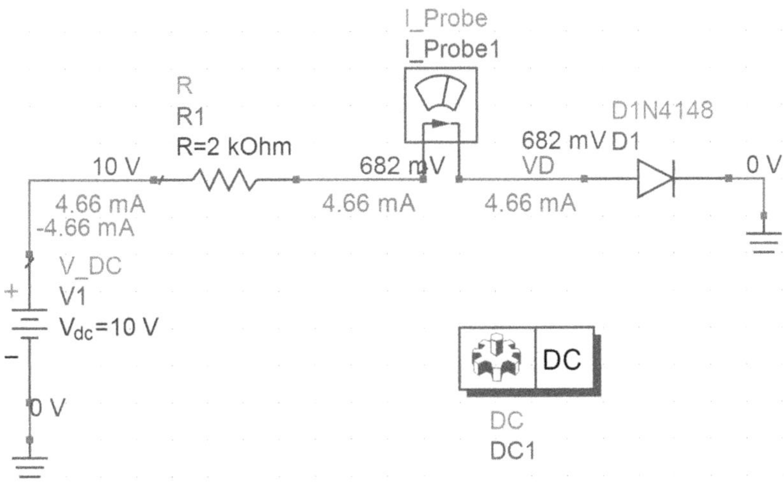

FIGURE 2.14 Increase of R causes a reduction of the current and a slight reduction of the diode's voltage.

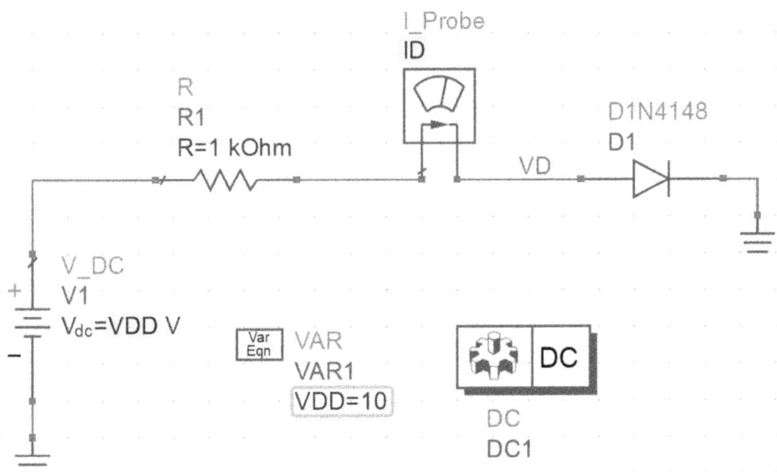

FIGURE 2.15 Pre-sweep preparations.

Small changes (see Figure 2.15): Name of probe changed to I_D. The DC supply voltage became variable V_{DD}, with a default value of 10 V.

We want to plot I_D as function of the independent variable V_{DD}. Then we want to plot I_D as a function of the dependent variable V_D.

One method of doing single-parameter curving (in DC simulation) is to use the sweeping ability of the DC simulation controller. See Figure 2.16.

Double-click the DC simulation block in Figure 2.15. Fill in the dialog box for Sweep. As can be observed from the Sweep dialog box, there will be a total of 201 V_{DD} points, ranging from 0 to 20 at increments of 0.1. That is, 201 DC simulation runs, accumulating all the output data.

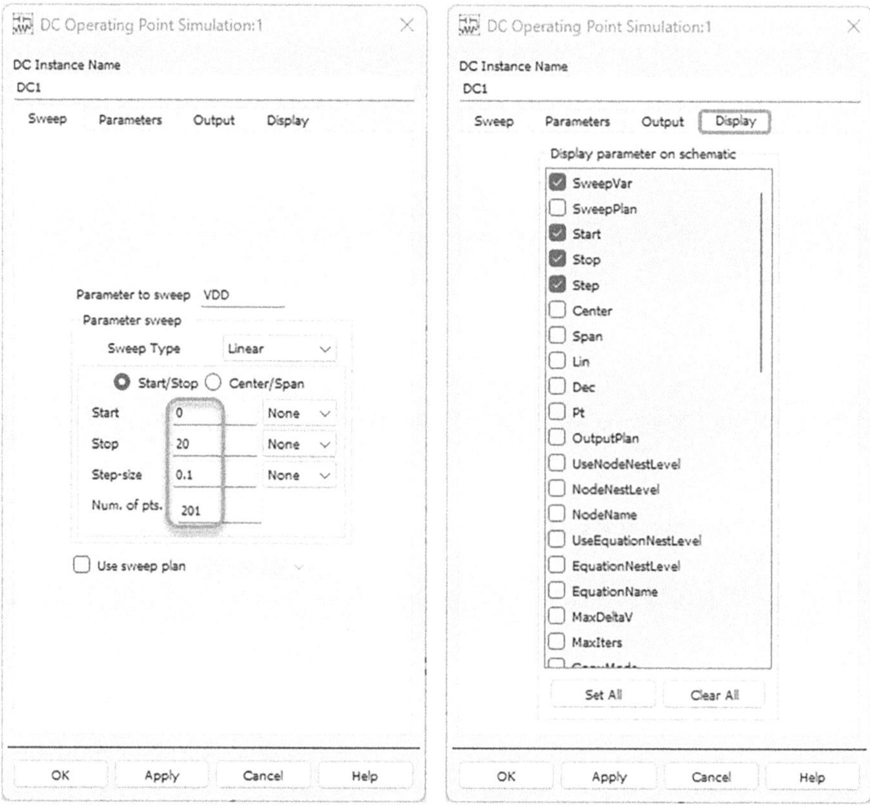

FIGURE 2.16 Sweep information and sweep display information.

Select SweepVar, Start, Stop, and Step in the dialog box under the display tab. This will make these parameters visible in the schematic page. See Figure 2.16.

See Figure 2.17. Voltage and current swept data will be generated. However, the plots do not have to be only with respect to the swept variable (V_{DD} in this case). Any variable can provide the X axis, and any variable can be the Y axis.

What do we expect to see? The diode has a cut-in voltage of nearly 0.5 V. Therefore, as seen in Figure 2.18, for low supply voltage, the diode will be OFF. Then the diode turns ON and has a voltage near 0.7 V.

Following the usual Insert > Plot routine, we choose this time the >> Add vs. >> option. First, we pick up ID.i. Then we picked up V_{DD} from the offered menu.

We see in Figure 2.18 that $I_D \approx 0$ for $V_{DD} < 0.5$ V. Why is the plot in Figure 2.18 linear? $(V_{DD} - V_D)/R_1 \approx V_{DD}/R_1$, which is linear in V_{DD}.

Repeating the Insert > Plot steps, we can get the diode's I_D vs. V_D exponential curve. The Y-axis unit is [A]. With a resistor of 1 kΩ and V_{DD} that can run up to 20 V, the highest curve point was $V_D = 0.763$ V and $I_D = 19$ mA, as seen in Figure 2.19.

Changing the sweep range of V_{DD} from 0–20 V to −120–20 V allows us to see the full reverse bias region as well as the diode's breakdown region, with a breakdown voltage of −100 V of the D1N4148 diode (Figure 2.20).

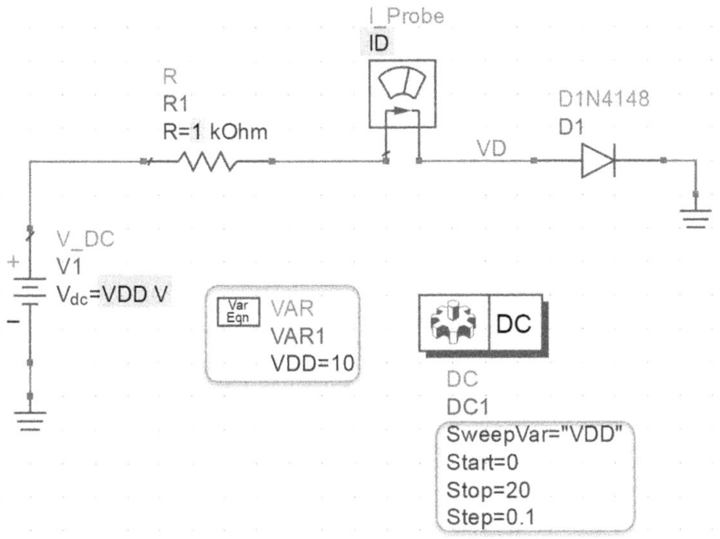

FIGURE 2.17 A ready-to-run diode curves sweep study schematic.

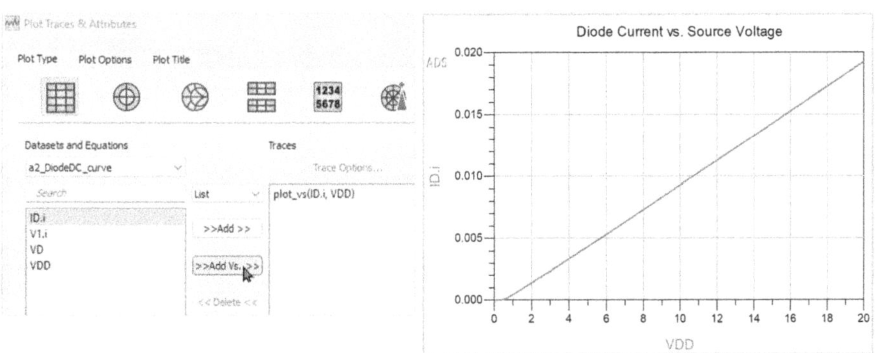

FIGURE 2.18 Diode's current vs. supply voltage V_{DD}.

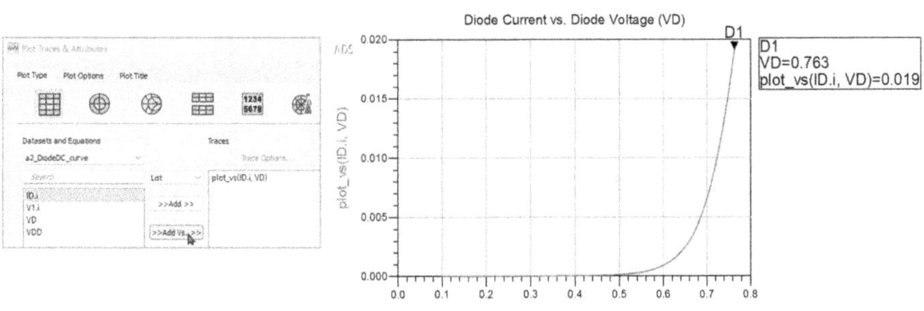

FIGURE 2.19 Diode's current vs. diode's voltage.

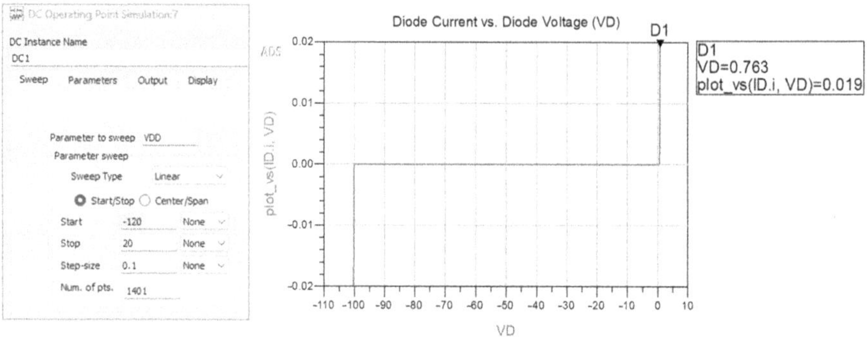

FIGURE 2.20 Diode's I–V curve going from breakdown, through reverse bias, and ending at forward bias.

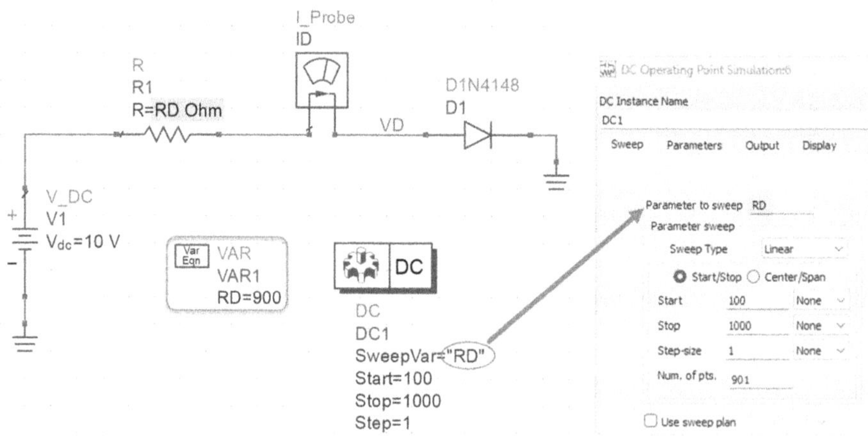

FIGURE 2.21 DC simulation with resistor value sweeping.

Example 2.1 (continued)

Let us use the original values: DC supply of V = 10 V and R = 1 kΩ. Find a new value of R for which the diode's voltage becomes $v_d = 0.8$ V.

We can sweep the value of R (named RD in Figure 2.21) and estimate, with the aid of markers, its needed value from the V_D vs. R_D plot.

The resulting VD vs. RD plot is shown in Figure 2.22. By placing a marker m1 and carefully moving it along the plot until a VD = 0.8 V value indicates that we need RD = 286 Ω.

See Section 2.4 for "How to move a marker" hints.

Finally, let us demonstrate the temperature sweep. In this experiment, the circuit does not change, only the ambient temperature. See Figure 2.23.

In the DC simulation block, choose the Sweep column and the Parameter to Sweep field. Type "temp" to indicate that the sweeping will be of the temperature. See Figure 2.24.

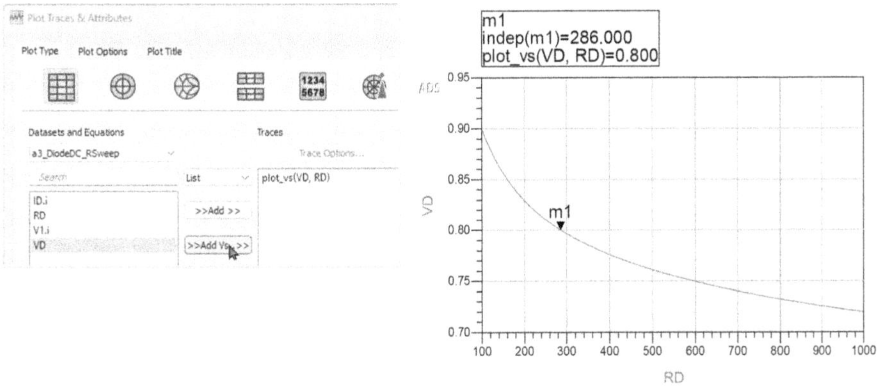

FIGURE 2.22 Results of the resistor value sweep.

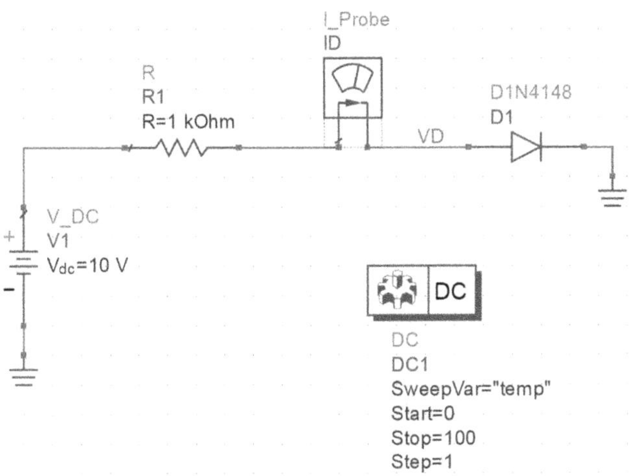

FIGURE 2.23 Temperature sweep of the resistor–diode circuit.

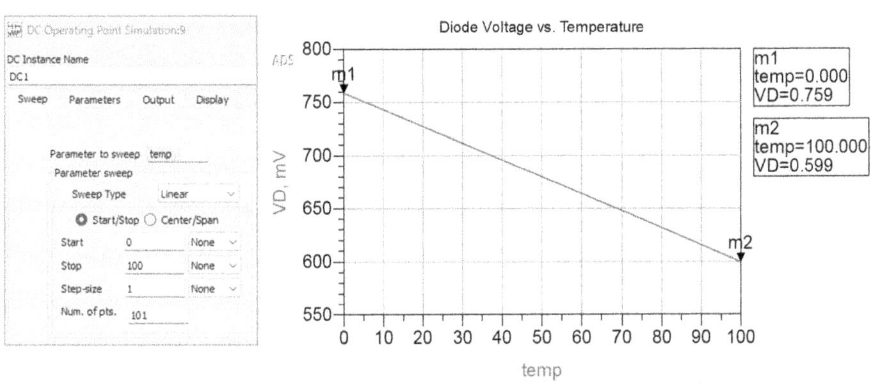

FIGURE 2.24 (Left) Setting of the temperature sweeping dialog box. (Right) The resulting plot of VD vs. temperature.

"temp" is a built-in ADS variable. The temperature's unit is [°C]. So, sweeping from 0°C to 100°C is like placing the tested circuit in conditions that run from freezing cold to boiling hot.

In this simple resistor and diode circuit (Figure 2.23), which elements are temperature sensitive?

Recall the thermal voltage parameter $V_T = KT/q$ in the diode's current vs. voltage equation. Also recall the pure silicon's intrinsic concentration n_i, which depends strongly on the temperature.

ADS resistors do have thermal coefficients; however, their effect on the resistance is of secondary importance. In this circuit, the diode is the most temperature-sensitive element.

The diode's voltage V_D appears to be decreasing linearly with temperature.

$V_D = 0.76\,V$ at $T = 0°C$ and $V_D = 0.6\,V$ at $T = 100°C$.

So, a diode embedded in a V+R+D circuit can serve as a linear temperature sensor. It is probably not the best temperature sensor, as its thermal sensitivity is only $(0.76 - 0.6)/100 = 1.6$ mV/°C. This sensor definitely needs an accompanying voltage level shifting circuit. Some thermistors (which are resistors that are specially constructed to be strongly temperature dependent) have a large temperature coefficient.

2.4 ADS TIPS AND TRICKS 2A: MARKER HINTS

When we place a marker, we can drag it to a location using the mouse. On occasions, if we have high resolution (many points), it may be hard to get exactly where we want with the mouse. We use the left/right arrow keys for finer movement.

We may sometimes want more significant digits for your marker's results. We double-click on the marker to bring up the properties window. Then we select the format tab and increase the number of "significant digits". See Figure 2.25. You may want to experiment with the format as well. See Figure 2.25.

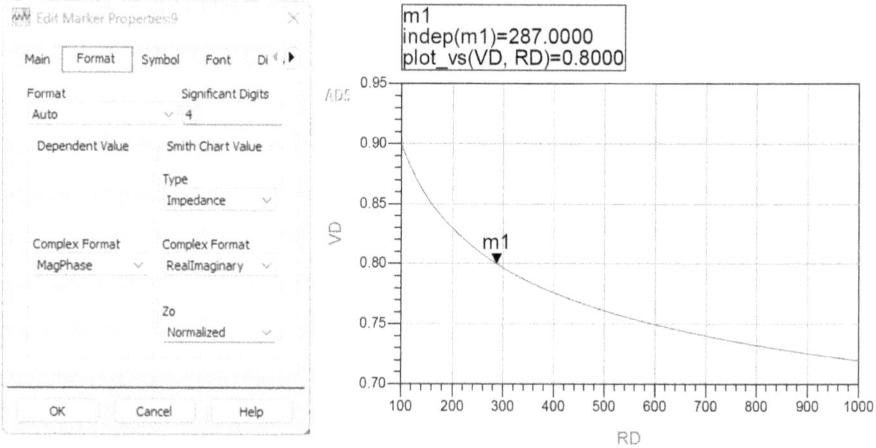

FIGURE 2.25 Increase markers significant digits.

2.5 ADS TIPS AND TRICKS 2B: STAY ORGANIZED

Many newcomers to electronics and simulation learn the hard way that keeping your work organized pays off many times sooner than later.

In the new workspace that we created for this part of the book, we decided to name our circuits with a prefix using the letter "a+num" in sequence as we created the different circuits. This has the fortunate effect of listing the circuits sequentially in the ADS main window folder view. It makes it easy to find what we are looking for. Note that we gave names that are meaningful, that is, no need to open a circuit to know what is inside or guess what is inside. See Figure 2.26.

2.6 DIODE RECTIFIERS: HALF-WAVE RECTIFIERS

A diode in a DC circuit is either ON all the time or OFF all the time. Not so whenever a diode's nodes are exposed to time-varying voltage signals. In such an event, a diode may be ON for some of the signal's values and OFF for the other signal's values.

A **half-wave voltage rectifier** is one of the simplest diode circuits to help us study diode behavior in the time domain, and it has real-life applications. What a half-wave rectifier does is to eliminate parts of the input voltage signal that are either negative (as shown in Figure 2.27) or positive (as shown in Figure 2.28). The output of the half-wave rectifier is ideally equal to the input voltage, without the input signal portions that got erased. "Ideally" is if we use the diode's open-circuit-short-circuit ideal model. In reality, we need to take into account the diode's cut-in voltage and its voltage drop while conducting.

Using first a diode's ideal model, we see from Figure 2.27 that whenever $v_{in}(t) > 0$, the diode is ON, and because we assume that it works like a short circuit $v_o(t) = v_{in}(t)$. Whenever $v_{in}(t) < 0$, the diode is OFF and therefore there is no current flowing through R, hence $v_o(t) = 0$.

For example, if $v_{in}(t) = A \cdot \sin(2\pi t/T)$, then (ideally) $v_o(t) = A \cdot \sin(2\pi t/T)$ for the first half of each period: $nT < t < nT + T/2$, $n = 0, 1, 2, \ldots$, and $v_o(t) = 0$ for the second half of each period: $nT + T/2 < t < (n+1)T$, $n = 0, 1, 2, \ldots$.

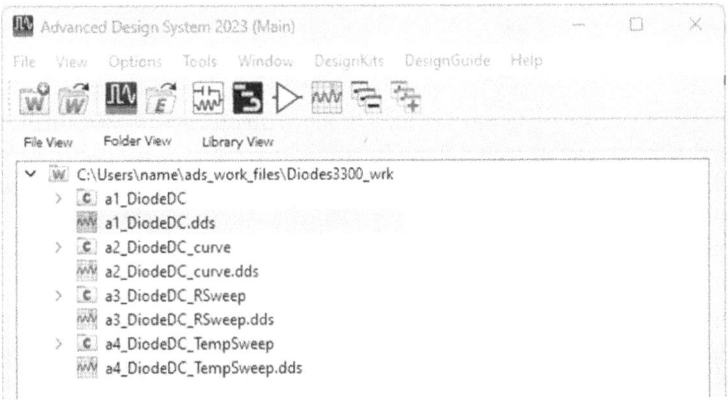

FIGURE 2.26 Using workspace cell names to stay organized.

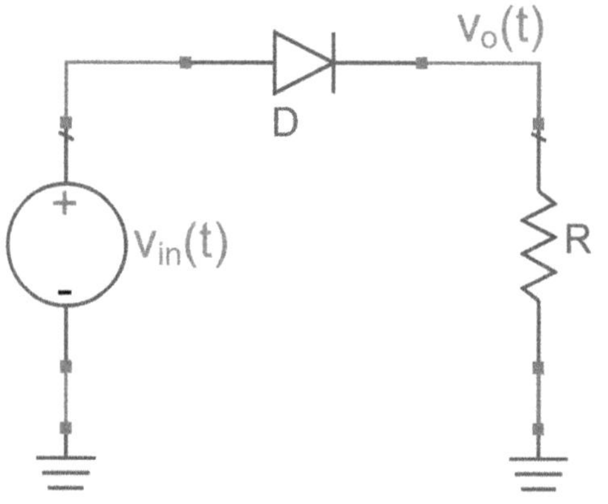

FIGURE 2.27 A half-wave rectifier that removes negative-valued parts of the input voltage.

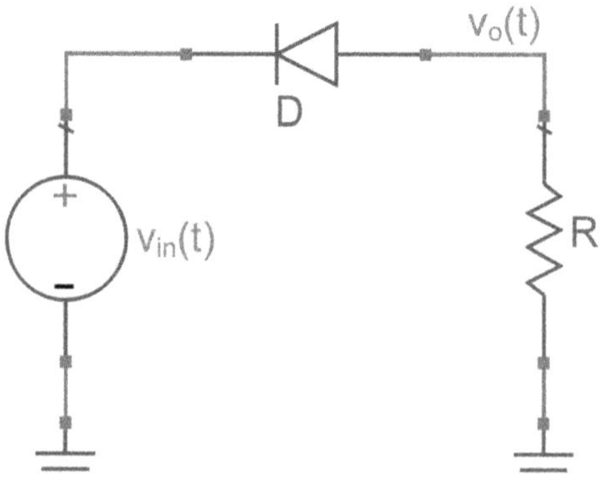

FIGURE 2.28 A half-wave rectifier that removes positive-valued parts of the input voltage.

In the above example, taking into account the diode's exact behavior (as well as its practical model), the diode will not turn ON till the AC input signal becomes larger than the diode's cut-in voltage of about 0.5 V. So, as seen in Figure 2.27 showing the simulation schematic of a half-wave rectifier driven by an AC voltage, we expect the positive part of $v_0(t)$ to occupy less than half of the input signal's period, and its peak voltage to be by one diode voltage (which is approximately 0.7 V) less than the peak input voltage. More explanations about Figure 2.27 are given shortly.

In order to eliminate positive portions of the input voltage signal and retain its negative-valued portions, all we have to do is to flip the diode, as shown in Figure 2.28.

In Figure 2.28, the diode is OFF whenever the input voltage is positive, and is (ideally) ON whenever the input voltage is negative.

A **full-wave rectifier** can be designed to fold up negative parts of the input voltage, rather than eliminating them, as is done by the half-wave rectifier. It can also be designed to fold down positive parts of the input instead of eliminating them.

Before explaining how full-wave rectifiers are built, let us mention that any rectifier (be it either half wave or full wave) is an important part of a bigger circuit designed

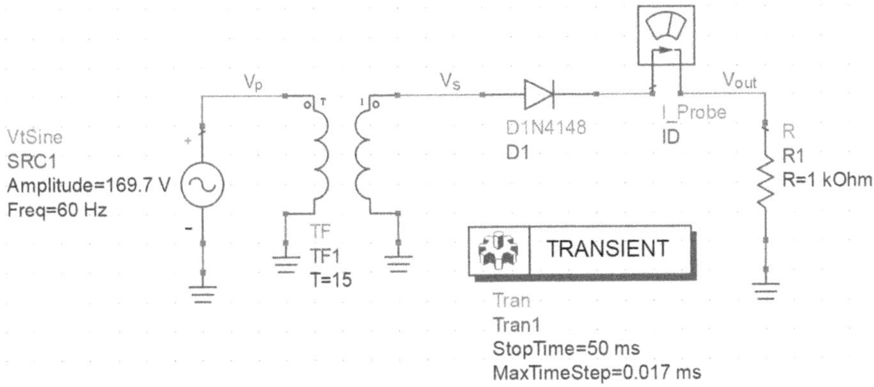

FIGURE 2.29 A half-wave rectifier fed from the electricity line via a regular transformer.

to convert an AC line voltage (in the US: Amplitude of 169.7 V, frequency 60 Hz) to a DC voltage of a desired size, subsequently acting as a DC voltage supply to an electronic circuit. Therefore, the input voltage (such as $v_{in}(t)$ shown in Figures 2.27 and 2.28) may often come from the secondary side of a transformer that has its primary side connected to the electricity line. A regular transformer is featured, for instance, in the ADS schematic shown in Figure 2.29.

2.7 ADS IDEAL TRANSFORMERS; TRANSIENT SIMULATION OF A HALF-WAVE RECTIFIER

Let us add a few explanations to the above schematic. A regular transformer TF may be found in the components library. The transformer is ideal (that is, it has no power losses). The user has to specify the transformer's turning ratio T. In the schematic shown in Figure 2.29, the name V_p stands for the transformer's primary voltage (a 60-Hz AC voltage with an amplitude of 120 Vrms, which is 169.7 V peak amplitude). V_s is the name given to the transformer's secondary voltage. Its amplitude for a turning ratio of T = 15 is 169.7/15 = 11.27 V. The transient simulation parameters are based on the line period of T = 16.67 ms (apologizing for giving the same name to the transformer turning ratio and the line voltage period). A StopTime of 50 ms allows for three periods to be viewed. The MaxTimeStep is taken as T/100 = 16.67/100 ≈ 0.17 ms.

The transformer is not so essential in these simulations. Its action can be emulated by a single VtSine source equal to $V_s(t)$.

The simulation results of the above schematic are shown in Figure 2.30. Notice how V_{out} comes out smaller than V_s, due to the diode's voltage drop, whenever the diode is ON.

In Figure 2.31, we show the same circuit, but with a flipped direction for the diode, resulting in a negative rectified voltage.

So, if rectifying a signal is considered Step One toward AC to DC conversion, then positive rectifying is intended to create some positive DC voltage. Negative rectifying is intended for creating a negative DC voltage.

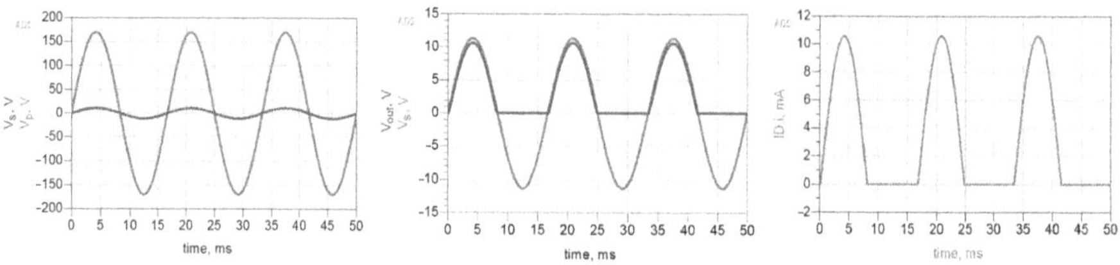

FIGURE 2.30 (Left) Transformer's primary and secondary voltages. (Middle) Rectifier's positive output voltage, compared to the transformer's secondary voltage. (Right) The half-rectified diode's current.

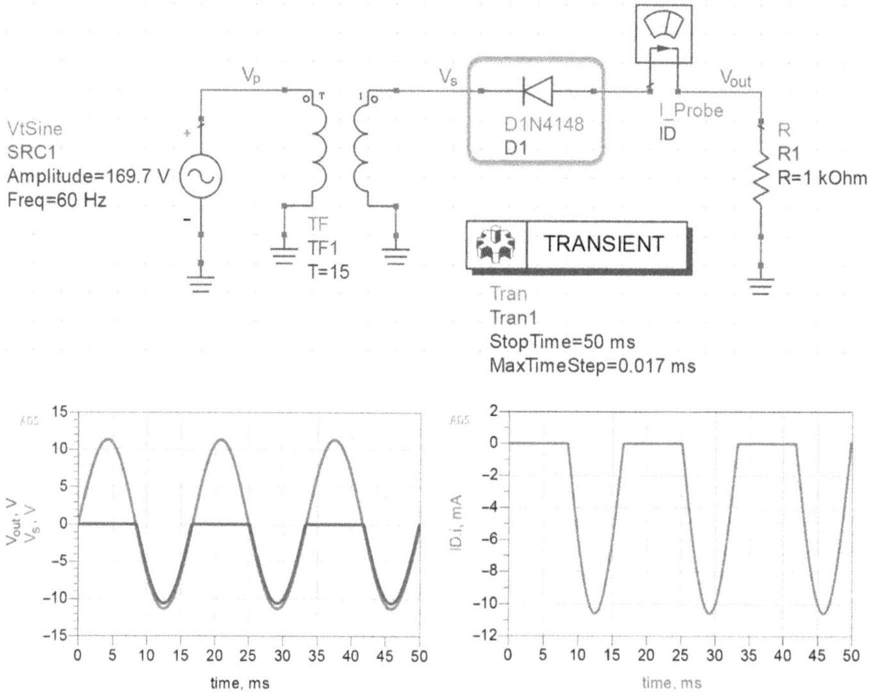

FIGURE 2.31 Negative rectified voltage. Up: Rectifier circuit with a flipped direction for the diode. Down left: The negatively rectified output voltage, compared to the transformer secondary signal. Down right: The negatively rectified diode's current.

2.8 FULL-WAVE RECTIFIERS

In a full-wave rectifier, instead of eliminating the positive (or negative) part of the rectifier's input voltage, we make such signal parts fold over to the opposite side.

Let us explain two methods of full-wave rectifying: One that uses two diodes and a special transformer, called a "center-tapped transformer". The second method uses a regular transformer, but it requires using four diodes.

The transformer's secondary-side center tap divides the secondary coil into two equal parts. The two voltages $V_{s1}(t)$ and $V_{s2}(t)$, shown in Figure 2.32 (left), add up to

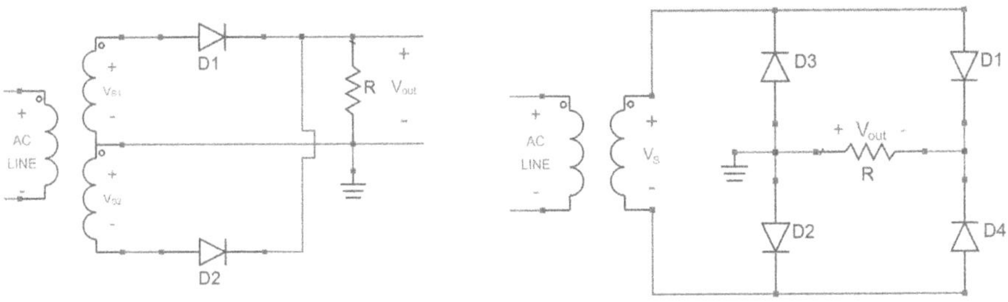

FIGURE 2.32 Two methods for full-wave rectifying. (Left) Center-tapped transformer and two diodes. (Right) a regular transformer and four diodes forming a "diode bridge".

the full secondary voltage $V_s(t)$. The grounding of the center tap causes, at any time, any one of the two diodes to have an input voltage that is the negative of the voltage input that feeds the other diode. In other words, whenever D_1 is ON, D_2 must be OFF, and vice versa. Furthermore, a diode that is ON sends a current to the load R, which goes in the same direction as the current sent to R when the other diode becomes ON. The two current signals complement one another and, by that, create the voltage folding effect.

Unlike the half-wave rectifier and the center-tapped transformer full-wave rectifier, where some spot of the, secondary transformer becomes grounded, in the diode-bridge full-wave rectifier the secondary side of the transformer is kept floating. The diodes are arranged so that at any half-cycle of the AC secondary voltage, two of the diodes are ON, whereas the other two diodes are OFF. Whenever the upper node of the secondary coil is at a positive voltage, diodes D_1 and D_2 are ON. Whenever the upper node of the coil is at a negative voltage, diodes D_3 and D_4 are ON. See Figure 2.32 (right) for two alternating Kirchhoff paths: GND → D_2 → Secondary (from bottom to top) → D_1 → R → GND, or GND → D_3 → Secondary (from top to bottom) → D_4 → R → GND. In both paths, the current signal flows into the resistor in the same direction, creating voltage folding.

In both full-wave rectifying methods, the voltage drop of the resistor is the rectifier's output signal. For a center-tapped transformer method the output voltage is by one diode voltage drop smaller than half of the transformer secondary voltage.

For a diode-bridge method, the output voltage is by two diode voltage drops smaller than the full secondary voltage.

2.9 ADS: CENTER-TAPPED TRANSFORMER AND ITS EMULATION; TRANSIENT SIMULATION OF FULL-WAVE RECTIFIERS

First, let us create a rectifier based on a center-tapped transformer and two diodes. Searching for a center-tapped transformer may proceed in the component's palette or in the component library, as shown in Figure 2.33 regarding the TF3 component.

The turn ratios are: $TR_1 = 1/T_1$ and $TR_2 = 1/T_2$. A turns ratio *less than 1* describes a transformer in a step-down configuration. The ends that are in phase are identified by a small open circle on the schematic symbol. The connection between the two pins and ground in the secondary and equal turn ratios make this transformer a center-tapped ideal transformer.

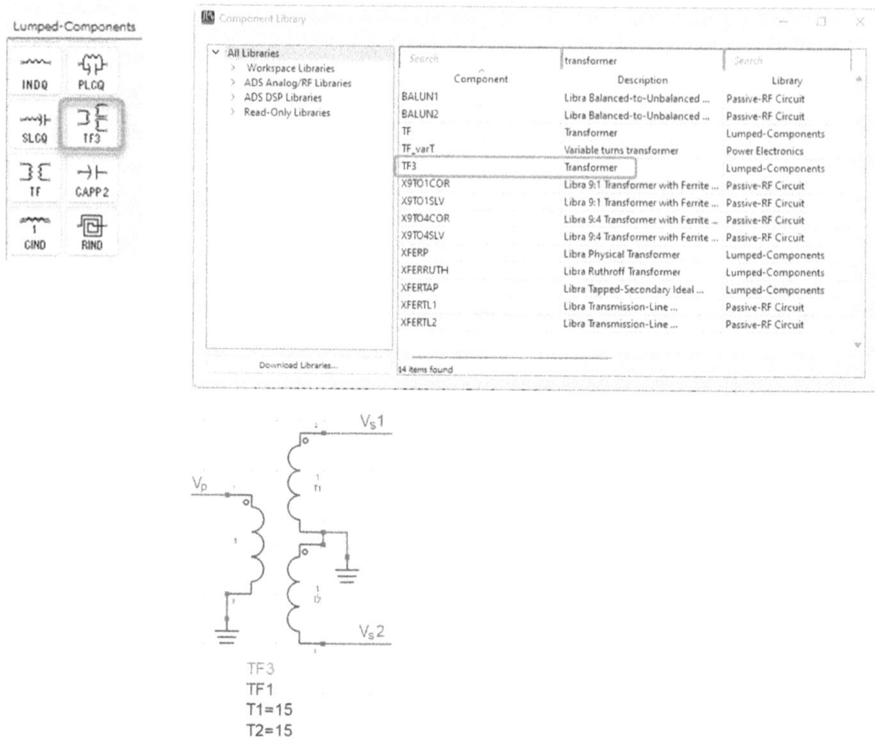

FIGURE 2.33 TF3 is an ideal center-tapped transformer, included in ADS "Lumped Components".

The complete schematic of a full-wave rectifier with a center-tapped transformer is given in Figure 2.34. The simulation results are in Figure 2.35.

Section 2.12 shows an approach by which the center-tapped transformer is emulated by two VtSine sources in series, emulating the secondary windings of the transformer.

Note that $V_{out}(t)$ is smaller by one diode voltage from any of the secondary voltages $V_{s1}(t)$ or $V_{s2}(t)$. The diode currents complement one another. The load R_l "sees" both contributions.

Let us conclude this section with a simulation of a full-wave diode-bridge rectifier (Figure 2.32).

Note (in Figure 2.36) that because the transformer's secondary is floating, there is a need to create the secondary voltage by means of an ADS measurement equation.

In Figure 2.37, note that $V_{out}(t)$ is smaller by two diode voltages from the input signal $V_s(t)$.

2.10 A RECTIFIER FILTER

The process of converting AC voltage to DC will be studied more deeply in a subsequent section. The conversion process consists of three simultaneous steps: Rectifying the AC input, filtering of the rectified voltage, in order to create a DC

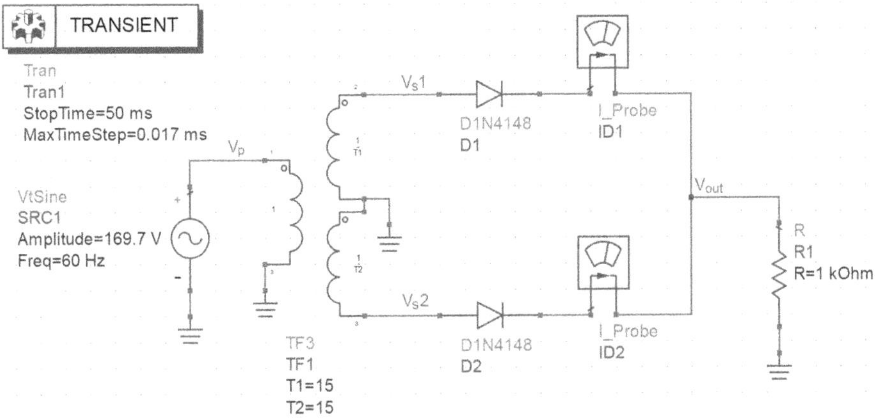

FIGURE 2.34 A complete full-wave rectifier with center-tapped transformer.

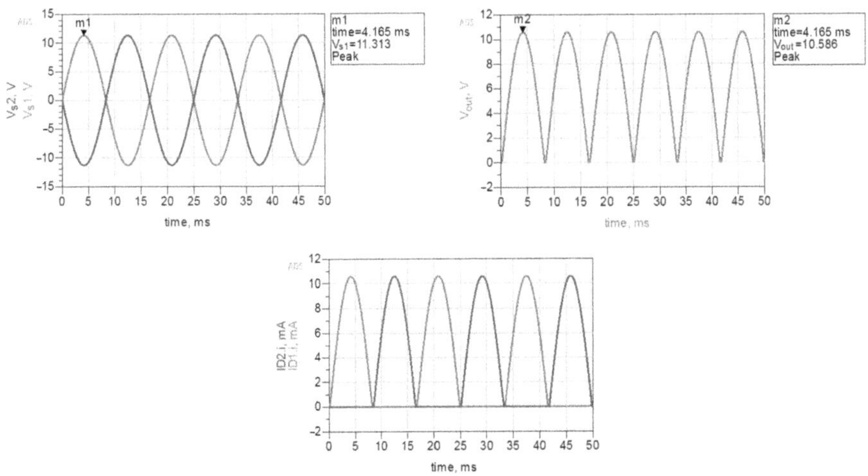

FIGURE 2.35 Resulting voltages and currents in the full-wave rectifier.

voltage with small ripples, and ripples reduction and DC voltage regulation done with a Zener diode.

The filtering of a rectified voltage is obtained by connecting a large capacitor in parallel to the rectifier's resistive load. See Figure 2.38, which demonstrates a filter added to a half-wave rectifier.

In Figure 2.38, the input voltage $v_s(t)$ represents some given transformer's secondary AC voltage with a peak amplitude of v_{sp} and a period of T.

In the US system T = 1/60 = 16.67 ms. The peak amplitude v_{sp} depends on the choice of some desired transformer's turns ratio.

Let us explain (for Figure 2.38) how the filtering of the rectifier's output signal works.

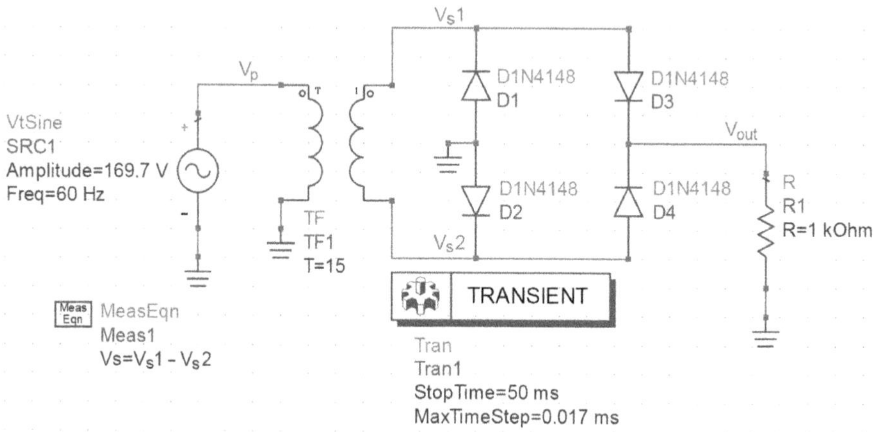

FIGURE 2.36 Complete circuit of a full-wave rectifier with a diode bridge.

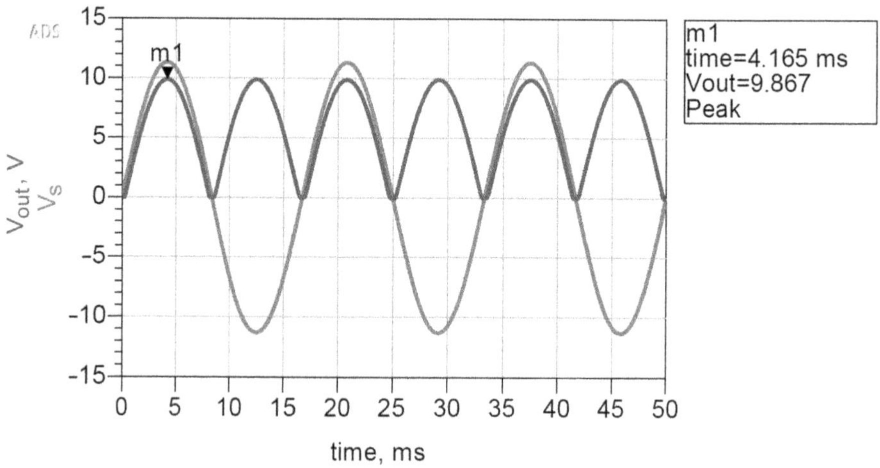

FIGURE 2.37 Full-wave rectifier with diode-bridge input and output voltages.

We assume that the value of R is in the hundreds of Ω to a few kΩ range, and that the value of C is in the tens to hundreds μF range. Let us further assume that the ON resistance of the diode is small (in the range of a few tens of Ω).

In the positive part of $v_s(t)$ the diode is ON, and C charges relatively quickly, due to the small ON resistance of the diode. Therefore, the output voltage $v_{out}(t)$ follows the rising AC input $v_s(t)$, slightly lagging behind.

As the AC input reaches beyond its peak v_{sp} and begins to head down, the diode cuts off because its anode voltage becomes smaller than its cathode voltage, which is tied to the capacitor.

The capacitor, that is charged to a peak voltage v_{cpeak}, that is close to v_{sp} minus one diode voltage, begins to discharge exponentially through R. As $v_s(t)$ begins to rise again, and $v_{out}(t)$ continues to fall down, at some point the diode turns ON again, and

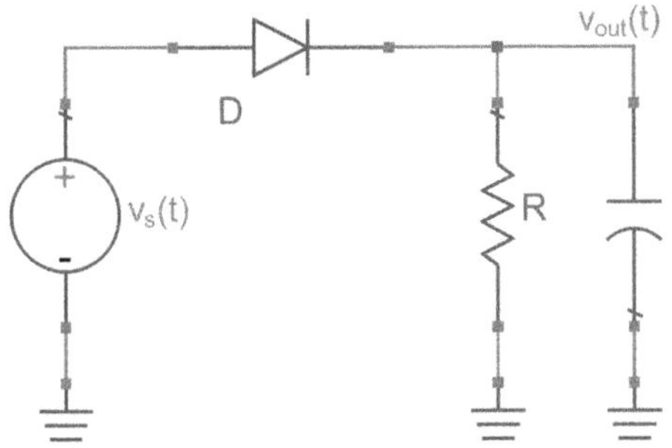

FIGURE 2.38 Filtering of a half-wave output voltage.

again the capacitor charges quickly to the value of v_{cpeak}. So forth periodically. If the time constant RC is much larger than the input's period T, then V_r, the peak-to-peak amplitude of $v_{out}(t)$, which is equal to $v_{cpeak} - v_{cmin}$, is rather small. The signal $v_{out}(t)$ then resembles a DC signal with small ripples added on top.

Let us show a mathematical derivation that is valid during times at which the capacitor's voltage decays through R. Let us denote by $t = t_1$ the time at which the diode cuts off and $v_c(t)$ begins to fall. Let us denote by $t = t_2$ the time at which the diode turns ON again. Then $v_c(t) = v_{cpeak} \cdot \exp[-(t - t_1)/RC]$. At $t = t_2$, $v_c(t)$ reaches a minimum value v_{cmin}: $v_c(t_2) = v_{cmin} = v_{cpeak} \cdot \exp[-(t_2 - t_1)/RC]$.

The time difference $t_2 - t_1$ is less than T. Assuming that T << RC implies that $t_2 - t_1 \approx T$, and that $\exp(-T/RC) \approx 1 - T/RC$. In other words, during one period T of the input voltage, we see only the linear beginning of the very long exponential decay. The shape of the periodic rippled DC output of the rectifier's filter, during one period, includes a fast sinusoidal rise from v_{cmin} to v_{cpeak}, followed by a much longer linear decay from v_{cpeak} to v_{cmin}.

The above approximations result in an approximate simple design formula (2.3a) for the half-wave rectifier's filter:

$$V_r = v_{cpeak} - v_{cmin} \approx v_{cpeak} \frac{T}{RC} \tag{2.3a}$$

$$V_r = v_{cpeak} - v_{cmin} \approx v_{cpeak} \frac{T}{2RC} \tag{2.3b}$$

The filtering of a full-wave rectifier is done by connecting a large capacitor C in parallel to R. However, in a full-wave rectifier, the rectifier's output voltage has a frequency that is double that of the half-wave rectifier's frequency. Therefore, the design formula (2.3b) is derived exactly as that for the half-wave rectifier's filter, assuming that $t_2 - t_1 < T/2$.

We call equations (2.3) "design formulas" because V_r/v_{cpeak} is usually specified (say 0.01, 0.02, 0.05, …), T is known, and RC needs to be determined. As Example 2.3, let T = 16.67 ms, and let the ratio between the peak-to-peak ripples and the peak capacitor's voltage be specified to be no larger than 5%. For a half-wave rectifier, we use equation (2.3a) to obtain RC ≥ 333.4 ms. Say that R = 1 kΩ, then C ≥ 333.4 μF.

For a full-wave rectifier, by equation (2.3b), we see that C can be half as much of the value needed for a half-wave rectifier. This is sometimes one of the reasons for choosing a full-wave rectifier over a half-wave one.

2.11 ADS TRANSIENT SIMULATION OF A RECTIFIER FILTER; DATA DISPLAY EQUATIONS

Example 2.4

Let us first simulate a half-wave rectifier's filter, taking some arbitrary "predesign" values for R = 1 kΩ and C = 1 μF. See Figure 2.39 (up). The result shown in Figure 2.39 (down) clearly shows that the choice of RC was way too small – in half a period the capacitor became almost fully discharged.

Recalling the design formula: $V_r/V_{cpeak} \approx T/RC$, if we want V_r/V_{cpeak} to be no larger than 0.05 (as in Example 2.3), then $RC \geq T/0.05 = 16.67/0.05 = 333.4$ ms.

R must not be too large. Let it be equal to R = 10 kΩ. Then C = 33.34 μF. Let us implement the design that we just carried out. See Figure 2.40. Results are shown in Figure 2.41.

In order to check whether or not the peak-to-peak ripples are no larger than 5% of v_{cpeak}, let us learn how to "Zoom to Designated Area".

First select the plot (the one shown in Figure 2.40). Click the Zoom to Area button on the toolbar (See Figure 2.41). The Autoscale Plot dialogue pops up (as seen in Figure 2.42).

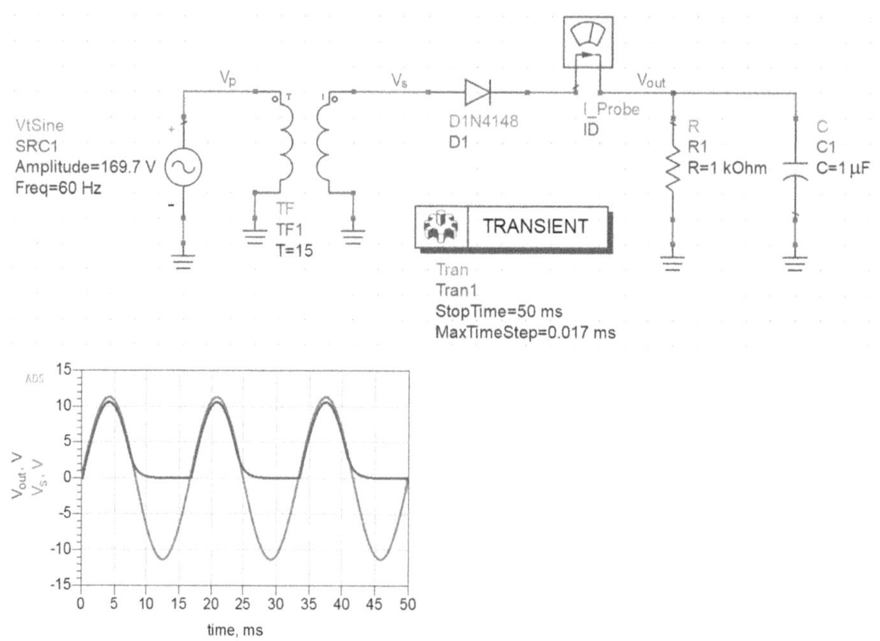

FIGURE 2.39 Mistaken choice of a too small RC = 1 ms. $V_{out}(t)$ decays too fast.

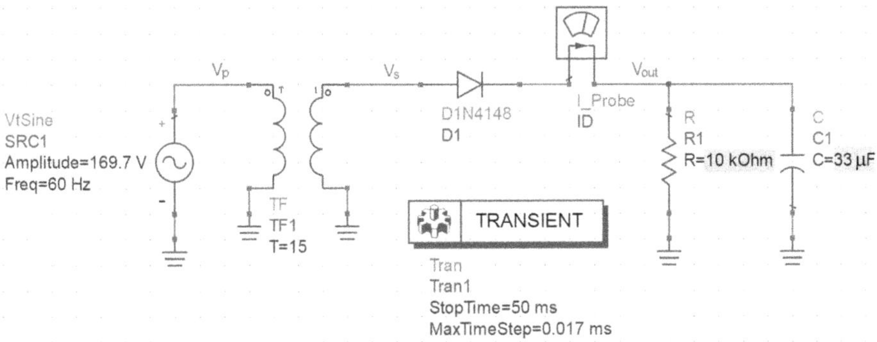

FIGURE 2.40 Half-wave rectifier with a correctly designed filter to create a DC with no more than 5% ripples.

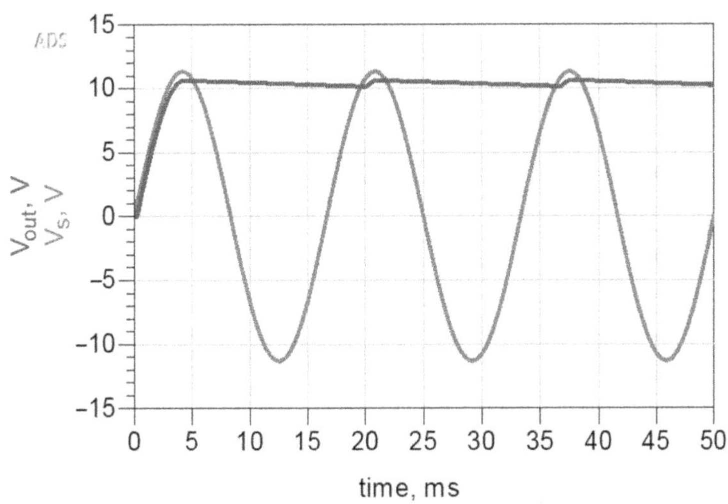

FIGURE 2.41 DC and ripple output voltage of the rectifier's filter.

One can then, as a second step, draw a rectangle for the desired zoom area, seen in Figure 2.43.

The resulting area that is zoomed is shown in Figure 2.44. We can now bring peak and valley markers in order to calculate V_r and see if the design has been successful.

Once we place the peak and valley markers on the plot, we use the ADS feature that allows us to use the markers in the data display equation to do computations for ripple and percent-ripple.

The peak-to-peak amplitude of the rippled V_{out} is:

$V_r = V_{max} - V_{min} = m_1 - m_2 = 0.484\,V$ and $\%V_r = V_r/V_{cpeak} = 0.484/10.627 = 4.6\%$.

So, yes, we have ripples of 4.6%, and the design was for ripples no bigger than 5%. Let us not forget that the rectifying filter design formulas are very approximate.

In summary, the rectifying filter design works the same way for the half-wave rectifier and the two types of full-wave rectifiers. There are three different design formulas:

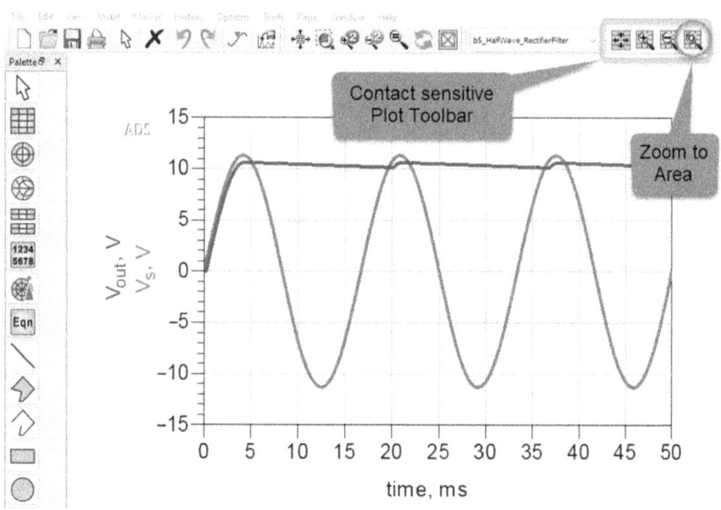

FIGURE 2.42 Initiation of zoom to area.

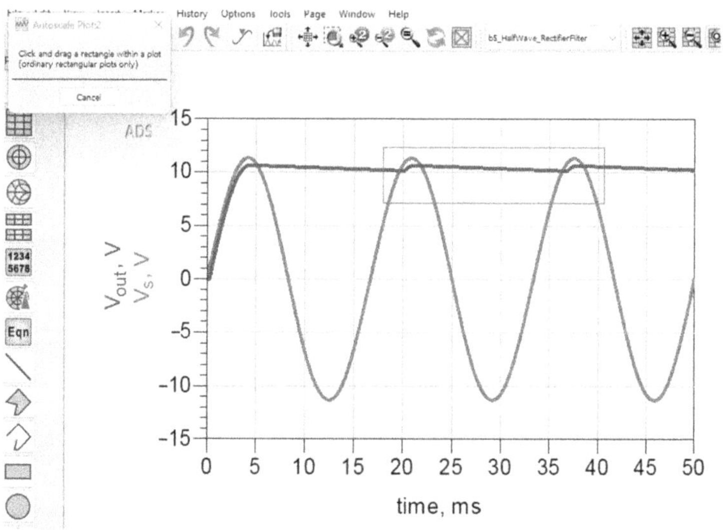

FIGURE 2.43 Selection of the area to zoom.

- Half-wave: $V_r/V_{cpeak} = T/RC$ where $V_{cpeak} = V_{speak} - V_d$.
- Full-wave center tap: $V_r/V_{cpeak} = T/2RC$ where $V_{cpeak} = V_{speak}/2 - V_d$.
- Full-wave bridge: $V_r/V_{cpeak} = T/2RC$ where $V_{cpeak} = V_{speak} - 2V_d$.

If we just specify the desired ratio V_r/V_{cpeak} then we don't even need to know the value of V_{cpeak}.

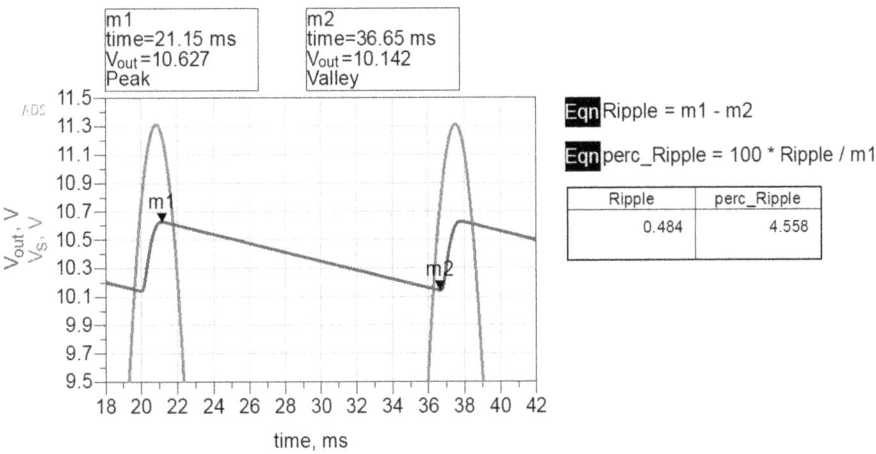

FIGURE 2.44 Measurements on simulation results and related calculations using equations.

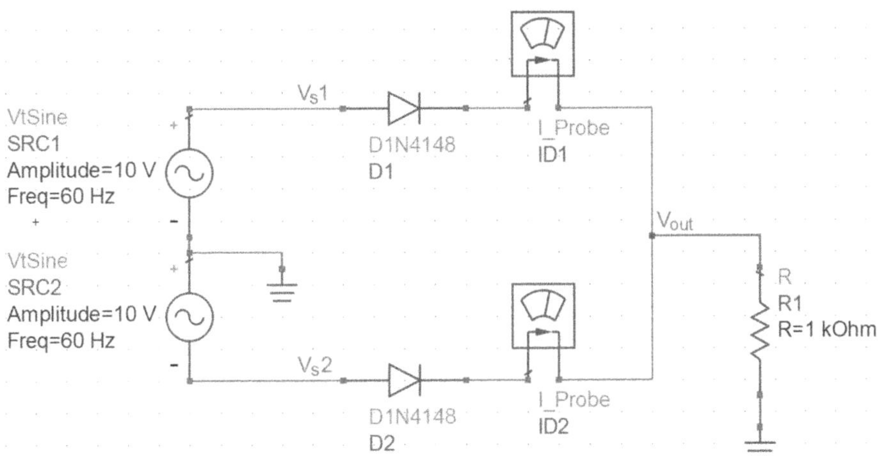

FIGURE 2.45 Center-tap emulation using two AC voltage sources.

2.12 ADS TIPS AND TRICKS 2B: EMULATION OF CENTER-TAPPED TRANSFORMERS

The ideal center-tapped transformer may be emulated using two AC sources, each one is half of the secondary voltage. See Figure 2.45.

2.13 ZENER DIODES AND THEIR PRACTICAL MODEL IN DC CIRCUITS

As the reverse bias voltage of a diode increases, so does the electrical field created over the diode's depletion region. As that field becomes intense, one of two effects (or even both) occurs that causes the diode to enter its breakdown region. One is the

Zener effect that is due to a large electrical field that may cause many donor atom electrons that reside in low energy bands to elevate to the conduction band [Sze1]. The other is the Avalanche effect involving free electrons, that due to a large field force, accelerate at high speed to possibly collide with silicon, donor or acceptor atoms and pry open covalent bonds to release a large number of free electrons and holes. The breakdown phenomenon, for most regular diodes, occur at a very large reverse voltage (Figures 2.2 and 2.20). Zener diodes, on the other hand, are deliberately fabricated to have their breakdown voltage happen at one of multiple standard values of a low reverse bias voltage. Such diodes may conduct from the cathode to the anode, while developing a voltage drop equal to the breakdown voltage, called the Zener voltage. These diodes, depending on their bias conditions, may still conduct from anode to cathode, like regular diodes. A Zener diode may be OFF if the bias voltage is too small.

Example 2.5

Note in Figure 2.46 the electrical symbol given to Zener diodes.

The value of the diode's Zener voltage V_z is often marked on the circuit diagram next to the diode. Let the Zener voltage in this example be $V_z = 4.7$ V and $R = 1$ kΩ. Let us consider several numerical values for the DC supply V_{DD}:

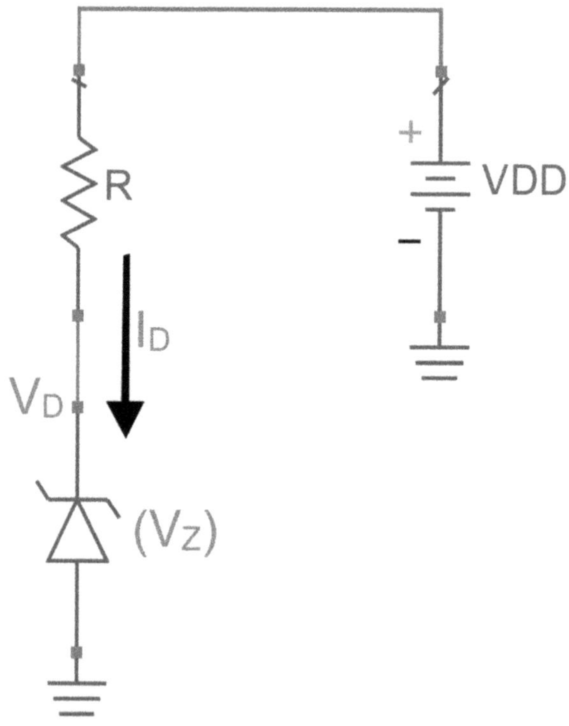

FIGURE 2.46 Illustration of the three possible states of a Zener diode: ON as Zener, ON as a regular diode, and OFF.

a. V_{DD} = 10 V: The diode conducts from cathode to anode. It develops a voltage $V_D = V_z = 4.7$ V. The diode's current (arbitrarily considered positive if the diode conducts in its breakdown mode) is $I_D = (V_{DD} - V_D)/R = 5.3$ mA.
b. $V_{DD} = -5$ V: The diode conducts from anode to cathode. It develops a regular ON voltage of a silicon diode: $V_D = -0.7$ V (using a diode practical model). The regular diode's current (with positive direction as defined in Figure 2.46) is $I_D = (V_{DD} - V_D)/R = [-5 - (-0.7)]/1 = -4.3$ mA.
c. $V_{DD} = 4$ V: This supply voltage is too low for turning ON a Zener diode of $V_z = 4.7$ V. The diode is OFF and can be considered an open circuit. $I_D = 0$ and $V_D = V_{DD} = 4$ V.
d. $V_{DD} = -0.4$ V: This supply voltage is too low for turning ON a regular diode. The diode is OFF. $I_D = 0$ and $V_D = V_{DD} = -0.4$ V.

In a DC circuit that may include regular and Zener diodes, resistors, and DC supplies, the solution approach is similar to the one taken for DC circuits with regular diodes only. Need to start with a plausible assumption about the state of each diode. Regular diodes (for all practical purposes) have two states. Zener diodes have three states each. We often use the "water flow analogy". The second step is to follow up on the initial assumption, replacing diodes that are ON with batteries with values equal to their voltage drops, and considering diodes that are OFF as open circuits. We then solve the circuit, finding all node voltages and components' currents. Finally, need to go diode by diode and see if the initial assumption can be considered justified.

2.14 SIMULATION OF DIODE AND ZENER DIODE DC CIRCUIT; ADS DESIGN KIT ZENER DIODE VS. EMULATED ZENER DIODE

Example 2.6

Let us start with a commercial Zener diode named BZX3v3 that is available in the EE design kit. It has a Zener voltage of 3.3 V (see Figure 2.47). Double-clicking on the diode does not produce any means to change the Zener BV (breakdown voltage); being a commercial part, it has a fixed model. This part is a model with a subcircuit in our library. We have created these models such that the library user can see the inner workings of the part model.

To see the part model, select it and from the toolbar click "Push into Hierarchy". See Figure 2.48.

The resulting deeper level of the hierarchy is shown in Figure 2.49.

The model is made of other components, hence the term "subcircuit model". The exact details of this model are beyond the scope of this book; suffice to say, it is quiet complex, and the user should not modify it.

Example 2.7

A Zener diode may be emulated using a DC voltage battery and two regular diodes. See the VSrcZener, D_1, and D_2 components in Figure 2.50. In order to create a 3.3-V Zener, we use a 2.6-VDC supply in series with a regular diode.

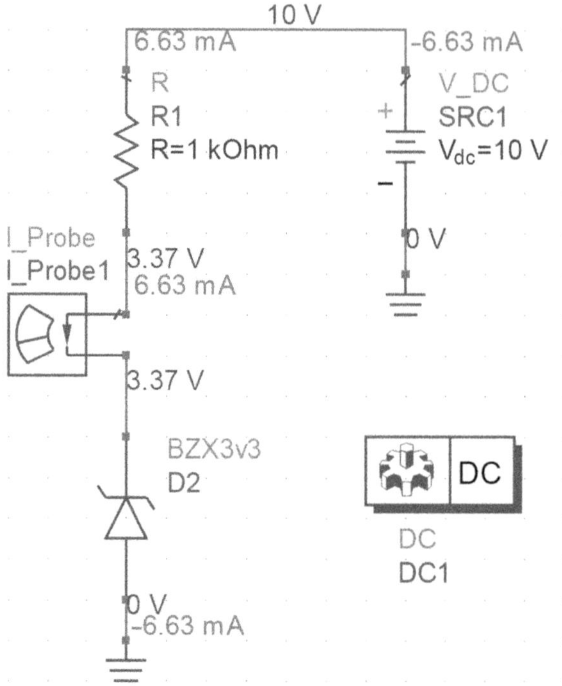

FIGURE 2.47 DC testing of the BZX3v3 Zener diode.

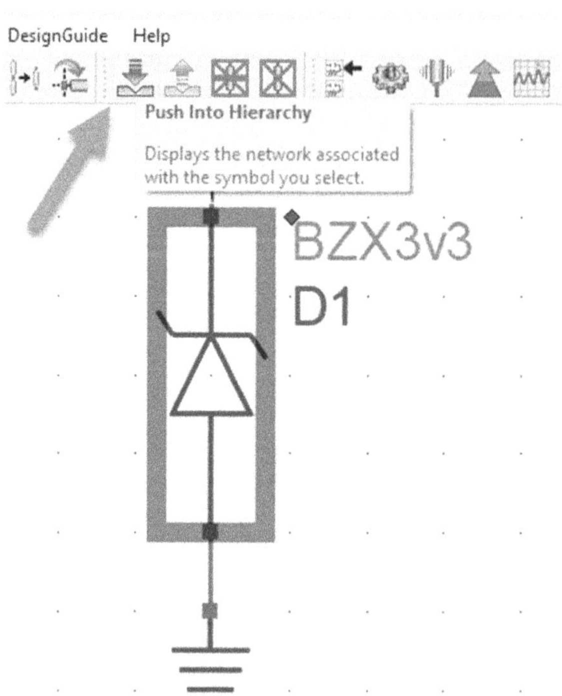

FIGURE 2.48 Push into hierarchy action.

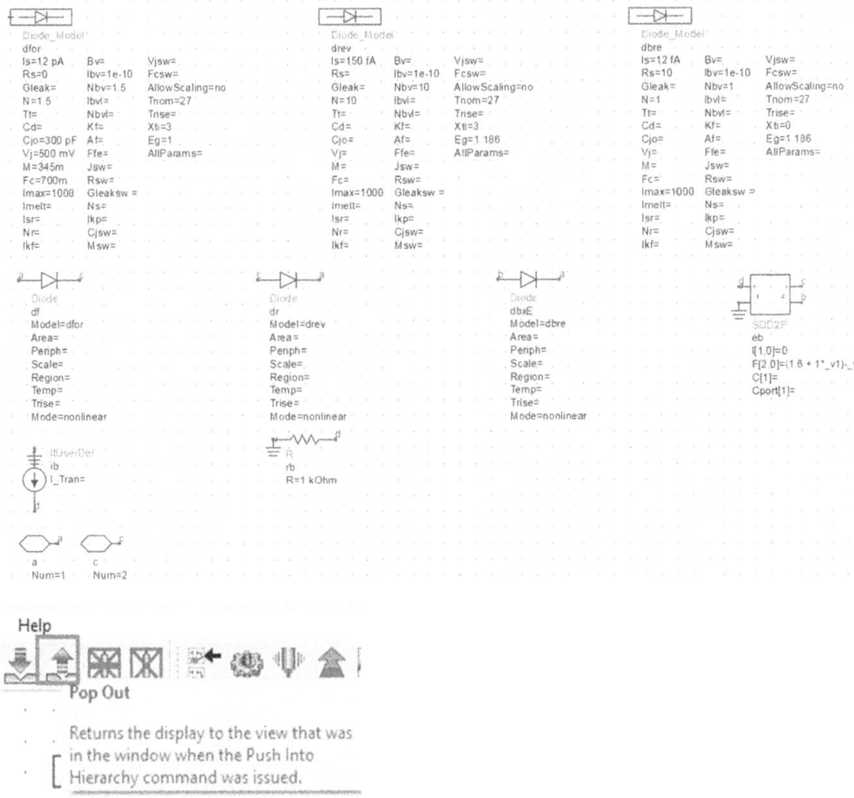

FIGURE 2.49 The subcircuit model of the BZX3v3 Zener diode (top) and returns the display to the view that was in the window when the Push into Hierarchy command was issued (bottom).

So, with a sufficiently large positive supply VD_1, the diode is ON. With $R_1 = 1$ kΩ, VD_1 should be at least 4.3 V or so. The voltage V_z becomes the sum of the regular diode voltage and the battery voltage. We see that it came to be 3.3 V. It is coincidental that diode D_1 developed a voltage of exactly 0.7 V.

If VD_1 is sufficiently negative (at least −0.5 V), then D_1 is OFF but D_2 is ON. The Zener works like a regular diode.

See Figure 2.51 for a schematic setup for obtaining the Zener diode I–V curves. Let us vary the supply voltage, denoted as V_{DD}. In order to observe both ON states of the Zener diode, let V_{DD} vary from some negative value to some positive value. We shall watch V_z and I_z each as a function of V_{DD}, and plot the Zener diode I–V curve.

In the V_z vs. V_{DD} plot in Figure 2.52, we see that for $V_{DD} > 3.5$ V (or so, right of the marker m_2), the Zener diode's voltage is around 3.3 V, as designed.

Whenever $V_{DD} < -0.5$ V (or so, left of marker m_1), the Zener diode is like a regular diode, developing a voltage of around −0.7 V. The minus sign is because the Zener current flows in the opposite direction (i.e., we still satisfy Kirchhoff current and voltage laws).

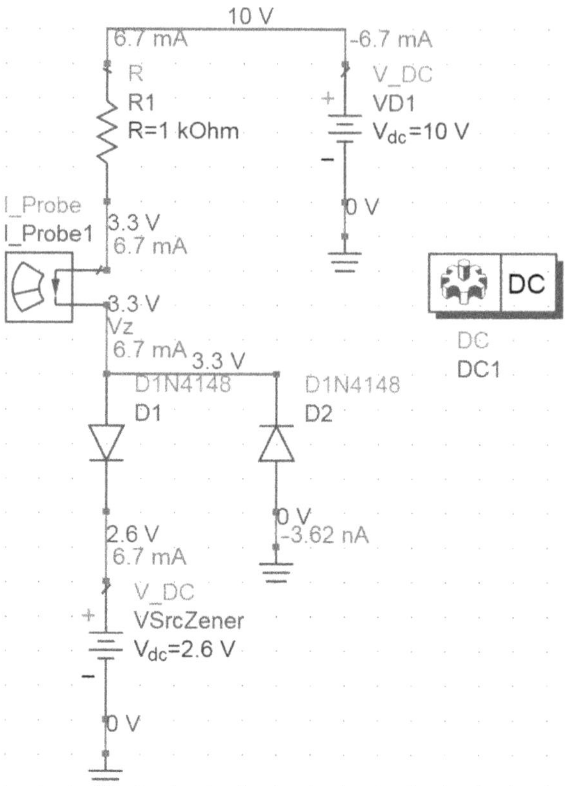

FIGURE 2.50 3.3 V Zener diode emulation with a 2.6-V battery and two regular diodes.

What is that linear curve of V_z vs. V_{DD} for the in-between voltages (between marker m_1 and m_2)?

In the I_z vs. V_{DD} plot, we see that wherever V_{DD} is between −0.5 and 3.5 V, the Zener diode is OFF. That is, $I_z \approx 0$. Therefore, the node voltage V_z becomes equal to V_{DD} (as there is no current).

Note: Figure 2.52 plot type is known as a stacked plot, and we use line markers. For details on how to create, this plot see Section 2.17.

The sweeping of V_{DD} allows us easily to obtain the I–V curve of the Zener diode. In Figure 2.53 we see it inverted, because in our schematic, the current is defined as positive whenever the Zener is in breakdown. So, on the left we see the exponential curve of a regular diode's I vs. V curve. For each current value, there is a distinct diode voltage value, and all these values are around −0.7 V.

On the right of Figure 2.53, we see the I–V curve as the Zener is working in a Zener mode. Again, there are distinct (I, V) points, and all the Zener voltage values hover around the nominal Zener breakdown voltage of 3.3 V.

From now on, let us consider V_z and I_z at the Zener mode (on the right) to be positive. It is an arbitrary current sign definition.

As facilitated by the schematic, shown in Figure 2.54, let us focus now on the positive I_z and V_z, at the vicinity of the Zener operation, shown in Figure 2.55 (left). We vary the supply VD_1 from 3 V (where we expect the Zener diode to be OFF) to 20 V (where we expect V_z to slightly exceed the nominal Zener voltage of 3.3 V).

In order to find the Zener diode's incremental resistance r_z (the theory of which is introduced in the subsequent Section 2.15), and denoted here as RZener), we need to compute the slope at each point along the Zener's I–V curve. We use the ADS differentiation function "diff". See the diff syntax in the MeasEqn block in Figure 2.54. The function diff performs a finite difference approximation of the derivative: It subtracts the values of two data points and divides by the independent variable difference between the above data points.

The diff function can be used only with respect to an independent variable. In our example, we can differentiate V_z with respect to V_{DD}, or we can differentiate $I_{z.i}$ with respect to V_{DD}. V_{DD} is the independent variable in the particular sweep parameter simulation that we did.

What if we want to differentiate with respect to a dependent variable? (as we need to do to find the diode's incremental resistance r_z: Derivative of V_z with respect to $I_{z.i}$).

So, the idea is to take a ratio of two derivatives: dy/dx = (dy/dt)/(dx/dt), say that t is an independent variable.

Another syntax rule about diff: diff(data,1) means that the vector of differentiation results is augmented by one element at the end, in order to have the two data vectors (say dVz/d(V_{DD}) and V_{DD}) be of the same length. Otherwise, we write diff(data,0).

The results for r_z as a function of V_z are shown in Figure 2.55(right). The Zener diode is almost OFF at V_z that is below 3.2 V, and therefore r_z value is relatively large (hundreds of ohms).

The more interesting region, where V_z is near 3.3 V does not show clearly – resistance values are relatively small. Need to zoom into the vicinity of V_z = 3.3 V. This is done by changing the Start sweeping value from V_{DD} = 3 to 8 V. The results are shown in Figure 2.56.

We finally see I_z vs. V_z and r_z vs. V_z at the range of V_z values that is relevant to the voltage regulator design that we shall study soon. The units of r_z and I_z (not shown in Figure 2.55) are [Ω] and [A], respectively.

Thus far, the Zener diode has been emulated using two diodes and a DC source model, as explained earlier. This gives us the flexibility to do experiments at any Zener voltage. It has been shown that the commercial diode we started with and this model match fairly closely.

Of course, if one has a commercial model as we do have several Zeners in our EE design kit, one should prefer those for accuracy and to match the real Zener much more closely. However, our emulated model has proven very useful, and we will continue to use it. The next logical step was to add this model to our library by first converting it into a subcircuit. The subcircuit will need parameter V_z and a symbol to make it possible to instantiate in a schematic. The flexibility of specifying any Zener voltage as parameter V_Z was added to the model (Figure 2.57). For details see Section 2.18.

2.15 ZENER DIODE'S VOLTAGE FLUCTUATIONS

We recall that in a regular diode, that is forward biased, we often opt for a practical assumption that v_D = 0.7 V regardless of the diode's current. Of course, the diode's exact i_D vs. v_D curve is exponential. The same happens for a Zener diode: We often assume that for many practical purposes the Zener diode's voltage drop is equal to the nominal Zener voltage, regardless of the size of the diode's breakdown current. However, the exact Zener diode's I_z vs. V_z curve is exponential as well. It means that

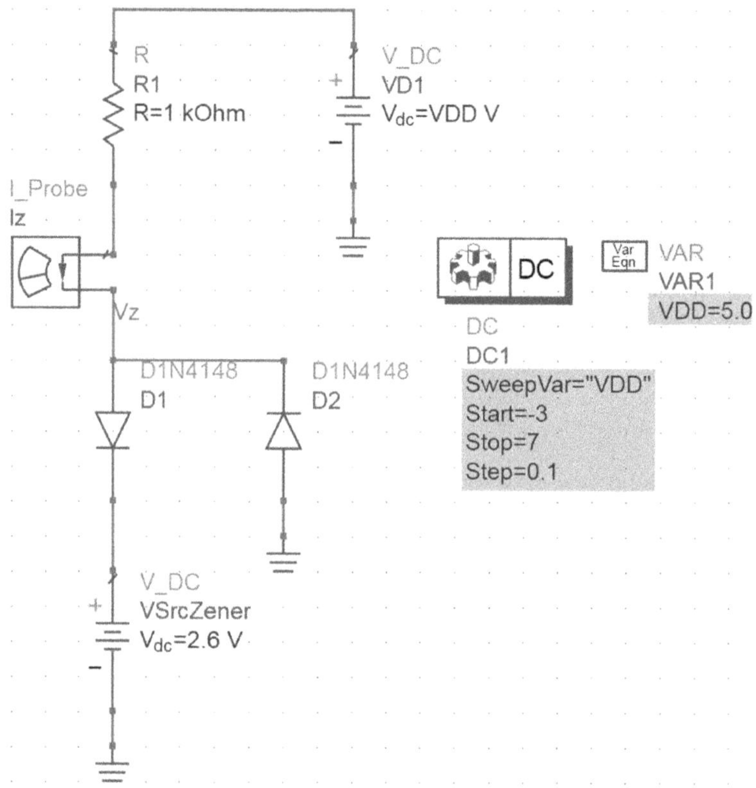

FIGURE 2.51 Schematic to show the three regions of a 3.3-V emulated Zener diode.

the diode's **nominal Zener voltage**, $V_z = V_{znom}$, happens at a specific current value, called the diode's **nominal current** $I_z = I_{znom}$. The slope $\mathbf{r_z = dV_z/dI_z}$, of the current–voltage exponential curve is called the Zener diode's **incremental resistance**. The steeper the I_z vs. V_z curve, the smaller the incremental resistance r_z, the better the Zener diode. A small value of r_z at a selected point along the current–voltage curve means that, at that operating point, possible sizable diode current changes only create small changes in the Zener diode voltage.

Figure 2.58 illustrates a typical AC to DC voltage conversion. The circuit consists of a half-wave rectifier, with a capacitor filter, and a Zener diode voltage regulator. The same can be done using any of the full-wave rectifier circuits. The Zener diode's voltage V_z is the output of this circuit. The load circuit is any circuit that needs V_z to act as a DC supply. The load current I_L may be positive (as shown) or negative. Note the triple role played by resistor R: A load resistor for the rectifier, a voltage discharge resistance for the filter capacitor C, whenever the rectifier's diode cuts off, and a current driver of the Zener diode.

There are two error sources that may cause I_z to fluctuate, and consequently, for V_z to fluctuate as well. In the circuit of Figure 2.58 $v_c(t)$ is typically a DC voltage with ripples. The ripples cause the resistor's current to change, and that may be the first cause of I_z fluctuations. The load current $I_L(t)$ (if pointing into the load circuit)

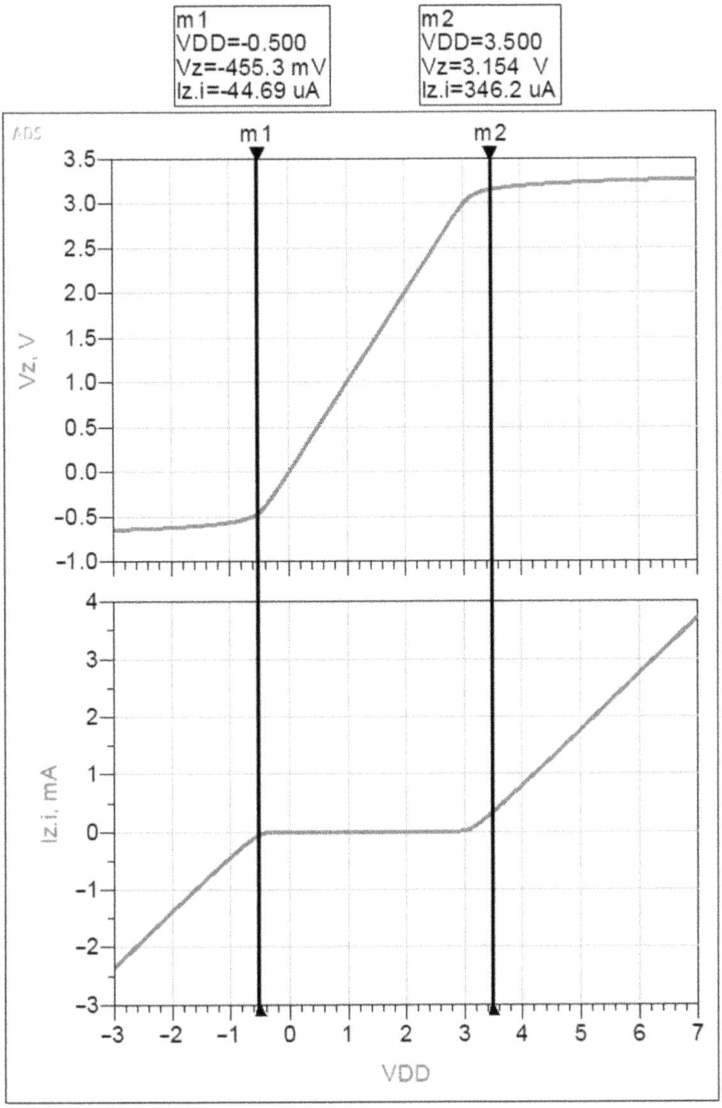

FIGURE 2.52 Simulation results of Figure 2.51: V_z and $I_{z.i}$ as functions of the independent variable V_{DD}.

"steals" some of the resistor's current away from the Zener diode. This is a second source of potential V_z fluctuations.

Let us analyze first the effects of the filtered rectifier's voltage ripples. Assume initially that the Zener diode load is disconnected. That is $I_L = 0$. Let us refer to the simplified circuit shown in Figure 2.59.

By Ohm's law: $I_z(t) = [V_{in}(t) - V_z(t)]/R$. Voltage changes $\Delta V_{in}(t)$ cause changes $\Delta I_z(t)$ and $\Delta V_z(t)$. Then, by the definition of the incremental resistance r_z of a Zener diode, we obtain a relationship that reminds us of a "voltage division": $\mathbf{\Delta V_z(t) \approx \Delta V_{in}(t) \cdot r_z/}$

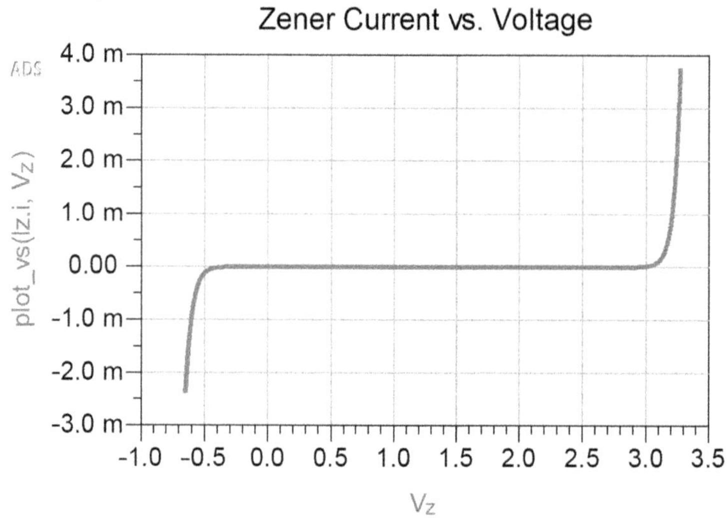

FIGURE 2.53 I–V plot of the Zener diode showing both ON regions and the OFF region.

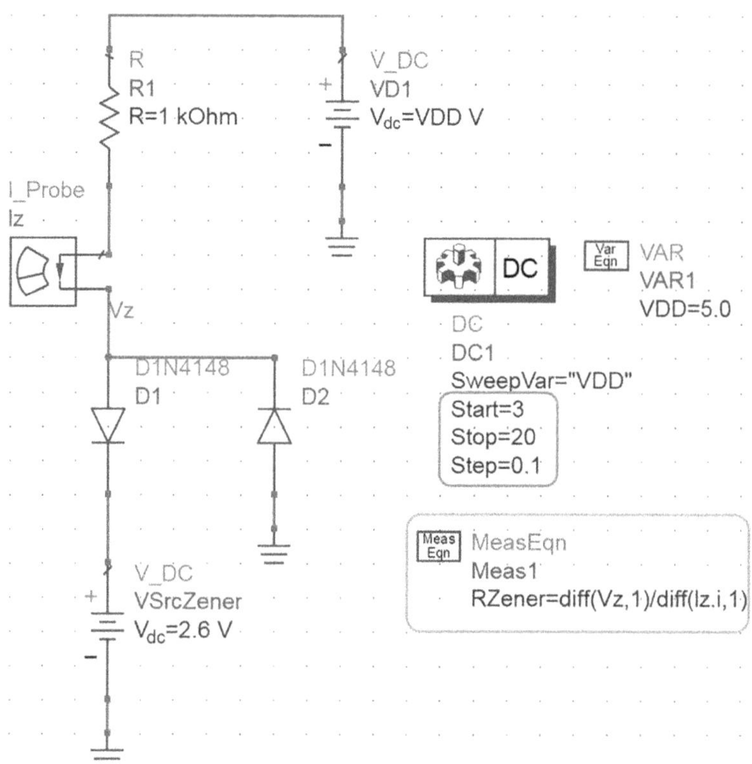

FIGURE 2.54 Schematic for plotting the I–V curve of the Zener diode, focusing on the Zener ON region. The schematics also lay the ground for measuring the diode's incremental resistance.

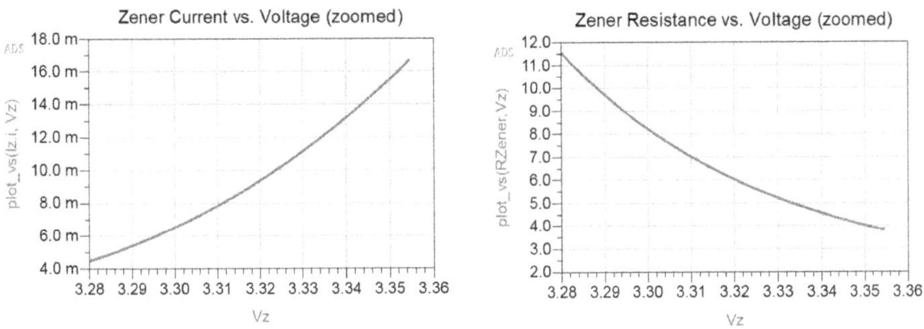

FIGURE 2.55 (Left) The I–V curve of the Zener in the vicinity of the Zener's nominal Zener voltage. (Right) Plot of Zener's incremental resistance vs. V_z.

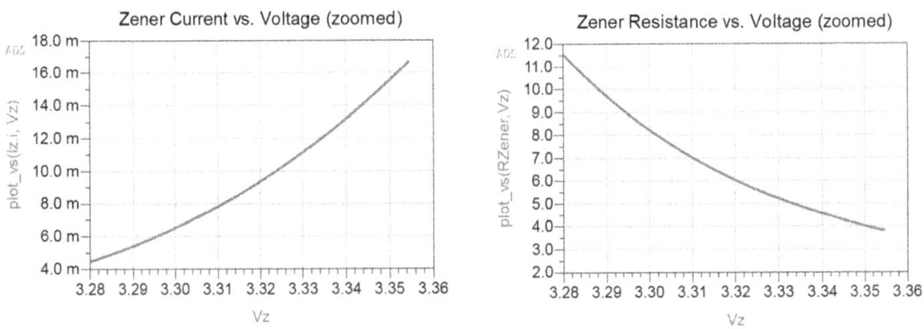

FIGURE 2.56 A revision of Figure 2.55 resulting from a modification of the sweeping range.

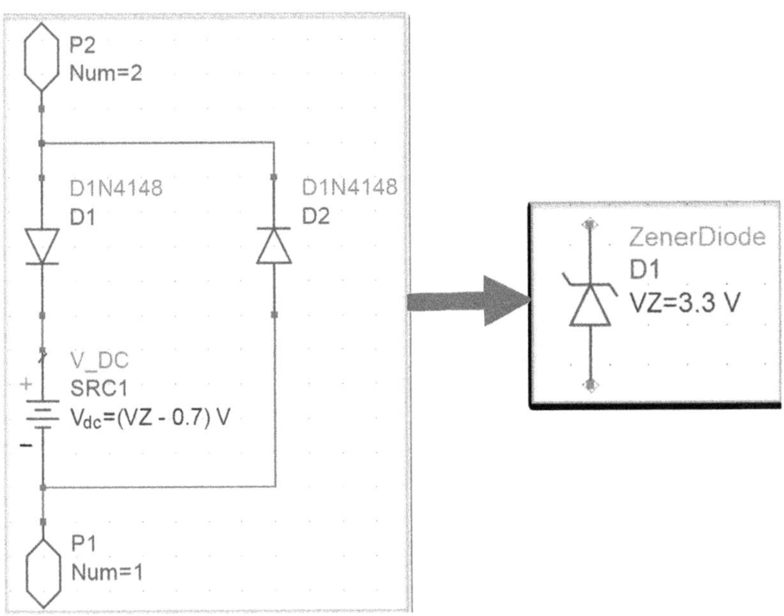

FIGURE 2.57 Parameterized model along with a symbol creates a new library component.

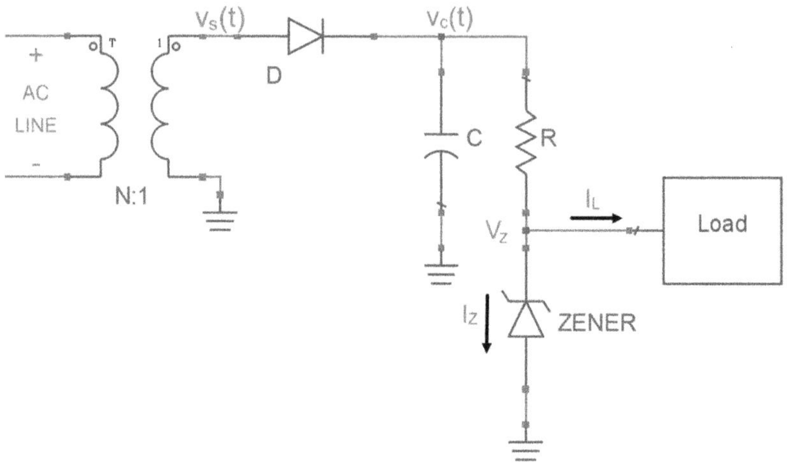

FIGURE 2.58 AC to DC with a half-wave rectifier with a filter and a voltage regulated by a Zener diode.

($R + r_z$). The ratio $\mathbf{\Delta V_z/\Delta V_{in} = r_z/(R + r_z)}$ is called the **Line Regulation**. The smaller the line regulation, the better the Zener diode's regulated output DC voltage. The Line Regulation is often expressed in [mV/V] units.

Example 2.8

A 5V Zener diode has its nominal voltage $V_z = 5$V at a nominal current of $I_z = 10$mA. At this operating point, the diode's incremental resistance is $r_z = 10\ \Omega$. If I_z changes by +10%, by what percentage will V_z change? How much is the line regulation if R = 1 kΩ? If V_{in}(t) changes by ±5% with respect to a 10 VDC level, and R = 1 kΩ, by how much will V_z change?

Solution: The first answer needs the definition of r_z. That is $r_z \approx \Delta V_z/\Delta I_z$. I_z increases by 10% which is a 1 mA change. Therefore, (using mA and kΩ units) $\Delta V_z \approx \Delta I_z \cdot r_z = 1 \cdot 0.01 = 0.01$ V. This 10 mV Zener diode voltage change is 0.2% of the nominal Zener voltage of 5V.

The line regulation equals 10/(1,000 + 10) = 0.0099 [V/V] = 9.9 [mV/V]. That is, a 1-V change of V_{in} creates a 9.9-mV change in V_z.

The third answer also needs the line regulation formula. With V_{in} ripples amplitude of 0.5V, $\Delta V_z \approx \pm 0.5 \cdot (10/(1{,}000 + 10) = \pm 0.00495$ V.

Referring again to Figure 2.59, let us explore the second source of V_z changes, the one related to changes in the load current I_L. The overall change in V_z is a superposition of the two effects – input voltage ripples and load current variations. Assuming that v_c(t) is a perfect DC voltage, we see that $\Delta V_z \approx -\Delta I_L \cdot (r_z || R)$. The **Load Regulation** is therefore defined as $\mathbf{\Delta V_z/\Delta I_L = -(r_z || R)}$, and its typical unit is [mV/mA]. The minus sign indicates that if I_L (going into the load circuit) increases, V_z decreases because I_z decreases.

Another useful formula allows us to estimate the instantaneous value of I_z(t) for any instantaneous V_{in}(t) value if R and r_z are given. A "zero-order approximation"

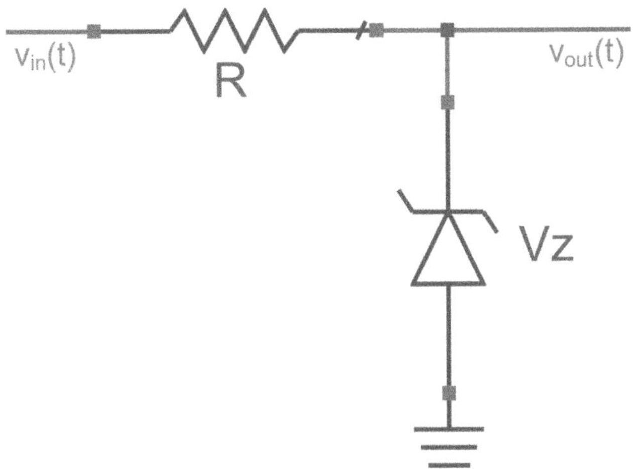

FIGURE 2.59 A Zener diode fed from a rippled voltage $V_{in}(t)$. $V_{out}(t)$ is the Zener diode's voltage.

for $I_z(t)$ is obtained from the Zener diode practical model, taking the Zener diode voltage to be equal to its nominal voltage: $I_z(t) \approx (V_{in}(t) - V_{znom})/R$. A slightly more accurate "first-order approximation" of $I_z(t)$ needs us to linearize the I_z vs. V_z curve at the nominal Zener voltage and current: $V_z \approx V_{z0} + r_z \cdot I_z$. Substitution of the nominal point V_{znom} and I_{znom} values, and knowing r_z, allows calculation of the linearized model intercept parameter V_{z0} (a value that always comes out slightly less than the nominal Zener diode voltage).

The resulting "first-order approximation" of $I_z(t)$ is $\mathbf{I_z(t) \approx [V_{in}(t) - V_{z0}]/(R + r_z)}$.

Example 2.8 (Continued)

In Figure 2.59, if $V_{in}(t)$ changes by ±5% with respect to a 10-VDC level, and R = 1 kΩ, what is the peak of the Zener diode's current I_z? What is the peak of the Zener diode's voltage V_z? What is the load regulation of the circuit shown in Figure 2.59?

Solution: Whenever $I_L = 0$, the ripples of V_{in} are synchronized with the ripples of I_z and V_z, and all have the same shape (fast sinusoidal rise followed by slow linear decay). The peak of the input voltage is $V_{inpeak} = 10.5$ V. The zero-order approximation of $I_{zpeak} \approx (10.5 - 5)/1 = 5.5$ mA. The zero-order approximation of V_z is, of course, 5 V, and it has no ripples.

If we linearize the I_z vs. V_z curve at the nominal values (10 mA, 5 V), then $V_{z0} = 5 - 10 \cdot 0.01 = 4.9$ V. The first-order approximation of $I_{zpeak} \approx (10.5 - 4.9)/(1 + 0.01) \approx 5.54$ mA. The first-order approximation of the Zener diode's peak voltage is obtained from the linearized model: $V_{zpeak} \approx 4.9 + 0.01 \cdot 5.54 = 4.955$ V. Of course, with a Zener diode current that is near 5 mA, the Zener diode's voltage is lower than its nominal voltage of 5 V (that develops whenever the current is near 10 mA).

The load regulation is $-(10||1{,}000) = -9.9$ mV/mA, which means that every 1 mA of a load current I_L that is "stolen" from the Zener diode lowers its voltage by

9.9 mV. The load regulation, whenever we use r_z of the Zener's nominal point, is a feature of that nominal point and the driving resistor R.

2.16 ADS: DC SIMULATION OF ZENER DIODE CIRCUITS; PARAMETER SWEEP AND DOUBLE SWEEP ANALYSIS

An interesting **Example 2.9** (suggested in [Sedra1]) illustrates the effects created by a parallel load resistance attached to a Zener diode. The example calls for a 6.8-V Zener, created by the new Zener model described previously. See Figure 2.60.

Unlike Zener diode textbook problems, where we had one good point on the I_z vs. V_z curve, here we just have the whole curve itself. We also have a curve for r_z vs. V_z. In the new Zener diode part that we created, we just changed to $V_Z = 6.8$ V. The curves created by the schematic of Figure 2.60 are shown in Figure 2.61. The curves are a preparation to Example 2.9.

We see that the r_z curve is unsatisfactory, as the desired V_z range is not visible. We took a curve of V_z vs. V_{DD} in order to change the V_{DD} sweep range. In order to avoid the I_z vs. V_z "knee", let V_{DD} start at 8 V (yielding V_z of nearly 6.75 V). We reran the V_{DD} sweep, this time from 8 to 20 V. See Figure 2.62 for the revised results.

The curves are needed for the example that we are about to study. From Figure 2.62, curves we learn that at the nominal Zener voltage of $V_z = 6.80$ V the Zener's nominal current is $I_z = 6.4$ mA. The incremental resistance r_z, measured at $V_z = 6.80$ V is 8 Ω

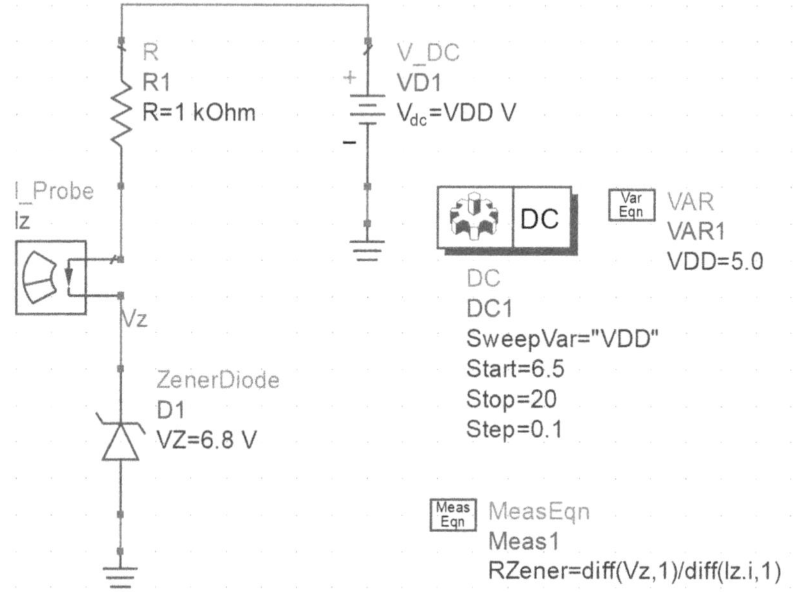

FIGURE 2.60 A study of the I–V curve of a 6.8-V Zener and its incremental resistance curve.

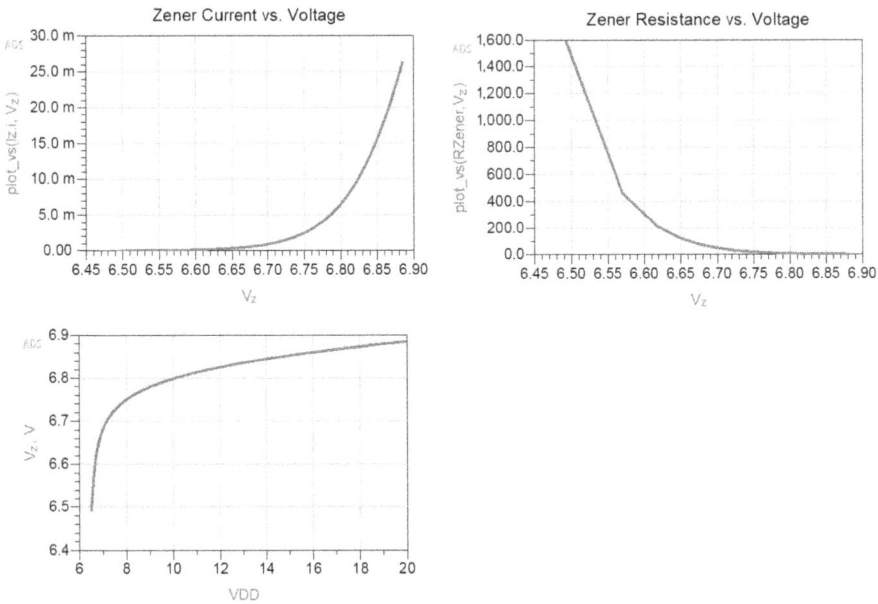

FIGURE 2.61 6.8 V Zener diode curves.

(approximately). Linearizing the I_z vs. V_z curve at $V_z = 6.80\,\text{V}$ yields the linearized model intercept parameter $V_{z0} = 6.75\,\text{V}$.

Example 2.9 Part 1

See Figure 2.63. With no input voltage ripples and with no-load current, what is the exact Zener diode's voltage V_z?

It is coincidental that with $V_{DD} = 10\,\text{V}$ and $R = 500\ \Omega$, the current Iz came to be the exact Zener nominal current of 6.4 mA, yielding the nominal Zener voltage of 6.80 V.

Example 2.9 Part 2

With an input voltage ripple amplitude of 1 V, what is the amplitude of the Zener diode's voltage?

One may be tempted to run transient simulation emulating the input voltage ripples by a VtSine source. A simpler approach is to continue to sweep V_{DD} within DC simulation. The schematic is shown in Figure 2.64. The sweeping results of V_z vs. V_{DD} are shown in Figure 2.65.

The DC sweep simulation ranges from $10 - 1 = 9\,\text{V}$ to $10 + 1 = 11\,\text{V}$. The result in Figure 2.65 predicts peak-to-peak V_z amplitude of $6.814 - 6.780 = 0.034\,\text{V} = 34\,\text{mV}$. Approximately $\pm 17\,\text{mV}$. The approximated formula, which uses "voltage division" between r_z and R predicts $\pm 1(8/508) = \pm 15.75\,\text{mV}$.

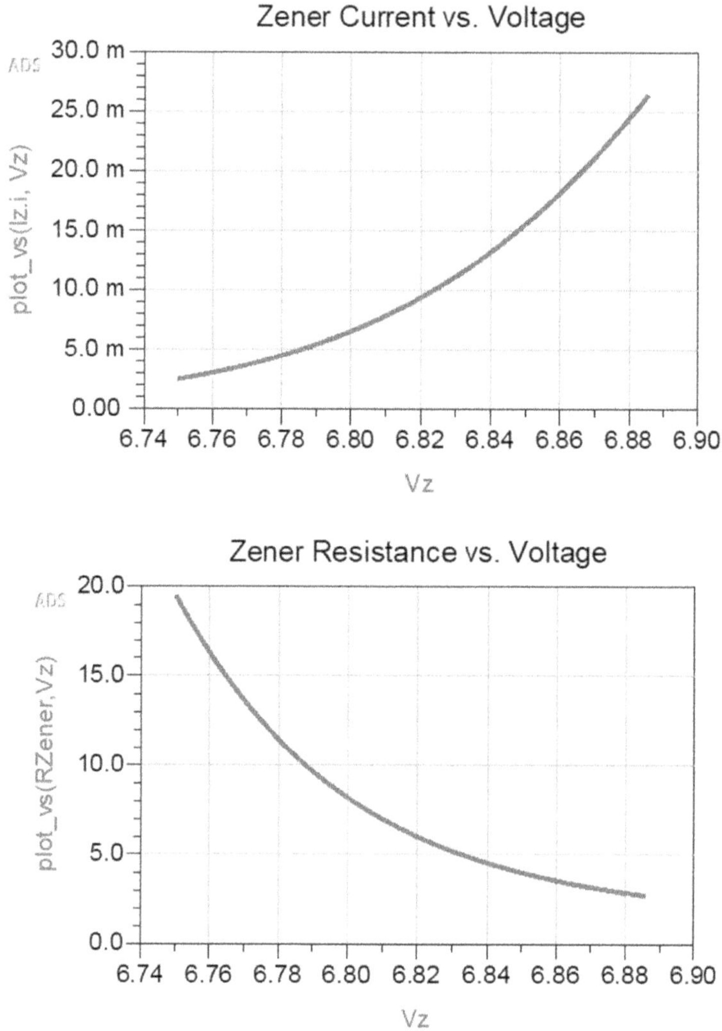

FIGURE 2.62 Revised sweep study of the 6.8 V Zener diode curves.

Example 2.9 Part 3

With no supply ripples but with a load resistance of $R_L = 2$ kΩ, by how much does the Zener diode's voltage change, compared to the no-load case?

See Figure 2.66.

Exact DC simulation shows that I_z is reduced by $I_L = 3.38$ mA. As a result, V_z dropped from 6.80 to 6.76 V (40 mV drop).

Let us compute the drop of V_z by the approximation formula (which is the definition of r_z):

Zero-order estimate of I_z drop: $\Delta I_z = -I_L = -V_z/R_L = -6.8/2 = -3.4$ mA. Thus $\Delta V_z \approx \Delta I_z \cdot r_z = -3.4 \cdot 0.008 = -0.0272$ V $= -27.2$ mV.

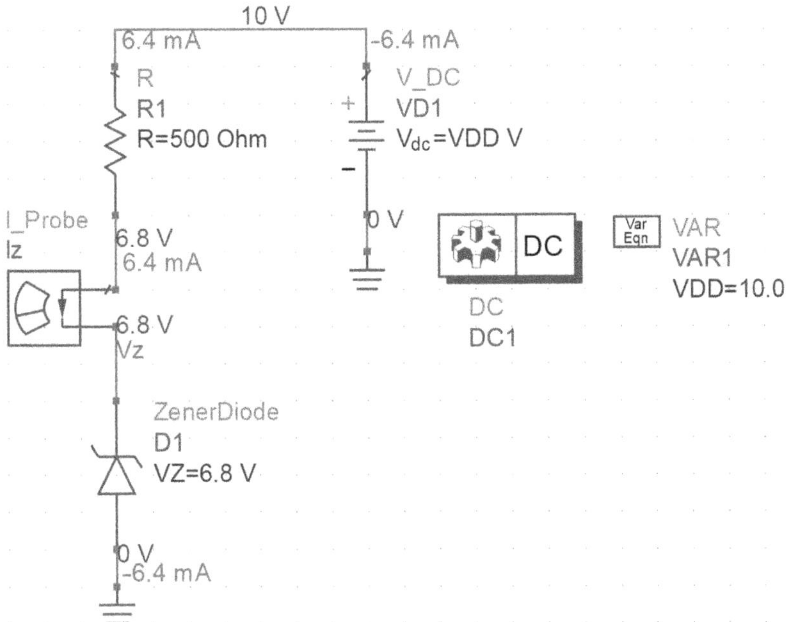

FIGURE 2.63 Schematic for solving Example 2.9 Part 1.

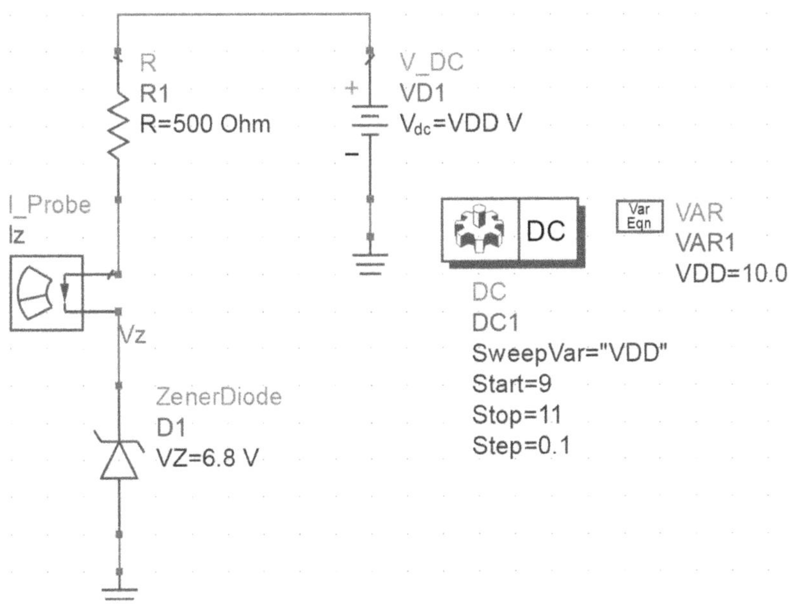

FIGURE 2.64 Study of a 6.8-V Zener, with no load, driven by a rippled input voltage.

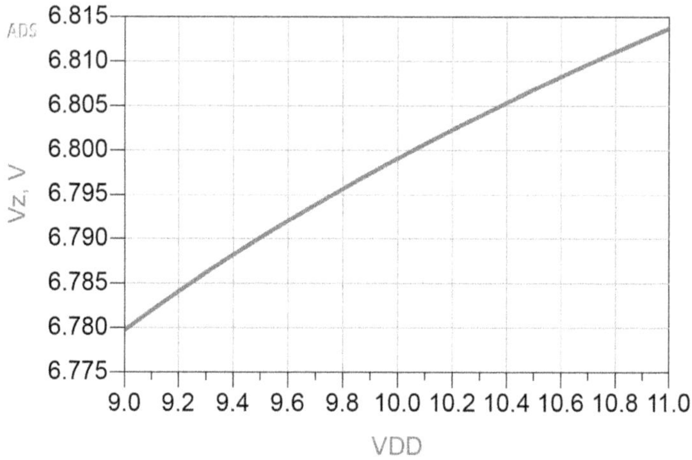

FIGURE 2.65 A DC Simulation sweep study of voltage fluctuations of a 6.8-V Zener diode, with no load, driven from a rippled voltage source.

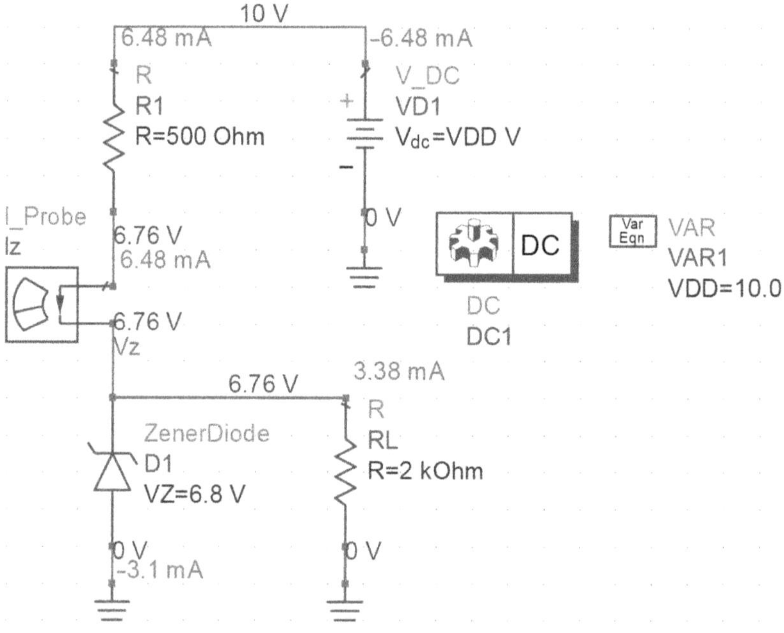

FIGURE 2.66 6.8 V Zener loaded by a 2-kΩ resistor.

Calculation by the r_z definition formula was not accurate because I_z changed significantly, and the linearized r_z value is no longer accurate.

Example 2.9 Part 4

What happens if R_L steals too much current from the Zener?

We see by DC simulation (Figure 2.67) that with $R_L = 500\ \Omega$, the Zener diode cuts off. The 10 VDC supply drives two 500 Ω resistors in series. Hence, a current of 10 mA and a V_z voltage of 5 V, which comes from R_L.

Example 2.9 Part 5

As we vary R_L, where is the borderline? In other words, what is the smallest value of R_L for which the Zener diode is still ON?

An easier version of Part 5 would be to eliminate the V_{in} ripples. Say, we let $V_{in} = V_{DD} = 10\ V = \text{const.}$

In such a case, we just sweep one variable R_L, and such sweeping can be done from within the DC simulation engine, as we have seen before.

Here we shall vary both V_{DD} and R_L. See Figure 2.68.

Let us explain more about **Double Sweep** and how to solve Part 5 of Example 2.9. Whenever we sweep more than one parameter, we must use **parameter sweep** simulation (found in the palette, or in the components library browser, by typing "par…").

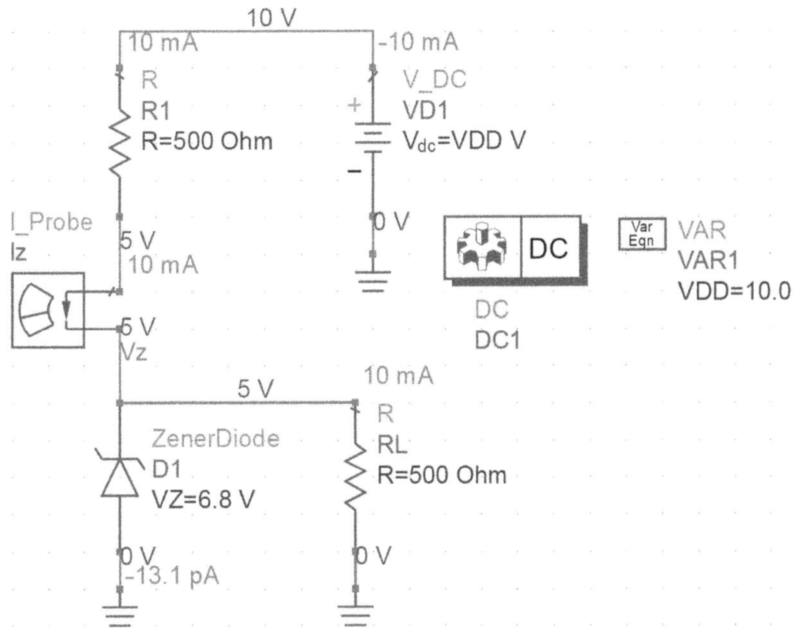

FIGURE 2.67 An R_L that steals too much current from the Zener, forcing it to cut off.

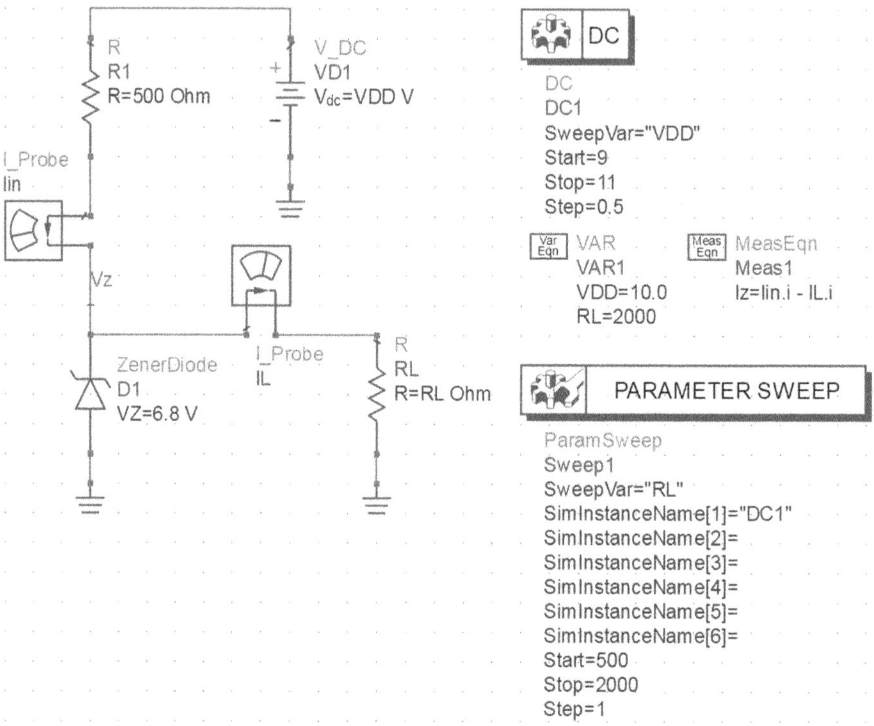

FIGURE 2.68 Schematic for a double sweep of two parameters, using DC simulation and parameter sweep. The sweeping study provides an answer to the question regarding the minimum load resistance that still keeps the Zener diode ON.

One of the sweeps may still be done by DC simulation. Alternatively, we may do all the sweeps via parameter sweep simulation.

With a double parameter sweep, ADS needs to know what the order is for doing the sweeps: The first (primary) sweep parameter determines the X axis for the plots. The secondary sweep determines the number of plots, using the same X and Y axes as in the primary simulation.

Looking for "borderline" means looking for I_z as it becomes 0.

The Zener diode current I_z is the difference between I_{in} (coming from the supply) and I_L that goes to the load. Recall: Current probe reading is marked with ".i" suffix. Of course, we could have put a current probe in series with D_1. It was just as simple to create a measurement equation.

Let us inspect more closely the settings of the parameter sweep (Figure 2.69).

For the primary sweep (R_L, in our case), we take many values along the X axis, so that we can determine the borderline resistance as accurately as possible. For the secondary sweep (V_{DD}), we take a few representative values; otherwise, the plots will appear too dense.

The double sweep results are shown in Figure 2.70.

The highest curve is for $V_{DD}=11$ V, whereas the lowest is for $V_{DD}=9$ V. Considering the swept V_{DD} value, the worst case is $V_{DD}=9$ V (the minimum value), dictating that $R_{Lmin}=1{,}300\ \Omega$.

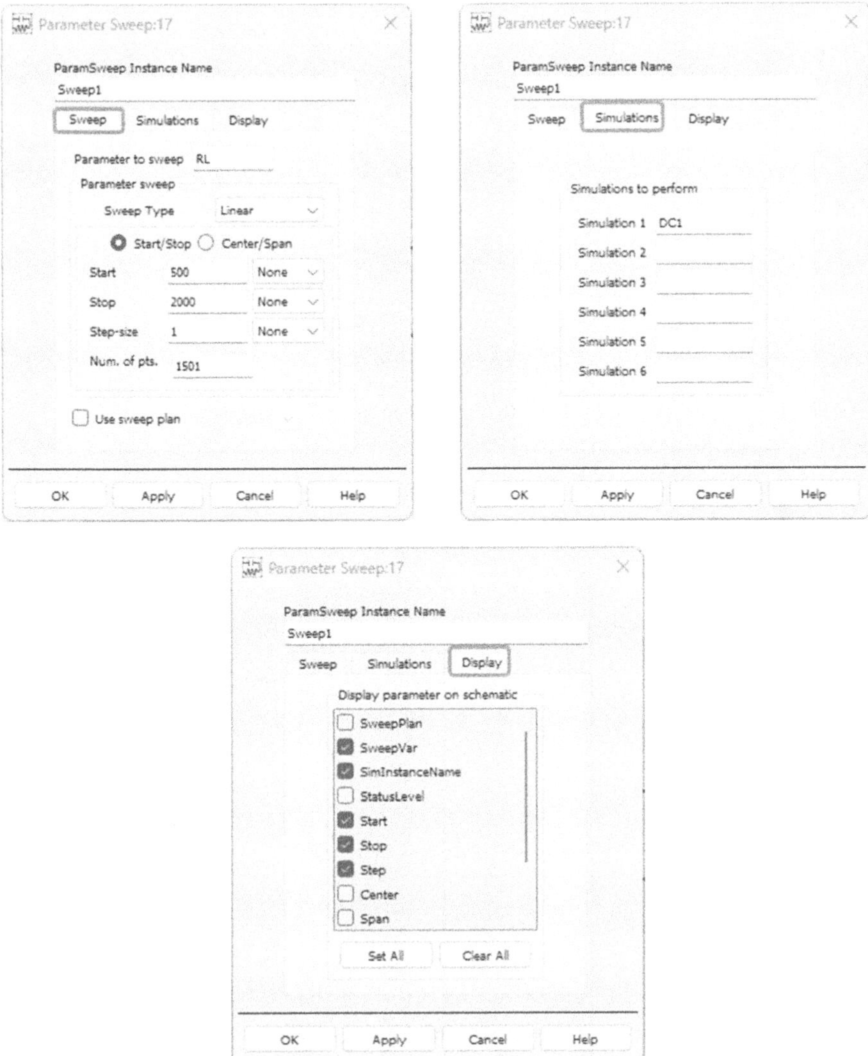

FIGURE 2.69 The settings of the parameter sweep.

2.17 ADS TIPS AND TRICKS 2D: STACKED RECTANGULAR PLOTS AND LINE MARKERS

Stacked rectangular plots (shown in Figure 2.71) are great for comparing several parameters that all depend on the same variable for example, V_{DD}. To instantiate one of these plots in the data display, select the stacked plot icon from the palette and drag it over to the display area (see Figure 2.71, left side). The rest of the setup is similar to the regular rectangular plot. The difference is in the rendering of the data; each trace is plotted separately (middle of Figure 2.71).

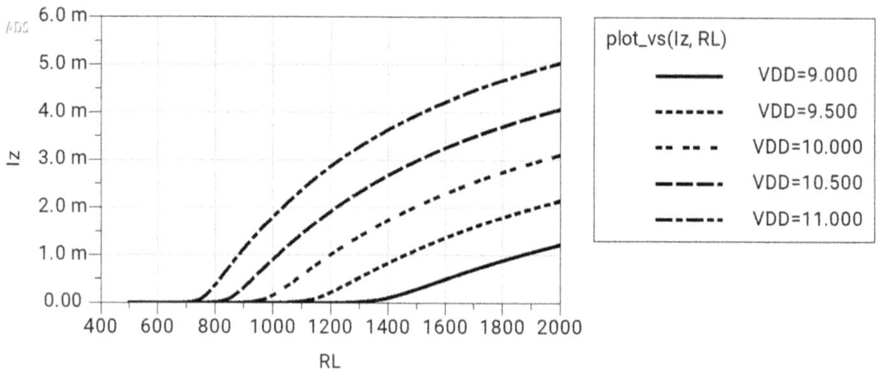

FIGURE 2.70 Curves of I_z vs. R_L with V_{DD} as a parameter.

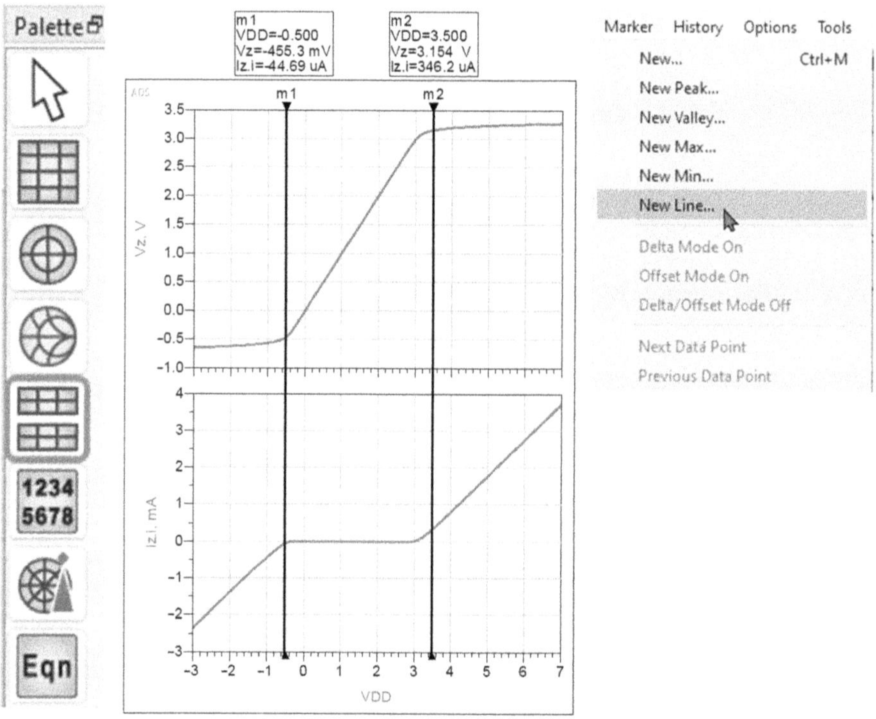

FIGURE 2.71 Stacked plots and line markers.

Line markers are the perfect type of marker whenever one wants to display values of parameters with the same independent variable on a rectangular type of graph. To add a line marker from the toolbar, select Marker > New Line … and click on the rectangular plot to instantiate the marker. The line readout will show the values of all the parameters in a single location.

2.18 ADS TIPS AND TRICKS 2E: ZENER MODEL CREATION

We start by creating a Zener diode schematic. On this schematic, we capture the Zener equivalent model and add pins for the cathode and anode of the Zener diode component. See Figure 2.72 (left and middle).

In the same cell, we add a symbol view, typically just called "symbol" as shown in Figure 2.72 (right).

On the symbols page, we need to capture the desired symbol, in our case, a Zener diode symbol. We drew a triangle and a few lines on the symbol page and then added ports.

See ADS documentation for more details on creating symbols [Keysight1]. Make sure the cathode and anode pin numbers match those of the circuit model.

On the schematic page, we select File > Design Parameters … (see Figure 2.73 (left)). We fill in a description. The Component Instance Name of D is what each diode instance starts with, here D as in "D1" and so on.

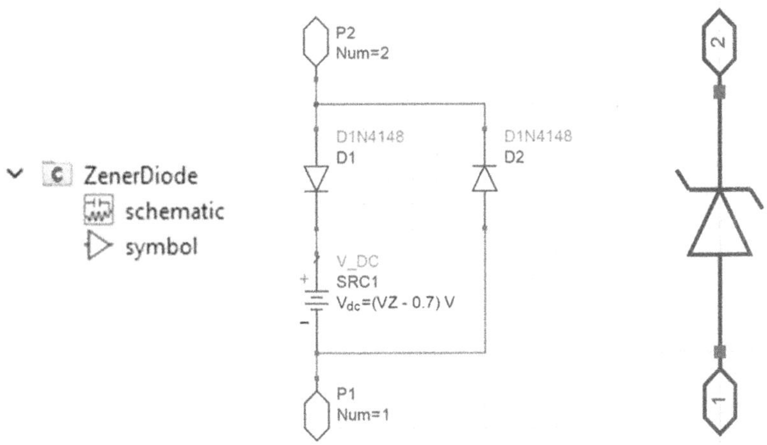

FIGURE 2.72 Zener model creation: (left) Cell, (middle) equivalent circuit model schematic view, and (right) symbol view.

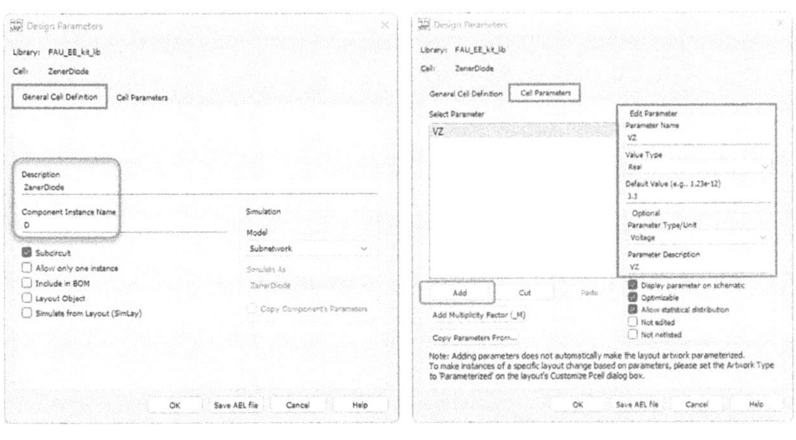

FIGURE 2.73 Zener model creation: Cell definition (left) and assign cell parameter (right).

The parameter name (see Figure 2.73 (right)) is obvious and creates a variable V_Z that we used in our model. We assign Value Type, Default Value and Unit and then click the Add button.

The above steps now lead to a part that can be instantiated in any schematic, and varying the value V_Z emulates a Zener diode of the value.

We stress again that this is an equivalent simple model. For more accurate models, we always use vendor models whenever available. If no vendor model is available, this model may prove very useful.

2.19 THE AC TO DC CONVERSION DESIGN PROCESS

Let us refer again to Figure 2.58 (repeated here for convenience):

Figure 2.58: AC to DC conversion, with a half-wave rectifier and filter, and a Zener diode voltage regulation.

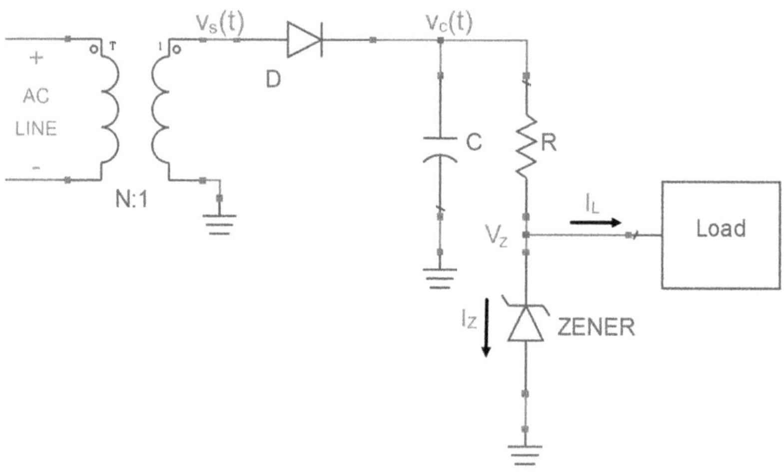

FIGURE 2.58 AC to DC with a half-wave rectifier with a filter and a voltage regulated by a Zener diode.

Let us also remind the reader of the equations (discussed earlier) that will be relevant to the AC to DC conversion process:

$$V_r = v_{cpeak} - v_{cmin} \approx v_{cpeak}\frac{T}{RC} \quad \text{Half-wave rectifier} \tag{2.3a}$$

$$V_r = v_{cpeak} - v_{cmin} \approx v_{cpeak}\frac{T}{2RC} \quad \text{Full-wave rectifier} \tag{2.3b}$$

$$v_{cpeak} = v_{speak} - v_d \quad \text{Half-wave rectifier} \tag{2.4a}$$

$$v_{cpeak} = \frac{v_{speak}}{2} - v_d \quad \text{Center-tapped rectifier} \tag{2.4b}$$

$$v_{cpeak} = v_{speak} - 2v_d \quad \text{Diode-bridge rectifier} \tag{2.4c}$$

$$V_z \approx V_{z0} + I_z r_z \quad \text{Linearization } I_z \text{ vs. } V_z \tag{2.5a}$$

$$I_z \approx \frac{v_C - V_{z0}}{R + r_z} \quad \text{Linearized } I_z \text{ for } I_L = 0 \tag{2.5b}$$

$$\text{Line-regulation} = \frac{r_z}{R + r_z} \tag{2.6a}$$

$$\text{Load-regulation} = -r_z \parallel R \tag{2.6b}$$

We shall demonstrate the design process whenever a half-wave rectifier is employed. The approach is similar if any of the full-wave rectifiers are to be used. Therefore, the relevant equations to the half-wave rectifier case are equations (2.3a), (2.4a), and (2.5)–(2.6). Equations (2.3) are related to the voltage ripples of the rectifier's filter. Equations (2.4) relate the peak capacitor's rippled voltage to the peak of the signal coming from the transformer. Equations (2.5) are related to the linearized model of Zener's I_z vs. V_z curve at the nominal Zener diode's point.

The first and foremost design specification of this DC supply circuit is the desired nominal DC output voltage value. This specification dictates the Zener diode that needs to be selected, which has such a nominal Zener diode voltage V_{znom}.

A good Zener diode voltage regulator is expected to exhibit very small output voltage ripples. That is, it needs to have a small line regulation and a small load regulation (see equation 2.6). Both are achieved if the incremental resistance r_z of the Zener diode is as small as possible, and in particular, it needs to be much smaller than the resistance R that drives the Zener diode.

Another important design specification is the supply's ability to handle all expected load currents. The designer must know whether the load current I_L goes into the load circuit, or comes from the circuit, or that both scenarios are possible. All that we need to know are the maximum I_{Lmax} and minimum I_{Lmin} load currents. A frequently occurring minimum load current is the no-load current: $I_{Lmin} = 0$.

One of the design specifications is introduced to assure that the Zener diode's current I_z never fall below a specified minimum current value I_{zmin}. I_{zmin} is a fraction of the nominal Zener current I_{znom}. It may be selected as $I_{znom}/3$, $I_{znom}/2$ or similar values.

In order to assure that I_z never falls below I_{zmin}, an upper limit constraint on R will be derived. This design formula for R requires the following data: Zener diode's linearized model (about the nominal point) parameters r_z and V_{z0}, the constraint value I_{zmin}, the load maximum current I_{Lmax}, and V_{cmin}, the dip value of the capacitor's ripples. The resulting constrained value of R is typically in the order of magnitude of hundreds of Ω.

As part of the design process, a limit is set on the size of the capacitor's voltage ripples. Such a limit dictates the maximum value of the time constant RC (see equation 2.3). Due to the relatively small value of R, the capacitor C often attains large values (in the order of magnitude of hundreds of μF). In an effort to reduce C, the designer may opt for a full-wave rectifier, needing only half as much of C needed for a half-wave rectifier. In addition, keeping in mind the approximations used for developing equation (2.3), the design formula for RC, a relaxation of the C ripples specification may often become a conveniet remedy, allowing C to be reduced.

Prior to deriving the design formula for R, equation (2.5b) needs to be revised to include the effects of the load current I_L. Starting with Kirchhoff's and Ohm's laws $I_R = I_z + I_L = \frac{v_c - V_z}{R} = \frac{v_c - (I_z r_z + V_{z0})}{R}$ yields the revised equation (2.5b): $I_z = \frac{v_c - V_{z0} - I_L R}{R + r_z}$. Setting a lower limit I_{zmin} on I_z implies in the worst case that $\frac{v_{cmin} - V_{z0} - I_{Lmax} R}{R + r_z} \geq I_{zmin}$. Hence, the design formula:

$$R \leq \frac{v_{cmin} - V_{z0} - I_{zmin} r_z}{I_{Lmax} + I_{zmin}} \tag{2.7}$$

Given the Zener diode with a nominal Zener voltage of V_{znom} (at a nominal current of I_{znom}), we can determine the minimum allowed Zener diode's current I_{zmin}, and find the incremental resistance r_z and the other parameter V_{z0} of the Zener's linearized model (equation 2.5a). The maximum load current I_{Lmax} is provided in the design specifications.

So the only question is how to find the capacitor's ripple dip voltage v_{cmin}. We have to (somewhat artificially, it should be admitted) add a capacitor's ripple amplitude specification (equation 2.3a). Once we do it, v_{cmin} can be determined from v_{cpeak} (which is known from the transformer's secondary voltage and the rectifying method (equation 2.4).

We now apply equation (2.7) to determine R. Then we head back to equation (2.3a) to determine C.

Example 2.10

With the aid of ADS, design a circuit that converts the US line voltage to a 10 VDC supply. The load circuit requires a load current I_L that may run between 0 and 15 mA. The Zener diode's current must never fall below $I_{zmin} = 5$ mA. Assume that only regular transformers are available, coming with the following standard turns ratios: 15, 12, 10, 8, and 5.

Solution: We decide arbitrarily to prefer a half-wave rectifier (Figure 2.58), over the more complicated full-wave rectifier with a diode bridge.

Let us first resolve the selection of a transformer. We keep in mind that the amplitude of the transformer's secondary voltage must exceed the specified DC supply level by a few volts, at least, to allow for the rectifier's diode voltage drop, the ripples of the capacitor's voltage, and the resistor's voltage drop that assures that a sufficient current goes to the Zener diode. A turns ratio of 15 creates the following secondary amplitude: 169.7:15 = 11.31 V, which is too small. A turns ratio of 12 creates a secondary amplitude of 14.14 V, which is probably okay. We choose a turns ratio of 10, getting $V_{smax} = 16.97$ V, a comfortable margin for 10 V.

The 10 V Zener diode needs to be tested (as shown in Section 2.15) in order to find the needed diode parameters. As we shall see soon, the found parameters are: $I_{znom} = 6.5$ mA, $r_z = 8.2\ \Omega$, $V_{z0} = 9.947$ V. Once we know the nominal current, we normally determine I_{zmin}. Here though, I_{zmin} has been specified to be 5 mA and we adopt it, seeing no conflict with the diode's I–V curve.

Equation (2.7) helps us select a resistor R that will not be too large. The only missing entry is V_{cmin}, the value of the dip point of the capacitor's ripples. Note that no specification has been given regarding the upper limit of the capacitor's peak-to-peak ripple amplitude. The designer is free to select any reasonable bound, say 3%, referring to equation (2.3a). The peak of the ripples signal is lower by one diode voltage than the secondary voltage amplitude. That is, $V_{cmax}=V_{smax}-V_d=16.97-0.7=16.27\,V$. With the 3% ripples, we find that $V_{cmin}=V_{cmax}\cdot 0.97=15.78\,V$.

Now we can find the upper bound of R, using equation (2.7).

$$R \le \frac{V_{cmin}-V_{z0}-r_z I_{zmin}}{I_{Lmax}+I_{zmin}} = \frac{15.78-9.947-0.0082.5}{15+5} \approx 0.290\ k\Omega$$

With R = 290 Ω, let us find C that accomplishes the capacitor's voltage ripples of 3%. $V_r/V_{cmax} \approx T/RC \le 0.03$ we obtain C ≈ 1.9 mF. C came out too large but let us leave it like that, for the time being. If needed, C can be halved if we replace the half-wave rectifier with a full-wave rectifier. It may also become smaller if we allow the capacitor's ripples to be 5%–10%, instead of 3%.

Time to test the above design in ADS.

2.20 ADS-BASED DESIGN OF AC TO DC CONVERSION; LOAD CURRENT EMULATION; NONZERO SIMULATION START TIME

Example 2.10 (Continued)

The starting schematic (Figure 2.74) features the full rectifier, filter, and Zener driving circuit. It has temporary arbitrary values for R and C (R = 500 Ω, C = 100 μF). It also features an ItSine current source that emulates I_L. For I_L that may fluctuate between $I_L=0$ and $I_L=15\,mA$, the ItSine source has a DC level of $I_{DC}=7.5\,mA$ and an amplitude of 7.5 mA. The frequency (200 Hz) of the load current emulator is arbitrary, chosen so that it will not be an integer multiple of the 60 Hz line frequency. That is, in the upcoming simulation $I_L(t)$ and $V_c(t)$ are not synchronized.

This is an all-purpose schematic that can be used for any such design. If we want $V_{out}=10$ VDC, then we need to choose a Zener diode with nominal Zener voltage of $V_z=10\,V$. Be sure that the current probe that measures I_z, does exactly that (and not I_R or I_L). There may be designs where I_L may come from the supply (source) or sometimes go into the supply (sink). We always have to account for the "no-load" possibility ($I_L=0$).

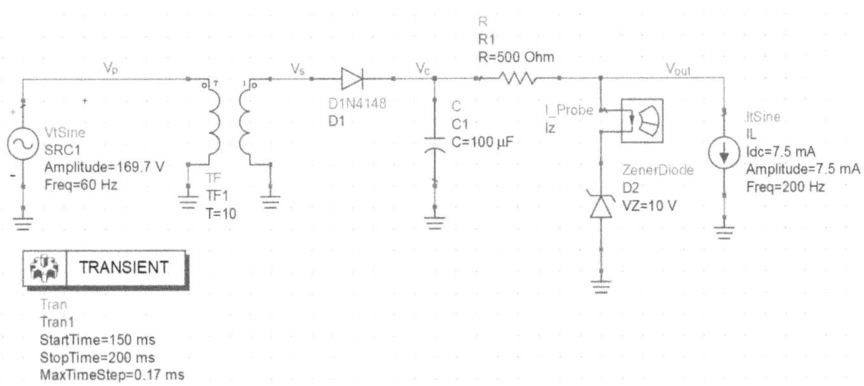

FIGURE 2.74 Starting schematic of the 10 VDC supply.

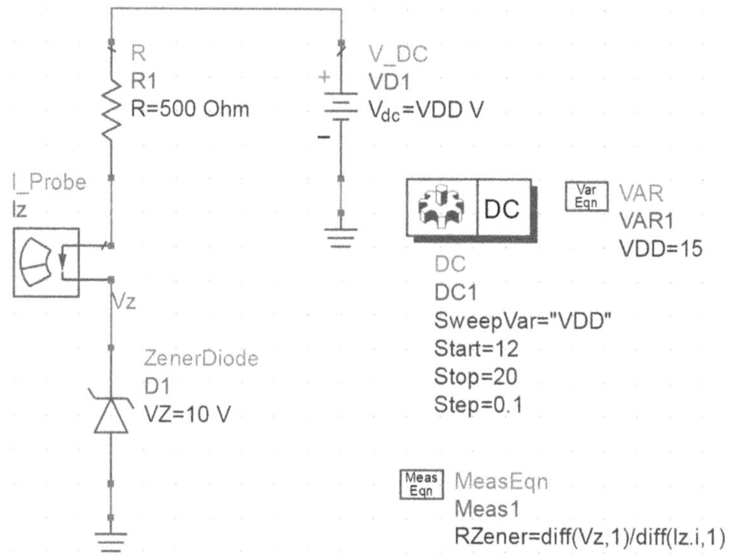

FIGURE 2.75 Schematic for testing the 10 V Zener diode I_z vs. V_z and r_z vs. V_z.

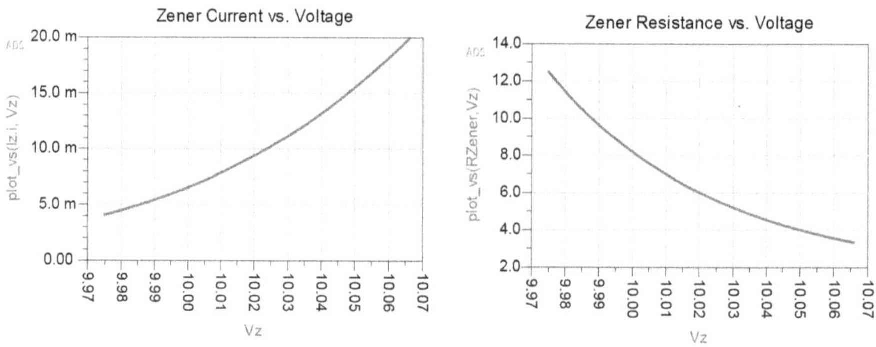

FIGURE 2.76 We see that for $V_z = 10\,V$, I_z is approximately $I_z = 6.5\,mA$. Around this operating point the incremental resistance is approximately $r_z = 8.2\ \Omega$. We can now compute $V_{z0} = 9.947\,V$.

Due to the new Zener voltage (now 10 V, compared to 3.3 and 6.8 V that we used in an earlier section), we need to redo the I_z vs. V_z curve and the r_z vs. V_z curve (Figure 2.75).

The resulting curves are provided by Figure 2.76.

The design conducted in Section 2.19 resulted in the values $R = 290\ \Omega$ and $C = 1.9$ mF. See Figure 2.77.

Note (in Figure 2.77) that the transient simulation runs from 150 to 200 ms (instead of 0–50 ms) in order to avoid watching the transient effect on the scale of the resulting steady-state signals.

For example, assuming that $V_c(0) = 0$, a simulation that starts at $t = 0$ will have the rise of $V_c(t)$ from zero to the ripple values as the dominant effect for setting the scale of the $V_c(t)$ vs. t plot.

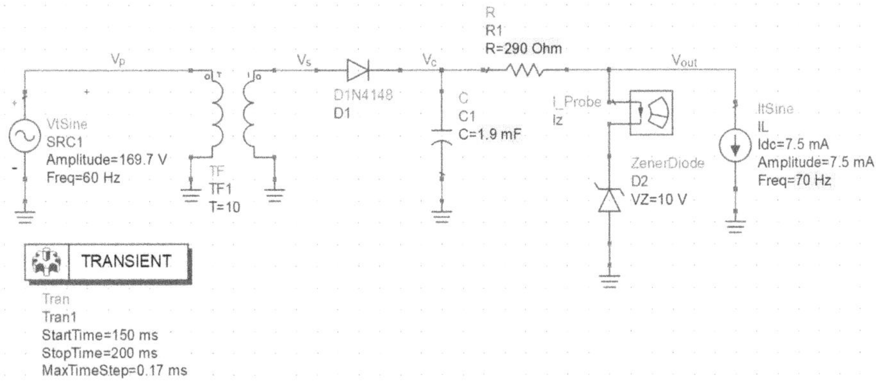

FIGURE 2.77 Schematic that displays the final design values is awaiting the plots of $V_{out}(t)$, $V_c(t)$, and $I_z(t)$ to check if all the specifications have been met. The load current emulator features a new frequency of 70 Hz (for no special reason).

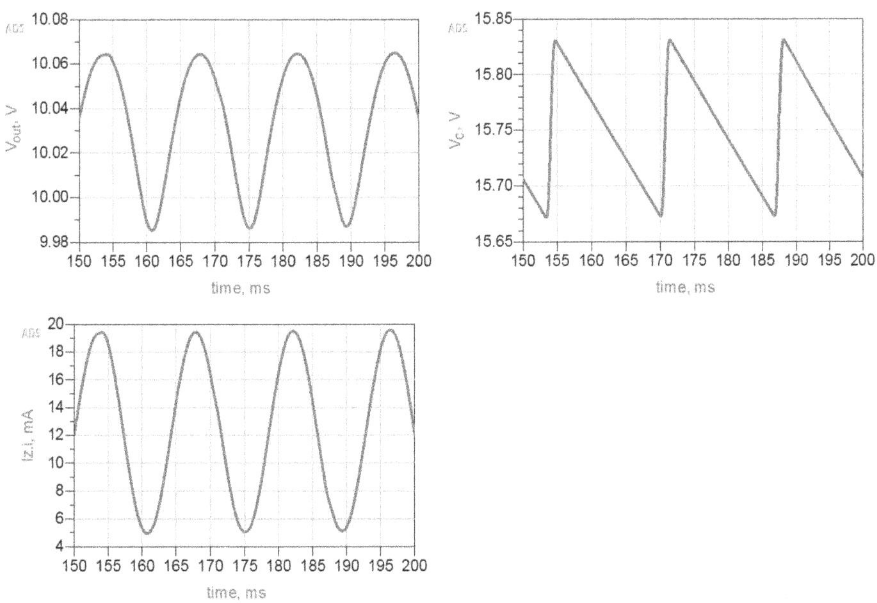

FIGURE 2.78 10 VDC supply resulting signals: $V_z(t)$, $I_z(t)$ and $V_c(t)$.

On the other hand, a belated onset of the simulation plot skips the transient, leaving only the ripples to determine the plot scale. The simulation results are shown in Figure 2.78.

We see that the selection of $R = 290\ \Omega$, according to the design formula, assures that $I_z(t)$ never falls below $I_{zmin} = 5$ mA. We also see that $V_{out}(t)$ is around 10 V, but there are peak-to-peak fluctuations of $10.063 - 9.985 = 0.078$ V. That is, there are 78 mV peak-to-peak ripples of the 10 VDC supply output voltage.

Finally, the peak-to-peak ripple amplitude of $V_c(t)$ is $V_r = 15.83 - 15.675 = 0.155$ V, which is slightly less than 1%. The design formula (2.3a) is too conservative. We can now reduce C significantly (as shown in Figure 2.79).

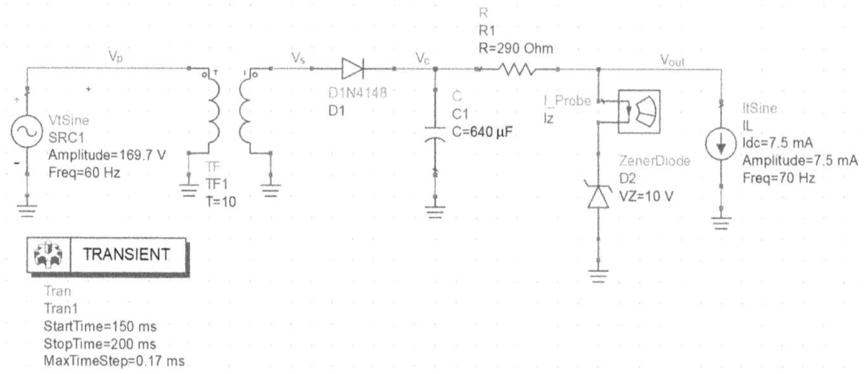

FIGURE 2.79 The final design, featuring a more reasonable C value.

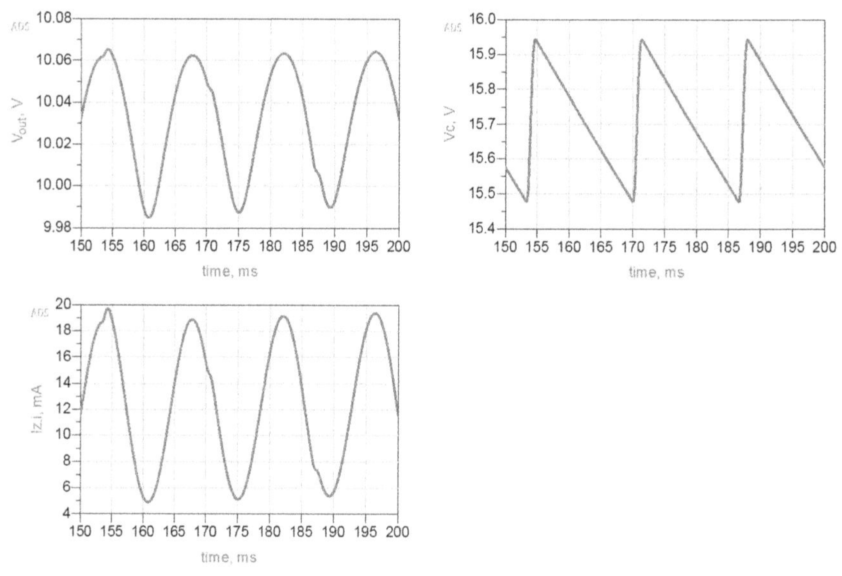

FIGURE 2.80 Simulation results of the final design.

The new set of results is shown in Figure 2.80. The "funny shapes" are because the 60 Hz voltage and the 70 Hz load current are not synchronized.

This time the peak-to-peak amplitude of $V_c(t)$ ripples is 2.8%. The ripples of $V_{out}(t)$ increased slightly to 80 mV ptp. This is because the $V_c(t)$ ripples are larger.

2.21 DIODE LIMITERS

Diodes can provide protection to critical circuit components, preventing such components from being exposed to electrical signals that may be too large. This can be done by signal limiting circuits [Sedra1]. A limiter may be implemented by a circuit that contains one or more diodes (or Zener diodes), resistors, and DC voltage supplies. Let us start by studying several simple limiters:

Example 2.11

Simple One Diode Voltage Limiters (Figure 2.81).

Whenever $V_{in}(t)$ is negative (or positive below the cut-in voltage), the diode is OFF and we have that $V_{out}(t) = V_{in}(t)$. As $V_{in}(t)$ increases, the diode turns ON and from there on $V_{out}(t)$ begins to be limited by a one diode voltage (around 0.7 V), regardless of how large $V_{in}(t)$ may be. Extensions of the idea are presented in Figures 2.82 and 2.83.

A signal $V_{in}(t)$ that turns ON a limiting diode becomes clipped at the diode voltage.

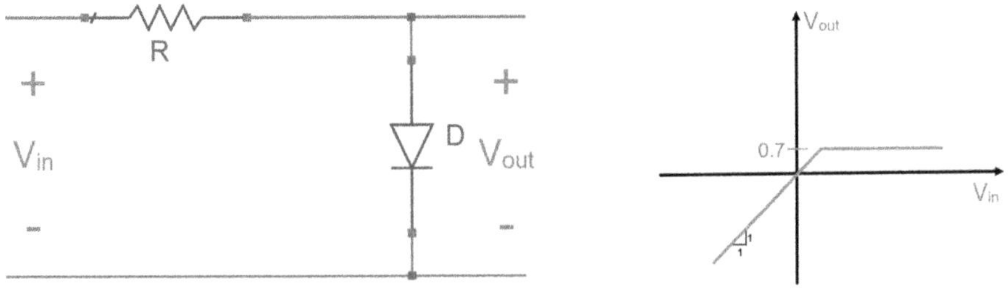

FIGURE 2.81 An R+D circuit that limits the upper part of an input signal to one diode voltage.

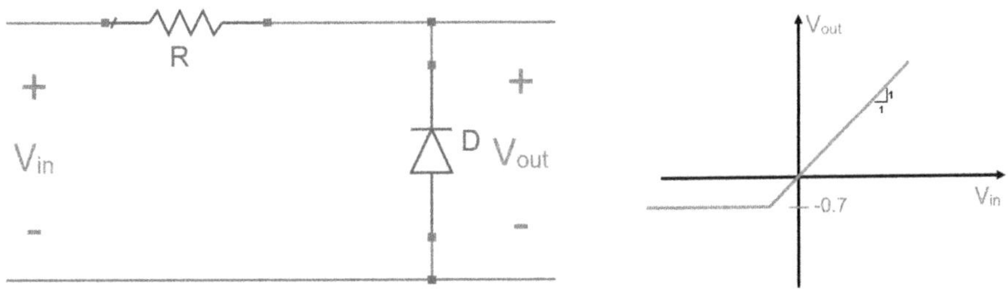

FIGURE 2.82 Turning the diode around creates a lower limit by one diode voltage.

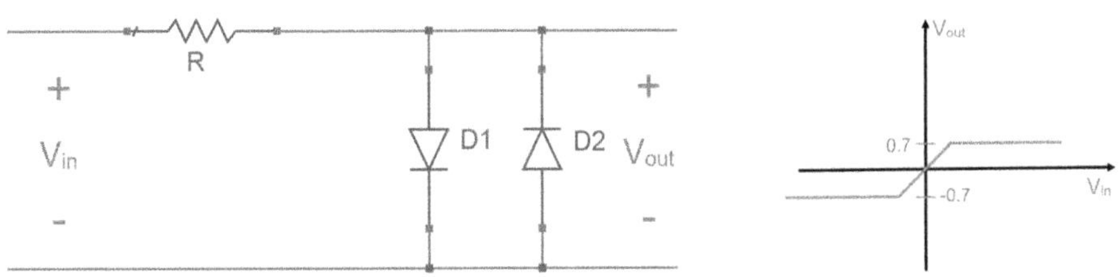

FIGURE 2.83 Upper and lower one diode voltage limits. $V_{out}(t) = V_{in}(t)$ whenever the two diodes are OFF simultaneously.

Example 2.12

Creation of arbitrary voltage limiting levels.

Figure 2.84 illustrates an upper limit of 5.7 V, created by placing a diode and a 5 VDC supply in series. The diode is OFF whenever the input voltage is lower than the set limit.

Practically, though, batteries may often be replaced by Zener diodes of appropriate nominal voltages. A current path consisting of two Zener diodes placed face-to-face (or back-to-back) creates independent upper and lower voltage limits, as shown in Figure 2.85.

In Figure 2.85, the Zener diodes have Zener voltages of V_{z1} and V_{z2}, respectively. The value V_{z2} is selected in order to create an upper limit of $V_{z2}+0.7$, whenever Zener diode Z_2 is ON as a Zener and Zener diode Z_1 is ON as a regular diode. Likewise for the lower limit $-(V_{z1}+0.7)$.

Say, for example, that we want a limiter that has an upper limit of 6 V and a lower limit of −4.5 V, then Z_1 should have a nominal Zener voltage of 3.8 V, and Z_2 should have a Zener voltage of 5.3 V.

Example 2.13

Most op-amps have a built-in "do-nothing" subcircuit that turns on only when the op-amp's output becomes short-circuited, or becomes connected to a large load.

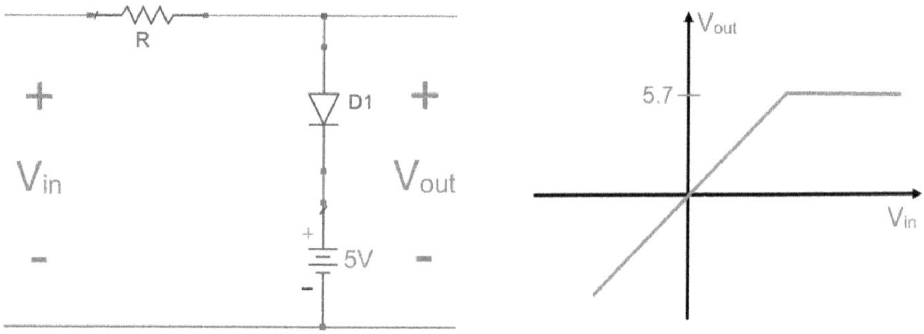

FIGURE 2.84 Upper limit increase with the help of a DC battery.

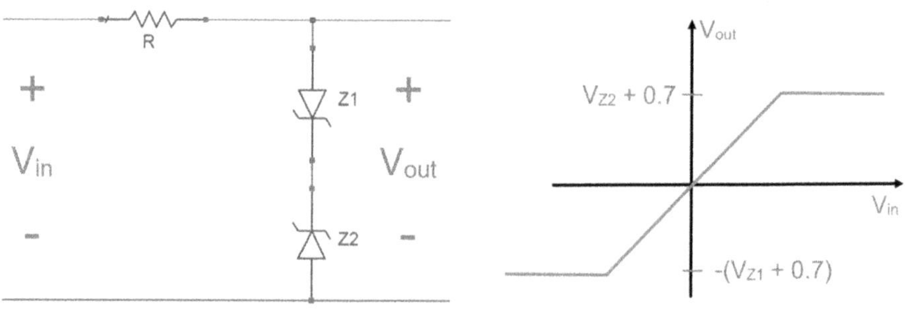

FIGURE 2.85 Two Zener diodes with opposite orientations and different nominal Zener voltages create user-defined upper and lower voltage limits.

Some op-amps, on the other hand, may be damaged whenever the input voltage becomes larger than the op-amp's supply voltages. The op-amp may be protected as shown in Figure 2.86 (right). Likewise, digital circuits that receive an input voltage that exceeds the chip's DC supply level (left).

2.22 ADS: DC AND TRANSIENT SIMULATIONS OF LIMITERS; BACK-TO-BACK EMULATED ZENER DIODES

Example 2.14

To observe an input–output curve of a voltage limiter, we shall use voltage sweeping via DC simulation. Then we shall follow up with a transient simulation to actually see how the limited output looks in the time domain. Refer to Figure 2.87.

Say that we need a limiter that has a 6.5-V limit on the positive side, and −4 V limit on the negative side.

We can take the Zener diode emulation again, using a DC supply and two regular diodes. What should be the DC supply values?

Let us consider the Zener voltage selection shown in Figure 2.87. The selection was based on the fact that for V_{in} that runs between 6.5 and −4 V, both Zener diodes are OFF, and $V_{out}=V_{in}$. The V_{in} sweep result shown in Figure 2.88 seems a little off: The lower limit looks to be around −4.7 V, and the upper limit looks to be around 7.2 V. What went wrong?

We recall that our design formula told us that each Zener voltage needs to be the sum of the battery voltage and one diode voltage. However, it is wrong whenever we put Zener diodes back-to-back. Studying the internal model of an emulated Zener diode (left side of Figure 2.87) reveals that there are two diode voltages added to each battery, not one. Let us revise the design by revising the battery selections, implemented by choosing another appropriate value for V_Z.

The simulation input–output curve for the circuit of Figure 2.89 is shown in Figure 2.90.

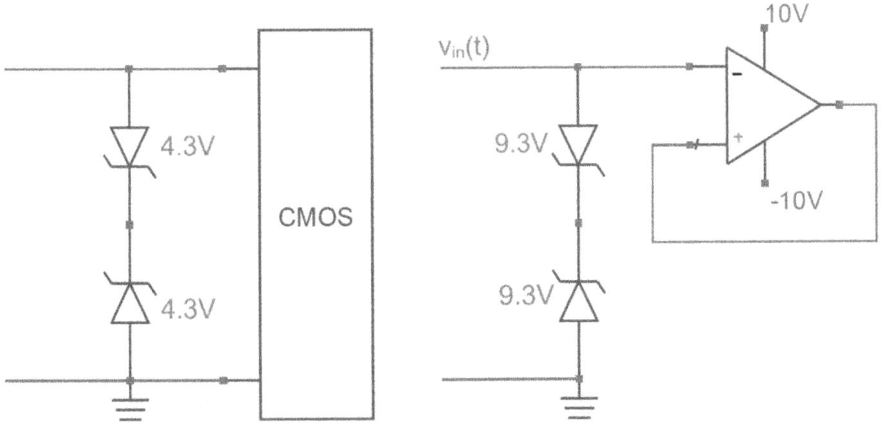

FIGURE 2.86 Protection of an op-amp and digital circuits from excessive input voltages.

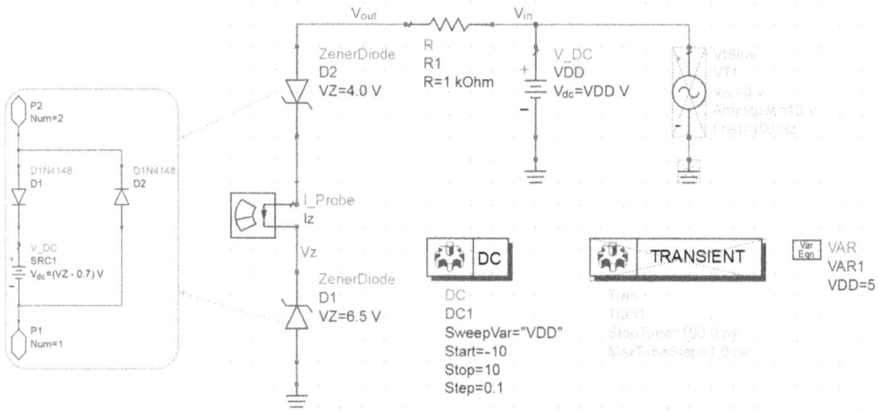

FIGURE 2.87 (Left) Zener diode emulation circuit. (Right) Schematic of a voltage limiter made of two Zener diodes.

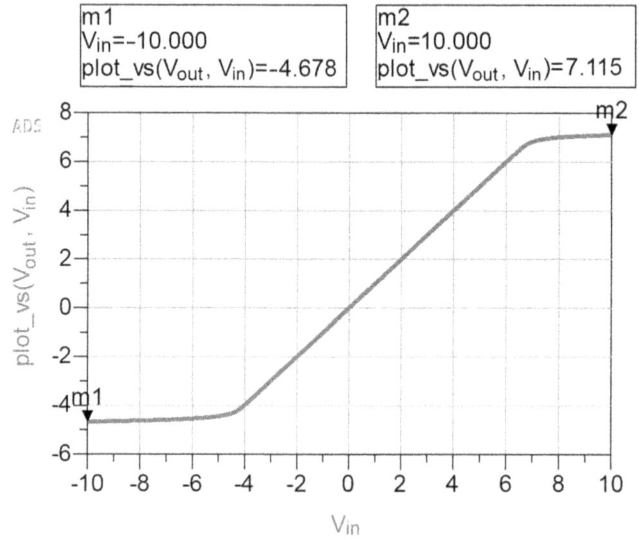

FIGURE 2.88 An incorrect input–output limiter curve produced by the circuit of Figure 2.87.

This is better!

For the time simulation, let us deactivate the DC simulation, the DC supply, and the VAR block. Instead, let us use VtSine signal (of any frequency) of amplitude large enough to be able to observe both limiting actions. See Figure 2.91.

We see in Figure 2.92 that V_{in} and V_{out} overlap whenever there is no current through the Zener diodes. Whenever the Zener diodes conduct, the limiting action takes place.

Ideally, we like the limiting levels to be straight horizontal, but let us recall that Zener diode voltages are not really constant.

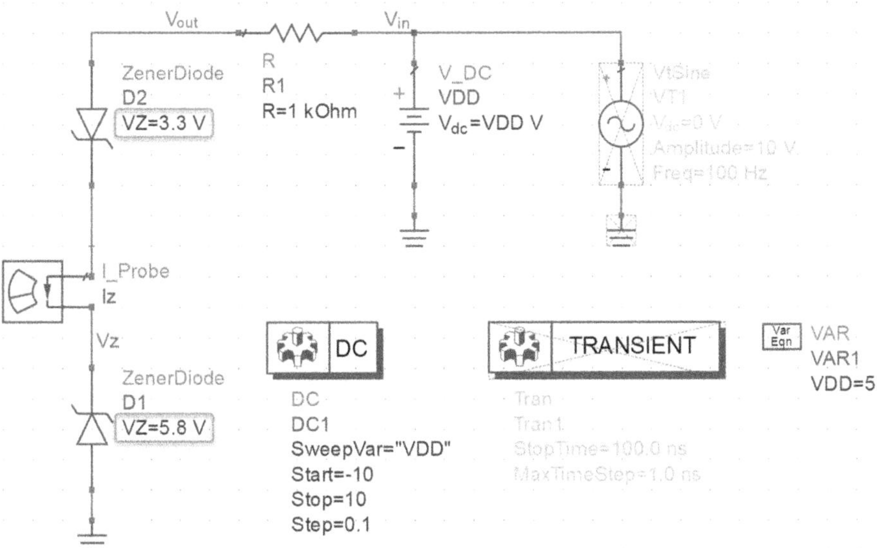

FIGURE 2.89 Revised Zener diodes limiter circuit.

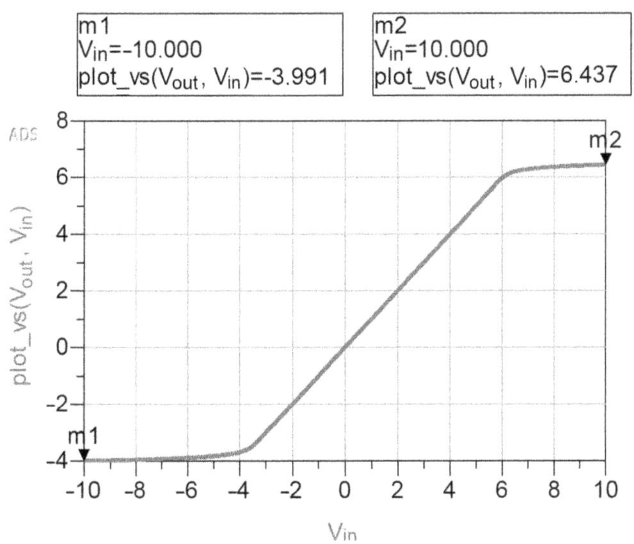

FIGURE 2.90 Corrected input–output limiter curve.

Example 2.13 Revisited

Let us see next how to protect a 741 op-amp from the effects of excessive input signals.

The 741 is, in general, well protected, via an automatic current limit circuit, from events where the output terminal becomes short-circuited to ground. What may damage an op-amp are input signals with amplitudes larger than the op-amp DC supply voltages.

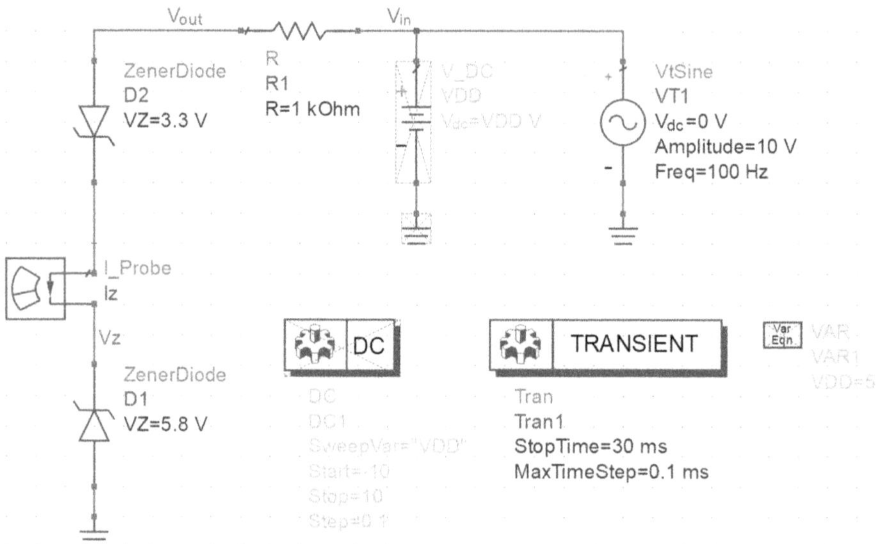

FIGURE 2.91 Transient simulation of the limiter circuit.

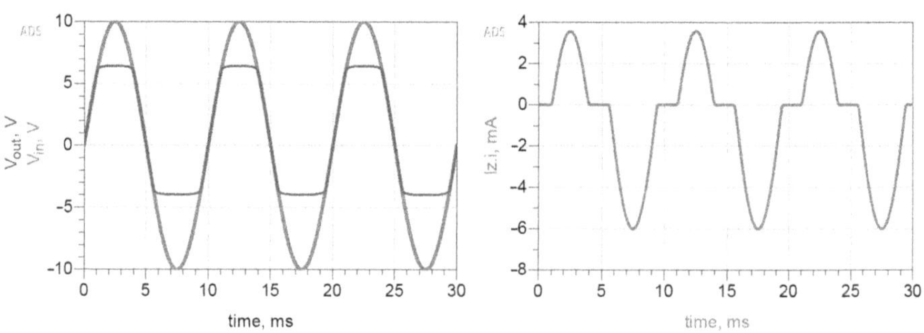

FIGURE 2.92 Time signals of the limiter circuit as the voltage limits are tripped.

In the example brought in Figure 2.93 we look at a comparator with supplies of +10 and −10 V.

Say that we limit the input signal not to exceed 9 or −9 V, which explains why we choose a DC battery of 7.6 V for each limiting. That is, for the Zener component, we choose $V_Z = 8.3$ V.

Whenever the signal V_{in} is inside the limit levels, the two back-to-back Zener diodes are both OFF. It is a do-nothing protection circuit. The limiter kicks in when the input signal becomes large (Figure 2.94).

First, let us try an input amplitude of 6 V and observe a normal comparator function. Then we shall take a 15-V amplitude.

In Figure 2.95, we shall observe what happens when the limiter is disabled.

ADS issued a warning for leaving some components unconnected.

The larger-than-supply levels input signal causes an incorrect operation of the comparator circuit, as seen in Figure 2.95 (right).

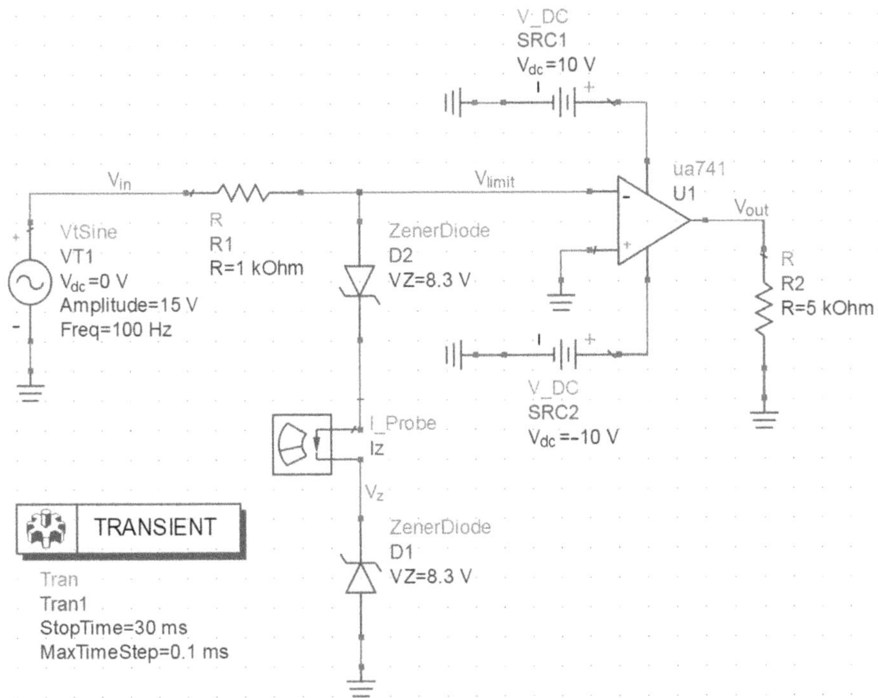

FIGURE 2.93 Zener limiter as a protector of a 741 op-amp.

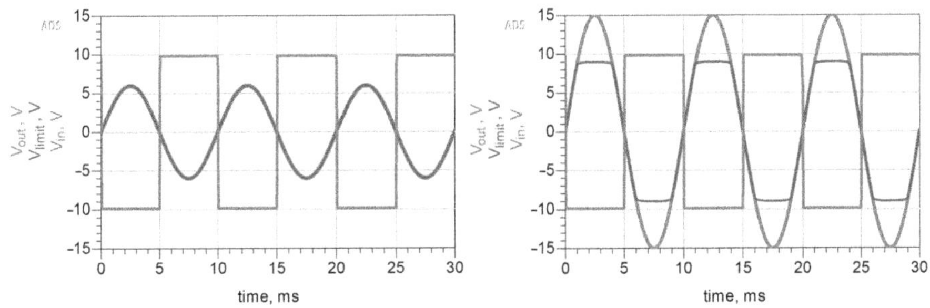

FIGURE 2.94 (Left) Comparator operation when the limiter is OFF due to a low enough input signal. (Right) Comparator and Limiter operation when the input signal is large enough to turn the limiter ON.

2.23 PROTECTION AGAINST AN INDUCTOR SUDDEN DISCONNECT

An inductor's current cannot jump (in the same way that a capacitor's voltage cannot suddenly jump). What happens if an inductor conducts current (and thus holds some energy) and all of a sudden, the current path becomes disconnected? We refer to [Horowitz1].

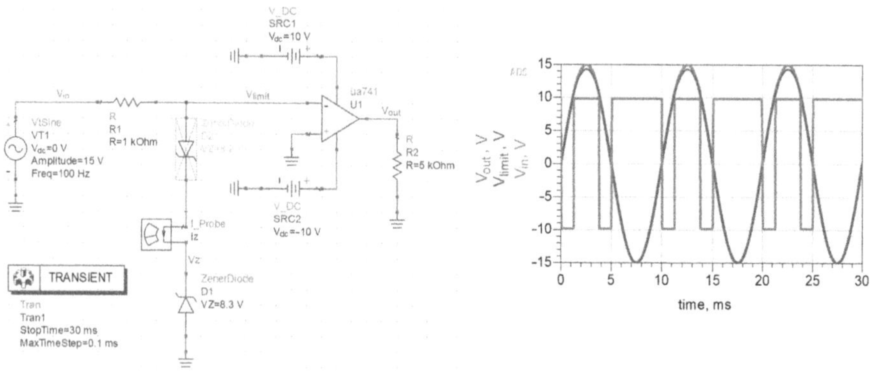

FIGURE 2.95 (Left) Comparator with a disconnected limiter circuit. (Right) Incorrect comparator operation.

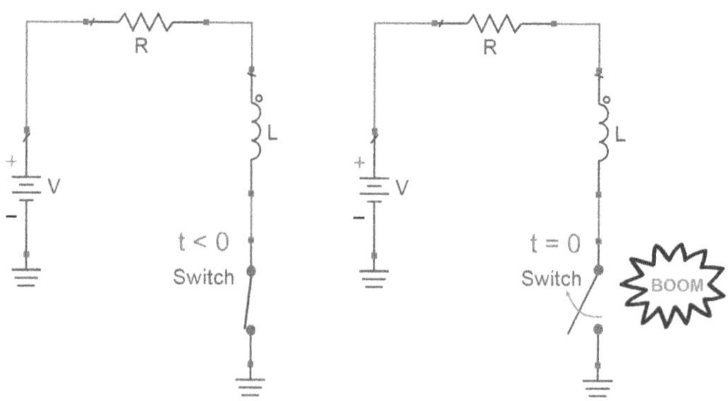

FIGURE 2.96 Illustrating the inductor's current path that becomes disconnected.

Look at Figure 2.96 (left) in which an inductor L is connected via a resistor R to a DC battery V. The switch, assumed to be normally closed, emulates a wire to ground that may undergo a sudden disconnect.

At t = 0 the switch opens up, and the energy stored in the inductor, at negative times, must be discharged. Mathematically speaking, and as seen from the current–voltage equation of any inductor: $v_L = L \cdot di_L/dt$, the sudden wire disconnect creates a large negative voltage spike.

An arc is created as the inductor tries to maintain current continuity, even through the air.

A simple protection involves a reverse-biased ("do-nothing") diode added across the inductor (Figure 2.97). The diode "gets into action" if the switch opens up, allowing the inductor to discharge its stored energy.

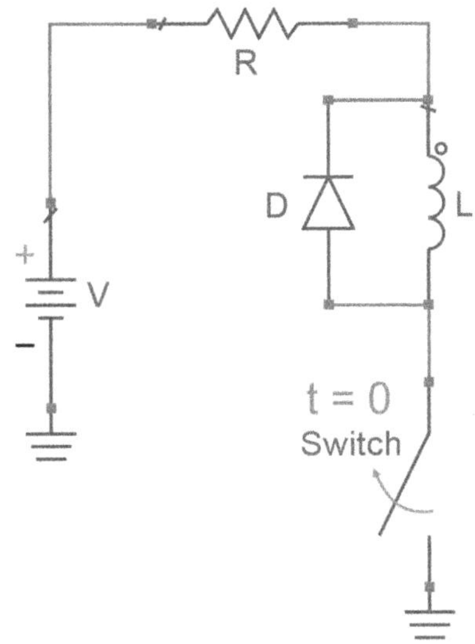

FIGURE 2.97 Diode is OFF almost all the time. Whenever the inductor's voltage becomes negative, the diode turns ON, allowing the inductor's energy to be discharged via an L–D circuit.

2.24 ADS TRANSIENT SIMULATION OF AN INDUCTOR DISCONNECT; VTPULSE, VTSTEP SIGNALS, AND VOLTAGE-CONTROLLED SWITCH

Example 2.15

An inductor L=1 mH and a series resistor R=1 kΩ, have an inductor's current charging time constant of L/R=1 μs. Let us assume an imperfect step signal input VtStep, and a voltage-controlled switch SwitchV. The switch is closed whenever the control voltage V is 0 V (or less). The switch becomes open if V exceeds 1 V. SWITCHV1 is controlled by a VtPulse (of levels 0 V and 2 V). The pulse kicks in at t=10 μs and returns to 0 at t=20 μs. See Figure 2.98.

Need to clear the SwitchV component model field, as shown, otherwise the SwitchV needs a "model card". No further action from the user. Model cards will be studied along with MOSFET transistors in Chapter 4.

The transient parameters are selected StopTime=30 μs based on the events created by the pulse control signal, and MaxTimeStop=1 ns, based on the time constant of the RL circuit.

The simulation results are displayed in Figure 2.99.

We see $I_L(t)$ charges, then it becomes zero when the switch opens up, and then it charges again. The problem is $V_L(t)$. At the disconnected instance, dI_L/dt is negative and huge. It creates a negative spark of thousands of V. Such sparks can be damaging to the coil and to nearby circuits.

As seen in Figure 2.100, as long as the circuit works normally (current flowing from left to right), the diode is OFF. The diode comes into protective action as the current path breaks down. The diode provides the inductor with a path to discharge its stored energy.

The simulation results of the schematic of Figure 2.100 are shown in Figure 2.101.

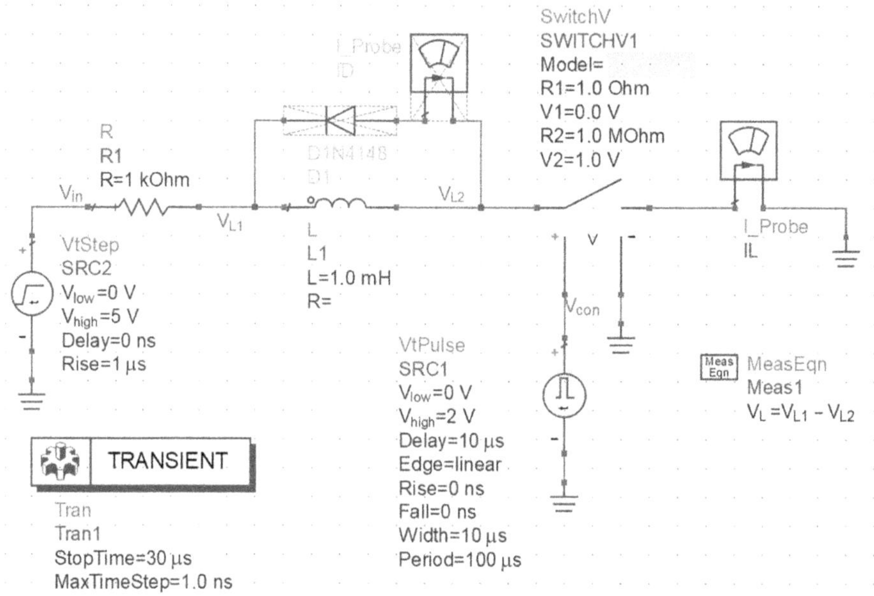

FIGURE 2.98 A schematic to simulate Figures 2.96 and 2.97.

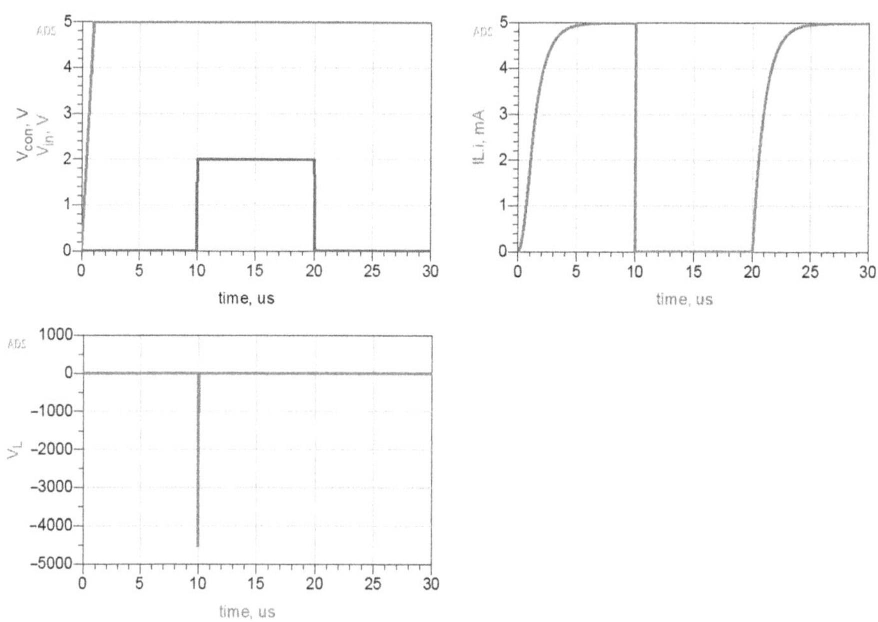

FIGURE 2.99 The circuit experiences a sudden inductor disconnect.

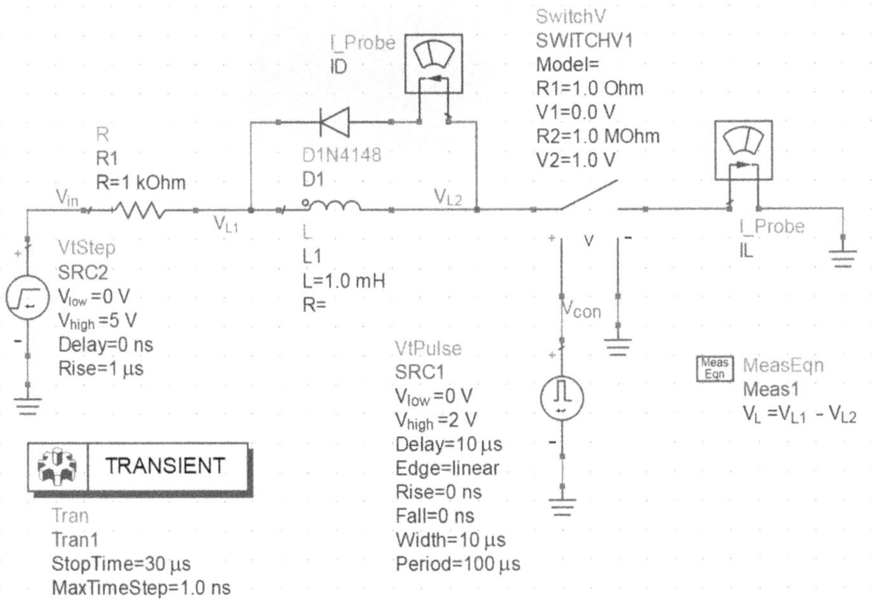

FIGURE 2.100 Current disconnect of an inductor with a protection diode.

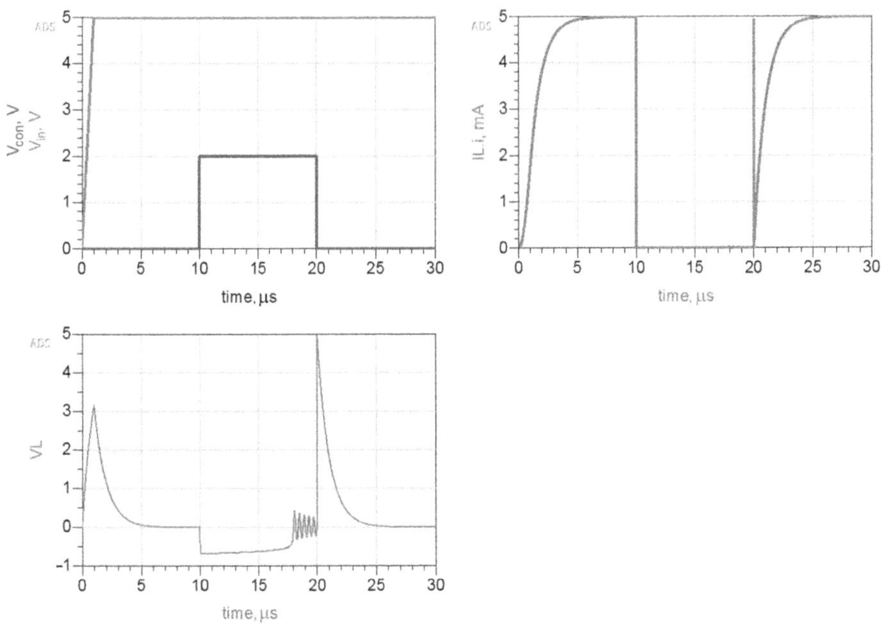

FIGURE 2.101 The effect of the protection diode on the inductor's voltage.

2.25 DIODE–CAPACITOR CLAMPING CIRCUITS

As we earlier discussed, the filtering of a rectified signal, the curious combination of a capacitor and a diode connected in series played an important role. When the diode is ON, the capacitor charges with a small time constant, determined by the diode's small ON resistance. Whenever the diode turns OFF the capacitor holds on to its charged voltage. The circuit of Figure 2.102 illustrates a capacitor and a diode driven by a discontinuous voltage signal, a periodic rectangular pulse.

We see that as a result, the input voltage $V_{in}(t)$ undergoes a DC voltage level shifting upward by an amount equal to the voltage that the capacitor is charged to. Let us explain in more detail how the circuit works (referring to [Sedra1]).

Assume that at $t=0$ the input voltage equals −6 V and the capacitor C is fully discharged. Then at $t=0$ the diode turns ON, and C is charged very fast to a voltage of 6 V (polarized as + on the diode's side and − on the input signal's side). After this brief charging, the diode turns OFF and the capacitor holds on to its charged voltage. At the time instance in which $V_{in}(t)$ jumps from −6 to 4 V the diode stays OFF and the circuit's output voltage $V_{out}(t)$ jumps to 10 V (the sum of the new input voltage level of 4 V and the capacitor's charged voltage of 6 V). As $V_{in}(t)$ jumps back from 4 to −6 V, $V_{out}(t)$ has to jump from 10 to 0 V because C still holds on to its charged voltage of 6 V. This time around the diode doesn't even turn ON, not even briefly.

Switching the diode's direction causes the capacitor to charge to 4 V, and this time it is the upper level of $V_{in}(t)$ that gets clamped to zero. See Figure 2.103. These clamping actions are independent of the value of C, as long as the diode–capacitor charging time constant is small.

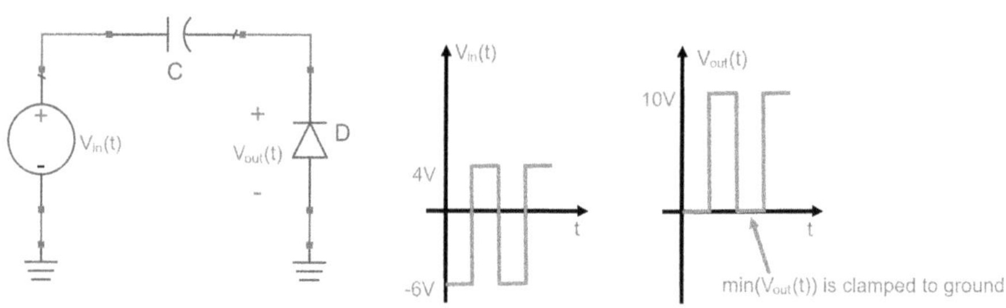

FIGURE 2.102 Lower clamping: The lowest level of a periodic pulse is clamped to a 0 V level.

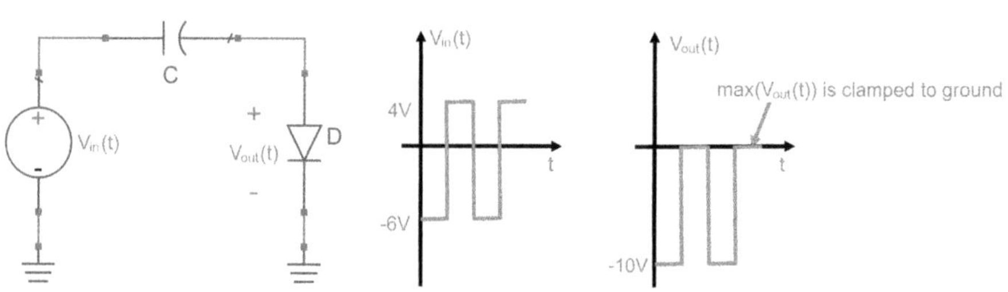

FIGURE 2.103 Upper clamping: The upper level of the input signal is clamped to zero.

So, what is the purpose of voltage clamping, also known as "DC Level Restoration"?

Note that in the above two examples $V_{in}(t)$ has a (mean) DC level of −1 V, whereas $V_{out}(t)$ has a DC level of 5 V (whenever the lower level of $V_{in}(t)$ got clamped to 0 V) or −5 V (whenever the upper level of $V_{in}(t)$ got clamped to 0 V).

There are many voltage signals in which the DC level carries important information. Consider, for instance, a pulse-width-modulation (PWM) amplifier. The output voltage signal of a PWM amplifier is a pulse wave with levels (0, V) or (0, −V) and a frequency f = 1/T with a period T that is much smaller than the time constant of the circuit fed by the PWM signal (say, a DC motor). Because f is large, all the harmonics 1, 2, … of the PWM signal are small. The only significant harmonic is the 0 harmonic (DC). So, the PWM amplifier works like a power amplifier with a tunable voltage gain of K. The PWM signal may be modulated by varying its duty cycle (defined here as the portion of T in which the signal is nonzero). The larger the duty cycle, the larger K.

If, for some reason, the PWM signal encounters an AC coupling to its load, then the coupling capacitor blocks the PWM signal's originally intended DC level, replacing it with 0. The clamping of one of the PWM voltage levels to zero is a way of recovering the original operation of the PWM amplifier.

2.26 ADS: SIMULATION OF CLAMPING CIRCUITS

Figure 2.104 features a schematic that can illustrate multiple transient demonstrations somewhat similar to the ones discussed in Section 2.25: Lower clamping of a pulse signal, upper clamping of a pulse signal, and even a clamping of an AC signal. Each run necessitates the disabling of certain components in the schematic. Note the altered input pulse signal: (6, −4) V in place of (4, −6) V.

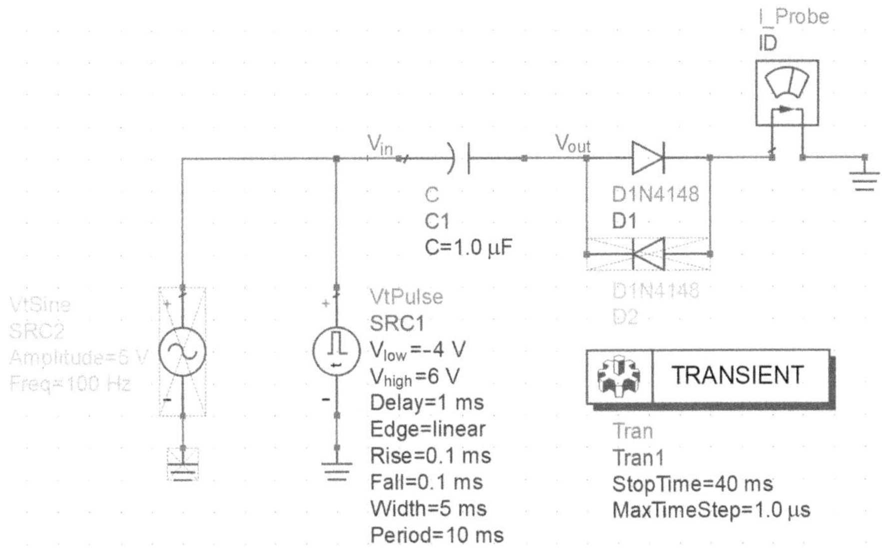

FIGURE 2.104 Schematic for multiple clamping actions, starting with upper clamping.

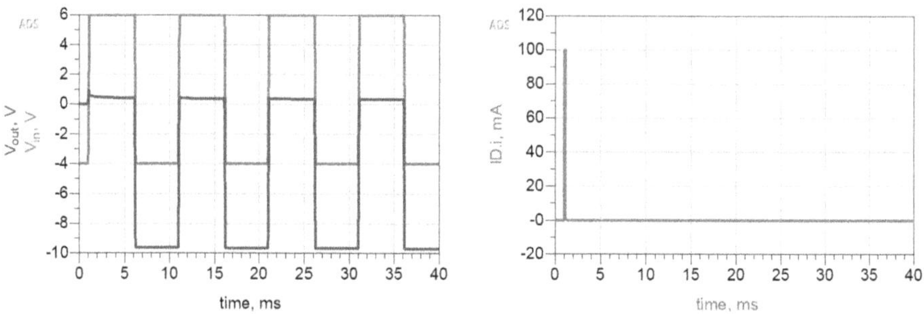

FIGURE 2.105 Upper clamping of a pulse signal.

Example 2.16

By disabling D_2 in Figure 2.104 and the AC source, the above schematic serves for simulating the circuit for upper clamping of a pulse signal. Results are shown in Figure 2.105.

We notice that most of the time there is no current. The charging of the capacitor to 6 V is very fast. The diode turns ON when V_{in} jumps from −4 to 6 V, and V_{out} nears zero. The diode quickly turns OFF, and C holds on to its 6 V charged voltage. That is why V_{out} jumps to around −10 V when V_{in} changes from 6 back to −4 V.

Example 2.17

Lower clamping is shown in the schematic of Figure 2.106, in which diode D_1 is disabled and diode D_2 becomes enabled. We added MeasEqn to measure the voltage across the capacitor.

The results are shown in Figure 2.107. There is an initial 1 ms delay of the VtPulse source. The current is practically zero.

Whenever the input source is nonideal, the clamping phenomenon may happen after some transient period. In Figure 2.108, the clamping circuit is fed by a voltage source with source resistance $R_S = 1\ k\Omega$. In a real voltage source, there is always an output resistance that is usually small. Notice that in our circuit, this has the effect of "slowing" things down. What we mean is that the circuit time constants will increase, allowing our setup to show the effects of charging of the capacitor more clearly. The effects of R_S on V_{out} are minimal. We clearly see the impact on the capacitor charging. The current peak has not changed, but it is smaller due to the current limiting effect that R_S has, otherwise, the current tends to zero in a steady state.

The addition of the resistor R_S did not change the steady-state response which is in line with expectations (Figure 2.109).

Further comments on the capacitor voltage. This is beyond our scope. We include the discussion in Appendix A (of this section) for the interested reader. In the transient simulation, the capacitor is initially treated as an open circuit and hence a voltage of 4 V at time = 0 s. What would happen if the capacitor's initial voltage were 0 V instead? Essentially, we want the *initial condition* of the capacitor voltage to be 0 V. See Section 2.27 for the discussion.

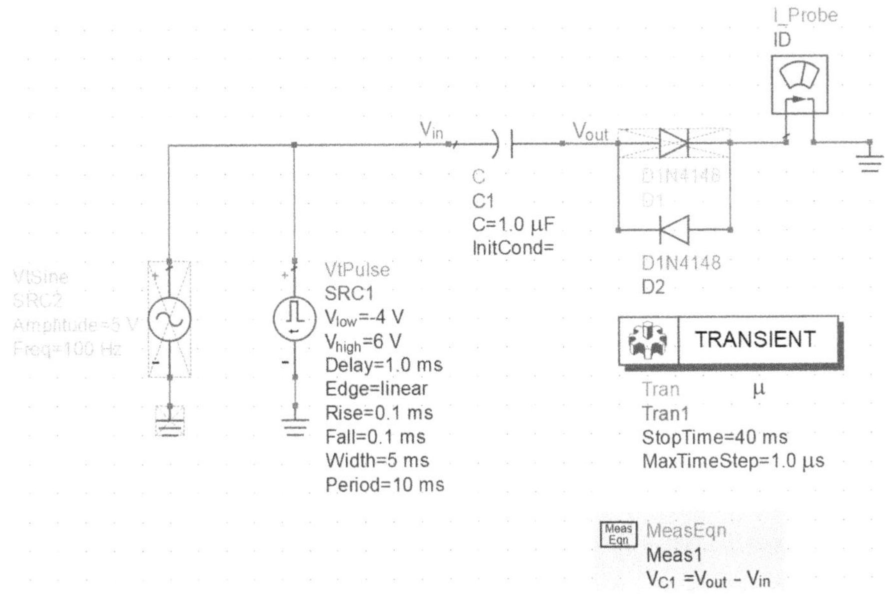

FIGURE 2.106 Lower clamping of a pulse signal, showing also the capacitor's voltage.

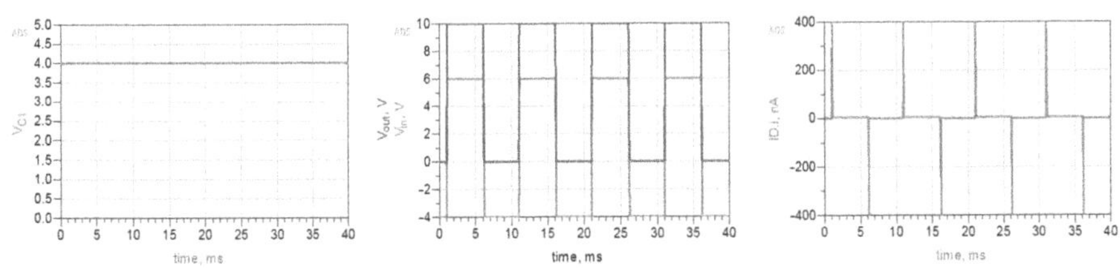

FIGURE 2.107 Capacitor's voltage and other results for a low clamping of a pulse signal.

Example 2.18

Clamping of an AC signal is demonstrated in Figure 2.110.

As seen in Figure 2.111, the clamping is not perfect and the effect of the diode's ON voltage is visible.

2.27 ADS TIPS AND TRICKS 2F: SIMULATION INITIAL CONDITIONS

Time domain simulators like SPICE, ADS, and others, assume that at time = 0 s capacitors behave as open circuits. We know from the physics of a capacitor that this is an assumption that, in many cases, is not true. We aim to show what happens if the default simulator assumption is changed, and what happens to the time response of the basic diode–capacitor clamping circuit. Is the assumption of an "open circuit capacitor" at t = 0 reasonable?

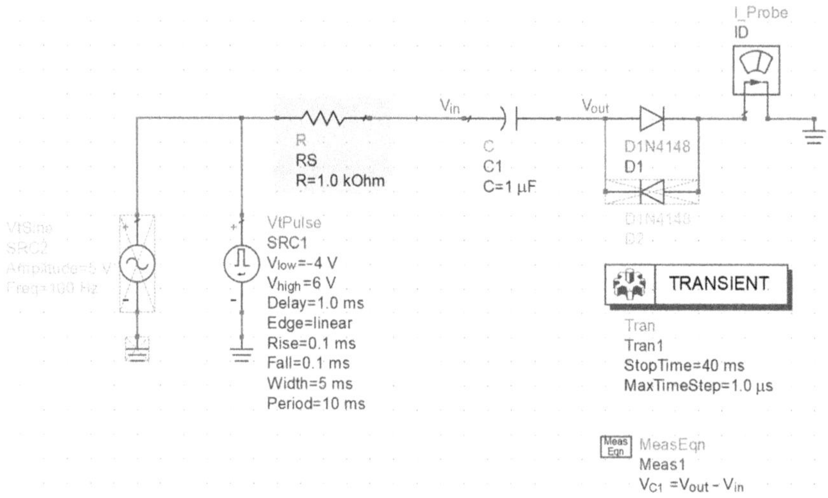

FIGURE 2.108 A clamping circuit fed by a nonideal source.

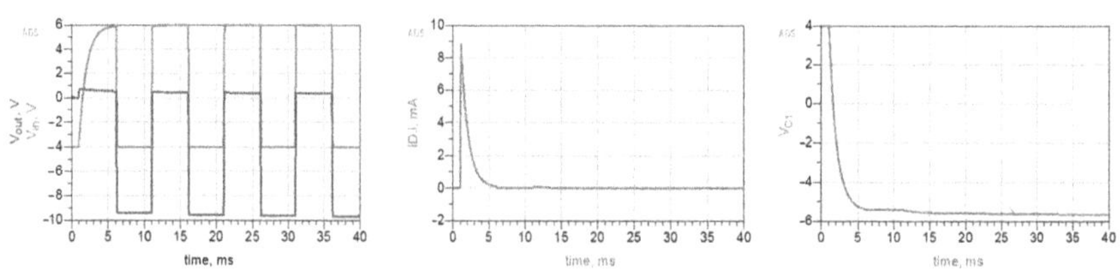

FIGURE 2.109 Performance of the clamping circuit fed by a nonideal source.

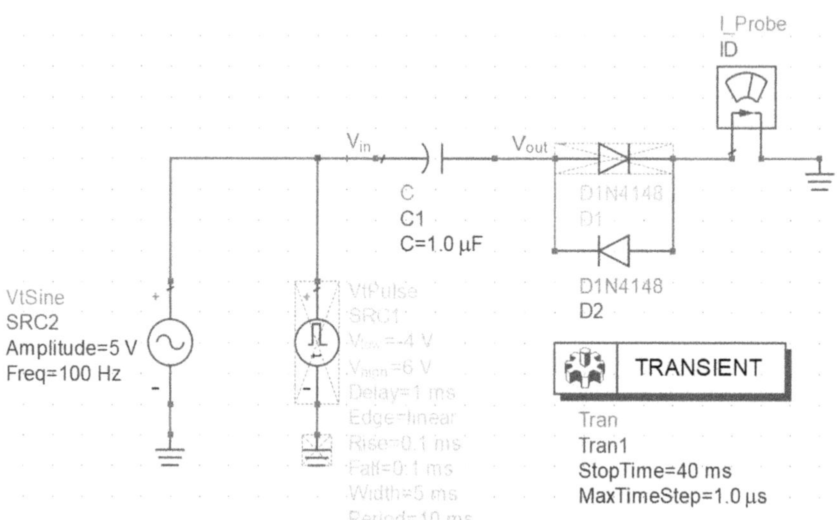

FIGURE 2.110 Clamping of an AC signal.

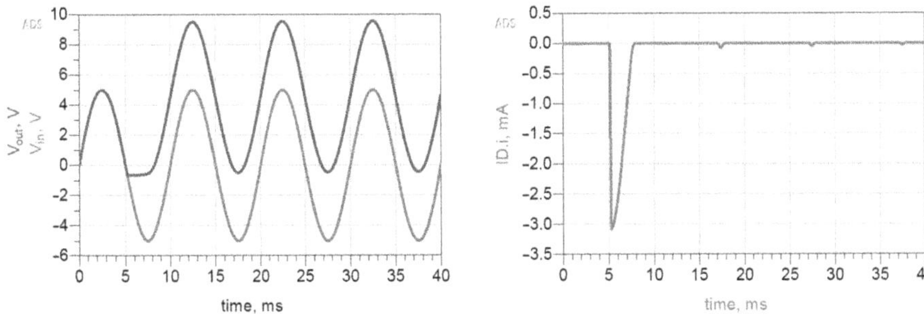

FIGURE 2.111 Low clamping of an AC signal.

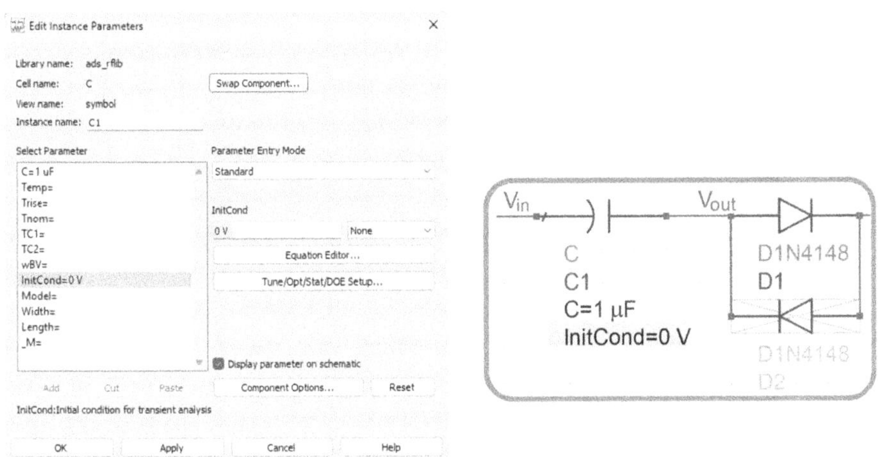

FIGURE 2.112 Setting InitCond = 0 V and making it visible on the schematic.

Starting from the schematic of Figure 2.104, the first step is to set up the capacitor and the simulator to include the initial conditions. For the capacitor, we look at the parameters seen in Figure 2.112.

The simulator, ADS transient, must be made aware that in the analysis, it must consider the initial conditions specified. This is done by opening the transient simulator controller properties window and selecting the convergence tab (Figure 2.113 (left)).

We select the "Use user-specified initial conditions", which is unselected by default. We also went to the display tab and made this selection visible on the schematic (Figure 2.113 (right)).

Figure 2.114 is the revised schematic.

From the results in Figure 2.115 the voltage VC_1 at $t = 0$ does indeed start at 0 V. There is some initial current and V_{out} transient response near $t = 0$ s, which was not there before, but it is brief. Significantly for our test circuit, the steady-state response is still the same, regardless of the initial condition. Readers may wonder why a simulator would allow users to specify initial conditions if the steady state is not affected.

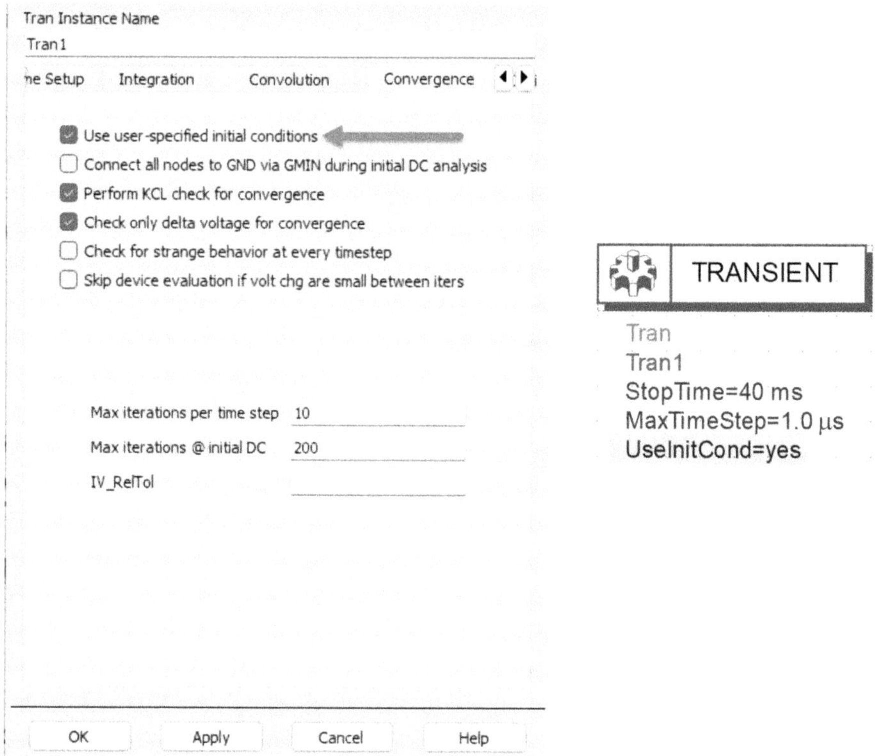

FIGURE 2.113 Transient simulator controller properties.

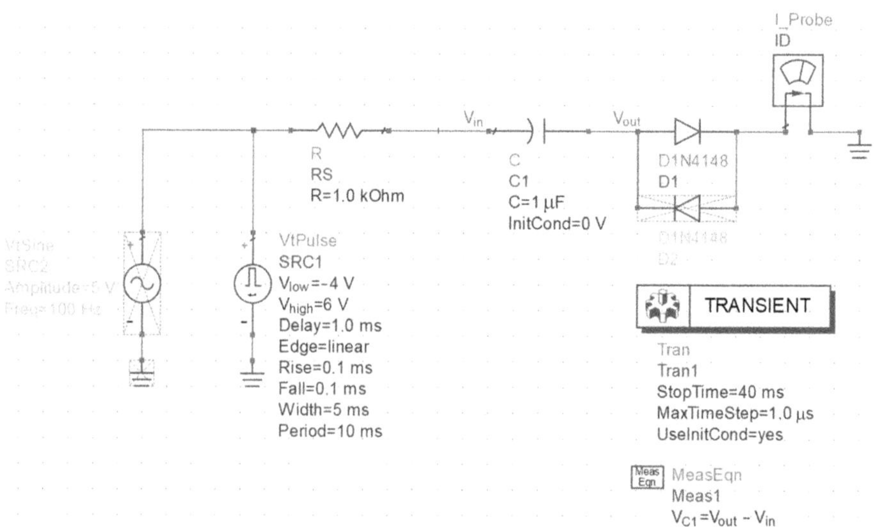

FIGURE 2.114 Clamping circuit with enforced zero initial condition of a capacitor's voltage.

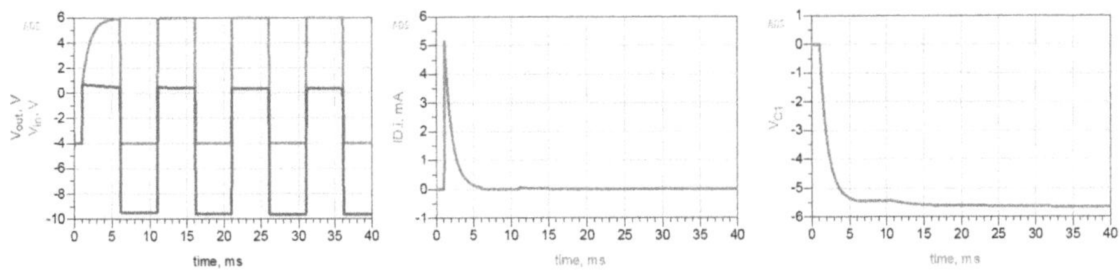

FIGURE 2.115 Results of the circuit of Figure 2.114.

In general, one cannot always assume, without *a priori* knowledge, that the steady state is not affected by the initial condition. There are circuits, such as digital latch circuits, that can be shown to have different steady-state responses to different initial conditions.

REFERENCES

[Gray1] Paul R. Gray, Paul J. Hurst, Stephen H. Lewis and Robert G. Meyer, *Analysis and Design of Analog Integrated Circuits*, 4th Edition, John Wiley & Sons, 2001.

[Horowitz1] Paul Horowitz and Winfield Hill, *The Art of Electronics*, 2nd Edition, Cambridge University Press, 1989.

[Keysight1] *Keysight Advanced System Design (ADS) Release 2025*, Keysight Technologies, Inc. https://www.keysight.com/us/en/products/software/pathwave-design-software/pathwave-advanced-design-system.html

[Sedra1] Adel S. Sedra and Kenneth C. Smith, *Microelectronic Circuits*, 7th Edition, Oxford University Press, 2015.

[Sze1] S.M. Sze, *Semiconductor Devices: Physics and Technology*, 2nd Edition, John Wiley & Sons, 2002.

CHAPTER 2 PROBLEMS – DIODES

Problem 2.1: Regular Diodes and Zener Diodes in DC

In the right circuit, let $v_i = -10\,V$ (Figure 2.116).

1a: Solve both circuits (above). First, make assumptions about each diode (ON? OFF?). After solving each circuit, justify your assumptions. Compare to an ADS solution.

1b: In the left circuit, replace the 180 Ω resistor with a 1-kΩ resistor. Solve again.

1c: In the left circuit, replace D_2 with a 1.8 V Zener diode, the anode of which is connected to ground. Solve again.

Problem 2.2: Regular Diodes and Zener Diodes in DC

In the given DC circuit of Figure 2.117 needs to assume a practical model for each diode. The Zener diode has a nominal Zener voltage of $V_z = 5\,V$.

2a: Assume that all the diodes are ON (including the Zener diode, which is ON as a Zener). Show that it is true.

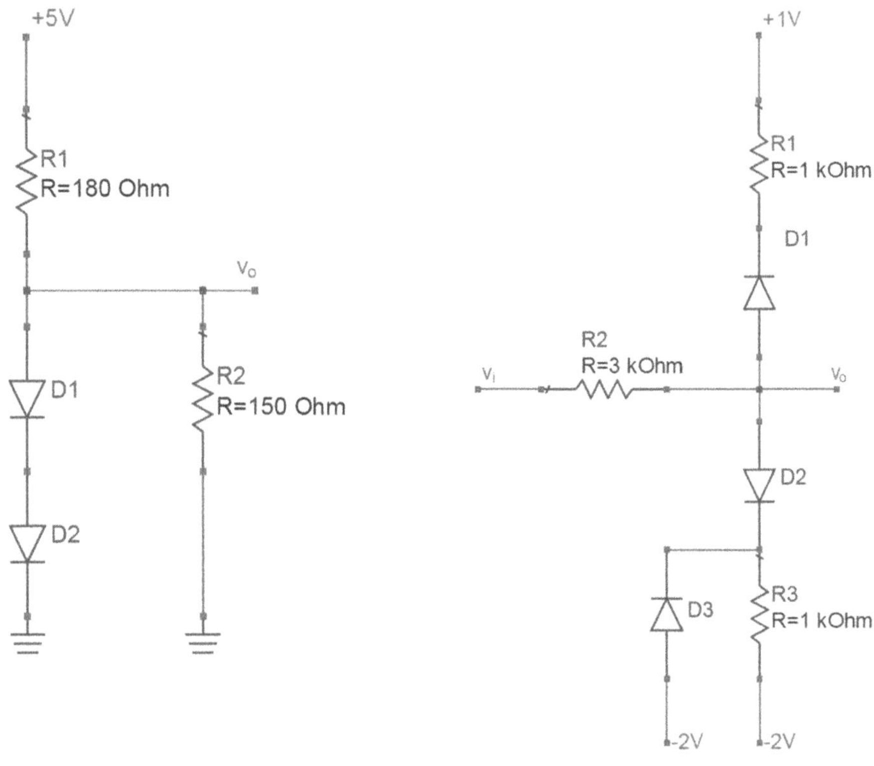

FIGURE 2.116 Problem 2.1 circuits.

Hint: Set a Kirchhoff current law equation for the node voltage v_1, solve it, and find all the circuit currents. In particular, find the Zener diode's current and its direction.

2b: Compare an ADS simulation of the circuit to your manual calculations of Problem 2.2a. Look critically at every current and every node voltage. Explain any differences between the simulation and calculations.

2c: What happens if R_4 changes from $R_4 = 5$ kΩ to $R_4 = 2$ kΩ? Which diodes turn OFF and why? Again, find v_1 and the currents. Compare to an ADS simulation.

Hint: Whenever you assume that a diode is off, don't remove it in the simulation. It has to stay. The simulation will show if the diode is OFF by displaying a very small current that goes through the diode.

2d: Go back to the circuit of Problem 2.2a. This time, connect the anode of D_3 to a DC voltage source $V_a = 10$ V. Analyze the circuit: Which diodes are ON and which ones are OFF? Is the Zener diode ON, and if so, is it ON as a Zener or as a regular diode? Compare to an ADS simulation.

2e: Go back to the circuit of Problem 2.2a. Replace $R_3 = 6$ kΩ by $R_3 = 100$ Ω. Analyze the circuit. Are all the diodes OFF? If so, why? Compare to an ADS simulation.

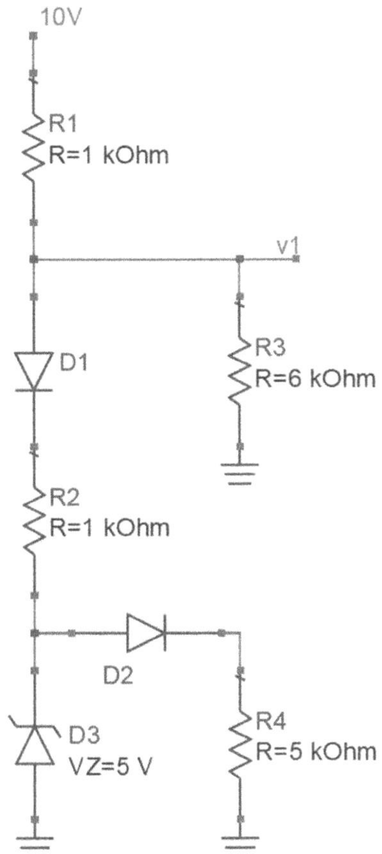

FIGURE 2.117 Problem 2.2 circuit.

Problem 2.3: Regular Diode and Zener Diode in DC Circuit

Use practical models in your manual calculations. ADS, on the other hand, uses exact models. Name the cathode voltage v_1 and the anode voltage v_2 (Figure 2.118).

3a: In the circuit (above), show that diode D_1 is OFF. In your ADS verification, leave the diode D_1 in its place – don't disconnect it.

3b: Can you make D_1 turn ON? You are allowed to modify just one of the resistors (of your choice). Explain how to do the design systematically (not by trial-and-error).

3c: Replace diode D_1, any way you want (between v_1 and v_2), with a 2.7-V Zener diode. You may modify one resistor so that the Zener diode conducts as a Zener. Again, explain your method. No trial-and-error.

Problem 2.4: Rectifier and Filter

4a: The input signal V_{in} needs to be converted to a rectified+filtered signal, with ripple amplitude of less than 1.1 V. For that purpose, a half-wave rectifier and filter

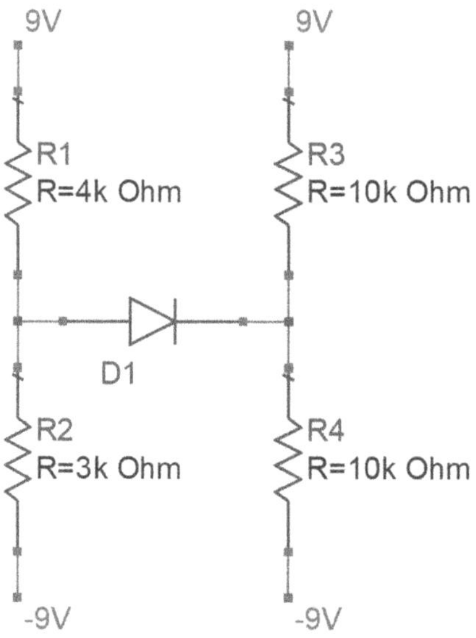

FIGURE 2.118 Problem 2.3 circuit.

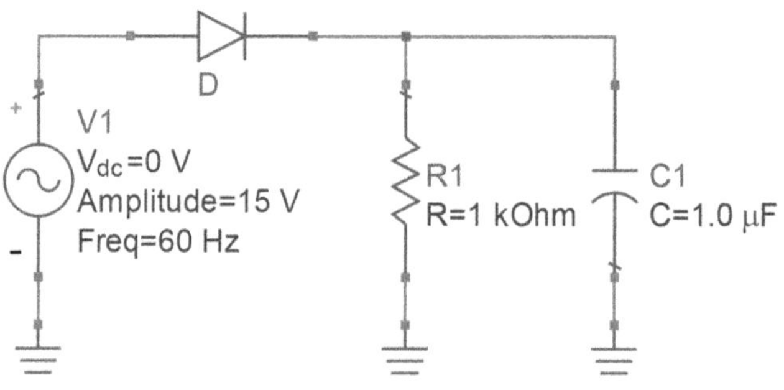

FIGURE 2.119 Problem 2.4 circuit.

are used. How large should R and C be, and how long does the conversion take? (Don't use capacitor values larger than 200 μF, or resistor values larger than 1 MΩ (Figure 2.119).

Also be aware that a too large value for R may cause the diode's current to be very small, and it may degrade the diode's performance. Demonstrate that your design is successful by measuring the ripples' peak-to-peak on ADS with markers. Also, compare the calculated and simulated values of V_{cmax} and V_{cmin}.

4b: Repeat (Problem 2.4a) using a full-wave rectifier consisting of a diode bridge. Can you use the same resistor value as in Problem 2.4a and half of the capacitor's value?

Problem 2.5: AC to DC Conversion

In this problem, the line voltage is (120 Vrms or 169.7 V peak amplitude, 60 Hz). The load connected to the output DC voltage V_z is resistive: $R_{Lmin} \leq R_L \leq 11$ kΩ.

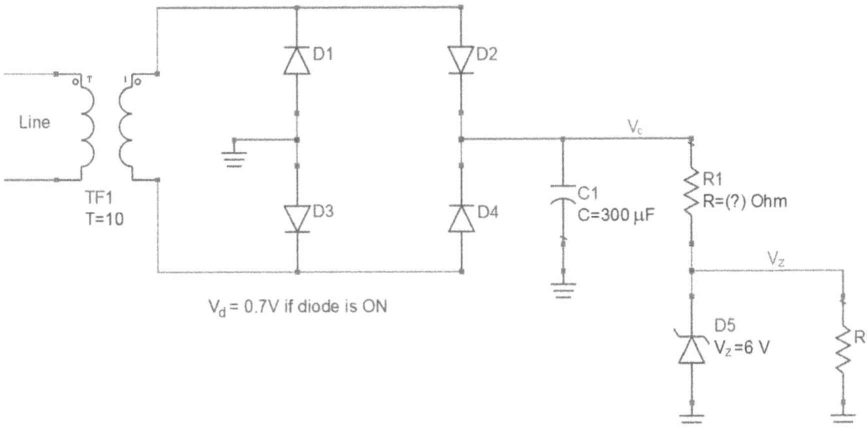

FIGURE 2.120 Problem 2.5 circuit.

The Zener diode parameters are {$V_z = 5\,V$ if $I_z = 1.2\,mA$, $r_z = 15\ \Omega$, I_z must not fall below $I_{zmin} = 0.6\,mA$.} *Note*: As you implement in ADS the emulation of the Zener diode, you need to measure the correct nominal current I_{zn} that develops the nominal Zener voltage, and also measure the correct value of r_z at that point. The new nominal current affects the value of I_{zmin}, taken as one half of the nominal current (Figure 2.120).

5a: Find the peak of the capacitor voltage V_c. Your calculations should consider the transformer action and the type of rectifier. *Hint*: When you simulate this part of the circuit, assume initially that $R = 1\ k\Omega$ and $C = 300\ \mu F$.

5b: Let $C_1 = 300\ \mu F$ (as shown) and let $R = 180\ \Omega$. Find the peak-to-peak amplitude of the ripples of $V_c(t)$. Plot this voltage and mark its voltage and time parameters. Use ADS markers for your measurements.

5c: Assume that for $R = 180\ \Omega$, we have that whenever $I_L = I_{Lmax}$ the Zener diode current may fall to $I_z = I_{zmin} = 0.6\,mA$ (or the new I_{zmin}, found for the Zener diode that you built). What is the value of I_{Lmax} and consequently, what is the permitted value of R_{Lmin}?

Problem 2.6: AC to DC Conversion

6a: Assume that a sinusoidal signal that has an amplitude of 26 V and a frequency of 440 Hz is the total secondary-side voltage of a center-tapped transformer. Model that transformer in ADS by simply using two VtSine sources in series with ground in between, each VtSine source has an amplitude of 13 V. Using D1N4148 diodes, design a **full-wave rectifier** and filter, to obtain a (very crude) "DC voltage" with ripples that are no bigger than 5% (ripples' peak-to-peak amplitude is considered with respect to the peak capacitor's voltage). Compare your design to the results obtained in the ADS simulation, and comment on any differences.

6b: Create in ADS an 8.1 V Zener diode. Measure its voltage–current curve. At what current value does the Zener diode have a voltage that is exactly equal to its nominal voltage? Also plot the Zener diode's incremental resistance r_z as a function of V_z. What is the value of r_z at the nominal Zener voltage?

6c: Use your ADS 8.1 V Zener diode, created in (Problem 2.6b), to design a voltage regulator for the rectified and filtered voltage that was created in (Problem 2.6a) in order to create a "clean" 8.1 VDC voltage. Can you observe any ripples on the Zener diode voltage? Note that in this design, no data was provided about the load current. The Zener diode minimum allowed current I_{zmin} must be no smaller than 1/3 of the Zener nominal current.

6d: Need to redesign R and C based on some load current data and minimum Zener diode current spec. Assume that the filter's maximum voltage ripples are as in (Problem 2.6a) above. The load current may vary between 0 and 11 mA. You may represent it by an ItSine source hooked up in parallel to the Zener diode – make the current go from the Zener diode node to ground with DC level of 5.5 mA, an amplitude of 5.5 mA, and a frequency of 100 Hz. [*Precaution*: As you hook up the ISIN source representing the load current, make sure that this load takes away current from the Zener diode and does not add current to the Zener diode]. The design should ensure that the Zener diode's current never falls below I_{zmin} (specified in Problem 2.6c). Choose and explain your choice of R, using the inequality design formula, and, once you know what R is, revise C as necessary to keep the same filter ripples as before. Then demonstrate in ADS that the specs are met.

Problem 2.7: AC to DC Conversion

In Figure 2.121, let $N=8$, $V_z=5.2\,V$ at $I_{znom}=6.5\,mA$ with $r_z=10\ \Omega$, $I_{zmin}=I_{znom}/2$, capacitor's voltage ripples of 5% and $I_{Lmax}=8.4\,mA$. Your calculations may differ from the ADS results because the Zener diode may have different parameters. Elaborate on these differences.

7a: Find R and C, explaining every step.

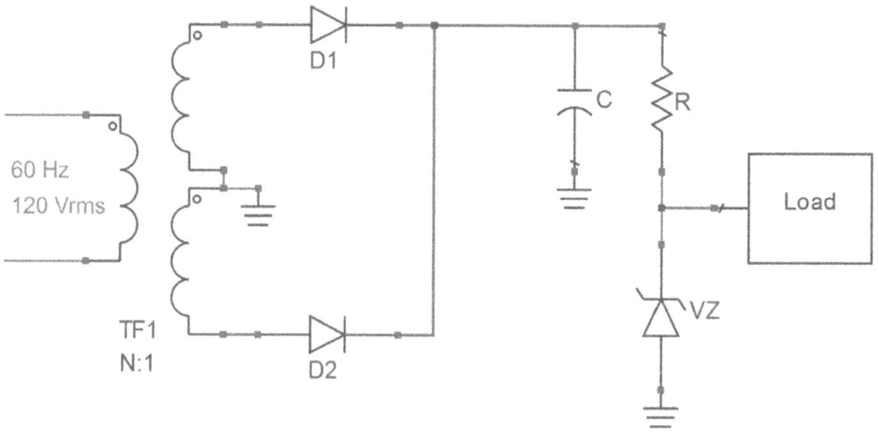

FIGURE 2.121 Problem 2.7 circuit.

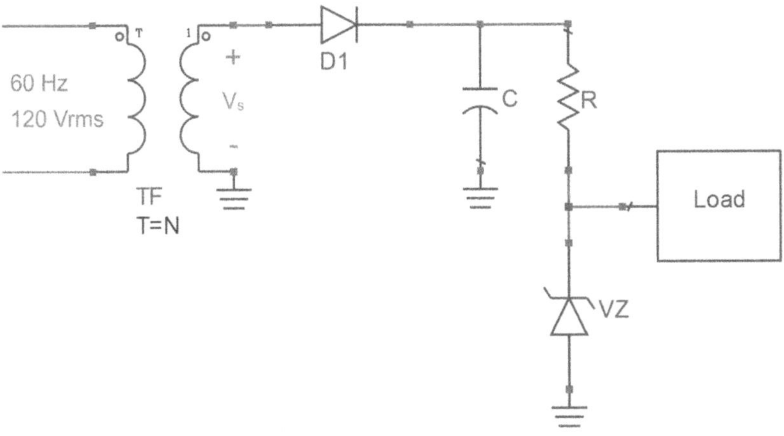

FIGURE 2.122 Problem 2.8 circuit.

7b: For the design of (Problem 2.7a), if the Zener diode is replaced by one with the same parameters except for $r_z = 8\ \Omega$, by what percentage will the line and load regulations change? Is it good or bad? Explain every step. No ADS verification is needed in this part.

Problem 2.8: AC to DC Conversion

Let: $N = 10$, $V_z = 7.5$ V at $I_{znom} = 6.4$ mA with $r_z = 20\ \Omega$, $I_{zmin} = I_{znom}/3$ and capacitor's voltage ripples are of 5%. The load regulation is equal to −17 mV/mA. See Problem 2.7 comments regarding possible differences with ADS (Figure 2.122).

8a: Find R, C, and the maximum load current I_{Lmax}. Explain every step.

8b: When there is no load, what is the peak Zener diode current? Explain every step.

Problem 2.9: AC to DC Conversion

Let: $N = 9$, $V_z = 6.8$ V at $I_{znom} = 19$ mA with $r_z = 12\ \Omega$ and $I_{zmin} = I_{znom}/3$. The line regulation is equal to 18 mV/V. As in Problems 2.7 and 2.8, regarding comparison with ADS (Figure 2.123).

9a: Find R, C, and the maximum load current I_{Lmax}, if we know that at no load, the dip Zener current equals 16 mA. Explain every step. [*Explanation*: I_{zmin} is the specified minimum current at full load operation. It is smaller than the no-load dip current.]

9b: For a load current of $I_{Lmax}/2$, what is the Zener diode's voltage?

Problem 2.10: AC to DC Conversion

Need to create a V = 7.5 VDC supply from the US line voltage (170 V peak amplitude, 60 Hz frequency). Two 10:1 turns ratio transformers are available, one with a

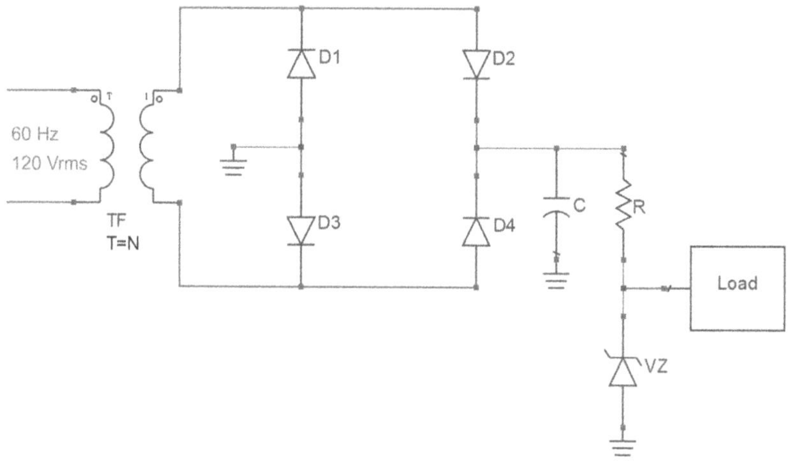

FIGURE 2.123 Problem 2.9 circuit.

center tap and another without the center tap. The DC supply feeds a load circuit that can draw up to $I_{Lmax} = 3\,mA$.

10a: Obviously, you need a 7.5-V Zener diode. Create one in ADS (using a DC battery and two regular diodes) and run tests on this emulated Zener diode to find its parameters: Nominal current I_{znom}, Minimum allowed current I_{zmin}, and the incremental resistance r_z.

10b: Set your own spec for the capacitor's voltage ripples (say 5%, or any other figure that you choose). Design the resistor R (that feeds I_z into the Zener) and the capacitor C. In your ADS verification, don't show the transformer – just emulate it by creating a suitable AC secondary voltage. Test your design and measure the following performance indicators:

The minimum value that $I_z(t)$ reaches.
The ptp ripples of the supply V.
The ptp ripples of the capacitor's voltage.
The resulting line regulation.
The resulting load regulation.

Measurement of some of the above indicators may require that you set the circuit in certain special settings. For example, disconnect I_L when you measure the line regulation.

3 Basic BJT Circuits

3.1 BJT IN DC CIRCUITS: WHAT IS A BIPOLAR JUNCTION TRANSISTOR?

We refer the reader to [Sedra1], [Gray1], and [Sze1]. A bipolar junction transistor (BJT) is a three-terminal semiconductor device consisting of two adjacent p–n junctions. A BJT has three regions named **emitter**, **base**, and **collector**. There are two types of BJT transistors: **PNP** and **NPN**. In a PNP transistor, the emitter is p-type, the base is n-type and the collector is p-type. Likewise, in an NPN transistor, the emitter is n-type, the base is p-type, and the collector is n-type. The base region is "sandwiched" between the emitter and collector regions (Figure 3.1).

An alert reader may wonder how a BJT can work at all, consisting of two back-to-back diodes in series. Indeed, what distinguishes a BJT from a series connection of two opposite polarity diodes, which are permanently OFF, is the narrowness of the base region (Figure 3.2). This is why, under bias conditions that assure active mode, most carriers (free electrons or free holes) emitted from the emitter region into the base region end up being collected into the collector region.

Figure 3.3 illustrates the electrical symbols of NPN and PNP transistors, as well as the directions of the three BJT currents. The three arrows that show the directions of I_E, I_C, and I_B are not part of the electrical symbols. Each BJT symbol features a single arrow that indicates the direction of I_E.

In both types of BJT transistors, by Kirchhoff current law, the emitter current equals the sum of the collector and base currents.

In each type of BJT, there are two p–n junctions, the **base–emitter** (BE) and the **base–collector** (BC) junctions. Each of the junctions may be forward- or reverse-biased, hence the BJT has four modes of operation. The most important mode of operation is **active mode**, in which the BE junction is forward-biased while the BC junction is reverse-biased. In a **saturation mode**, both junctions are forward-biased. In a **cutoff mode**, both junctions are reverse-biased. The fourth mode of operation, called reverse active mode, in which the BC collector junction is forward-biased and the BE junction is reverse-biased, is rarely encountered.

In a BJT that operates in active mode, the majority carriers diffuse from the emitter region into the base region. These are free holes for a PNP BJT and free electrons

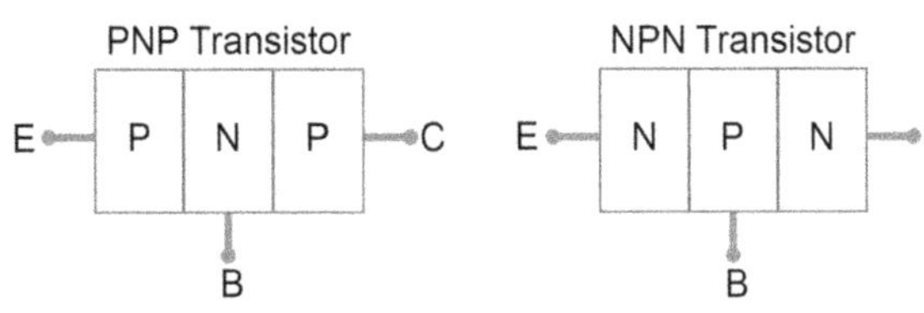

FIGURE 3.1 The two types of BJT transistors.

DOI: 10.1201/9781003606772-3

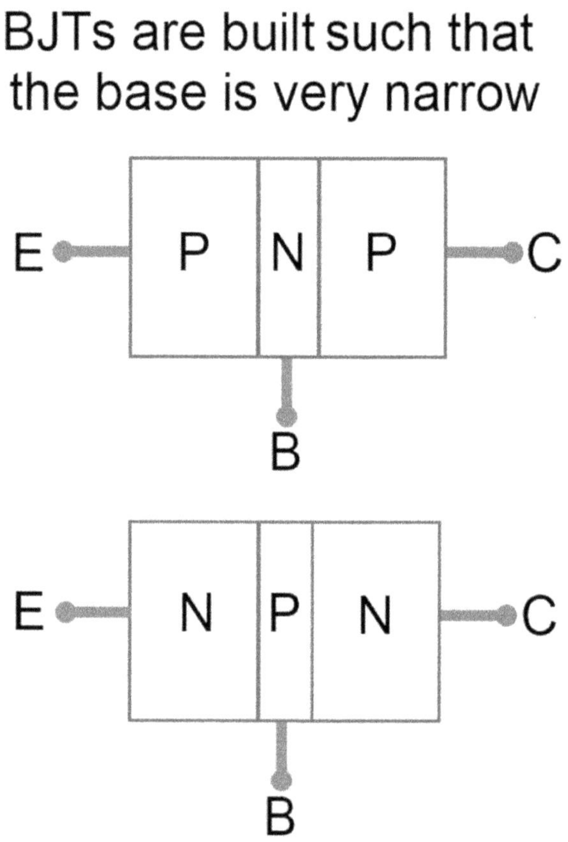

FIGURE 3.2 Illustration of a narrow base.

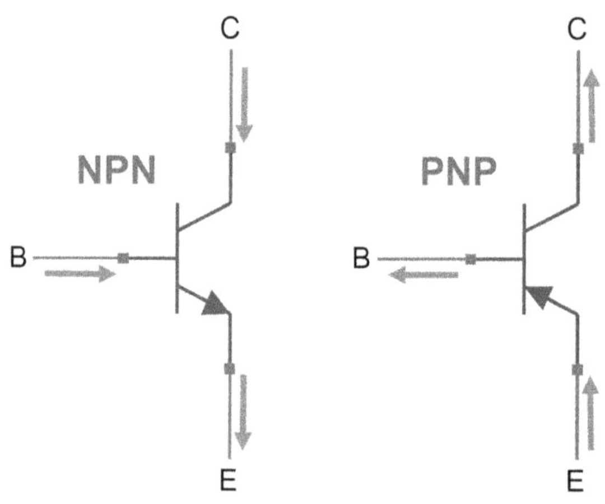

FIGURE 3.3 BJT current directions and BJT electrical symbols.

for an NPN BJT. The large majority of these carriers are swept into the collector region. That is, the **collector current I_C** is only slightly lower than the **emitter current I_E**. The **base current I_B**, which is the difference between I_E and I_C, is accounted for by base majority carriers that diffuse from the base to the emitter, and by hole–electron recombinations taking place at the base region.

A BJT that operates in **active mode** has the following **properties**:

i. The BE junction operates as a regular diode that is forward-biased. We often use a "practical model" of $\mathbf{V_{BE} \approx 0.7\,V}$ (for an NPN BJT) or $\mathbf{V_{EB} \approx 0.7\,V}$ (for a PNP BJT), where V_{BE} stands for $V_B - V_E$ and $V_{EB} = V_E - V_B$. Recall, in a silicon diode, that is in forward bias, the p-type side of the p–n junction is approximately 0.7 V above the n-type side of the junction.
ii. $\mathbf{I_C = \beta I_B}$, where β is large (with a value that may range between 100 and 200). In approximate manual analysis, or in a manual design of a circuit that contains BJT transistors, we often assume that β is a given constant (say, $\beta = 100$). Actually, β strongly depends on W, the effective width of the base region. W may vary significantly due to variations in the length of the depletion region of the reverse-biased BC region. These changes in W cause β to fluctuate.
iii. $\mathbf{I_C = \alpha I_E}$ where $\boldsymbol{\alpha = \beta/(\beta + 1)}$. Because β is large, α is very close to 1. In approximate calculations, we often assume that in active mode $\mathbf{I_C \approx I_E}$.

A BJT that operates in **saturation mode** has the following **properties**, and it is important to note that properties (ii) and (iii) of the active mode (listed above) are no longer valid:

i. The BE junction operates as a regular diode that is forward-biased. We often use a "practical model" of $\mathbf{V_{BE} \approx 0.7\,V}$ (for an NPN BJT) or $\mathbf{V_{EB} \approx 0.7\,V}$ (for a PNP BJT).
ii. The BC junction operates as a diode that is forward-biased, where its ON voltage is slightly lower than that of the BE junction. We often assume that the ON voltage of the BC junction is constant (a sort of "practical model") and is given.
iii. Combining (i) and (ii) results in $\mathbf{V_{CE}} = V_C - V_E = \mathbf{V_{CEsat}} = \text{const}$ (for NPN BJT) and $\mathbf{V_{EC}} = \mathbf{V_{ECsat}} = \text{const}$ (for PNP BJT). A typical value for $|V_{CEsat}|$ may be between 0.05 and 0.1 V.
iv. Analyses of BJT circuits, in which the transistor is in saturation mode, almost always require using KCL: $\mathbf{I_E = I_C + I_B}$.

A BJT that operates in **cutoff mode** has the following **property: $\mathbf{I_E = I_C = I_B = 0}$.**

The easy DC circuit of **Example 3.1**, shown in Figure 3.4, illustrates typical approximated solution approaches, depending on the transistor's mode of operation.

We shall vary V_B and observe its effect on the BJT mode of operation. As with a solution of the diodes and resistors DC circuit, we often start with an assumption about the BJT's mode of operation. We then follow up on that initial assumption till it can be justified, or till a contradiction is encountered.

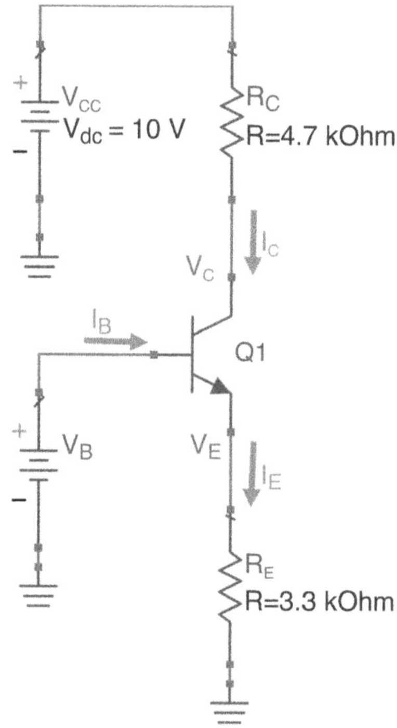

FIGURE 3.4 Example 3.1: A DC circuit featuring an NPN BJT, resistors, and batteries.

In the solution, below let us assume $V_{BE}=0.7\,V$, whenever the BE junction is forward-biased, $\beta=100$ and $V_{CEsat}=0.1\,V$.

Example 3.1 Part 1: $V_B=3\,V$

Let us assume active mode.

The solution proceeds as follows: $V_E=V_B-V_{BE}\approx2.3\,V$ by property (i). Then $I_E\approx V_E/R_E\approx0.7\,mA$ by Ohm's law. By property (iii) $I_C\approx I_E=0.7\,mA$. We can now find V_C: $V_C=V_{CC}-I_C\cdot R_C\approx6.7\,V$, and check the validity of the active mode assumption by comparing V_C to V_B: $V_C=6.7\ V>V_B=3\,V$. Junction BC is reverse-biased, as the n-type collector voltage is larger than the p-type base voltage. Finally, $I_B=I_C/\beta\approx7\ \mu A$.

Example 3.1 Part 2: $V_B=5\,V$

Let us assume again active mode. A calculation, similar to that of 3.1 Part 1, yields $V_E=4.3\,V$, $I_E=1.3\,mA$, and $I_C\approx1.3\,mA$. This time $V_C\approx3.9\,V$, which contradicts the active mode assumption ($V_C<V_B$). We change the assumption to saturation mode, but there is no need to restart the solution from the beginning. The calculation of $I_E=1.3\,mA$ is still correct, but property (iii) of active mode is no longer valid, so $I_C\neq I_E$. The clue how to proceed is V_{CEsat}, $V_c=V_E+V_{CEsat}=4.4\,V$, leading to $I_C=(V_{CC}-V_C)/R_C\approx1.19\,mA$. Finally $I_B=I_E-I_C=0.11\,mA$.

Example 3.1 Part 3: $V_B = 0\,V$

Let us assume cutoff mode. As $I_C = 0$ we have that $V_C = V_{CC}$, and as $I_E = 0$ we have $V_E = 0$. We now observe that $V_C > V_B$, implying that the CB junction is reverse-biased. In addition, $V_B = V_E = 0\,V$, implying that the BE junction is reverse-biased. Assumption is justified.

3.2 ADS: BJT DC SIMULATION: EXACT SOLUTION VS. APPROXIMATED SOLUTION; BJT CURVES AND DOUBLE-SWEEP CURVES

Let us start with an ADS simulation of Example 3.1 (Figure 3.5). Recall that in the manual solution of this example, we used some arbitrary assumptions, such as $\beta = 100$. It is of interest to compare the approximate manual calculations to the exact DC solution provided by ADS.

For the simple DC simulation, we shall just annotate the node voltages and component currents (Figure 3.6). The latter does not require the use of current probes. However, we do need the current probes for the subsequent sweep studies.

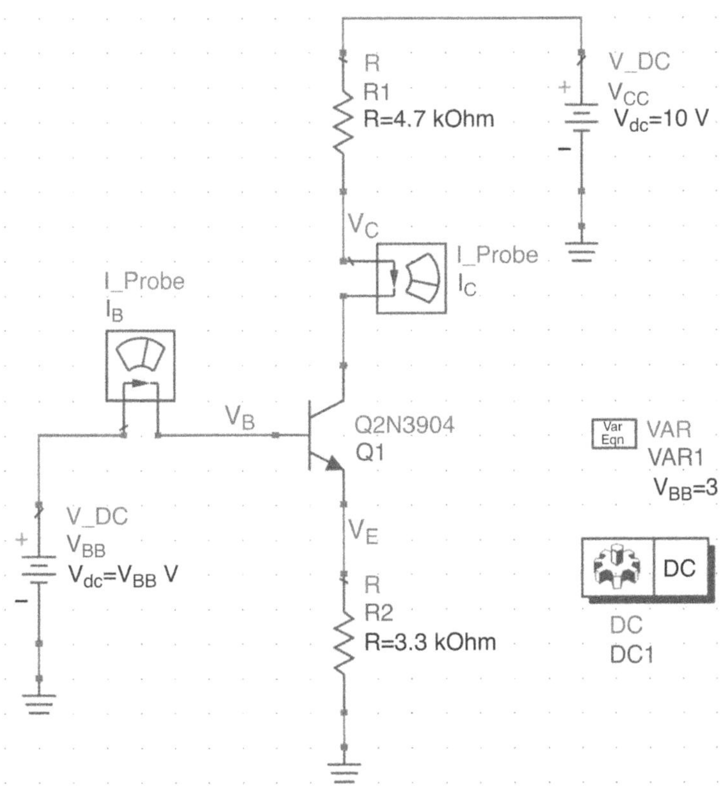

FIGURE 3.5 ADS schematic of Example 3.1.

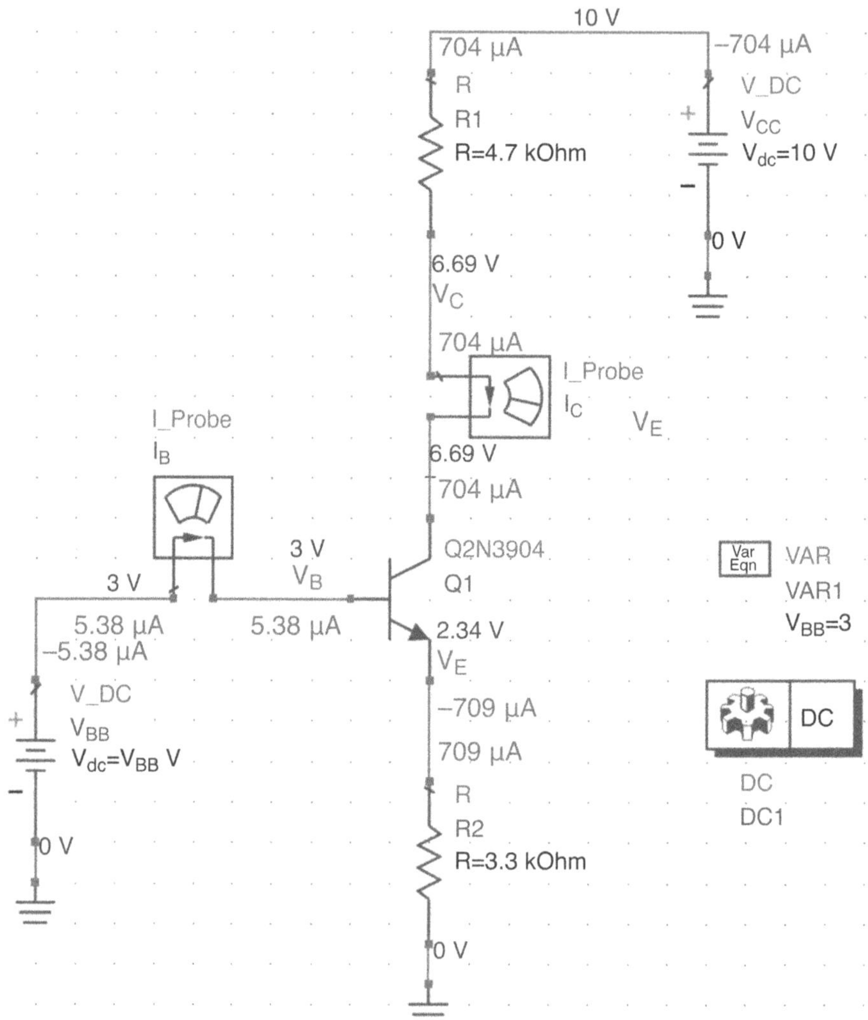

FIGURE 3.6 Exact DC currents and voltages of Example 3.1.1.

We got more or less the same voltages and currents as in the manual calculations. The assumption of $V_{BE}=0.7\,V$ is contrasted against the actual $V_{BE}=3-2.34=0.66\,V$. The assumption of $\beta=100$ is compared against the actual value of $\beta=I_C/I_B=704/5.38=130.86$.

In the calculations, we assumed $I_C=I_E$. Here $I_E=0.709\,mA$ and $I_C=0.704\,mA$. The CB junction is indeed reverse-biased as $V_C=6.69\,V$ and $V_B=3\,V$. The BJT is therefore in active mode. The calculated currents and voltages came very close to the actual values, even though the assumed β value was tens of percent off.

The DC solutions for other V_B cases are shown in Figure 3.7.

For $V_B=4\,V$, featured in Figure 3.7 (left), the BJT is still in active mode. Now: $V_{BE}=0.67\,V$ and $\beta=133.16$. The new $V_{CB}=1.29\,V$ can be compared to $V_{CB}=3.69\,V$

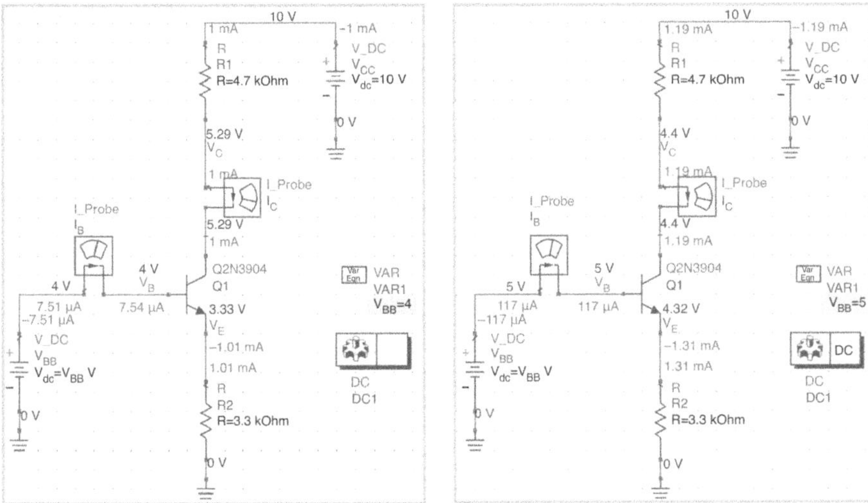

FIGURE 3.7 DC solutions, for Example 3.1.1, with other V_B values.

for VBB = 3 V. We see that as V_B increases, V_{CB} decreases, as the BJT comes nearer to saturation mode.

For VBB = 5 V in Figure 3.7 (right), the BJT is in saturation mode. Notice the big difference between I_C and I_E. We measure $V_{CEsat} = 4.4 - 4.32 = 0.08$ V. Here $V_{BE} = 0.68$ V.

The BE junction continues to be forward-biased. The β ratio is meaningless outside the active mode.

We just saw that for $V_B = 4$ V the BJT is in active mode, whereas for $V_B = 5$ V the BJT is in saturation mode. Obviously, the borderline is somewhere in between these two voltage levels. How do we pinpoint that transition moment?

Figure 3.8 features the schematic of Example 3.1, which lays the ground for several sweep studies.

Let us first explore how "constant" V_{BE}, β and V_{CB} are (as functions of the base voltage V_{BB}). We use the measurement equation MeasEqn to create these variables. See Figure 3.8.

Figure 3.9 features the curves of V_{BE} vs. V_{BB} and V_{CB} vs. V_{BB}.

The linear curve V_{BE} vs. V_{BB} (for V_{BB} between 0 and 0.5 V) is meaningless, as the BJT is cutoff. The BJT turns ON at $V_{BB} \approx 0.5$ V, and we see that V_{BE} slowly, but steadily, increases from 0.496 to 0.706 V.

At the BJT cutoff region $V_{CB} = V_{CC} - V_B$ (a meaningless linear segment).

For $V_{BB} > 0.5$ V, the reverse voltage V_{CB} of the BC junction varies dramatically.

There is a linear drop of V_{CB} that continues beyond $V_{CB} = 0$. The corner point at which V_{CB} becomes approximately equal to the cut-in V_{BC} voltage of the BC junction is the borderline between active and saturation modes. Precise measurements show that the corner is at $V_{BB} = 4.8$ V and then $V_{CB} = -0.554$ V.

As we have seen, for the V_{BB} range of 0–0.5 V, the BJT is in cutoff mode. In Figures 3.8 and 3.10, we use MeasEqn to define beta as a ratio of two currents.

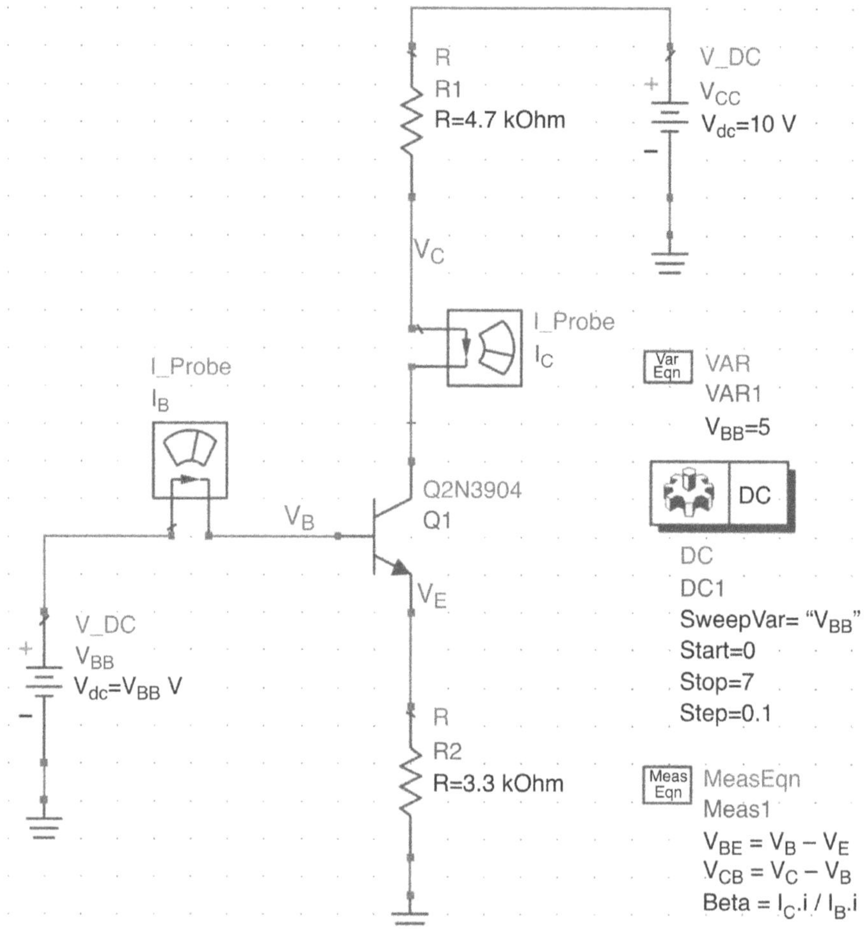

FIGURE 3.8 Schematic for sweep studies of Example 3.1.1 as V_{BB} varies.

However, beta becomes nonsensical for zero currents. In Figure 3.10, we just skipped the cutoff mode by starting the sweep of V_{BB} at $V_{BB}=0.5$ V. The β curve is shown in Figure 3.11.

Near BJT cutoff (0.5 V $\leq V_{BB} \leq$ 0.8 V), beta is relatively small (running between 50 and 96). Throughout the active mode region (0.8 V $\leq V_{BB} \leq$ 4.7 V), beta is not constant at all. It increases from 96 to around 133. The corner point where beta drops drastically (as it becomes meaningless) is a clear indication of the borderline between active and saturation modes.

We conclude this section with a double-sweep study of a BJT to generate a family of curves I_C vs. V_{CE} with I_B as a parameter. See Figure 3.12.

We observe in Figure 3.13 that in active mode, the larger I_{BB}, the larger $I_{C.i}$. Ideally, we expect I_C to be constant, that is, independent of V_{CE}. Ideally, I_C depends only on I_B.

Practically, though the curves of I_C vs. V_{CE}, are slanted. It is called "the Early Effect".

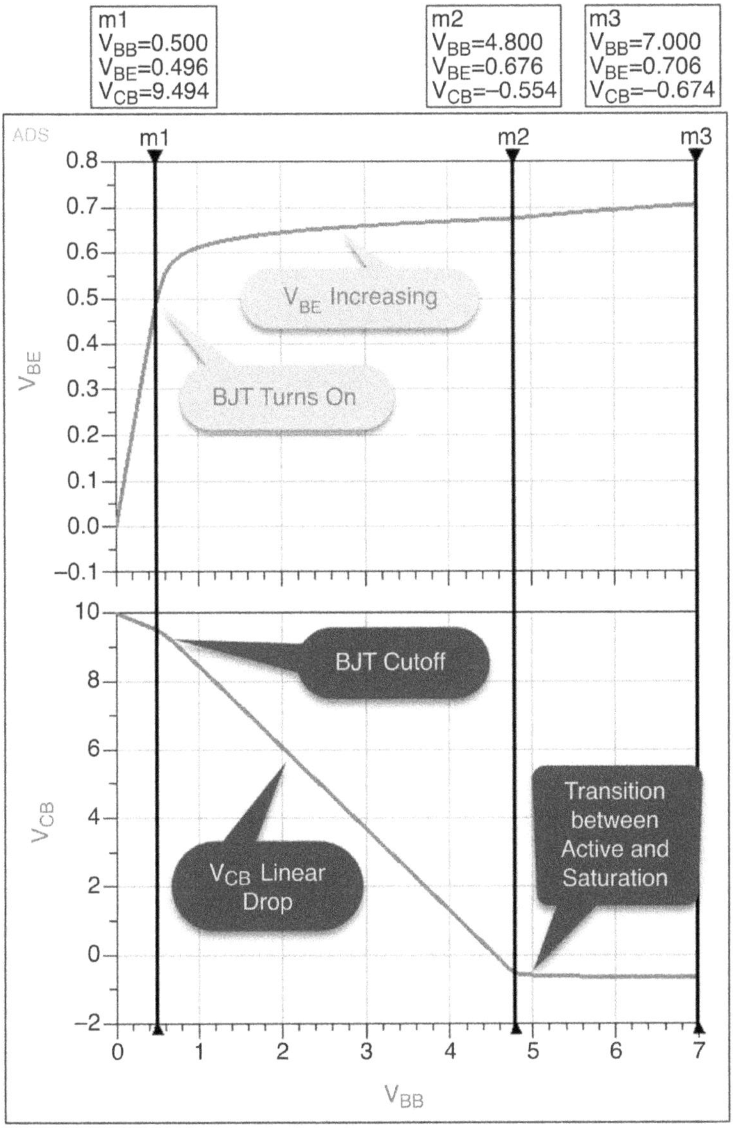

FIGURE 3.9 Sweep study results of Example 3.1: V_{BE} vs. V_{BB} and V_{CB} vs. V_{BB}.

The curves of Figure 3.13 can be improved by including a color code and a legend. See Figure 3.14. In a black and white plot, we use a different pattern for each plot.

The reason that I_C has a slight dependence on V_{CE} is because that changes in V_{CE} (even large ones, from one plot to another) create changes in V_{CB}, the reverse bias voltage of the BJT CB junction. V_{CB} changes affect the length of the CB junction depletion region, thus affecting the effective base width W, and with it the gain β.

A double-sweep requires two independent variables. Here, these are V_{CC} (which in Figure 3.12 is equal to V_{CE}) and I_{BB}. V_{CE} is swept by the DC simulation engine.

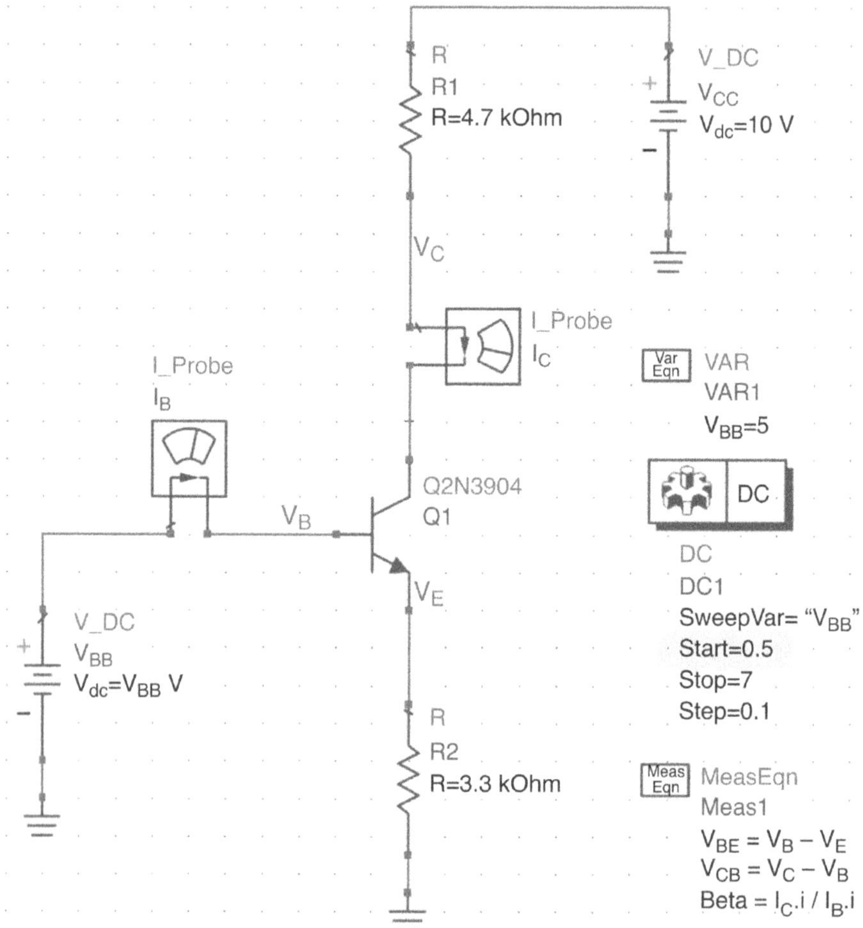

FIGURE 3.10 A slight change in the sweep range for V_{BB} to enable a more meaningful curve of β vs. V_{BB}.

I_{BB} is swept by the parameter sweep engine. The V_{CE} sweep is determined to be the "first sweep" by the parameter sweep block. See SimInstanceName[1] = "DC1". Consequently, the sweep of I_{BB} is the "second sweep". In a double-sweep study, the first (primary) sweep typically has many points. Each of the I_C vs. V_{CE} curves is created by taking many V_{CE} points. The secondary sweep variable has only a few points. This is done for good visibility of the curves. Double-sweep curves must not be too dense.

3.3 BJT CURRENT DRIVERS; STRONG VS. WEAK β-DEPENDENT CURRENT DRIVER DESIGN

A BJT can serve as an interface between an op-amp and a load, driving a current to the load that is too large for the op-amp to deliver on its own. See **Example 3.2** in Figure 3.15.

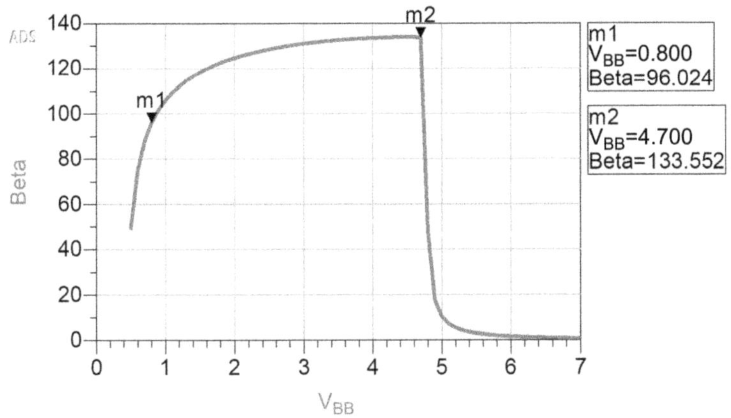

FIGURE 3.11 Curve of β over the active mode and saturation mode regions.

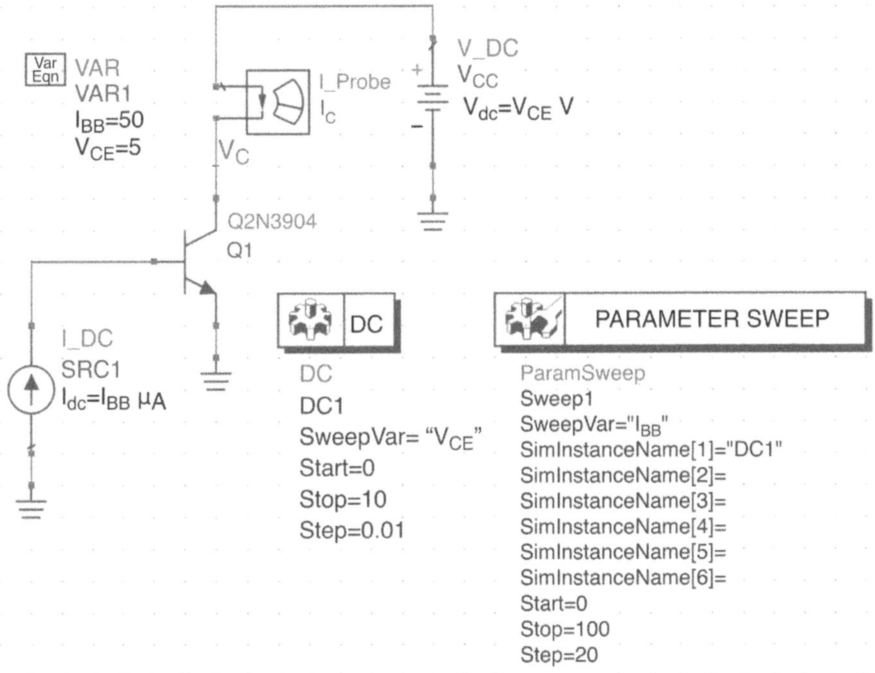

FIGURE 3.12 A double-sweep study of BJT current–voltage curves.

We see that in Figure 3.15 (left), the light-emitting diode (LED) turns ON for a positive comparator output voltage V_o and is OFF for a negative op-amp's output voltage. The current $I_L = (V_o - V_{LED\text{-}ON})/R$ needs to be large enough for the LED to develop its full light intensity. For a GaAs LED the ON voltage is approximately twice larger than the ON voltage of a regular silicon diode. R can be chosen to make I_L large, but such a current may be larger than the maximum output current of the op-amp.

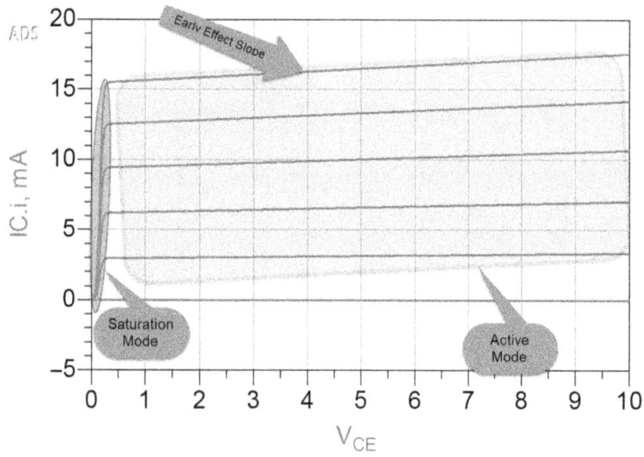

FIGURE 3.13 Plot of I_C vs. V_{CE} parameterized by I_{BB}.

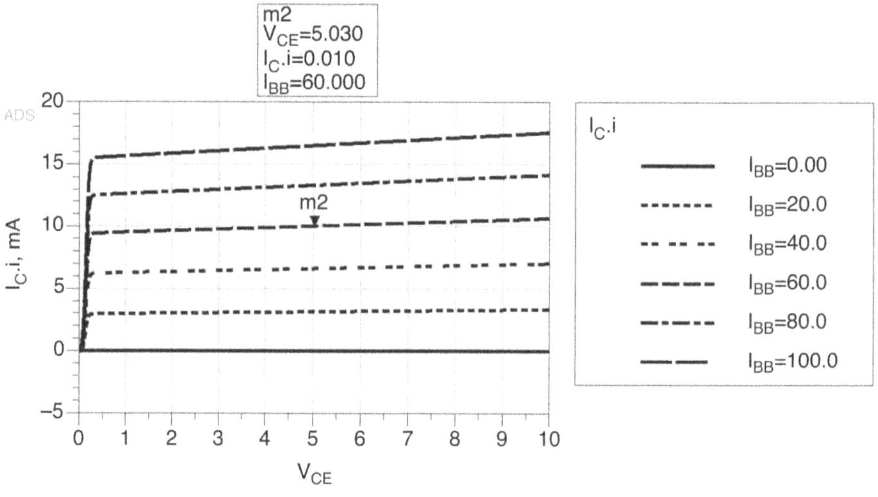

FIGURE 3.14 Plot of I_C vs. V_{CE} parameterized by I_{BB}, with a legend. Note that, based on the schematic in Figure 3.12, the unit of IBB is μA.

In Figure 3.15 (right), I_L is I_C of a BJT and the op-amp only has to provide the current I_B. Assuming that the BJT operates in active mode, for the positive output voltage of the op-amp, we obtain that $\mathbf{I_L = \beta \cdot (V_o - V_{BE})/R_B}$, and a specific choice of R_B provides the needed I_L value for some assumed constant value of β. The BJT is in active mode if (approximately) $V_{CC} - V_{LED\text{-}ON} \geq V_{BE}$, which is almost always true given that V_{CC} is large enough. If we, as a second design possibility place the LED between emitter and ground, we just need to modify R_B to obtain the same I_L current value for the same assumed β value: $\mathbf{I_L = \beta \cdot (V_o - V_{LED\text{-}ON} - V_{BE})/R_B}$. In the second design possibility, the BJT is in active mode if the collector voltage $V_{CC} \geq V_{LED\text{-}ON} + V_{BE}$, the base voltage.

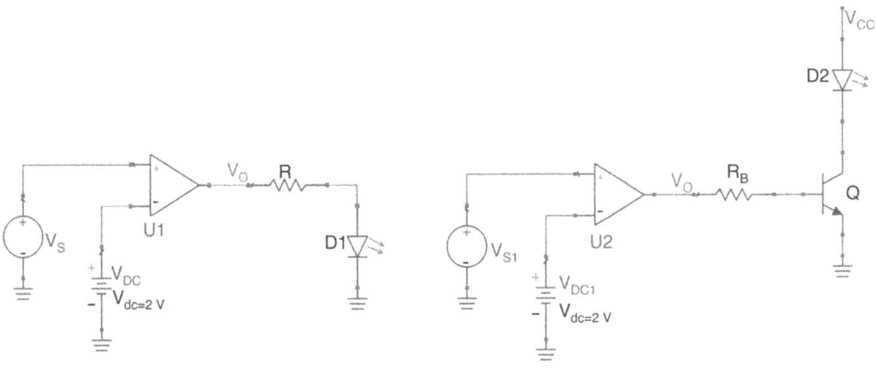

FIGURE 3.15 (Left) An op-amp comparator connected directly to an LED. (right) An op-amp comparator connected to an LED through a BJT current driver.

In both design formulas, the input voltage of the BJT is the comparator's output voltage V_o. Both designs feature what is called a "**strong β-dependency**". That is, the resulting I_L is proportional to β. If, for instance, due to temperature variations or a BJT replacement, β changes by, say, 20%, which is not uncommon, the LED current I_L will also change by exactly the same percentage. Strong β-dependent design is generally considered undesired if we aim at obtaining a specific load current.

So, can we make the current driving design less β-dependent? Yes. Figure 3.16 shows the correct architecture of such a BJT LED driver. It involves two design parameters: R_B and R_E.

Let us develop an expression for I_E, that is, valid only if the BJT is in active mode. $I_B = (Vo - I_E R_E - V_{BE})/R_B = I_E/(\beta + 1)$. The resulting expression for I_E is $\mathbf{I_E = (V_o - V_{BE})/[R_E + R_B/(\beta + 1)]}$.

In active mode $I_L = I_C \approx I_E$ and we select a specific I_L value by choosing the value of the denominator in the I_E's expression. I_L is made weakly dependent on β if $R_E \gg R_B/(\beta + 1)$. If R_E is taken too large, the BJT may glitch into saturation mode, which stops the current driving.

Example 3.2

Let the input be $V_o = 10\,V$ and the desired load current $I_L = 30\,mA$. Assume $V_{BE} = 0.7\,V$ and $\beta = 100$, and $V_{LED\text{-}ON} = 1.5\,V$. The resulting denominator of I_E becomes 310 Ω. How shall we split it between R_E and $R_B/(\beta+1)$?

First attempt: $R_E = 250$ Ω and $R_B/(\beta+1) = 60$ Ω (which implies that $R_B = 6$ kΩ). We note that $V_B = 8.2\,V$ and $V_C = 8.5\,V$. The BJT is in active mode. If β changes from 100 to 120, I_L will change from 30 to 31 mA (a change of 3.33%). The ratio of R_E to $R_B/(\beta+1)$ is about 4:1, not exactly "?". The design is "moderately weak β-dependent".

An unsuccessful second attempt: $R_E = 280$ Ω and $R_B/(\beta+1) = 30$ Ω (which implies that $R_B = 3$ kΩ). If β changes from 100 to 120, I_L will change from 30 to 30.5 mA (a change of 1.67%). However, we note that $V_B = 9.1$ V and $V_C = 8.5\,V$. The BJT is in saturation mode, and as a result, I_L becomes much smaller than the desired value of 30 mA. One should not be too greedy.

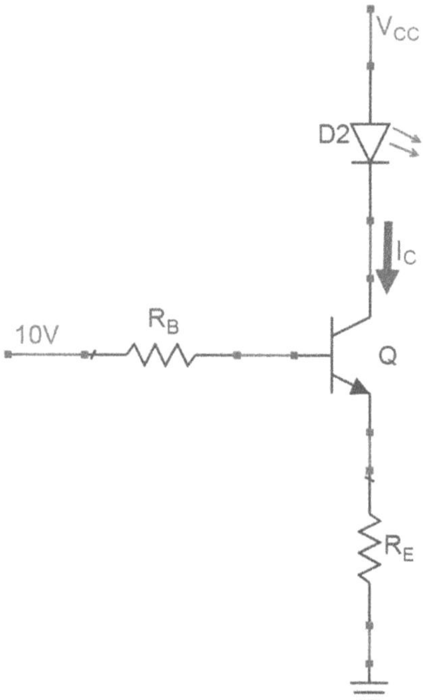

FIGURE 3.16 BJT current driver, with an emitter resistor, capable of driving a weakly β-dependent load current.

3.4 ADS: DC AND TRANSIENT SIMULATION OF BJT CURRENT DRIVERS; RED LED COMPONENT

Example 3.3

Our first simulation schematic, in Figure 3.17, is a circuit that resembles the "strong β-dependent" design shown in Figure 3.15 (right).

As in Example 3.2, a DC current of $I_C \approx 30\,\text{mA}$ must flow in the LED. We use an EE kit library component named LEDRed. It is a GaAs LED with an ON voltage that is around 1.8 V. We also assume that $V_{in} = 5$ VDC and it is not changeable. The LED may also be emulated by two regular D1N4148 diodes in series.

Of course, in order for a BJT to drive a sizable current, it must be in active mode.

The design involves a selection of a resistor R_B. Let us assume $V_{BE} = 0.7\,\text{V}$ and $\beta = 100$. The calculation goes: $V_B = V_{BE} = 0.7\,\text{V}$ because $V_E = 0$. $I_B = (V_{in} - V_B)/R_B = (5 - 0.7)/R_B$. We want $I_C = 30\,\text{mA}$, which means that we want $I_B = I_C/\beta = 0.3\,\text{mA}$. Therefore, we need $R_B = 14.33\ \text{k}\Omega$. Let us conduct a DC simulation and see if the design is accurate.

See Figure 3.18 for the schematic and for the DC annotation results. We observe that the BJT is in active mode: $V_C = 8.08\ \text{V} > V_B = 0.779\,\text{V}$.

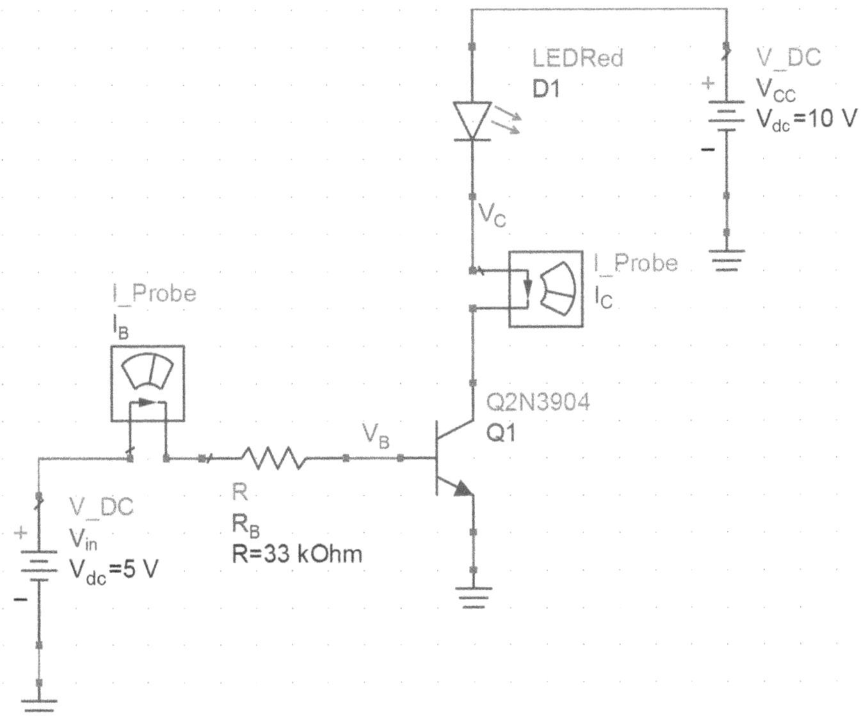

FIGURE 3.17 LED driven by a BJT. It is a strong β-dependent design.

We further see that the design is off by a large margin, as I_C came out to be 44.1 mA (instead of 30 mA). Is it because V_{BE} is not exactly 0.7 V (it is 0.779 V) and β is not exactly 100? Let us see what β really is: $\beta = I_C/I_B = 44.1/0.295 = 149.5$.

What we used in the design was 33.1% smaller than the true coefficient. Can we tweak the design using more accurate V_{BE} and β values?

$(5 - 0.779)/R_B = 30/149.5$ yielding the new value $R_B = 21$ kΩ. Let's check again the annotated results (Figure 3.19): $I_C = 32.2$ mA, closer but still off. This time $V_{BE} = 0.767$ V and $\beta = 32.2/0.202 = 159.4$.

Obviously, these parameters vary significantly between $I_C = 44$ mA and $I_C = 32$ mA. One can do a few more tweaking iterations and eventually converge to the desired design result. Let us instead demonstrate a computer-aided design. See Figure 3.20.

Let us sweep, the resistor R_B value (say between 10 and 30 kΩ). We plot I_C vs. R_{BB} and obtain the needed R_B value, without any computational tweaks. See Figure 3.21.

We see (with the aid of a marker) that for $I_C = 30$ mA the resistor R_B needs to be $R_B = 22.8$ kΩ.

Let us observe the strong β-dependency of the design by sweeping the temperature (as shown in Figure 3.22). Both V_{BE} and β are temperature-dependent. Results are shown in Figure 3.23.

We see that the desired current value $I_C = 30$ mA occurs at room temperature. It can vary by tens of percent as the temperature varies. The curve is linear, so we can estimate the temperature sensitivity of the resulting I_C:

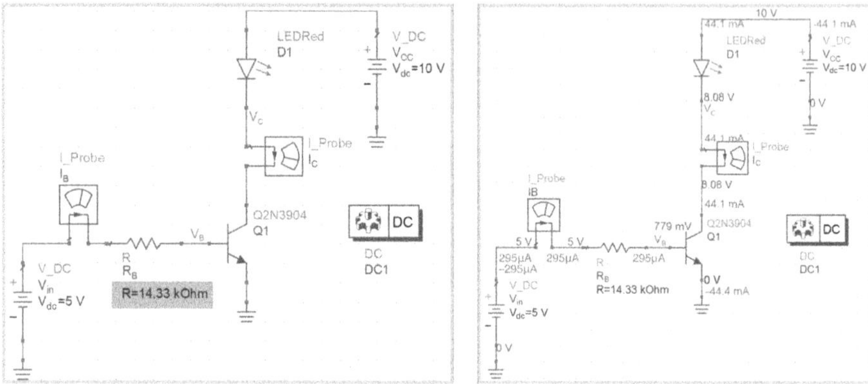

FIGURE 3.18 Current driving design based on an assumption β = 100.

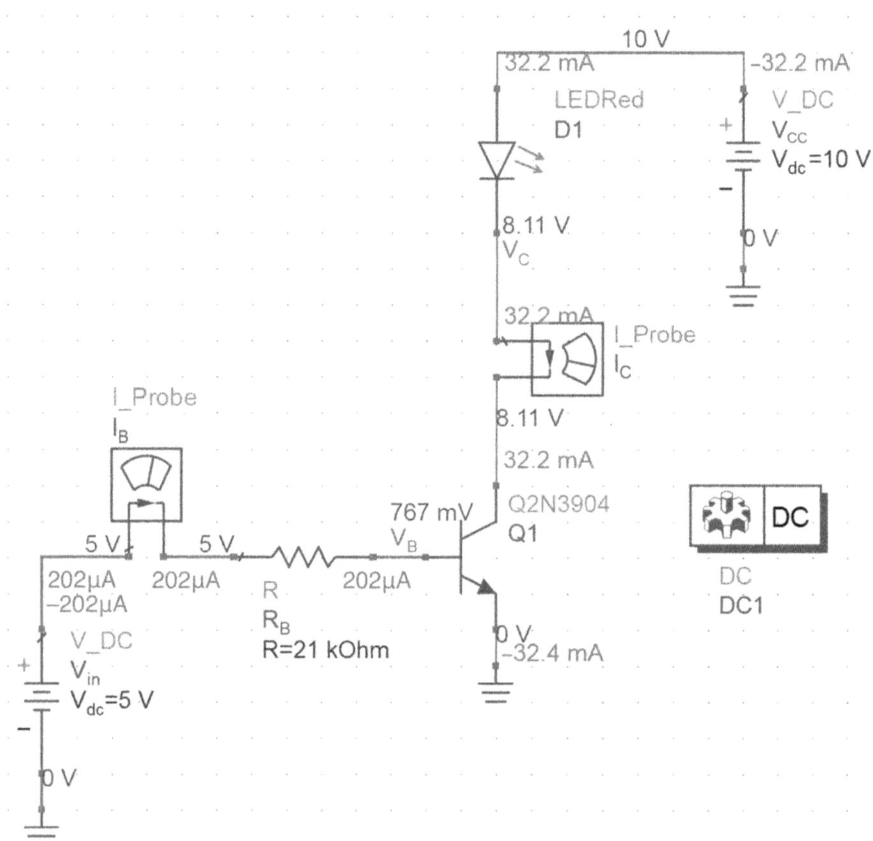

FIGURE 3.19 DC annotation results of a tweaked design.

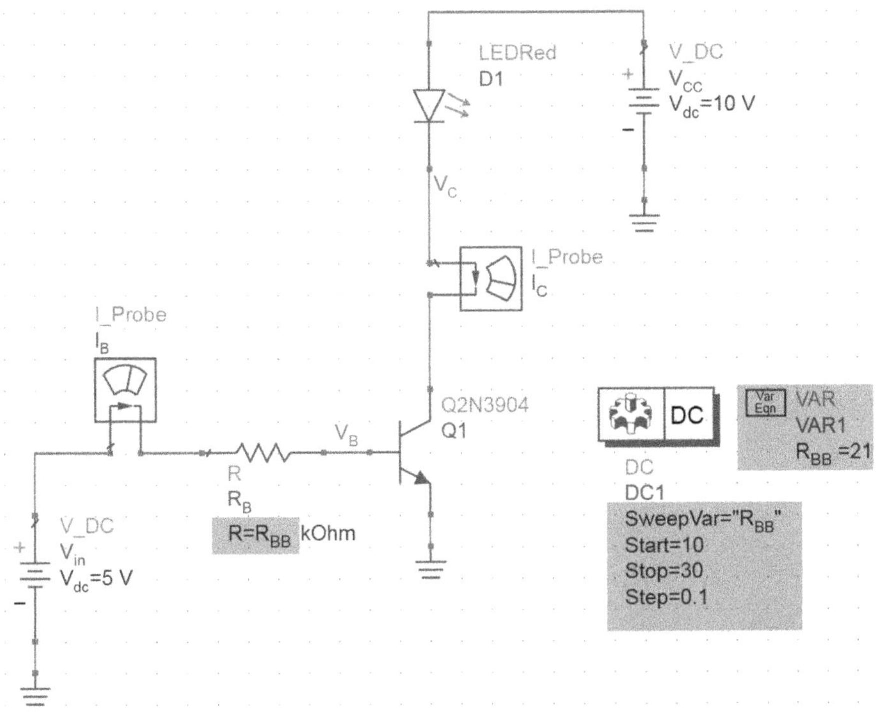

FIGURE 3.20 Design tweaking by means of a resistor sweep.

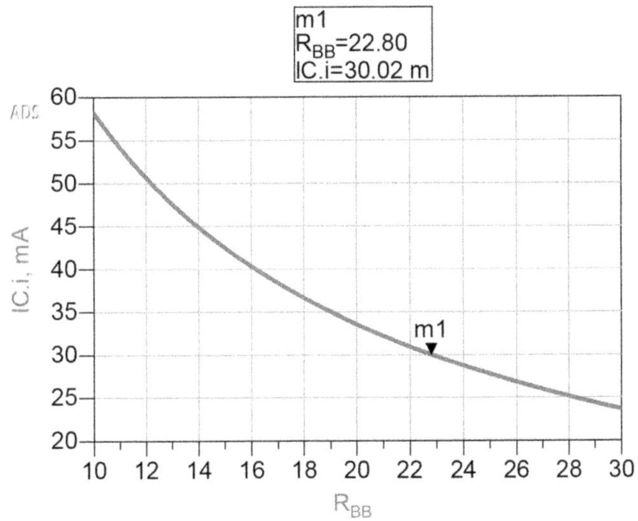

FIGURE 3.21 Result of the resistor sweeping study.

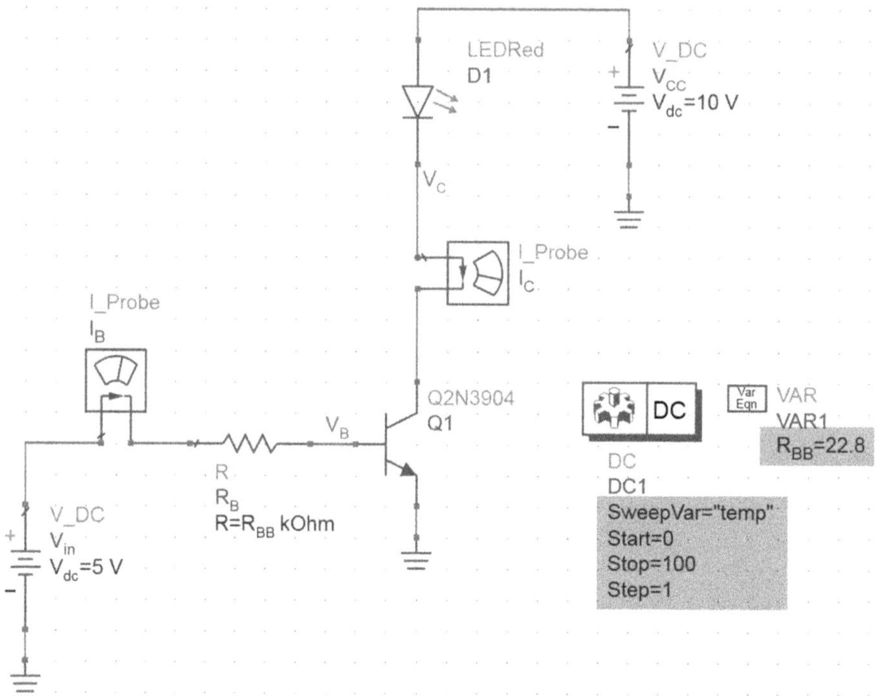

FIGURE 3.22 Temperature sweep study of the final design (between 0°C and 100°C).

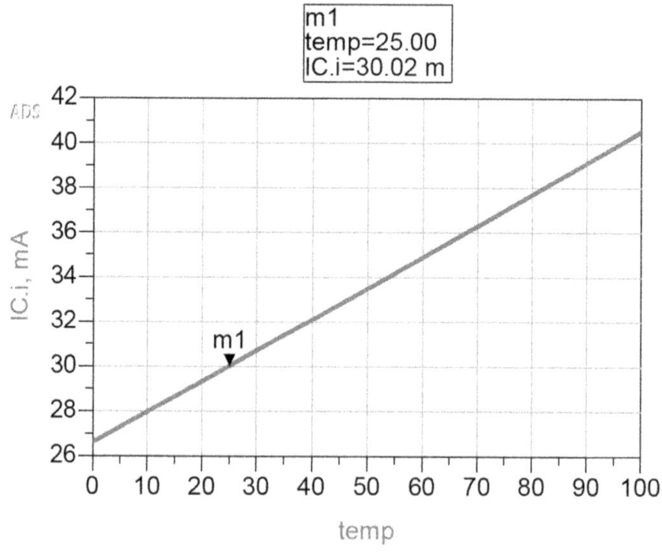

FIGURE 3.23 Temperature sweep result.

$T_{sensitivity} = (41 - 27)/(100 - 0) = 0.14$ mA/°C.

At the cost of one more resistor R_E, we can dramatically improve the design. There will be two design parameters: R_B and R_E. See Figure 3.24.

Presently, the values of R_B and R_E are arbitrary and have nothing to do with the resulting design. Let us refer to the previous Section 3.3 (specifically for the formula for I_E and the numerical Example 3.3). The formula for I_E is repeated here:

$$I_E = \frac{V_{in} - V_{BE}}{R_E + \frac{R_B}{\beta + 1}}$$

A design in which $V_{in} \gg V_{BE}$ and $R_E \gg R_B/(\beta + 1)$ will become "weakly dependent" on β and consequently, weakly dependent on temperature.

There are infinitely many pairs of resistors, R_E and R_B, that can be found to meet a desired I_C value. However, if R_E is too large, the BJT may enter a saturation mode.

Let us start with $I_E = 30$ mA, $V_{BE} = 0.7$ V, and $\beta = 100$, approximately, and $R_E = 0.5$ kΩ. The only unknown is R_B, and it comes out to be negative. Let us lower R_E to $R_E = 0.1$ kΩ. Now: $R_B = 4.3$ kΩ. Let us check the design in the DC simulation. See Figure 3.25.

We see in Figure 3.26 that with the crude approximations for V_{BE} and β, the result is off. Let us tweak: We see that I_C needs to be reduced by a factor of 30/32.8. Therefore, the denominator of I_E (which is 100 + (4,300/101) = 142.6 Ω) needs to be increased by a factor of 32.8/30 to 156 Ω. Can we simply modify R_E to be

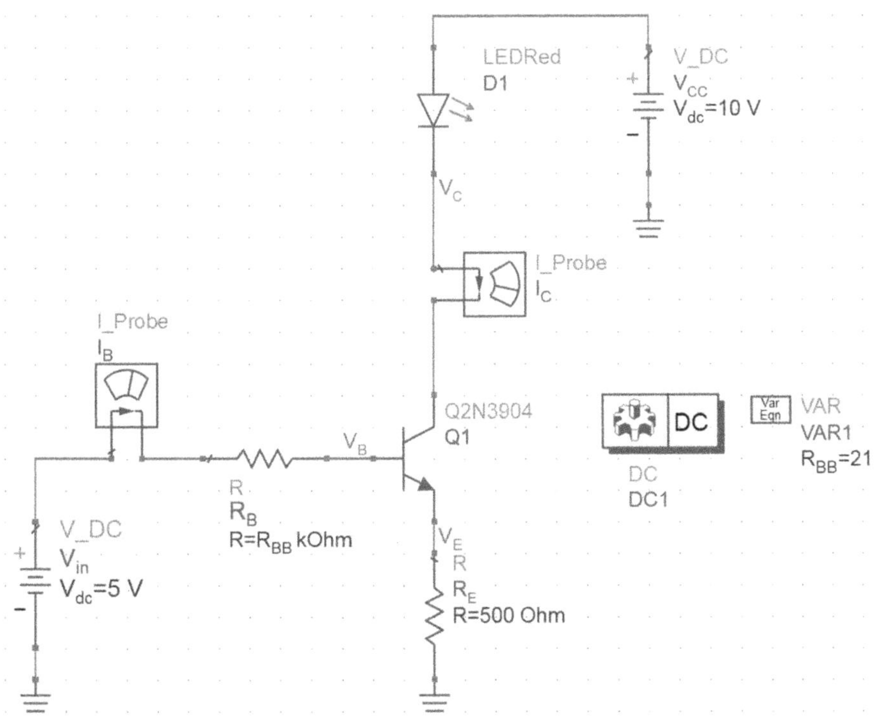

FIGURE 3.24 Current driver with two resistors.

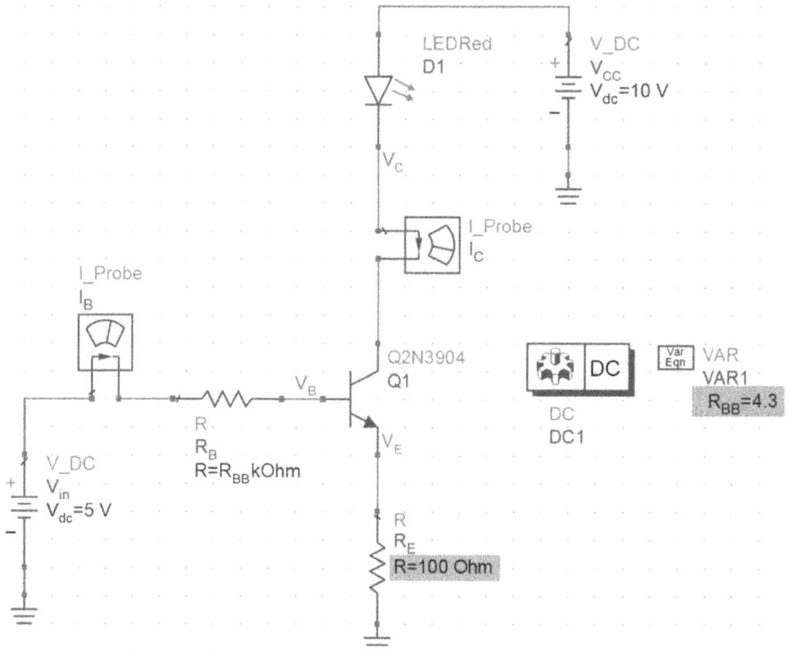

FIGURE 3.25 First iteration in the design of a weakly β-dependent design.

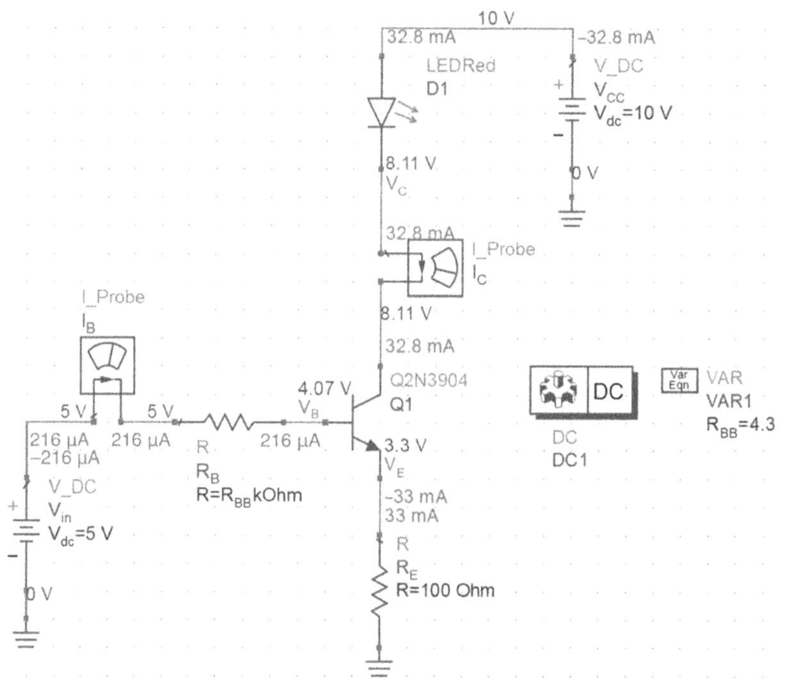

FIGURE 3.26 Simulation results of the first design iteration.

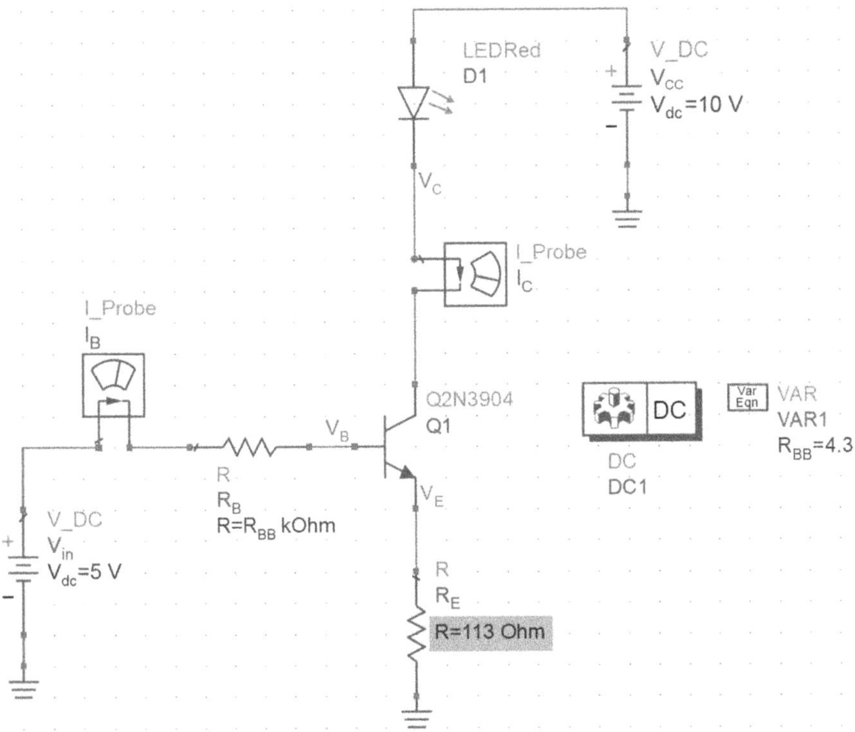

FIGURE 3.27 Second iteration in the design of the two-resistor current driver.

$100+(156-142.6)=113\ \Omega$ and keep R_B unchanged? Let us pursue the modifications (Figure 3.27), but watch V_C and V_B to ensure that active mode is maintained.

We see (Figure 3.28) that the new I_C is $I_C=29.9$ mA. We are almost there.

BJT is in active mode as $V_C=8.41\ V>V_B=4.17$ V. In a weak β-dependency design, it is easier to tweak.

Let us leave the design as is, and check temperature-dependency.

The temperature sensitivity is $(32.4-28.8)/100=0.036$ mA/°C.

Example 3.4

Finally, let us demonstrate a design in which V_{in} is not constant. As an example, consider an **optical communication transmitter** design. The $V_{in}(t)$ communicated information is converted to a current $I_C(t)$, and consequently, to the light intensity created by the LED. In a wireless communication system, the receiver may include a phototransistor that matches the wavelengths of the LED light signal.

We want the signal $I_C(t)$ to mimic the input voltage signal $V_{in}(t)$. Using a weakly β-dependent current driver, and recalling the formula $I_C \approx (V_{in}-V_{BE})/[R_E+(R_B/(\beta+1)]$, $I_C(t)$ may follow $V_{in}(t)$, if $V_{in}(t)$ is large enough (compared to V_{BE}) and R_E is large enough compared to $R_B/(\beta+1)$. Just need to assure that for the entire range of $V_{in}(t)$ values, the BJT never leaves the active mode region.

Figure 3.30 demonstrates a mistaken design. We use the current driver with weak β-dependency, which we discussed earlier. $V_{in}(t)$ is a 1-kHz AC signal with an amplitude of 5 V.

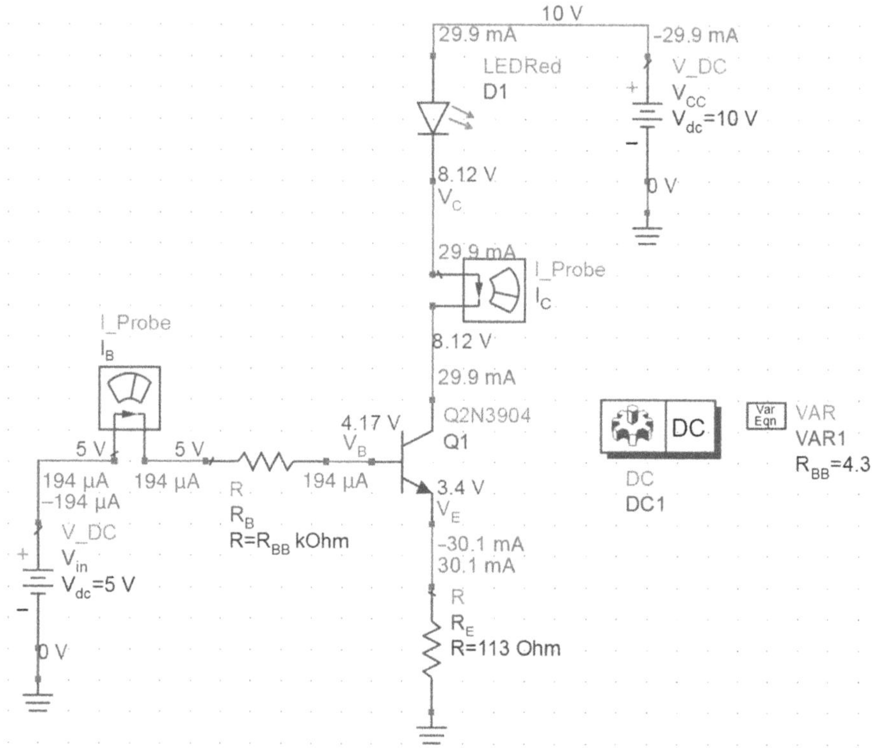

FIGURE 3.28 Tweaked DC results.

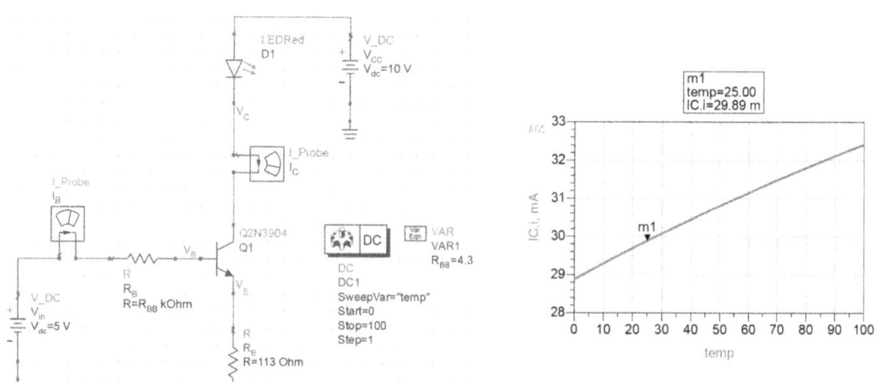

FIGURE 3.29 Temperature sensitivity of the weak β-dependency design.

We want $I_C(t)$ to be AC of the same frequency. What is wrong with the design is that whenever $V_{in}(t) < 0$, the BJT cuts off, as seen in Figure 3.31.

Let us observe the two signals (Figure 3.31).

We see that whenever $V_{in}(t)$ is positive $I_C(t)$ is sinusoidal with an amplitude of 30 mA. Whenever $V_{in}(t)$ is negative $I_C(t) = 0$ because the BJT is OFF.

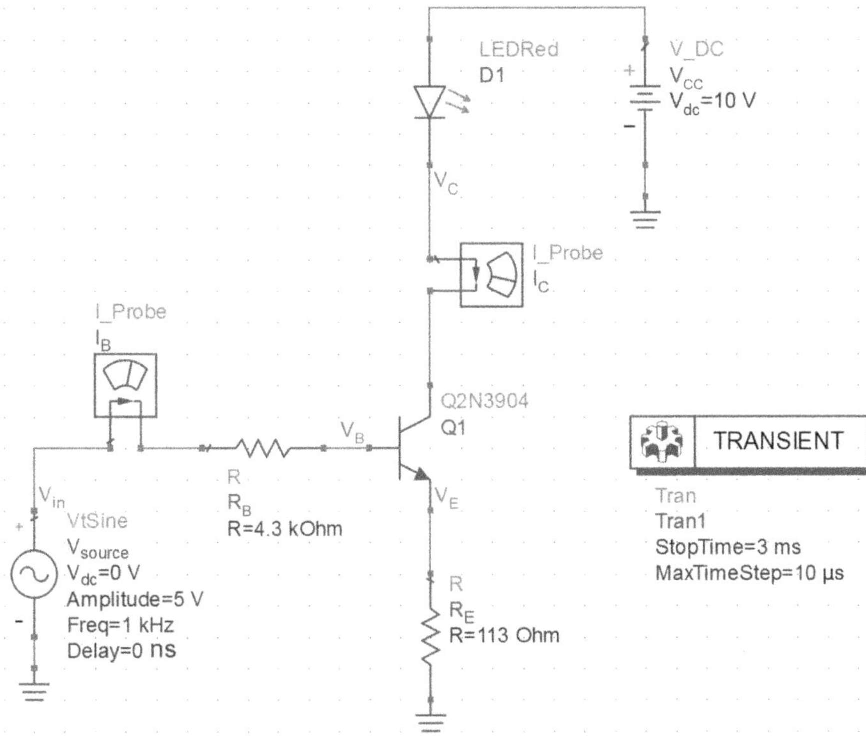

FIGURE 3.30 Mistaken design of an optical communication transmitter.

In order to correct the design, all we have to do is let $V_{in}(t)$ have some DC level, so that even when $V_{in}(t)$ is minimum, that minimum will be above zero, allowing the BJT to be ON all the time. The question is: Will the BJT stay in active mode when V_{in} is maximum?

Figure 3.32 illustrates an input voltage DC level of 5V, which is insufficient. Need to assure that $V_{in}(t)$ is above the cut-in voltage of the BJT B–E junction (which is around 0.5V). That is why there was a clipping.

The results shown in Figure 3.33 seem okay. The only question now is – will the LED sustain a peak current of almost 70mA? If not, say that I_{LEDmax} = 50mA, then we need to lower the amplitude of $V_{in}(t)$ so that we don't exceed the maximum allowed current value.

3.5 BJT PUSH–PULL AMPLIFIER

A current driver, as we have seen in Sections 3.3 and 3.4, creates a prescribed unidirectional current. A push–pull amplifier, as its name suggests, can create a large current in a bidirectional manner, feeding the current into a variety of load sizes. In Figure 3.34, that shows the amplifier in its basic form, the load resistor R_L can have any value (subject to some limitations) and its current $i_L(t)$ can flow into R_L ("push") or from the load ("pull").

The push–pull amplifier is basically a power amplifier. It consists of a power NPN BJT Q_N, the collector of which is connected to a positive DC voltage supply V_{CC}, and

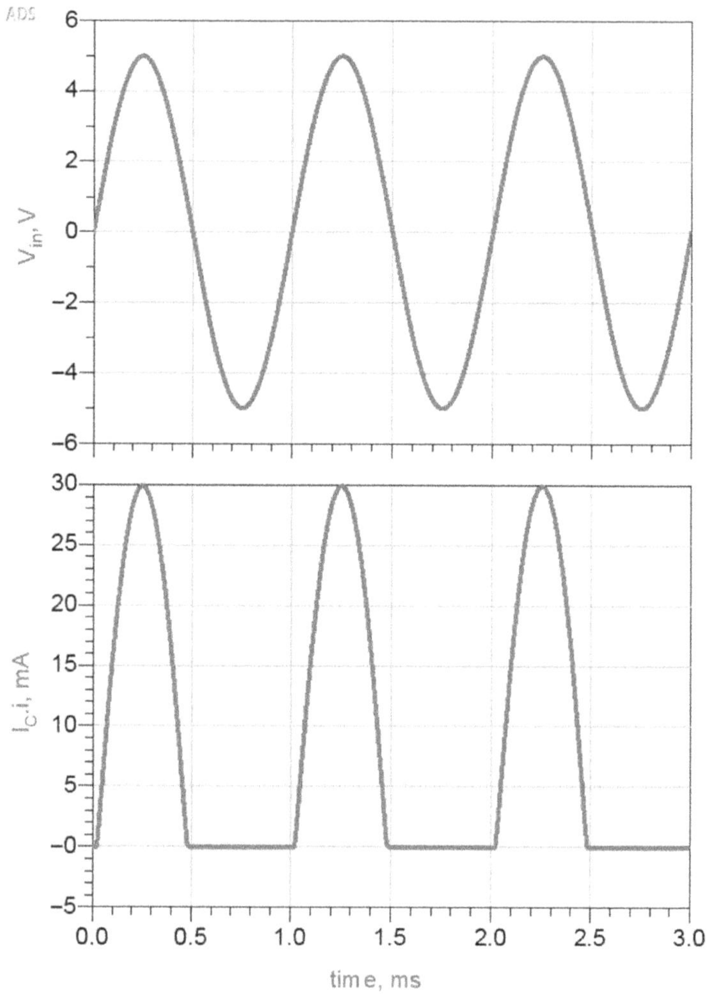

FIGURE 3.31 $V_{in}(t)$ and $I_C(t)$ of the mistakenly designed optical transmitter.

a power PNP BJT Q_P, the collector of which is connected to a symmetric negative supply $-V_{CC}$. **The two BJTs are connected emitter-to-emitter and base-to-base**.

Let us explore how the amplifier operates. Whenever Q_N is ON, the output voltage $v_o(t)$ is by V_{BEN} lower than the input voltage $v_{in}(t)$. Likewise, whenever Q_P is ON, the output voltage $v_o(t)$ is by V_{EBP} larger than $v_{in}(t)$. Obviously, **Q_N and Q_P cannot possibly be ON simulateneously**. For example, if $V_{in}=5$ VDC, and we assume that both BJTs are ON, then $V_{EN}=4.3$ V, whereas $V_{EP}=5.7$ V, but $V_{EN}=V_{EP}=v_o$, a contradiction. Whenever **$v_{in}(t)$ is positive** and larger than the cut-in voltage of junction BE_N **Q_N is ON** whereas Q_P must be OFF. Whenever **$v_{in}(t)$ is negative** and its absolute value is

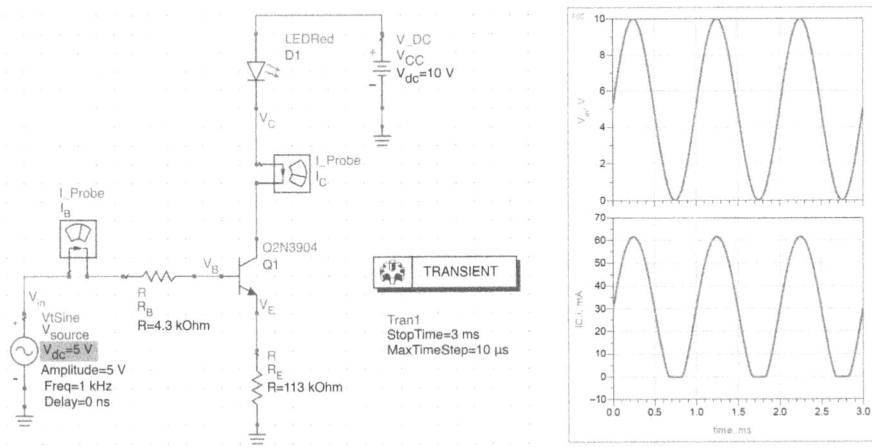

FIGURE 3.32 $I_C(t)$ clips at the bottom because the DC level of $V_{in}(t)$ is not large enough.

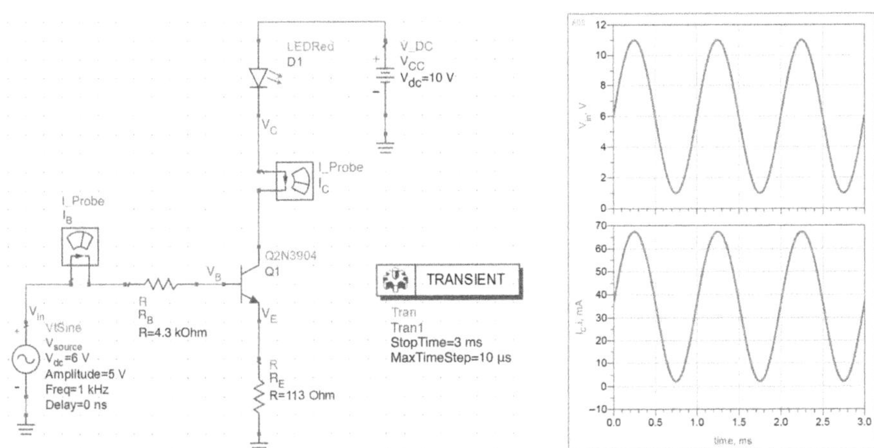

FIGURE 3.33 Corrected results after increasing the input's DC level to 6 V.

larger than the cut-in voltage of junction EB_P, **Q_P is ON** while Q_N is OFF. If we limit $v_{in}(t)$ to never exceed $\pm V_{CC}$, then whenever any of the BJTs is ON, it is in active mode.

We see that the amplifier's output voltage, as it mimics the shape of the input voltage, is slightly smaller (in absolute value) than the input voltage. The load current, however, coming from the emitter of either BJT, is much larger than the respective base current. Hence, the **push–pull amplifier amplifies power**. So, a smaller power that is delivered to the amplifier's input turns into a much larger power delivered to the load. Where is that extra energy coming from? It comes from the two DC supplies. The average power delivered to the load is always smaller than the average

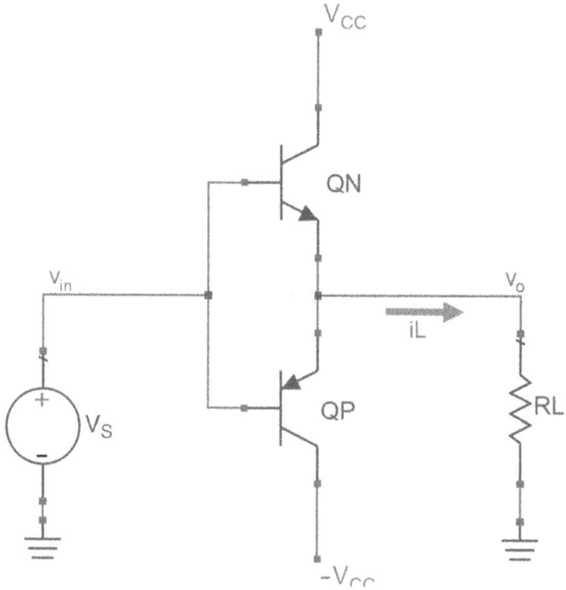

FIGURE 3.34 A push–pull amplifier consisting of NPN and PNP BJTs, connected emitter-to-emitter and base-to-base. The DC supplies are typically symmetrical.

power drawn from the batteries. The ratio between the average power delivered to the load and that drawn from the DC supplies is called the **amplifier's efficiency**. The energy difference is acounted for by the average power dissipated by the BJTs in the form of heat.

So, the BJTs cannot be both ON at the same time, but can they be OFF simultaneously? Yes, unfortunately. If $|v_{in}(t)|$ is no larger than the BE junctions' cut-in voltages (which are around 0.5 V for positive inputs and −0.5 V for negative inputs), both BJTs are OFF and then $v_o(t) = 0$. This is called the **crossover distortion**. Such a distortion can be corrected in several ways [Sedra1], one of which involves an op-amp. See Figure 3.35.

The op-amp, in Figure 3.35, works most of the time as a voltage buffer. This happens whenever $v_{oa}(t)$ is large enough to turn ON one of the BJTs. By that, a negative feedback is established. On the other hand, whenever $v_{oa}(t)$ is positive, but smaller than the cut-in voltage of the Q_N BJT, the feedback path is broken and the op-amp begins to work as a comparator, but only for a very brief time. The comparator's output shoots toward the positive saturation level, but as soon as the cut-in voltage is reached, the feedback is reestablished. Likewise, if v_{oa} is negative and small. End result: The crossover distortion becomes almost completely wiped out. There is, in addition, an unexpected bonus reward: The V_{BE} difference between $v_o(t)$ and $v_{in}(t)$ vanishes due to the op-amp's buffer action.

Finally, what does it take to design a push–pull amplifier?

Assuming that the supplies $\pm V_{CC}$ and the load resistor R_L are given, the only design choice is to select each of the BJTs to have the maximum expected average dissipation power P_{Davmax}. The BJT's maximum average power dissipation depends on the shape of $V_{in}(t)$ and on its amplitude.

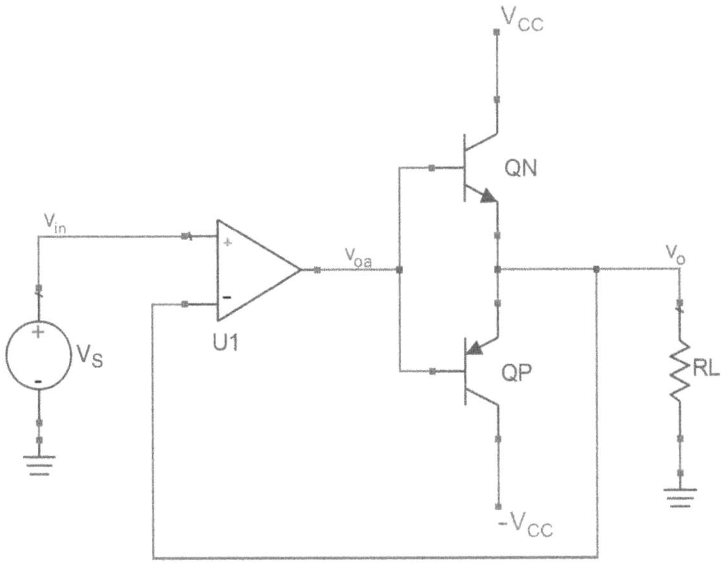

FIGURE 3.35 Correction circuit for the crossover distortion.

Example 3.5

In which we neglect the effects of the crossover distortion on average power calculations:

Part 1: $V_{out}(t)$ is AC with an amplitude of A and a frequency f. *Results*: The average power delivered to the load is $P_{Lav}=A^2/2R_L$. The average power drawn from the two supplies $P_{Sav}=2V_{CC}A/\pi R_L$. Hence, the efficiency of the amplifier is $P_{Lav}/P_{sav}=\pi A/4V_{CC}$. Differentiating $P_{Dav}=P_{Sav}-P_{Lav}$ with respect to the output voltage amplitude A yields $A_{opt}=2V_{CC}/\pi$. This is the worst case A, for an AC input voltage, at which the warming of the BJTs is the largest. Substituting the value of A_{opt} into P_{Dav} yields $P_{Davmax}=2V_{CC}^2/\pi^2R_L$. All the above results are independent of f.

Part 2: $V_{out}(t)$ is a symmetric square wave with amplitudes of $\pm$A and a frequency f. *Results*: The average power delivered to the load is $P_{Lav}=A^2/R_L$. The average power drawn from the two supplies $P_{Sav}=V_{CC}A/R_L$. Hence, the efficiency of the amplifier is $P_{Lav}/P_{sav}=A/V_{CC}$. Differentiating $P_{Dav}=P_{Sav}-P_{Lav}$ with respect to the output voltage amplitude A yields $A_{opt}=V_{CC}/2$. This is the worst case A, for a square wave input voltage (and also for a DC input), at which the warming of the BJTs is the largest. Substituting the value of A_{opt} into P_{Dav} yields $P_{Davmax}=V_{CC}^2/4R_L$. All the above results are independent of f and the duty cycle of $V_{out}(t)$.

We make the following observations from the above results:

a. The larger the amplitude of the input, the larger is the amplifier's efficiency.
b. The worst case BJT dissipated power happens at some mid-value of A, depending on the shape of the input signal.
c. The selection of the power BJTs should be done based on the results of a symmetric square wave input signal (which are also the same results for a DC input voltage).
d. Numerical example of the above BJT selection approach. Let $V_{CC}=10\,V$ and $R_L=10\,\Omega$. The maximum power dissipation split between the two BJTs,

based on AC input, is around 1 W. The maximum power dissipation split between the two BJTs, based on a square wave input, is around 1.25 W. If the input voltage is DC, then the dissipation power falls entirely on one of the BJTs, then each BJT needs to be capable of dissipating 2.5 W.

e. A power BJT selection is done based on the DC input BJT's maximum power dissipation.

3.6 ADS: TRANSIENT SIMULATION OF A PUSH–PULL AMPLIFIER; SWEEP STUDY TO FIND THE MAXIMUM POWER DISSIPATION AND MAXIMUM AMPLIFIER EFFICIENCY; AVERAGE POWER AND EFFICIENCY CALCULATIONS; DATA DISPLAY EQUATIONS

The following schematic explores the various time signals that exist in a push–pull amplifier.

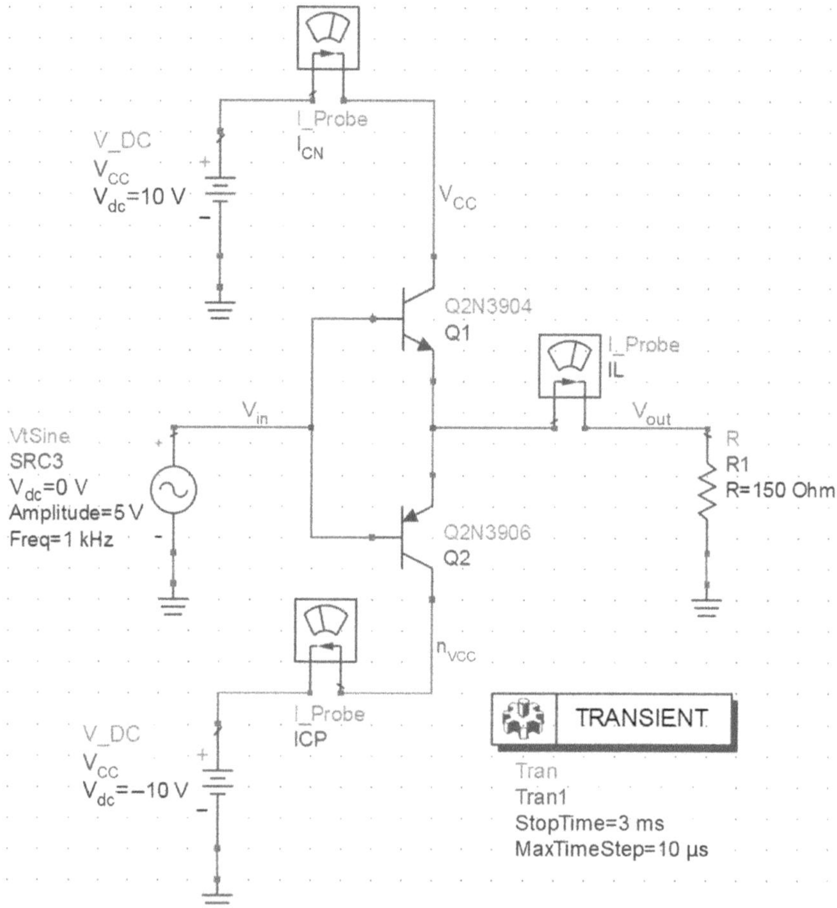

FIGURE 3.36 Basic schematic of a push–pull amplifier.

Don't forget to flip horizontally the PNP transistor, so that you can connect the emitter-to-emitter. Figure 3.37 shows the various time signals.

We can see how the NPN and PNP emitter currents complement one another in providing the load current. We can see the crossover distortion and the one-diode voltage difference between V_{in}(t) and V_{out}(t).

All the average powers of interest are created in measurement equations. See Figure 3.38. For example: $I_{L.i} \cdot V_{out}$ is the instantaneous power signal dissipated on R_L. The "mean" command computes average of a time signal. It generates a single number. We shall observe all these figures in a table.

Transient stop time must be large enough to allow more accurate mean calculations.

In efficiency calculations, we take a precaution to ensure that the two average supply powers don't subtract one from the other in the denominator. It sometimes happens due to "delivered" and "absorbed" sign conventions. The "abs" command was added to assure that P_minus_avg is positive (Figure 3.39).

Let us check conservation of energy. The total delivered power is that drawn from the DC supplies. We add the power delivered to the load to the power dissipated as heat in each BJT (Figure 3.40).

The subject of data display equations is revisited in Sections 3.7 and 3.8.

Next, let us explore two questions that involve sweep studies: (i) At what AC input amplitude does the NPN BJT dissipate maximum power?, and (ii) At what AC input amplitude does the amplifier have maximum efficiency?

Both questions (above) can be answered by the same parameter sweep study, at which we vary the amplitude of V_{in}, where the DC level of V_{in} is kept zero. The same two questions can be asked about DC V_{in}. As we vary V_{in} we set its amplitude to zero, so the VtSine becomes a DC voltage source with variable voltage. See Figure 3.41.

See Figure 3.42 for the simulation results of the swept AC input amplitude.

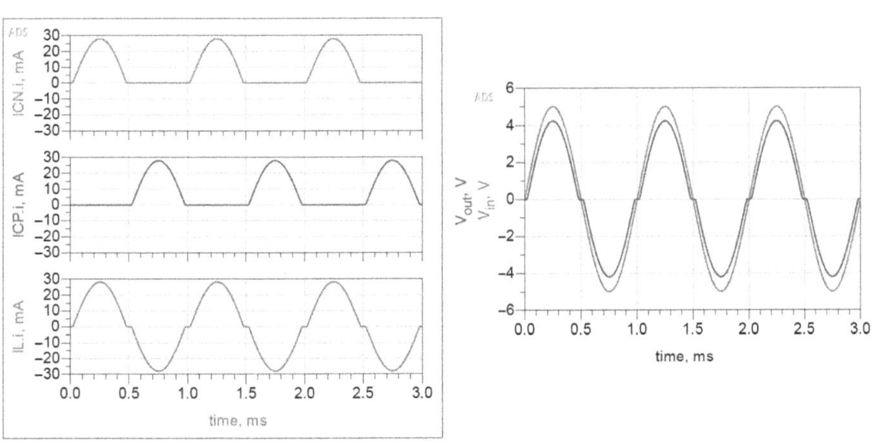

FIGURE 3.37 Time signals of a push–pull amplifier.

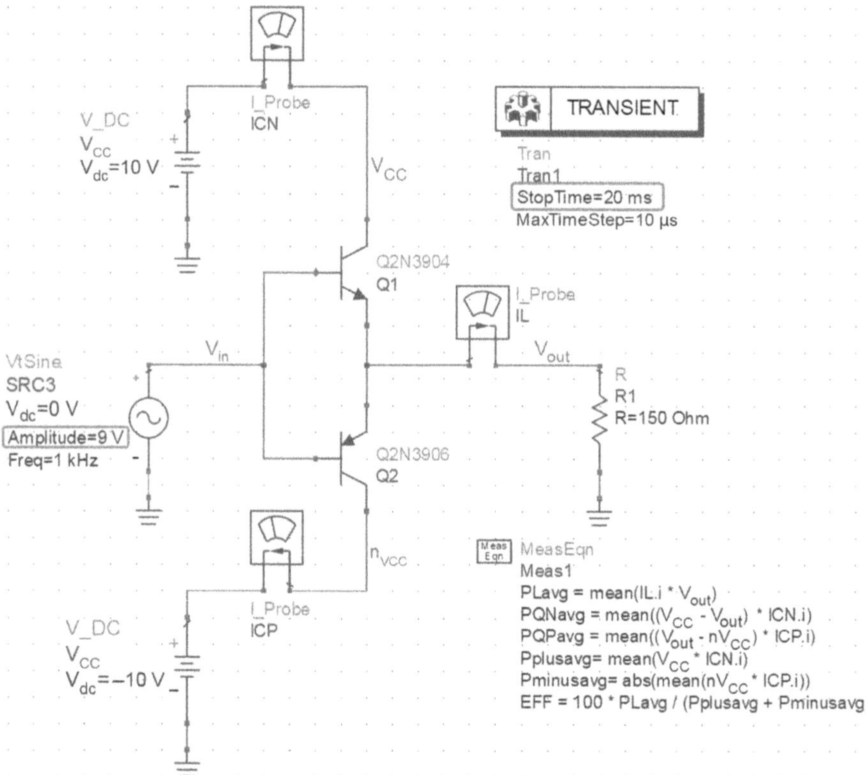

FIGURE 3.38 Push–pull amplifier schematic for average power calculations.

The average power plots are supposed to be parabolic. We observe that, in general, our simulation results show a parabolic "noisy" response. We need to improve the resolution of the mean measurements. This is done by reducing the MaxTimeStep of the transient simulation parameter, as seen in Figure 3.43.

The average power looks much better now. There is, however, "no free lunch" here.

The reduction in step time means the simulator must work harder and longer to give us these results. Taking 1,000 samples per period of a signal is often considered "overkill", but not here, though. For the DC sweep, we have to do a few changes. See Figure 3.44.

The DC input results are presented in Figure 3.45.

3.7 ADS TIPS AND TRICKS 3A: EQUATIONS

ADS allows for versatile equation handling, from simple variable assignment to the use of complex functions. While a complete treatment of this subject is beyond our scope, we wish to provide some introductory information. The reader is always encouraged to see ADS documentation for further details on the subject (Figure 3.46) [Keysight1].

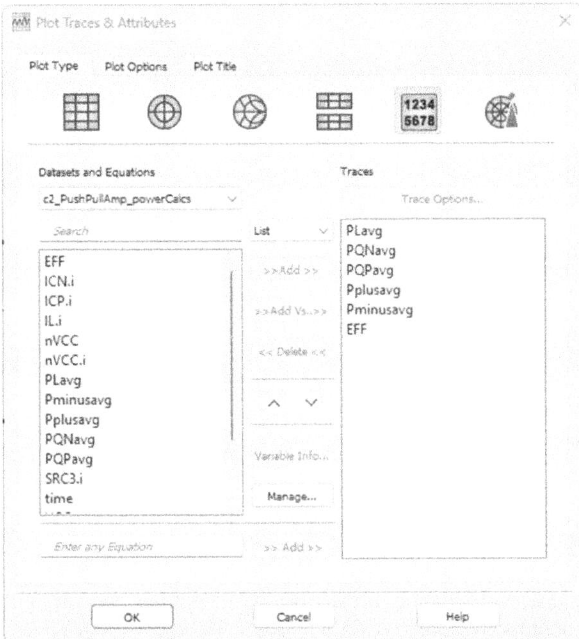

PLavg	PQNavg	PQPavg	Pplusavg	Pminusavg	EFF
0.197	0.054	0.055	0.153	0.152	64.676

FIGURE 3.39 Creation of a table for the calculated average powers. The units of the average power figures are [W]. The unit of efficiency EFF is %.

Eqn PDC = Pplusavg + Pminusavg

PDC
304.7 m

Eqn Pdelivered = PLavg + PQNavg + PQPavg

Pdelivered
306.2 m

FIGURE 3.40 Data display equations. The drawn and delivered powers are more or less equal.

3.8 ADS TIPS AND TRICKS 3B: DATA DISPLAY EQUATION

ADS Eqn, a.k.a data display equations, perform simple or involved mathematical operations on any data available in datasets. Eqns allow the user to process data and display results in plots or column listings (tables).

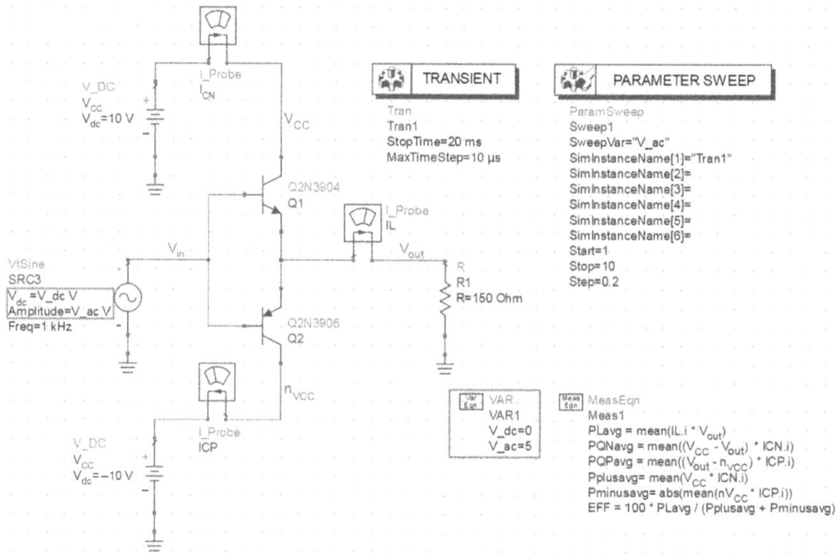

FIGURE 3.41 Transient simulation of the push–pull amplifier. We vary either the amplitude of the AC input or its DC level.

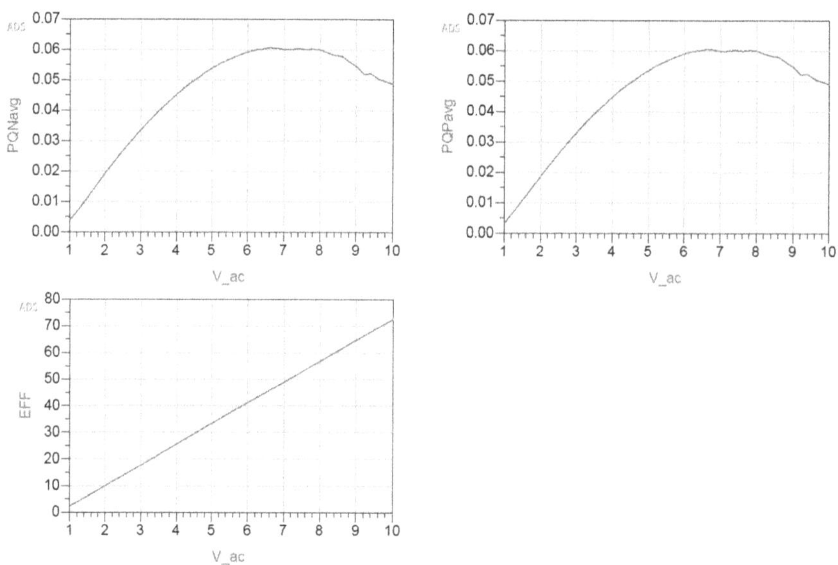

FIGURE 3.42 Simulation results for the swept amplitude of the AC input voltage.

Step 1: Drag and drop an Eqn onto the data display. Then enter equation dialogue will pop up. See Figure 3.47.

Step 2: Type in a variable name on the left side of the equality sign, and an expression with ADS functions, operators, and variables in the data set. See Figure 3.48.

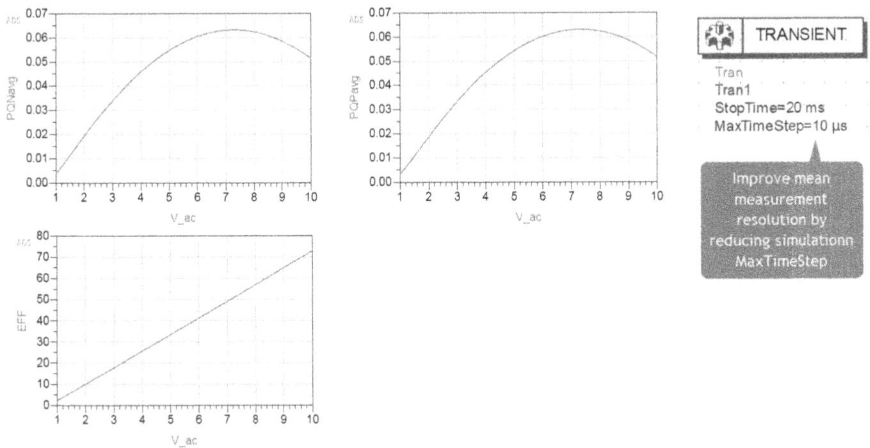

FIGURE 3.43 Simulation results for the swept amplitude of the AC input voltage (version 2).

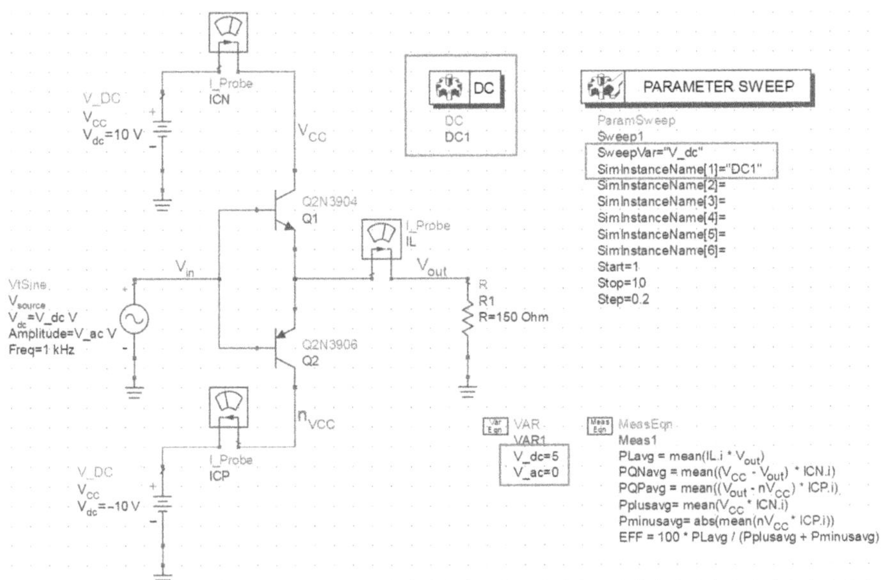

FIGURE 3.44 Push–pull amplifier with sweep of the DC input.

Step 3: Click OK when done. The equation shows up. See Figure 3.48.

Step 4: To plot or list the value of the Eqn, you must select "Equations" from the plot traces and attributes under datasets and equations. See Figure 3.49.

Note: The same procedure is used if you want to plot the EQN values.

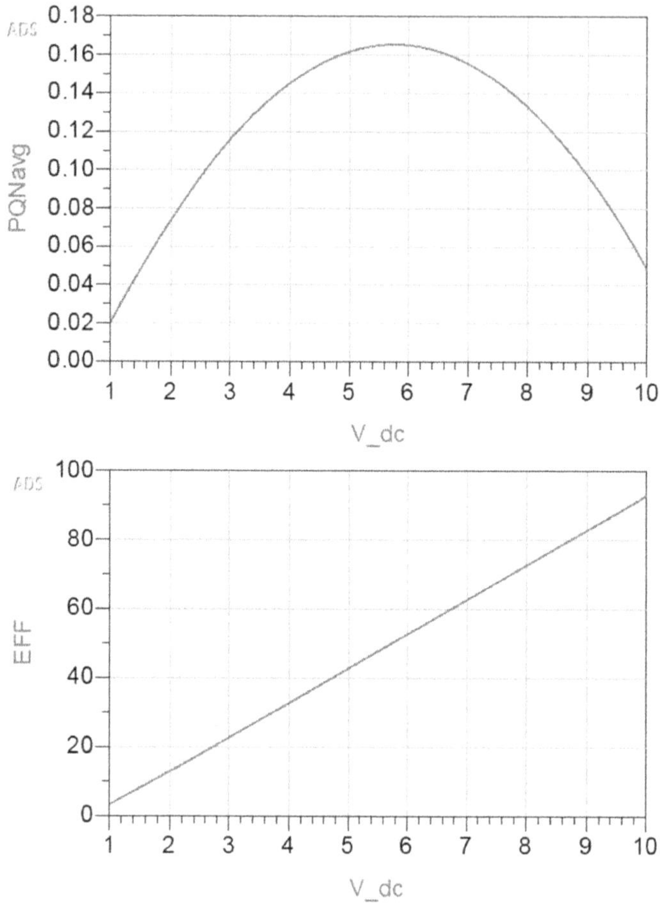

FIGURE 3.45 Simulation results of the push–pull amplifier driven by a variable DC input voltage.

- **VAR**: set up pre-simulation evaluated before simulation, defined on schematic

  Use on schematic for initializing variables. VARs are used for assigning parameter values and initializing sweep parameters. It can use built in ADS functions.

  VAR
  VAR1
  V_{CC}=5.0

- **MeasEqn**: set up pre-simulation evaluated post or during simulation, defined on schematic

  Use on schematic pre-simulation and it is available in the dataset after simulation. Frequently used for Yield Specs and Optimization Goal. It uses node names and ADS built in functions.

  MeasEqn
  Meas1
  Av = mag(Vout)/mag(Vin)

- **Eqn:** post-simulation evaluation and use, defined on data display page

  Use for calculations in the data display. Use node voltages, ADS built in functions, and any data available in a dataset.

  Eqn Pdelivered = PLavg + PQNavg + PQPavg

FIGURE 3.46 Explanation of ADS equation type and examples.

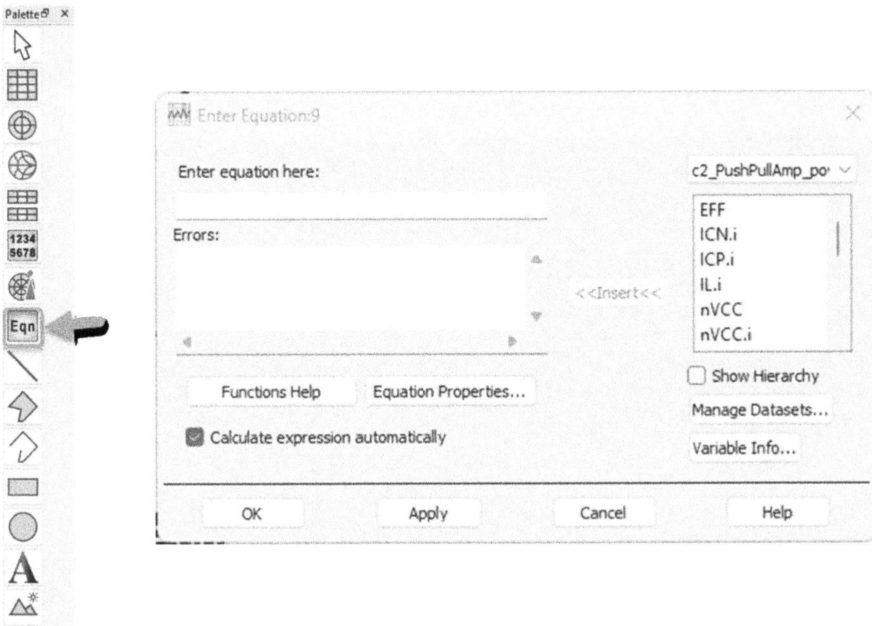

FIGURE 3.47 ADS data display equations palette icon and dialogue box.

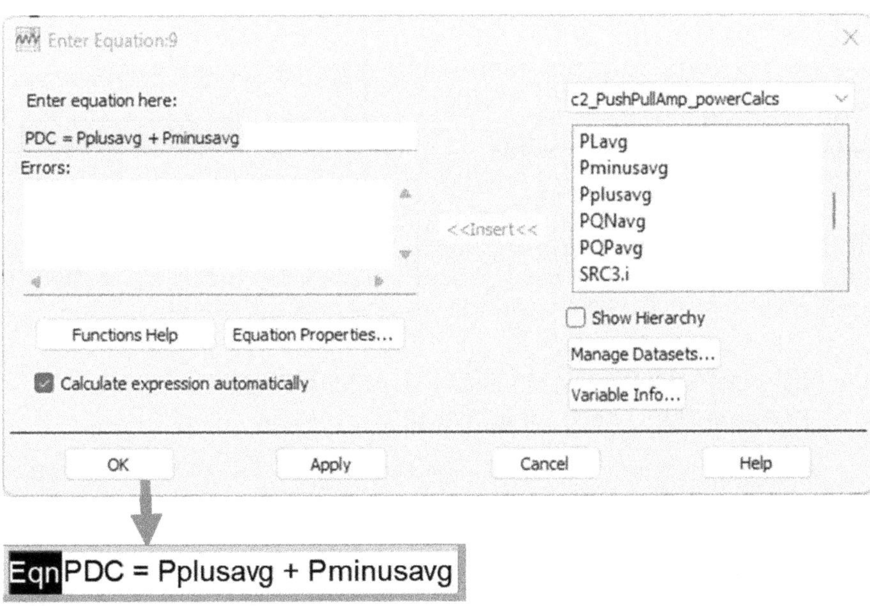

FIGURE 3.48 Creating and editing ADS data display equations.

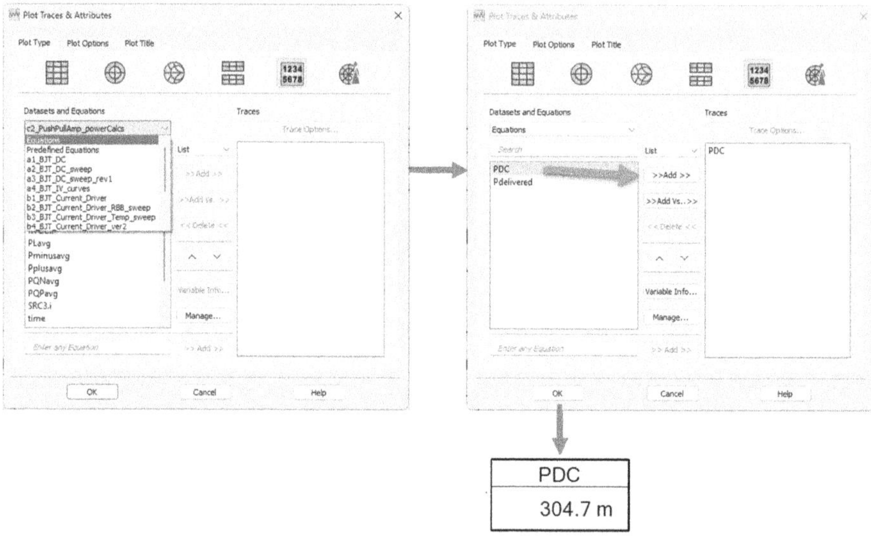

FIGURE 3.49 Listing or plotting ADS data display equations.

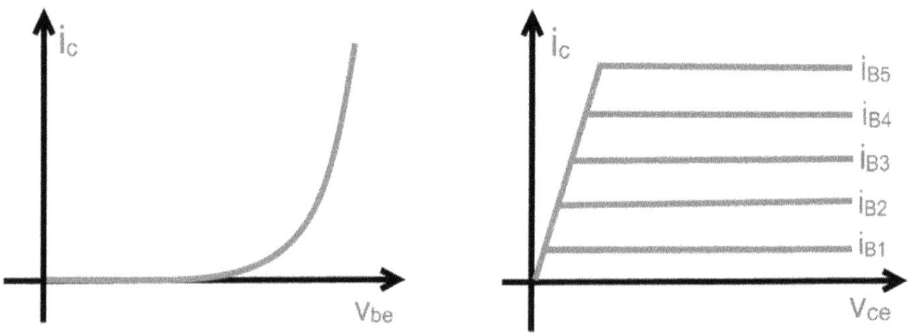

FIGURE 3.50 (Left) The exponential i_C vs. v_{BE} diode curve. (right) Approximate i_C vs. v_{CE}, with i_B as a parameter. In active mode, i_C is approximately independent of v_{CE}, and the BJT acts as a current-controlled current source. In saturation mode, v_{CE} is small.

3.9 BJT AMPLIFIER'S SMALL-SIGNAL ASSUMPTION, PARAMETERS, AND DIAGRAMS

BJT is a highly nonlinear device, as observed from the various transistor curves (Figure 3.50).

How can a BJT become the backbone of a linear voltage amplifier? Answer: The amplifier will be shown to be linear only if the input voltage signal is sufficiently small.

The BJT must have a preestablished **DC operating point $Q = \{I_C, I_B, V_{BE}, V_{CE}\}$** that is deep inside the BJT's active mode. "**Q point**" stands for "Quiescence", meaning that even if there is no input signal to amplify, the BJT has to stand ready to amplify small-signals, by having DC conditions shown in Q.

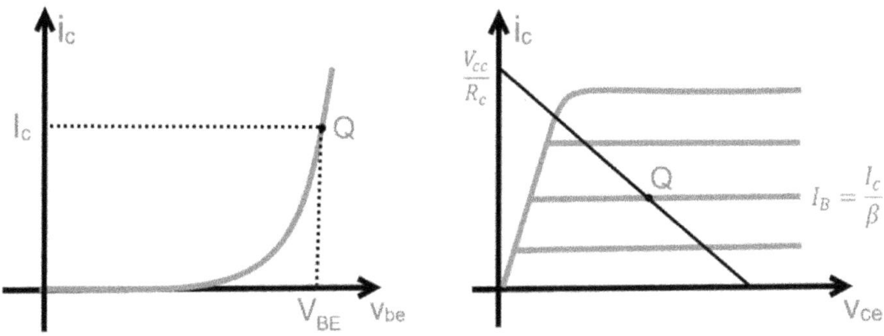

FIGURE 3.51 Two views of the amplifier's Q Point.

Figure 3.51 illustrates the DC operating point within the BJT curves.

"Deep inside the active mode region" means that Q must lie sufficiently far away from the borders of cutoff and saturation modes. The establishment of a Q point is called **Biasing of the Amplifier**. There are multiple circuit architectures that enable the creation of a desired Q point, along with nice properties of the biasing circuit, such as weakly β-dependent behavior, providing the amplifier with low sensitivity to temperature variations. Recall Section 3.3.

Two DC circuits, shown in Figure 3.52, demonstrate biasing architectures of NPN BJT. Voltage amplifiers can also be created around a PNP BJT (as shown in the end-of-chapter problems).

Every voltage and current of the amplifier is a mix of a DC and a signal: Every signal rides on top of a DC level. If indeed the signals are small enough, and therefore the amplifier operates linearly, as will be shown, then there is a superposition of the DC currents and voltages and the respective signal-related currents and voltages.

In order to better explain concepts related to small-signals in the BJT, let us adopt the following notation for each node voltage and component current in any BJT voltage amplifier [Sedra1]. Say, the collector current. Let I_C (capital–capital) be the DC collector current, i_c (lowercase–lowercase) be the small-signal collector current, and $i_C = I_C + i_c$ be the total collector current (lowercase variable and capital subscript). The exponential relationship between the DC I_C and DC V_{BE} (that is, $I_C = I_S \cdot \exp(V_{BE}/V_T)$) holds also for the total values: $i_C = I_S \cdot \exp(v_{BE}/V_T)$. $V_T = KT/q$ is the thermal voltage. In room temperature, it equals approximately 25 mV. This implies that the following small-signal relationship $i_c = I_C \cdot \exp(v_{be}/V_T)$. If $\mathbf{|v_{be}(t)| = V_T}$, we say that the amplifier meets the **small-signal assumption**. We then approximate the exponential function by its first two power series terms, yielding the small-signal relationship $\mathbf{i_c \approx (I_C/V_T)v_{be}}$. We define the **transconductance gain** small-signal parameter as $\mathbf{g_m = I_C/V_T}$. This gain is the main vehicle in creating voltage amplification for small input voltage signals. For BJT transistors, due to the steep BE junction collector current vs. BE forward voltage curve, typical values of g_m are in the order of magnitude of tens of mA/V.

More small-signal relationships are readily found with the help of the i_c vs. v_{be} relationship:

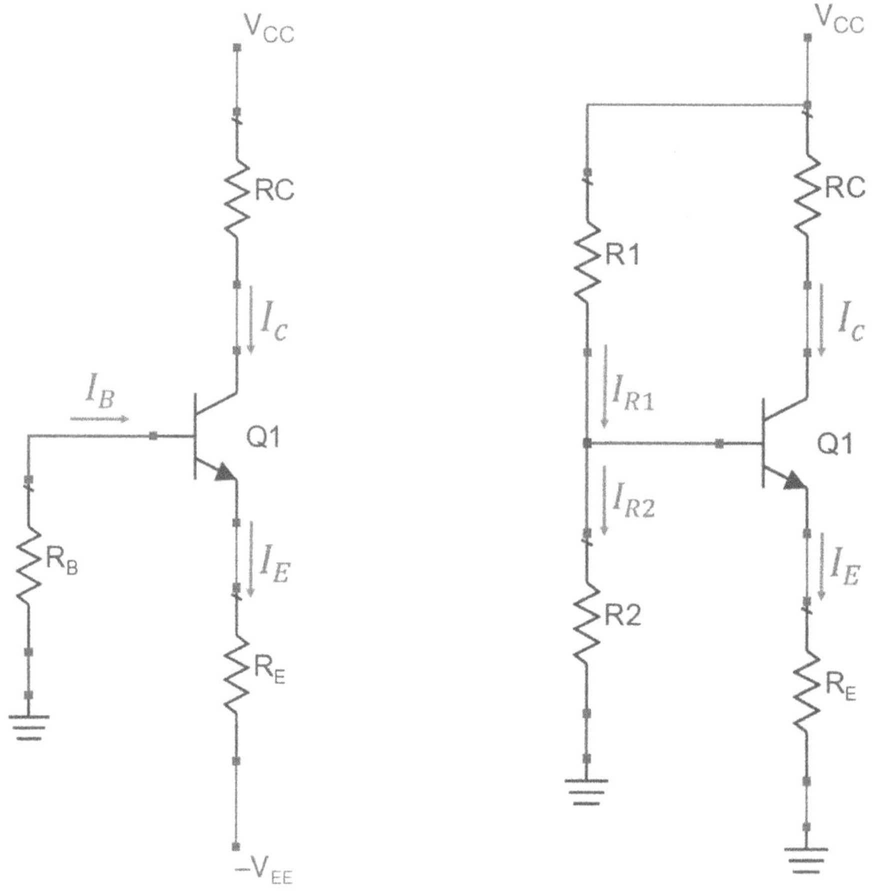

FIGURE 3.52 (Left) Dual supply NPN BJT biasing. (right) single-supply NPN BJT biasing.

If we neglect β variations, then $\mathbf{i_b \approx i_c/\beta}$ and $\mathbf{v_{be}/i_b = \beta/g_m}$. We define $\mathbf{r_\pi = \beta/g_m}$. Typical order of magnitude of the r_π small-signal parameter is a few kΩ. Also $\mathbf{v_{be}/i_e \approx \alpha/g_m = r_\pi/(\beta+1)}$. We define $\mathbf{r_e = r_\pi/(\beta+1)}$. Typical values of r_e are tens of Ω. What do we do with all these small-signal relationships?

The BJT **small-signal diagram** is a circuit that obeys all the small-signal relationships (above). Some (not all) of the small-signal parameters appear as components or gains in such circuits. There are several small-signal diagrams (all equivalent mathematically), each utilizes some of the small-signal parameters. See Figure 3.53 (left) for one such circuit, known as the **hybrid-π BJT small-signal diagram.**

There are other diagrams that feature β as well as r_e. See [Sedral]. All the small-signal parameters depend on the DC current I_C, established at the amplifier's Q point.

Let us explain briefly the small-signal component r_o shown in Figure 3.53 (right). Recall Figure 3.50 in which we show an approximate family of curves of I_C vs. V_{CE}. In the active mode zone, the current plots (of Figure 3.50) are horizontal, indicating

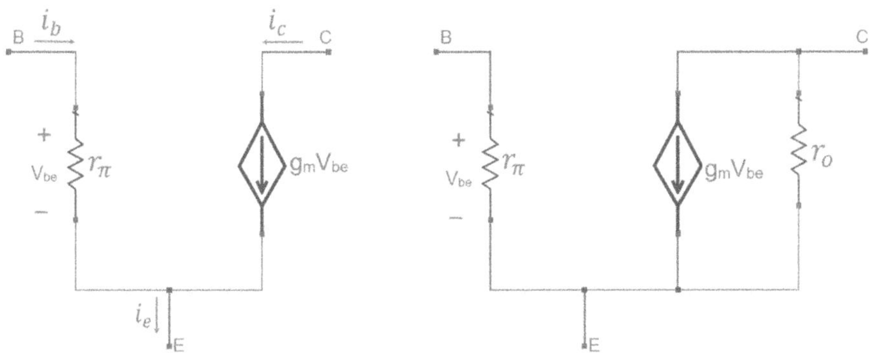

FIGURE 3.53 Hybrid-π BJT small-signal diagram featuring the r_π and g_m parameters (left) and an additional small-signal parameter r_o (right).

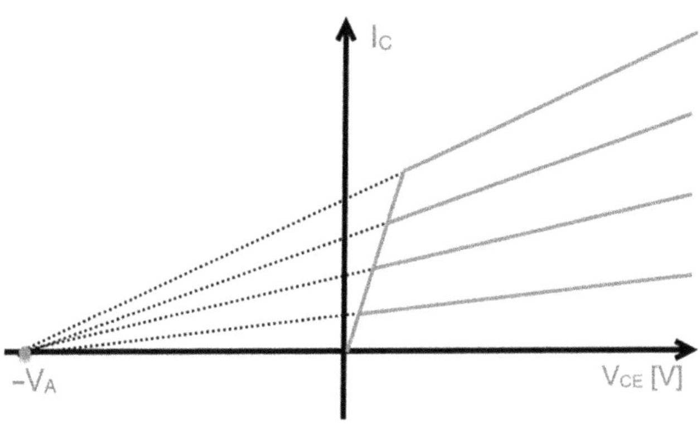

FIGURE 3.54 The early effect: In the active model, the I_C vs. V_{CE} curves are slightly slanted. Each plot of I_C vs. V_{CE} is parameterized by I_B.

that I_C is independent of V_{CE}. In reality, there is a small dependence of I_C on V_{CE}, as sketched in the exaggerrated Figure 3.54.

All the I_C vs. V_{CE} active mode lines meet at a point $V_{CE}=-V_A$. V_A known as the "**early voltage**" is a parameter of the BJT. Typical values of V_A are many tens of V. The reason for the small dependence of I_C on V_{CE} is that changes in V_{CE} cause changes in the reverse bias voltage V_{CB}, causing changes in the width of the CB junction depletion region, which cause changes in the width of the BJT's base region, and consequently, there are changes in the β parameter of the BJT. The ensuing small-signal parameter is $\mathbf{r_o=V_A/I_C}$ and its order of magnitude is many tens to a few hundreds of kΩ.

So, an analysis of a BJT amplifier, if the small-signal assumption is met, consists of a superposition of DC analysis and small-signal analysis. The DC analysis is done first. It verifies that the BJT operates in active mode. It establishes the Q point DC voltages and currents. The current I_C is used, prior to the small-signal analysis, to find all the small-signal parameters.

Small-signal analysis of a BJT amplifier involves the construction of a **small-signal diagram** of the amplifier. The starting point is the small-signal diagram of the BJT, like the one that is presented in Figure 3.53 (right). We then have to find what amplifier circuit components are connected to each of the BJT's small-signal diagram's terminals. This will be demonstrated for a common emitter (CE) amplifier explained in Section 3.10.

3.10 COMMON EMITTER AMPLIFIER

There are many architectures of a CE amplifier as seen in the end-of-chapter problems. The one shown in Figure 3.55 is based on a single NPN BJT and a single DC supply V_{CC}. There are BJT amplifiers that utilize PNP BJTs and/or dual supplies. CE amplifiers can provide large small-signal voltage amplification, in the order of magnitude of tens to more than a hundred.

Let us first introduce the CE amplifier's circuit components. The DC biasing of the BJT, shown in Figure 3.55, is done by the supply V_{CC}, two resistors R_{B1} and R_{B2} that create a voltage for the BJT's base, and most importantly, a resistor R_E that is crucial in determining the DC emitter current. This biasing setup was shown in Figure 3.52. There are three large capacitors: A coupling capacitor that prevents the small-signal voltage source v_{in}, from affecting the BJT's bias point. Another coupling capacitor

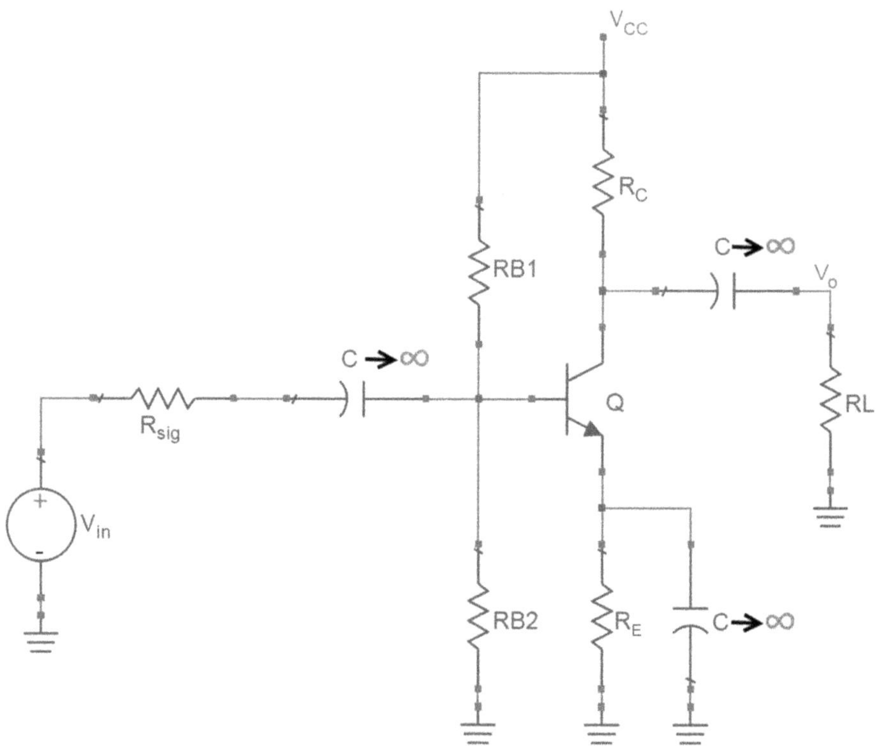

FIGURE 3.55 A single-supply NPN CE amplifier.

that prevents the load resistance R_L from affecting the DC bias point. A third large capacitor, called a bypass capacitor, short-circuits R_E to prevent it from participating in the small-signal operation of the amplifier. Typical values of each large capacitor are around 100 μF.

The resistors R_{B1} and R_{B2} work approximately like a voltage divider (as long as I_B is much smaller than the currents flowing in these resistors, which happens whenever these base resistors are selected to be typically in the order of magnitude of tens of kΩ). As a result, $V_B \approx V_{CC} \cdot R_{B2}/(R_{B1} + R_{B2})$ and $I_E \approx (V_B - V_{BE})/R_E$. If the BJT operates in active mode, then $I_C \approx I_E$. Active mode is then verified by checking if approximately $V_C \geq V_B$, where $V_C = V_{CC} - I_C \cdot R_C$.

The small-signal diagram of the above CE amplifier (shown in Figure 3.56) is constructed from the CE implementation diagram (Figure 3.55) by setting the BJT small-signal diagram and adding components, following three construction rules:

a. DC voltage supplies are considered short-circuited.
b. DC current sources are considered open-circuited.
c. Large capacitors are considered short-circuited.

The reason for rules (a) and (b) is to avoid double-counting of each supply effect. All DC supplies participate in the amplifier's DC analysis, and therefore should not participate in the small-signal analysis. Rule (c) implies that at the input signal's frequencies, the capacitor's impedance is very small, compared to other resistances that exist in the implementation circuit.

Let us explain how the amplifier's small-signal diagram is constructed based on the BJT small-signal diagram and the CE amplifier's implementation circuit. The starting point is the BJT's small-signal diagram (shown bracketed in Figure 3.56). Now, let us apply rules (a)–(c) (above) to the CE amplifier's circuit of Figure 3.55.

The emitter terminal is short-circuited to ground because the bypass capacitor short-circuits the resistor R_E.

In Figure 3.55, the BJT's collector terminal is connected to V_{CC} by R_C and is connected to R_L via a coupling capacitor. V_{CC} is considered a short circuit to ground. The coupling capacitor is considered short circuit. This is how R_C and R_L appear between the collector and the ground.

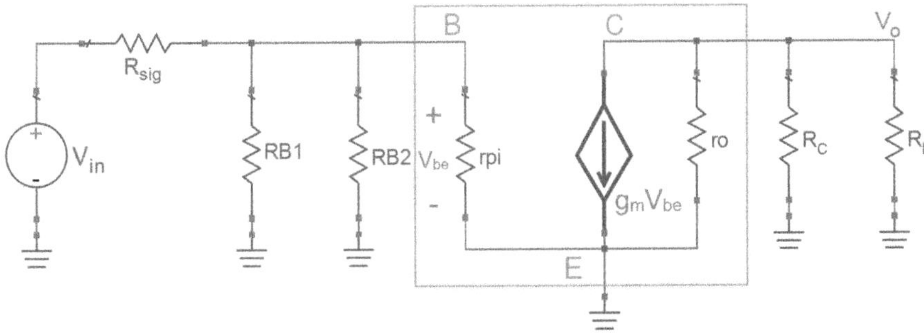

FIGURE 3.56 Small-signal diagram of a CE amplifier.

For the same reason that R_C is connected between collector and ground, resistor R_{B1} is connected between base and ground, in parallel to the other base resistor R_{B2}. The input coupling capacitor is considered short-circuited between the base terminal and the input signal's internal resistance R_{sig}.

What is somewhat amazing about the amplifier's small-signal diagram (shown in Figure 3.56) is that it serves all CE amplifiers' biasing architectures: single supply, dual supply, NPN, PNP, etc. Therefore, formulas that are obtained by performing small-signal analysis (such as small-signal voltage and current gains, small-signal input and output resistances) are valid in all versions of the CE amplifier. The only architectures that should be excluded are ones that feature a small resistance between the emitter and R_E, a resistor that is not bypassed. Such a resistor that shows up between emitter and ground in the amplifier's small-signal diagram is called a **degeneration resistor** and more will be said about it further on in the book (See Section 5.1). For the time being, let us assume that the emitter is at ground in all basic CE amplifiers.

Let us derive a few small-signal formulas to help us better understand the ADS simulation that will follow. In the diagram shown in Figure 3.56, the signal source v_{in} and its internal resistance R_{sig} are not part of the CE amplifier. Likewise, the load resistance R_L is not part of the amplifier. The CE amplifier interfaces between the small-signal voltage source and the load. R_{sig} and R_L do affect the voltage gain result, but let us consider R_{sig}, R_L, and the amplitude of $v_{in}(t)$ as given, unchangeable by the CE amplifier's designer.

The input resistance R_{in} of the CE amplifier is the equivalent resistance seen from the input source "looking into" the amplifier. It equals

$$R_{in} = \frac{v_b}{i_{in}} = R_B \parallel r_\pi \quad \text{where } R_B = R_{B1} \parallel R_{B2} \tag{3.1}$$

See Figure 3.56. The voltage v_b denotes the small-signal voltage at the base of the BJT. We have that $v_{be} = v_b$ because $v_e = 0$. The current i_{in} denotes the small-signal current flowing from the voltage signal source to the amplifier. The signal voltage $v_{be}(t)$ is an attenuated version of the input voltage signal $v_{in}(t)$ through voltage division between R_{in} and R_{sig}: $v_{be}(t) = V_{in}(t) \cdot R_{in}/(R_{in} + R_{sig})$. We recall that the BJT "small-signal assumption" is $|v_{be}(t)| = V_T$. It practically implies that the amplitude of $v_{be}(t)$ (positive and negative) should be no larger than 10–15 mV.

We further recall that $v_{be}(t)$ controls the small-signal diagram's dependent current source $i_c(t)$ via the transconductance gain g_m. The amplified small-signal output voltage $v_o(t)$ is created by $i_c(t)$ and the parallel resistors located between the collector and ground. The resulting small-signal voltage gain is thus:

$$A_V = \frac{v_o}{v_{in}} = -\frac{R_{in}}{R_{in} + R_{sig}} \cdot g_m \left(R_C \| R_L \| r_o \right) \tag{3.2}$$

The negative sign of A_V indicates that the CE amplifier is inverting. In the amplifier's small-signal diagram the direction of the current flowing in the dependent current source, is opposite to the positive polarity of the voltage drop of the parallel collector resistors.

3.11 ADS: DC AND TRANSIENT SIMULATION OF A CE AMPLIFIER: DC BIASING, DC CURRENT TWEAKING, MEASUREMENT OF THE SMALL-SIGNAL VOLTAGE GAIN AND INPUT RESISTANCE; QUALITATIVE CHECK OF THE SMALL-SIGNAL ASSUMPTION

Example 3.6

The following single-supply NPN CE amplifier example (Figure 3.55) was initially constructed by the selection of arbitrary, yet reasonable, component values. No deliberate design, that will be studied in Section 5.2. Let us explain the amplifier in more detail.

We chose a value of 100 µF for all the large coupling and bypass capacitors. It is large enough and okay, practically. For example, with such a choice, the low frequency range of the amplifier (in which the capacitors may not be considered short circuit) is expected to lie below 100 Hz. A choice of C = 1 mF (which lowers the borderline of the CE amplifier's low frequency range to about 10 Hz) will prompt a warning by the ADS program due to the "too large" capacitance.

We shall do DC annotations. The current probe is needed for input resistance calculations later on (Figure 3.65).

For DC simulation, the circuits that are to the left of C_1 and to the right of C_2 are irrelevant. We shall focus on Q_1, R_{B1}, R_{B2}, R_C, and R_E only.

Can we estimate the DC currents and voltage purely by inspection? Let's try:

Assuming that R_{B1} and R_{B2} form an "almost-voltage-divider", we estimate V_B to be around 5 V. We then estimate V_E to be around 4.3 V. We further estimate $I_C = I_E = 1$ mA, which leads to an estimated V_C to be around 7 V, which justifies the active mode assumption.

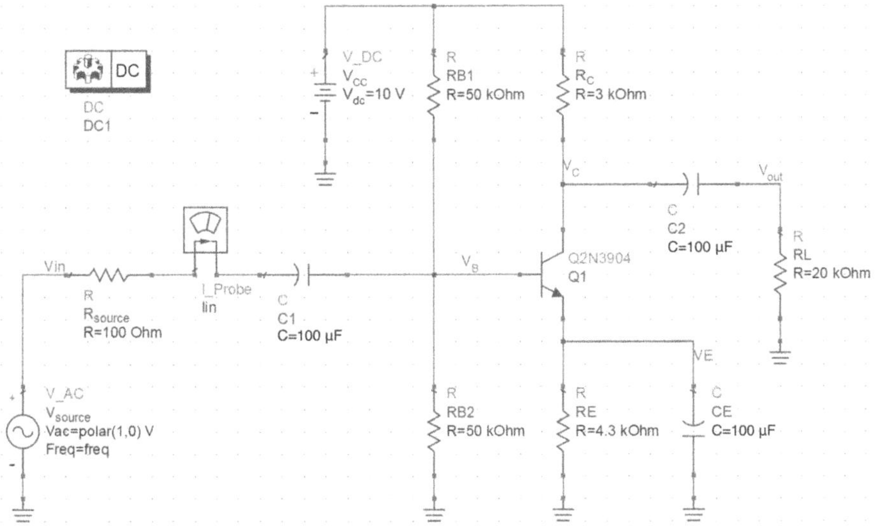

FIGURE 3.57 Single-supply NPN CE amplifier example.

We estimate I_B to be around 10 μA, and IR_B to be around 100 μA. It more or less justifies the "almost-voltage-divider" assumption. Let us compare to the exact values by running the DC simulation. See Figure 3.58 for the DC results.

We see that the voltage divider assumption is slightly off: exact $V_B = 4.82$ V compared to the estimated $V_B = 5$ V. Consequently, the current I_C is slightly off: 959 μA vs. the estimated 1 mA. Exact $V_C = 7.12$ V. Not bad compared to the estimated $V_C = 7$ V.

We see that the exact $I_B = 7.14$ μA is smaller than the expected $I_B = 10$ μA, which makes sense, because we assumed $\beta = 100$, but obviously, the actual β is a lot larger.

The IR_B currents are more or less around 100 μA, as estimated. Still, $IR_{B1} = 104$ μA and $IR_{B2} = 96.4$ μA, compared to $I_B = 7.14$ μA, constitute an "approximated voltage divider", and not "almost-voltage-divider".

Suggested exploration activity: Increase R_{B1} and R_{B2} from 50 (each) to 100 kΩ (each) and later 200 kΩ (each). Observe how the "voltage divider" assumption gradually deteriorates.

How do we tweak the value of I_C?

The main idea is that R_E is the dominant factor that influences I_C. Why?

You may recall the discussion about the weak β-dependent current driver and similarly, the exact formula for I_E: $I_E = (V_{BB} - V_{BE})/[R_E + R_B/(\beta + 1)]$ where V_{BB} is the voltage divider voltage, and R_B is the parallel connection of R_{B1} and R_{B2}. If $R_E \gg R_B/(\beta + 1)$, then the voltage divider approximation $I_E \approx (V_{BB} - V_{BE})/R_E$ works well.

Let us check: $R_E = 4.3$ kΩ, $R_B = 25$ kΩ, $R_B/(\beta + 1) = 250$ Ω. Is 4.3 kΩ much larger than 250 Ω? Yes.

Let us return to the task of tweaking the current I_C. Refer again to Figure 3.58. We want to increase I_C from 959 to 1,000 μA. If we decrease R_E by a factor of 1,000/959, will I_C come much closer to the desired 1 mA value?

Let us revise RE, and then repeat the simulation: $RE_{new} = RE_{old} / (1{,}000/959) = 4.1$ kΩ. Going beyond the first decimal to the more exact RE value RE=4.1237 kΩ is, of course, not significant.

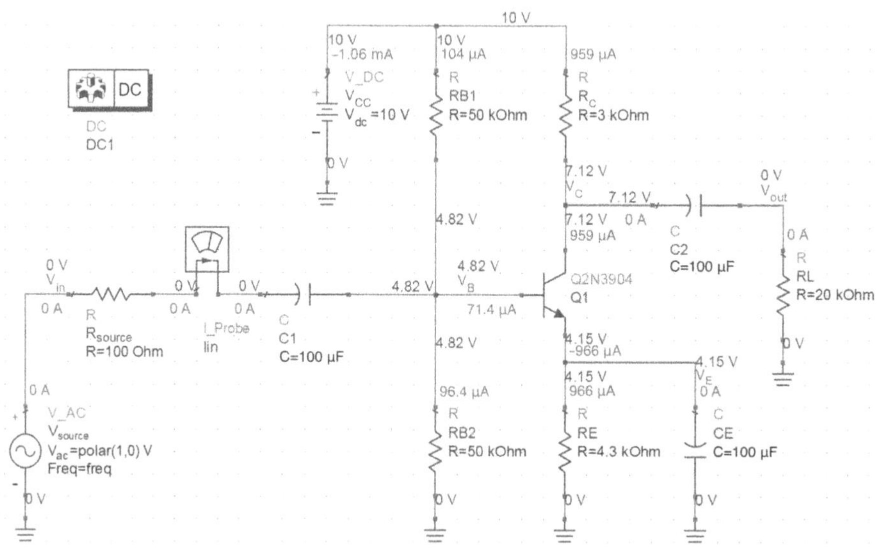

FIGURE 3.58 DC results of the CE amplifier example.

Let us see how well the tweaking of R_E from 4.3 to 4.1 kΩ works. It does indeed (Figure 3.59).

Next, we do transient simulation of the CE amplifier (Figure 3.60). We shall watch the small-signal input and the amplified small-signal output. We shall calculate the small-signal voltage gain by measuring the peak-to-peak (ptp) output amplitude. We shall observe how the output signal becomes more and more distorted as we increase the input signal's amplitude. Finally, we shall measure the

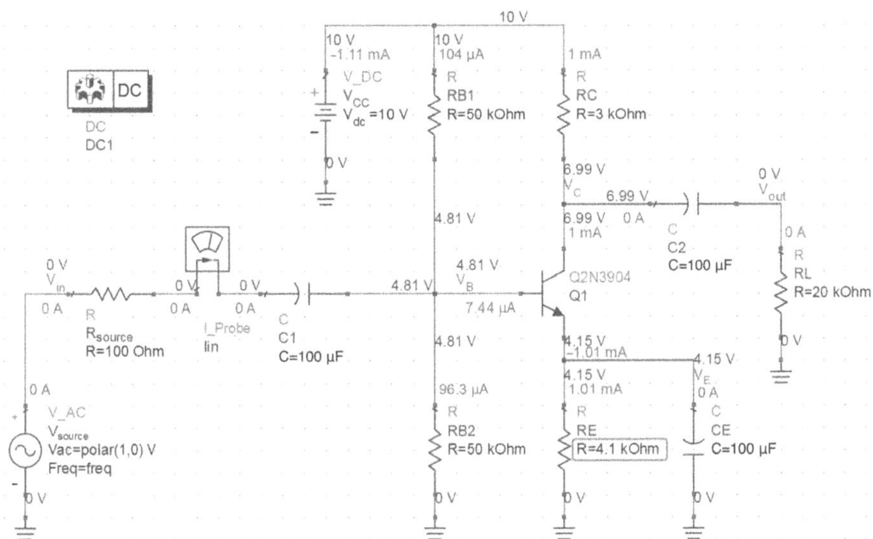

FIGURE 3.59 CE amplifier example after the tweaking of the DC current.

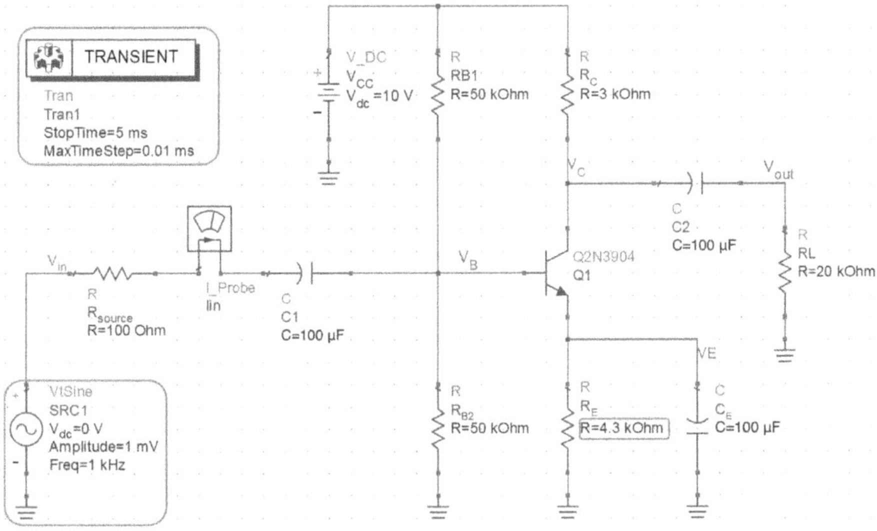

FIGURE 3.60 Transient simulation analysis of the CE amplifier example.

amplifier's input resistance by dividing the ptp base voltage amplitude by the ptp of the incoming small-signal current.

We use the pre-tweaking biasing that produced $I_C = 0.959$ mA.

Are the 100 μF capacitors short-circuited at f=1 kHz? Let us calculate the capacitor's impedance at such f. C=100 μF → $Z_c = 1/(j\omega C) = 1/(j2\pi \cdot 1{,}000 \cdot 100e{-}6) = -j1.59\ \Omega$. So, it is not a short circuit exactly, but compared to other resistors in the diagram each capacitor is practically a short circuit at f=1 kHz.

Later in this section, we shall demonstrate what happens to the small-signal voltage gain if we remove the bypass capacitor, named CE.

Let us run a simple transient simulation. Notice how small the input amplitude is (1 mV). Later, we shall increase it gradually.

The voltages V_B, V_E, and V_C all have some DC level, and on top of the DC level we have the AC small-signal. V_{out} has a DC of 0, because of C_2. Let us watch the simulation results (Figure 3.61).

Because of the large small-signal voltage gain, it is not a good idea to put V_{in} and V_{out} on the same axes. V_{in} is poorly visible. Need to plot each in its own scale. V_{in} is a pure AC signal, with a frequency of f=1 kHz and a ptp amplitude of 2 mV.

Let us measure the ptp amplitude of V_{out}, using peak and valley markers. See Figure 3.62.

Subtracting manually the V_{out} values included in each marker's box yields $ptp(V_{out}) = 180.4$ mV. Therefore, the small-signal voltage gain is $A_v = -ptp(V_{out})/ptp(V_{in}) = 180.4/2 = -90.2$. Notice that V_{out} is inverted compared to V_{in} hence, the negative sign in the A_v formula.

Note further that the manual calculations may be automated with the help of data display equations.

The names of the markers (m_1 and m_2) appear as variables in the output ptp calculation. These calculation results may be presented in a table.

In Figure 3.63 (obtained automatically as we opt not to close the data display window), the data display calculation process is repeated for V_{in} having an amplitude of 10 mV. The ptp amplitude of V_{in} is 0.02 V.

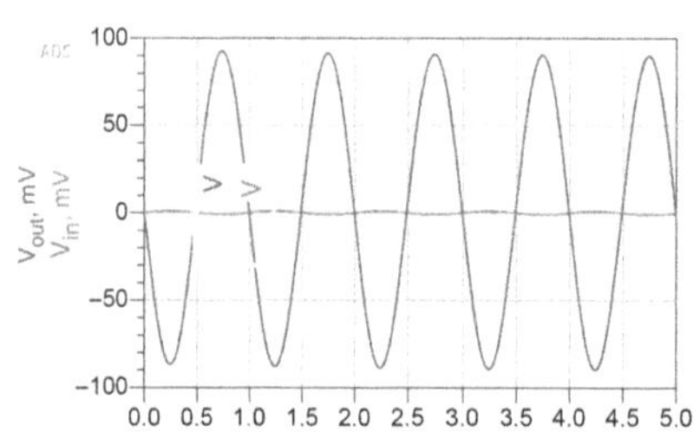

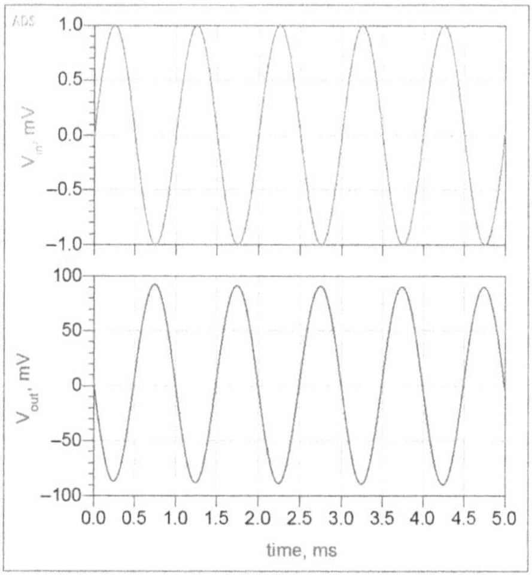

FIGURE 3.61 The CE amplifier's input and output small-signal voltages. (Left) Poorly visible V_{in}. (Right) Separate vertical scales for V_{in} and V_{out}.

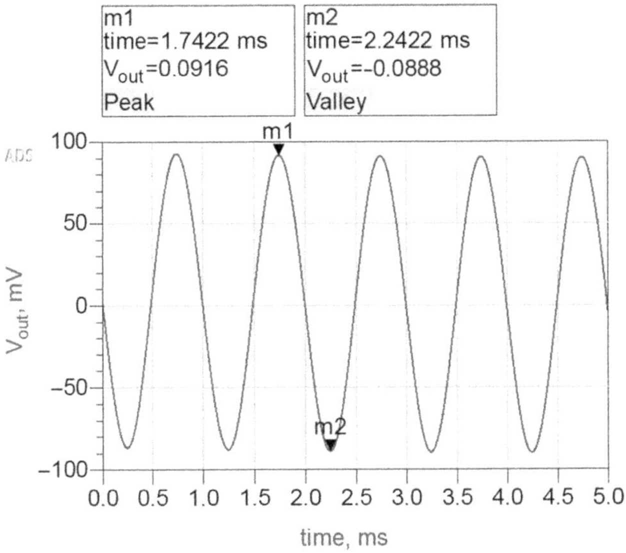

Eqn V_{out}_ptp = m1 - m2

Eqn V_{in}_ptp = 2e-3

Eqn Av = -V_{out}_ptp / V_{in}_ptp

V_{out}_ptp	Av
0.1804	−90.1903

FIGURE 3.62 Measurement of the ptp amplitude of V_{out} using markers, and post-processing equations in terms of the placed markers.

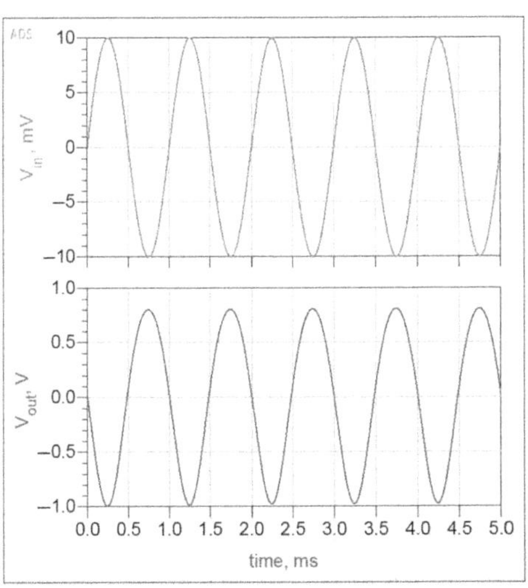

Eqn V_{out}_ptp = m1 - m2

Eqn V_{in}_ptp = 20e-3

Eqn Av = -V_{out}_ptp / V_{in}_ptp

V_{out}_ptp	Av
1.7861	-89.3057

FIGURE 3.63 Repetition of the measurement for V_{in} with a larger amplitude.

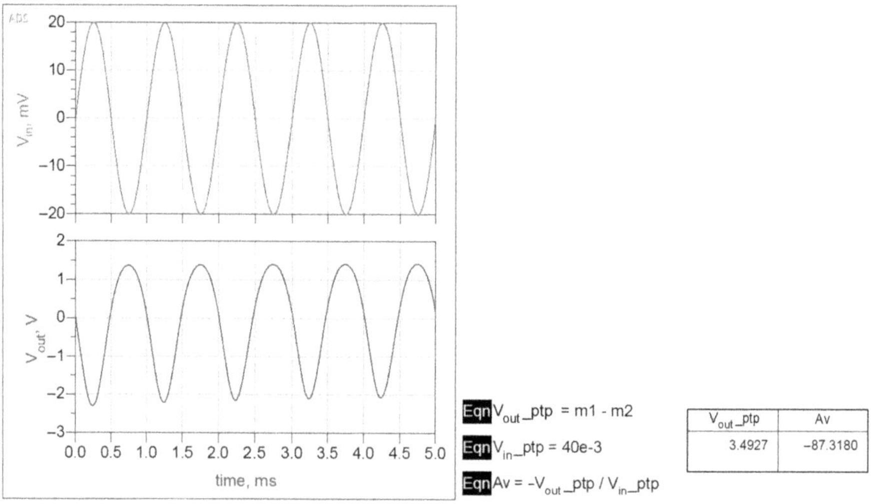

FIGURE 3.64 Simulation results for an even larger V_{in}.

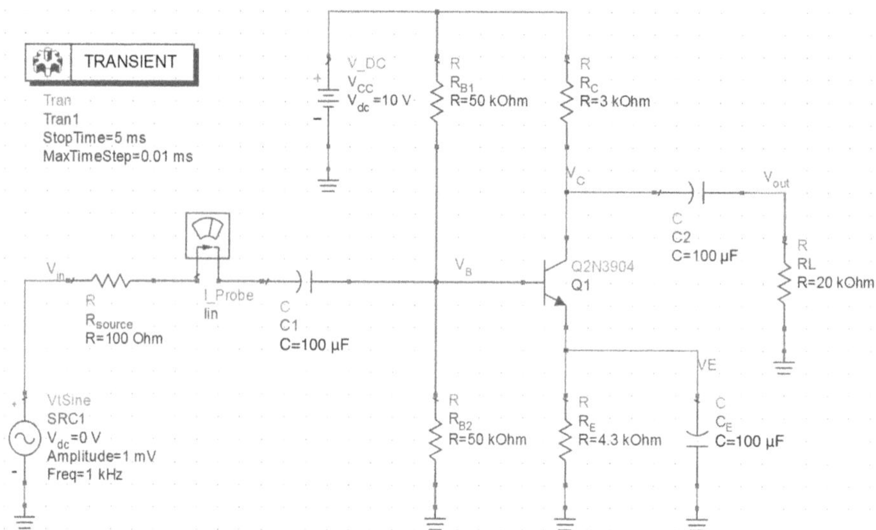

FIGURE 3.65 CE amplifier schematic for measuring R_{in}.

Notice the distorted shape of V_{out}.

V_{out} is no longer symmetric around 0. The positive part is fat and short, and the negative part is narrow and longer. The ptp amplitude of V_{out} is (by inspection) about 1.8 V. Therefore, the voltage gain still stays around −90.

With an input amplitude of 20 mV, V_{out} becomes more distorted (Figure 3.64).

Now the voltage gain falls to around −87, but V_{out} is not really a pure AC. As we recall, in a linear amplifier, the shape of V_{out} must be the same as that of V_{in}.

Where do we draw the line? For which input amplitude do we deem the amplifier's performance "acceptable"? An "unacceptable" $V_{out}(t)$ means that we no longer meet the "small-signal assumption". We shall offer answers soon.

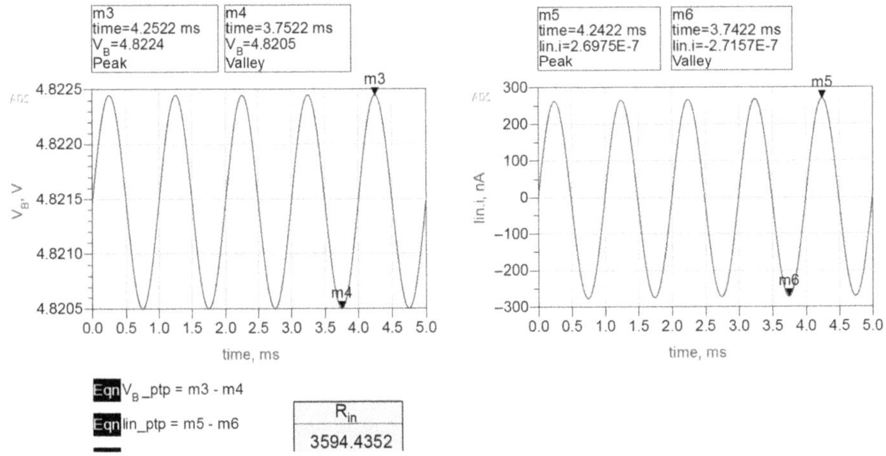

FIGURE 3.66 R_{in} computation.

Back to 1 mV V_{in} amplitude. The amplifier's input resistance R_{in} is determined from the voltage $V_B(t)$ and the incoming current signal $I(C_1) = I_{in,i}(t)$.

Let us use markers to calculate the ptp amplitude of each signal (Figure 3.66).

3.12 ADS: AC SIMULATION OF CE AMPLIFIER: MEASUREMENT OF THE SMALL-SIGNAL VOLTAGE GAIN AND INPUT RESISTANCE, VOLTAGE GAIN TWEAKING; HARMONIC BALANCE SIMULATION; QUANTITATIVE ASSESSMENT OF THE SMALL-SIGNAL ASSUMPTION

As we recall from our op-amp low-pass filter frequency response (in Chapter 1), AC simulation allows us to observe the frequency response of the amplifier. We can see the voltage gain as a function of the frequency f. The CE amplifier is a wide-band amplifier: It is desired that the voltage gain stay constant for a large range of input frequencies. We shall also observe the input resistance as a function of f. AC simulation, of a CE amplifier, is based on the small-signal diagram of the amplifier (Figure 3.56). The circuit response is therefore insensitive to the input signal's amplitude.

In Figure 3.67 note the special AC input voltage signal source. In the AC simulation block, we set the frequency to be in log scale. Each frequency decade (such as from 1 to 10 Hz, or from 100 kHz to 1 MHz) has the same X-axis length. In our simulation, the plots will cover a range of eight decades!

ADS runs impedance analysis. Therefore, every voltage and every current is a complex variable. We create measurement equations that make sense only in AC simulation. A_v is the ratio of magnitudes of frequency responses (of V_{out} and V_{in}). Likewise, the definition of R_{in}.

Figure 3.68 presents the magnitude frequency response of the voltage gain. As the simulation run ends, we do: Insert Plot → Select a rectangle → Ask for A_v vs. freq → OK.

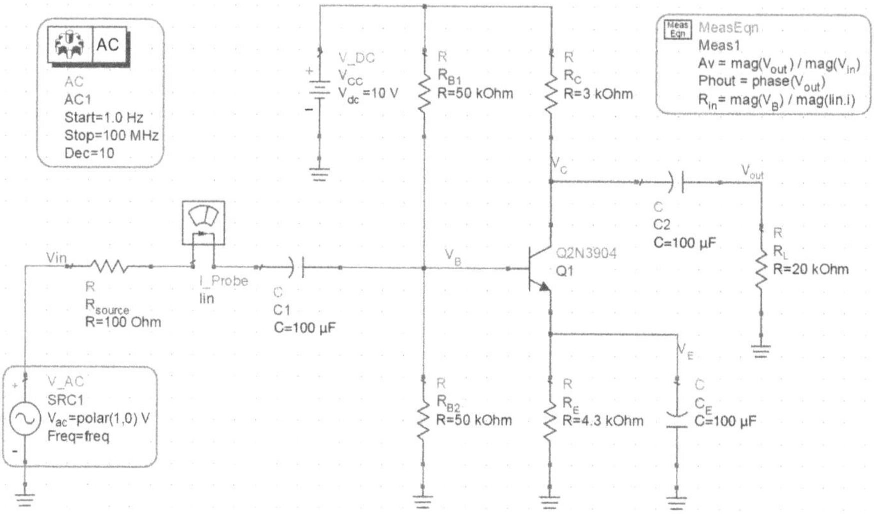

FIGURE 3.67 CE amplifier schematics for AC simulation.

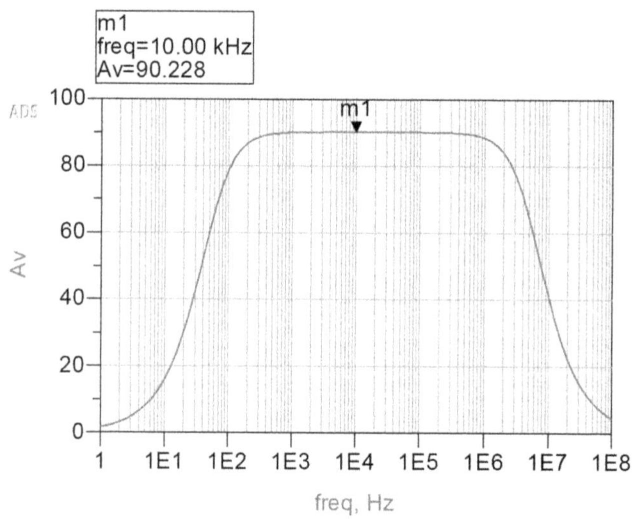

FIGURE 3.68 Magnitude frequency response of the CE amplifier.

Now choose "Plot Options" → For the X axis, choose "log", and for the Y axis leave it as "linear".

The useful (i.e., constant gain) region stretches from around 300 Hz to 1 MHz. This frequency range is called "**the mid-band frequency range of the amplifier**". The magnitude voltage gain at the mid-band frequency range is approximately 90.2.

In log scale, the frequency f = 0 (DC) is at $-\infty$. We avoid DC here so that the plot is meaningful throughout the entire frequency range. What causes the gain to fall at low frequencies is the large capacitors (C_1, C_2, and C_3). At DC, the gain is zero. DC input signals are blocked from entering the CE amplifier.

What causes the voltage gain to fall at high frequencies are the small parasitic capacitors created inside the BJT itself (like the capacitance of the PN junction depletion region), which will be covered in Chapter 6.

What happened to the negative sign of the gain?

In theory, the *negative sign* in an AC simulation means there is a 180° phase difference between the output voltage and the input voltage. Since our voltage sweep kept the input constant at zero degrees phase, it is okay to just look at the phase of V_{out} (Figures 3.69).

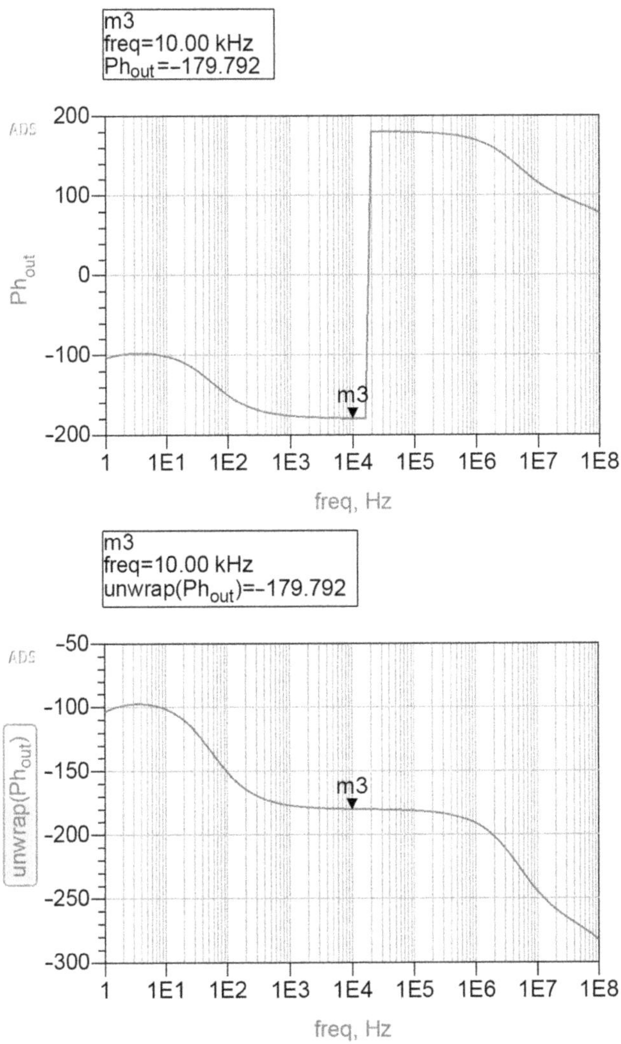

FIGURE 3.69 Two equivalent phase plots of the frequency response of the CE amplifier.

The phase over frequency will wrap around once the phase gets to −180°. In AC simulation to see a continuous phase, we use the unwrap() function.

Notice (in Figure 3.69 (down)) that in the mid-band frequencies, the output phase is more or less 180°.

We repeat the plotting process, this time for R_{in} vs. freq, where the frequency axis is kept on a log scale (Figure 3.70). The "mid-band" range for R_{in} is between 100 Hz and 100 kHz, more or less. At that range R_{in} is around 3.6 kΩ.

For voltage amplifiers, the larger R_{in}, the better.

Let us demonstrate two more frequency domain CE amplifier analysis activities.

In order to tweak the voltage gain, or simply to change it to another value, the main tool is the RC resistor. See Figure 3.67. We shall explore how it is done.

We shall also demonstrate what happens to the voltage gain and to R_{in} if we remove the bypass capacitor CE.

To study the voltage gain tweaking, let us refer again to Figures 3.67 and 3.68. In the CE amplifier circuit diagram, in Figure 3.67, we see that $R_C = 3$ kΩ and $R_L = 20$ kΩ. In the magnitude frequency response shown in Figure 3.68, we see that the magnitude of the voltage gain is 90.226. Say that we wish to modify the voltage gain from −90.226 to −95. How?

Recall the CE amplifier's A_v formula (3.2): $A_v = -g_m[R_{in}/(R_{in} + R_{source})](R_C \| R_L \| r_o)$.

The tweaking idea is that, if we don't change R_E, R_{B1}, and R_{B2}, then the Q point stays unchanged. Consequently, I_C stays the same, which means that the BJT small-signal parameters stay unchanged. Therefore, in the A_v formula, the only changeable parameter is R_C.

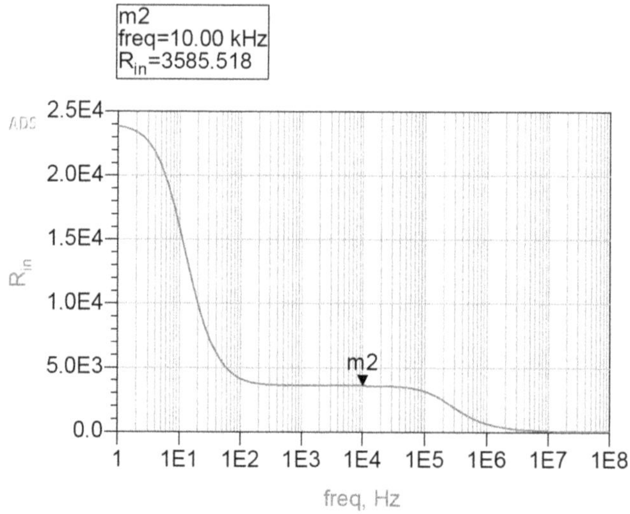

FIGURE 3.70 Frequency response plot of the CE amplifier input resistance.

Recall that R_C does not influence the Q point, as long as the BJT is in active mode. Let us increase $R_C||R_L||r_o$ by a factor of 95/90.226, and then find R_C. From Figure 3.58 (before we tweaked I_C), we see that $I_C = 0.96\,mA$. We assume that the early voltage is $V_A = 100\,V$, yielding an approximate value $r_o = 104\ k\Omega$. The tweaking calculation goes as follows:

$$Old(R_C||R_L||r_o) = 3||20||104 = (3^{-1} + 20^{-1} + 104^{-1})^{-1} = 2.55\ k\Omega.$$
$$New(R_C||R_L||r_o) = 2.55 \cdot 95/90.226 = 2.68\ k\Omega.$$

Now let us solve the equation: $RC||20||104 = 2.68$.

Manipulation of inverses is the fastest way to handle such equations:

$$R_C = (2.68^{-1} - 20^{-1} - 104^{-1})^{-1} = 3.2\ k\Omega.$$

Let us apply this new value of R_C and watch the new gain (Figure 3.71).

The new Av came to be −95.213. Why not an exact result of −95.0?

Several potential reasons: In the calculation of the new R_C, we went to the second decimal. In the implementation of the new R_C, we went to the first decimal. Also, for calculation, of r_o we assumed $V_A = 100\,V$, which is inaccurate. For Q2N3904 BJT, the early voltage is $V_A = 74\,V$.

Let us now demonstrate the role played by the bypass capacitor. See Figure 3.72 for the CE amplifier's schematic, with the bypass capacitor disabled.

The role of CE is to short circuit R_E in the amplifier's small-signal circuit. If R_E is not short-circuited, the effect on the voltage gain is catastrophic. The gain becomes approximately, $A_v = -R_{in}/(R_{in} + R_{source}) \cdot (R_C||R_L||r_o)/R_E$.

Finally, let us conduct a quantitative assessment of the small-signal assumption, using a new simulation engine called harmonic balance.

What are harmonics?

Fourier series theory teaches us that every periodic signal, of frequency f, is equal to an infinite sum of DC and pure AC signals, of frequencies f, 2f, 3f, … (see [Gable1]).

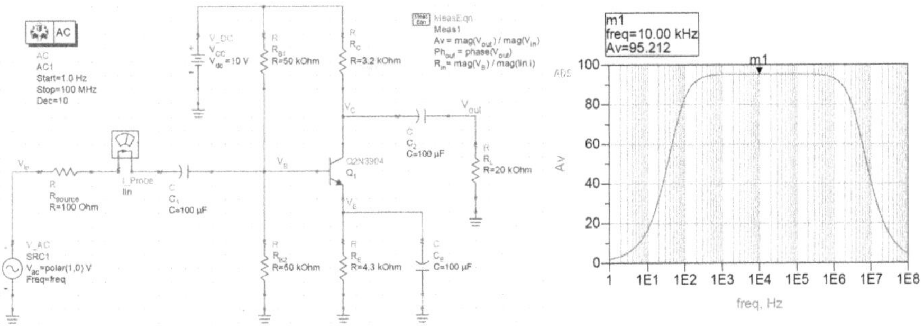

FIGURE 3.71 Voltage gain tweaked RC resistor leading closer to a desired gain.

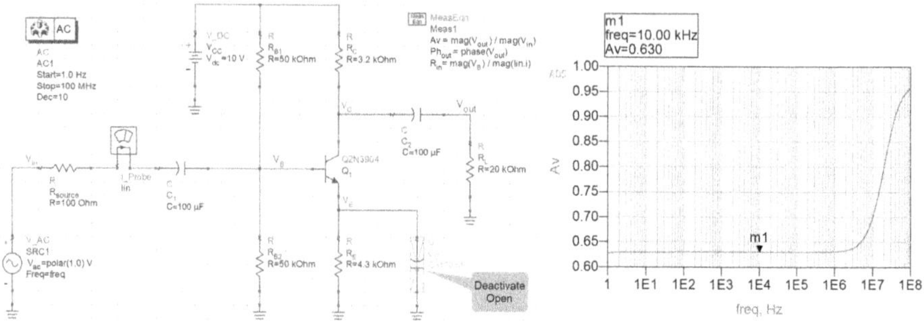

FIGURE 3.72 Whenever "forgetting" to include a bypass capacitor, resistor RE participates in determining the CE amplifier's voltage gain, making it very small.

The AC component of frequency f is called the "main harmonic", or "the first harmonic". The AC component of frequency 2f is called "the second harmonic", and so on.

There are infinitely many harmonics, but as their number rises, their magnitude becomes smaller and smaller. In ADS, the Fourier series analysis is done with a **harmonic balance** simulator.

What can be said about the distorted output voltage as the input amplitude increases?

We saw (in the transient simulation study; Figures 3.62–3.64) that as the amplitude of the AC V_{in} increases, V_{out} starts to look less and less like AC. Note that V_{out} still has the same frequency f as that of V_{in}. The distorted V_{out} has harmonics.

An "acceptable AC V_{out}" is a signal in which the first harmonic is much larger than any one of the other harmonics. **An "unacceptable AC V_{out}" may be arbitrarily defined as a signal in which at least one of the side harmonics becomes larger than 10% of Harmonic 1.**

In the harmonic balance simulation, we need to specify the frequency of harmonic 1. We need to specify the number of harmonics that we consider important (say N=5). That is, we allow ADS to truncate all the higher harmonics. The program computes all the desired harmonics, so we can quantitatively compare with each other. Because of the truncation, the computation is not accurate. We can increase N, from one simualtion to the next. As long as there are significant changes in the results, we may need to keep increasing N, until the results become insensitive to N.

In Figure 3.73, we show how the CE amplifier's schematic is prepared for harmonic balance simulation.

We continue to use the VtSine source, which is generally considered a time domain source. We keep it since we are familiar with it and since it works well with harmonic balance. There is also a voltage source called V_1Tone that can also be used and is set up slightly differently. See Figure 3.74.

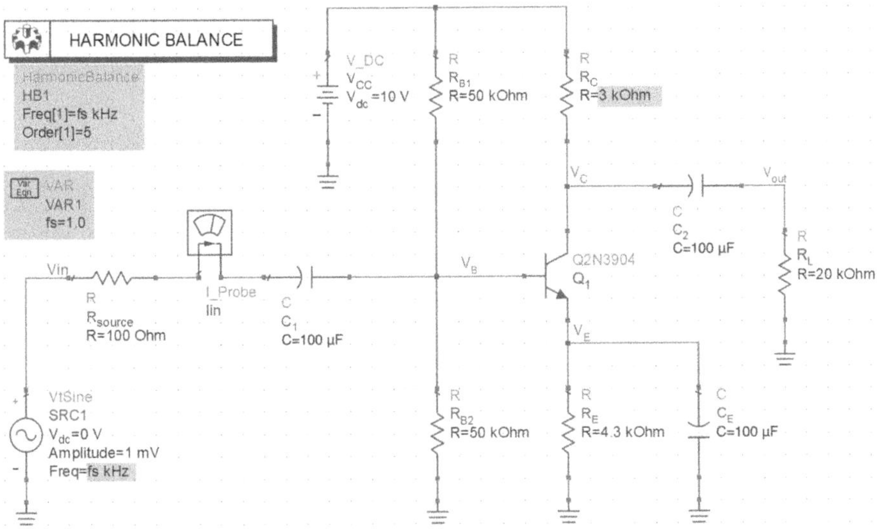

FIGURE 3.73 CE amplifier schematic harmonic balance settings.

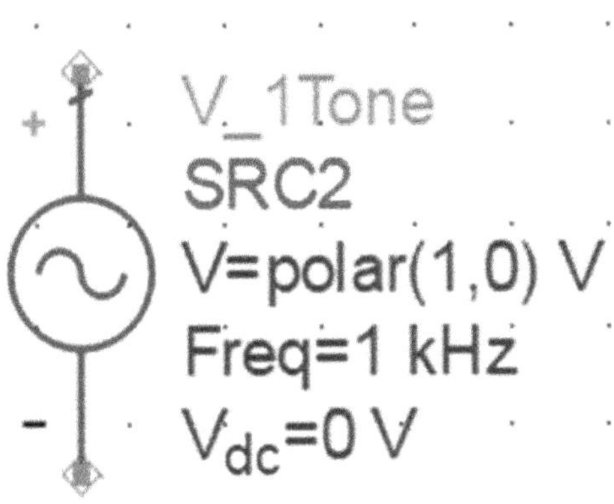

FIGURE 3.74 V_1Tone voltage source.

See Figure 3.75 for the dialog box for setting the harmonic balance simulation choices:

Harmonic 1 is 1 kHz. We shall look at the first five harmonics. Like in AC simulation, all the computed variables are complex numbers. See Figure 3.76 for an "ADS inquiry" about the user's preferred data display and the resulting plot.

Alternatively, since HB produces discrete data points at the harmonics, it may be convenient to look at the results in table form. Since we already have a plot, it is simple to convert the plot to a table, as shown in Figure 3.77.

The simulation of Figure 3.77 ran with an input amplitude of 1 mV.

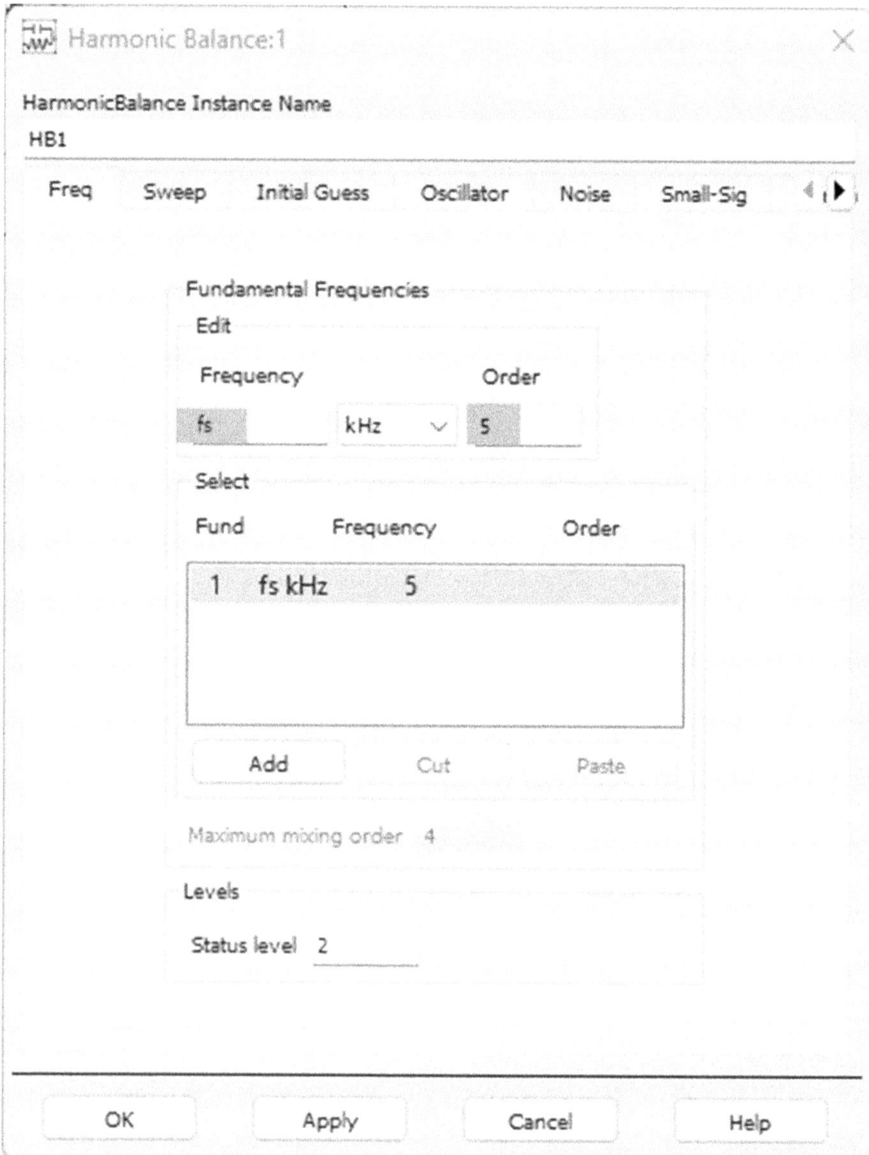

FIGURE 3.75 Harmonic balance simulation settings dialog box.

We see that harmonic 1 of V_{out} is 90 mV. Harmonic 2 of V_{out} is 753.9 μV, which is around 0.84% of harmonic 1. Harmonics 3 and 4 are much smaller.

Obviously, for an input amplitude of 1 mV, the signal V_{out} is very acceptable. It almost has no distortions at all.

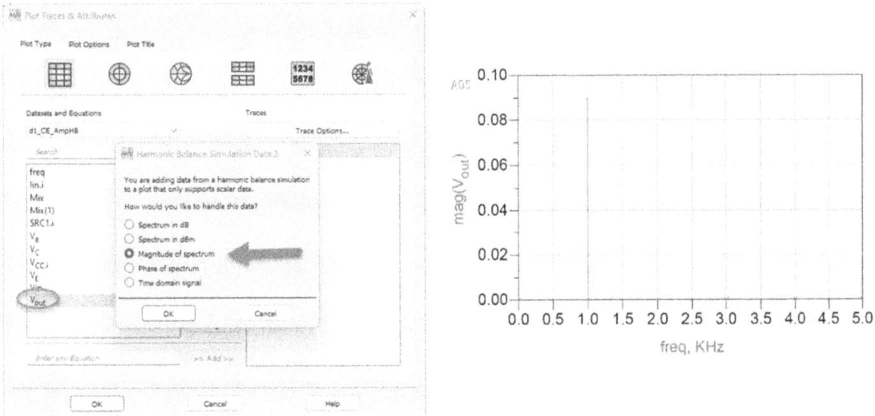

FIGURE 3.76 Data display selection and the mag(V_{out}) harmonics.

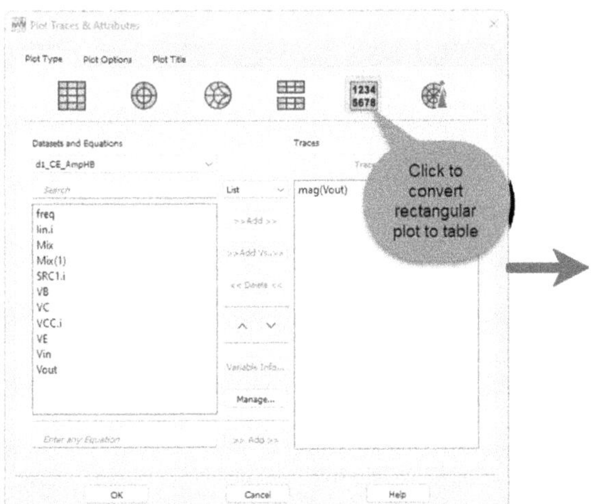

freq	mag(V_{out})
0.0000 Hz	0.0000
1000.0 Hz	90.07 m
2.000 kHz	753.9 u
3.000 kHz	3.559 u
4.000 kHz	7.460 n
5.000 kHz	36.81 p

FIGURE 3.77 Table format for the harmonics data.

Let us repeat the simulation, with input amplitudes of 10 and 15 mV (Figure 3.78).

Harmonic balance simulation can produce the time signals from the harmonics data (Figure 3.79).

We see that, compared to transient simulation, it is not a bad reconstruction of V_{out}. Alternatively, we may create a measurement equation using the ts() function of V_{out}.

Note: Harmonic balance is a steady state simulator, so no initial transient response is seen.

freq	$mag(V_{out})$
0.0000 Hz	0.0000
1000.0 Hz	886.8 m
2.000 kHz	73.94 m
3.000 kHz	3.499 m
4.000 kHz	74.70 u
5.000 kHz	3.443 u

Simulation with V_{in} amplitude of 10mV:
Harmonic 1 of V_{out} = 886.8 mV
Harmonic 2 of V_{out} = 73.94 mV (8.4% of harmonic 1)
Harmonic 3 of V_{out} = 3.5 mV (0.4% of harmonic 1).
So V_{out} is still acceptable.

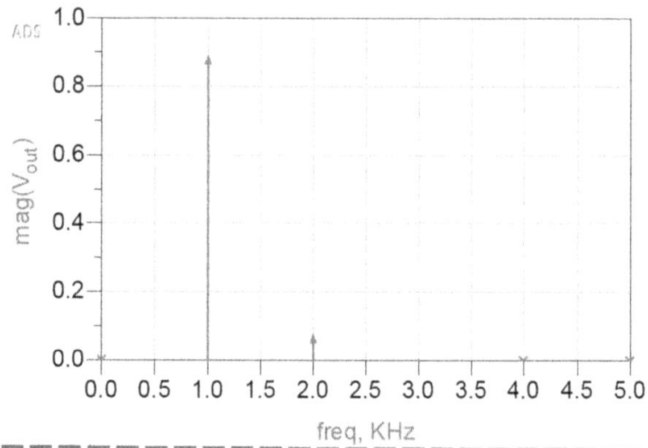

freq	$mag(V_{out})$
0.0000 Hz	0.0000
1000.0 Hz	1.305
2.000 kHz	162.4 m
3.000 kHz	11.56 m
4.000 kHz	378.4 u
5.000 kHz	24.04 u

V_{in} amplitude is 15mV:
Harmonic 1 of V_{out} = 1.305V
Harmonic 2 of V_{out} = 0.162V (12.4% of harmonic 1!)
Harmonic 3 of V_{out} = 0.012V (0.9% of harmonic 1)
Now, V_{out} is unacceptably distorted.

So, the "small-signal assumption" is valid for a V_{in} amplitude that is no larger than (possibly) 12-13mV.

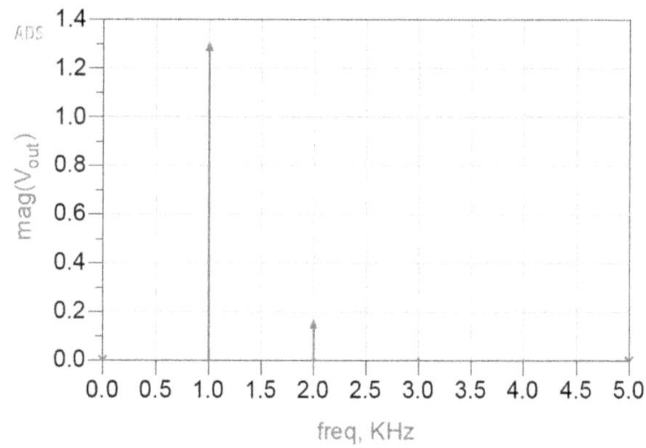

FIGURE 3.78 HB Simulation results (and analysis comments) for input amplitudes that are around the borderline of the small-signal assumption.

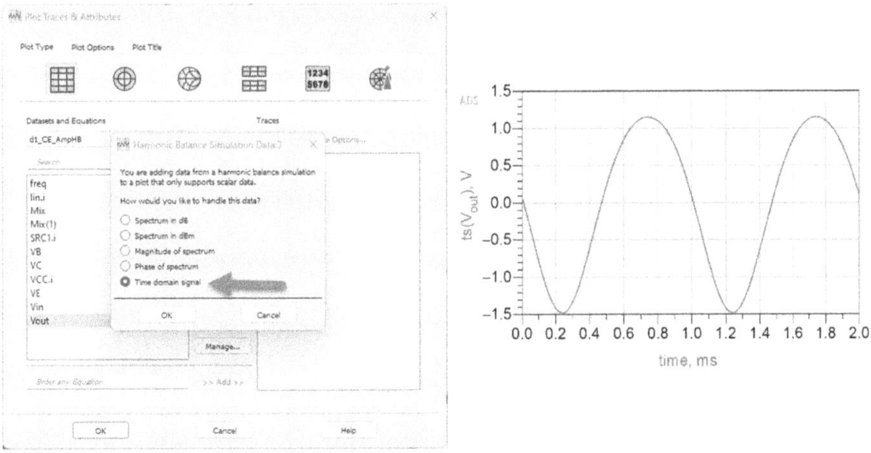

FIGURE 3.79 Generation of time domain signals in HB simulation.

REFERENCES

[Gable1] Robert A. Gable and Richard A. Roberts, *Signals and Linear Systems*, 3rd Edition, Wiley, 1986.

[Gray1] Paul R. Gray, Paul J. Hurst, Stephen H. Lewis and Robert G. Meyer, *Analysis and Design of Analog Integrated Circuits*, 4th Edition, John Wiley & Sons, 2001.

[Keysight1] *Keysight Advanced System Design (ADS) Release 2025*, Keysight Technologies, Inc. https://www.keysight.com/us/en/products/software/pathwave-design-software/pathwave-advanced-design-system.html

[Sedra1] Adel S. Sedra and Kenneth C. Smith, *Microelectronic Circuits*, 7th Edition, Oxford University Press, 2015.

[Sze1] S.M. Sze, "Semiconductor Devices: Physics and Technology", 2nd Edition, John Wiley & Sons, 2002.

CHAPTER 3 PROBLEMS – BJT

Problem 3.1: BJT DC Circuits

Analyze the four circuits below, first analytically and then verify using ADS DC simulation (with simple DC annotations of voltages and currents). You have to solve each circuit – that is, find the status of the BJT (active, saturation, or cutoff mode) and then find all the node voltages and all the currents, so that you can justify the BJT state assumption.

Whenever you solve *manually* always make the following simplifying assumptions:

- For NPN BJT $V_{BE}=0.7\,V$ (if the BE junction is forward-biased),
- For PNP BJT $V_{EB}=0.7\,V$ (if the EB junction is forward-biased),
- For NPN and PNP assume that $\beta=100$ (only if the BJT is in active mode)
- For NPN BJT (if it is in saturation mode) $V_{CEsat}=0.1\,V$
- For PNP BJT (if it is in saturation mode) $V_{ECsat}=0.1\,V$
- For NPN and PNP BJT, the borderline between active and saturation modes is $V_{CB}\approx 0$
- For NPN and PNP BJT in active mode $\alpha\approx 1$

In ADS simulations, use Q2N3904 for NPN BJT and Q2N3906 for PNP BJT from the ADS design kit. The above assumptions are gross approximations:

- $|V_{BE}|$ equals 0.7 V at one specific value of I_E. In general, these voltages vary slightly around 0.7 V and can typically be anywhere between 0.65 and 0.75 V.
- The actual value of β for each BJT is often much larger than 100. The parameter β actually varies not-so-slightly (as the reverse bias voltage $|V_{CB}|$ varies).
- The actual value of the saturation mode V_{CE} for the above two transistors is actually close to 0.05 V and is not exactly constant.

Compare the hand-calculation results to the results obtained by simulation. You may sometimes be amazed at how close the computation results come to the true results, despite the gross approximations.

1a: The circuit of Figure 3.80 is relatively easy, as one of the BJT currents is given, (i) Assume active mode and follow up with all the calculations proving that indeed the assumption is justified, (ii) Change R_C from 4 to 20 kΩ – does the BJT mode change from active to saturation? Show the calculations. (Actually, part (ii) is not so easy for manual solution and requires a solution technique like the one practiced in Problem 3.1c.)

1b: The circuit of Figure 3.81 is relatively easy as well, as it happens whenever V_B or V_E are given (here, it is $V_E=V_{EE}$), (i) Again assume active mode, find I_B and then show that the assumption makes no sense, (ii) Next, assume saturation mode and complete the solution, (iii) If we want the BJT to be in active mode and have a current of $I_C=1\,mA$ is it enough to just find another value for R_B?

1c: The circuit of Figure 3.82 is considerably more difficult, because none of the BJT voltages or currents are given, (i) Assume active mode and consider V_E as the unknown that must be found first: Express I_B and I_E in terms of the unknown voltage V_E and set an equation with one unknown – find V_E and then show that the solution leads to a contradiction, (ii) Assume saturation mode and again set V_E as the only unknown – express each of the BJT currents in terms of V_E and solve, (iii) Suggest changes in R_B and/or R_E in order to bring the BJT to active mode with a current of $I_C=2\,mA$.

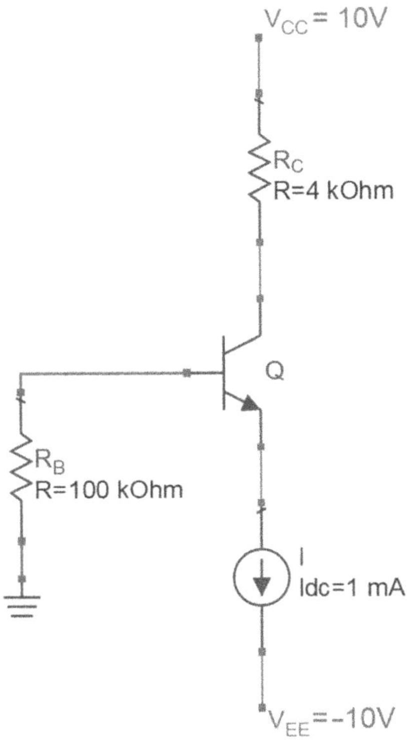

FIGURE 3.80 BJT DC circuit.

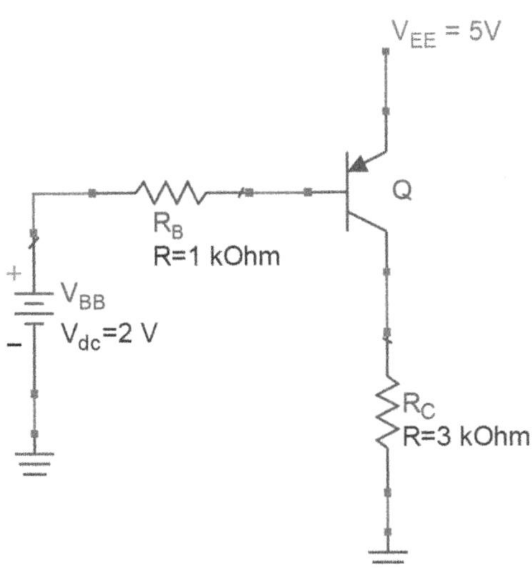

FIGURE 3.81 BJT DC circuit.

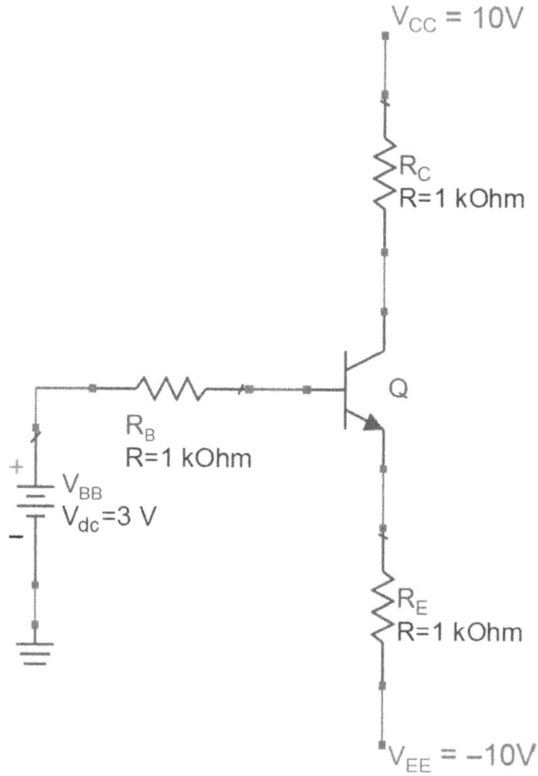

FIGURE 3.82 BJT DC circuit.

1d: The circuit of Figure 3.83 is not easy for active mode calculations, but it is easy for saturation and cutoff mode calculations. Do: (i) run the three sets of calculations (active, saturation, and cutoff modes) and show that the BJT is in cutoff mode, (ii) What changes are needed to be made in the circuit in order to bring the BJT to operate in active mode? Is it enough just to lower V_{BB}, or is it also necessary to increase R_B? What happens if V_{BB} becomes negative?

Problem 3.2: BJT DC Circuits

Consider the following BJT DC circuit. For the NPN BJT, let $V_{BE}=0.7\,V$ (if BE junction is ON), $\beta=100$, and $V_{CEsat}=0.1\,V$. Likewise (same numbers) for the PNP BJT (Figure 3.84).

2a: Explain why Q_1 is in active mode. What is the value of I_C?

2b: Engineer A claims that "by decreasing R_E by 50% (in Problem 3.2a) it is possible to increase I_C by a certain factor". Engineer B argues that "it would not be enough to change only R_E. R_C needs to be changed as well". Who is right? Explain by showing the relevant computations. If indeed I_C needs to be increased, what increase factor should it be?

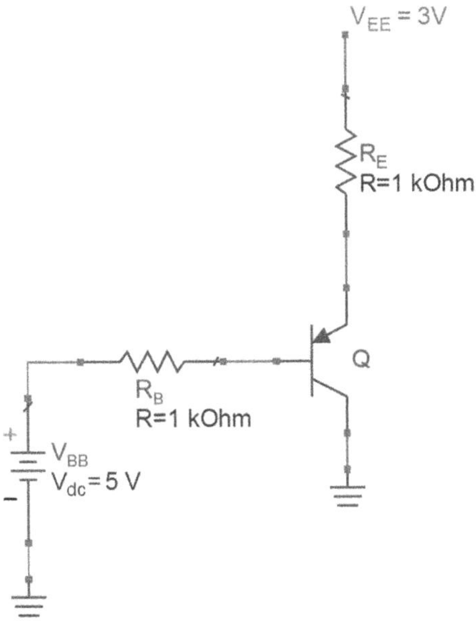

FIGURE 3.83 BJT DC circuit.

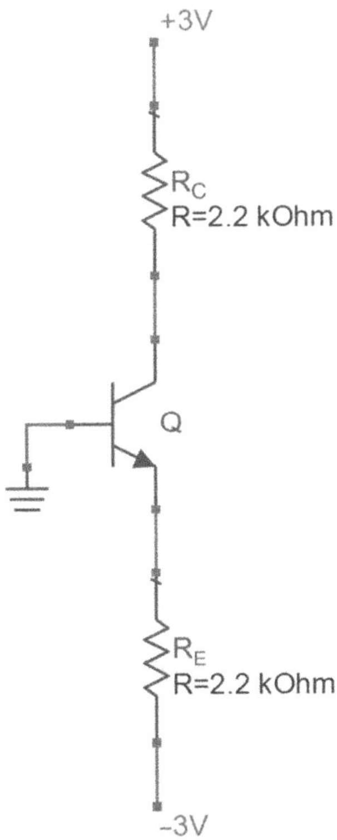

FIGURE 3.84 BJT DC circuit.

2c: In the circuit of Figure 3.85, choose a value for R_E such that the PNP BJT is in active mode and $I_C = 1.1\,mA$.

2d: Choose value for R_E (in Problem 3.2c) such that the BJT is in saturation and $I_B = 0.8\,mA$. (*Hint*: Let V_E be an unknown – set up a KCL equation).

Problem 3.3: BJT DC Circuits

Let NPN BJT be Q2N3904. For your convenience, continue to use {$V_{BE} = 0.7$ when BE junction is ON, $\beta = 100$ and $V_{CEsat} = 0.1$ V} even though these are not the datasheet parameters of the 2N3904 transistor. Let $R_{B1} = R_{B2}$ and each has a resistance value in the range of tens of kΩ. Let R_C and R_E have resistance values in the range of a few kΩ.

3a: Choose all the resistors in the circuit of Figure 3.86 so that $I_C = 1\,mA$ and $V_{CE} = 3\,V$. For your convenience, assume that R_{B1} and R_{B2} create an "almost-voltage-divider". Then you can easily determine V_B and V_E and choose R_E to set the desired I_C spec, and then set R_C to meet the V_{CE} spec. Of course, don't forget to justify the active mode assumption and the "almost-voltage-divider" assumption. Once the circuit is solved (i.e., all the node voltages and all components' currents are found), compare the voltages and currents to the ADS results. Can you tell what the true

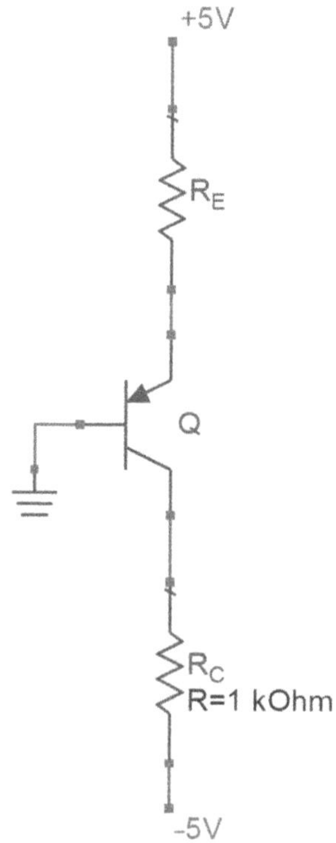

FIGURE 3.85 BJT DC circuit.

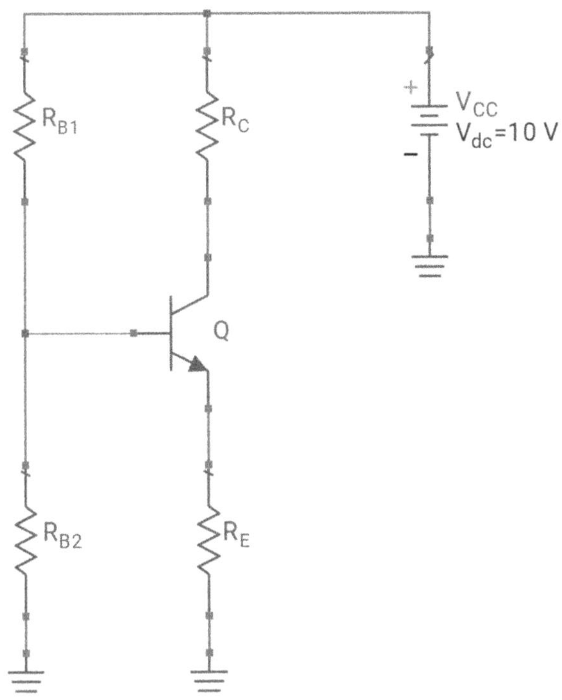

FIGURE 3.86 BJT DC circuit.

parameters of the BJT are? Does it really matter for the design to use more accurate datasheet parameters?

3b: Now, change one of the R_B resistors to drive the BJT into saturation mode. Once you violate the active mode assumption, while still keeping the BE junction forward-biased, solve the circuit that now operates in saturation mode.

3c: In your design of (Problem 3.3a), offer a change of one of the R_B resistors in order to drive the BJT into cutoff mode. Solve the circuit in cutoff mode.

Problem 3.4: BJT Current Drivers and Push–Pull Amplifiers

Same BJT parameters as in Problem 3.2.

4a: A GaAs LED shown in Figure 3.87 has a forward voltage of 1.4 V. The two regular diodes in series (shown below) emulate a single LED. You may also use the LED contained in the design kit. We want the BJT to be in active mode and the LED average current to be $I_C = 26\,\text{mA}$.

Accomplish the desired I_C by two methods:

Method A: Keep R_E constant (at the value of 200 Ω) and find a suitable value of VDC@VS (the DC level of the signal source). What should VDC@VS be? Justify that the BJT is in active mode.

Method B: Keep VDC@VS constant (at the value of 5 V) and find a suitable value of R_E. What should R_E be? Justify that the BJT is in active mode.

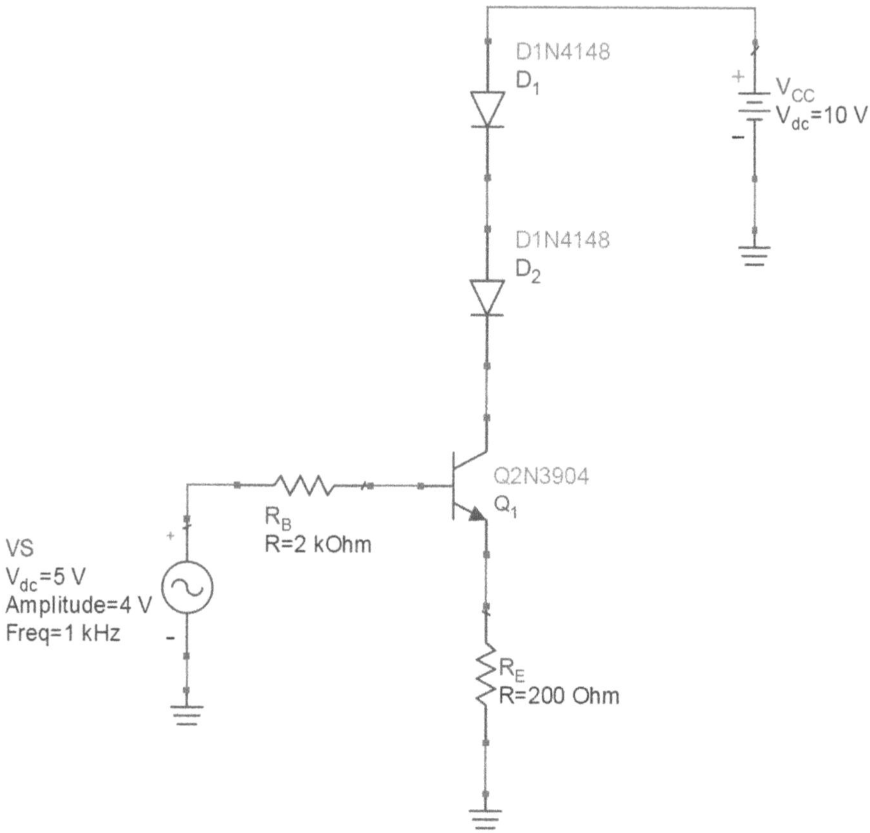

FIGURE 3.87 BJT current driver.

4b: If for the push–pull amplifier in Figure 3.88 $v_I = 3\,V$, what is V_E, and what is the mode of each BJT? *Hint*: The conducting BJT should be treated like a current driver.

4c: In the circuit of Figure 3.88, choose a value for a DC voltage source v_I such that BJT Q_2 is in saturation and $I_B = 1\,mA$. What is the I_C of each BJT?

4d: Replace the DC voltage source v_I in Figure 3.88 by a 150 Hz AC voltage source $v_i(t)$. What should be the amplitude of $v_i(t)$ if both BJTs dissipate maximum heat? What is the largest amplitude of $v_i(t)$ for which the average load power is zero?

Problem 3.5: BJT Small-Signal CE Voltage Amplifiers

In Section 3.4, we covered in detail the analysis of an NPN BJT CE amplifier that utilizes a single-supply DC biasing. In this problem, you will meet two additional CE amplifier architectures – one that is based on dual DC supplies, combined with a *DC current source*, and the other that utilizes a *PNP* BJT with single-supply biasing.

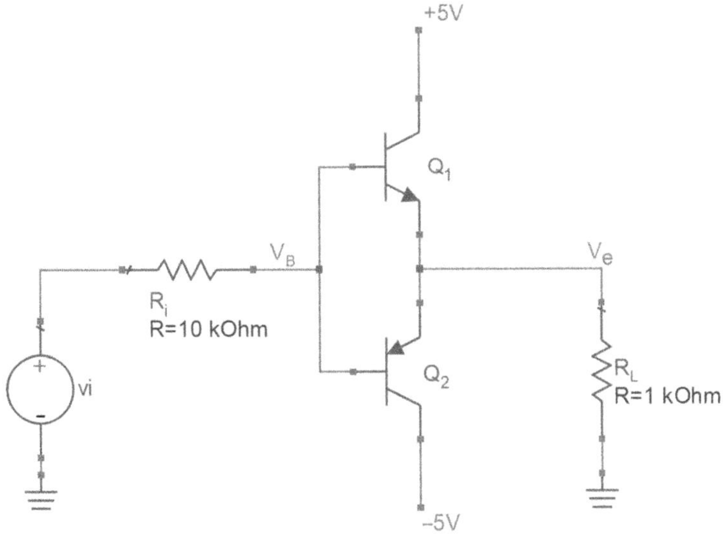

FIGURE 3.88 Push–pull amplifier.

In both problems (below) you need to manually find the amplifier's Q point (i.e., the DC values of I_C and V_{CE}), the BJT small-signal parameters (r_e, r_π, g_m, and r_o) the drawing of the CE amplifier's small-signal diagram, the amplifier's input resistance R_{in}, and the amplifier's small-signal voltage gain $A_V = v_o/v_s$.

In the ADS simulation, verify the DC conditions and measure R_{in} and A_V (using both simulation modes – transient and AC simulations).

The supply voltages all involve 10 V voltages (that is $V_{CC} = V_{EE} = 10\,V$). In both problems, the signal source internal resistance is $R_{sig} = 100\ \Omega$, and the load resistance is $R_L = 15\ k\Omega$. The large capacitors are assumed infinite (for hand-calculations) and in ADS take each to be equal to $C = 100\ \mu F$.

In both problems, let $R_C = 3.5\ k\Omega$.

5a: In the circuit of Figure 3.89, let $I = 1\,mA$ (and later in Section 4.4, we shall discuss how such DC current sources are implemented using transistor circuits, in an architecture known as "current mirrors") and let $R_B = 100\ k\Omega$.

5b: In the circuit of Figure 3.90 let $R_{B1} = R_{B2} = 30\ k\Omega$ and let $R_E = 4.3\ k\Omega$.

Problem 3.6: BJT CE Amplifiers

Consider the NPN CE amplifier of Figure 3.91. The BJT parameters are {$V_{BE} = 0.7\,V$ if junction is ON, $\beta = 120$ and $V_A = 85$ V}. Assume that I flow into a negative voltage supply $-V_{EE} = -10\,V$. Let the thermal voltage be $V_T = 25\,mV$. Current source I flows into some suitable and irrelevant negative DC supply.

6a: For $V_{CC} = 10\,V$, $R_{sig} = 0$, $I = 0.3\,mA$, $R_C = 9\ k\Omega$ and $R_L = 22\ k\Omega$ find the amplifier's input resistance and voltage gain. No need to draw the small-signal diagram.

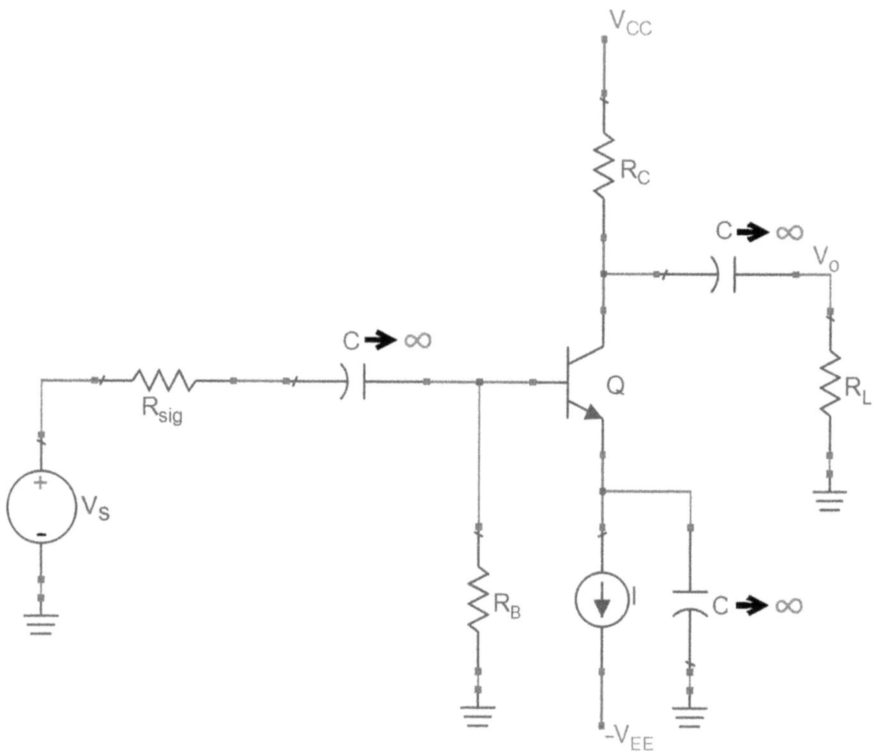

FIGURE 3.89 NPN CE amplifier with current source bias.

6b: In the amplifier in Figure 3.91, show how to tweak the absolute voltage gain down by 5.

Problem 3.7: BJT CE Amplifiers

Let the PNP BJT Q in Figure 3.92 be 2N3906. For your convenience, use {V_{EB} = 0.7 V if the EB junction is ON, β = 100, and early voltage V_A = 100 V} even though these are not the true parameters of the 2N3906 transistor.

The internal resistance of the input signal source $v_{sig}(t)$ is R_{sig} = 200 Ω. The load resistance is R_L = 10 kΩ. You cannot change the source resistance or the load. Let R_{B1} = R_{B2} = 40 kΩ. Let R_E = 3.9 kΩ and let R_C = 3 kΩ.

7a: Find the DC voltages and currents of the amplifier, the BJT's small-signal parameters, and draw the small-signal diagram of the amplifier. Is the active mode assumption true? Do your assumptions about V_{EBP} and β introduce large errors in the DC calculations?

7b: Find the amplifier's small-signal input resistance R_{in} and the small-signal voltage gain. When comparing to ADS, use AC simulation. Explain the differences between your manual results and the ADS results.

7c: Use ADS to find the maximum allowed amplitude of $v_{sig}(t)$, assuming that no side harmonic exceeds 10% of the main harmonic. Assume that $v_{sig}(t)$ is AC with a

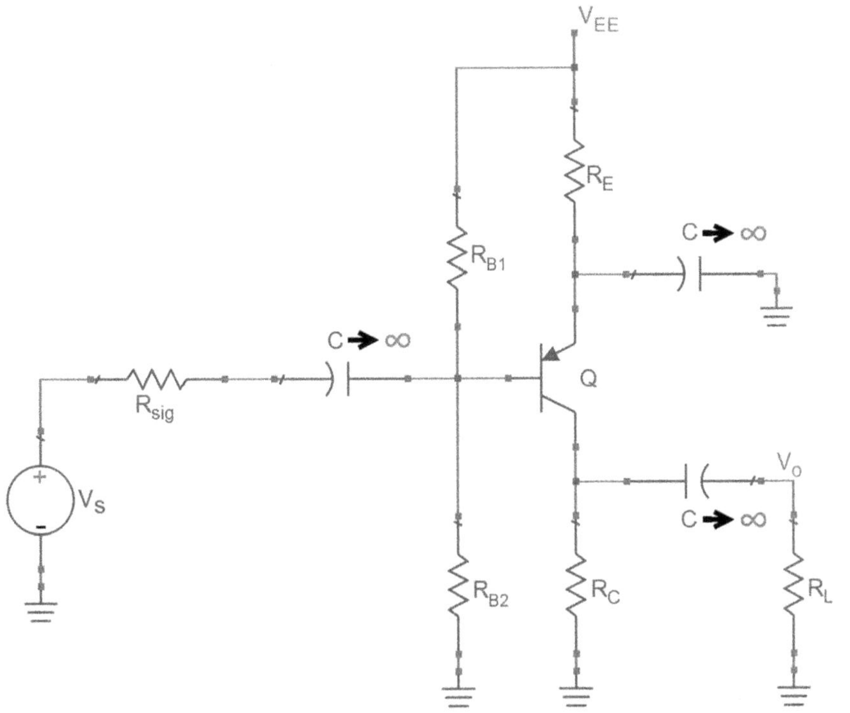

FIGURE 3.90 PNP CE amplifier with single-supply biasing.

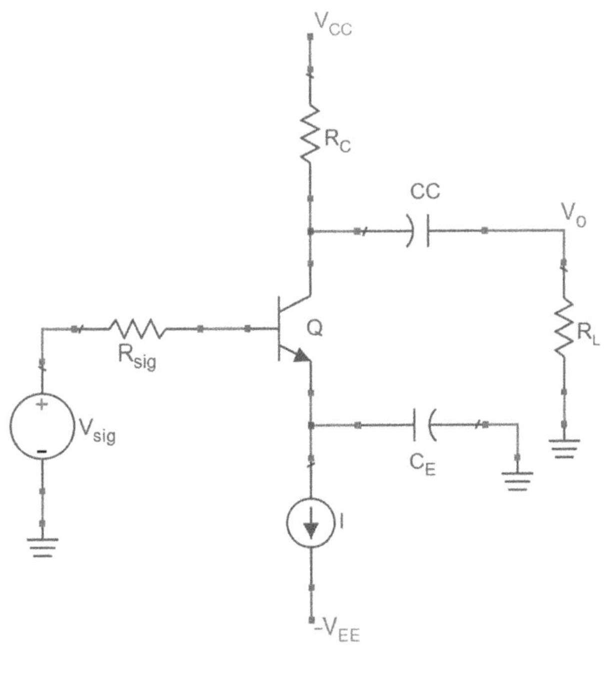

FIGURE 3.91 NPN CE amplifier with DC current source biasing.

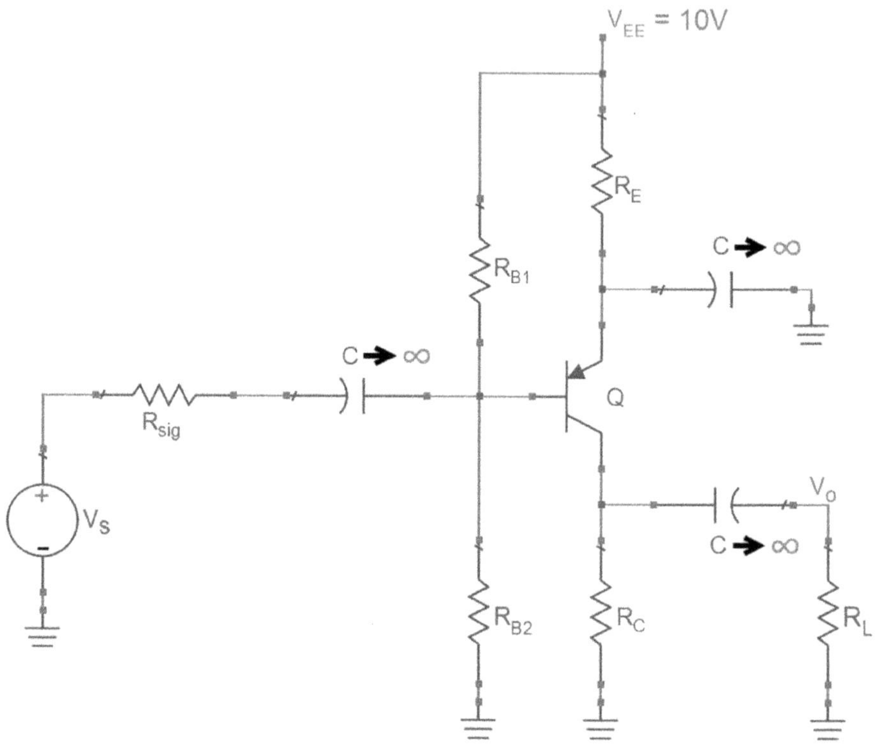

FIGURE 3.92 PNP CE amplifier.

frequency of 1.5 kHz. Use harmonic balance simulation. For the maximum allowed $v_{sig}(t)$ amplitude plot, the distorted output voltage $v_o(t)$.

7d: Modify one of the resistors in the circuit of Figure 3.92 in order to increase the voltage gain by 20%. *Hint*: It would be the simplest to tweak the only resistor that has nothing to do with setting the DC current I_C.

4 Basic MOSFET Circuits

4.1 INTRODUCTION TO MOSFET TRANSISTORS: WHAT IS A MOSFET?

Refer to [Gray1], [Razavi1], and [Sedra1]. MOSFET stands for Metal-Oxide-Semiconductor Field-Effect Transistor. Unlike the three-terminal BJT, MOSFET has four terminals: source (S), drain (D), gate (G), and substrate (B), with some similarities and differences of duties to those of the BJT terminals. A source functions similarly to an emitter, and a drain functions similarly to a collector. Both types of transistors are primarily aimed at acting as **controlled current sources** (a current that consists of carriers that flow from E to C in a BJT or from S to D in a MOSFET). Whereas the base of the **BJT** provides **current control** of the E–C flow of carriers, the gate of a **MOSFET** provides **voltage control** of the S–D flow of carriers.

There is a substrate that needs to reassure that **all p–n junctions**, residing within the MOSFET, **stay reverse-biased**. This will be revisited in a few more paragraphs. In many MOSFET circuits (but not all!), the source and the substrate are shorted, and thus such a transistor becomes a three-terminal device.

Every MOSFET transistor typically operates between a single positive supply (denoted as V_{DD}) and ground. The value of V_{DD} is dictated by the CMOS technology that is pursued.

As there are two types of BJT transistors, NPN and PNP, there are two types of MOSFET transistors – **NMOS** and **PMOS**. See Figure 4.1 for electrical symbols used for the two MOSFET types, operating as four-terminal or three-terminal devices.

In a four-terminal MOSFET, it is customary to mark the source terminal and indicate an arrow to denote the substrate. In a three-terminal device, the arrow is placed at the source (now short-circuited to the substrate), and it shows the current direction.

In an NMOS, the current consists of free electrons moving from the source to the drain if a path (called a **channel**) is created. Therefore, the current, denoted as I_D, flows from drain to source. In PMOS, the current consists of free holes flowing

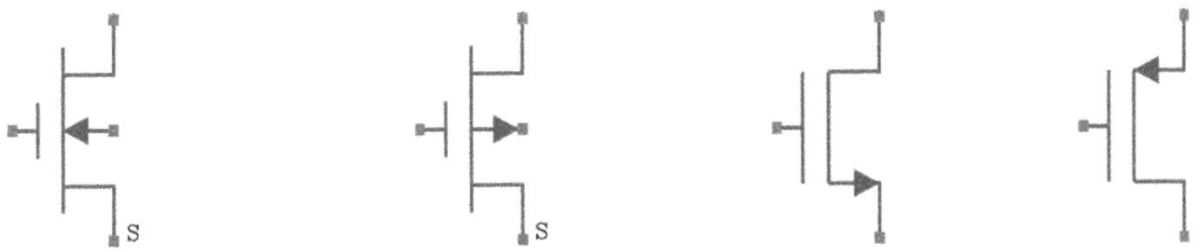

FIGURE 4.1 (from left to right) Four-terminal NMOS, four-terminal PMOS, three-terminal NMOS (in which the current leaves the source), and three-terminal PMOS (with current entering the source).

DOI: 10.1201/9781003606772-4

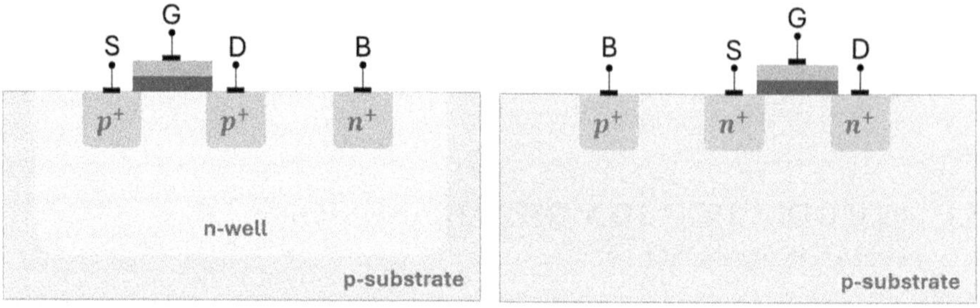

FIGURE 4.2 Side view of PMOS (left) and NMOS (right) fabrication.

from the source to the drain, and so flows I_D. The gate oxide (see Figure 4.2) is an insulator. Therefore $\mathbf{I_G = 0}$. That is, there is no DC current flowing in or out of the MOSFET gate.

"Discrete" BJT transistors populate printed circuit boards (and students' prototype circuits built on breadboards).

MOSFET microelectronic circuits, on the other hand, are intended to be parts of "systems on a chip". Figure 4.2 illustrates side views of NMOS and PMOS transistors residing on the same wafer (along with hundreds, or maybe thousands, of other similar transistors).

The + superscripts, in Figure 4.2, denote regions of high doping levels. We observe that the p-type substrate and its terminal (B) serve all NMOS transistors on the chip. PMOS transistors are built inside n-wells, which allows every PMOS transistor to have its own B terminal. We further see the gate contact of each transistor and the insulating SiO_2 gate oxide layer residing between each gate and the substrate, above the "future channel" that will bridge the drain and source regions, whenever the MOSFET turns ON. There are multiple p–n junctions (shown in Figure 4.2), such as the one created by the NMOS n-type drain and the p-type substrate, or the PMOS p-type source and the PMOS n-well n-type substrate.

All p–n junctions must be reverse-biased. This is done in NMOS transistors by connecting the p-substrate to the lowest available voltage, typically 0 V. In PMOS transistors, the n-substrate is connected to the most positive voltage, typically V_{DD}.

Unlike a BJT transistor, in which a nonzero current must be created by one (or possibly two) p-n junctions turning forward-biased, in a MOSFET, carriers can only flow (from S to D) along a thin channel created underneath the gate region: In an NMOS transistor, a sufficiently positive gate-to-source voltage $V_{GS} = V_G - V_S$ causes inversion of the region underneath the gate oxide, turning it from p-type to n-type. The channel connects electrically the source to the drain. In PMOS, a sufficiently negative gate-to-source voltage causes an inversion of the region underneath the gate oxide from n-type to p-type.

The reason for the semiconductor type inversion of the substrate region at and near the surface bordering the gate oxide is because of a capacitor formed by the gate metal (or polysilicon) and substrate "plate", and the gate oxide insulator sandwiched in between the plates. More accurately, and it doesn't change the explanation, there are two capacitors in series (the oxide capacitor and a substrate depletion region capacitor) that create, by voltage division, the voltage at the oxide–substrate interface. In an NMOS, an applied positive gate voltage creates a positive charge on the gate-side "plate", resulting in a negative charge created on the substrate-side "plate".

In an NMOS, the negatively charged near-oxide substrate happens due to (i) free holes near the oxide surface, repelled inwards into the substrate, which exposes the negatively charged static acceptors, and (ii) free electrons that are attracted from the source and drain regions into the substrate region that is near the surface underneath the gate oxide. Likewise, for PMOS. Let us not forget that in a PMOS, V_S is larger than V_G. The negative V_{GS} creates a positive charge on the substrate side, which eventually creates the p-type channel. The inversion process of the substrate region that borders the oxide is gradual. A threshold voltage $V_{GS}=V_{TH}$ is the voltage V_{GS} for which the concentration of free holes and free electrons in the interface region becomes equal. NMOS transistors are fabricated to have a positive V_{TH}, and PMOS transistors are fabricated to have a negative V_{TH}. In an NMOS, the larger the overdrive voltage $V_{GS}-V_{TH}$, the larger is the free electron concentration of the n-channel connecting the source to the drain. Similarly, in a PMOS, the more negative the overdrive voltage $V_{GS}-V_{TH}$, the larger the free hole concentration of the p-channel.

In BJT, the emitter and collector currents consist primarily of diffused carriers. In MOSFET, on the other hand, the current consists primarily of drifted carriers. That is, the current flowing through the thin channel is affected in part by the lateral electrical field created by the applied voltage $V_{DS}=V_D-V_S$, which causes carriers to move from S to D, and in part by the vertical electrical field created by the overdrive voltage $V_{GS}-V_{TH,}$ which determines the channel's electrical properties.

Figure 4.3 illustrates a partial view of a MOSFET with a rectangular geometry. Not shown are the contacts attached to the metal layers and the contact for the substrate terminal. The gate oxide layer and the metal above it form a rectangle of length L and width W. L and W are the layout-drawn dimensions of the MOSFET gate.

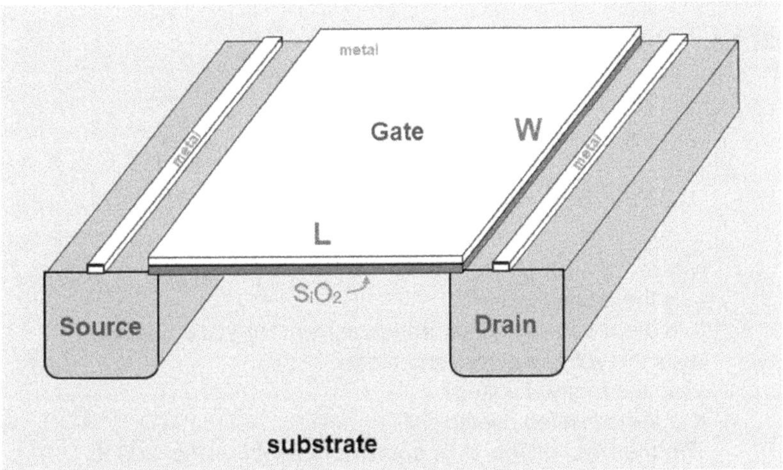

FIGURE 4.3 Partial 3D view illustration of an idealized single MOSFET.

Ideally, L becomes the length of the channel; however, as illustrated in Figures 4.2 and 4.3, due to lateral diffusion during fabrication, the effective L is slightly smaller than the drawn value.

Over the years, as L scaled down from one CMOS technology to the next, so did the supply voltage V_{DD}. For instance, for L = 2 µm, the supply was $V_{DD} = 5$ V, whereas for L = 700 nm technology, the supply voltage shrank to $V_{DD} = 3.3$ V.

4.2 ADS: MODEL CARDS FOR SEMICONDUCTOR DEVICES

Let us start with a brief discussion of modeling. A device model is a set of mathematical constructs used to predict the behavior of that part under varying conditions.

Models range from simple descriptions of a passive component, such as a resistor, shown in Figure 4.4, to extremely complex sub-circuits that represent the foundry process and technology of semiconductor devices.

Users interact with the models through the device parameters available to them, for example, the resistance of a resistor. If our resistor has a value of 100 Ω, then the mathematical expression of the model becomes (as seen in Figure 4.4):

A diode model has a more involved model equation that relates to the voltage and current, as seen in Figure 4.5.

The device modelers typically concern themselves with many effects that affect the performance of the device.

For example, in ADS, a diode is modeled by the equivalent circuit shown in Figure 4.6.

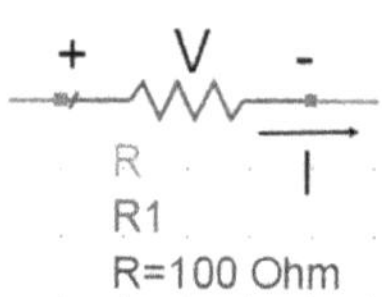

$$V = I * R = 100 * I$$

FIGURE 4.4 A 100 Ω resistor's model.

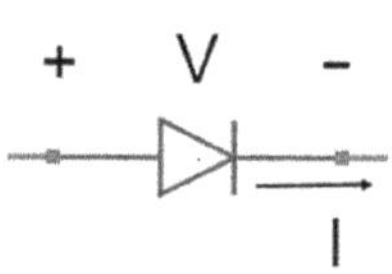

$$I_D = I_S\left(e^{\frac{V_D}{nV_T}} - 1\right)$$

where
I_D is the diode current
I_S is the reverse-bias saturation current (or scale current)
V_D is the voltage across the diode
V_T is the thermal voltage
n is the emission coefficient
The thermal voltage V_T is approximately 26 mV at 300 K,

FIGURE 4.5 A diode model.

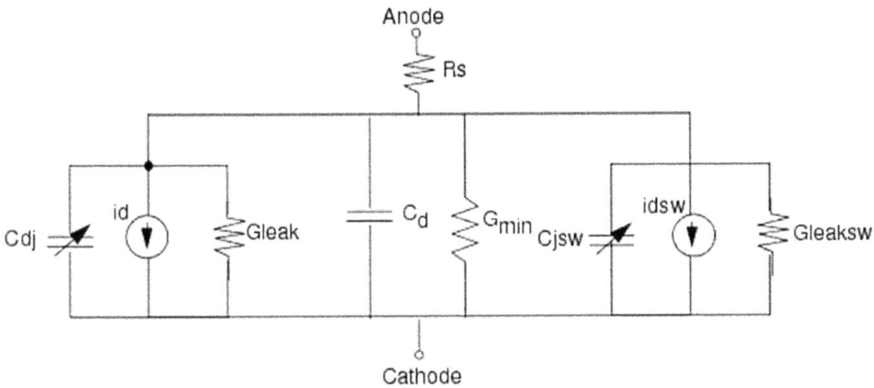

FIGURE 4.6 A circuit that represents ADS model of a diode.

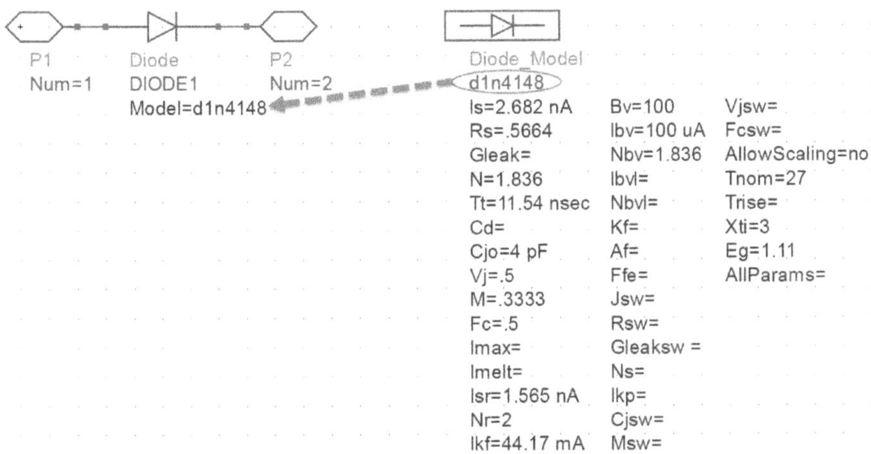

FIGURE 4.7 FAU EE KIT model card of a D1N4148 diode.

The circuit of Figure 4.6 is "created" and used by the simulator in the background, unseen by the circuit designer. To deal with this model complexity with many parameters, CAD tools like ADS use model cards when simulating semiconductor devices. For example, in the FAU EE KIT, the D1N4148 is modeled as shown in Figure 4.7.

Notice, in Figure 4.7, how the DIODE1 part model parameter points to the diode's Model Card via its unique name "d1n4148". This is a basic step needed to model semiconductor devices. Device modelers still may have a need to extend certain models.

One may find models that, in addition to the basic devices, have other circuit components. An example of this is the Zener Diode model that we discussed earlier (in Section 2.3). Other semiconductor devices too, like BJT or MOSFET, use model cards, as shown in Figure 4.8.

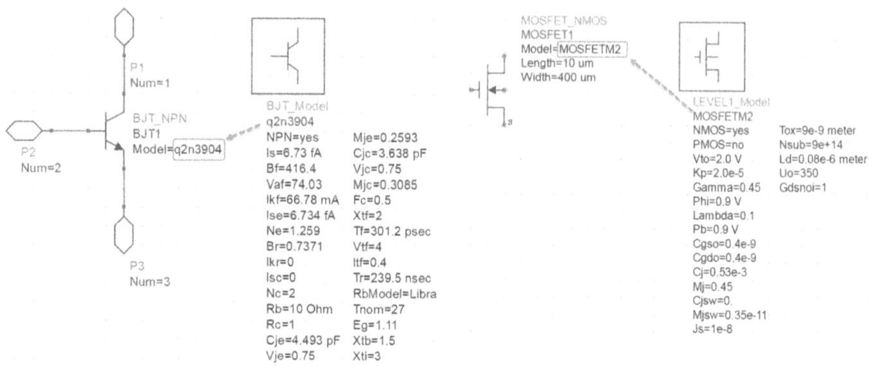

FIGURE 4.8 ADS FAU EE KIT model card of a Q2N3904 NPN (left) and an ADS Level 1 model of an NMOS (right).

4.3 MOSFET LEVEL 1 MODEL

The earliest MOSFET mathematical model, known as the Level 1 model, has been valid for L that is no smaller than 2 μm. The supply voltage for that earliest CMOS technology was $V_{DD} = 5\,V$.

The Level 1 model assumes an abrupt formation of the MOSFET channel. That is, an NMOS is in cutoff if $V_{GS} \leq V_{TH}$ and is ON if $V_{GS} > V_{TH}$, where V_{TH} is the threshold voltage parameter. For PMOS, where both V_{GS} and V_{TH} are negative, the MOSFET is OFF if $V_{GS} \geq V_{TH}$, and it abruptly turns ON if $V_{GS} < V_{TH}$. The gradual inversion process, which gives rise to subthreshold current, is covered by more advanced MOSFET models that were introduced for technologies smaller than 2 μm.

If both the source and drain voltages are grounded, $V_S = V_D = 0$ and $|V_{GS}| > |V_{TH}|$, the resulting channel has a uniform thickness. In NMOS, a positive V_{DS} causes the channel to be tapered, thicker on the source side and narrower on the drain side. For PMOS, a negative V_{DS} causes the channel to be tapered, thicker on the source side.

A MOSFET has two different ON conditions: triode mode and saturation mode. In triode mode, all along the tapered channel, the channel's thickness is finite. In saturation mode, the tapered channel pinches off on its drain side. That is, the channel's thickness becomes zero at a point that borders the drain.

What does a pinched-off channel physically mean? Is it "good" or "bad"?

Recalling that a channel is characterized by the existence of free carriers, such as in an n-channel, which has more free electrons than free holes. At a pinched-off channel point, there are no carriers. It doesn't mean that the pinch-off point stops the current I_D from flowing. All the carriers, flowing from the source, get swept up by the drain. The physical difference between the two conduction modes is that in a triode mode, there is a strong influence on the current I_D by the lateral electrical field created by the voltage V_{DS}. In triode mode, both V_{DS} and V_{GS} determine I_D. In saturation mode, the influence of V_{DS} on I_D is ideally zero. Saturation mode is actually "good" for creating a voltage-controlled current source: that is, I_D ideally depends on V_{GS} only.

Let us first explain the transistor's bias conditions that cause each of the above conduction modes. Let us assume an X-axis going along the channel: At x=0, the channel borders the source. At x=L, the channel borders the drain. **A Level 1 model's second assumption is that V(x), the voltage along the channel, varies linearly from $V(0)=V_S$ to $V(L)=V_D$.** For an n-channel to have a finite thickness, the following condition must hold for every value of x (ranging from 0 to L): $V_G - V(x) \geq V_{TH}$. As $V_D > V_S$, the worst-case scenario in a **triode mode** dictates a condition $V_G - V_D \geq V_{TH}$. Equivalently: $\mathbf{V_{DS} \leq V_{GS} - V_{TH}}$. Symmetrically for a PMOS (along with the negative voltage values). **PMOS triode mode requires $V_{DS} \geq V_{GS} - V_{TH}$.** In other words, whenever $|V_{DS}|$ becomes equal to the absolute value of the overdrive voltage $|V_{GS} - V_{TH}|$, the channel pinches off and the MOSFET is on the borderline between saturation and triode modes. As $|V_{DS}|$ becomes larger than $|V_{GS} - V_{TH}|$, the MOSFET goes deeper into saturation mode, and ideally the current does not change, as long as $|V_{GS} - V_{TH}|$ does not change.

The idealized current–voltage relationship in triode mode is shown below in equation (4.1), and the idealized current–voltage relationship in saturation mode is brought in equation (4.2). Equation (4.2) results instantly upon substituting the triode–saturation borderline condition $V_{DS} = V_{GS} - V_{TH}$ into equation (4.1). In other words, the current I_D changes continuously from one conduction mode to the other. V_{DS} at the borderline is called the "**saturation voltage**".

$$I_D = \mu C_{ox} \frac{W}{L}\left[(V_{GS} - V_{TH})V_{DS} - \frac{1}{2}V_{DS}^2\right] \tag{4.1}$$

$$I_D = \frac{1}{2}\mu C_{ox} \frac{W}{L}(V_{GS} - V_{TH})^2 \tag{4.2}$$

The coefficient μ is the **mobility** of the channel carriers. Electron mobility is denoted as μ_n, and hole mobility is denoted as μ_p. C_{ox} is the capacitance per area of the oxide capacitor. In ADS (as in any SPICE code), we often denote $\mu C_{ox} = K_p$, regardless of the MOSFET being NMOS or PMOS. K_p for NMOS has a different value than K_p for PMOS due to the different mobilities:

Let us outline briefly how equation (4.1) is derived [Sedra1]:

A. By Kirchhoff's Current Law, the current I_D is constant along the channel.
B. At any point x along the channel, the current is $I_D = \dfrac{dQ(x)}{dt} = \dfrac{dQ(x)}{dx} \cdot \dfrac{dx}{dt}$.
C. Q(x) is the charge in an infinitesimal slice of the channel, located at x, and has a cross-section area of W·dx.
D. It is assumed that $Q(x) = C_{ox} \cdot W \cdot dx\left[V_G - V(x) - V_{TH}\right]$.
E. The current I_D is a drift current. The carrier's velocity is $\dfrac{dx}{dt} = \mu \cdot E(x)$.
F. At any point x, the electrical field is $E(x) = \dfrac{dV(x)}{dx}$.
G. Substituting D, E, and F into B yields $I_D dx = \mu C_{ox} W\left[V_G - V(x) - V_{TH}\right]dV(x)$.
H. The expression G is integrated on the left side from x=0 to x=L, and on the right side from $V(0)=V_S$ to $V(L)=V_D$. The result (after some easy manipulation) is equation (4.1).

Equations (4.1) and (4.2) serve both NMOS and PMOS transistors. One just needs to adopt the correct mobility μ and threshold voltage. No other changes are needed. In PMOS, for instance, both V_{GS} and V_{TH} are negative.

In addition to the above idealized current–voltage equations, there are versions of the Level 1 model of a MOSFET that account for three secondary effects that cause I_D to deviate slightly from the values predicted by equations (4.1) or (4.2) for some given gate-to-source and/or drain-to-source voltages: channel length modulation, lateral source and drain diffusion, and the body effect. Let us explain briefly. See [Razavi1].

4.3.1 Channel Length Modulation Effect

$$I_D = \frac{1}{2}\mu C_{ox}\frac{W}{L}(V_{GS} - V_{TH})^2(1 + \lambda|V_{DS}|) \tag{4.3}$$

In saturation mode, as V_{DS} increases with respect to the saturation voltage, the point x at which the channel pinches off moves closer to the source (x = 0). This effect causes the effective channel length L to become smaller than its drawn value. The multiplicative term $1 + \lambda|V_{DS}|$ in equation (4.3) is a correction term. Notice that the L that is used in equation (4.3) is still the drawn value. Electronic circuit simulators attach the channel length modulation correction term also to the triode mode current–voltage equation, even though the effect happens in saturation mode only. This is for better continuity between the two modes. LAMBDA is a Level 1 model parameter.

4.3.2 Lateral Diffusion Effect

In the CMOS fabrication process of MOSFET circuits, ion implantation fabrication steps (to create n-wells, source and drain regions, etc.) may cause some slight damage to the silicon crystal structure, damage that is mostly corrected by an annealing process step. During the annealing thermal process, some side diffusion of dopants causes symmetric overlap between the source and gate on one side of the gate, and between the drain and gate on the other side of the gate, as shown in Figure 4.2. The effect is modeled mathematically by replacing L, in equations (4.1) and (4.2), with $L - 2L_D$. L_D is the lateral diffusion length parameter of a MOSFET Level 1 model.

4.3.3 The Body Effect

The Level 1 parameter V_{TO} is the value of the threshold voltage V_{TH} whenever the source and the substrate are short-circuited. That is, whenever $V_{SB} = 0$. There are many CMOS amplifier configurations in which the source must not be grounded or be connected to V_{DD}, whereas the substrate can and should. The difference between V_S and V_B causes V_{TH} to deviate from V_{TO}. This is called the body effect. Skipping the relevant semiconductor physics [Sze1], the MOSFET Level 1 body effect mathematical relationships are given by equations (4.4) and (4.5).

$$V_{TH} = V_{TO} + \gamma\left(\sqrt{2\Phi_F + V_{SB}} - \sqrt{2\Phi_F}\right) \tag{4.4}$$

where

$$\gamma = \frac{\sqrt{2q\varepsilon_{si}N_{sub}}}{C_{ox}} \qquad \Phi_F = \frac{KT}{q}\ln\left(\frac{N_{sub}}{n_i}\right) \tag{4.5}$$

The body effect coefficient γ (GAMMA) depends on the dielectric constant of silicon and on the doping level of the substrate (N_{sub}). The PHI parameter depends on the thermal voltage and the intrinsic concentration of free electrons and holes in undoped silicon. In NMOS, the voltage V_{SB} is sometimes positive, causing V_{TH} to increase as V_{SB} gets larger. In PMOS, V_{SB} is often negative, causing V_{TH} to become more negative as V_{SB} gets more negative.

The MOSFET Level 1 model further includes the transistor's high-frequency capacitances and the small-signal parameters. These issues are deferred to later on in this chapter.

4.4 MOSFET DC CIRCUITS

Three brief examples (presented in [Sedral]) serve to illustrate the manual solution method of MOSFET DC circuits. In a manual solution, we often adopt the idealized current–voltage relationships (4.1) and (4.2) and choose to leave the more computationally intensive secondary effects to the simulation. We often have to guess the MOSFET's conduction mode and follow up on the assumption until it is justified or until it results in a contradiction. The quadratic nature of equations (4.1) and (4.2) often leads to quadratic equations, of which one or both solutions may be false.

Example 4.1

For the circuits shown in Figure 4.9, the NMOS parameters are $V_{TH}=2\,V$, $K_p=20$ [μA/V²], and L = 10 μm (for both Example 4.1a and 4.1b) with W = 400 μm in Example 4.1a and 100 μm in Example 4.1b. In Example 4.1a, we need to select resistors R_D and R_S so that {$I_D=0.4\,mA$, $V_D=1$ V}. In Example 4.1b, we need to select R for $I_D=0.4\,mA$.

The reader should disregard the unrealistic features of Example 4.1 (the dual positive–negative supplies, the shorted B and S at a voltage that is not the most negative, etc.) and focus on the solution method.

Solution

In Example 4.1a, we first notice that both V_G and V_D are given, which allows us to check the triode vs. saturation modes inequality: $V_D=1$ $V_D>V_G-V_{TH}=0-2=-2\,V$. Note that V_S cancels out on both sides of the inequality. The NMOS, after meeting the saturation mode inequality, still needs to meet the conduction inequality $V_{GS}>V_{TH}$.

In saturation mode, by equation (4.2), I_D and V_{GS} must be consistent. The specified I_D value dictates the VGS value needed. The current voltage equation, after substituting the given variables and coefficients, is $0.4=(1/2)\cdot 20\cdot 10^{-3}(400/10)(V_{GS}-2)^2$, or equivalently $1=(V_{GS}-2)^2$. The solution $V_{GS}=1\,V$ is false, as it falls below the threshold voltage. The correct solution is $V_{GS}=3\,V$. The design concludes with $V_S=-3\,V$, and then $R_S=5$ kΩ. The resistor $R_D=10$ kΩ is known from the design specifications.

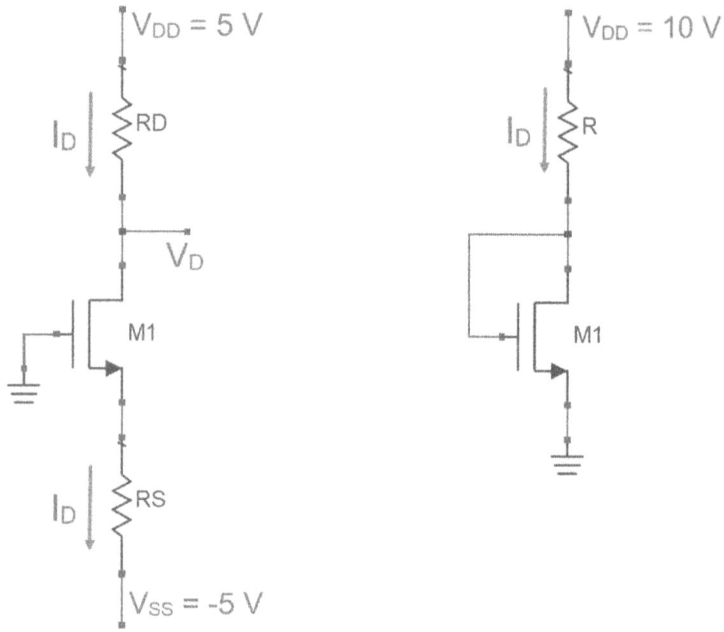

FIGURE 4.9 Circuits of Example 4.1a (left) and Example 4.1b (right).

Example 4.1b is very similar to Example 4.1a. We notice that because $V_G = V_D$ and because V_{TH} is positive, the shorting of the gate and the drain leads to saturation mode. Equation (4.2) yields two solutions: $V_{GS} = 4\,V$ (correct) and $V_{GS} = 0\,V$ (incorrect). The resulting resistor is $R = 15$ kΩ.

Example 4.2

For the circuit shown in Figure 4.10, the PMOS parameters are $V_{TH} = -1\,V$ and $K_p(W/L) = 1$ [mA/V²]. You need to choose R_{G1}, R_{G2}, and R_D so that the PMOS operates in saturation mode, $I_D = 0.5\,mA$, and $V_D = 3\,V$.

Solution

Saturation mode is specified. Let us substitute the given values into equation (4.2): $0.5 = (1/2) \cdot 1[V_{GS} - (-1)]^2$, yielding solutions $V_{GS} = 0$ (false) and $V_{GS} = -2\,V$ (correct). Now, $V_G - V_S = V_G - 5 = -2$ resulting in $V_G = 3\,V$. $V_G = 3 = V_{DD}[R_{G2}/(R_{G1} + R_{G2})] \rightarrow 5/3 = 1 + R_{G1}/R_{G2}$. We can take $R_{G1} = 2$ MΩ and $R_{G2} = 3$ MΩ. "Why MΩ"? will become clearer in Section 4.8. By the specified I_D and V_D, we have $R_D = 6$ kΩ.

Example 4.3

See Figure 4.11. Solve for following two conditions: (4.3a) NMOS parameters are $V_{TH} = 1\,V$ and K_p (W/L) = 1 [mA/V²]. We need to solve the circuit's currents and voltages. (4.3b) Same, but let us choose $R_D = 15$ kΩ.

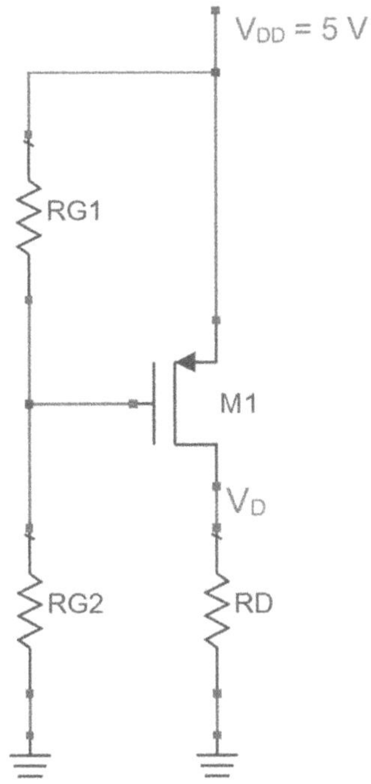

FIGURE 4.10 Example 4.2 circuit.

Solution 4.3a

We observe that $I_G=0$ and $V_G=5\,V$. Let us assume saturation mode (for no better reason than the simpler current equation).

I_D is unknown, and so is V_S; however, $V_S=I_DR_S$. Therefore, a quadratic equation for I_D may be formed from equation (4.2) as follows:

$$I_D=(1/2)\,KP(W/L)(V_G-I_DR_S-V_{TH})^2$$

substituting the known values leads to:

$$(1/2)1(5-6I_D-1)^2-I_D=0$$

$$18I_D^2-25I_D+8=0$$

The two solutions are $I_D=0.89\,mA$ and $I_D=0.5\,mA$; both look plausible. Let us make a second assumption: $I_D=0.89\,mA$. Then $V_S=6I_D=5.34\,V$; hence, a conduction inequality contradiction $V_G-V_S=5-5.34<V_{TH}=1\,V$. Let us retract one step and make a second assumption that $I_D=0.5\,mA$. Now $V_S=3\,V$, and $V_G-V_S=5-3=2>V_{TH}=1\,V$. Now the conduction inequality is met, but we are not done yet. We need to check the saturation mode inequality: $V_D=V_{DD}-I_DR_D=10-0.5\cdot 6=7>V_G-V_{TH}=5-1=4\,V$. After justifying both assumptions, the rest of the solution is easy.

Solution 4.3b

It starts as the solution for Example 4.3a. Again, we assume saturation mode, and again we have to choose $I_D=0.5\,mA$. This time though, this current solution fails the saturation mode inequality: $V_D=V_{DD}-I_DR_D=10-7.5<V_G-V_{TH}=5-1$.

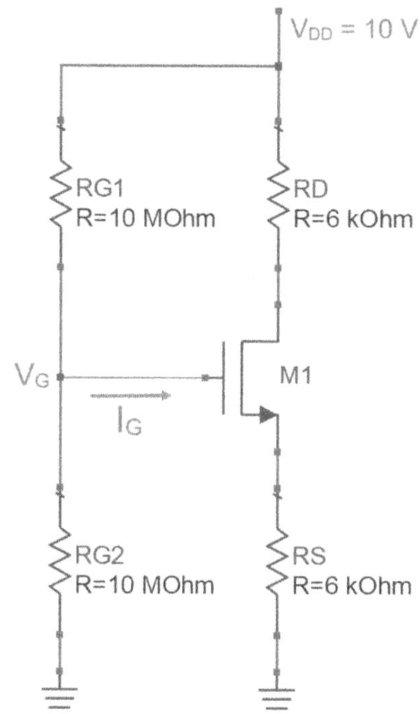

FIGURE 4.11 Example 4.3 circuit.

As a result of the contradiction, we need to go all the way back to the first assumption. This time we take triode mode. We substitute $V_D = 10 - 15I_D$ and $V_S = 6I_D$ into the triode mode current equation: $I_D = 1 \cdot [(5 - 6I_D - 1) \cdot (10 - 15I_D - 6I_D) - (1/2) \cdot (10 - 15I_D - 6I_D)^2]$, resulting (with some effort) in the quadratic equation $94.5I_D^2 - 65I_D + 10 = 0$. The solutions are $I_D = 0.455$ mA and $I_D = 0.232$ mA. $I_D = 0.455$ mA meets both inequality conditions.

In the following ADS simulation, we shall compare the above manual solutions of Examples 4.1 and 4.2, which utilize idealized current equations, to solutions where the MOSFET features some second-order effects. A study of Example 4.3 may be useful before tackling some of the end-of-chapter problems.

4.5 ADS: MOSFET INSTANCE COMPONENTS AND MODEL CARDS; LEVELS OF COMPLEXITY OF A MOSFET LEVEL 1 MODEL; COMPARISON BETWEEN SIMPLIFIED MANUAL SOLUTIONS AND ADS SOLUTIONS THAT ADDRESS SECOND-ORDER EFFECTS

See Figure 4.12 for locating MOSFET components. First, let us look at "instance components" for MOSFET devices. An instance component is the one included in the circuit implementation, but the component is like an empty shell.

For a MOSFET instance to work, we need to attach a MOSFET model card to it. Notice in Figure 4.13 (left) that in NMOS, the substrate arrow points inwards. In PMOS, the substrate arrow points outwards.

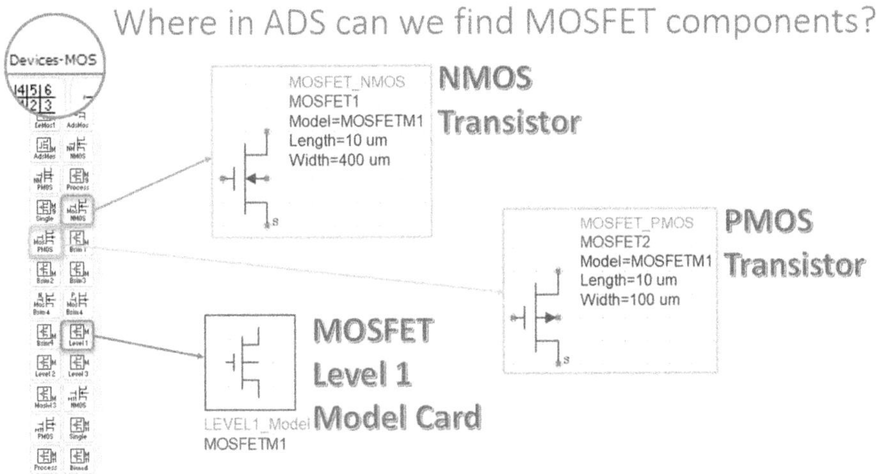

FIGURE 4.12 Devices-MOS components palette.

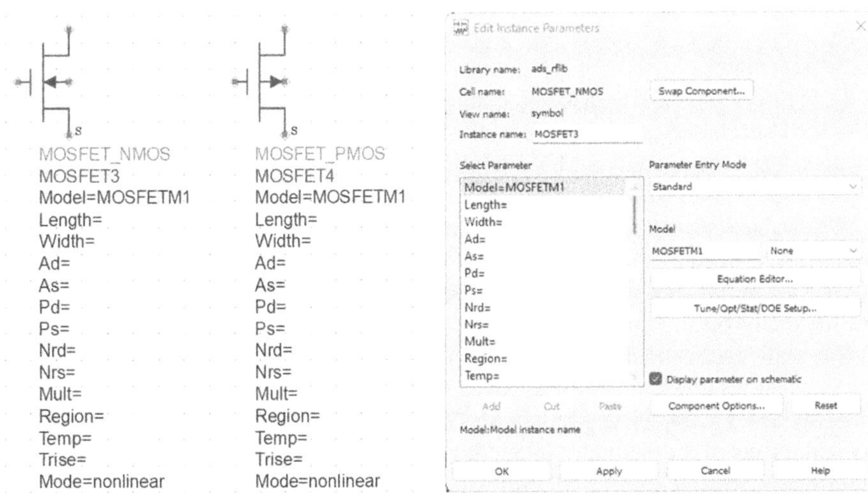

FIGURE 4.13 NMOS and PMOS instance components (left) and parameters dialog box (right).

We will undisplay parameters that are not needed for this simulation, as shown on the right side of Figure 4.13. When one double-clicks on the NMOS component, the dialog box in Figure 4.13 (right) opens up.

The parameters A_d and A_s are areas of the drain and source, respectively. P_d and P_s are perimeters of the drain and source, respectively. These are important if we want to account correctly for the high-frequency capacitance effects in a MOSFET small-signal amplifier (Sections 4.7 and 4.8).

Model is the name of the model card to be used. We like to keep this parameter displayed.

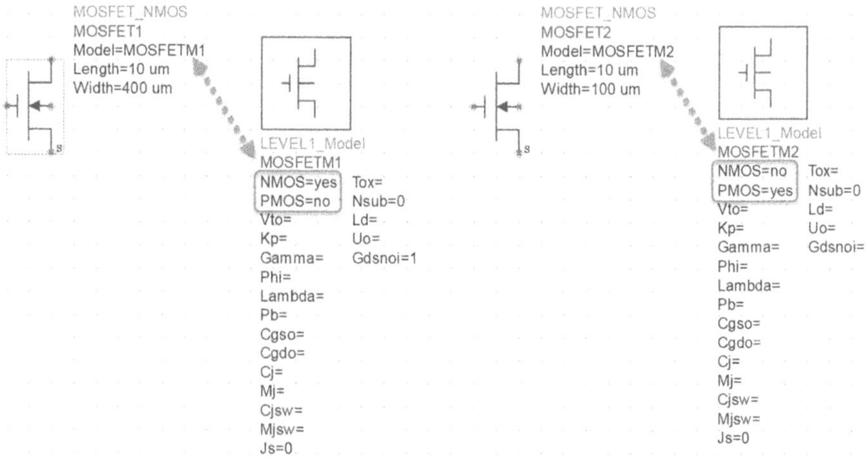

FIGURE 4.14 Instance components and model cards.

Length is L and width is W. We leave them displayed as well and add values and units, such as 5 um ("u" stands for "μ"). Let us de-select everything else from display on the schematic.

In Figure 4.14, see the instance components (first from left and third from left) after the unnecessary parameters are de-selected not to be displayed.

Note that even though a MOSFET has drain-source symmetry, the designated source node is marked by the letter "s".

We can have several model cards, but each part instance (e.g., "M1") uses a unique model, recognized by the model's unique name (e.g., "MOSFETM1").

On each model card, we make sure to select between NMOS and PMOS correctly. See the red boxes in Figure 4.14.

We shall start by looking only at Level 1 MOSFET models. These are valid for channel lengths down to $L = 2\,\mu m$. No further down!

A MOSFET component needs a MOSFET model card. In order to compare the manual results of the examples in Section 4.4 to simulation results, let us first create the simplest possible Level 1 MOSFET model.

The most primitive Level 1 model has values for V_{TO} and K_p only.

In order to simulate body effect, channel length modulation, and side diffusion effect, we need to add nonzero GAMMA, PHI, LAMBDA, and LD parameters, as suggested in [Razavi1].

MOSFET devices have many different models available. The use of a particular model is dependent on the technology that the foundry process uses.

The schematic in Figure 4.15 has two Level 1 model cards. We will initially use MOSFETM1 for the MOSFET1 transistor. Model MOSFETM1 has no body effect (as Gamma = 0), no channel length modulation effect (as Lambda = 0), and no side diffusion effect (as LD = 0).

In Example 4.1a, it is given that $K_p = \mu_n C_{ox} = 20\ \mu A/V^2$. In Level 1 models, the parameter K_p has units of A/V^2. That is why we set it to $K_p = 2.0e^{-5}$.

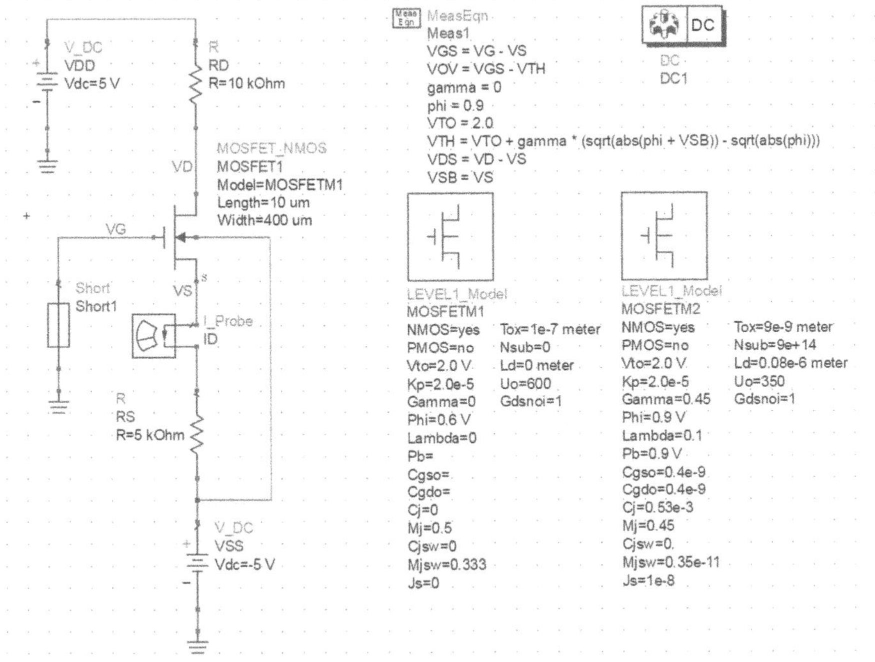

FIGURE 4.15 Schematic of Example 4.1a (from Section 4.3).

TABLE 4.1
Simulation DC Results for Example 4.1a, Using the Most Primitive Level 1 Model Card

freq	$I_{D.i}$	V_D	V_{DS}	V_G	V_{GS}	V_{OV}	V_S	V_{SB}	V_{TH}
0.0000 Hz	400.0 uA	1000.0 mV	4.000	0.0000 V	3.000	1.000	−3.000 V	−3.000	2.000

Note that the substrate is connected to the most negative voltage (which in this example is $V_{SS} = -5$ V).

The reason for adding *Short1*, which does not change the fact that $V_G = 0$, is to be able to mark the name V_G at the gate node. Without *Short1*, we would be renaming a ground node, which ADS does not allow (Figure 4.15).

We see that $I_D = 400$ uA and $V_D = 1000$ mV (as designed, seen in Table 4.1).

Even though V_{SB} is nonzero, which was supposed to create the body effect, the threshold voltage $V_{TH} = V_{TO}$ because Gamma = 0. There is no channel length modulation because Lambda = 0. As a result, there is a perfect match between the manual calculations and the simulation.

In the next run, let us adopt the MOSFETM2 model shown fully in Figure 4.15 (which is a Level 1 NMOS model suggested in Prof. Razavi's book [Razavi1]). This time, the design values will be influenced. Example 4.1a is a design example.

The manual solution neglected second-order effects. Whenever such effects exist, the design values of R_D and R_S must be tweaked.

Note that in Figure 4.16, we not only update the model card's parameters, but we also update the MeasEqn for the data display VTH post-processing calculation.

The updated DC results are shown in Table 4.2.

As a result of the nonzero value of V_{SB} ("Body Effect"), the actual V_{TH} is now 2.244 V (deviating from $V_{TO} = 2$ V).

V_D and V_S deviate significantly from the calculated values. $V_D = 1$ V was specified, but we got $V_D = 1.248$ V.

Due to the strong channel length modulation effect (LAMBDA = 0.1 1/V), the current I_D, calculated originally according to the simple saturation mode I_D equation and designed to be equal to 0.4 mA, now has a multiplicative correction term $(1 + \text{LAMBDA} \cdot V_{DS})$ causing it to fall to 375 uA. Of course, I_D is also affected by the new value of V_{TH}.

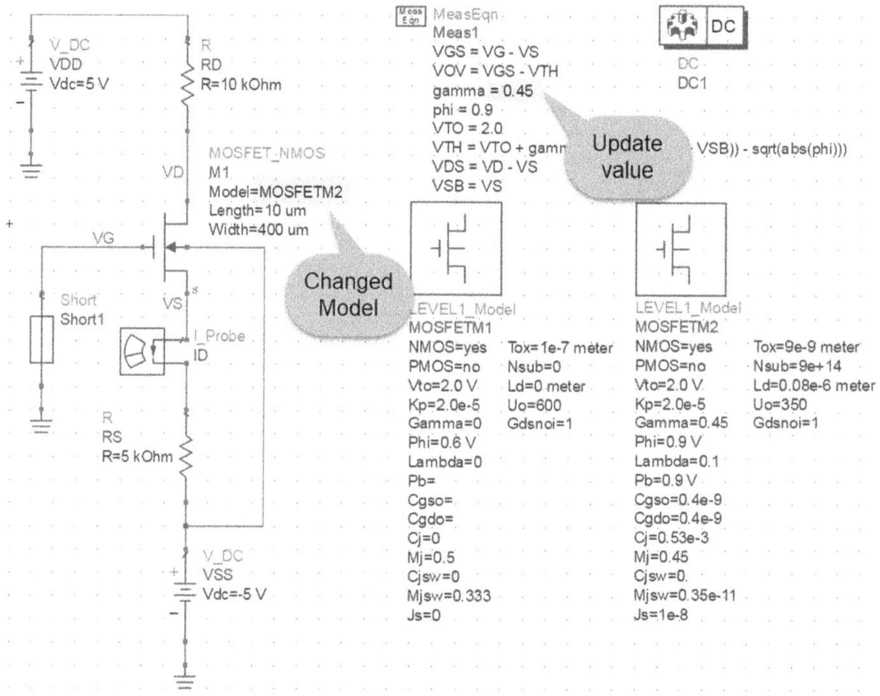

FIGURE 4.16 Changing the model card and updating its parameters.

TABLE 4.2
Example 4.1a DC Results Updated to Include Secondary Effects

freq	$I_{D.i}$	V_D	V_{DS}	V_G	V_{GS}	V_{OV}	V_S	V_{SB}	V_{TH}
0.0000 Hz	375.2 uA	1.248 V	4.372	0.0000 V	3.124	880.0 m	−3.124 V	−3.124	2.244

L_D and LAMBDA each cause I_D to rise. An increase of V_{TH} causes I_D to be reduced.

In this example, since I_D became smaller, it is obvious that body effect dominated the other two error sources.

The tweaking of R_D and R_S is not shown in this example.

The schematic for Example 4.1b is shown in Figure 4.17, again featuring the full Razavi NMOS model. The simulation results for the Level 1 NMOS model, which contained no second-order effects, and for the one that does, are shown in Table 4.3.

There is no body effect in Example 4.1b, as the source and substrate are connected. Example 4.1b, with Razavi's model, does feature channel length modulation as well as lateral diffusion, each contributing to the increase of I_D, compared to the simplified design. Again, tweaking of R is left as an exercise to the reader.

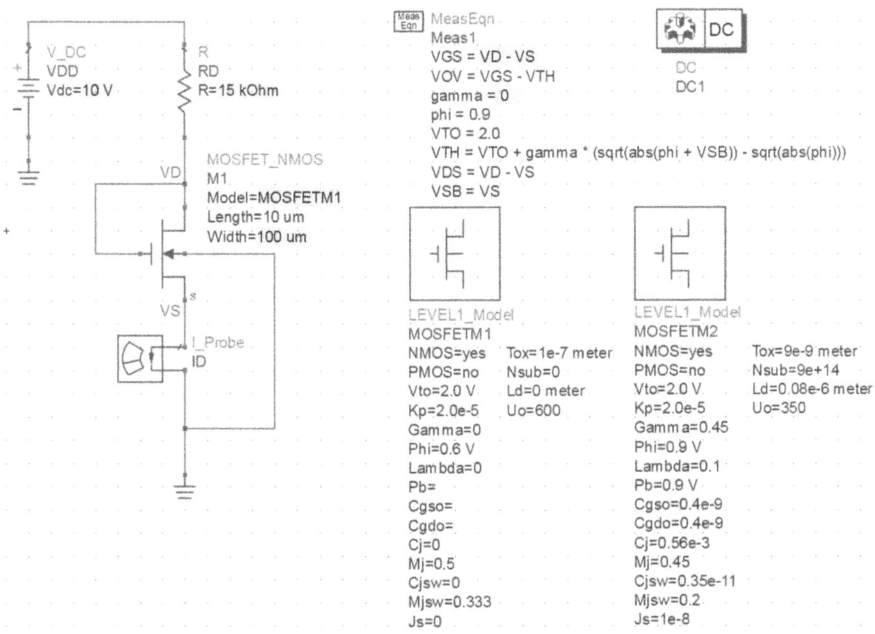

FIGURE 4.17 Example 4.1b schematics.

TABLE 4.3
DC Simulation Results, for Example 4.1b, Validating the Manual Solution (Upper Table) and Featuring Second-Order Effect (Lower Table)

freq	$I_{D.i}$	V_D	V_{DS}	V_{GS}	V_{OV}	V_{TH}
0.0000 Hz	400.0 uA	4.000 V	4.000	4.000	2.000	2.000

freq	$I_{D.i}$	V_D	V_{DS}	V_{GS}	V_{OV}	V_{TH}
0.0000 Hz	418.0 uA	3.731 V	3.731	3.731	1.731	2.000

Figure 4.18 features the resulting diagram of the simplified design of Example 4.2 (of Section 4.4).

Example 4.2, solved in Section 4.4, used a combined PMOS parameter $K_p(W/L) = 1\ mA/V^2$. For the simulation, we assumed $W/L = 20$. Therefore, $K_p = 5e^{-5}$ [A/V^2]. Also given $V_{TO} = -1$ V, and it is equal to V_{TH} because there is no body effect in this problem. Model MOSFETM1 neglects the channel length modulation and the side diffusion effects. Model MOSFETM2 was suggested by Razavi as a realistic Level 1 model for PMOS transistors.

The simulation DC results are shown in Table 4.4.

With LAMBDA and L_D effects, the current I_D increased from the designed I_D of 0.5 mA to I_D of 0.567 mA. V_D is off too.

Notice that V_{DS}, which is now less negative (−1.598 V, compared to −2 V), *still maintains a saturation mode* for the PMOS, as −1.598 V is more negative than $V_{GS} - V_{TH} = -2 - (-1) = -1$ V.

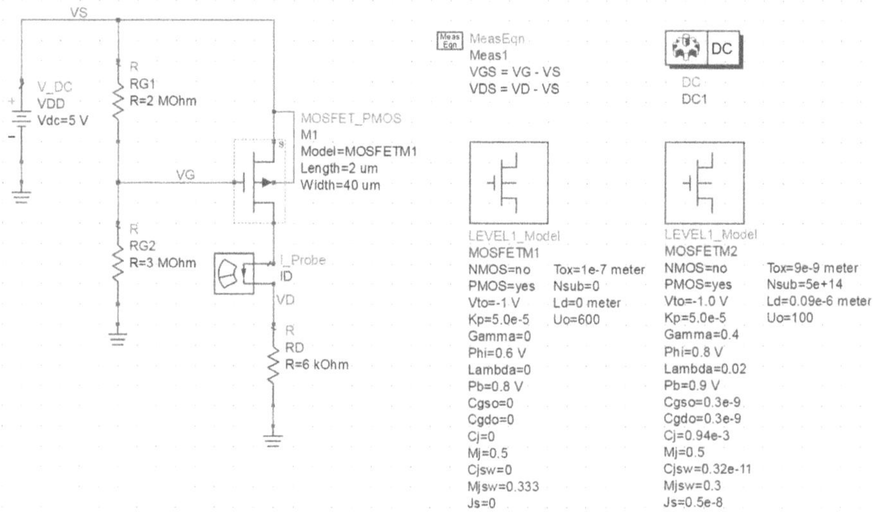

FIGURE 4.18 Schematic of Example 4.2, featuring the full PMOS Level 1 Razavi model.

TABLE 4.4
DC Simulation Results of Example 2 (PMOS): Idealized Values Neglecting Second-Order Effects (Upper Table), Results That Include Channel Length Modulation and Side Diffusion (Lower Table)

freq	$I_{D.i}$	V_D	V_{DS}	V_G	V_{GS}	V_S
0.0000 Hz	500.0 uA	3.000 V	−2.000	3.000 V	−2.000	5.000 V

freq	$I_{D.i}$	V_D	V_{DS}	V_G	V_{GS}	V_S
0.0000 Hz	567.0 uA	3.402 V	−1.598	3.000 V	−2.000	5.000 V

4.6 MOSFET CURVES, SATURATION VS. TRIODE MODES, THE BODY EFFECT, CHANNEL LENGTH MODULATION EFFECT

Figure 4.19 shows the schematic for running a simple DC simulation voltage sweep, creating the curve of I_D vs. V_{GS} for a fixed value of V_{DS}. We use a MOSFET model card based on Razavi's Level 1 NMOS model [Razavi1].

Leaving K_p blank means that the simulator will calculate it when needed. In the DC examples covered in Section 4.4, we used the default value for K_p as $K_p = 2.0e^{-5}$ A/V^2. We see that $W/L = 10$. Therefore, $K_p(W/L) = 0.2$ mA/V^2.

What do we expect to see in this sweep study?

As we vary V_{GS}, the NMOS is in cutoff as long as $V_{GS} < V_{TH}$. Because $V_{BS} = 0$, there is no body effect. Therefore, $V_{TH} = VTO$. We have $V_{DS} = 5\,V$, as seen in Figure 4.19. Obviously, $V_{DS} > V_{GS} - V_{TH}$, which means that once the NMOS turns ON, it will be in saturation mode (Figure 4.20).

Something is wrong. We expect (without the Level 1 effects) that $I_D \approx K_p(W/L)\cdot(V_{GS} - V_{TH})^2$. Say that $V_{GS} - V_{TH} = 2\,V$. Then with $K_p(W/L) = 0.2$ mA/V^2, we expect $I_D \approx 0.2\cdot 4 = 0.8\,mA$. What we actually observe for $V_{GS} = 2.7\,V$ is I_D close to 4 mA. What went wrong?

Theoretically, $K_p = \mu_n C_{ox}$. Razavi's model specifies a mobility of $U_0 = 350$. C_{ox} depends on T_{ox} (the thickness of the gate oxide layer). If K_p is left blank, the U_0 and T_{ox} parameters dominate and determine the effective K_p. If we write $K_p = 2.0e^{-5}$, even though it is the default value, this K_p overrides the effect of T_{ox} and U_0 in the simulation.

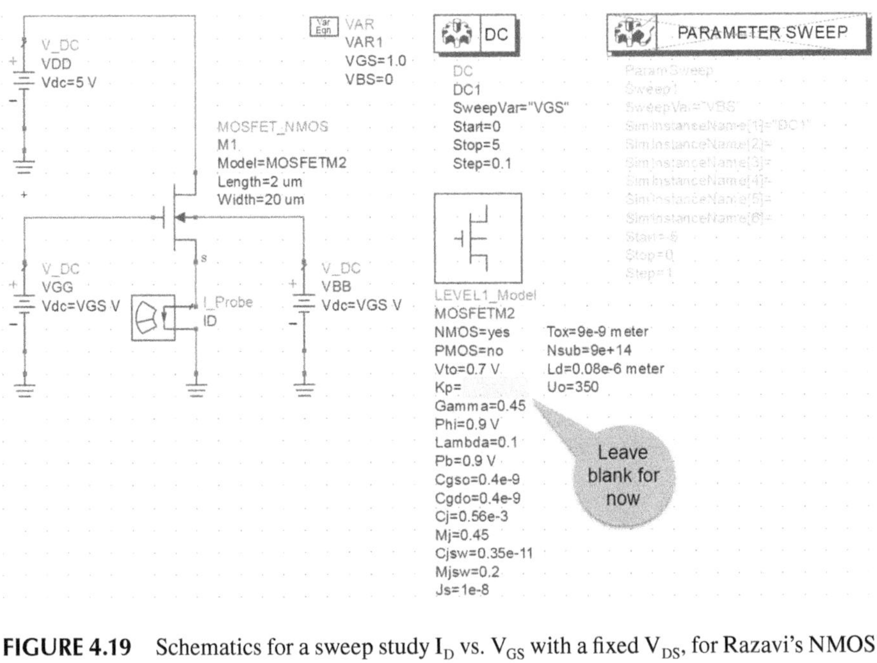

FIGURE 4.19 Schematics for a sweep study I_D vs. V_{GS} with a fixed V_{DS}, for Razavi's NMOS Level 1 model.

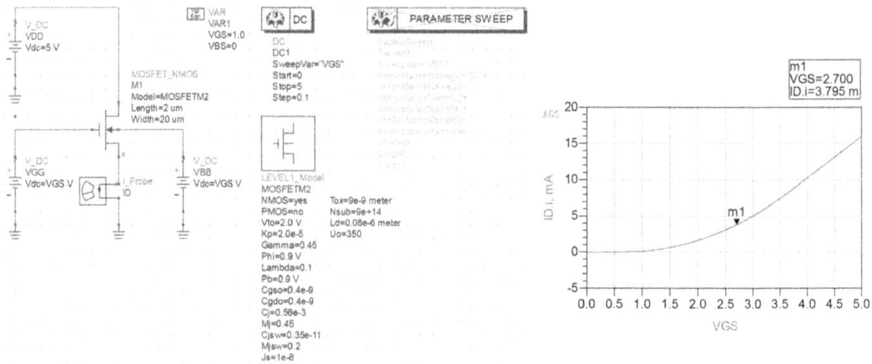

FIGURE 4.20 I_D vs. V_{GS} for a fixed V_{DS} (first attempt).

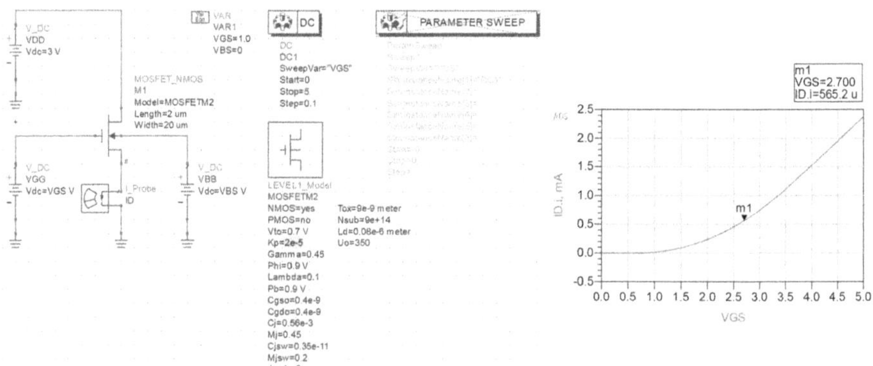

FIGURE 4.21 I_D vs. V_{GS} for a fixed V_{DS} (second attempt).

Figure 4.21 shows the corrected situation:

With K_p being specifically included, the results of I_D vs. V_{GS} now make sense.

We see that whenever $V_{GS} \le V_{TH} = 0.7\,V$, $I_D = 0$ (the MOSFET is OFF). As we enter the saturation mode, the graph of I_D vs. V_{GS} becomes a parabola.

Can we see the triode mode region?

With $V_{DS} = 5\,V$, and V_{GS} swept between 0.7 and 5 V, we always have $V_{DS} > V_{GS} - V_{TH}$ (saturation mode).

We introduce one small change: $V_{DS} = 3\,V$ (in place of 5 V). See Figure 4.22.

Now, saturation mode is when $V_{DS} \ge V_{GS} - V_{TH} \rightarrow 3 \ge V_{GS} - 0.7$. The borderline is $V_{GS} = 3.7\,V$.

So, for $V_{GS} > 3.7\,V$, the NMOS is in triode mode. In triode mode, I_D depends on V_{GS} linearly. In saturation mode, I_D depends on V_{GS} quadratically.

Next, let us do a double sweep. See Figure 4.23. The primary sweep is I_D vs. V_{GS} (as before). The secondary sweep (and hence fewer points) is V_{BS}. For an NMOS, V_B must be at the "most negative" voltage. Here, because $V_S = 0$, we must take only negative values for V_{BS}. The primary variable V_{GS} is swept by the DC engine.

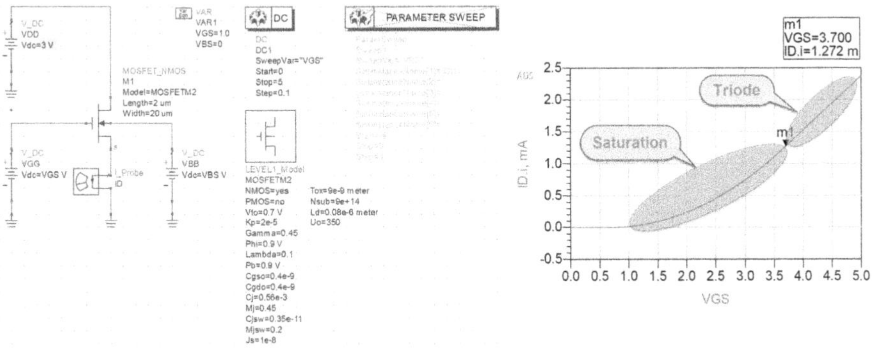

FIGURE 4.22 I_D vs. V_{GS}, showing both MOSFET conduction modes.

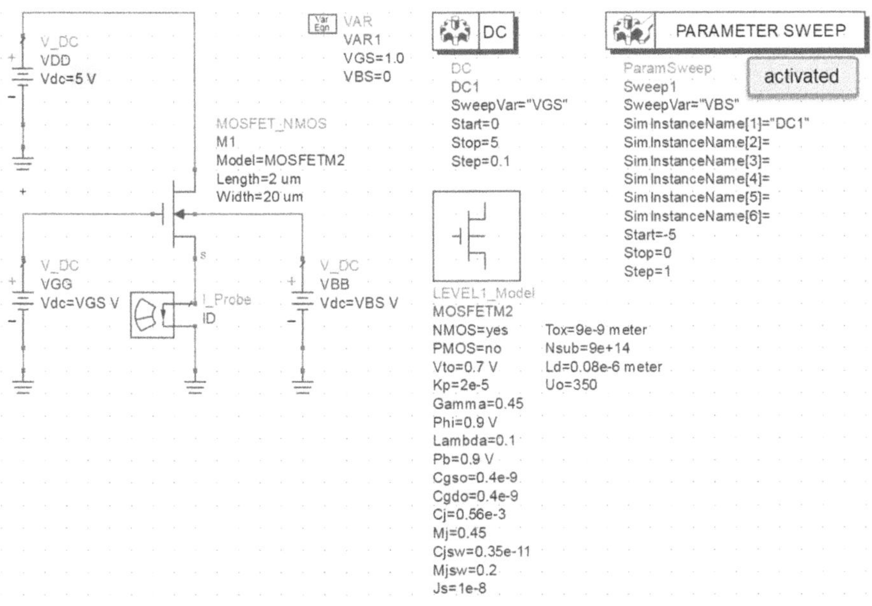

FIGURE 4.23 Schematic for a double sweep study of an NMOS.

The secondary variable V_{BS} is swept by the Parameter Sweep block. Note that V_{BS} starts at -5 V, increments by 1, and stops at 0. Results are shown in Figure 4.24.

We expect, according to the V_{TH} formula, to see that as V_{SB} becomes larger, V_{TH} becomes larger. Indeed, that is what we see in these results. Note that in the simulation we swept V_{BS}, which is $V_{SB} = -V_{BS}$.

Zooming into the area of the results for lower currents allows us to get a sense of the values of V_{TH} as V_{SB} changes. See Figure 4.25.

Let us now go back to a single sweep looking at I_D vs. V_{DS} for some fixed V_{GS}. See Figure 4.26. The resulting curve is presented in Figure 4.27.

The slope of the I_D vs. V_{DS} curve, in the saturation mode region, is huge because LAMBDA = 0.1 is huge in 2 μm technology.

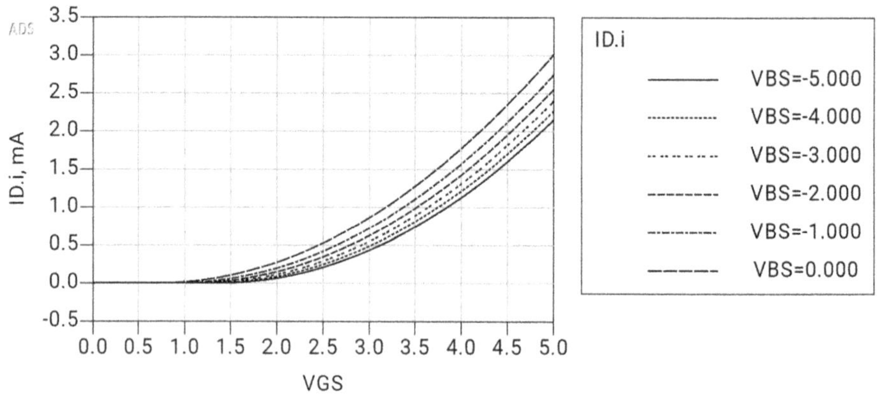

FIGURE 4.24 Multiple I_D vs. V_{GS} curves parameterized by V_{BS}.

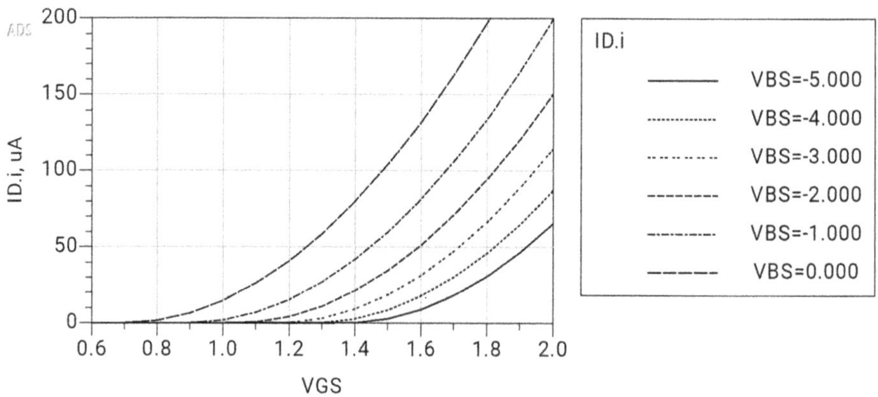

FIGURE 4.25 Zooming in the low current region of the I_D vs. V_{GS} curves parameterized by V_{SB}. The zooming in is done by reducing the sweep range for V_{GS}.

Normally, in Level 1 models, LAMBDA is smaller by an order of magnitude. Let us repeat the simulation for LAMBDA = 0.01. Figure 4.28 is more like it. In saturation mode, I_D is "almost independent" of V_{DS}. That is, the curve is almost parallel to the X-axis.

The MOSFET in saturation mode acts like a current source.

We conclude this section with a double-sweep study (primary = V_{DS}, secondary = V_{GS}) to show how V_{GS} controls I_D in saturation mode (Figures 4.29 and 4.30).

The larger the $V_{GS,}$ the larger the I_D becomes in saturation mode. All saturation mode I_D vs. V_{DS} lines are slanted. All the lines meet at the large negative voltage, sometimes called "Early Voltage". In MOSFET, the Early Voltage is $V_A = 1/$ LAMBDA. Notice the parabolic curves where the MOSFET is in triode mode.

For $V_{DS} < 1$ V (more or less) the curves are almost straight lines: Voltage-controlled resistor curves.

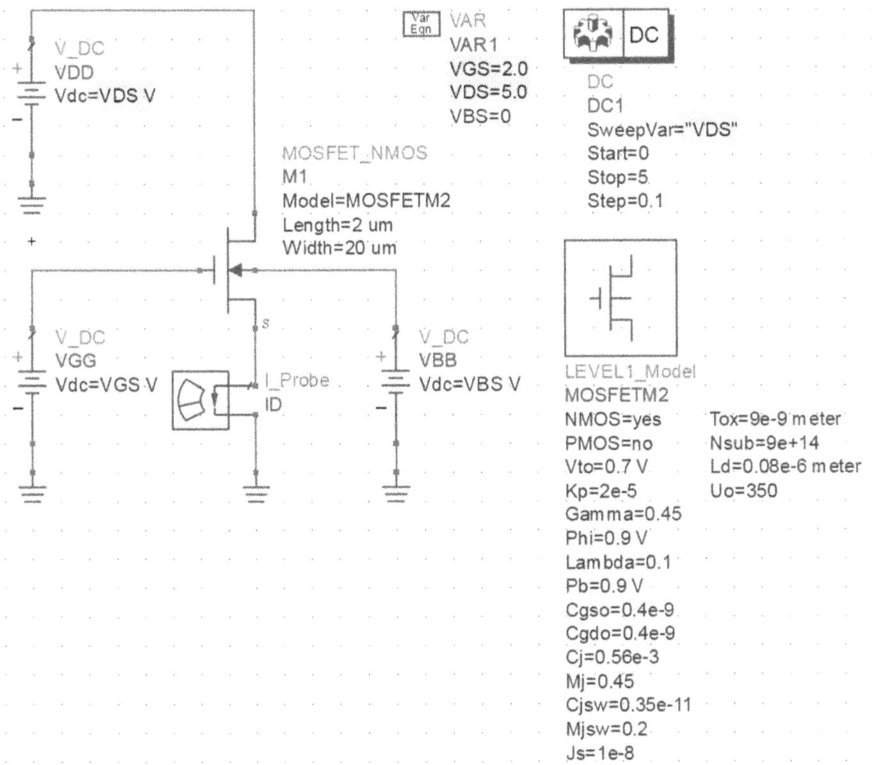

FIGURE 4.26 Schematic for a single sweep of I_D vs. V_{DS}.

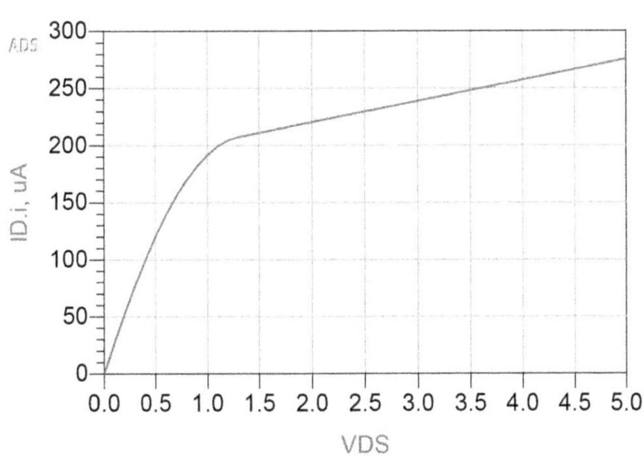

FIGURE 4.27 I_D vs. V_{DS} for $V_{GS}=2$ V. LAMBDA = 0.1.

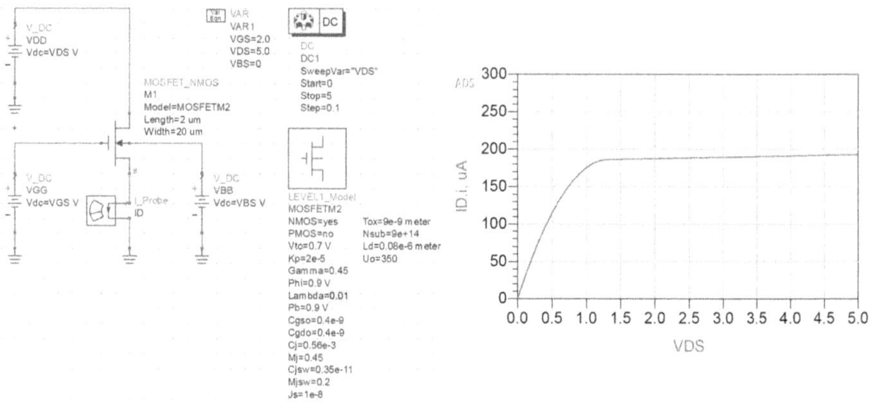

FIGURE 4.28 I_D vs. V_{DS} for $V_{GS} = 2$ V. LAMBDA = 0.01.

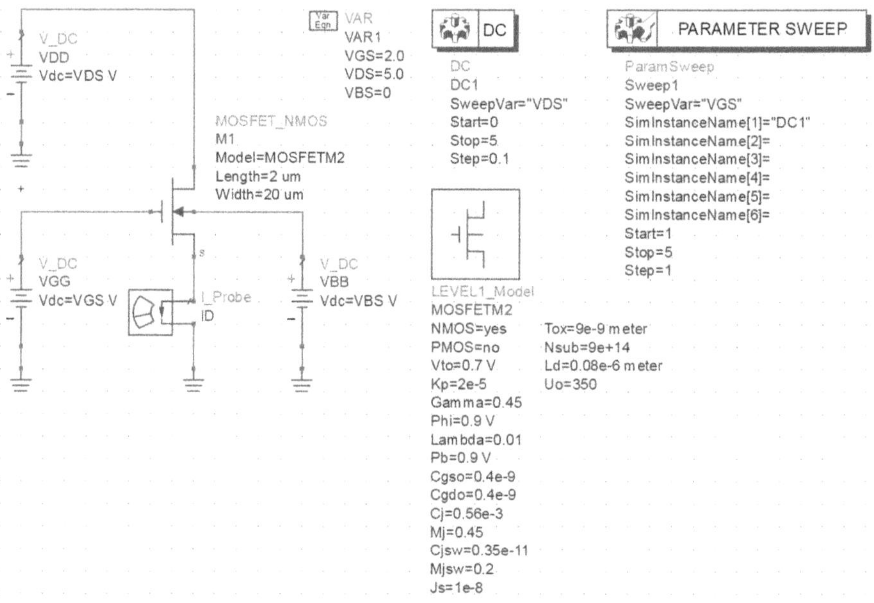

FIGURE 4.29 Schematic for sweeping V_{DS} and V_{GS}.

4.7 BASICS OF SMALL-SIGNAL MOSFET AMPLIFIER

Some basics of MOSFET small-signal performance can be illustrated with the aid of the primitive common-source (CS) amplifier with resistive load shown in Figure 4.31. Refer to [Sedra1]. For many basic single-stage MOSFET amplifiers, see [Murmann1].

Again, let us denote DC variables by capital letters (name and subscript), small signal variables by lowercase letters, and total signals (DC and small signal) with a lowercase letter for the variable and a capital letter for the subscript. For example, $v_{GS} = V_{GS} + v_{gs}$ or $i_D = I_D + i_d$.

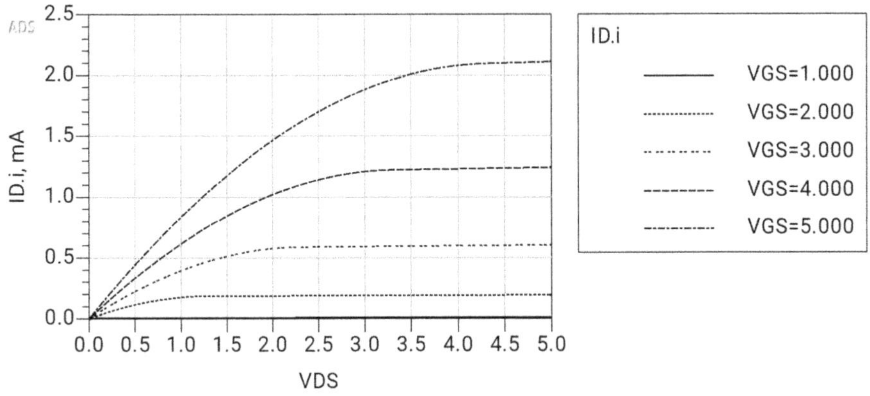

FIGURE 4.30 I_D vs. V_{DS} with V_{GS} as a parameter.

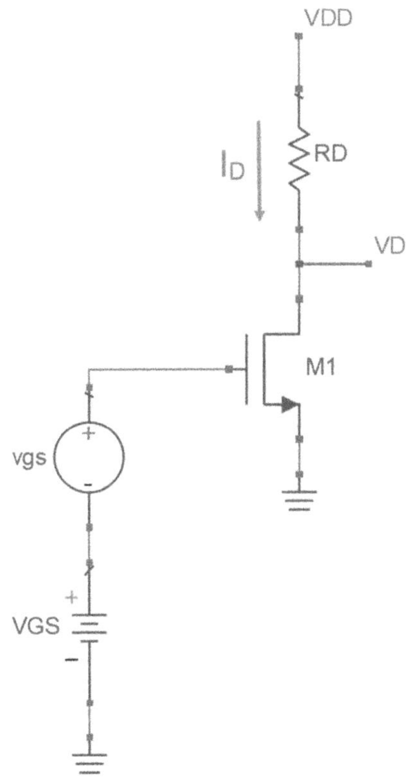

FIGURE 4.31 CS amplifier with R_D load illustrative circuit.

We assume that the above NMOS operates in saturation mode.

First, let us neglect the Level 1 model second-order effects. The idealized current–voltage equation (4.2), shown earlier for DC operation, works exactly the same for total signals.

$$i_D = \frac{1}{2}\mu_n C_{ox}\frac{W}{L}(v_{GS} - V_{TH})^2 \tag{4.6}$$

Substituting each total signal of equation (4.6) with a representation as a sum of its DC and small-signal component provides an expression (4.7) for the small signal i_d as a function of the small-signal input voltage v_{gs}:

$$i_d = \mu_n C_{ox}\frac{W}{L}(V_{GS} - V_{TH})v_{gs} + \frac{1}{2}\mu_n C_{ox}\frac{W}{L}v_{gs}^2 \tag{4.7}$$

The first term of i_d is linear in terms of v_{gs} and the second term is quadratic in v_{gs}. For the small-signal amplifier to operate linearly, the first term must be much larger than the second term, leading to the **MOSFET small-signal assumption** equation (4.8) (an expression that works for NMOS and PMOS). The small-signal assumption is a condition on the size of the input voltage's amplitude.

$$|v_{gs}(t)| \ll 2|V_{GS} - V_{TH}| \tag{4.8}$$

If the MOSFET obeys the small-signal assumption, thus operating approximately as a linear small-signal amplifier, it has an approximated transconductance gain g_m given in equation (4.9).

$$g_m \approx K_p\frac{W}{L}(V_{GS} - V_{TH}) \tag{4.9}$$

We keep in mind that with g_m defined as in equation (4.9), the small-signal diagram's dependent current source $g_m v_{gs}$ (shown in Figure 4.32) works the same for NMOS and PMOS. For any discussion of g_m as a standalone entity, the definition of equation (4.9) for PMOS must include the absolute value of the DC overdrive voltage.

The simplest small-signal diagram for a MOSFET is shown in Figure 4.32.

Equation (4.9) combined with the idealized saturation mode current equation (4.2) creates two more expressions for g_m; all are mathematically equivalent.

$$g_m \approx \frac{2I_D}{V_{GS} - V_{TH}} \tag{4.9a}$$

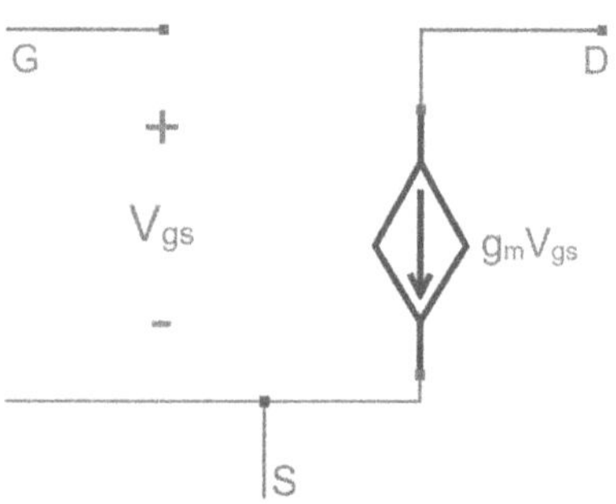

FIGURE 4.32 The simplest small-signal diagram of MOSFET consisting of a single small-signal parameter.

$$g_m \approx \sqrt{2K_p \frac{W}{L} I_D} \tag{4.9b}$$

The channel length modulation effect adds a resistor r_o to the small-signal diagram, as shown in Figure 4.33.

From equation (4.3), a differentiation of I_D with respect to V_{DS} provides an expression for r_o, shown in (4.10):

$$r_o = \frac{1}{\lambda I_D} \tag{4.10}$$

Inclusion of λ in the current equation, as in equation (4.3), necessitates modifications in the g_m expressions. In equation (4.6), there is a need to include a multiplicative correction $1+\lambda V_{DS}$. In equation (4.9b), the same correction term is included in the denominator inside the square root expression. Equation (4.9a) remains unchanged despite the nonzero value of λ, making it users' favorite expression of g_m.

We refer the reader to Razavi's book for a discussion of the MOSFET's most inclusive Level 1 small-signal diagram. See Figure 4.34. The diagram also considers the body effect and high-frequency capacitance effects.

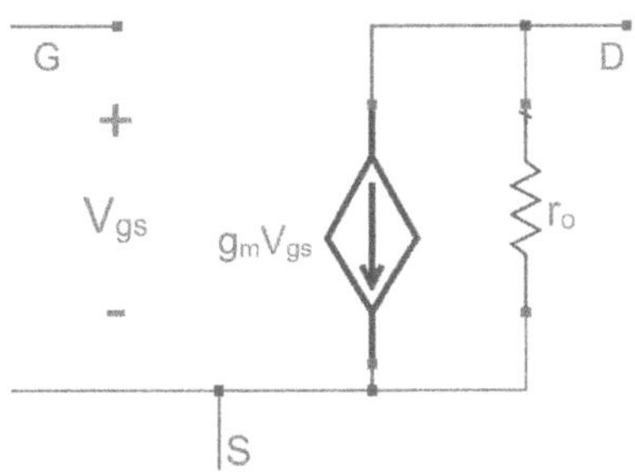

FIGURE 4.33 MOSFET small-signal diagram that includes the channel length modulation effect.

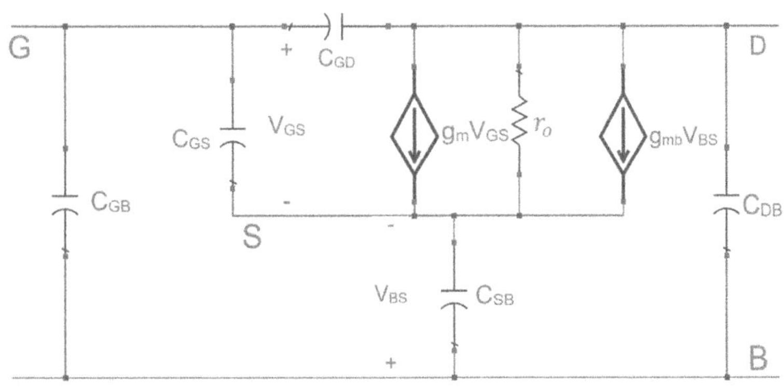

FIGURE 4.34 The full Level 1 MOSFET small-signal diagram [Razavi1].

The transconductance g_{mb} depends on g_m and the DC voltage V_{SB}. For more discussion, see [Razavi1].

Of the five capacitors, shown in Figure 4.34, C_{GB} is negligible in saturation mode.

C_{SB} and C_{DB} are both bias voltage dependent and depend on the area and perimeter of the drain or the source, respectively.

The overlap capacitances C_{GD} and C_{GS} are constant.

4.8 ADS DESIGN AND SIMULATION OF A CS AMPLIFIERS; COMPARISON WITH AND WITHOUT MOSFET HIGH-FREQUENCY CAPACITANCE EFFECTS; VOLTAGE GAIN TWEAKING OF CS AMPLIFIER WITH R_D LOAD; DRAIN AND SOURCE AREA AND PERIMETER AND EFFECT ON AMPLIFIER'S BANDWIDTH

Example 4.4

Let us create a schematic circuit of the NMOS CS amplifier with an R_D load, given in Figure 4.31. See Figure 4.35. Let us choose some arbitrary component and source values: $V_G = 2$ V (this is also V_{GS}, as the source is at ground) and $R_D = 2.5$ kΩ.

We shall compare two MOSFET Level 1 models (that have the same V_{TH}, Lambda, and W/L):

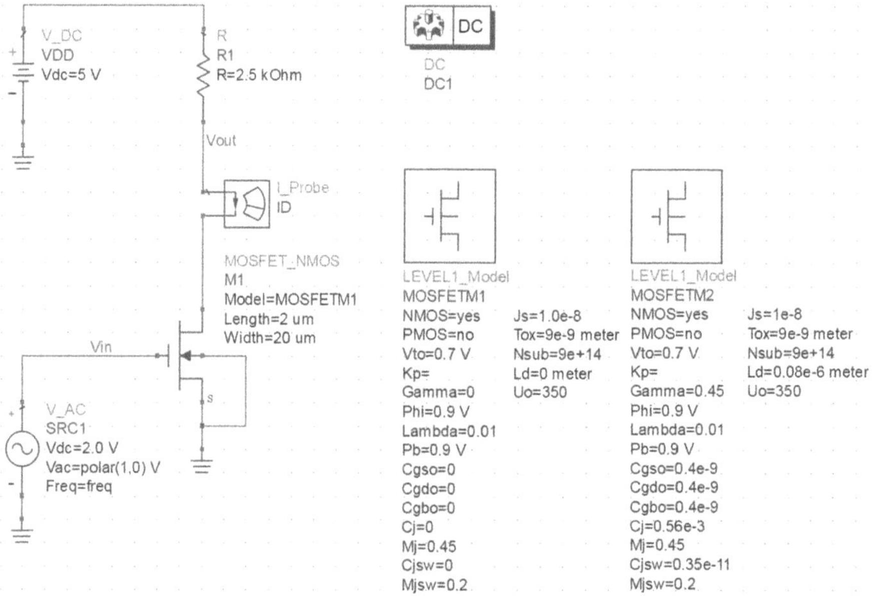

FIGURE 4.35 Schematic of the CS amplifier of Figure 4.31.

MOSFETM$_1$ has parameters as in MOSFETM$_2$, but all capacitance effects, body effect, and L_D effect were nulled.
MOSFETM$_2$ follows Razavi's suggestion for a realistic NMOS Level 1 model.

Both models have the same nonzero $\lambda = 0.01$ [1/V].

We set W/L = 10. All drain and source areas and perimeters are initially assumed to be zero.

In Figure 4.36, the DC simulation controller is replaced by the AC simulation controller. Yet, we use DC schematic annotation, but as seen in Figure 4.36, there is no DC controller in the schematic. How can this be?

In modern simulators like ADS, a DC simulation is always performed by the other simulators (e.g., AC simulation) before their analysis is performed. This allows for things like topology checks (e.g., DC convergence). In the case of AC analysis, it provides the information for the small-signal equivalent circuit created in the background.

DC annotation to the schematic proceeds in the same way, just with no DC controller.

Back to **Example 4.4**. We see in Figure 4.36 that with both models the NMOS is in saturation mode:

Model MOSFETM$_2$: $V_D = 1.86\ V > V_G - V_{TH} = 2 - 0.7 = 1.4\ V$
Model MOSFETM$_1$: $V_D = 2.1\ V > V_G - V_{TH} = 2 - 0.7 = 1.4\ V$.

With model MOSFETM$_2$ (Razavi), $I_D = 1.26$ mA, whereas with model MOSFETM$_1$, $I_D = 1.16$ mA. Why are the I_D currents different? MOSFETM$_2$ has an L_D effect, while model MOSFETM$_1$ does not. L_D slightly lowers the effective L. Therefore, whenever the L_D effect is present, the aspect ratio W/L becomes effectively larger, which increases I_D, and consequently the small-signal voltage gain.

In Figure 4.36, let us estimate the K_p parameter. You may recall that we specified U_0 and T_{ox} (and left K_p empty). If we directly specify K_p, it overrides U_0 and T_{ox}. We include both the L_D and LAMBDA effects, of course, because this time we are after a precision calculation of a parameter.

$$\text{MOSFETM}_2 : I_D = 0.5 \cdot K_p \cdot \frac{20}{2 - 2 \cdot 0.08}(2 - 0.7)^2(1 + 0.01 \cdot 1.86) = 1.26 \text{ mA}$$

$$\text{MOSFETM}_1 : I_D = 0.5 \cdot K_p \cdot \frac{20}{2}(2 - 0.7)^2(1 + 0.01 \cdot 2.1) = 1.16 \text{ mA}$$

In both models, we get the same K_p parameter: $K_p = 0.135$ mA/V^2.

In Figure 4.37, we see that the CS amplifier with the NMOS model MOSFETM$_2$ has a larger voltage gain (of 4.688). The gain begins to fall around f = 200 MHz due to MOSFET high-frequency capacitance effects. The other CS amplifier, which has model MOSFETM$_1$ (no capacitances and $L_D = 0$), has slightly lower gain (of 4.333) and an infinite bandwidth (BW).

Why is one gain larger than the other? Recalling some of the equations (4.9) g_m equations. If $V_{GS} - V_{TH}$ stays constant, and R_D stays the same, then the larger the I_D, the larger the g_m, and the larger the voltage gain.

A strategy for increasing the voltage gain by a factor of K is as follows:

- Keep both I_D and R_D constant.
- Decrease the overdrive voltage by a factor K.
- Increase W/L by factor K^2.

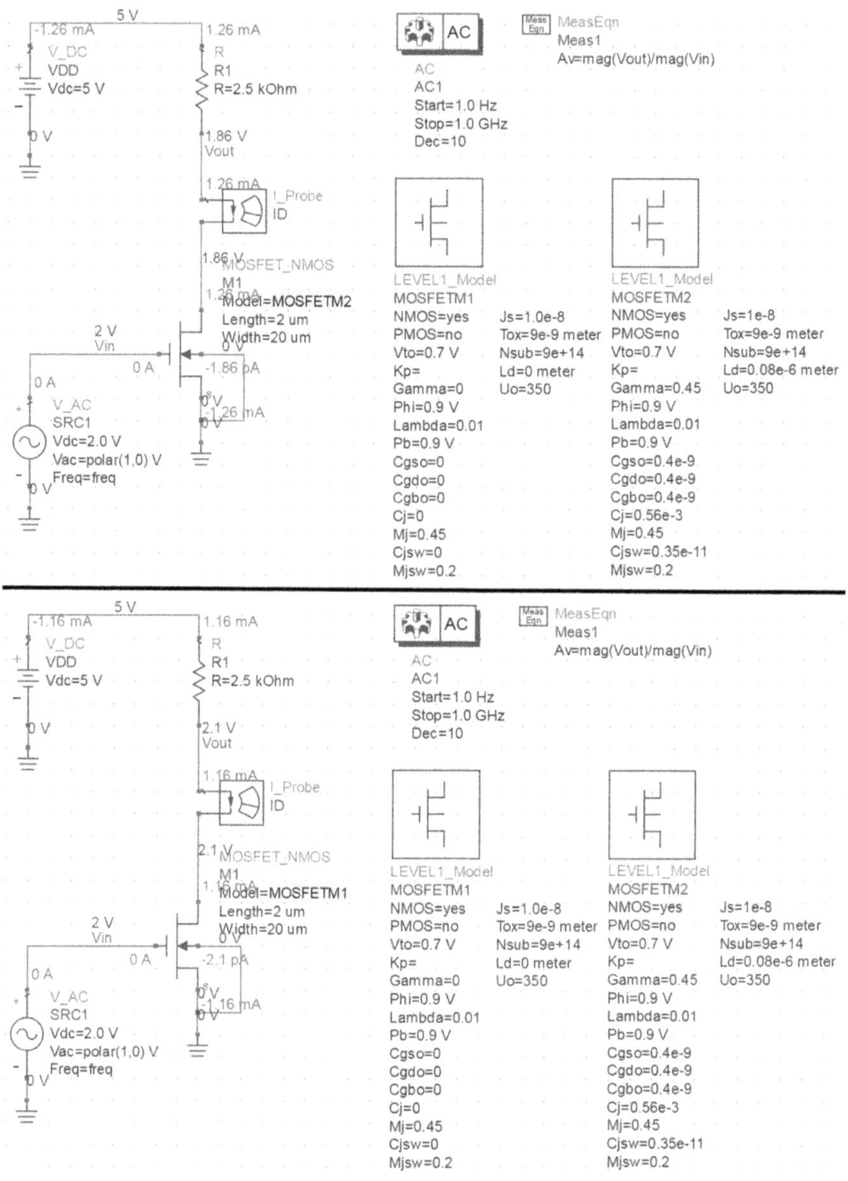

FIGURE 4.36 DC results: Model MOSFETM2 (up) and Model MOSFETM1 (down).

Let us explain the strategy using idealized saturation current equations (4.2) and (4.9) for g_m. Since the voltage gain absolute value more or less equals $g_m \cdot R_D$, g_m must increase by a factor of K. Note that I_D can be written in terms of g_m: $I_D = (1/2) g_m(V_{GS} - V_{TH})$. If g_m increases by a factor of K and I_D stays unchanged, then the overdrive voltage must decrease by a factor K. By the g_m expression, if g_m needs to increase by a factor of K and the overdrive voltage decreases by a factor of K, then the only way to raise g_m by a factor of K is to raise W/L by a factor of K^2.

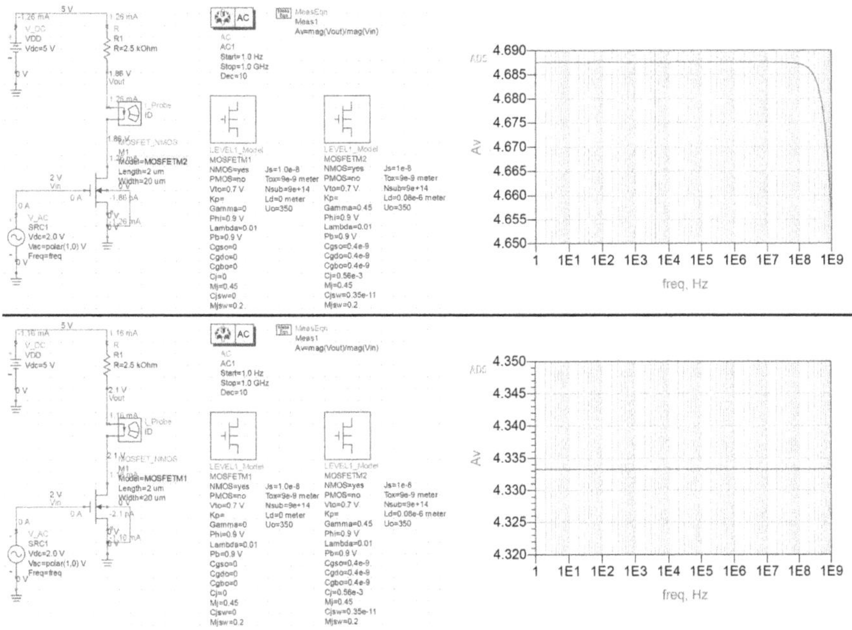

FIGURE 4.37 Comparison of gain frequency response of the two NMOS models.

Example 4.5

The gain-increasing strategy will next be demonstrated on the CS amplifier of Example 4.3, with the NMOS model MOSFETM2 that had a voltage gain of −4.688. Say that we wish the new gain to be −30.

- Then $K = 30/4.688 = 6.4$.
- $\text{Overdrive}_{old} = V_{G,old} - V_{TH} = 2 - 0.7 = 1.3\,V$.
- $\text{Overdrive}_{new} = \text{Overdrive}_{old}/K = 1.3/6.4 = 0.203\,V$.
- $V_{G,new} = \text{Overdrive}_{new} + V_{TH} = 0.203 + 0.7 = 0.903\,V$.
- $(W/L)_{old} = 10$ (assuming no L_D effect). With L_D effect $(W/L)_{old} = 20/(2 - 2\cdot 0.08) = 10.87$.
- $(W/L)_{new} = (W/L)_{old}\cdot 6.4^2 = 409.6$ (with no L_D effect) or $(W/L)_{new} = (W/L)_{old}\cdot 6.4^2 = 10.87\cdot 40.96 = 445.2$.
- $W_{new} = 409.6\cdot 2 = 819\,\mu m$ (approximating $L_D = 0$), or $W_{new} = 445.2\cdot 1.84 = 819.2\,\mu m$ (with nonzero L_D).

Note that L_D has no effect on the gain enhancement method that involves decreasing V_G and increasing W.

We applied the new V_G and new W, inspecting first the DC solution. We see (Figure 4.38) that I_D stayed more or less the same: $I_D = 1.25\,mA$.

Next, let us see if the new V_G and W values indeed enhanced the voltage gain (Figure 4.39).

Indeed, the gain enhancement method works. Notice the reduction in the amplifier's BW. The enhanced amplifier voltage gain A_v begins to fall around

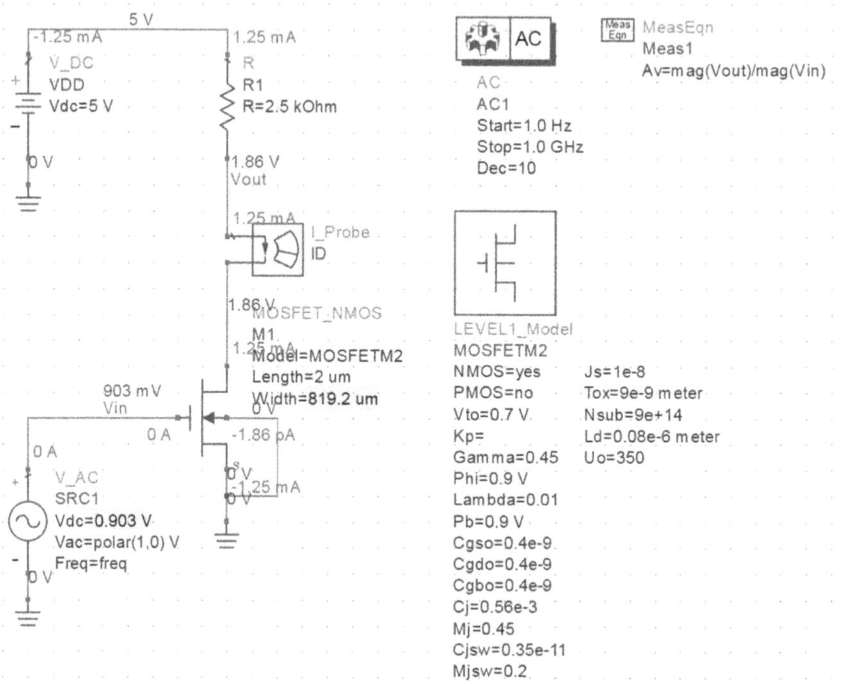

FIGURE 4.38 Schematic of the circuit that implements the voltage gain enhancement strategy.

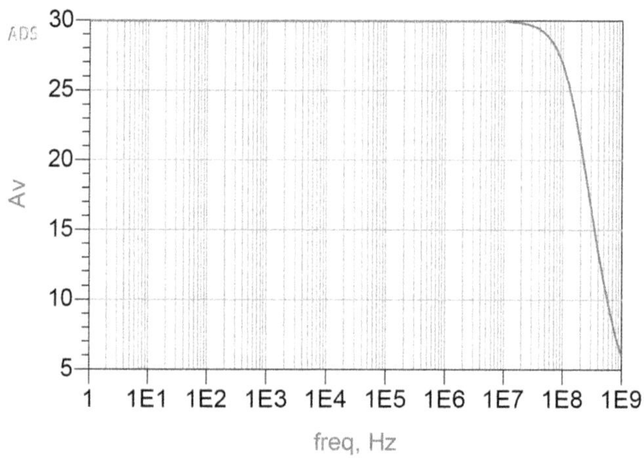

FIGURE 4.39 The resulting enhanced gain.

f=20 MHz (compared to 200 MHz that we had prior to implementing the gain enhancement). Why?

It is because the MOSFET is now much bigger, which increases the size of the high-frequency capacitances. Actually, we are missing several capacitors of Figure 4.34 because of taking $A_D = A_S = 0$ and $P_D = P_S = 0$, as happens by default.

Next, we shall add nonzero drain's as well as source's areas and perimeters. It will cause further reduction in the amplifier's BW.

Before we move on, let us talk briefly about the concept of BW. As a figure of merit, BW indicates the frequency range for which the design is "useful". The BW is typically specified as the frequency range at which, for a constant amplitude input signal, the output power of the signal drops by half (a.k.a the 3 dB BW). In a voltage amplifier like the CS, the BW is specified as the frequency range at which the voltage gain is $0.707 \cdot A_{vmax}$ (Figure 4.40).

In the ADS data display page of the Figure 4.40 schematic, we perform a simple calculation for the 3 dB BW and display the result in a table. We then place a marker and move it to approximately the 3 dB BW gain magnitude calculated and read BW from the X-axis. BW ≈ 199.5 MHz.

In the updated schematic (of Figure 4.41), the units of L and W were changed from μm to meters (m) to support the calculation of A_d and A_s, which have units of m^2, and to support the calculation of P_d and P_s with m units.

Example 4.6

The "bandwidth_func(db(Av),3" function measures the −3 dB frequency. That is, the frequency f at which the gain falls to $1/\sqrt{2}$ of the midband gain. We shall simulate the CS amplifier that has enhanced gain (Figure 4.38), with results showing in Figure 4.42.

The MOSFET parameters cgso and cgdo activate overlap capacitances whenever L_D is nonzero.

The parameter cgbo activates a depletion capacitance between G and B. According to Razavi's book, C_{GB} is negligible in saturation and triode modes due to the "masking from the gate" provided by the channel.

Tox influences the MOSFET oxide capacitance.

All the above capacitance effects exist regardless of the drain and source dimensions. They all play a role in determining the BW of 200 MHz (measured in Figure 4.42).

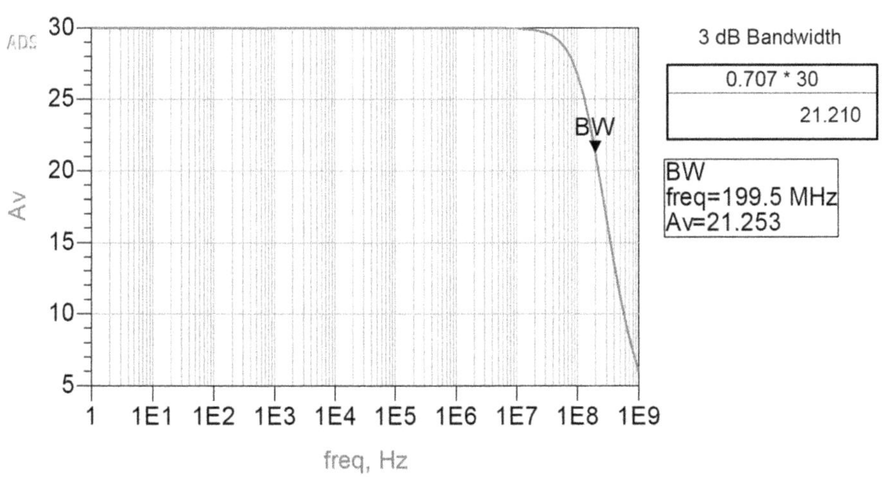

FIGURE 4.40 Manual BW measurement.

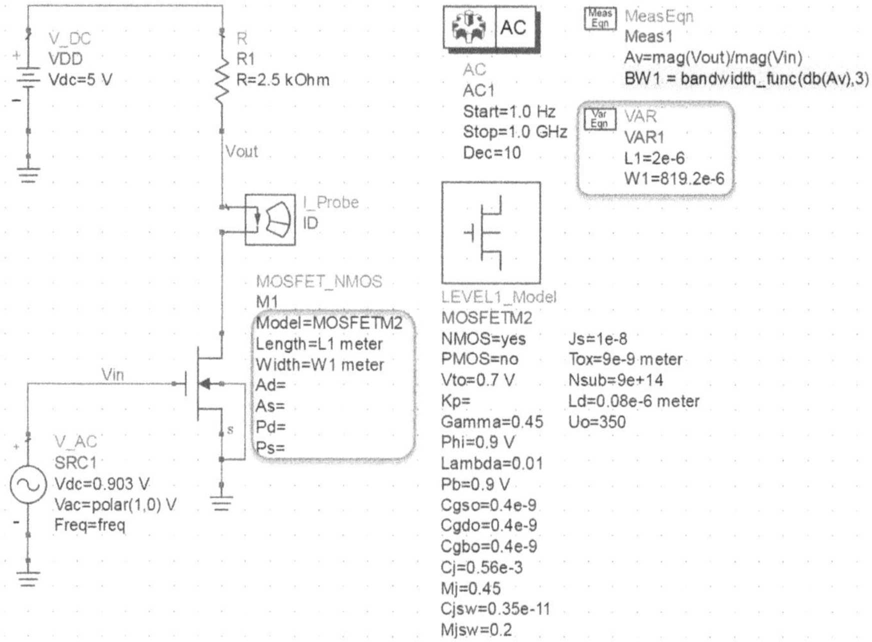

FIGURE 4.41 Updated schematic for measuring CS amplifier BW.

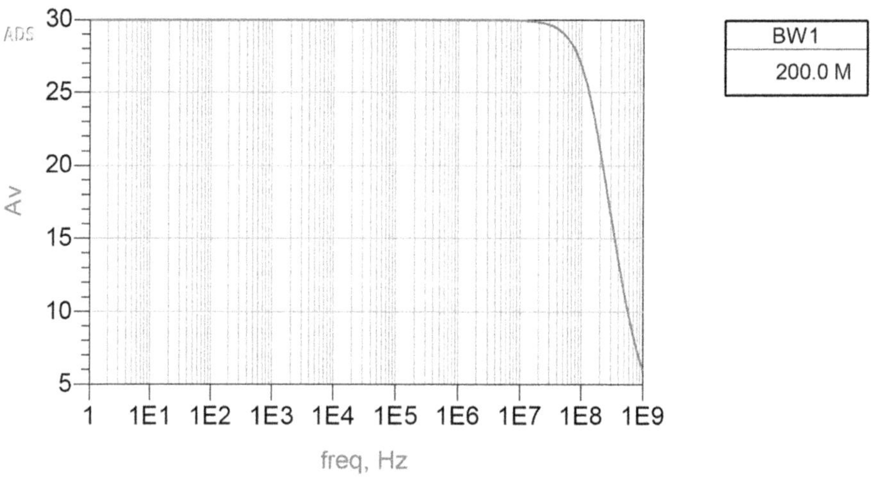

FIGURE 4.42 The starting CS BW for zero drain and source areas and perimeters.

Next, we consider a rectangular geometry for the MOSFET transistor (Figure 4.43):

A MOSFET transistor viewed from above (Figure 4.3) reveals a rectangular gate region (with dimensions L_1 by W_1). We assume that the drain and source regions are rectangles with the same dimensions (L) as that of the gate. Nonzero areas of

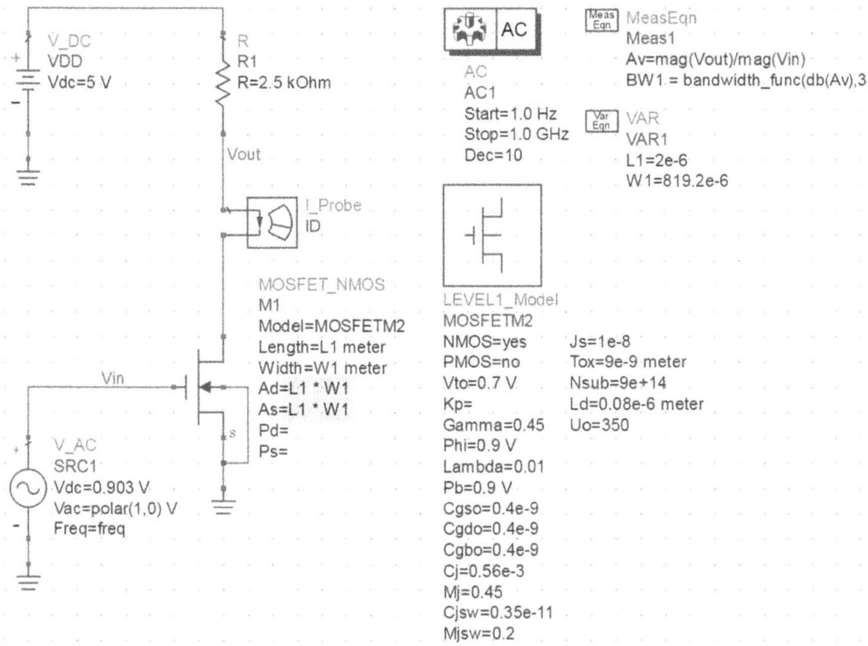

FIGURE 4.43 Inclusion of drain and source areas, based on a rectangular MOSFET geometry assumption.

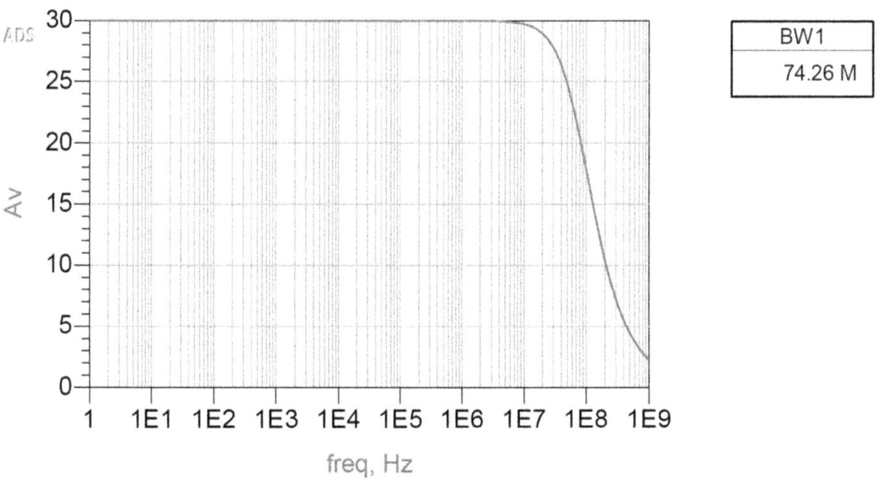

FIGURE 4.44 Frequency response, considering the source and drain areas.

the drain and source, combined with the cj and mj parameters, are responsible for some depletion capacitances between D and B and between S and B.

The additional capacitance effects (that depend on the drain and source areas) cause a reduction of the amplifier's BW from 200 to 74.26 MHz. See Figure 4.44.

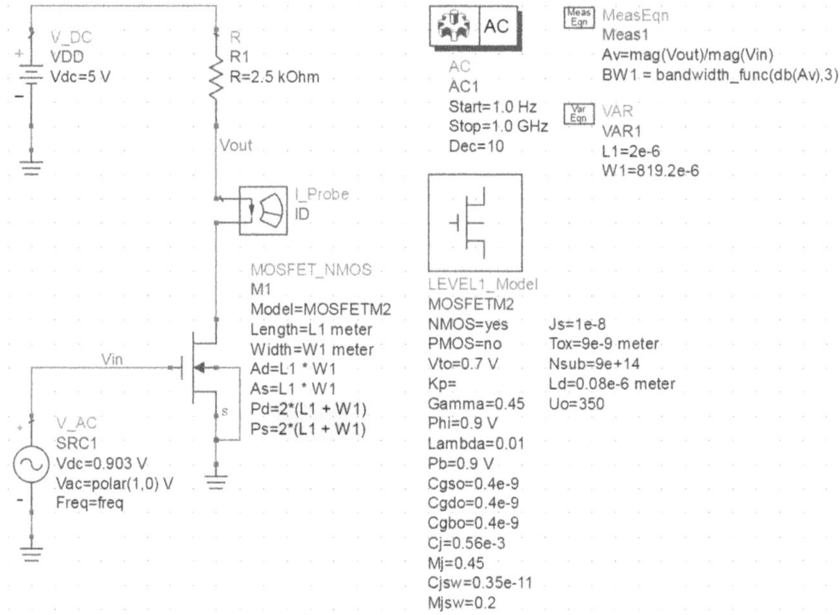

FIGURE 4.45 Including both drain and source areas and perimeters, assuming rectangular geometry.

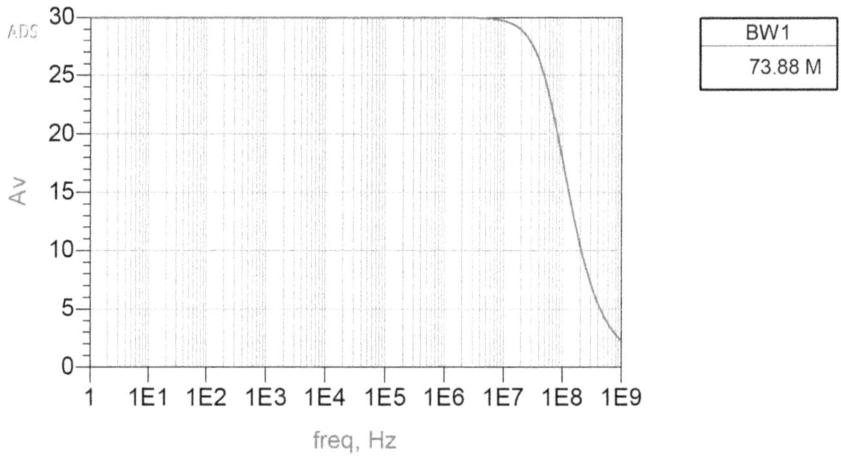

FIGURE 4.46 Frequency response that considers drain and source areas and perimeters, assuming rectangular geometry.

Now, let us add the perimeter length of the drain and source regions (Figure 4.45). The nonzero perimeter, combined with the cjsw and Mjsw parameters, creates additional depletion capacitance effects.

It further reduces the BW of the amplifier, as seen in Figure 4.46.

With all the MOSFET Level 1 high-frequency capacitances included, the correct amplifier's BW, based on the Level 1 model, is 73.88 MHz.

4.9 BASIC MOSFET CURRENT MIRROR CURRENT SOURCES

Some basic principles of simple MOSFET current mirrors can be demonstrated by the circuits shown in Figure 4.47. See [Sedra1] and [Razavi1].

It is assumed that the MOSFET transistors **M_1 and M_2** (in Figure 4.47 (left)) **are identical (except possibly for their aspect ratios)**. The transistors have the same K_p, V_{TH}, and other model parameters. M_1 and M_2 are connected gate-to-gate and source-to-source, so, "by structure", both have the same V_{GS} voltage.

The purpose of a DC current mirror is to create a DC output current I_o based on an input reference current I_{ref}. As is shown in Figure 4.47 (right), **multiple independent current mirrors may be constructed with a single reference current**. M_1, M_2, and M_3 are identical, but with different aspect ratios (as will be discussed soon), different output currents (such as I_{o2} and I_{o3}) may be realized.

The reference side of the mirror, recognized by the diode connection of the MOSFET, carries I_{ref}, whereas the output side of the mirror carries the current I_o. In Figure 4.47, transistor M_1 operates in saturation mode because of the short circuit between its drain and gate. Transistor M_2 may operate in any of the ON modes, depending on the load that I_o is connected to. For each current mirror to work correctly as a DC current source, the respective **output transistor must operate in saturation mode**.

The simplest implementation of I_{ref} can be done by means of a positive supply V_{DD} and a resistor R, connected between V_{DD} and the drain of the reference transistor. Such an implementation may suffer from V_{DD} fluctuations and temperature variations of R. The current source I_{ref} is often implemented as a standalone circuit, and it is carefully regulated. A typical system on a chip may need a large number of DC current sources, but it needs only one reference current source.

In Figure 4.47 (left), let us assume that M_2 operates in saturation mode, an assumption that must be checked and justified once the load of M_2 becomes known. Let us further neglect MOSFET Level 1 second-order effects. In the ratio I_o/I_{ref} of the

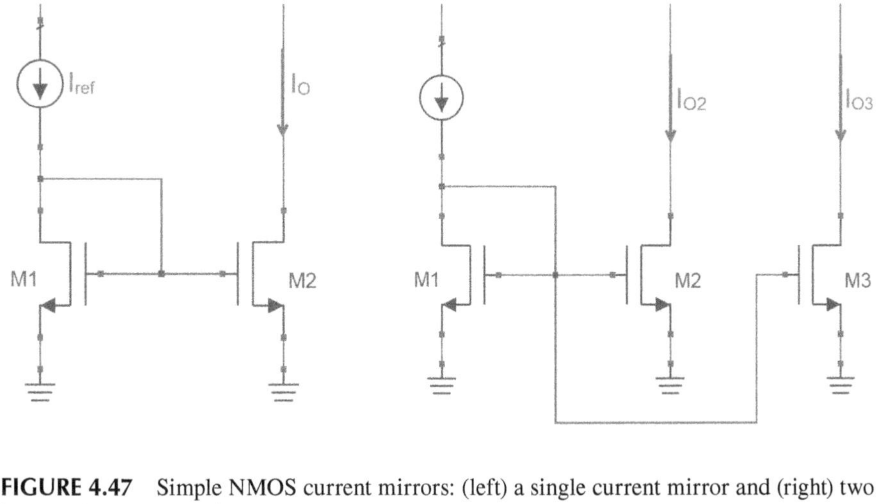

FIGURE 4.47 Simple NMOS current mirrors: (left) a single current mirror and (right) two independent current mirrors.

respective currents, using the idealized equation (4.2), we note that the K_p parameter and the overdrive voltage of each current cancel one another. We are left with the **"ratio of aspect ratios" simplified design formula**:

$$\frac{I_o}{I_{ref}} = \frac{\frac{1}{2}K_{P2}\left(\frac{W}{L}\right)_2 (V_{GS2} - V_{TH2})^2}{\frac{1}{2}K_{P1}\left(\frac{W}{L}\right)_1 (V_{GS1} - V_{TH1})^2} = \frac{\left(\frac{W}{L}\right)_2}{\left(\frac{W}{L}\right)_1} \tag{4.11}$$

On the reference side of the mirror, due to the saturation mode, the selected $V_{GS1} = V_{DS1} = V_{GS2} = V_{GS}$ and $(W/L)_1$ must be consistent with the value of the current I_{ref}. The load circuit that is fed by I_o creates some V_{DS2} value. V_{DS2} must satisfy the saturation mode condition for M_2. In general, V_{DS1} and V_{DS2} may be significantly different. As a result, and even though $\lambda_1 = \lambda_2 = \lambda$, the channel modulation effect introduces an error to the equation (4.11) design formula, as shown in equation (4.12):

$$\frac{I_o}{I_{ref}} = \frac{\left(\frac{W}{L}\right)_2 (1 + \lambda_2 V_{DS2})}{\left(\frac{W}{L}\right)_1 (1 + \lambda_1 V_{DS1})} \tag{4.12}$$

A simple NMOS current mirror produces Io that goes from the load back into the current source. This type of current source is known as a "**sink**". A current source that produces a DC current that goes from the current source into the load is called "**source**". A **source current mirror is made of identical PMOS transistors**, as shown in Figure 4.48 (left).

In a circuit that contains many current mirror current sources, some NMOS mirrors and some PMOS mirrors, all the NMOS transistors need to have the same parameters, and all the PMOS need to have the same parameters, though not necessarily similar to those of the NMOS group of mirrors.

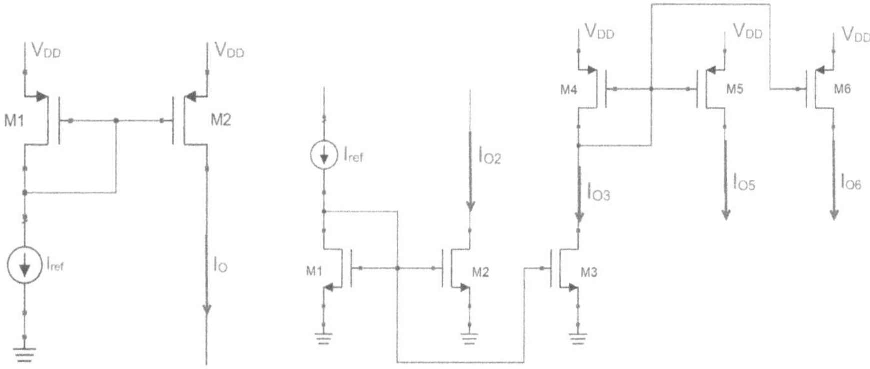

FIGURE 4.48 (Left) Simple PMOS current mirror current source. (Right) Current steering from NMOS mirrors to PMOS mirrors.

The single I_{ref} that serves all the mirrors may be inserted within the NMOS mirrors or within the PMOS mirrors. The action of transferring a current from the NMOS group to the PMOS group, or vice versa, is called "**Current Steering**". It is demonstrated in Figure 4.48 (right).

In the current steering that is demonstrated in Figure 4.48 (right), the steered current is I_{o3}, which becomes a reference current for M_4 that serves all the PMOS sources. The example features a single sink (I_{o2}) and two sources (I_{o5} and I_{o6}).

4.10 ADS: STUDY OF SIMPLE CURRENT MIRRORS

Example 4.7

In the example of Figure 4.49, let us imagine that we are given an external (to the chip) temperature and supply variations stabilized DC current source $I_{ref} = 100$ μA. We wish to build a sink current source of $I_{D2} = 300$ μA. We created a mirror consisting of two NMOS transistors. We let the nominal design be $(W/L)_2/(W/L)_1 = 3$.

The annotated DC results are shown in Figure 4.50.

The actual output current I_{D2} has an error. It is $I_{D2} = 310$ μA instead of the intended $I_{D2} = 300$ μA. Why?

The current error is due to V_{DS1} and V_{DS2} not being equal. In Figure 4.50, we see that $V_{DS1} = 1.07$ V and $V_{DS2} = 4.38$ V. Therefore, because LAMBDA is nonzero, there is a multiplicative error of $(1 + \lambda V_{DS2})/(1 + \lambda V_{DS1})$.

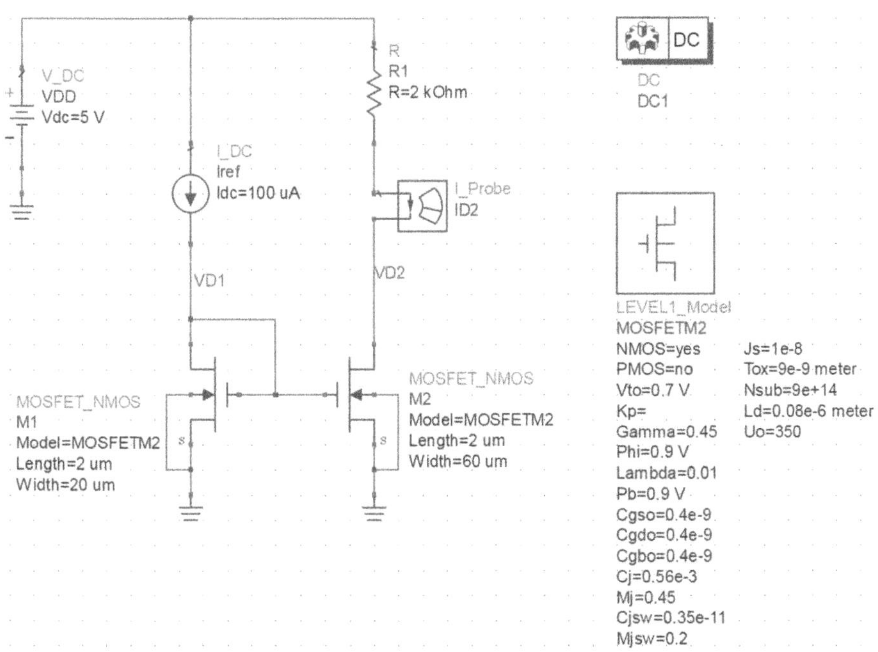

FIGURE 4.49 Simple NMOS current mirror.

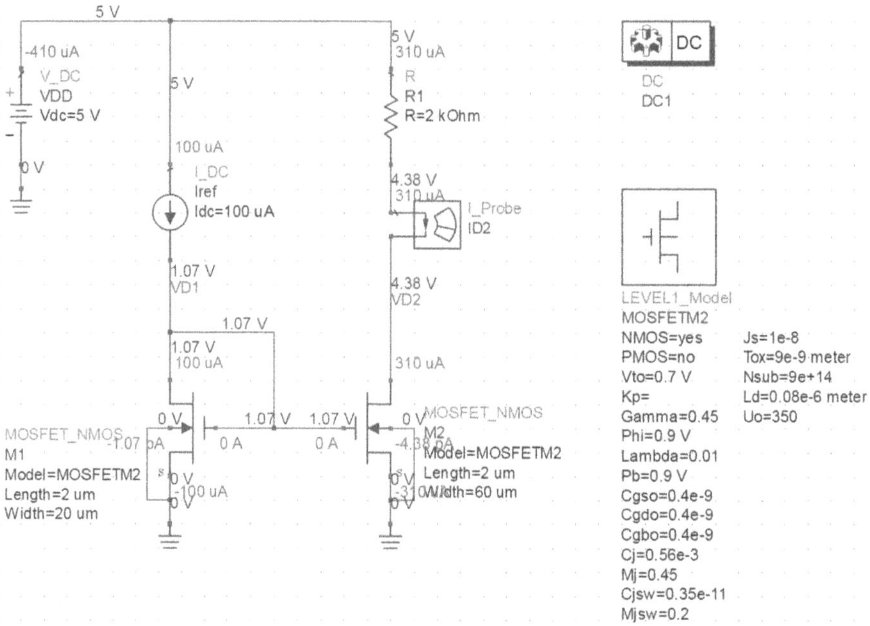

FIGURE 4.50 DC results of the simple NMOS mirror intended to provide a sink current of $I_{D2}=300\ \mu A$.

Resistor R_1, along with the supply V_{DD}, represents a load served by the output current I_{D2}. For different values of R_1, transistor M_2 will experience different V_{DS} values, and as a result, the current I_{D2} will vary.

For large values of R_1, transistor M_2 may go from saturation mode to triode mode (Figure 4.51).

Let's conduct two sweep studies.

We shall first sweep R_1. Then, for $R_1=2\ k\Omega$, let us sweep W_2 in order to tune $I_{D2}=300\ \mu A$.

In Figure 4.52, the unit of R_1 is kΩ. The unit of I_{D2} is μA. A good DC current source stays constant for any load resistance. We see that I_{D2} varies as R_1 increases, due to the V_{DS} and LAMBDA errors.

Does the corner (in the plot of Figure 4.52) represent a borderline between saturation and triode mode? Let us check:

$V_{GS2}=1.07\ V$, $I_{D2}=0.3\ mA$ for $R_1=15\ k\Omega$, $V_{DS2}=V_{DD}-I_{D2}\cdot R_1=5-0.3\cdot 15=0.5\ V>V_{GS2}-V_{TH2}=1.07-0.7=0.37\ V$. It is close to the borderline, though.

In the triode mode, the current mirror source completely stops functioning, as we see in Figure 4.52.

The schematic of Figure 4.53 is for a W_2 sweep for a fixed load resistor of $R_1=2\ k\Omega$.

The V_{DS} error caused I_{D2} to become larger than what was intended in the approximated design (that was based on the ratio of aspect ratios) because V_{DS2} was larger than V_{DS1}. We see from the W_2 tuning results in Figure 4.54 that W_2 needs to be slightly reduced from 60 to 58 μm. Of course, this tuning is good only for $R_1=2\ k\Omega$. For another load, we need to re-tune.

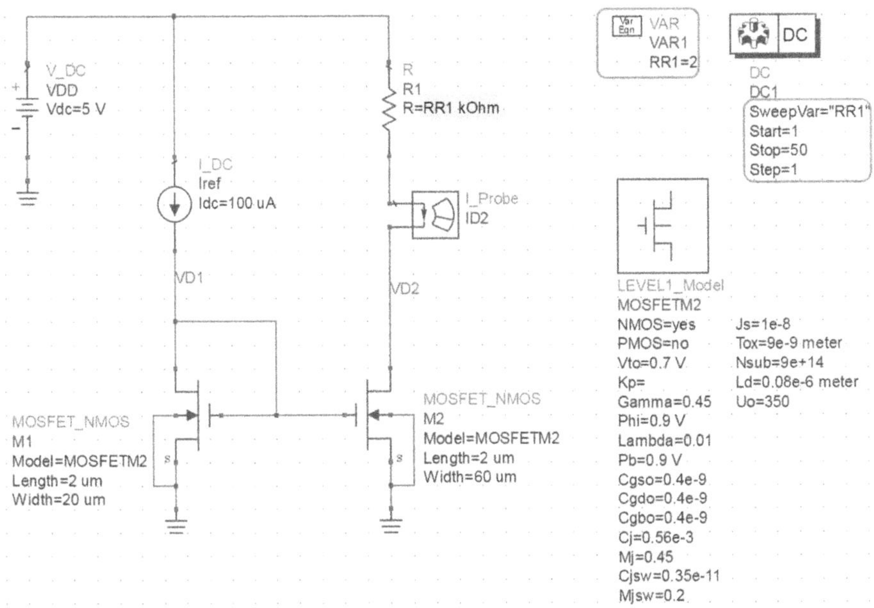

FIGURE 4.51 Simple NMOS mirrors with a sweeping of the load resistance.

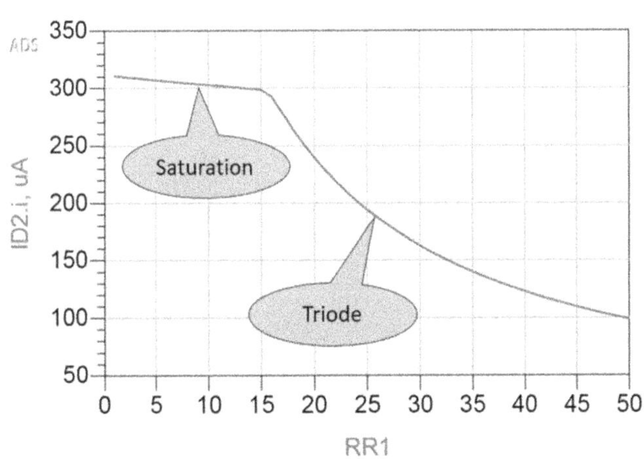

FIGURE 4.52 R load sweeping in a simple NMOS mirror.

Example 4.8

Next, in Figure 4.55, we shall demonstrate a current steering example.

All the mirrors need just one reference current, $I_{ref} = 100\ \mu A$.

Because, here, I_{ref} is tied to the NMOS mirrors, we need to transfer a current from the group of NMOS mirrors to the group of PMOS mirrors. Say that the transfer current is $I_{D3} = 100\ \mu A$.

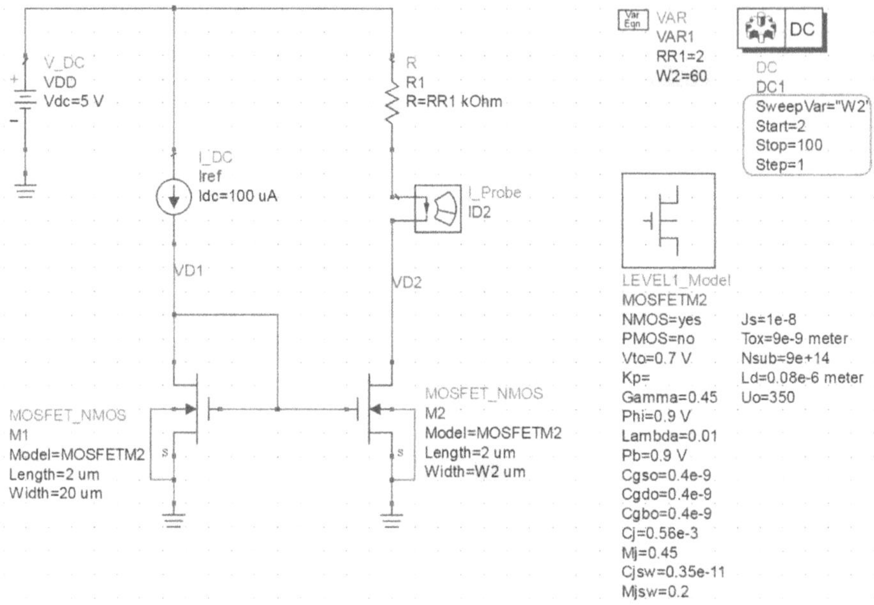

FIGURE 4.53 Simple NMOS mirror with a W sweep of the output transistor.

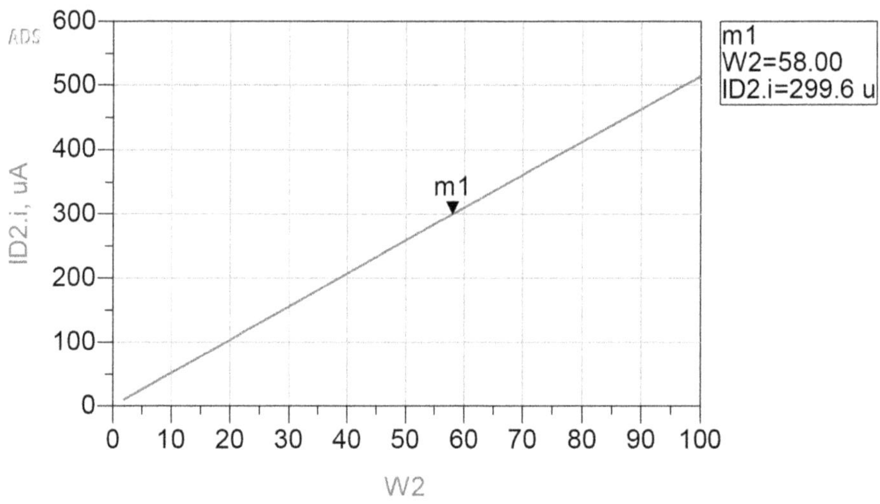

FIGURE 4.54 Mirror's output current as function of W of the output transistor, for a fixed R load.

The schematic shown in Figure 4.55 is constructed by expanding the schematic shown in Figure 4.49. In addition to the NMOS mirror that sinks 300 μA from R_1, let us add a PMOS mirror that sources 400 μA into another load, R_2.

The annotated DC results of the circuit of Figure 4.55 are displayed in Figure 4.56.

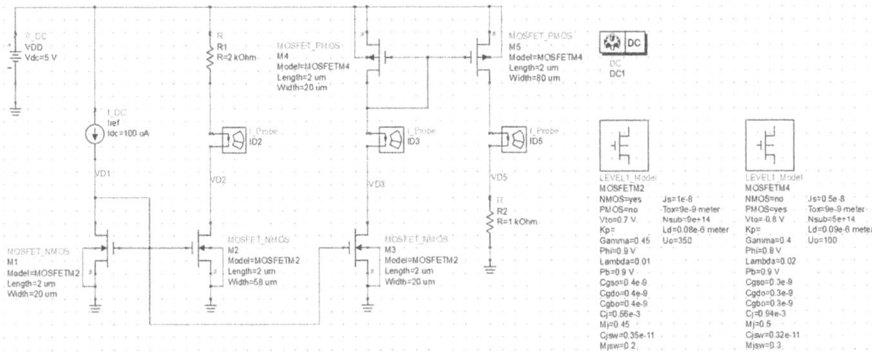

FIGURE 4.55 Current steering from NMOS mirrors to PMOS mirrors.

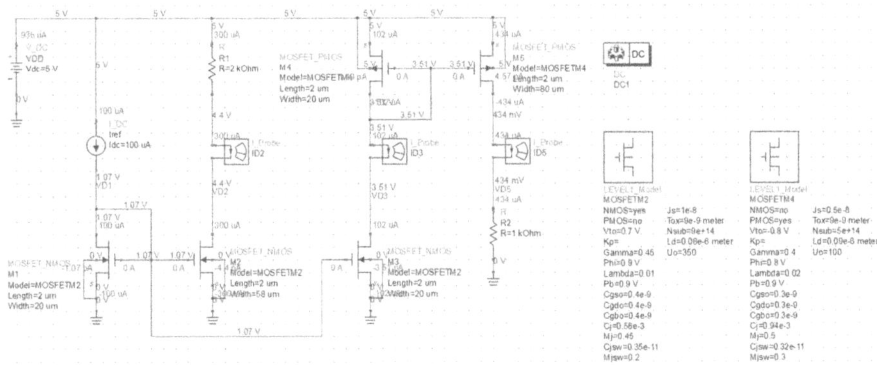

FIGURE 4.56 DC results of the current steering example.

The NMOS mirror (M_1, M_2) is already tuned to provide a sink current of $I_{D2}=300$ μA. The second NMOS mirror (the one that provides the current ID3 steering to the PMOS group) was designed to deliver $I_{D3}=100$ μA.

We note in Figure 4.56 that there is a slight error.

As seen in Figure 4.56, the second NMOS mirror actually delivers $I_{D3}=102$ μA. Does it matter?

The source current I_{D5} was designed for $I_{D5}=400$ μA. The resulting current was $I_{D5}=434$ μA, indicating a large error.

There are two sources of this error: the error in the steering current I_{D3} and the LAMBDA error between PMOS M_4 and M_5.

We can simply tune the output transistor M_5 in order to fix the error. See Figures 4.57 and 4.58.

Moving the marker (in Figure 4.57) with the keyboard has a resolution that is determined by the sweep step size. We see that a good choice for W_5 is $W_5=74$ μm (replacing the design value of $W_5=80$ μm).

In conclusion, it didn't matter that we left $I_{D3}=102$ μA, as the tuning of W_5 covered both sources of error (LAMBDA mismatch and steering current error).

Actually, there is one more source of error, and it is the L_D effect.

The design allows for many degrees of freedom. That is, there are many ways to distribute the (W/L) of each stage in order to get the desired output current.

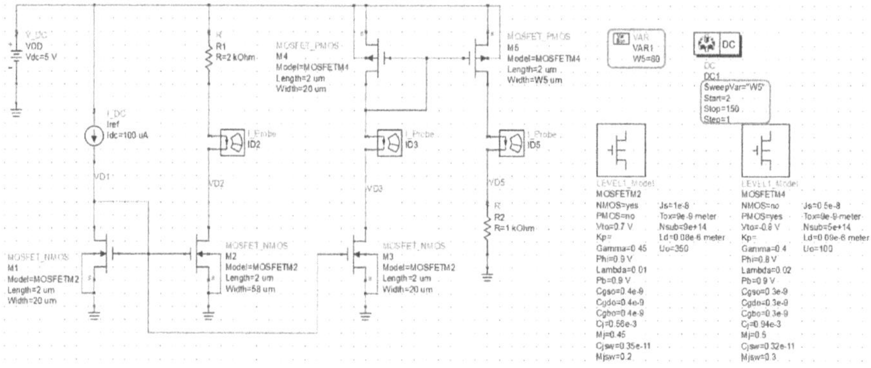

FIGURE 4.57 Current steering example with W tuning of the source output transistor.

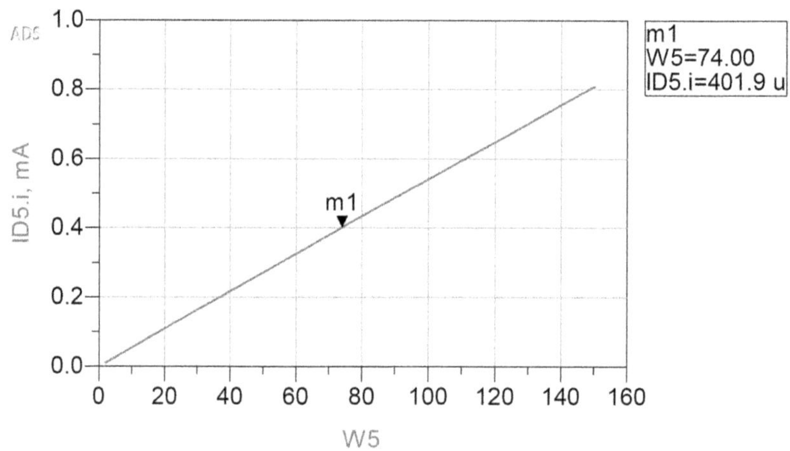

FIGURE 4.58 Tuning plot of the output transistor of a PMOS mirror.

The LAMBDA mismatch error is significant and causes the current source to become very non-ideal.

There are current mirrors that are more sophisticated and more resilient to V_{DS} mismatch errors [Razavi1].

REFERENCES

[Gray1] Paul R. Gray, Paul J. Hurst, Stephen H. Lewis and Robert G. Meyer, *Analysis and Design of Analog Integrated Circuits*, 4th Edition, John Wiley & Sons, 2001.

[Murmann1] Boris Murmann, *Analysis and Design of Elementary MOS Amplifier Stages*, The Modular Series of Microelectronic Device & Circuit Design, R. T. Howe and C. G. Sodini, Series Editors, NTS Press, 2013.

[Razavi1] Behzad Razavi, *Design of Analog CMOS Integrated Circuits*", 2nd Edition, McGraw-Hill, 2017.
[Sedra1] Adel S. Sedra and Kenneth C. Smith, *Microelectronic Circuits*, 7th Edition, Oxford University Press, 2015.
[Sze1] S.M. Sze, *Semiconductor Devices: Physics and Technology*, 2nd Edition, John Wiley & Sons, 2002.

CHAPTER 4 PROBLEMS – MOSFET

Comments About the Level 1 K_p Parameter

In manual calculations, use any K_p that you need. If a choice of K_p is inconsistent with the given parameters T_{ox} (oxide layer thickness) and U_0 (carriers' mobility), K_p overrides.

Below are two different calculations of K_p from t_{ox} and U_0. The results are different, possibly because of different values for T_{ox}. You may use either one if needed.

Permittivity of $S_i = \varepsilon_{ox} = 3.51 \cdot 10^{-13}$ [F/m]

Calculation A: With the given oxide thickness T_{ox}, the capacitance per unit area becomes $C_{ox} = 3.903 \cdot 10^{-7}$ [F/m^2], the same for NMOS and PMOS.

The NMOS K_p coefficient is therefore 136.6 [μA/V^2]. In ADS, the unit of K_p is [A/V^2]. For PMOS, K_p is 3.5 times smaller than that of an NMOS.

Calculation B: From the given $T_{ox} = 9e^{-6}$ [cm], one can find that $C_{ox} = 4.13e^{-5}$ [mF/cm^2] for both transistors. We know that $K_p = U_0C_{ox}$. Therefore, for NMOS, we have $K_p = 1.45e^{-2}$ [mA/V^2], and for PMOS we have $K_p = 4.13e^{-3}$ [mA/V^2]. In the SPICE code for each model card, because the units of K_p must be [A/V^2], simply let NMOS $K_p = 1.45e^{-5}$ and PMOS: $K_p = 4.13e^{-6}$.

Problem 4.1: MOSFET DC Circuit

For manual calculations, consider the following NMOS parameters: $K'_n = 0.1$ mA/V^2, $V_{TH} = 0.9$ V and $\lambda = 0$. For ADS simulation, use the table of Level 1 parameters (taken from Razavi's book). The problems are based on [Sedra1]. Comment on the differences between the manual calculation results and the simulation results. See Figure 4.59.

1a: Let W/L = 10. We want the transistor to be in *Saturation* mode with $I_D = 0.9$ mA and $V_{DS} = 2.1$ V. What should be the values of R_D and R_S? [*Hint*: Find first the required value of V_{GS}].

1b: Now we want the transistor to be in *Triode* mode with $I_D = 0.4$ mA and $V_{DS} = 0.5$ V. What should be W/L, R_D, and R_S? [Hint: Find first an inequality condition for V_{GS}. Then pick up any suitable specific value for V_{GS}. You will then be able to determine the rest].

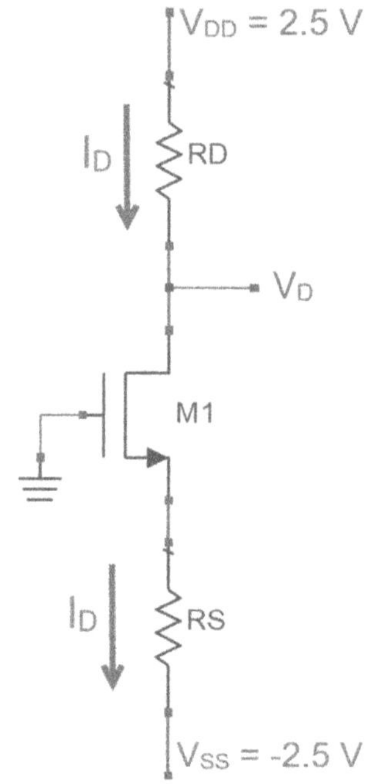

FIGURE 4.59 Problem 4.1.

Problem 4.2: MOSFET DC Circuits

See the circuits of Figure 4.60.

For manual calculations, the PMOS is characterized by threshold voltage $V_{TH} = -1$ V, $K'_p(W/L) = 1.2$ mA/V^2 and $\lambda = 0$. The NMOS is characterized by threshold voltage $V_{TH} = 1$ V, $K'_n(W/L) = 1.1$ mA/V^2 and $\lambda = 0$. For each problem, compare your manual results to the ADS simulation results based on Razavi's book model and explain the deviations. The problems build on [Sedra1].

2a: Find the labeled voltages in the Figure 4.60 (left) circuit diagram. State and justify the transistor mode of conduction.

2b: Replace the DC current source (in Problem 4.2a) by a resistor R_S so that $I_D = 2$ mA. What is R_S?

2c: In the circuit of Figure 4.60 (right), what happens when we replace $R_{G2} = 10$ MΩ by $R_{G2} = 500$ kΩ? [*Hint*: See how small V_G becomes. Can you assume cutoff mode and then justify such an assumption?].

2d: What happens (in the diagram of Problem 4.2c) if we let $R_S = 6$ kΩ become $R_S = 0$? Show first why it is that if you assume that the MOSFET is in saturation mode, and follow up on that assumption, you reach a contradiction.

2e: Assume triode mode (in the circuit of Problem 4.2d). Set up the quadratic equation for I_D and solve it.

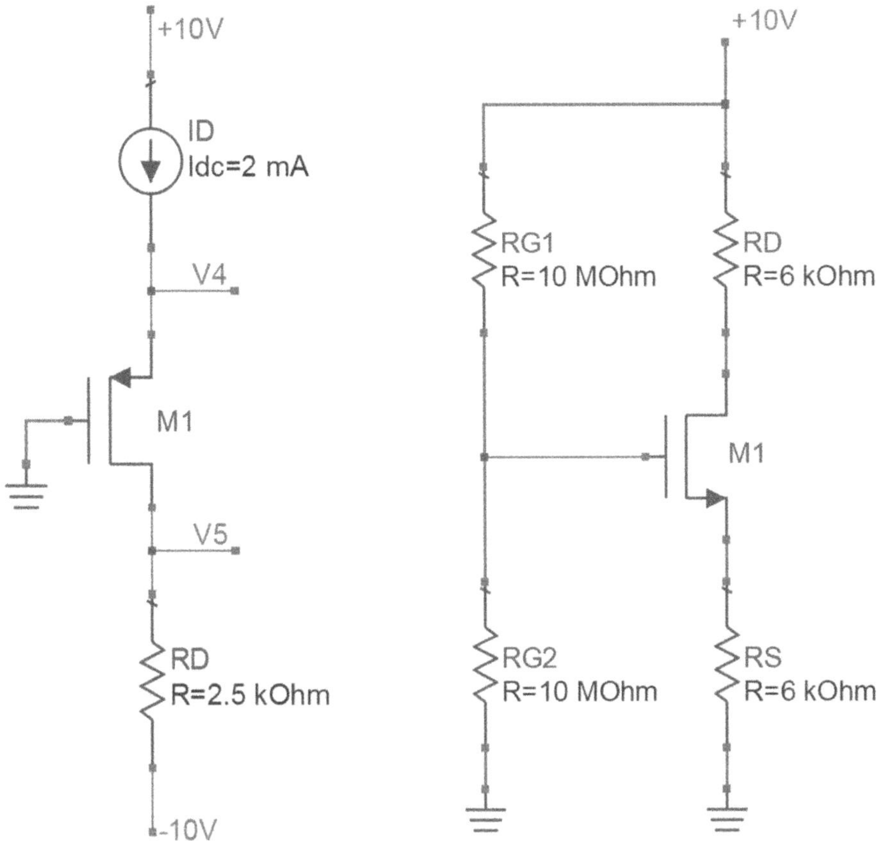

FIGURE 4.60 Problem 4.2.

Problem 4.3: MOSFET DC Circuit

The parameters of the PMOS transistor (below) for manual calculations are $K_p' = 0.14\ \text{mA/V}^2$, $V_{TH} = -1.2\,\text{V}$, $W/L = 11$, and $\lambda = 0$. For ADS simulation consider two Level 1 models: the most simplistic, using K_p and V_{TH} only and nullifying all the second-order effects vs. the full Level 1 model offered in Razavi's book. See Figure 4.61.

3a: Find the current I_D and the mode of the transistor (saturation? triode? cutoff?). Explain every step of your solution.

3b: In Problem 4.3a (above), how would you change V_G in order for the circuit to be in triode mode, having the same I_D as in Problem 4.3a? Analyze the new circuit.

Problem 4.4: MOSFET DC Circuits

The problem builds on two problems offered in Razavi's book. Analyze each of the circuits (in Figure 4.62), explaining how each circuit behaves for different DC V_X

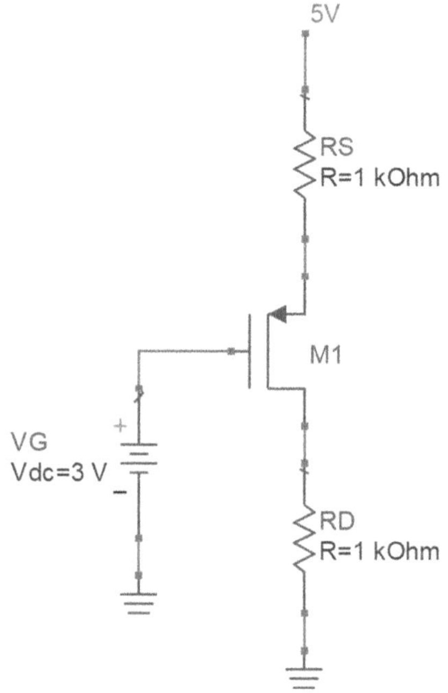

FIGURE 4.61 Problem 4.3.

values, as V_X varies from 0 to 5 V. Pay attention to the following questions, solving each analytically and supporting the solutions with ADS simulations:

4a: For what V_X values is the MOSFET in saturation mode? For what V_X values is the MOSFET in triode mode? For what values of V_X is the transistor in cutoff mode?

4b: For what values of the varying supply V_X do the transistor's drain and source reverse their roles? (Comment: Recall that in an NMOS I_D flows from drain to source. In a PMOS, I_D flows from source to drain).

4c: Is there a body effect in any of the MOSFET transistors? If so, what is the threshold voltage V_{TH} as a function of V_X?

Do the above analysis for channel width $W=20\ \mu m$ and channel length $L=2\ \mu m$.

Problem 4.5: CS Amplifier with R_D Load

The problem analyzes the MOSFET CS Amplifier with R_D load (Figure 4.63). In your manual calculations, use the simplistic Level 1 model and compare it in simulations to the fuller Level 1 model.

The CS amplifier (shown below) uses a PMOS transistor that has $K'_p = 0.15\ mA/V^2$, $V_{TH}=-0.9\ V$ and $\lambda=0.01$ [1/V]. The bias voltage is $V_G=3\ V$. We also have $V_{DD}=5\ V$ and $R_D=3\ k\Omega$. Use these parameters in both manual calculations and ADS simulations.

5a: Neglect the effect of nonzero λ on the amplifier's DC conditions. We want $I_D=0.3\ mA$ and the PMOS to be in saturation mode. What should be the value of

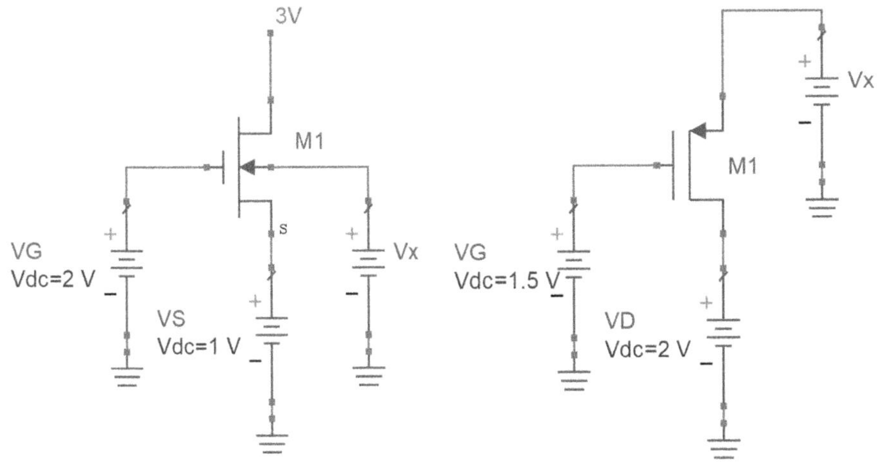

FIGURE 4.62 Problem 4.4.

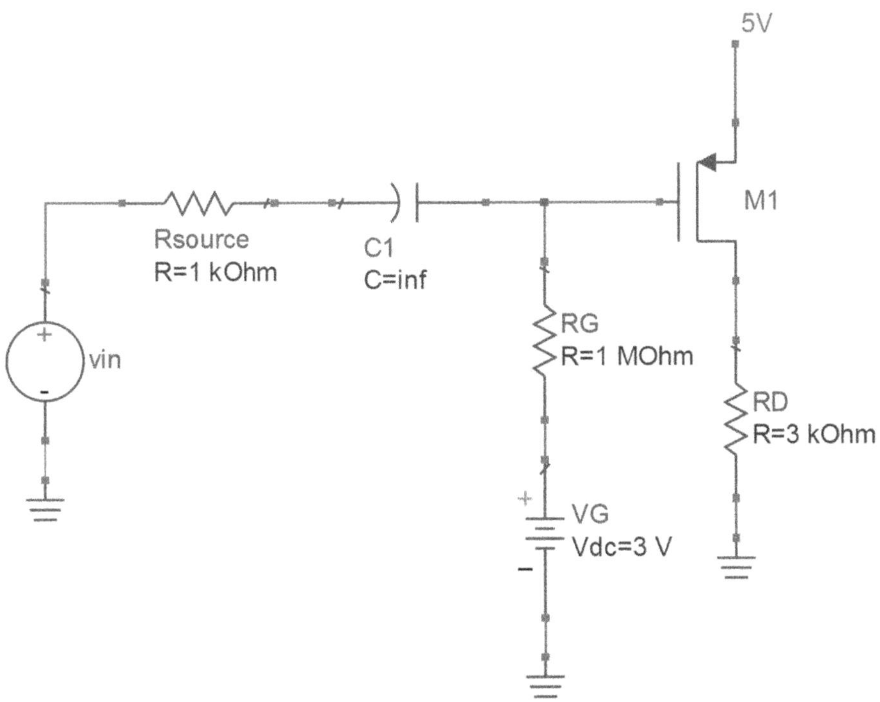

FIGURE 4.63 Problem 4.5.

W/L? Draw the small-signal diagram and compute R_{in} and the voltage gain (considering the effect of λ).

5b: Now we want to modify V_G and W/L such that the voltage gain obtained in Problem 4.5a becomes nine times larger, without changing any other circuit parameters and without changing I_D. Can it be done? If so, specify how. *Hint*: For PMOS the overdrive voltage $V_{GS} - V_{TH}$ is negative.

Problem 4.6: CS Amplifier with R_D Load

See Figure 4.64.

The NMOS model parameters are $V_{TH} = 0.85\,V$, $K'_n = 0.1\,mA/V^2$, $\lambda = 0$. Other given component values are $V_{DD} = 5\,V$, $R_D = 2.2\,k\Omega$, $R_L = 20\,k\Omega$, $R_{sig} = 20\,k\Omega$, and $R_G = 1\,M\Omega$.

6a: Let the NMOS aspect ratio be $W/L = 19$. Let $V_G = 1.4\,V$. Explain why it is that the NMOS conducts at all. What is I_D? Explain why it is that the NMOS is in saturation mode.

6b: Find the small-signal parameters of the NMOS and draw the small-signal diagram of the CS amplifier.

6c: Find the amplifier's input resistance R_{in} and its small-signal voltage gain $A_V = v_o/v_{sig}$.

6d: Let $v_{sig}(t)$ be an AC voltage signal with an amplitude of 20 mV and a frequency of $f = 400\,Hz$. Write an expression for $v_o(t)$.

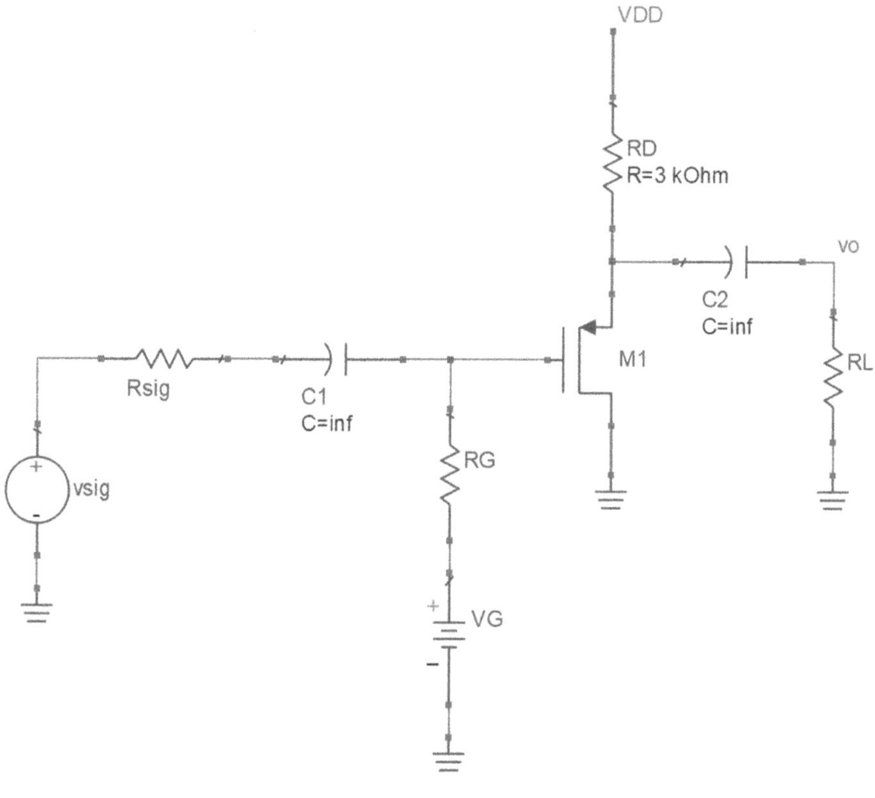

FIGURE 4.64 Problem 4.6.

Problem 4.7: CS Amplifier with R_D Load

The problem in Figure 4.65 builds on a problem from [Sedra1].

7a: For the NMOS shown in the figure (below), assume K′W/L = 2 mA/V^2, V_{TH} = 1 V and λ = 0.01 [1/V]. Analyze the amplifier by finding (i) its DC Operating Point (I_D, V_{DS}), (ii) the MOSFET small signal parameters, (iii) drawing the small signal diagram of the full amplifier, (iv) finding the small-signal input resistance R_{in} and the small-signal voltage gain v_{out}/v_{sig}, and (v) tweaking the amplifier in order to raise the voltage gain by a factor of 8.

7b: Replace the NMOS by a PMOS that has K′W/L = 2 mA/V^2, V_{TH} = −1 V, and λ = 0.02 [1/V] and rearrange the amplifier: Flip the PMOS so that its source connects to V_{DD} via the 3 kΩ resistor, the drain connects to ground via the 7.5 kΩ resistor, and the load connects to the drain via the coupling capacitor. Repeat the analysis of Problem 4.7a.

Problem 4.8: CS Amplifier with R_D Load

The parameters of the NMOS transistor (Figure 4.66) are $K'_n = 0.18\ \text{mA/V}^2$, V_{TH} = 1 V, W/L = 15, and λ = 0.01. In this CS amplifier R_D = 3.2 kΩ, R_L = 15 kΩ, and R_{sig} = 1 kΩ. *Hint*: Neglect λ in all DC calculations, but include r_o in the gain calculation.

8a: Find the DC current I_D and check that the NMOS is in saturation mode. If not, suggest some slight circuit modifications (R_D, R_{G1}, or R_{G2}) so that the MOSFET is in saturation mode.

8b: Find the MOSFET small-signal parameters and draw the small-signal diagram of the CS amplifier.

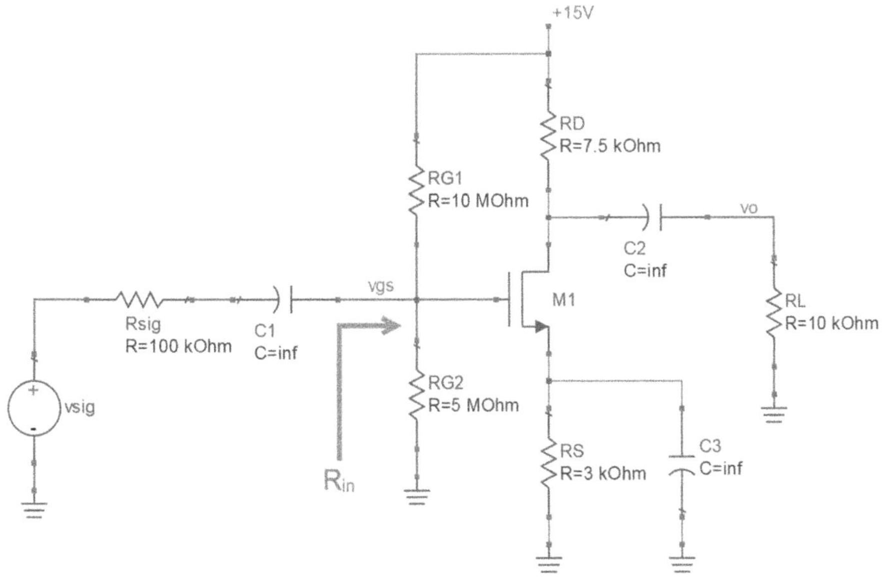

FIGURE 4.65 Problem 4.7

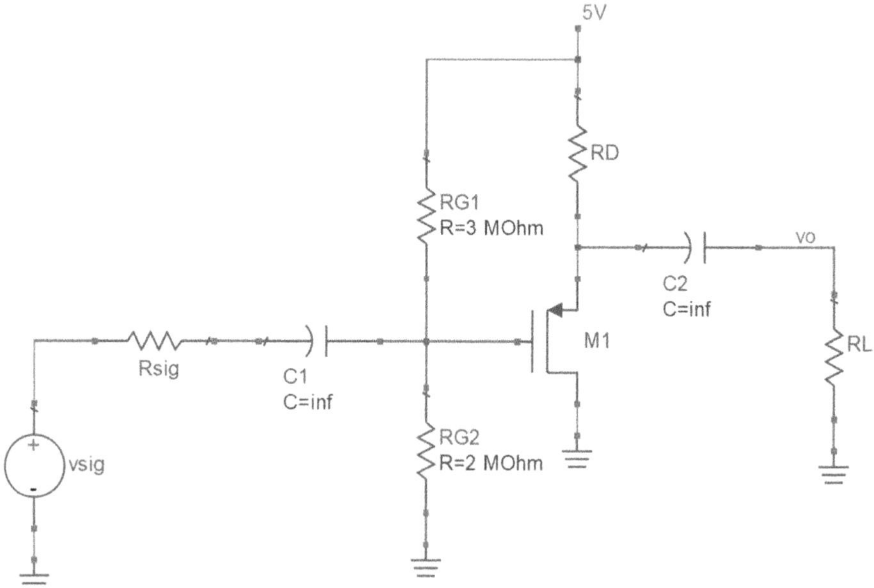

FIGURE 4.66 Problem 4.8.

8c: Find the amplifier's input resistance R_{in}, and the small-signal voltage gain A_V.

8d: There is a need to increase the voltage gain by a factor of 8. Suggest new values for R_{G1}, R_{G2} and W/L. Find the new input resistance and new voltage gain.

Problem 4.9: MOSFET Current Mirrors

You are given a stabilized reference DC current source $I_{ref} = 250\ \mu A$ connected to PMOS transistor M_1. The parameters of the MOSFET transistors are NMOS: $V_{TH,n} = 0.7\,V$, $K'_n = 0.13\ mA/V^2$, $\lambda_n = 0$, PMOS: $V_{TH,p} = -0.8\,V$, $K'_p = 0.1\ mA/V^2$, $\lambda_p = 0$. The transistor dimensions (L and W) must each be no smaller than $2\,\mu m$. Let all L parameters be equal, each to $2\,\mu m$. See Figure 4.67.

9a: Select W_1 so that $V_{GS1} = -2.5\,V$

9b: Select W_2 so that the source current $I_2 = 450\ \mu A$. What are the values of V_{GS2} (of the PMOS M_2) and of V_{DS3} (of the NMOS M_3)?

9c: Determine the value for W_3 (of the NMOS M_3) so that the current of M_3 is consistent with I_2.

9d: Determine W_4 so that $I_4 = 580\ \mu A$ (a sink current). Is M_4 in saturation mode?

Problem 4.10: MOSFET Current Mirrors

Consider the following NMOS current mirror circuit (Figure 4.68). M_1 and M_2 both have the following parameters: $\{K'_n = 0.15\ mA/V^2, V_{TH} = 0.9\,V, \lambda = 0, L = 2\ \mu m\}$. For $M_{1,}$ it is given that $W_1 = 25\ \mu m$, and for M_2 we have $W_2 = 35\ \mu m$. The voltage at the gate of both transistors is known to be $V_G = 1.4\,V$. Answer the following questions, explaining all the steps. Let $V_{DD} = 5\,V$.

What is the value of I_{ref}? What is the value of I_2? Are M_1 and M_2 in saturation mode? Why?

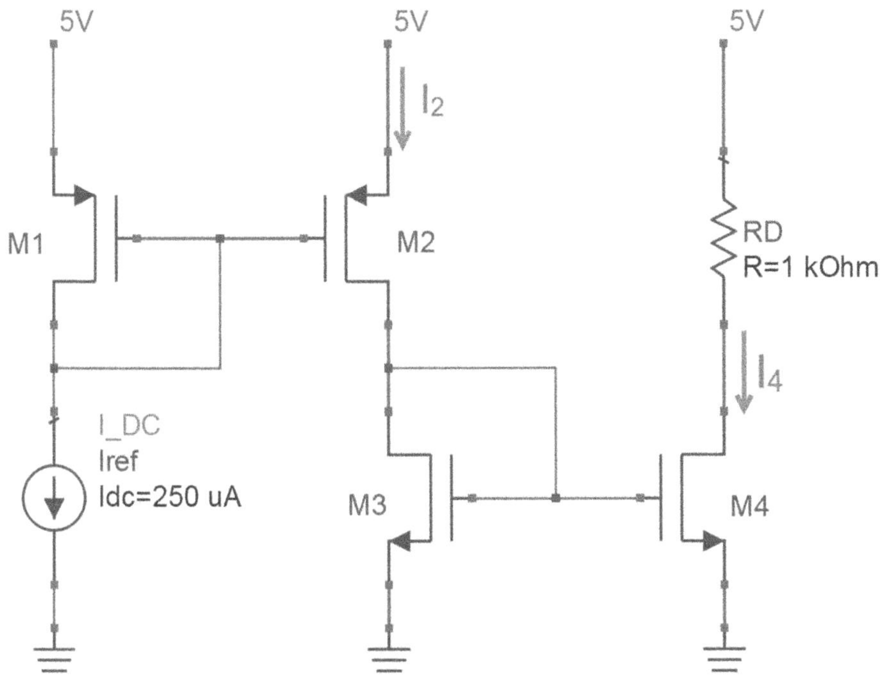

FIGURE 4.67 Problem 4.9.

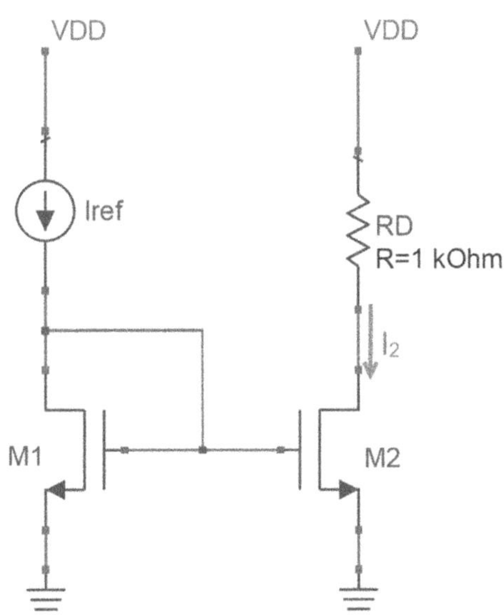

FIGURE 4.68 Problem 4.10.

5 Single- and Two-Stage BJT Amplifiers

5.1 BJT WITH A DEGENERATION RESISTOR

We may recall from Sections 3.10 to 3.11 (common emitter (CE) amplifier) the important role played by R_E, a resistor that is placed between the emitter and ground, in determining the BJT DC current I_E. R_E is bypassed by a large capacitor so that it does not lower the small-signal voltage gain of the CE amplifier. A degeneration resistor R_{ee} is a small resistor that is placed between the emitter and R_E (see Figure 5.1). R_{ee}

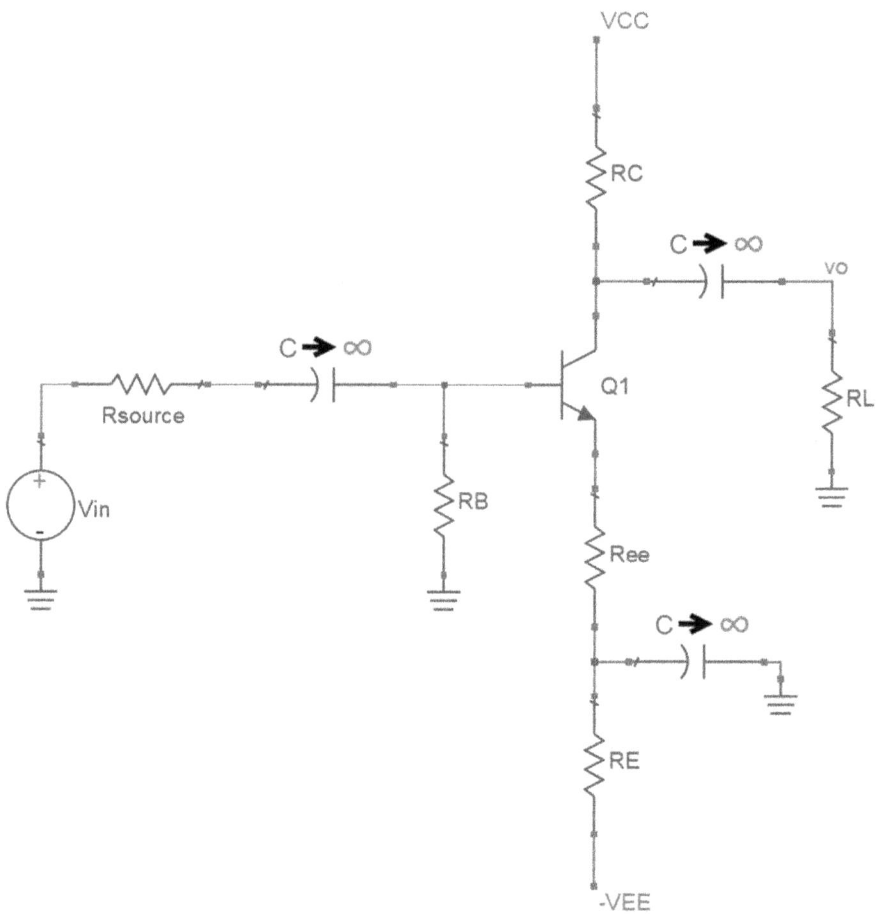

FIGURE 5.1 A dual-supply NPN CE amplifier featuring a degeneration resistor R_{ee}.

DOI: 10.1201/9781003606772-5

and R_E are connected in series for DC, but because $R_{ee} << R_E$, the degeneration resistor plays a minor role in the amplifier's DC analysis. For small signals, on the other hand, as R_{ee} is not bypassed, this small resistor has a profound effect on the CE amplifier's performance.

The reader may recall the CE amplifier's small-signal assumption. That is, the input signal's amplitude $|v_{be}(t)|$ must be no larger than 10–15 mV at room temperature. As we are about to see, **the main reason for incorporating a degeneration resistor is to be able to accommodate input signals that may have much larger amplitudes than that dictated by the small-signal assumption** (say 30–50 mV). The larger linear range of the amplifier does not come for free, though. It is a tradeoff: the input signal's amplitude and the amplifier's R_{in} may increase, but the achievable small-signal voltage gains decrease.

Figure 5.1 demonstrates one of the many architectures of a CE amplifier, including a degeneration resistor R_{ee}.

The small-signal diagram of the CE amplifier, shown in Figure 5.1, is shown in Figure 5.2.

Earlier (in Section 3.10), we made an interesting discovery: most CE amplifiers' architectures lead to the same small-signal diagram. These architectures include single- and dual-supply NPN or PNP CE amplifiers and CE amplifiers that use a DC current source to set the current I_C.

Therefore, an analysis of the diagram presented in Figure 5.2 creates practical approximated formulas that are applicable to many CE amplifier architectures. The formulas for A_v and R_{in} will generalize those obtained earlier, in conjunction with the small-signal diagram of Figure 3.56 that had no degeneration resistor. These were formulas (3.1) and (3.2).

The analysis of the diagram in Figure 5.2 starts with setting a KCL equation at the emitter node and neglecting the small-signal currents that flow through r_o and r_π with respect to the small-signal current flowing through the dependent source and R_{ee}

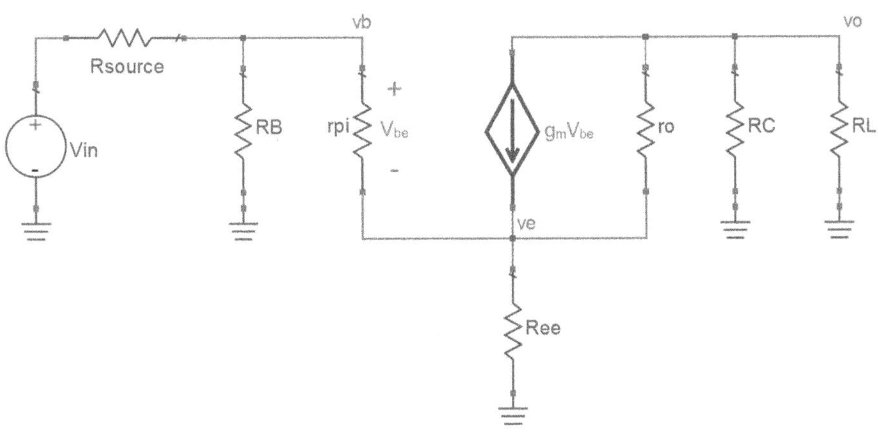

FIGURE 5.2 The small-signal diagram of the CE amplifier of Figure 5.1.

(and this is because $g_m \approx 1/r_e >> 1/r_\pi >> 1/r_o$). The approximate KCL calculation results in two relationships. One relates v_e to v_{be}: $\mathbf{v_e/v_{be} \approx R_{ee}/r_e}$. This leads to a second relationship, one that is between v_b and v_{be}: $\mathbf{v_{be}/v_b \approx r_e/(R_{ee}+r_e)}$. The latter, which looks like a "voltage division" between R_{ee} and r_e, holds the key to accommodating large input signal amplitudes.

The small signal $v_b(t)$ represents the input signal $v_{in}(t)$, as we shall discuss more soon. The small signal $v_{be}(t)$ is at the center of the amplifier's small-signal assumption $|v_{be}(t)| \leq |v_{be}|_{max}$, where $|v_{be}|_{max}$ may be a voltage level of, say, 10–15 mV at room temperature. Thus, from the above "voltage division" between R_{ee} and r_e, we obtain an inequality that guides us on how to choose R_{ee}: $\mathbf{|v_b|_{max}/|v_{be}|_{max} \leq 1 + R_{ee}/r_e}$.

Next, let us observe the effect of R_{ee} on the amplifier's input resistance, R_{in}. The small-signal input resistance R_{in} is the resistance "seen" from the input signal source v_{in}, "looking into" the amplifier. It can be measured by dividing the peak-to-peak amplitude of $v_b(t)$ by the peak-to-peak current $i_{in}(t)$, the current that goes through R_{source} and the input coupling capacitor (see Figure 5.1). $R_{in} = R_B \parallel R'_{in}$, where R'_{in} is the resistance defined by v_b and the current flowing through r_π. $R'_{in} \approx v_b/(v_{be}/r_\pi) \approx v_b/\left[v_b \cdot r_e/(R_{ee}+r_e)\right]/r_\pi = (\beta+1)\cdot(R_{ee}+r_e)$. Note that for $R_{ee}=0$ $R'_{in} = r_\pi$, and whenever $R_{ee} \neq 0$, a resistance of $(\beta+1)\cdot R_{ee}$ is added to r_π, causing an increase of R_{in}. The extended formula for R_{in} is therefore $\mathbf{R_{in} \approx R_B||[(\beta+1)\cdot(R_{ee}+r_e)]}$.

Coming back to the accommodation of larger small-signal voltage inputs, we have a voltage division between R_{in} and R_{source}: $v_b = v_{in} \cdot R_{in}/(R_{in}+R_{source})$. Consequently, $\mathbf{R_{ee}/r_e \geq (|v_{in}|_{max}/|v_{be}|_{max})\cdot(R_{in}/(R_{in}+R_{source}) - 1}$. R_{in} is the input resistance with R_{ee} included.

Example 5.1

If $|v_{in}|_{max} = 40$ mV, $R_{in} = 2$ kΩ, and $R_{source} = 500$ Ω, then $R_{ee}/r_e \geq 2.2$ guarantees that $|v_{be}|$ does not exceed 10 mV.

Let us finally discuss the extended formula for the small-signal voltage gain. In Figure 5.2, let us assume initially that $r_o = \infty$. The small-signal voltage output is $v_o \approx -g_m \cdot v_{be} \cdot (R_C||R_L) \approx v_{in} \cdot [R_{in}/(R_{in}+R_{source})] \cdot [r_e/(R_{ee}+r_e)] \cdot [-g_m(R_C||R_L)]$. The small-signal voltage gain is $A_v = v_o/v_{in} \approx -[R_{in}/(R_{in}+R_{source})] \cdot [(R_C||R_L)/(R_{ee}+r_e)]$. A better approximation that is consistent with the voltage gain formula obtained for $R_{ee}=0$ and for r_o included again is $\mathbf{A_v = v_o/v_{in} \approx -[R_{in}/(R_{in}+R_{source})]\cdot[(R_C||R_L||r_o)/(R_{ee}+r_e)]}$.

Example 5.2

As an example, to illustrate the tradeoffs encountered in using a degeneration resistor, let us compare two cases: In one $R_{ee}=0$, and in the other $R_{ee}=3r_e$, where $r_e = 25$ Ω (resulting from a DC current of $I_C = 1$ mA). We assume that $R_{source} = 500$ Ω, $R_B = 20$ kΩ, $R_C = 3$ kΩ, $R_L = 15$ kΩ, $r_o = 100$ kΩ, $\beta = 100$ and $|v_{be}|_{max} = 10$ mV. Table 5.1 summarizes the results.

The voltage gain reduces more or less by the same factor by which R_{in} increases, and the ability to accommodate large input signals increases. The calculations above assumed that $R_E >> R_{ee}$.

TABLE 5.1
$|v_{in}|_{max}$, R_{in} and A_v Comparison, with and Without R_{ee}

	Maximum Allowed Input Amplitude in mV	Small-Signal Input Resistance R_{in} in kΩ	Small-Signal Voltage Gain A_v
$R_{ee}=0$	12	2.22	−79.6
$R_{ee}=3r_e=75\ \Omega$	43	6.71	−22.7

5.2 DESIGN OF A CE AMPLIFIER EXAMPLE

The whole Section 5.2 is a CE amplifier design example.

Example 5.3

Let us consider only three design specifications in the design of a CE amplifier: small-signal voltage gain, small-signal input resistance, and accommodation of small input voltage signals with amplitudes that are no larger than a certain level. This is a simplified design task, as normally there may be many more specifications, such as bandwidth, linearity, and temperature resilience.

The design parameters may include a choice of amplifier architecture, a choice of specific BJT transistors, and a choice of component values for the circuit's resistors, capacitors, and voltage supplies. In the example discussed in this section, the architecture, the BJT transistor, and the capacitor values are all fixed. The remaining design parameters are DC supply and resistor values; all are shown with a question mark in Figure 5.3.

"Simple" designs are such that each specification may be addressed by means of one or two design parameters.

Designs in which each specification depends on multiple design parameters may require a highly simplified manual solution, followed by computer-aided design tweaking.

Figure 5.3 defines the design parameters in the example that we intend to solve. The CE amplifier's architecture is that of an NPN with a single supply.

Note that the source resistance R_{source} and load resistance R_L are not considered design parameters. These resistances are given. The designed amplifier is supposed to serve various pairs of voltage sources and load resistances. In this example, we shall assume the following fixed values: $R_{source}=1$ kΩ and $R_L=25$ kΩ. Later on, we shall consider tolerances for these components, as well as the tolerance of the amplifier's components.

The NPN transistor that we shall use to test the design is Q2N3904.

However, in the manual design we shall use approximate values $V_{BE}=0.7$ V when the BE junction is ON, {$\beta=100$ and $V_A=100$ V}. The input signal **$v_{in}(t)$, which we wish to accommodate, is known to have a maximum amplitude of 20 mV**. This is one of the design specifications.

The other two specifications are:

A small-signal voltage gain of $A_v=-20$ and a small-signal input resistance of $R_{in}\geq 20$ kΩ.

Other implicit design specifications, true for every CE amplifier design, are that the transistor operate in active mode and that the DC operating point be placed

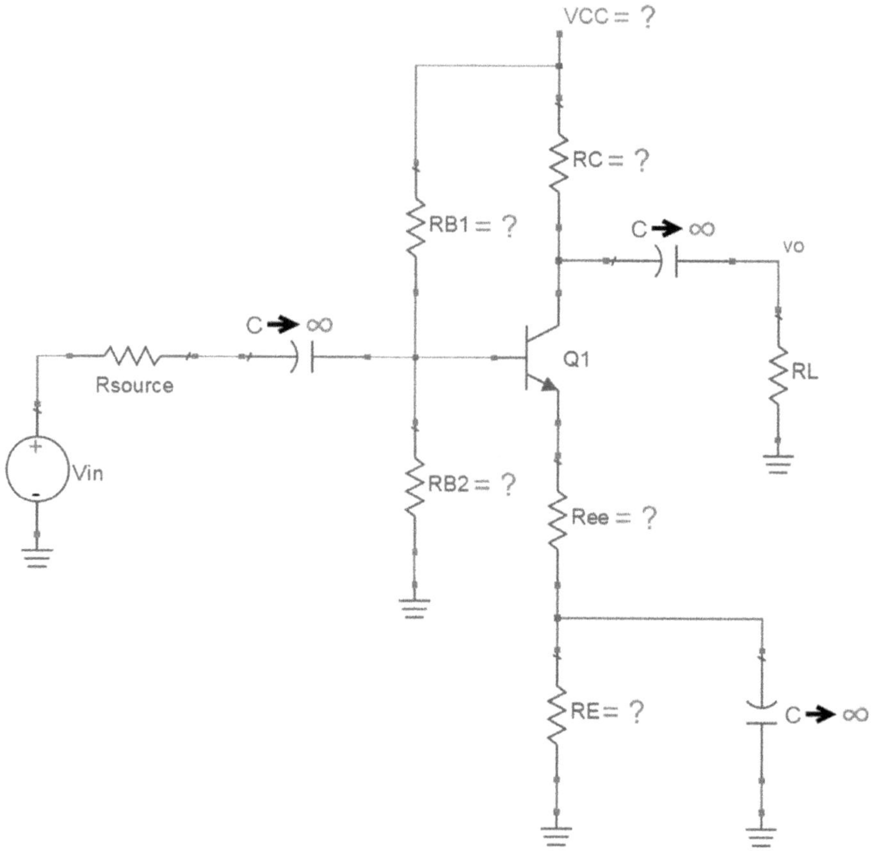

FIGURE 5.3 Single-supply NPN CE amplifier. The design parameters are the DC supply value and the base, collector, and emitter resistors.

"deep inside" the active mode region. We count in Figure 5.3 no less than six design parameters. There are therefore a few degrees of freedom. That is, some design parameters that may be determined arbitrarily.

We recall, from Section 5.1, that in order to accommodate a large input voltage amplitude, one needs to choose an appropriate degeneration resistor R_{ee}. This seems to be a good place to start the design. Engineers often leave some "safety margins" for meeting each design specification. This may be advisable for a second round of design calculations. Let us instead meet each specification with no margin, just to see if the design is feasible and the specifications do not conflict with one another. In addition, we shall approximate resistor values to the nearest first decimal. In the R_{ee}/r_e inequality, let us substitute $|v_{in}|_{max}=20\,\text{mV}$ and $R_{in}=20\ \text{k}\Omega$:

$R_{ee}/r_e \geq (|v_{in}|_{max}/|v_{be}|_{max})\ [R_{in}/(R_{in}+R_{source})]-1=(20/10)\ [20/(20+1)]-1\approx 0.905.$

Next, let us continue to interpret the specification for R_{in}, recalling that $R_{in}\approx R_B||[(\beta+1)\ (R_{ee}+r_e)]$, and that in the previous design step we assumed $R_{in}\approx 20$ kΩ. That is, $20\approx R_B||[101\cdot(R_{ee}+r_e)]$, where the resistances are all in kΩ.

We are now encountering the first design degree of freedom: We need to decide on how to select R_B and $R_{ee}+r_e$. For instance, we may take $R_B \approx 40$ kΩ and $100\cdot(R_{ee}+r_e) \approx 40$ kΩ. It is not obvious to see that $R_{ee}+r_e$ must not be taken to be unnecessarily large. Intuitively, by watching the voltage gain formula $A_v = v_o/v_{in} \approx -[R_{in}/(R_{in}+R_{source})]\ [(R_C||R_L||r_o)/(R_{ee}+r_e)]$, a large value for $R_{ee}+r_e$ may later require a large value for R_C, which may lead to an unnecessarily large value for the supply V_{CC}. On the other hand, recalling the "weak β-dependence" concept, we do not want R_B (and consequently the base resistors) to become too large. As a compromise, let us choose $R_{ee}+r_e \approx 300$ Ω and $R_B = 60$ kΩ. Let us briefly explain the calculation of R_B.

The solution of an equation $a||x = b$, where a and b are given constants and x is the unknown, $x = (b^{-1} - a^{-1})^{-1}$. Therefore, $R_B||(100(R_{ee}+r_e)) = 20$ has a solution $R_B = (20^{-1} - 30^{-1})^{-1} = 60$ kΩ.

Let us set aside R_B, for the time being, and calculate R_{ee} and r_e: $R_{ee}+r_e = 1.905r_e \approx 300$ Ω. Then $r_e = 157.5$ Ω and $R_{ee} = 142.5$ Ω. Let us choose $\mathbf{R_{ee} = 150\ \Omega}$. The parameter $r_e = 157.5$ Ω implies a DC current $I_C = V_T/r_e = 25/157.5 \approx 0.16$ mA.

For meeting the specification for the small-signal voltage gain, we note in the voltage gain formula that R_C is the only unknown and that $r_o \approx V_A/I_C = 100/0.16 = 625$ kΩ is negligible.

$A_v \approx -[R_{in}/(R_{in}+R_{source})] \cdot [(R_C||R_L||r_o)/(R_{ee}+r_e)] = -20 \approx -[20/(20+1)] \cdot (R_C||25)/0.3$. The equation for R_C is $R_C||25 = 6.3$, leading to $\mathbf{R_C \approx 8.4\ k\Omega}$.

We mentioned earlier that we have to be on the watch for inconsistent specifications. Imagine, for instance, that the voltage gain specification is changed from $A_v = -20$ to -100. Then the equation for R_C becomes $R_C||25 = 31.5$, which has no solution.

So far, we have found two design parameters: $R_{ee} = 150$ Ω and $R_C = 8.4$ kΩ. We also found $I_C = 0.16$ mA and $R_B = 60$ kΩ. What is left to be done?

We need to find R_E, R_{B1}, and R_{B2}, and the supply V_{CC}. We also need to verify that the BJT is in active mode and it has a Q-point that is not close to the cutoff or saturation mode borderlines.

The above last point leads us to a second degree of freedom in our design. The DC voltage V_{CE} must not be too small nor too large. Common selections of V_{CE} range from $V_{CC}/4$ to $V_{CC}/2$. How shall we allocate voltages from V_{CC} to R_C, R_E, and V_{CE}? One simple strategy is to take $V_{CE} \approx I_C R_C \approx I_E\cdot(R_E + R_{ee})$, assuming, of course, that in active mode $I_C \approx I_E$. In other words, let $V_{CC} = 3I_C R_C = 3\cdot 0.16\cdot 8.4 \approx 4$ V and $V_{CE} \approx V_{CC}/3$. In a discrete circuit (one that is built on a breadboard), one may opt for a supply voltage that is an integer multiple of 1.5 V. Say $\mathbf{V_{CC} = 4.5\ V}$, which we choose for this example. We then find R_E: $V_{CC} - I_C R_C = V_C = 4.5 - 1.344 = 3.156$ V $= V_{CE} + I_E\cdot(R_E + R_{ee})$. There are many choices for R_E. Let $V_E = I_E(R_E + R_{ee}) = 1.5$ V, and then $V_{CE} = 1.656$ V. The R_E calculation is $0.16\cdot(R_E + 0.15) = 1.5$, yielding $\mathbf{R_E \approx 9.2\ k\Omega}$. Let us check the active mode assumption. $V_E \approx 1.5$ V. Then $V_B \approx 2.2$ V $< V_C = 3.156$ V. Active mode is justified. The design concludes with finding the two base resistors, R_{B1} and R_{B2}. There are two equations with two unknowns resulting from the Thevenin equivalent circuit of V_{CC}, R_{B1}, and R_{B2}. The Thevenin voltage V_{BB} is solved from the I_E formula: $I_E = (V_{BB} - V_{BE})/[(R_E + R_{ee}) + R_B/(\beta+1)]$. Then $V_{BB} = V_{CC}R_{B2}/(R_{B1}+R_{B2})$ and $R_B = R_{B1}||R_{B2} = R_{B1}R_{B2}/(R_{B1}+R_{B2})$. Numerically, $I_E = 0.16 = (V_{BB} - 0.7)/[(9.2+0.15)+60/101]$, yielding $V_{BB} \approx 2.3$ V. The equations for the two base resistors are $2.3 = 4.5R_{B2}/(R_{B1}+R_{B2})$ and $60 = R_{B1}||R_{B2}$. From the first equation, $R_{B2}/(R_{B1}+R_{B2}) = V_{BB}/V_{CC} = 2.3/4.5$, and then from the second equation, $\mathbf{R_{B1}} = R_B V_{CC}/V_{BB} \approx \mathbf{117\ k\Omega}$. Finally, $\mathbf{R_{B2}} = \left(R_B^{-1} - R_{B1}^{-1}\right)^{-1} \approx \mathbf{123\ k\Omega}$. This design will be tested in ADS simulation in the next section.

5.3 DESIGN APPROXIMATIONS FOR A CE AMPLIFIER; ADS TRANSIENT AND HARMONIC BALANCE SIMULATION TESTING AND QUALITATIVE ASSESSMENT OF OUTPUT DISTORTIONS; COMPARISON OF DEGENERATED AND NON-DEGENERATED EMITTER RESISTANCE CE AMPLIFIERS; TIME SIGNAL PLOTS FOR HB SIMULATION RESULTS

As we recall, the design had three quantitative specifications (voltage gain, R_{in}, and accommodation of $|v_{in}|_{max}$), so that the design we just met the specs with apparently has no safety margins. We also used standard approximations such as $V_{BE}=0.7\,V$ and $\beta=100$. We need to test if the specs are met. We will test using DC, AC, transient, and HB simulations; each simulation type is done separately. The first testing in DC simulation is as shown in Figure 5.4. The annotated results are presented in Figure 5.5. Recall the design value of $I_C=0.16\,mA$.

We first want to see if the BJT is in active mode and if $I_C=0.16\,mA$. The BJT is indeed in active mode, as (from Figure 5.5) $V_C=3.09\ V \geq V_B=2.21\ V$.

$I_C=0.168\,mA$ (instead of 0.160 mA as designed). Does it matter? Do we need to tweak the DC current?

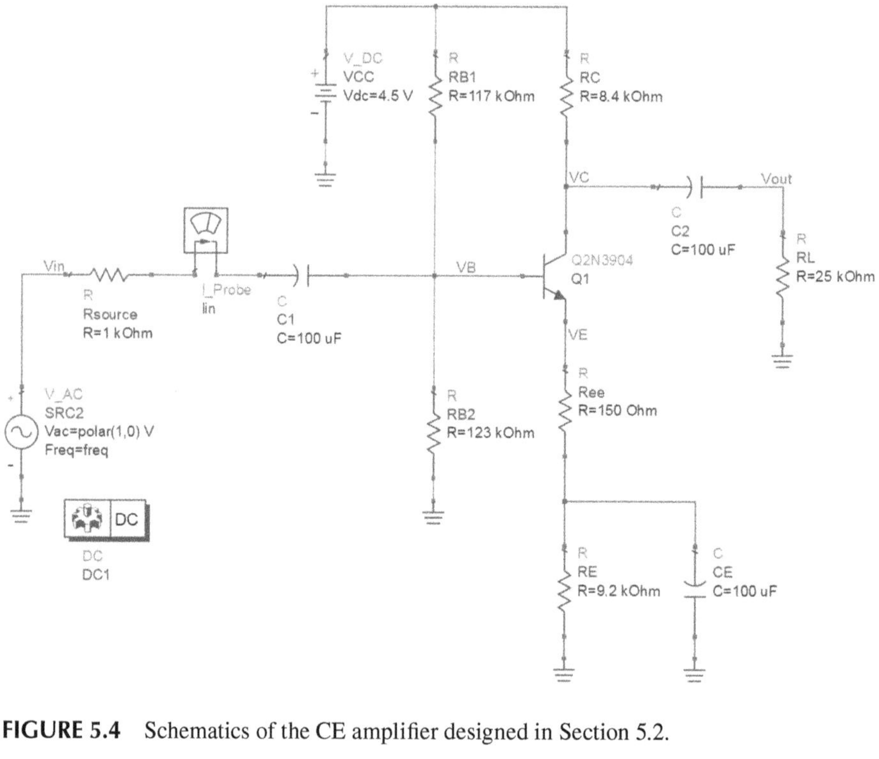

FIGURE 5.4 Schematics of the CE amplifier designed in Section 5.2.

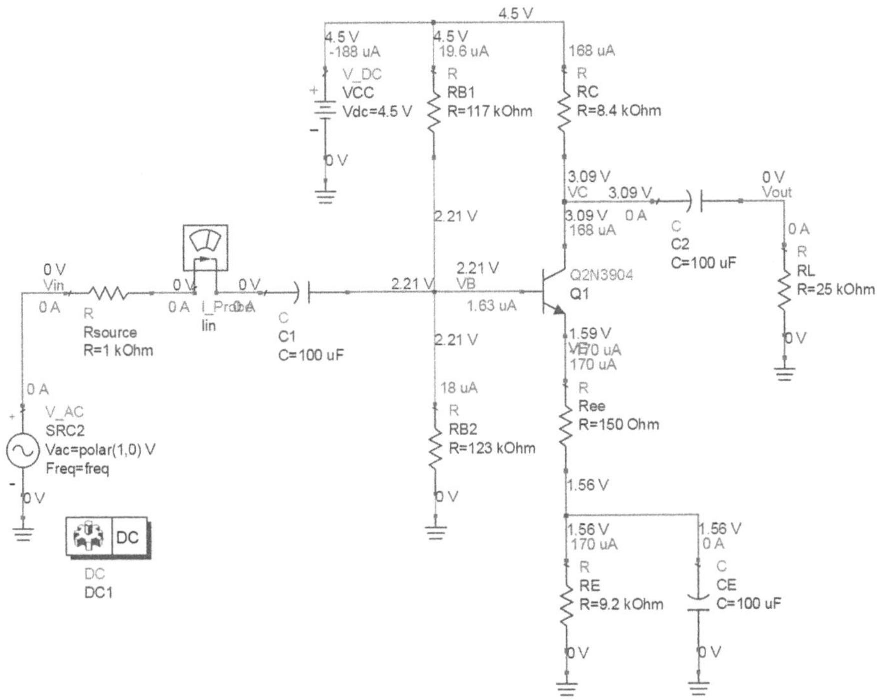

FIGURE 5.5 DC voltages and currents of the designed CE amplifier.

Let us not rush into bringing I_C back to the exact design value. We note that $V_{BE} = 2.21 - 1.59 = 0.62\,V$ (instead of 0.7 V). We also note that $\beta = 168/1.63 = 103$ (instead of 100).

Let's proceed with the testing, checking in AC simulation the voltage gain and the input resistance (Figure 5.6).

We check the frequency response of the CE amplifier. We are interested in the "mid-frequency band", where A_v and R_{in} are more or less constant. The frequency f (along the X-axis) will be taken on a logarithmic scale. Between 1 Hz and 100 MHz there are eight decades.

The current probe is needed to measure R_{in}. Simulation results are shown in Figures 5.7 and 5.8.

The voltage gain $|A_v| = 19.625$ is 1.875% smaller than the specified value $|A_v| = 20$. It is needed in general to tweak the gain, as shown in an earlier example (in Section 3.11). In this example, though, the focus is different.

For the tweaking of A_v, we shall use the R_C resistor. We keep I_C unchanged because this keeps all the small-signal parameters unchanged.

We could have used a more accurate BJT Early voltage of $V_A = 74\,V$ (for the Q2N3904 BJT), but because I_C is small, the effect of r_o on the voltage gain is very small ($r_o = V_A/I_C$).

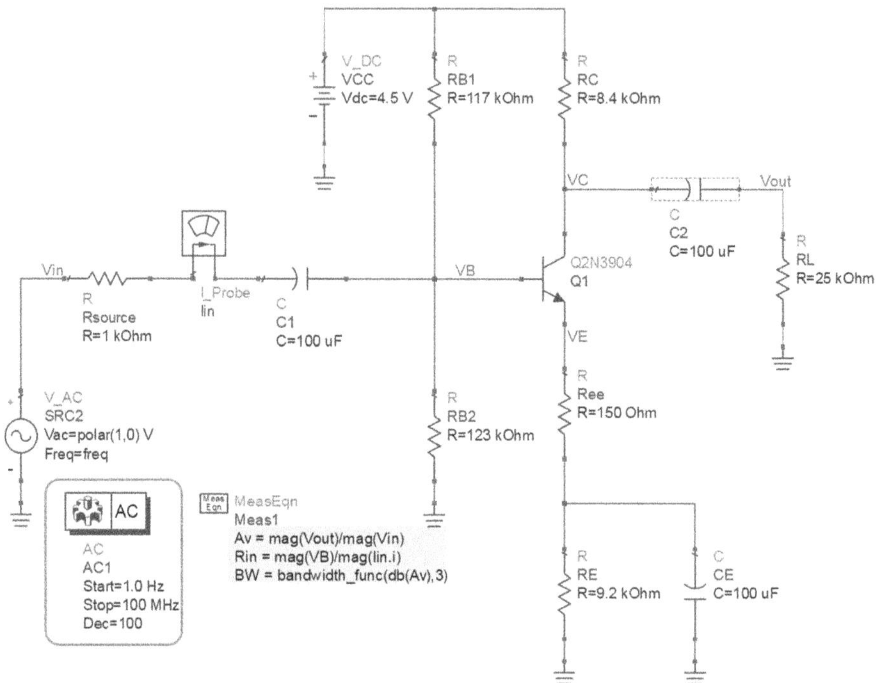

FIGURE 5.6 AC simulation testing of the designed CE amplifier.

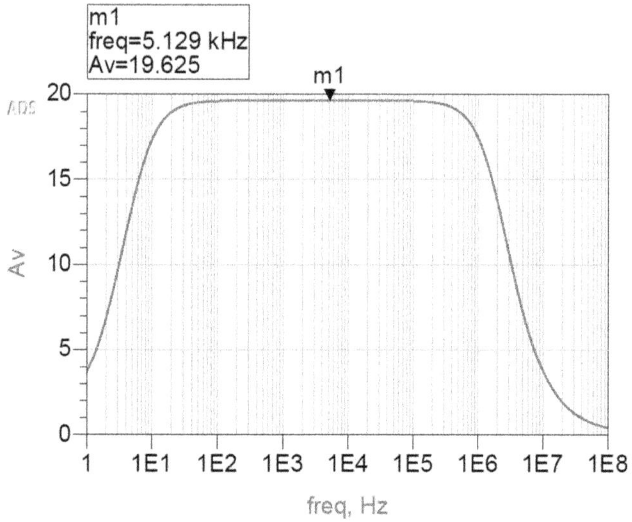

FIGURE 5.7 Voltage gain frequency response of the designed CE amplifier.

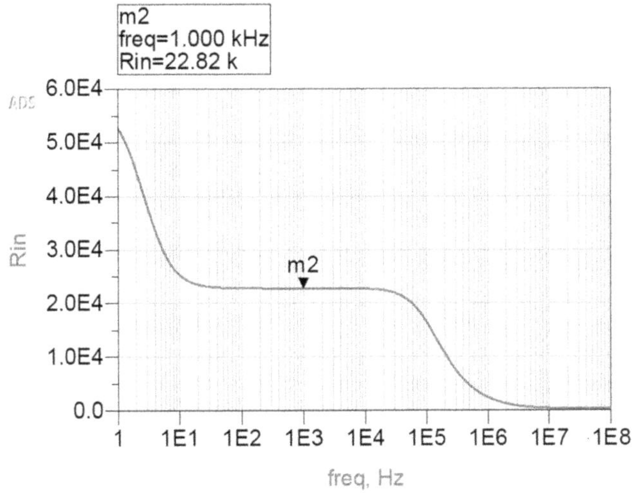

FIGURE 5.8 R_{in} frequency response of the designed CE amplifier.

The observed $R_{in} = 22.82$ kΩ exceeds the specified value of $R_{in} = 20$ kΩ because of our underestimating of β. The larger the R_{in} the merrier.

Even though the CE amplifier's bandwidth was not specified, we still included it in the formulas of MeasEqn (in Figure 5.6).

The BW of A_v can be measured manually from the A_v curve. With a cursor (or a marker), look for a gain of 0.707 of the mid-band gain. The frequency at which this gain is found is the −3 dB point. Here, of course, we can show the BW results from the BW formula in MeasEqn on the schematic.

We will use transient simulation to check the circuit's performance at different input signal amplitudes.

In Figure 5.9, note the Transient Simulation StartTime and StopTime parameters, selected to avoid $V_{out}(t)$ transient due to the large capacitors.

In Figure 5.10, we include a representative result for input amplitude of 20 mV.

Note the distortions of $V_{out}(t)$: Non-symmetric peak amplitudes, upper period (positive) wider signal than that at the lower periods (negative). What is considered "unacceptable"? We will discuss this in subsequent simulations.

In the harmonic balance simulation (Figure 5.11), we study the main signal and the harmonics of currents and voltages due to an AC input (or sometimes multiple AC inputs).

Harmonic 1 is the "main signal tone" chosen to be 1 kHz. "Order[]" specifies how many harmonics shall be used to approximate each signal in the circuit.

An arbitrary quantitative distortion assessment criterion is that if none of the harmonics has an amplitude bigger than 10% of the amplitude of the main tone, the signal is considered "not too distorted". Otherwise, it is deemed "too distorted".

Notice, in Figure 5.12, that for an input of 5 mV, we expect an output $|V_{out}| = |A_v| \cdot |V_{in}| = 5 \cdot 19.625 = 98.125$ mV.

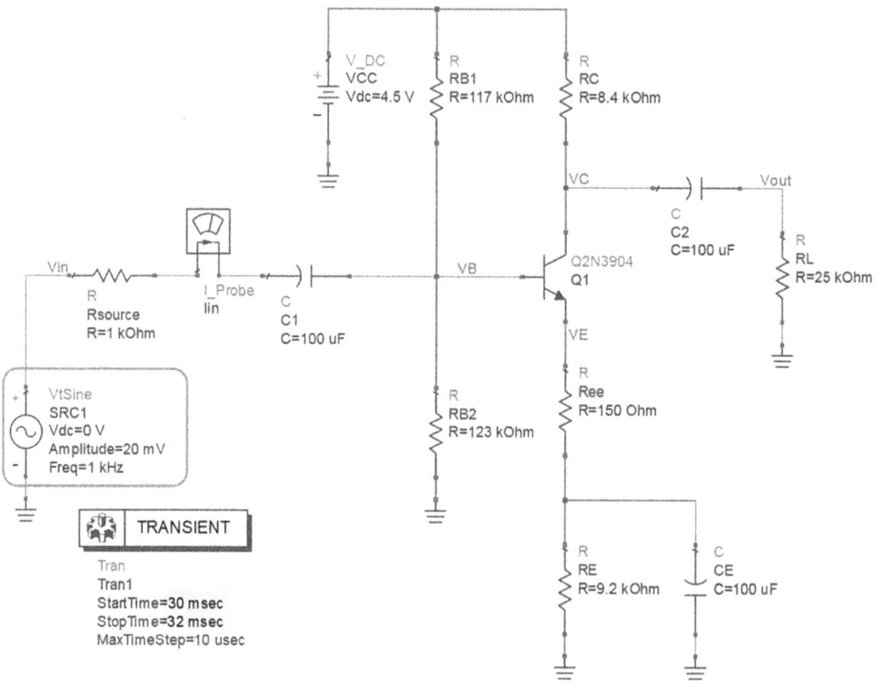

FIGURE 5.9 Transient simulation testing of the designed CE amplifier.

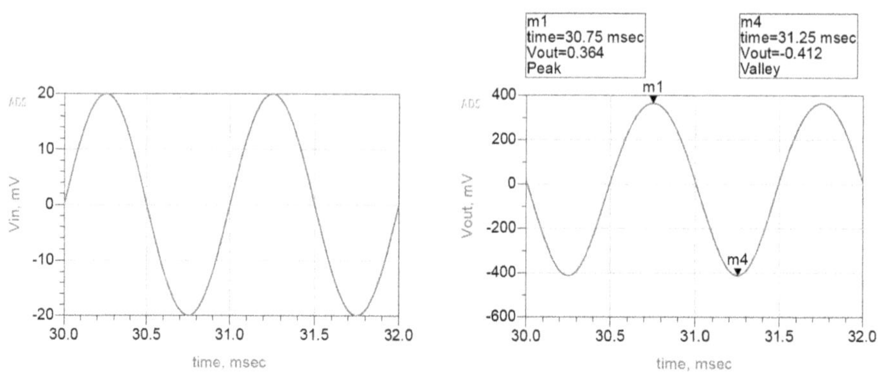

FIGURE 5.10 Transient testing of the CE amplifier at the maximum specified input signal amplitude of 20 mV.

The simulation results show 98.05 mV, close to our expected results. Note that some of the signal energy goes into the harmonics, which explains the small difference when compared to the expected results from the small-signal assumption. The distortion calculation shows 1.137% distortion. Thus, the output is considered not too distorted.

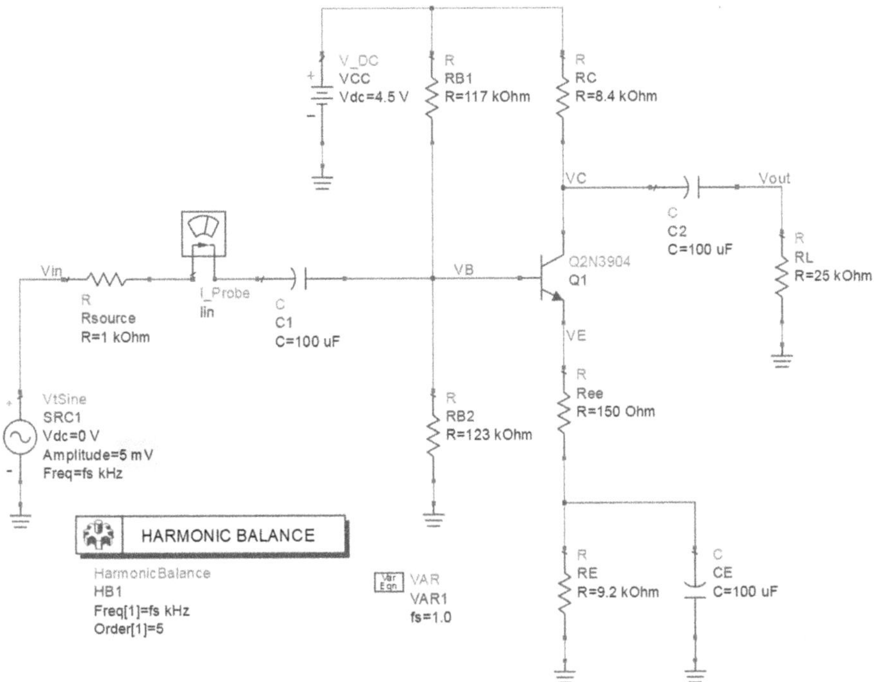

FIGURE 5.11 Harmonic balance simulation testing setup.

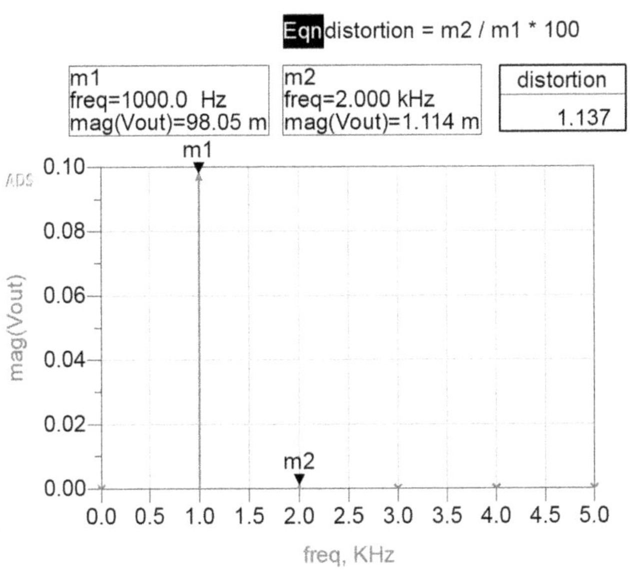

FIGURE 5.12 Harmonics of the CE amplifier output voltage for an input amplitude of 5 mV.

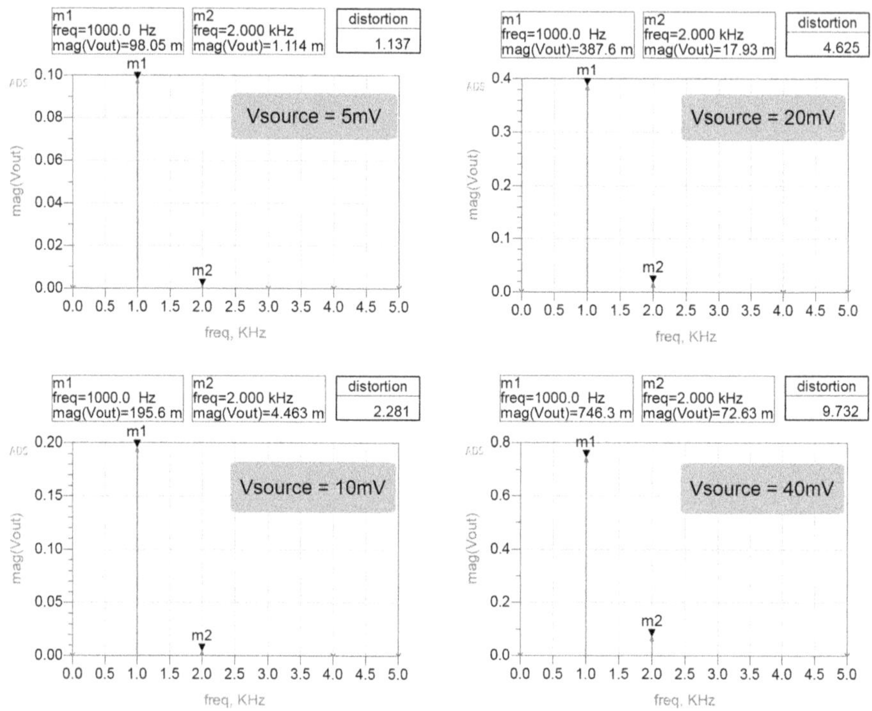

FIGURE 5.13 HB simulation results for four representative input amplitudes.

As shown in Figure 5.13, we tested the amplifier with several input signal amplitudes. The CE amplifier with a degenerative emitter resistor that we have been using does not show severe distortion until voltage levels greater than 40 mV.

We wish to compare an emitter-resistance-degenerated CE amplifier (as tested in Figures 5.11 and 5.13) with a non-degenerated CE amplifier. We shall look first at the distortion effects vs. input voltage amplitude. In the next section, we will look at statistical variability and yield.

Steps to obtain a non-degenerated CE with the same gain (−20) as a degenerated one, with minimal changes, are as follows:

1. Bypass R_{ee} (together with R_E). This keeps the DC current unchanged.
2. No need to consider the effects of r_o on the gain tweaking. It is truly negligible for $I_C = 0.16\,mA$ that we have in this example.
3. $R_{Cnew} \| R_L = (R_{Cold} \| R_L)[r_e/(R_{ee} + r_e)]$ brings down the gain of the non-degenerated stage to the desired level.
4. In this example, $R_{Cnew} = 3.73\ k\Omega$.

The non-degenerated CE amplifier is shown in Figure 5.14, along with the annotated DC results.

The DC results (in Figure 5.14) show that the value of the collector's current is unchanged. The collector voltage is now 3.87 V, which keeps the transistor in the active region.

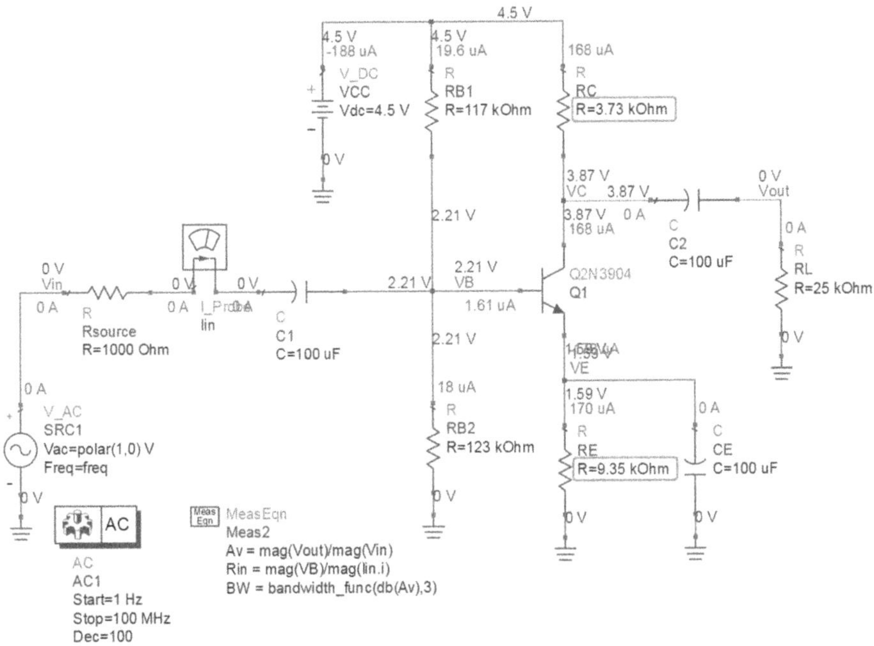

FIGURE 5.14 DC results of a non-degenerated R_E CE amplifier with the same gain as the designed CE amplifier.

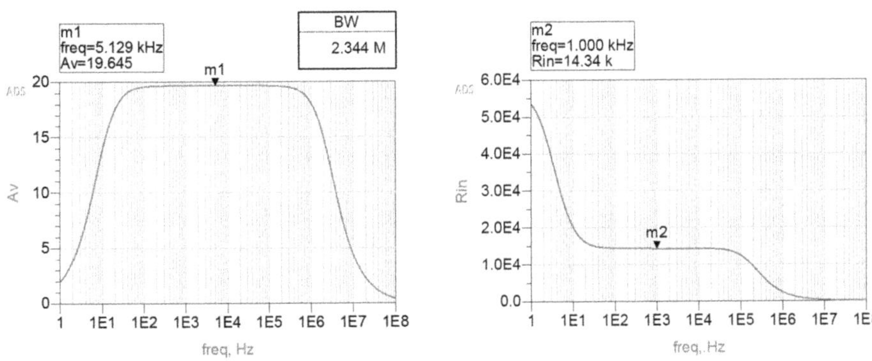

FIGURE 5.15 AC simulation results for the non-degenerated emitter resistance CE amplifier.

The voltage gain is the same, as observed in Figure 5.15, but the input resistance no longer meets the specification of 20 kΩ. It is only 14.34 kΩ, still high enough and of no concern for this example.

Figure 5.16 presents the schematic for a harmonic balance simulation of the non-degenerated CE amplifier.

Comparing Figures 5.13 and 5.17, the CE with a non-degenerated emitter resistor shows severe distortion at input voltage levels greater than 10 mV. This is

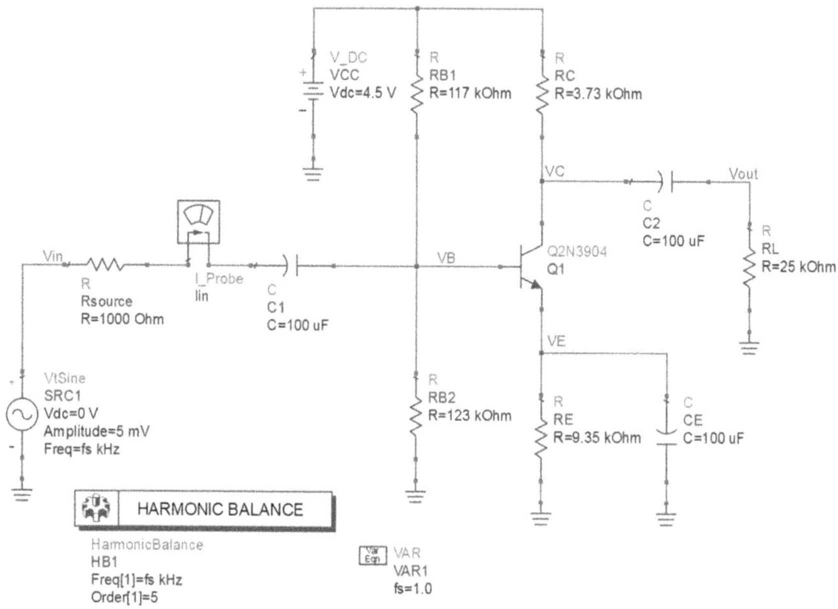

FIGURE 5.16 Harmonic balance distortion analysis of the non-degenerated CE amplifier.

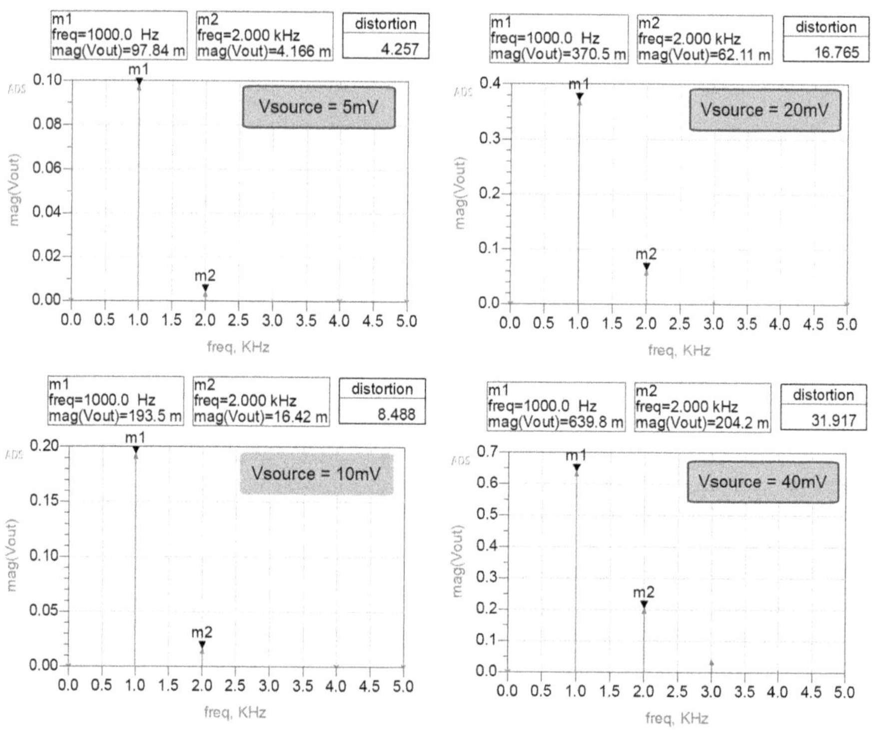

FIGURE 5.17 Harmonic balance simulation results for the non-degenerated CE amplifier.

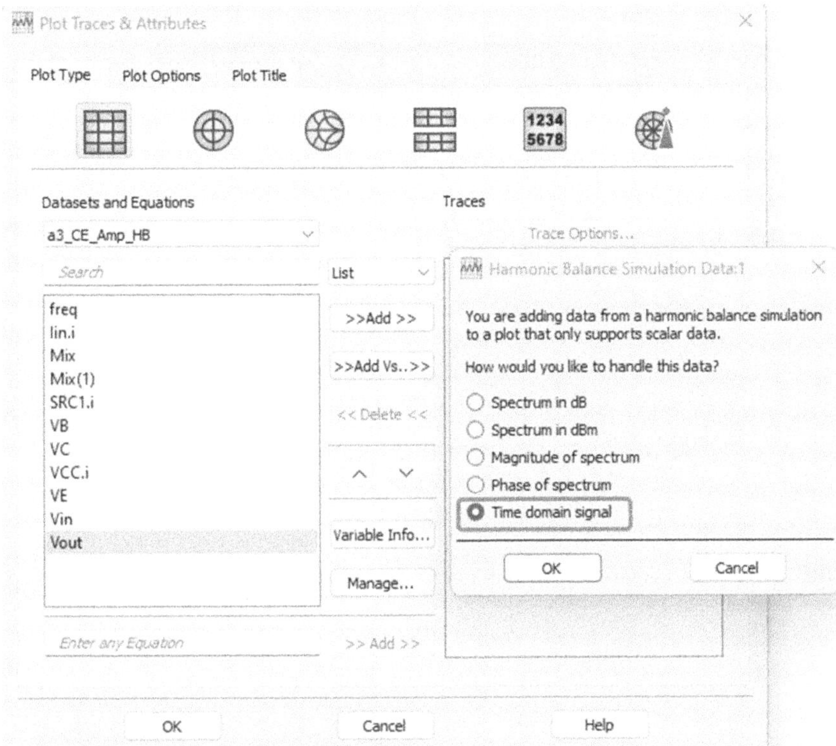

FIGURE 5.18 HB simulation dialog box for obtaining time signals from a finite set of harmonics.

four times smaller than the input signals that can be handled by the degenerated emitter design.

Harmonic balance simulation can generate very good approximations of the true time signals. Figure 5.18 shows the HB simulation dialog box and how to observe time signals.

Harmonic balance results are steady-state results. Within the limits of the order, used in the HB controller, the time domain signal can be calculated based on the HB results. One may use the ts() function to convert HB results from the frequency domain to the time domain or use the dialog box above.

The signal $V_{out}(t)$, shown in Figure 5.19, obtained in HB simulation based on five harmonics, is very similar to the $V_{out}(t)$ shown in Figure 5.10, done by transient simulation, which is based on many more harmonics.

We conclude this section with a comparison between the time signals of the degenerated and non-degenerated same-gain CE amplifiers. See Figures 5.20 and 5.21.

Figure 5.21 clearly shows the superior distortion performance of a CE with a degenerated emitter resistor over that of a non-degenerated emitter resistance CE amplifier.

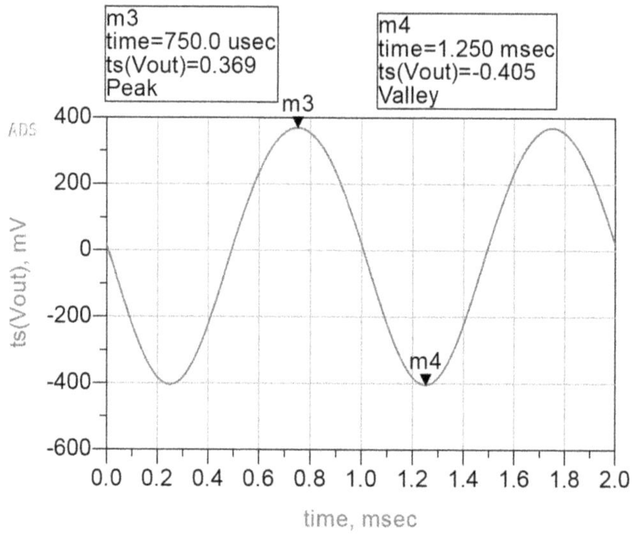

FIGURE 5.19 Recreation of $V_{out}(t)$ from the first five harmonics of V_{out} of the degenerated emitter CE amplifier.

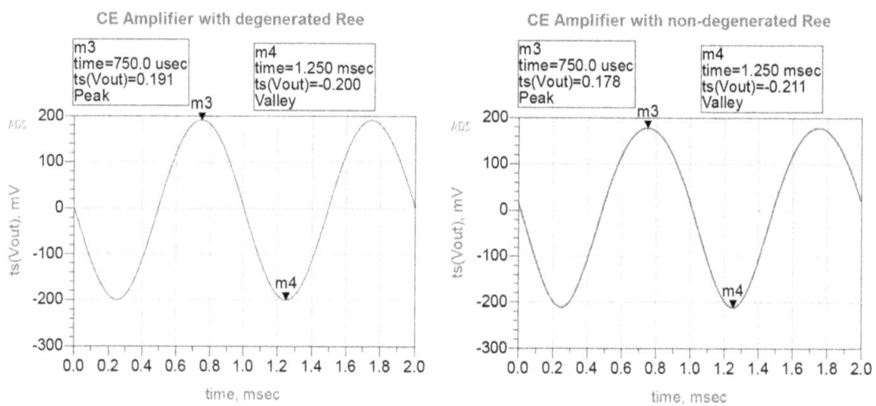

FIGURE 5.20 Comparison with V_{source} amplitude of 10 mV.

5.4 ADS YIELD ANALYSIS AND STATISTICAL DATA POST-PROCESSING

What is yield analysis? In some other electronic simulation programs, it is often called Monte Carlo analysis.

The design of any circuit initially assumes that all the components are infinitely precise. In the actual circuit implementation, we often approximate some component values to the nearest standard values. We also tweak the design to compensate for approximate parameter value assumptions. Yield analysis is done whenever we

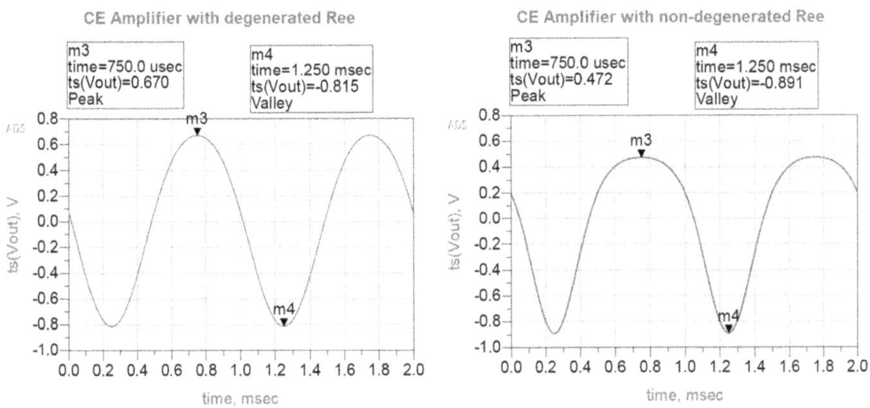

FIGURE 5.21 Comparison with V_{source} amplitude of 40 mV.

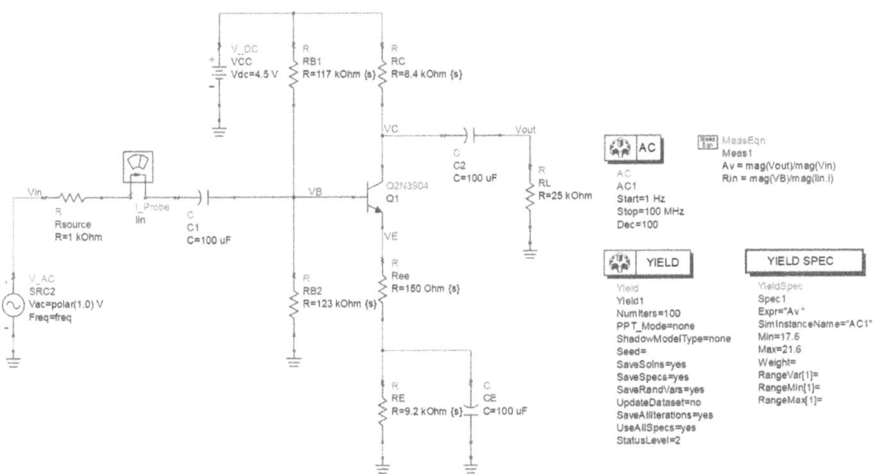

FIGURE 5.22 Schematic preparation for yield analysis of design Example 5.3.

consider components' tolerance. We do multiple simulation runs and assess the percentage of "successful runs".

Yield analysis should be combined with another simulation engine – DC, AC, transient or harmonic balance simulation. In the coming example we combine it with AC simulation. See Figure 5.22.

As seen in Figure 5.22, we need one yield analysis controller and at least one yield spec block.

Prior to running yield analysis, we need to make sure that the requirements we are interested in are met by the design and that we have an analysis setup that can perform the required measurements, such as gain and R_{in}. Recall from Section 5.3 that the following frequency response plots present the nominal results for this design.

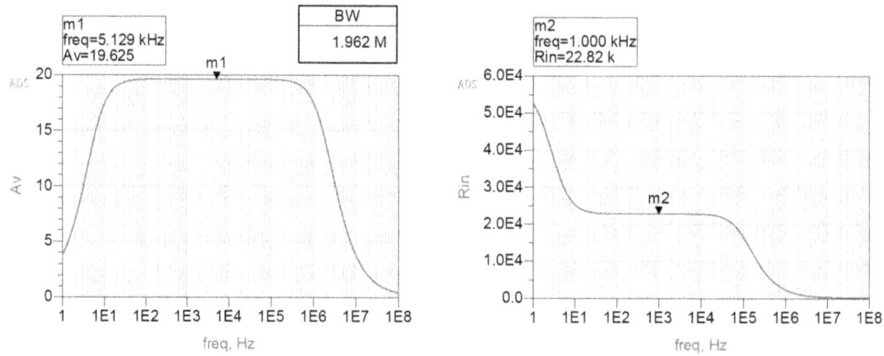

FIGURE 5.23 Nominal voltage gain and input resistance plots for Example 5.3.

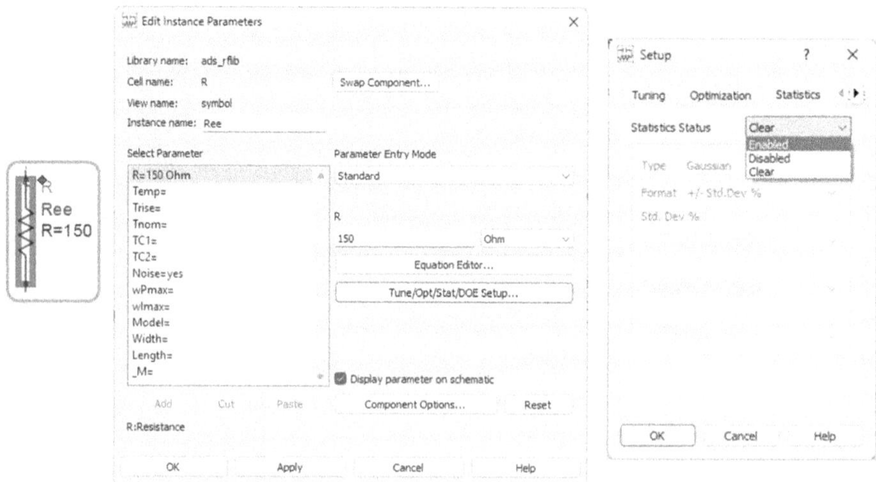

FIGURE 5.24 Adding tolerance values via the part properties dialog box.

Notice in Figure 5.24 (center) that in the part properties dialog, invoked by clicking on the part, as in Figure 5.24 (left), there is a **Tune/Opt/Stat/DOE** Setup … button. The **stat** portion will allow us to specify the part tolerances.

Under the **Statistics** tab the first action is the **Enable** statistics for this part. See Figure 5.24 (right).

As seen in Figure 5.25 (left), the **Type** pulldown menu has different types of probability density functions. For our example, we will use **Uniform**. In other examples later on, we shall sometimes use Gaussian.

A format must be selected for the statistics definition. See Figure 5.25 (center). We choose **±Delta %** to specify the tolerance in percentage.

The last action needed is to specify the tolerance range. In our example, we select "Delta %" to be 10. See Figure 5.25 (right), which means ±10%**.** Modern resistors are normally found at ±5% and even smaller. Indeed, manufacturers are encouraged

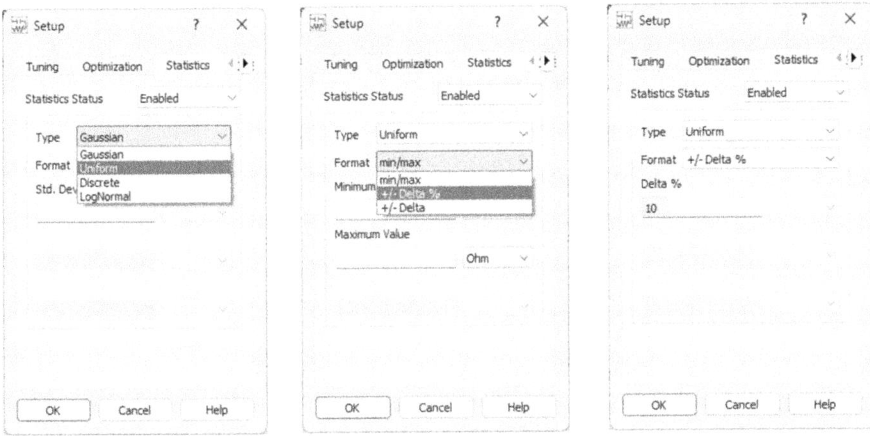

FIGURE 5.25 Selection of probability distribution and its relevant parameters.

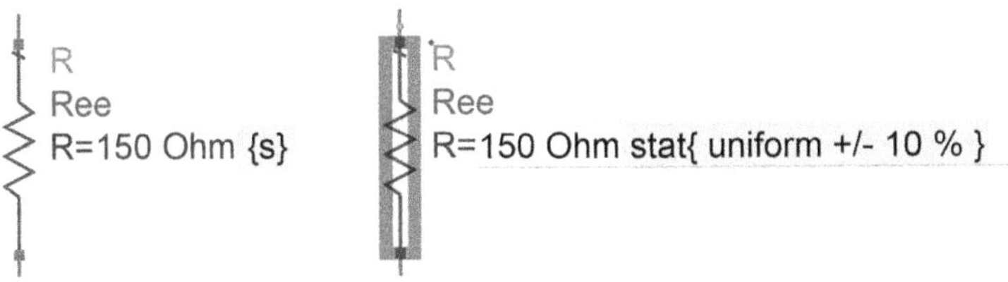

FIGURE 5.26 (Left) The {s} is shown in the schematic. (Right) The details about the distribution are shown only after clicking on {s}.

to reduce tolerance in their processes. Note that this is important since tighter tolerances lead to higher yields.

Once the statistics definition process is completed, and if the value is visible on the schematic, an **{s} symbol appears next to the component's value**. Look again at Figure 5.22. This symbol indicates that the part has defined statistics. This symbol is a shorthand notation. If we want to see more information, we need to click once on the {s} to view the details of the statistical setup for this parameter.

See Figure 5.26.

An alternative to the process of adding tolerance values, component by component, is shown in Figure 5.27. ADS provides the capability to set up statistics in a table format. This comes in very handy when many parameters need statistical setup or to view statistics setup for several parts. This is indeed the preferred way to define parts statistics.

The table, shown in Figure 5.27, is accessed through the upper toolbar "**Simulate**" tab, followed by clicking on "**Simulation Variables Setup ...**". The table serves users not only for assigning parts' statistics but also for optimization and other applications that will be studied later on.

The table, called from "Simulate" for a particular schematic, automatically lists all the schematic's component names, DC supply names, signal source names, and the

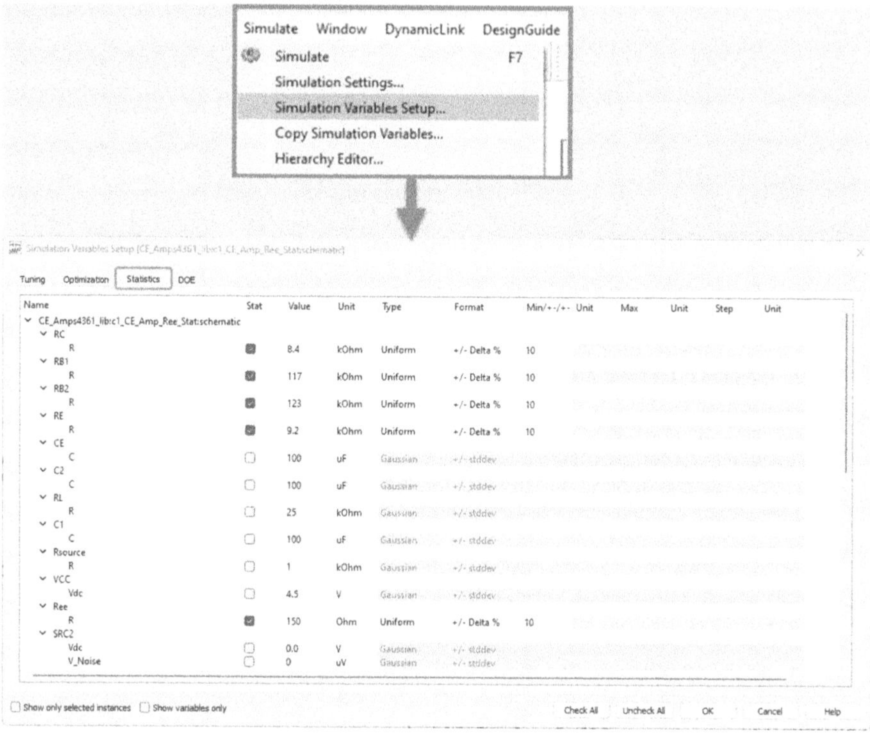

FIGURE 5.27 Table format Tune/Opt/Stat/DOE assignment.

sources' DC and noise parts, the nominal values of each, and the units. Figure 5.27 shows the parameters of Example 5.3.

The user then selects which components (and sources) need statistics to be assigned to them, under the stat column.

The statistics parameters (type, format, and value) are written by the user directly into the respective table columns.

The yeld simulation components may be found in the parts palette (Figure 5.28 (left)) or in the Components Library, like any other parts, as shown in Figure 5.28.

Yield simulation setup steps are shown in Figure 5.29. We get there by clicking on the yield component.

Under the **Setup** tab the default values, shown in Figure 5.29 (left) and (center), will work fine for our purposes. Make sure to set the **number of trials;** here, we use 100.

In the **Parameters** tab, Figure 5.29 (right), select the **Output Data** and **Output Data Control**, as shown. The **Seed** parameter of Figure 5.29 (right) presents users with two options:

Option 1: Leave the seed value unmarked. This instructs ADS not to use the same random values in the Monte Carlo analysis. From one run to the next, and from one Monte Carlo study to the next, everything is randomized.

Option 2: Choose any specific arbitrary seed number (such as 54,377 or 1079 or so on). A choice of a specific seed forces the random number generators to repeat all the

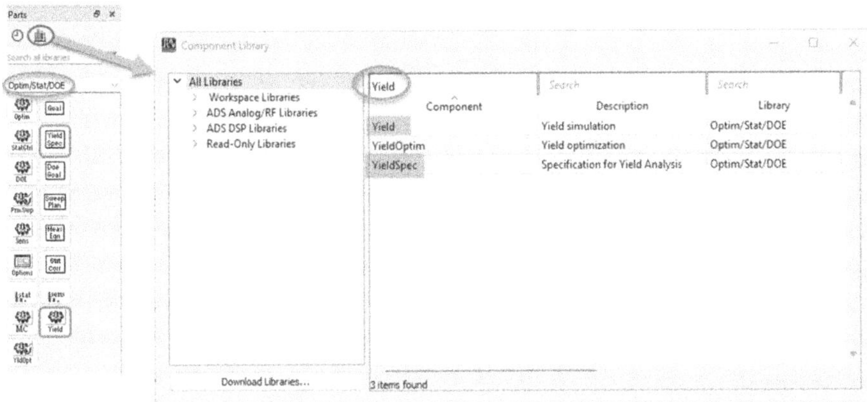

FIGURE 5.28 Yield components in the parts palette (left) and in the Component Library (right).

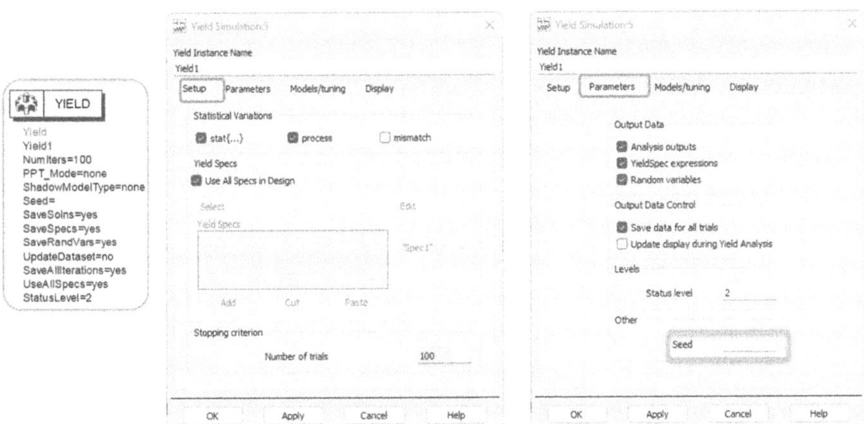

FIGURE 5.29 Yield simulation dialog box.

random sequences used in the first Monte Carlo study. The Monte Carlo study can be repeated with the same random sequences.

Option 1 is good for quality control, where we want all the parameters with statistics to be truly random.

Option 2 may be helpful for debugging a design. The designer can trace the combination of parameters that causes the design to fail to meet the specs.

In this book, we do not require a seed for our analyses, so each yield analysis may produce values that are different.

The yield specification setup and dialog box are shown in Figure 5.30.

For the yield specification an expression must be provided (**Expr**). In our case, we have a MeasEqn that measures the gain so that we can use it here. See Figure 5.22 for the gain formula.

It is important to specify which analysis measures the values in **SimInstanceName**, here we use "AC1".

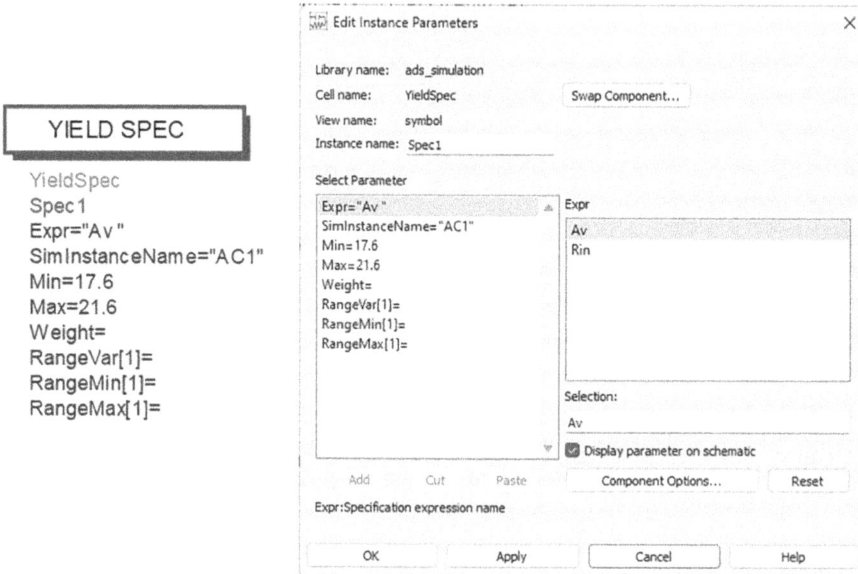

FIGURE 5.30 Yield spec dialog box and setup.

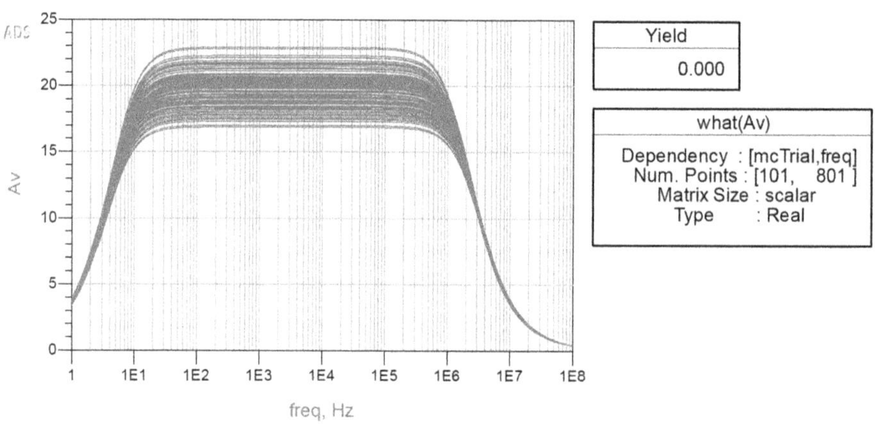

FIGURE 5.31 Yield analysis results: 100 gain frequency response plots and a yield figure.

We then must specify the range of passing values for the measurement. Assume, for instance, a ±10% range. We know that the magnitude of the gain is 19.6 (as shown on the left side of Figure 5.23), so we specify a range of 17.6–21.9 as our passing range and set yield spec min and Max to these values. Any of the 100 runs in which the gain falls between ±10% of the nominal gain is counted as "success". Any run in which the obtained gain falls outside the 10% boundaries is counted "failure". A designer may sometimes opt for tightening the boundaries of a "success", say ±5%.

Yield analysis results are shown in Figure 5.31. The **Yield variable is the number of successes expressed in % of the number of runs**. Note the use of the **what() function**. It takes a variable as an argument and returns a description of the variable content. Here, it tells us that the dependencies for Av are mcTrial and freq with 101 and 801 points, respectively. Further details on the what() function and data manipulation can be found later on in Chapter 5.

After running the yield analysis simulation, we see that the variable "yield" appears in the data display menu. We listed yield (as a table) and noted that it came out to be 0%.

However, since ADS allows us to plot all the traces from the yield analysis trials, we see that indeed many trials have the expected gain shape, and in the mid-band the gain meets the specs.

The problem is that ADS was instructed to evaluate the entire frequency sweep from 1 Hz to 100 MHz. Indeed, in this large range all trials do not meet the gain spec outside the mid-band range of frequencies. We need to specify the yield analysis to only include frequencies in the mid-band range. How do we do it?

One option is to use yield spec **RangeVar**.

From the what() function information, we note that "**freq**" is the variable ADS used to keep track of the analysis frequency. We can use freq in the RangeVar setup of the yield spec to limit the range of frequencies somewhere in the mid-band. Let us say we use 1–10 kHz (as shown in Figure 5.32).

A second option is to limit the entire AC analysis frequency range. The analysis Start and Stop frequencies for the yield analysis can also be limited in range. This option is typically saved for those cases in which the simulation runs slowly or saves large amounts of data, and one wants to save on data storage and manipulation time.

Let us first demonstrate the method that involves RangeVar in the yield spec (Figure 5.33).

The simulation results are shown in Figure 5.34. The results indicate that the yield is about 92%. It means that between f = 1 kHz and f = 10 kHz, the gain A_V is a "success" in 92 runs out of 100. Running the yield analysis several more times gave the following values of 93%, 89%, and 90%. Increasing the number of trials to 1000 gave a range of successes of 88.5%–89.5%. In general, the more iterations, the more accurate our yield results. However, this comes at the expense of longer simulation time and larger computer memory and storage requirements.

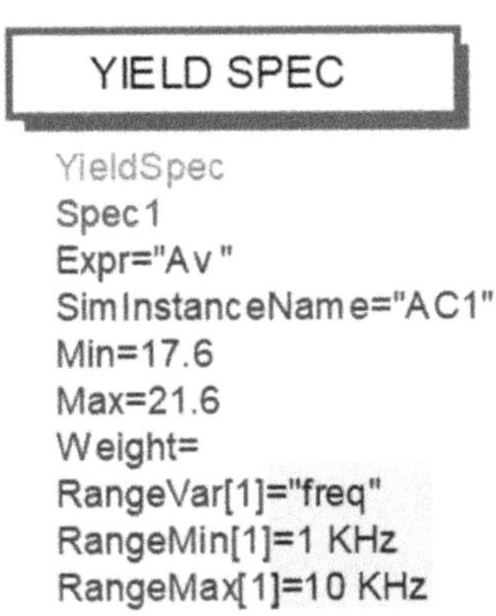

FIGURE 5.32 RangeVar settings in the yield spec.

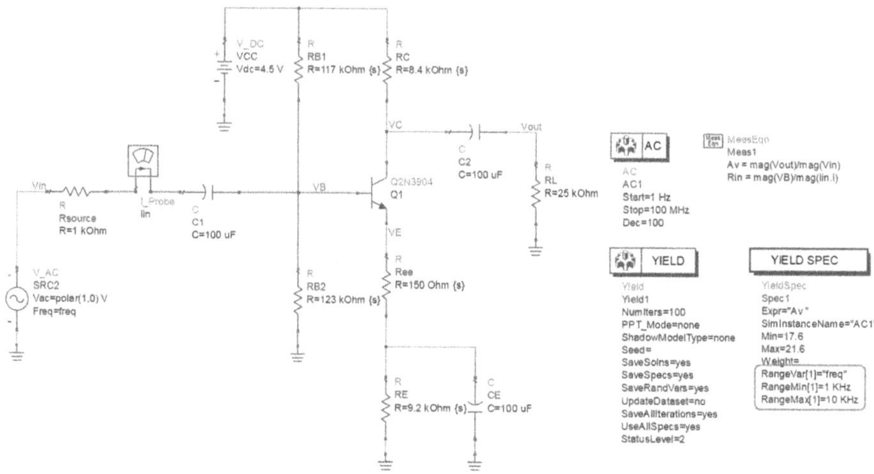

FIGURE 5.33 Revised yield spec setup prior to running yield analysis.

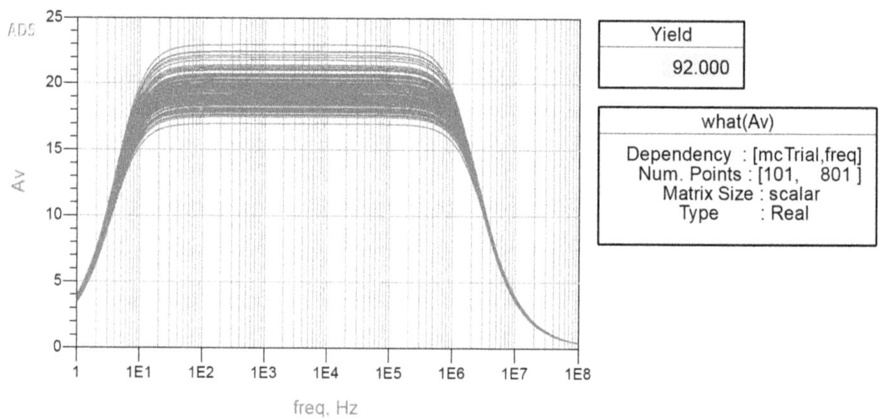

FIGURE 5.34 New yield analysis results after revising the gain inspection frequency range.

Rather than using a reduced frequency range on the AC controller (replacing the reduced frequency range offered by the yield spec), let us go to the extreme and use a **single frequency** in the AC simulation. Indeed, the gain is flat in the mid-band, so a single frequency in our case works fine. The setup for single frequency AC simulation is shown in Figure 5.35.

Single Frequency Analysis dialog box setup is shown in Figure 5.36.

"Single Frequency" is one of the modes of AC simulation. We can choose a representative frequency of the mid-band, such as f = 1 kHz. This option is computationally the fastest.

The results shown in Figure 5.37 indicate that the yield is about 89%, which agrees with what we saw before (using RangeVar).

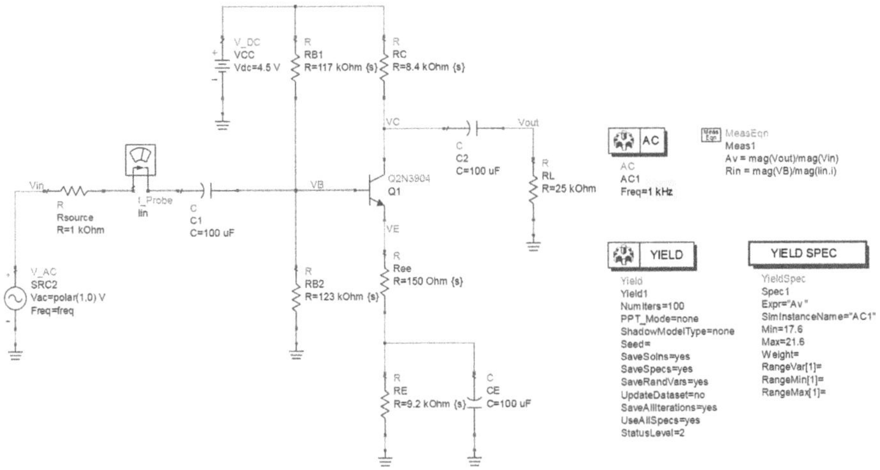

FIGURE 5.35 Yield analysis embedded in a single frequency AC simulation.

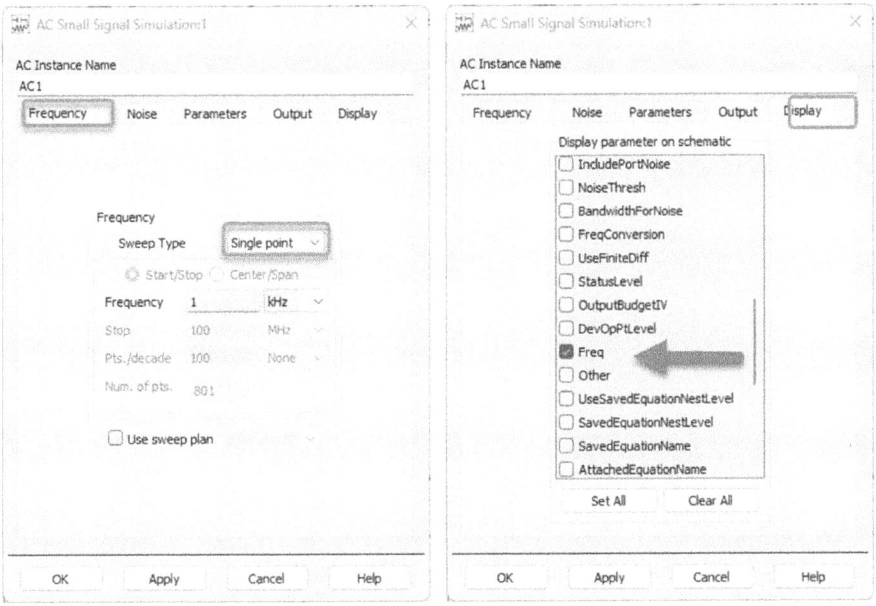

FIGURE 5.36 Dialog box for single frequency AC simulation.

The A_v plot does look strange, though. This makes sense since the results from the what(A_v) function clearly show that we only have a single frequency value, and this A_v plot has freq in the X-axis.

We can make this data more useful by using **data indexing** in ADS. Indexing allows us to select a portion of the data. In ADS indexes start at index 0, unlike some tools where 1 is the first index. We can say that ADS data arrays are zero indexed. More information about this can be found in Section 5.5.

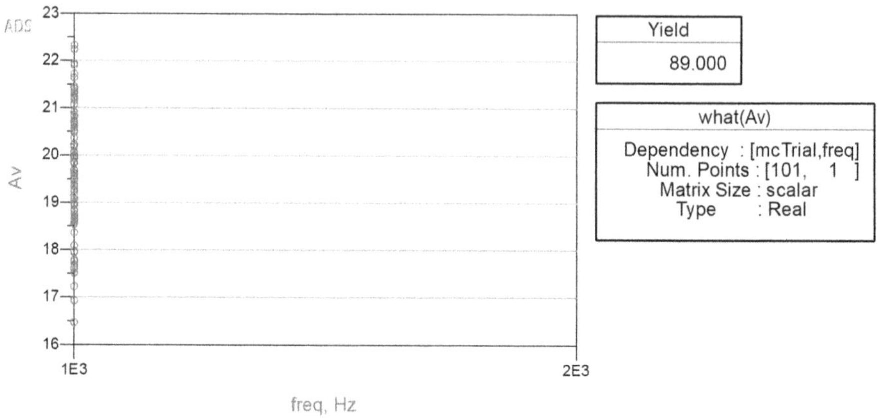

FIGURE 5.37 Results of the yield analysis with single-frequency AC simulation.

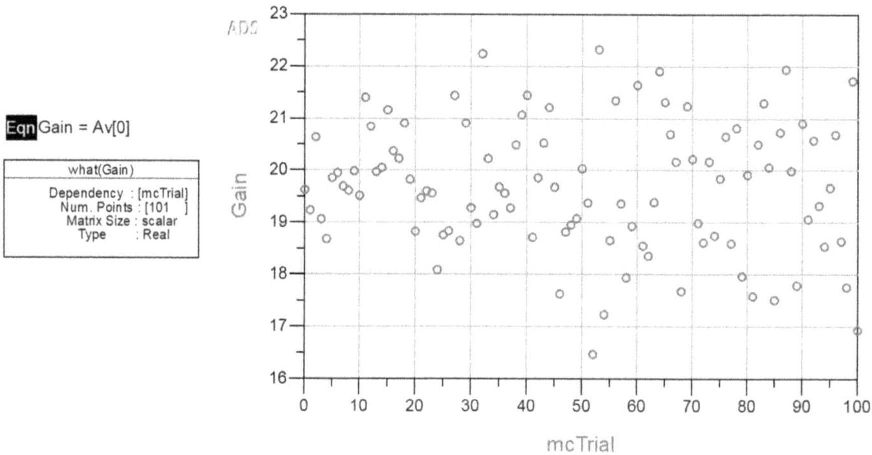

FIGURE 5.38 From the two-dependencies A_v variable to a single-dependency gain variable via zero indexing data display equation.

Note in Figure 5.38 that gain now has only mcTrial as a dependent variable, so the plot is more reasonable now that mcTrial is the X-axis.

When dealing with statistical data, it is often a good idea (or a requirement) to look at the mean and standard deviation. ADS provides such functions as seen in Figure 5.39 (left). Figure 5.39 (right) shows the presentation of the data by a **histogram**. Histograms will be further explained in Section 5.5.

Recalling the comparison in Section 5.3 between two CE amplifiers with the same voltage gain, one having a degenerated emitter and the other not, let us also compare the yields.

Figure 5.40 shows the non-degenerated CE amplifier set up for a yield analysis.

For a fair comparison of the two CE amplifiers, we will use the same parts' statistics, and once again, we will conduct a single-frequency AC simulation.

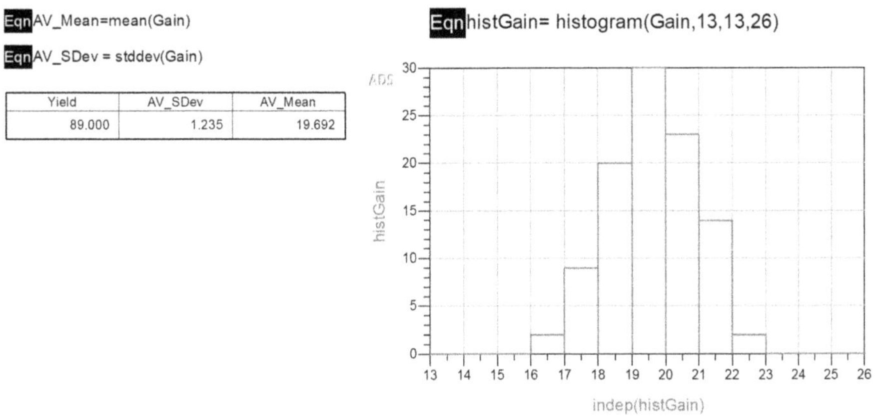

Yield	AV_SDev	AV_Mean
89.000	1.235	19.692

FIGURE 5.39 Computation of statistical functions using data display equations on a dependency-reduced random variable. The equations of Figure 5.39 follow the equation of Figure 5.38.

The two gain means, standard deviations, histograms, and yields are compared in Figure 5.41.

Quite remarkably, both designs have the same average gain.

The standard deviation for the non-degenerated case is higher, so gain values are more spread out. This greater spread is also observed in the histogram plots.

Finally, the Re-degenerated CE amplifier has a better gain yield than the non-degenerated design.

5.5 ADS: DATA DESCRIPTION AND MANIPULATION; TRACE EXPRESSION WHAT(X), ADS DATA REPRESENTATION, "FIND_INDEX()" COMMAND, DATA VECTORS, DATA DIMENSIONALITY REDUCTION, SCATTER PLOTTING OF YIELD ANALYSIS DATA, HISTOGRAM PLOTS

We have seen several analysis types, with sweeps and evaluation, such as yield analysis. Understanding how ADS saves this data is useful for our understanding of the results and to further process the data.

For this purpose, we use the what() function by entering it in a column listing and using the **trace expression what(x)**. To see how it is done, observe Figures 5.42 and 5.43.

In data display, we place a column listing.

We **>>Add >>** the variable that we wish information on, say $\mathbf{A_v}$.

We then click on **Trace Options ...**

The above steps are shown in Figure 5.42.

In the Trace Options window, we change the trace expression field by placing the variable inside the what() function (as shown in Figure 5.43).

Click OK to dismiss dialog windows. One should see a column listing like in Table 5.2.

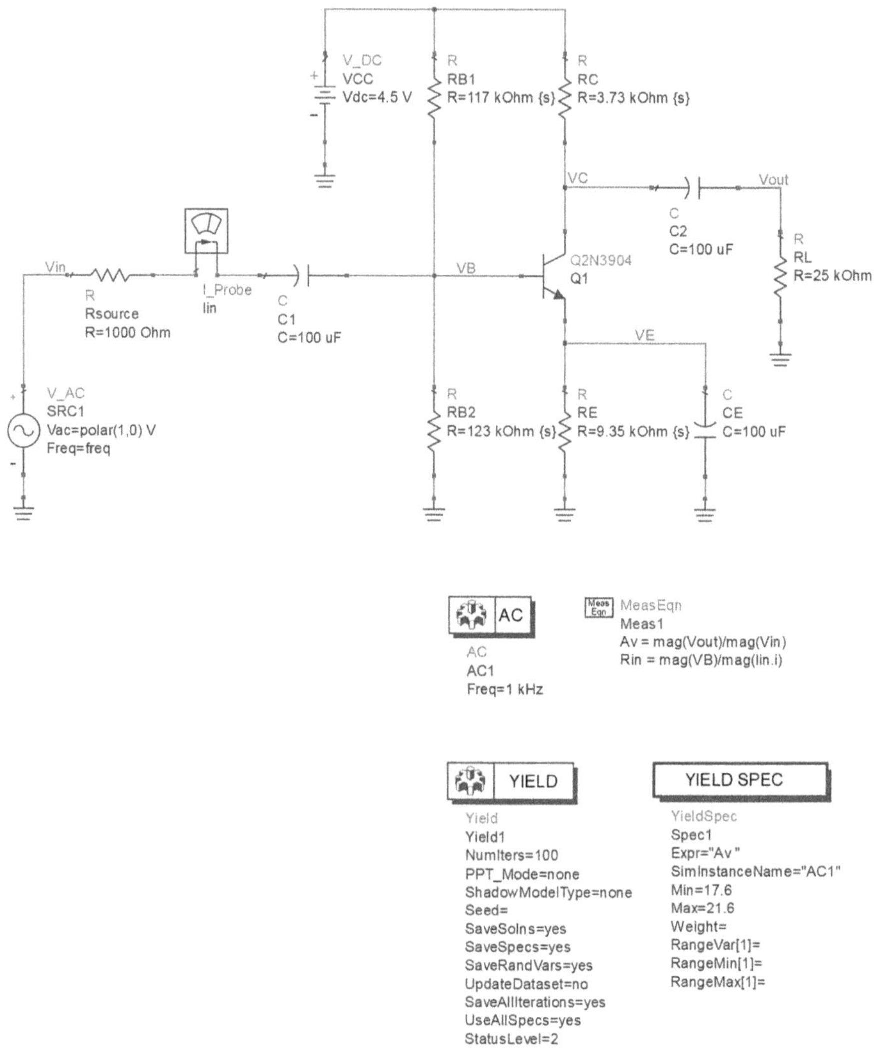

FIGURE 5.40 Yield analysis of a non-degenerated CE amplifier.

The important parts are dependency, num. points, and type. Matrix size is rarely of any importance or use. Subsequently, let us show a few data cases and some useful, and sometimes needed, data manipulation based on this information.

Example 5.4

Single simulation modes.

Three single simulation modes are presented in Figure 5.44.

The AC simulation is set up for eight decades of 100 points/decade, which results in 801 complex data points that depend on frequency.

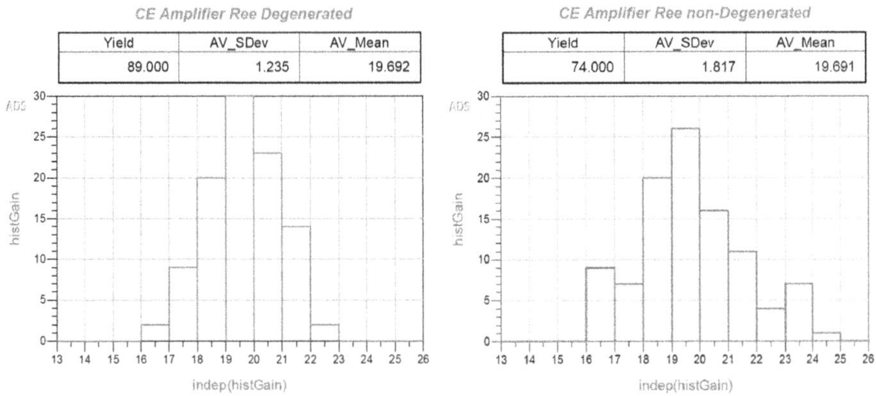

FIGURE 5.41 Statistical comparison of two CE amplifiers, one with emitter degeneration and one without, having the same voltage gain and the same component statistics.

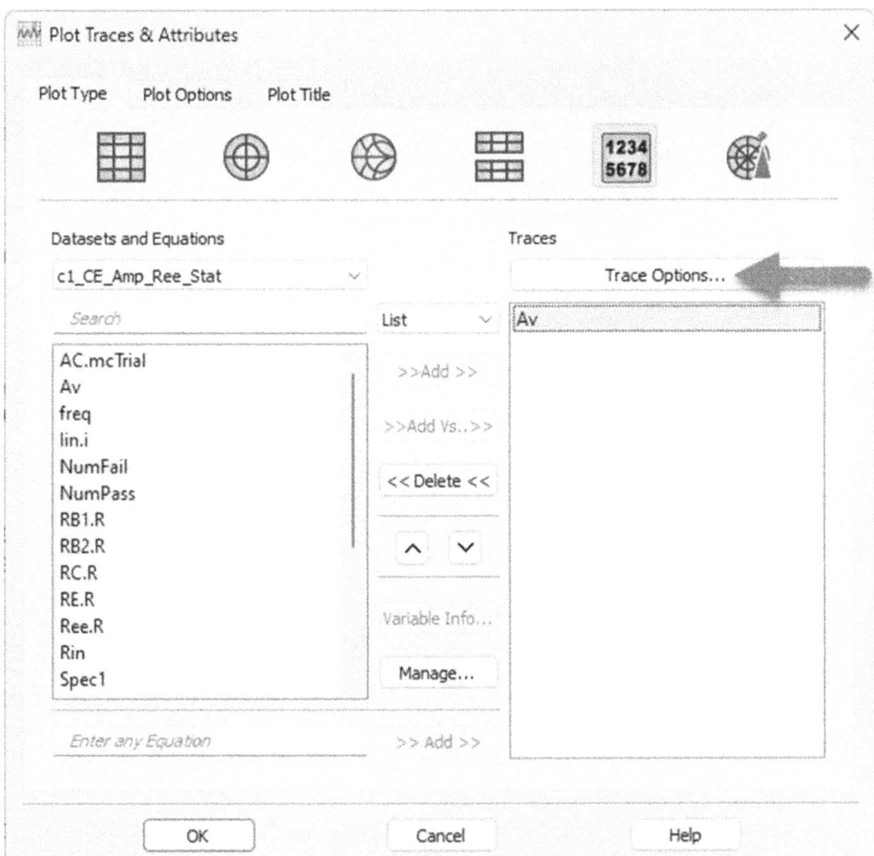

FIGURE 5.42 Working with the what() function within plot & attributes.

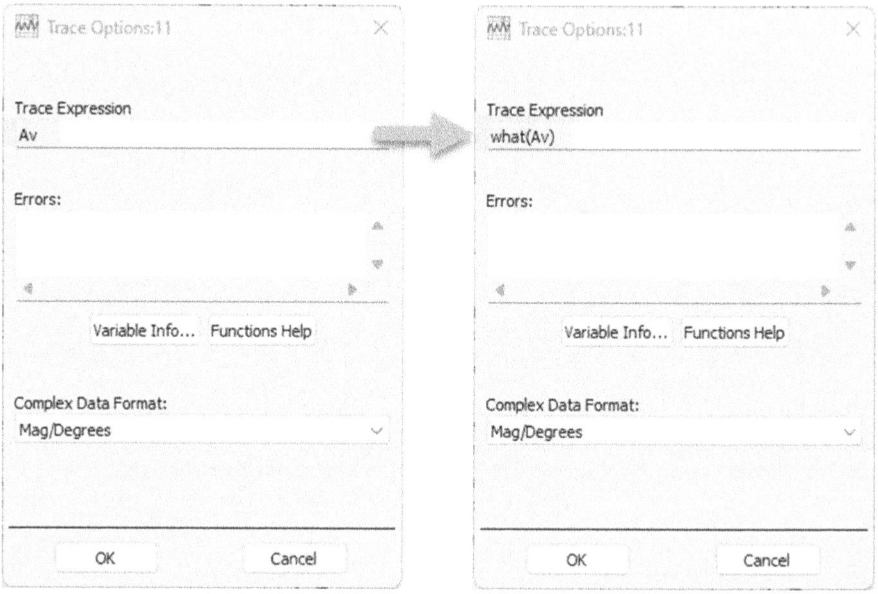

FIGURE 5.43 Specifying the trace expression what().

TABLE 5.2
What(A_v) Information

what(A_v)
Dependency : [mcTrial,freq]
Num. Points:[101, 801]
Matrix Size:scalar
Type :Real

The transient simulation is set up to save a total of 2 ms of data with a maximum time step of 10 μs. In this case, it resulted in 201 real data points that depend on time. Note: Transient can produce more points than dictated by the simulation interval and time step if it switches to time control methodology; further discussion of this is beyond our present scope.

The HB simulation (above) is set up for an order of 5. It results in six complex data points that depend on freq. Note that HB includes a DC value, the main signal frequency, and the harmonics 2–5 in this case.

Example 5.5

AC simulation together with yield analysis.

In addition to the AC analysis, we run a yield analysis that calls the AC simulation. This will add an additional dimension (dependency) to the variable data. In this particular example, it adds dependency mcTrial. See Figure 5.45. We set 100 trials but got 101.

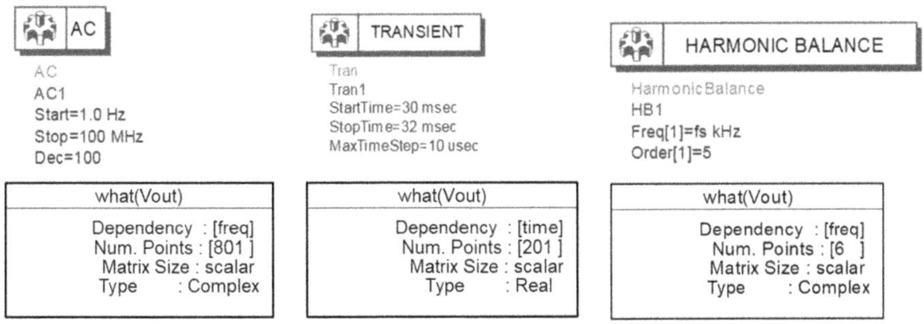

FIGURE 5.44 Data description for AC, transient, and harmonic balance simulations.

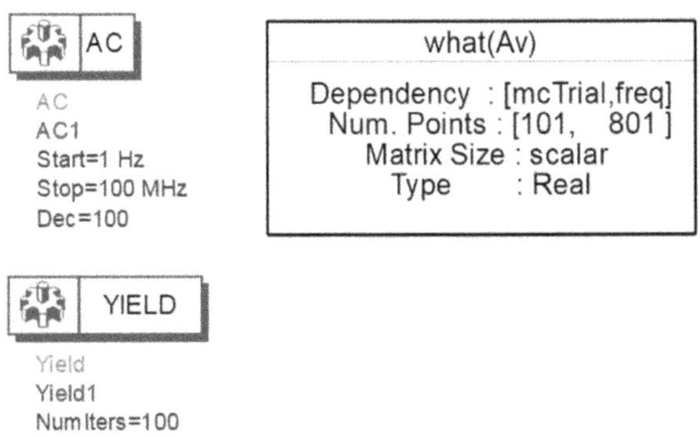

FIGURE 5.45 Data description for AC simulation combined with yield analysis. A_v is real.

This is because ADS will also run first a nominal analysis and saves' the results along with statistical runs. This initial run is not counted in yield calculation.

Example 5.6

Indexing of AC Simulation Data.

ADS allows indexing of data. For these purposes, we place [ ] after the variable and place the index (or indexes) of interest inside the brackets, for example, $V_{out}[0]$. ADS uses zero-based indexing, which means that indexes start at zero instead of one like other computational tools.

Let us stay with the AC data and say that we were interested in what the output voltage is at freq = 1 kHz.

As seen in Figure 5.46, we have 801 freq values, but which index is 1 kHz? To answer this question, we can use the "find_index()" function which has the following syntax: idx = find_index(data sweep, search value). It will find the closest index idx for a given search value.

Note that once we find idx (Figure 5.47), we can use it to index into freq data or any other data. To complete this example, let us find the value of V_{out} at 1 kHz. See Figure 5.48.

Vout is complex data shown here in magnitude/angle format.

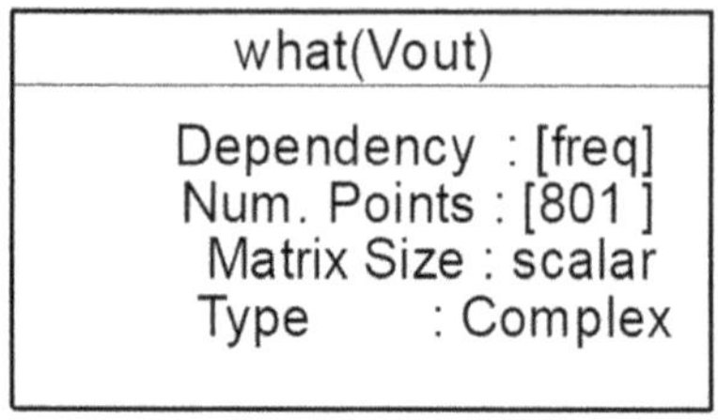

FIGURE 5.46 Data description for AC simulation. Finding V_{out}, which is complex.

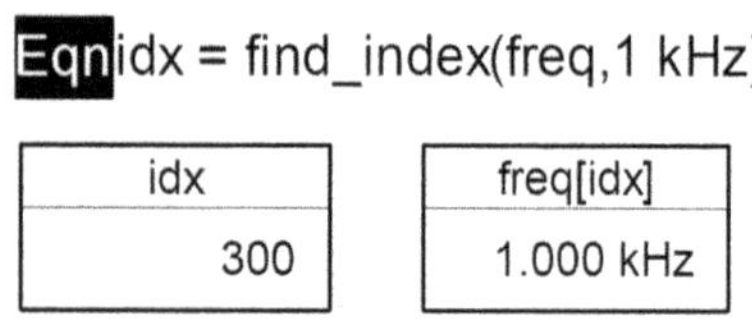

FIGURE 5.47 Data display equation to find an index from a freq data set.

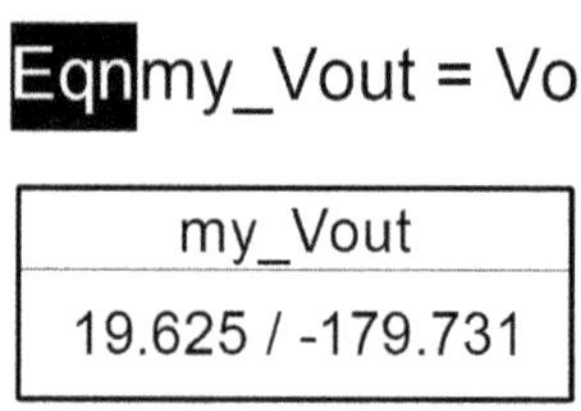

FIGURE 5.48 Data display equation to find a specific AC simulation voltage value, after finding the index.

Example 5.7

Creating a vector of data.

ADS can generate a vector of data using its vector notation. The notation is var1 = [start::increment::stop]. One can write var2 = [start::stop] in which case the increment equals 1. Simple examples are shown in Figure 5.49.

The double colon :: is used to form the vector. Be aware that the double colon is also used as a wildcard (select all), as will be seen soon.

ADS uses this notation to create more complex arrays of data.

ADS even has the capability to produce matrices using the {} brackets. This is, however, beyond the scope. For further information consult the ADS documentation [Keysight1].

Example 5.8

Indexing of transient simulation data.

Using transient simulation data, say that we want to plot V_{out} between 30.5 and 31.5 ms and then calculate the average across this time range.

```
Eqn var1= [10::0.5::15]
Eqn var2=[10::15]
```

var1	var2
10.000	10
10.500	11
11.000	12
11.500	13
12.000	14
12.500	15
13.000	
13.500	
14.000	
14.500	
15.000	

FIGURE 5.49 Synthesis of vectors.

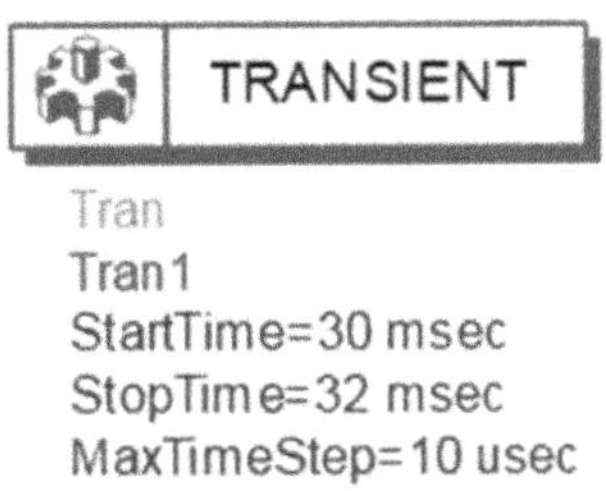

what(Vout)
Dependency : [time]
Num. Points : [201]
Matrix Size : scalar
Type : Real

FIGURE 5.50 Data description of transient simulation variables.

```
Eqn idx1 = find_index(time, 30.5 ms)
Eqn idx2 = find_index(time, 31.5 ms)
```

idx1	idx2
50	150

FIGURE 5.51 Data display equations to find indexes associated with specific time values.

See Figure 5.50. We have 201 time points, and we need to index from 30.5 to 31.5 ms. As a first step, let us get these two indexes. See Figure 5.51.

For the plotting we need to create two vectors of the same length: one for time and the other for V_{out}. The vector for V_{out} is also used for the mean calculation. See Figure 5.52.

Eqn exTime = time[idx1::idx2]

Eqn exVout = Vout[idx1::idx2]

FIGURE 5.52 Vectors creation for subsequent plotting and mean calculation.

Eqn exMeanVout = mean(exVout)

exMeanVout
-0.006

FIGURE 5.53 Data display equation for mean calculation.

The **mean** of the indexed data can be calculated using the **mean() function**, as in Figure 5.53.

To plot the requested signal segment, we use >>Add vs. >> as shown in Figure 5.54.

Example 5.9

Manipulation of two-dimensional yield analysis data.

Notice that on the A_v plot (of Figure 5.55) we placed a marker at mcTrial = 86 because this is the largest gain trial. The task is to extract this trace from the data and plot it against frequency. This data is two-dimensional. We need to select the data for only one index, mcTrial = 86, and all the freq indexes for this trial. The equation is as shown in Figure 5.56. Note that in the equations of Figure 5.56, we use the double-colon :: as a wildcard to select all frequency indexes. As we only used a single mcTrial index, the dimensionality of the data has reduced from 2 to 1. That is, maxTrialGain no longer has mcTrial dependency (Table 5.3).

We can now plot maxTrialGain using Add vs. maxTrialfreq, as shown in Figure 5.57.

Example 5.10

Scatter plot and histogram of yield analysis data.

In Example 5.9, we reduced the dimensionality of the AC simulation yield analysis data by removing mcTrial dependency, leaving only the freq dependency. The other way of reducing the dimensionality of such data is to remove the frequency dependency, by choosing a single frequency, and leaving the mcTrial dependency.

So let us say that we want the value of A_v for all mcTrials at the frequency of 1 kHz.

The solution steps are as follows:

- **Step 1:** We first need to find the index for 1 kHz in the freq data
- **Step 2:** we use the wildcard operator (::) to select all the mcTrials at this frequency.

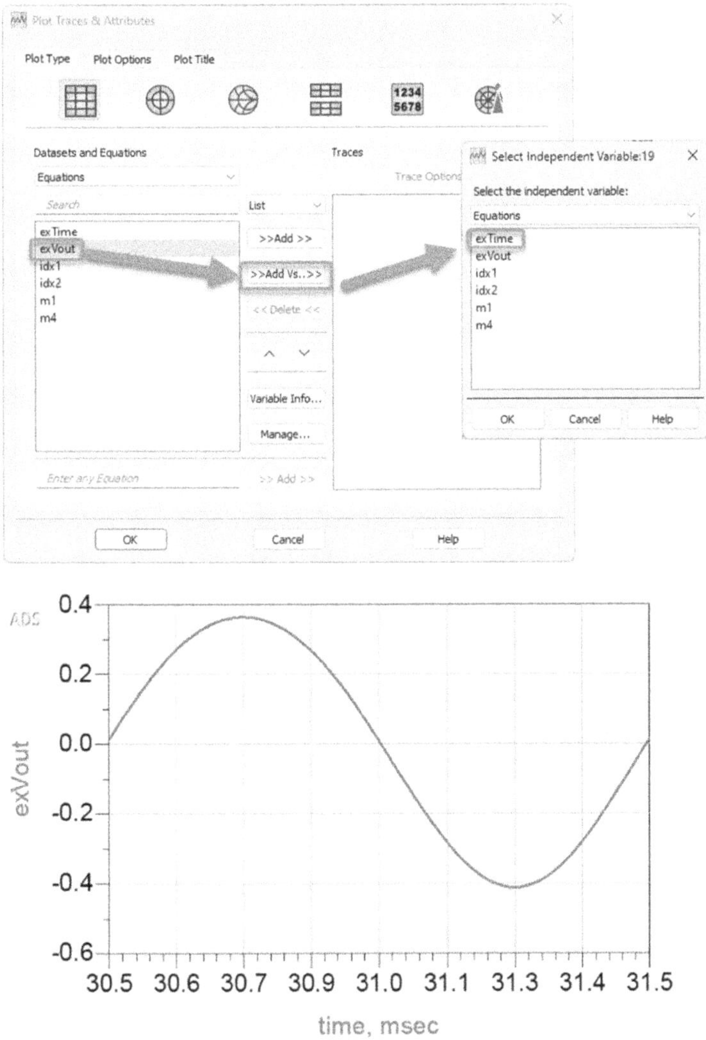

FIGURE 5.54 How to plot a segment of a time signal. Example 5.8.

In this example, we wish to plot the results in a scatter plot and in a histogram. It's a little tricky, though.

The first step is to understand the frequency dimensionality.

Note that what(freq) (in Figure 5.58) tells us that freq depends on [mcTrial,freq]. This is essentially saying that freq also depends on mcTrial. Yet we know that for every mcTrial, the freq values are the same.

Therefore, to search for a variable for a single freq value, we must first reduce freq dimensionality. This is why we use **freq[0,::]** in the idx_freq equation, that is, we reduce freq to a single dimension to search for the index of the value we want, knowing that all trials use the same frequencies.

Now that we have the frequency index of interest, we use it to reduce the dimensionality of A_v such that it only includes the mcTrial values at this freq, as was

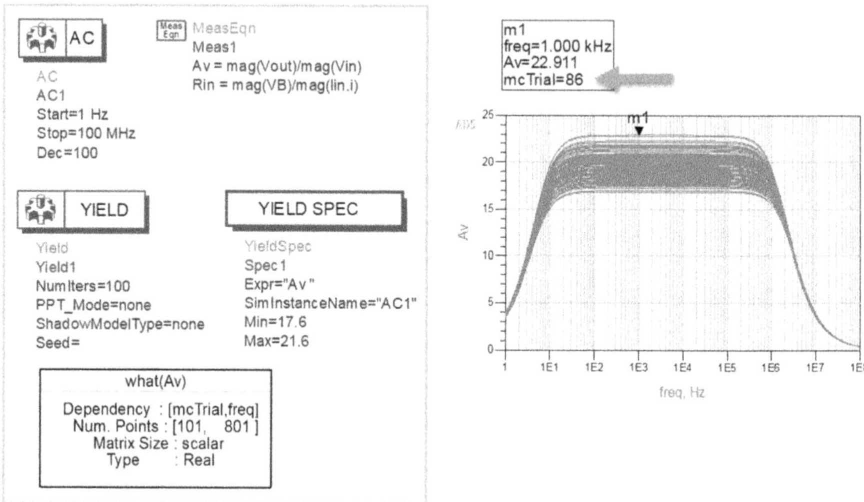

FIGURE 5.55 Yield analysis data to be manipulated in Example 5.9.

Eqn maxTrialGain = Av[86, ::]

Eqn maxTrialfreq = freq[86, ::]

FIGURE 5.56 Data display equations for extracting a single trial's curve.

TABLE 5.3
Data Description of the Picked Up Single Curve

what(maxTrialGain)
Dependency : [freq]
Num. Points:[801]
Matrix Size:scalar
Type :Real

done in the gain at 1 kHz equation, in Figure 5.58. The data is now ready to create both the scatter and histogram plots.

The scatter plot procedure involves plotting gain at 1 kHz and then selecting the **Scatter Trace Type** under **Trace Options ...**, as shown in Figure 5.59.

For the histogram plot, we use the ADS built-in histogram function that has the following syntax: **histogram(data,numberBins,minBin,maxBin).**

Experience tells us that setting the numBins (number of bins), minBin (minimum bin number), and maxBin (maximum bin number) is a matter of trial-and-error experimentation. Setting histogram bins information is also a matter of personal preference.

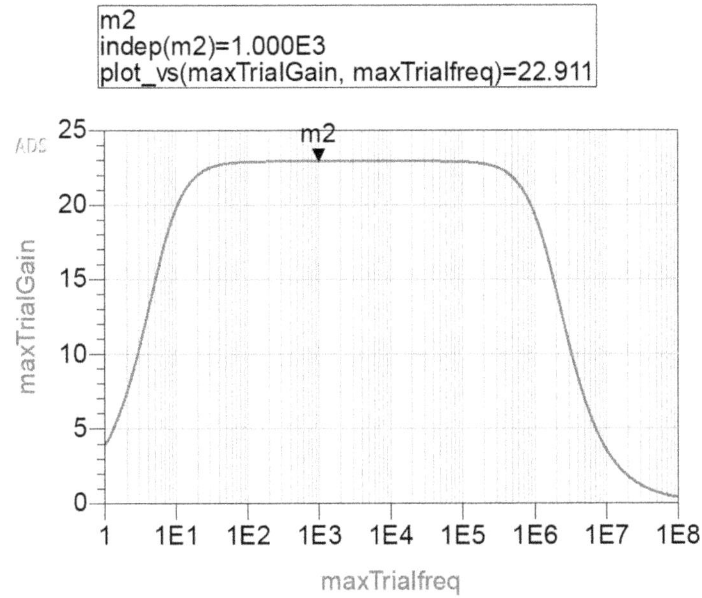

FIGURE 5.57 The maximum gain curve of all the trials. Example 5.9.

```
what(freq)
Dependency  : [mcTrial,freq]
Num. Points : [101,   801 ]
Matrix Size : scalar
Type        : Real
```

```
Eqn idx_freq = find_index(freq[0,::], 1 KHz)

Eqn gainAt1KHz = Av[::,idx_freq]
```

```
what(gainAt1KHz)
Dependency  : [mcTrial]
Num. Points : [101    ]
Matrix Size : scalar
Type        : Real
```

FIGURE 5.58 Voltage gain dimensionality reduction, to a single frequency, involves a tricky dimensionality reduction of freq itself.

In Example 5.10, we made the bins have a size of 1. We chose enough bins to cover slightly outside the range of values so that on either side (right or left) the data is zero to guarantee that we capture all the data. See the equation on top of Figure 5.60.

See the resulting Figure 5.61. The bin numbers (X-axis) are the same as the gain values (due to setting each bin's width to be 1). The Y-axis shows the number of trials contained in each bin.

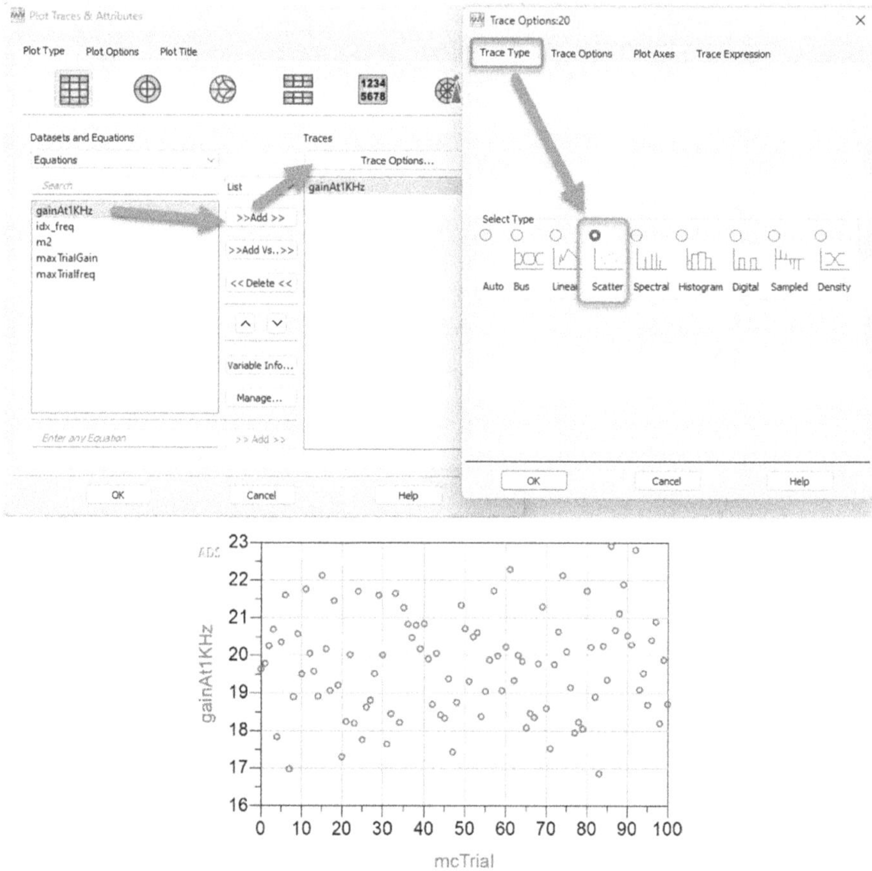

FIGURE 5.59 Plotting of a scatter type plot of the variable gain at 1 kHz.

5.6 EMITTER FOLLOWER AND THE YIELD OF A TWO-STAGE AMPLIFIER

An **emitter follower** (also known as a BJT common collector (CC) amplifier) has a small-signal **voltage gain of nearly 1**, a **large input resistance**, a **small output resistance** and a **large current gain**. As in a CE amplifier, in an emitter follower the small-signal input voltage is sent to the BJT base. Unlike a CE amplifier, though, the small-signal output voltage is taken from the BJT emitter. An emitter follower serves as a **voltage buffer** (connecting, for instance, a CE amplifier to a load). It also can serve as a power amplifier (called a Class A power amplifier).

In Figure 5.62, the input source v_s may represent the output voltage of a prior amplifier stage, such as the collector voltage of a CE amplifier. In all the emitter follower architectures, the collector is at ground in the small-signal diagram (Figure 5.63), hence the name CC.

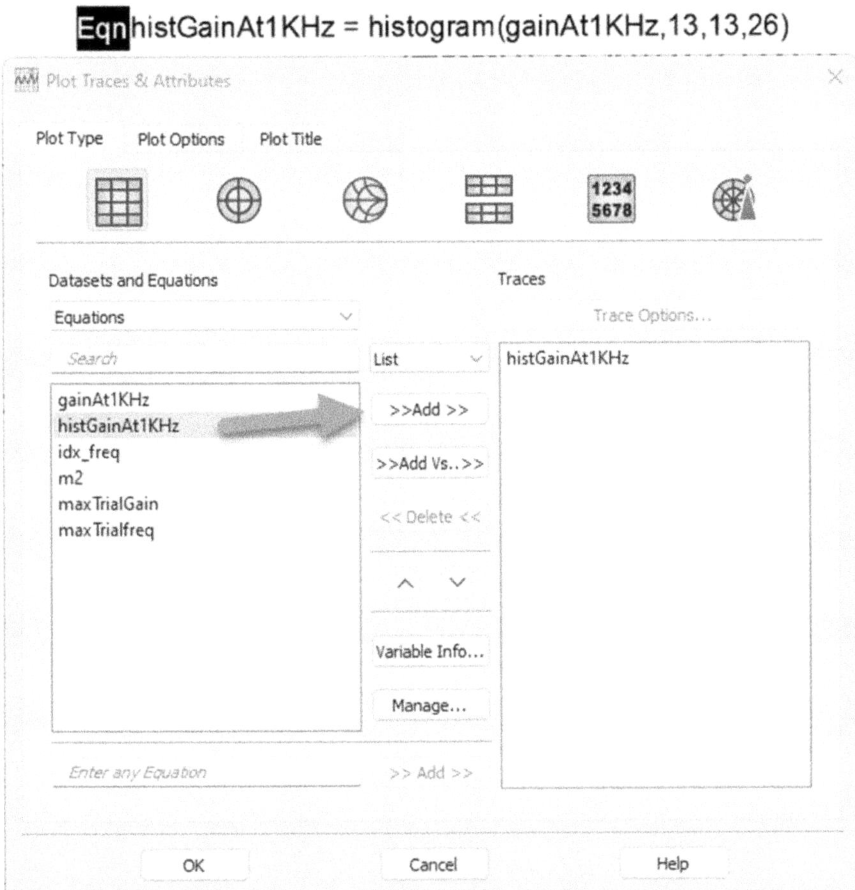

FIGURE 5.60 Plotting the histogram of the voltage gain and setting histogram parameters.

In Figure 5.63, referred to Figure 5.62 (upper row), $R_B = R_{B1} || R_{B2}$. In reference to Figure 5.62 (lower row), $R_E = \infty$.

The derivation of the small-signal input resistance R_{in} of an emitter follower is identical to that of R_{in} of a CE amplifier with a degeneration resistor R_{ee} (done in Section 5.1). In the R_{in} formula for the CE with emitter degeneration, we just replace R_{ee} by $R_E || r_o || R_L$ and then neglect r_e with respect to $R_E || R_L || r_o$. Therefore, $\mathbf{Rin \approx R_B || [(\beta + 1)(R_E || R_L || r_o)]}$. Let us name the R_{in} of the emitter follower as R_{incc} and note that it is typically much larger than the R_{in} of a CE BJT that has the same DC current.

To find the small-signal voltage gain of an emitter follower, we conduct a KCL calculation at the emitter node of Figure 5.63: $v_{be}/r_\pi + g_m v_{be} = v_e/(R_E || R_L || r_o)$, which leads, after neglecting $1/r_\pi$ with respect to g_m, to $v_e/v_b \approx g_m/[g_m + 1/(R_E || R_L || r_o)] = (R_E || R_L || r_o)/([r_e + (R_E || R_L || r_o)]$. Finally, $\mathbf{v_o/v_s} = (v_b/v_s)(v_e/v_b) \approx \mathbf{(R_{incc}/(R_{incc} + R_s)) \cdot (R_E || R_L || r_o)/(r_e + (R_E || R_L || r_o)) \approx 1}$.

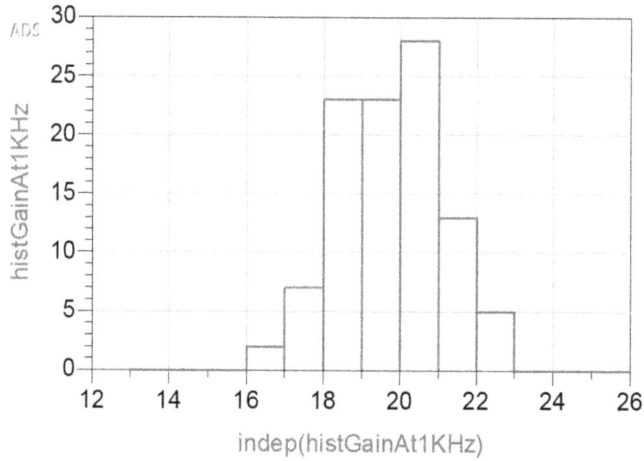

FIGURE 5.61 The resulting histogram of the voltage gain.

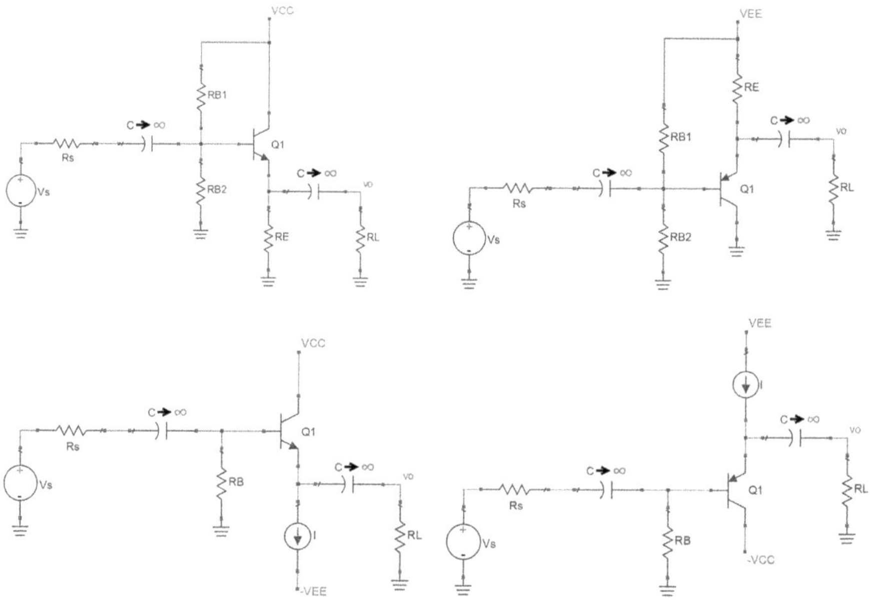

FIGURE 5.62 Four examples of emitter followers: NPN, PNP, single, and dual supply.

Next, let us discuss the voltage buffering ability of an emitter follower. Figure 5.64 illustrates how the CC block is connected between the CE block and the load resistance R_L.

The reader may recall the small-signal voltage gain formula of a CE amplifier (shown in Section 3.10 as well as in Section 5.1): $v_o/v_s \approx -[R_{in}/(R_{in}+R_s)]\cdot(R_C||R_L||r_o)/(R_{ee}+r_e)$.

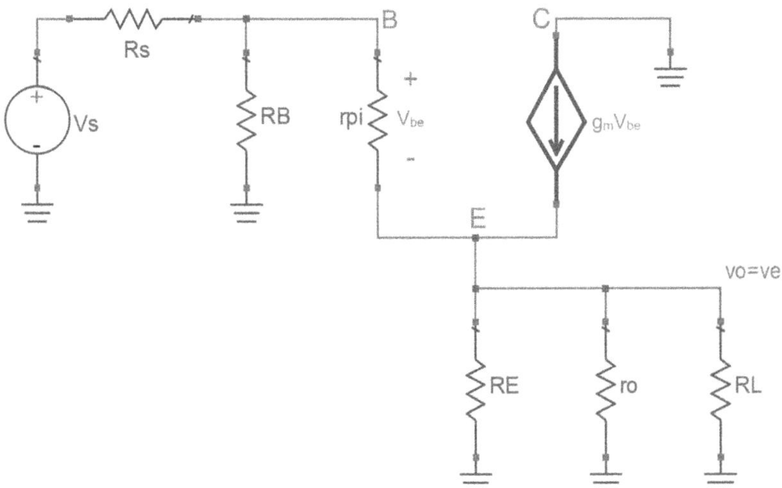

FIGURE 5.63 The small-signal diagram of all emitter follower architectures.

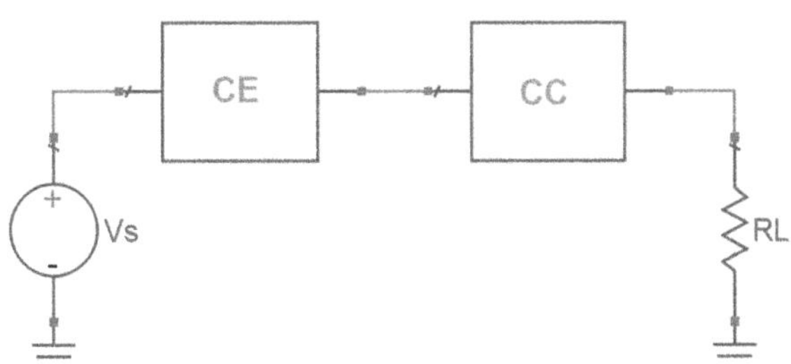

FIGURE 5.64 CE amplifier feeding into an emitter follower amplifier which feeds R_L.

The source voltage resistance R_s and the load resistance R_L are typically given and are not allowed to be changed, as the CE and emitter follower amplifiers are designed. Furthermore, sometimes a CE amplifier of a fixed voltage gain is expected to serve a variety of R_s and R_L values. We note from the CE voltage gain formula that as long as $R_{in} \gg R_s$ and $R_L \gg R_C$, changes in R_s and R_L have a secondary effect on the overall voltage gain. This is not always the case, especially considering R_L variations. A CE amplifier is designed to provide a desired voltage gain for some nominal value of R_L. Changes in R_L may cause significant changes in the amplifier's voltage gain if no buffer is used.

An emitter follower has a voltage gain of 1 for any value of R_L. It does not mean that placing it between a CE and its load R_L will make R_L effects on the voltage gain nonexistent. The CE amplifier was designed assuming that a nominal value of R_L is "seen" from the CE BJT collector. With an emitter follower buffer, the nominal R_L that was "seen" by the CE amplifier is replaced by R_{incc}. R_L does have a nonzero secondary effect on R_{incc}. A proper design of the CE → CC → R_L two-stage amplifier is to leave the CE design unchanged, as long as **the nominal value of R_{incc} equals the nominal value of R_L**. This is the only design requirement, which makes the design of the emitter follower relatively easy.

Example 5.11

Design of an emitter follower buffer.

Let the CE amplifier be the same as the one designed in Section 5.2 and tested in Section 5.3:

$V_{CC}=4.5$ V, $R_C=8.4$ kΩ, $R_E=9.2$ kΩ, $R_e=R_{ee}=150$ Ω, $R_1=R_{B1}=117$ kΩ, $R_2=R_{B2}=123$ kΩ, all capacitors are 100 μF, the nominal source resistance is $R_s=1$ kΩ, and the nominal load resistance is $R_L=25$ kΩ. Using the same V_{CC} for the emitter follower, we need to design $R_3=R_{B1cc}$, $R_4=R_{B2cc}$, and $R_5=R_{Ecc}$ (Figure 5.65).

The design of the emitter follower has many degrees of freedom. The resistor R_5 can be chosen arbitrarily; say $R_5=3$ kΩ. Let us also take $R_3=R_4$.

Now, we wish $R_{incc}=25$ kΩ (the nominal R_L value). A glance at the R_{incc} formula reveals that $R_B=R_{Bcc}=R_3||R_4$ dominates $(\beta+1)\cdot(R_{Ecc}||R_L||r_o)$, which is in the order of magnitude of hundreds of kΩ. Let us choose $R_3=R_4=50$ kΩ and prepare to tweak it up a little to make R_{incc} match the nominal R_L more exactly.

Final comment about the choice of R_5:

In a Class A amplifier, R_L is typically a lot smaller, requiring a larger signal current. The main design requirement is to ensure that the DC current of the Class A (emitter follower) amplifier is larger than the maximum expected negative R_L signal current peak; otherwise, the BJT cuts off. In our example R_L is large, and therefore R_5 can be large as well.

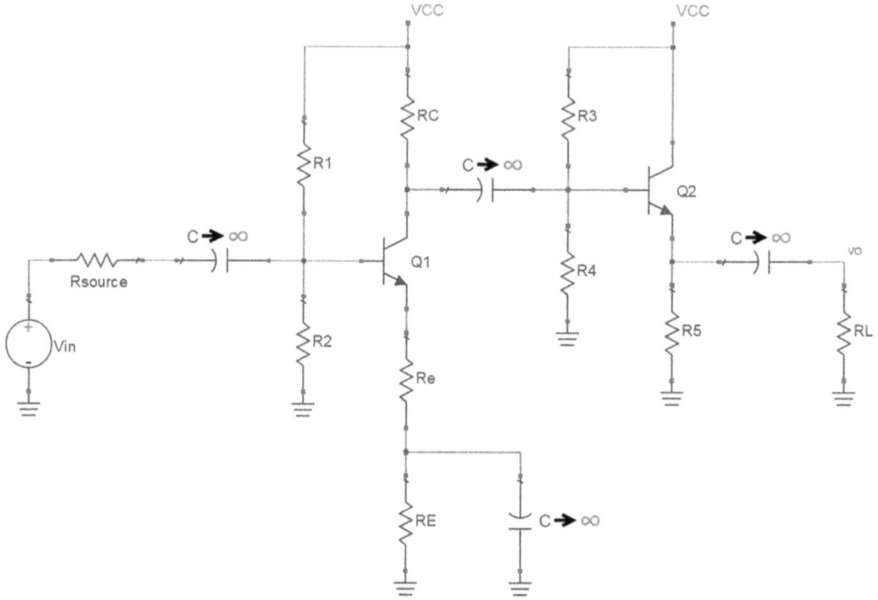

FIGURE 5.65 A single-supply CE amplifier followed by a single-supply emitter follower.

5.7 ADS: SENSITIVITY ANALYSIS AND YIELD ANALYSIS OF THE TWO-STAGE AMPLIFIER CONSISTING OF CE, EMITTER FOLLOWER AND RESISTIVE LOAD; YIELD ANALYSIS OF THE TWO-STAGE AMPLIFIER; SENSITIVITY ANALYSIS: SIMULATION VARIABLES SETUP FOR OPTIMIZATION, SENSITIVITY CONTROLLER AND A GOAL BLOCK, ANALYSIS OF SENSITIVITY ANALYSIS RESULTS; YIELD IMPROVEMENT BY TIGHTENING OF TOLERANCES

The measurement equations in Figure 5.66 provide the means to observe the voltage gain and R_{in} of each stage and the overall amplifier's gain.

Example 5.12

Say that we aim at overall $|A_v| = 20$.

Is $R_{incc} = R_L$? If not, do we need to tweak it?

Do we need to tweak the gain of the CE amplifier?

The simulation results of the schematic of Figure 5.66 are shown in Figure 5.67.

We see that the overall gain did not reach the goal. Let us do some tweaking of the design.

Let us first take care of our original specification for the amplifier's overall gain to be $|A_v| = 20$.

We note that the emitter follower's mid-band gain is 0.98 (and not 1 as expected). This gain lowers the overall voltage gain, unless we slightly increase the CE gain in anticipation of the second stage's gain.

For such tweaking, we note that the specified overall gain $A_{vreq} = A_{vCEnew} \cdot A_{vCC}$.

Therefore, $A_{vCEnew} = 20/0.98 = 20.4$. What load resistance should the CE amplifier "see" to achieve this gain?

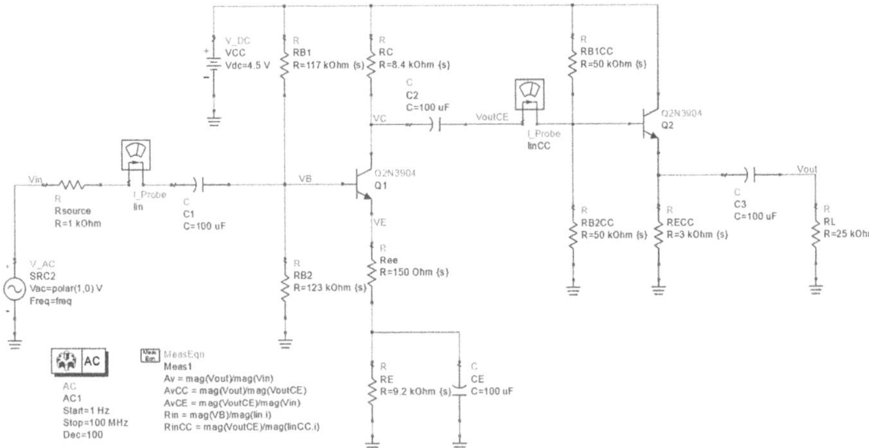

FIGURE 5.66 Yield analysis of a CE amplifier that is buffered from R_L by an emitter follower.

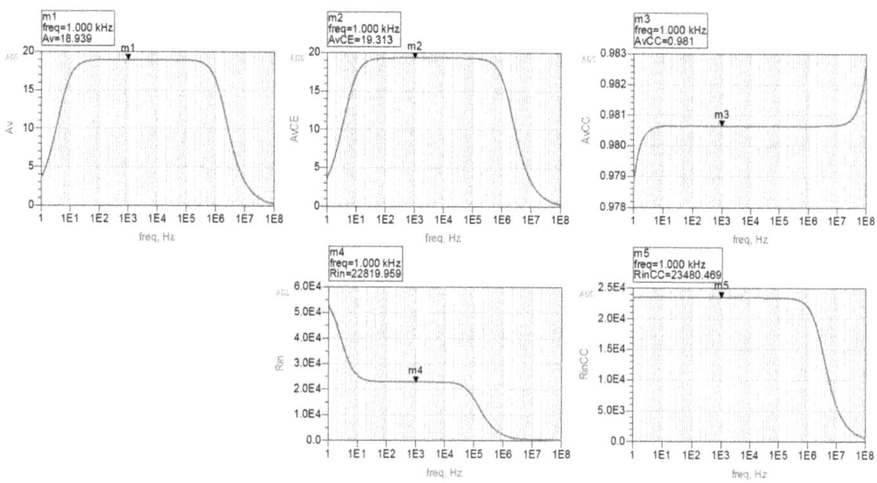

FIGURE 5.67 Simulation results of the CE+CC two-stage amplifier

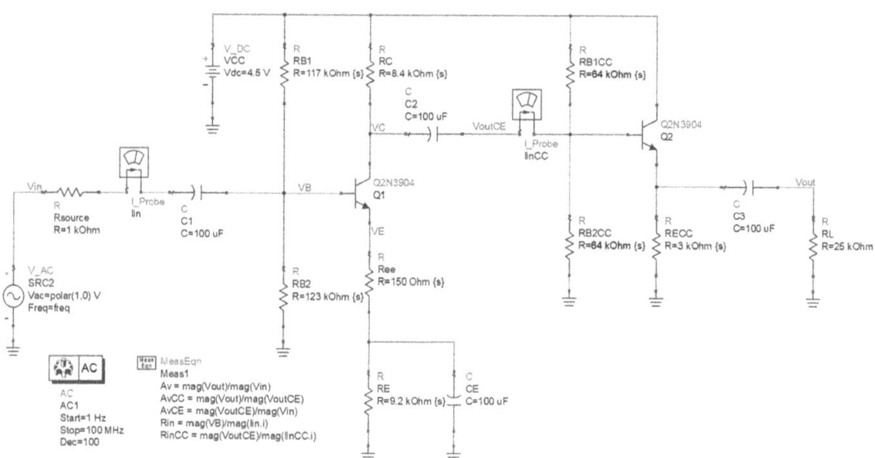

FIGURE 5.68 Tweaked the CE+CC design.

Ignoring r_o, we have $R_C||R_{Lnew}=(20.4/19.625)\cdot(R_C||R_{Lold})$, resulting in $R_{Lnew}=29.43$ kΩ. *Comment*: The reader is cautioned not to confuse R_L of the two-stage amplifier with "R_L" of the CE stage, which is R_{inncc} of Stage 2.

We then calculate the newly needed R_{BCC}: $R_{BCC}=50\cdot(29.43/23.48)=62.7$ kΩ. We shall use 64 kΩ since we ignored r_o.

Note that the CE amplifier retains its original design. The CE gain tweaking is done by the emitter follower's redesign. The tweaked design is shown in Figure 5.68.

The simulation results of the tweaked circuit are shown in Figure 5.69. We see that the two-stage amplifier meets its design specification.

Let us now perform a yield analysis of the tweaked two-stage amplifier. Note, in Figure 5.70, that R_{source} and R_L are fixed. No statistical properties are added to

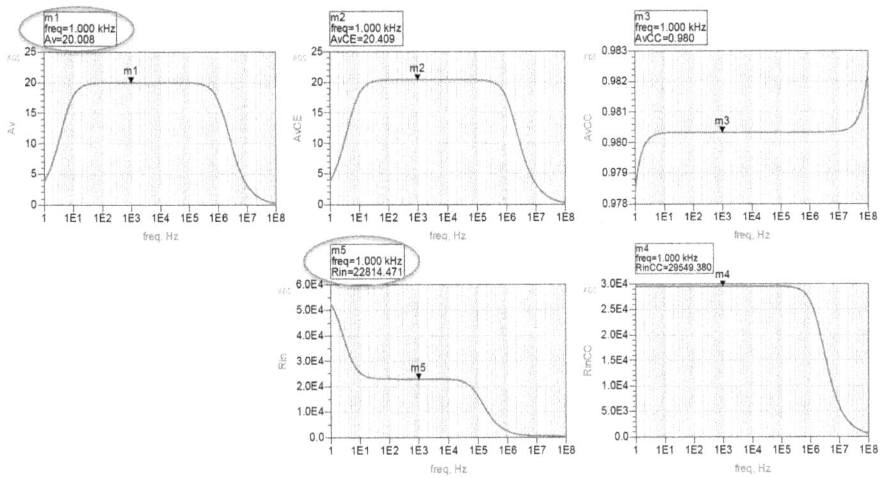

FIGURE 5.69 The tweaked results of the CE+CC amplifier.

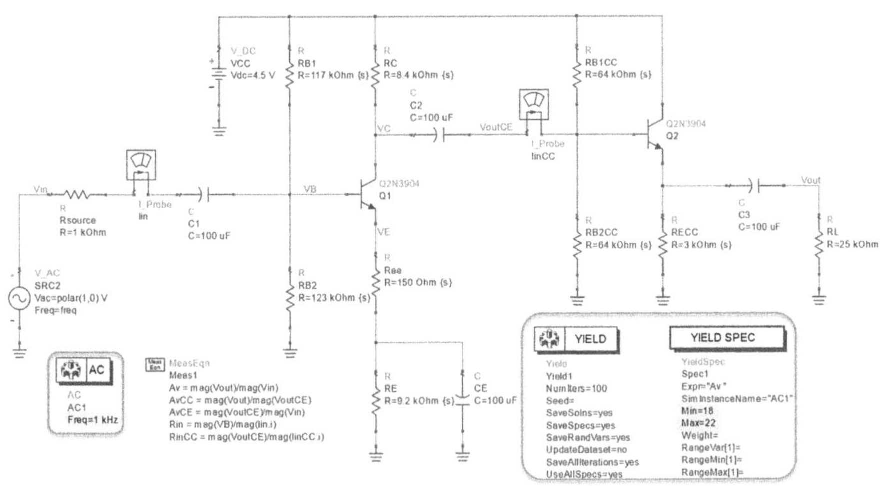

FIGURE 5.70 Single frequency AC simulation yield analysis of the two-stage amplifier.

them. The emitter follower is expected to improve the overall yield, whenever R_L fluctuates significantly.

So, in this example, as R_L is fixed, we expect the yield to be worse than that of the CE amplifier, simply because there are more random-valued components, and the effect of all adds up.

In Figure 5.71 we note a reduced yield of 83% (compared to the nearly 89% yield of the CE amplifier). Let us watch the fluctuations in certain performance Figure 5.72.

We see that R_{incc} varies between 26.5 and 32.5 kΩ. Not a bad buffering of R_L. The CE voltage gain varies from 18 to 23. So, what else influences the gain fluctuations?

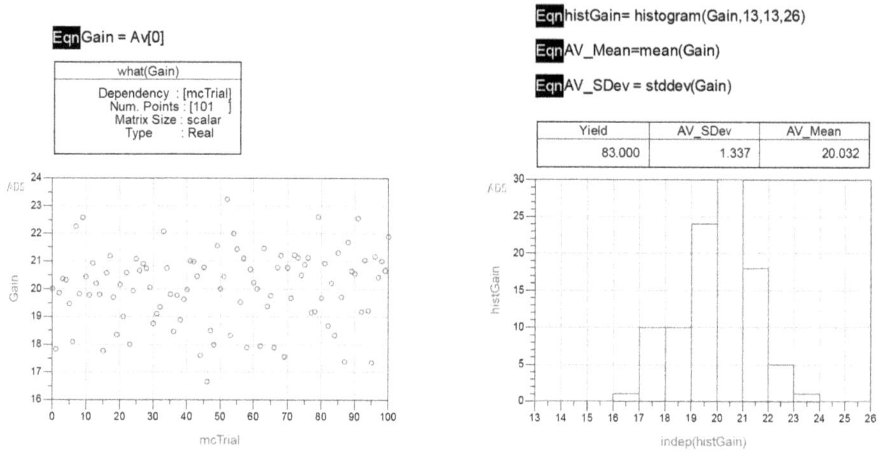

FIGURE 5.71 Yield analysis results of the tweaked CE+CC amplifier.

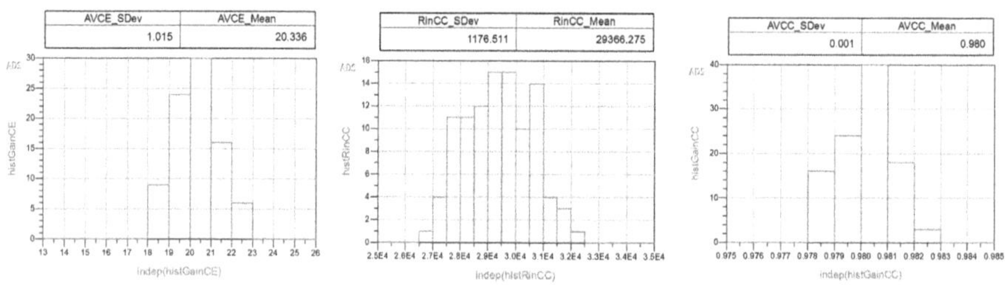

FIGURE 5.72 Histograms of the CE gain, R_{in}CC and CC gain, of the two-stage amplifier.

We note the A_{vCC} varies from 0.978 to 0.983. These are insignificant variations. So, it seems that the emitter follower works well as a buffer, and the additional components do not contribute much to the overall gain variations.

Example 5.13

We need to study the voltage gain sensitivity to each of the fluctuating components of the CE+CC+RL amplifier, as the above simulation results have not identified any culprits.

Sensitivity Analysis is part of the Optimizer Suite, as seen in Figure 5.73.

We need to set the "**Sens**" engine and a "**Goal**".

The goal is essentially similar to the previously used yield spec.

As part of working under the Optimizer, all the variable components (still marked each by {s}) will also be designated as optimized variables, marked each with {o}, as seen in Figure 5.75. Let us stress that no optimizing is expected to be done, when all that we conduct is sensitivity analysis tied to yield analysis.

Sensitivity analysis tests each changeable component, one at a time.

A component is varied by a certain small amount, and all other components are kept at their nominal values. The difference in circuit goal value, between the

FIGURE 5.73 Parts palette of Optim/Stat/DOE.

all-nominal circuit, and the one in which one component is varied, is assessed. The same is done for every one of the changeable components.

From the ADS documentation [Keysight1], sensitivities are approximated by finite differences:

$$Sp_i = \frac{\partial R(P_i)}{\partial P_i} \approx \frac{R(P_i^0) - R(P_i^+)}{P_i^0 - P_i^+}$$

Where $R(P_i^0)$ is the response evaluated at the nominal value, and $R(P_i^+)$ is the response due to the perturbation of the ith parameter.

The sensitivity analysis also outputs **normalized sensitivities.** Normalized sensitivities predict the percentage change in the response due to a 1% change in the design variable.

$$S_{Ni} = \left\{\left[R\left(P_i^0\right) - R\left(P_i^+\right)\right]/R\left(P_i^0\right)\right\}/\left\{\left[P_i^0 - P_i^+\right]/P_i^0\right\}$$

The sensitivity information is saved in the dataset using the goal instance name. Normalized sensitivities are saved with the name: norm_<goal_instance_name>.

We will tend to use normalized sensitivity values, as for the present design it is appropriate and easier to work with, as will be seen shortly.

The parameter optimization setup table is shown in Figure 5.74. The two-stage amplifier's schematic, setting the stage for a Sensitivity Analysis, is shown in Figure 5.75.

Note that the goal is the voltage gain A_v. All the components that got designation of {s} for the yield analysis also received a {o} designation for the Sensitivity Analysis.

A summary of Table 5.4 shows the normalized sensitivity of the goal to each of the components.

Meaning of the numbers: "If resistor RX varies by 1%, by how many % does A_v vary"? RX goes up 1%, and as a result, A_v may go up or down by a certain amount. For yield improvement, we care most about the absolute value of each normalized sensitivity.

From Table 5.4, A_v is most sensitive to R_C (of the CE amplifier). If R_C varies by 1%, A_v varies by 0.771%.

Second in line (after R_C) is R_{ee} of the CE amplifier: if R_{ee} varies by 1%, then A_v varies by the negative amount −0.487%.

Third in line: R_E of the CE amplifier. The change of 1% results in a change of A_v by −0.45%.

So, why are we doing a sensitivity analysis?

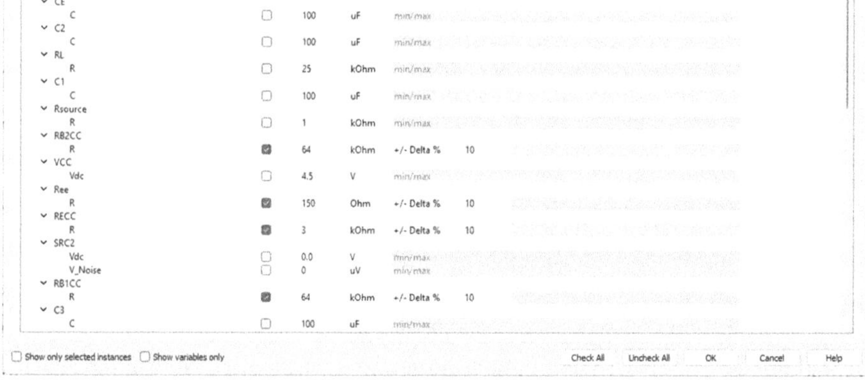

FIGURE 5.74 Sensitivity analysis parameter optimization setup.

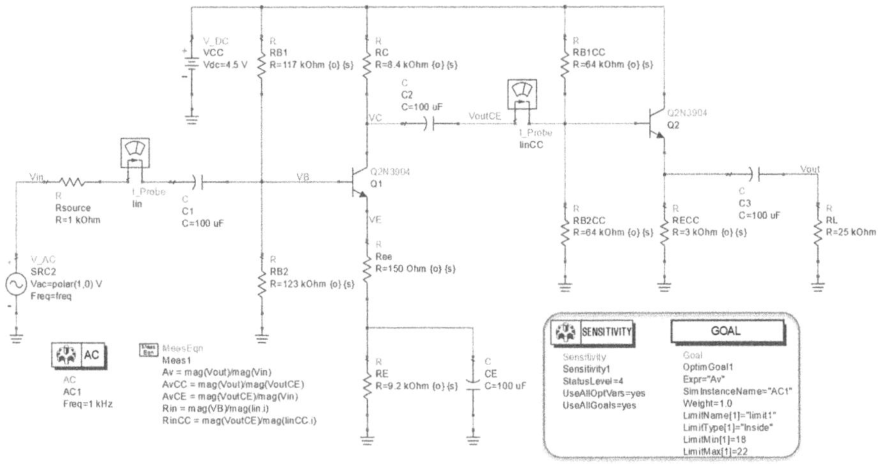

FIGURE 5.75 Sensitivity analysis schematics.

TABLE 5.4
Sensitivity Analysis Results

sensVariables	norm_OptimGoal 1
RB1.R	−0.329
RB1CC.R	0.086
RB2.R	0.317
RB2CC.R	0.115
RC.R	0.771
RE.R	−0.450
RECC.R	0.013
Ree.R	−0.487

In our yield analysis, we recall that all the resistors of the amplifier (CE+CC) each had a tolerance of ±10%. See Figure 5.74. Would the yield improve if all the resistors were replaced by resistors that have ±1% tolerance? Certainly.

Do we need to change all the regular resistors to precision resistors? Certainly not.

The sensitivity analysis points out the order by which we change a specific 5% (or 10%) resistor by a 1% resistor and observe the resulting yield (Figure 5.76).

We identified from the sensitivity analysis that the resistors highlighted in Figure 5.76 caused the most change in the voltage gain. Therefore, we reduced their tolerance to ±1%. We were feeling confident that these values would improve the yield, so to get a more accurate yield value, we increased the NumIters parameter to 1000.

Indeed, as seen in Figure 5.77, we were justified in changes to the three largest contributors to the gain variability, as evidenced by the 100% yield result for 1000 iterations.

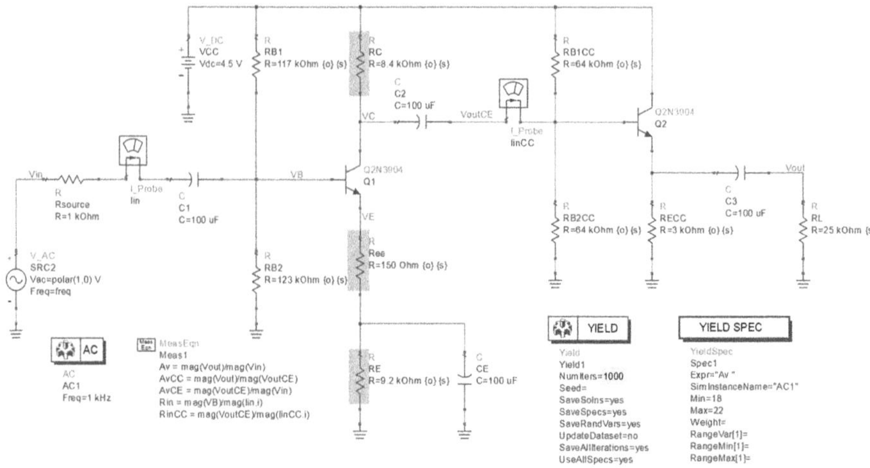

FIGURE 5.76 Yield analysis schematic of the two-stage amplifier after replacing three resistors.

Yield	AV_SDev	AV_Mean
100.000	0.599	19.981

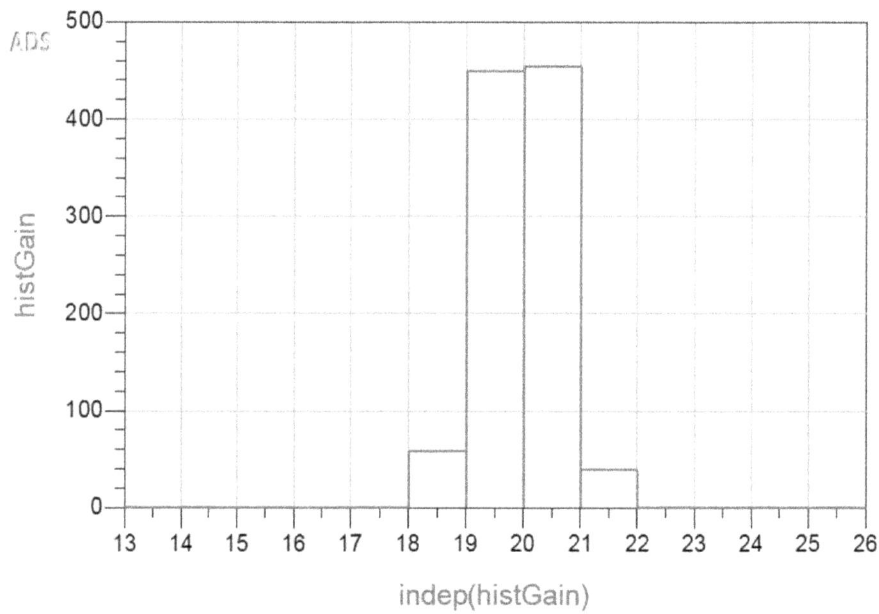

FIGURE 5.77 Yield analysis results of the two-stage amplifier after replacing three resistors.

Yield	AV_SDev	AV_Mean
100.000	0.589	19.992

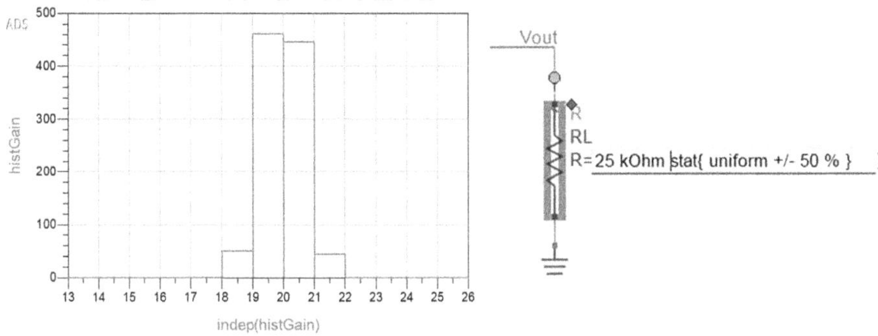

FIGURE 5.78 Yield analysis results in the presence of 50% variations in the load resistance R_L.

Notice the mean gain is pretty much 20, as required, and the standard deviation has reduced from 1.337 initially to 0.6.

One test still remains. We should test the CE+CC amplifier against variations in R_L. Let us run a test in which we allow R_L to vary ±25%.

What do we expect?

If the CC amplifier is doing a good buffering job, the yield (after placing the three 1% resistors) should not change. Indeed so.

In Figure 5.78, we show the yield analysis results in the presence of 50% R_L variations, and the yield stands at 100%. The CC buffer is performing as we had expected, and our design is tolerant to large load variations.

REFERENCE

[Keysight1] *Keysight Advanced System Design (ADS) Release 2025*, Keysight Technologies, Inc. https://www.keysight.com/us/en/products/software/pathwave-design-software/pathwave-advanced-design-system.html

CHAPTER 5 PROBLEMS – CE AMPLIFIER DESIGN WITH YIELD AND SENSITIVITY ANALYSIS

Problem 5.1: Design a CE amplifier with a small-signal voltage gain of $A_V = -36$ and input resistance $R_{in} \geq 14$ kΩ. Assume that (i) the amplitude of the input voltage signal is no larger than 33 mV and the signal source internal resistance is $R_{sig} = 20$ Ω, (ii) the load is $R_L = 16$ kΩ, and (iii) the amplifier works with a single DC voltage supply of V_{CC} that needs to be designed as well. Demonstrate the amplifier's performance in simulations (using ADS AC Analysis and Transient Analysis) – visit each of the specifications and show that the specification is met.

Hints/Requirements:

a. Try to design the circuit for the smallest possible value of V_{CC}. One of the ways to accomplish it is to make sure that r_e does not become too large. Specifically, when you meet the R_{in} requirement, you choose how much to allocate to the $(\beta+1)(R_e+r_e)$ term vs. the R_B term. Make $(\beta+1)(R_e+r_e)$ close to the desired R_{in} (say 17 kΩ), and that will make R_B larger.
b. In your initial design you may surely assume $V_T \approx 25$ mV, $\beta \approx 100$, $V_{BE}=0.7$ V, and $V_A=100$ V. Eventually you will see that V_T is actually closer to 26 mV, β is closer to 135, V_A is around 74 V, and even V_{BE} may be around 0.65 V. Do it one way or another; it doesn't matter. The design is amazingly insensitive to such assumptions. At the end, you will always have to tweak the gain (using modification of R_C) to the exact value of –36.
c. In your design, at the beginning, you find r_e based on meeting the R_{in} specification. At the end, test your design in AC simulation (to see the obtained voltage gain and R_{in}). Test in transient simulation at an input frequency of 2 kHz how distorted the output voltage is. In harmonic balance simulation, at the maximum input amplitude, watch $v_o(t)$ – the main harmonic of 2 kHz, and four to five side harmonics. Be sure that no side harmonic exceeds 10% of the main harmonic.

Problem 5.2: Consider your CE amplifier design of Problem 5.1. Assume that all amplifier's resistors have 5% tolerance (uniform distribution). Assume that R_{source} varies 5% (like the amplifiers resistors), but R_L varies 20% (above and below nominal, uniform distribution). Find the yield of the CE amplifier (assuming that a successful design is whenever the voltage gain is within ±5% of the nominal gain). Add an emitter follower and see if the yield improves. Run sensitivity analysis of the CE+CC circuit (without including sensitivities for R_{source} and R_L). How would you improve the yield to 100% or almost 100% with the minimum number of 1% resistors?

Problem 5.3: Each of the class's students will work on slightly different sets of design specifications. Let's denote the last four digits of your student number by WXYZ.

Design a CE amplifier with a small-signal voltage gain of $A_V=-20\text{-W}$ and input resistance $R_{in} \geq (7+X)$ kΩ. Assume that (i) the amplitude of the input voltage signal is no larger than (23+Y) mV and the signal source internal resistance is $R_{sig}=(500+Z00)$ Ω, where Z is the hundreds digit, (ii) the load is $R_L=(16+Y)$ kΩ, and (iii) the amplifier works with a single DC voltage supply of $V_{CC}=9$ V, which is fixed. Demonstrate the amplifier's performance in simulations (using ADS DC, AC, transient, and harmonic balance simulations) – visit each of the specifications and show that the respective specification is met.

Problem 5.4: Consider your CE amplifier design for Problem 5.3. Assume that each of the amplifier's resistors has 5% tolerance (uniform distribution). Assume that R_{sig} varies 5% (like the amplifiers resistors) and R_L too. Find the yield of the CE amplifier (assuming that a successful design is whenever the voltage gain is within ±5% of the nominal gain, or the "new nominal gain", if you had to lower the gain because of the fixed voltage supply).

Problem 5.5: Add an emitter follower to the design of Problem 5.4. Be sure that R_{in} of the emitter follower equals the nominal value of R_L. Include tolerances for the emitter follower's components. See if the yield improves. Run sensitivity analysis of the CE+CC circuit (without including sensitivity for R_{sig} and R_L). How would you improve the yield to 100% or almost 100% with minimum number of 1% resistors?

Hint: It would be easiest to select the two base resistors R_{BCC1} and R_{BCC2} of the emitter follower to be equal. Each needs to be slightly larger than 2·(needed $R_{in,\,cc}$). Assume a voltage divider for the two emitter follower base resistors and choose R_{ECC} to create a DC current $I_{C,\,CC}$ that is approximately 1 mA.

Additional Suggestion: As you strive to achieve perfect performance, don't forget to tweak the voltage gain of the CE+CC amplifier to the exact required nominal overall gain value.

6 Single- and Multistage Differential Amplifiers

6.1 BJT SIMPLE AND ADVANCED CURRENT MIRROR CURRENT SOURCES

In Sections 4.9 and 4.10, we studied simple MOSFET current mirrors. Simple BJT current mirrors, shown in Figure 6.1, have many similarities to MOSFET mirrors: Two transistors that are connected base-to-base and emitter-to-emitter and have identical parameters, NPN mirrors that generate sink current source, and PNP mirrors that generate source current sources and stabilized reference current going into a diode-connected transistor Q_1. There can be one (Q_2) or more (Q_2, Q_3, ...) output transistors. Current may be steered from the group of NPN mirrors to the group of PNP mirrors, or vice versa. Q_1 is automatically in active mode, and Q_2 needs to be in active mode, but whether or not it is depends on the circuit that I_o goes to.

There are some differences, though, between MOSFET and BJT mirrors.

For a perfect similarity between Q_1 and Q_2, we nominally have $I_o = I_{ref}$. If Q_2's BE junction area is K times the BE junction area of Q_1 (K being any positive real number), then nominally $I_o = KI_{ref}$.

Nominally, but not exactly. An identicality of Q_1 and Q_2, along with equal V_{BE} voltage of the two BJTs, makes $I_{E1} = I_{E2}$. See Figure 6.1. However, $I_o = I_{E2} \cdot \beta/(\beta + 1)$,

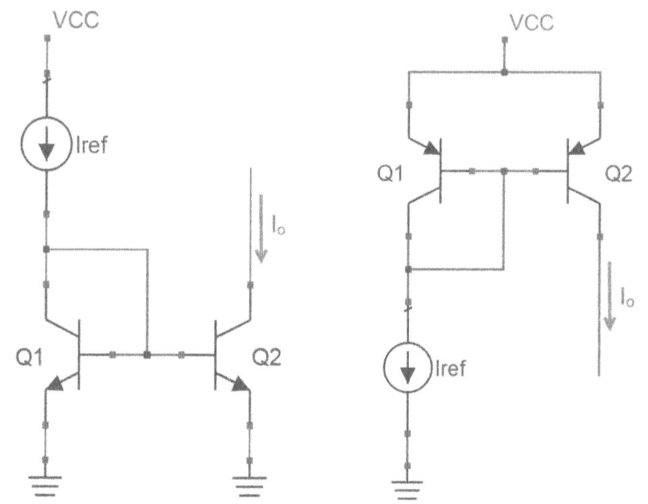

FIGURE 6.1 Simple BJT current mirror DC current source: (left) sink, (right) source.

DOI: 10.1201/9781003606772-6

whereas $I_{ref} = I_{C1} + I_{B1} + I_{B2} = I_{E1}(\beta + 2)/(\beta + 1)$. The β-dependent error between I_{ref} and I_o may reach a few %, and it is not the main error.

In addition to the β-dependent error between I_o and I_{ref}, there is an early effect error due to non-similar V_{CE1} and V_{CE2} voltages. See Figure 6.1.

Like in MOSFET mirrors, output BJT transistors that are connected at their collectors cause the respective output currents to add up. For example, say that we create two mirrors with Q_2 and Q_3 that are identical to Q_1 and short-circuit the collectors of Q_2 and Q_3; then, nominally, the combined output current I_o becomes twice the reference current.

Figure 6.2 shows the small-signal diagram for measuring Rout of the simple current mirror current source, which is the same for both the sink and source mirrors. I_{ref} is replaced by an open circuit. The circuit is driven by a small voltage signal v_x. As no signal reaches the left-hand side of the mirror, the v_{be} voltages are zero. Therefore, $\mathbf{R_{out} = r_{o2}}$. This is another similarity of simple MOSFET and simple BJT mirrors. For MOSFET mirror $R_{out} = r_o$ (where r_o relates to the MOSFET channel length modulation effect at the output transistor). For BJT mirror $R_{out} = r_o$ (where r_o relates to the BJT Early voltage effect in the output transistor).

Occasionally, a need arises to replace a simple current mirror with one called "an advanced current mirror current source". It has less output current error (with respect to the reference current) and larger output resistance. Let us briefly discuss the performance of one such advanced mirror, called a **Wilson Current Source**, shown in Figure 6.3. See [Sedral].

Again, by structure and by similarity of Q_1 and Q_2, the emitter currents I_{E1} and I_{E2} are equal. A straightforward comparison of I_{ref} and I_o leads to $I_o/I_{ref} \approx 1/(1 + 2/\beta^2)$, indeed a significant reduction of the β-dependent error. Moreover, R_{out} of the Wilson current source, without showing the detailed calculation, is drastically larger than that of a simple current mirror: $\mathbf{R_{out} \approx (\beta/2) \cdot r_{o3}}$. For proof, see [Sedral].

It happens because the diode-connected Q_1 degenerates the emitter of the output transistor Q_3.

6.2 BJT DIFFERENTIAL AMPLIFIERS

Recall the op-amp differential amplifier, discussed in Sections 1.17–1.19. A device that has two inputs and one output, where the output voltage should be proportional

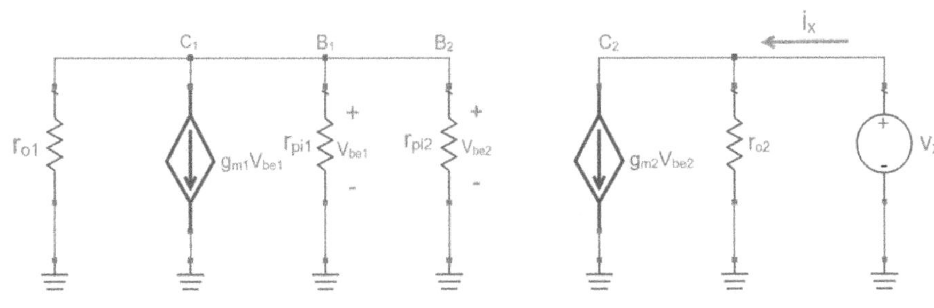

FIGURE 6.2 Output resistance measurement for a simple current mirror.

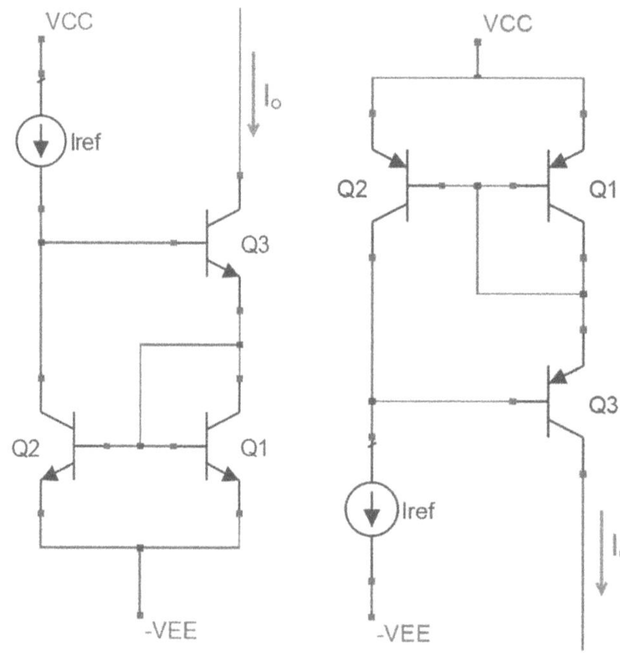

FIGURE 6.3 A Wilson current mirror current source: (left) sink, (right) source.

to the difference between the two input voltages, is called the **differential input voltage**. The output voltage should be insensitive to signals that are common to both inputs, called **common-mode voltages**.

In this section, we shall explore differential amplifiers implemented by BJT transistors. The theory is very similar to that of differential amplifiers built by MOSFET transistors. See [Sedra1] and [Razavi1].

The simplest BJT differential amplifier architecture is shown in Figure 6.4. The key to differential amplification is the symmetry between the left and right sides of the amplifier: The two BJTs are similar, must operate in active mode, and are tied to symmetric R_C resistors. The biasing of the circuit is done by a DC current source I, "tailing" the two tied emitters. The two input signals $v_{b1}(t)$ and $v_{b2}(t)$ are arbitrary, but their DC levels must be symmetrical. That is, the DC levels of the input signals are part of the amplifier's common-mode signal. These symmetric DC levels participate in the biasing of the BJTs.

The mathematical definitions of differential and common-mode input signals are, respectively, $\mathbf{v_d(t) = v_{b1}(t) - v_{b2}(t)}$ and $\mathbf{v_{CM}(t) = [v_{b1}(t) + v_{b2}(t)]/2}$.

Any one of the BJT collectors, or both, can serve as outputs of the amplifier.

Due to the symmetry of the amplifier, if $v_{b1}(t) = v_{b2}(t)$, then the DC current of each BJT equals I/2.

If the two input voltage signals are non-equal, then the BJT emitter currents $i_{E1}(t)$ and $i_{E2}(t)$ (DC and signal) become different as well. The BE junctions of the BJTs are diodes that presumably operate in forward bias. Recall the diode's exact model, discussed in Section 2.2. Therefore, the ratio of the emitter currents is an exponential

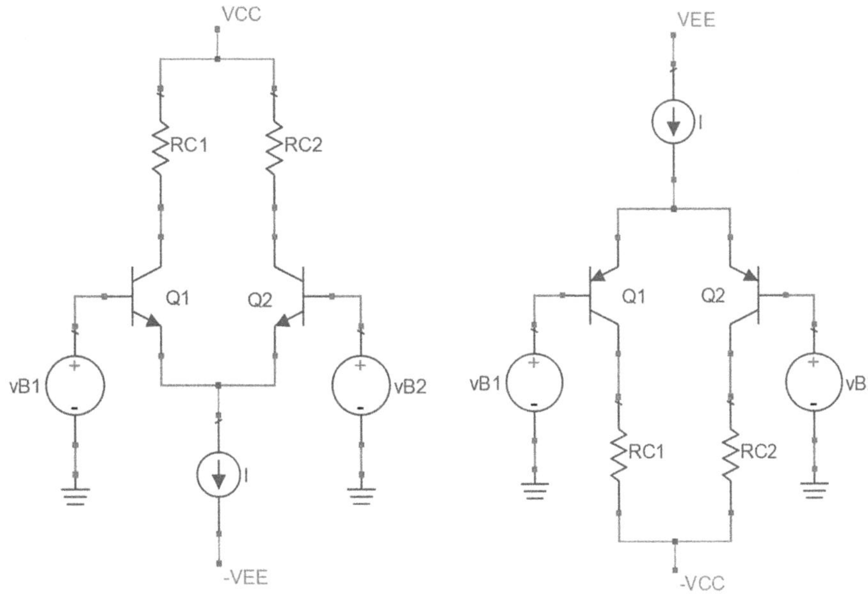

FIGURE 6.4 NPN and PNP differential amplifiers.

function of the differential input voltage: $i_{E1}(t)/i_{E2}(t) = \exp\{[v_{b1}(t) - v_{b2}(t)]/V_T\}$. The two currents must always add up to I: $i_{E1}(t) + i_{E2}(t) = I$. These two equations for the two unknowns $i_{E1}(t)$ and $i_{E2}(t)$ can be solved to yield $\mathbf{i_{E1}(t) = I/\{1 + \exp[-v_d(t)/V_T]\}}$ and $\mathbf{i_{E2}(t) = I/\{1 + \exp[v_d(t)/V_T]\}}$.

For the amplifier to work as a linear small-signal differential amplifier, the differential input voltage $v_d(t)$ must be small indeed. There is a "small-signal assumption" for BJT differential amplifiers, similar to what we witnessed in a CE amplifier (recall Section 3.9). If the differential input voltage is not small, the amplifier operates nonlinearly. At the extreme of such nonlinear operation, one of the BJTs may even cut off, and the other BJT "steals" the full bias current I. If the differential input voltage is small (compared with the thermal voltage V_T), then the exponential terms of the emitter currents hover around 1, meaning that **in a "small-signal operation" each BJT bias current is I/2**.

The differential input voltage's amplitude that, more or less, justifies a **small-signal assumption** is $\mathbf{|v_d(t)| \leq V_T/2}$. The starting point for a BJT differential amplifier small-signal analysis is the above nonlinear expressions for the emitter currents in which the exponential functions are replaced by their first two terms of the power series. This results in the following expressions for the total collector currents (DC plus small signal): $\mathbf{i_{C1}(t) \approx I/2 + g_m[v_d(t)/2]}$ and $\mathbf{i_{C2}(t) \approx I/2 - g_m[v_d(t)/2]}$, where the transconductance gain is $\mathbf{g_m \approx (I/2)/V_T}$.

The next step in understanding small-signal analysis of the differential amplifier lies in comparing Figures 6.4 and 6.5. In Figure 6.5, we can verify that the left-hand side inputs add up to $v_{b1}(t)$. Likewise, for $v_{b2}(t)$.

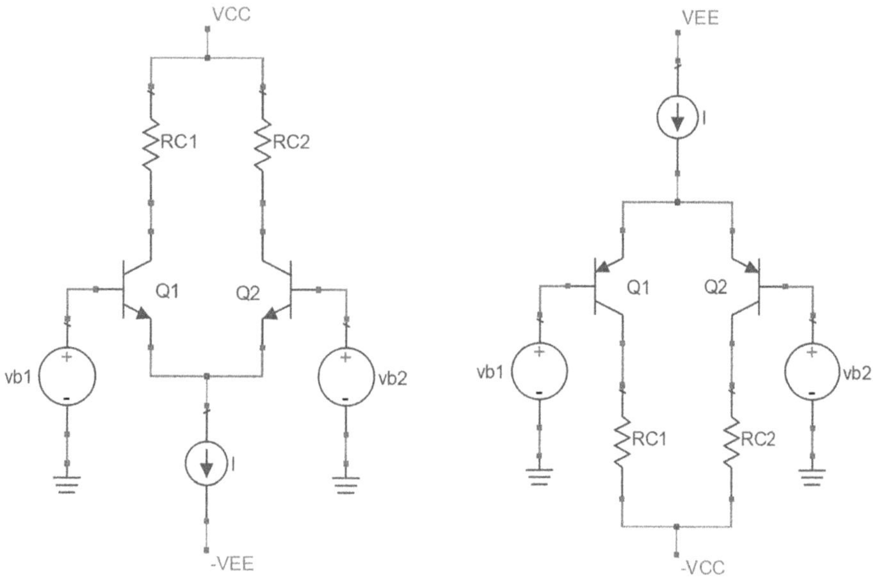

FIGURE 6.5 A NPN differential amplifier that is mathematically equivalent to the one shown in Figure 6.4 (left).

This equivalency is valid for both large and small signal operations. For a small-signal analysis though, as the amplifier is assumed to operate linearly, Figure 6.5 holds the key to a superposition of two analyses: A **Differential Mode Analysis** (in which we assume that $v_{CM}(t)=0$ and more about it when we implement it in ADS) and a **Common-Mode Analysis** (in which we assume that $v_d(t)=0$). Both amplifier modes are easy to analyze due to the symmetrical (or anti-symmetrical) input voltages, which allow us to benefit from the already-known results of CE amplifiers.

In a **differential mode analysis**, shown in Figure 6.6, the inputs $v_d(t)/2$ and $-v_d(t)/2$ are anti-symmetric. As a result, the emitter voltage is DC and independent of the input signals.

In the small-signal diagram (Figure 6.6) of the circuit shown in Figure 6.5 (under the assumption of zero common-mode input), the emitter becomes a "virtual ground", as seen by KVL equation in the loop that contains the differential input sources and the r_π resistors.

Consequently, as $v_e=0$, the small-signal diagram breaks into two separate "**half circuits**", each one a CE amplifier. Say that we continue the analysis using the left half-circuit (Figure 6.5), let the output voltage $v_{oL}(t)$ be taken from the left collector, and recall that a CE amplifier inverts, then $v_{oL}(t)=[v_d(t)/2]\cdot[-g_m(R_C||r_o)]$.

Similarly for the right half-circuit: $v_{oR}(t)=[-v_d(t)/2]\cdot[-g_m(R_C||r_o)]$.

Recall that we defined the differential input voltage as $v_{B1}(t)-v_{B2}(t)$. Therefore, we note that the left collector small-signal voltage (with respect to ground) is amplified with respect to the differential input voltage by a gain of $\mathbf{A_{VdL,SE}=-g_m(R_C||r_o)/2}$, and the right collector voltage small-signal is amplified by $\mathbf{A_{VdR,SE}=g_m(R_C||r_o)/2}$. Each collector's small-signal voltage (with respect to ground) is called a **single-ended**

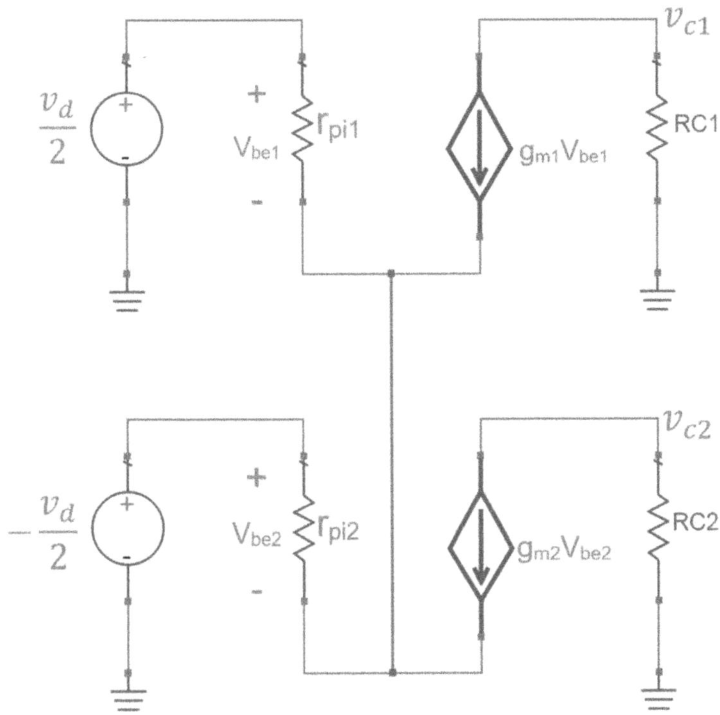

FIGURE 6.6 Small-signal diagram of a differential amplifier in differential mode. Resistors r_o are not shown.

output of the differential amplifier. The two single-ended gains have equal absolute value; one is inverting, and the other is non-inverting.

A third option, for an output of the differential amplifier, is a **differential output $v_{o,diff}$ (t) = v_{oR}(t) − v_{oL}(t).** The differential output differential voltage gain is $\mathbf{A_{Vd,diff} = g_m(R_C \| r_o)}$. It is twice as large as any of the single-ended voltage gains. The question is how we implement differential output. You may recall the op-amp linear combiner of Sections 1.11 and 1.12, which is irrelevant here, as we seek a more basic voltage subtraction method, done at the transistor level. Keep in mind that a BJT differential amplifier may be part of the internal circuit of an op-amp. One way of creating a differential amplitude is to have a cascade of differential amplifiers, as shown in Figure 6.7. In such a two-stage amplifier, the two outputs of stage 1 feed into the inputs of stage 2: $v_{inL2} = v_{oL1}$ and $v_{inR2} = v_{oR1}$. As stage 2 output voltages are sensitive to $v_{inL2} - v_{inR2}$, this arrangement implements a differential output in stage 1. The outputs of stage 2 are single-ended.

As is further seen from Figure 6.6, the **differential-mode input resistance**, relating v_d(t) to the loop current (flowing from $v_d/2$ to $-v_d/2$) is $\mathbf{R_{id} = 2r_\pi}$.

Note that in Figure 6.6, the input differential signal sources are assumed to be ideal. There are no load resistors R_L, which can be added symmetrically. There is no degeneration resistor R_{EE}.

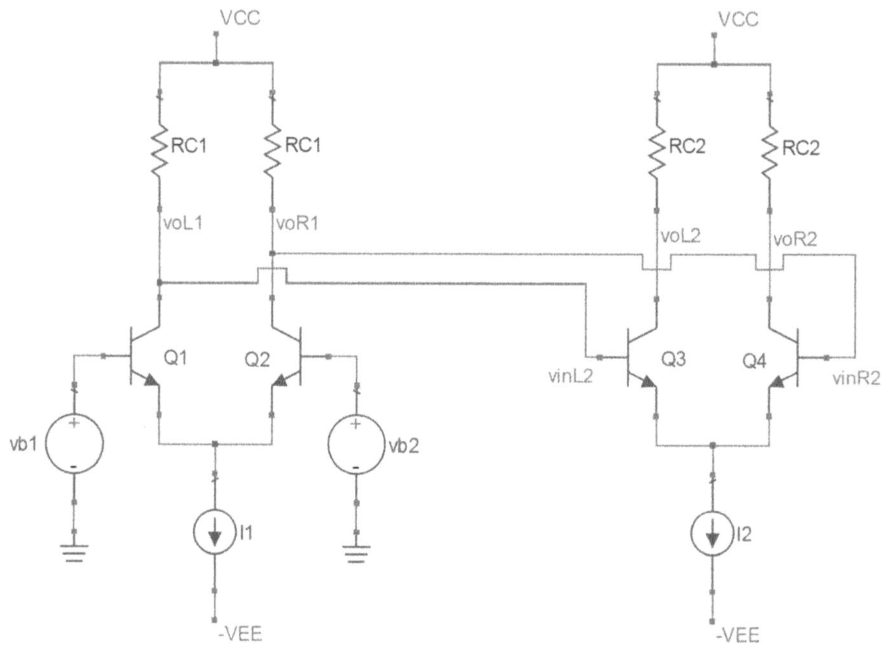

FIGURE 6.7 Cascade of BJT differential amplifiers.

There are several ways to extend the differential-mode differential voltage gain formula, similar to CE amplifier voltage gain formulas. Let us write down the extended differential output differential mode gain, keeping in mind that the respective single-ended gains are twice as low.

$A_{Vd,diff} = [R_{id}/(R_{id} + 2R_{source})]\cdot[(R_C||r_o||R_L)/(2(r_e + R_{EE})]$ where $R_{id} = 2(\beta + 1)(r_e + R_{EE})$. The ratio between the degeneration resistor R_{EE} and r_e is selected based on the maximum amplitude of $v_d(t)$ that we wish to accommodate. The needed equation is $v_{be}/(v_d/2) = (R_{id}/(R_{id} + 2R_{source}))\cdot(r_e/(r_e + R_{EE}))$. Like in CE amplifiers, we restrict $|v_{be}(t)|$ to be no larger than 10 mV.

In a **common-mode analysis**, the input $v_{CM}(t)$ is applied symmetrically to the BJT bases, as shown in the small-signal diagram of Figure 6.8. The resistor R represents the output resistance of the DC current source that feeds a current I into the emitters of the BJTs.

There is a complete symmetry in the circuit: DC current of each BJT is I/2. Small-signal parameters and R_C resistors of the BJTs are equal, respectively.

Let us omit the 1 and 2 subscripts from the small signal parameters and from R_C.

As $v_{be1} = v_{be2} = v_{be} = v_{CM} - v_e$, by using KCL at the emitter node, while neglecting $1/r_\pi$ with respect to g_m, we obtain $v_e \approx v_{CM}/(1 + r_e/2R)$ and $v_{be} \approx v_{CM}\cdot r_e/(2R + r_e)$. Consequently, $v_{c1} = v_{c2} = v_c \approx -v_{CM}\cdot R_C/2R$. The **common-mode differential output voltage gain** is $\mathbf{A_{VCMdiff} = (v_{c1} - v_{c2})/v_{CM} \approx 0}$. The **common-mode single-ended output voltage gain** is $\mathbf{A_{VCMSE} = v_c/v_{CM} \approx -R_C/2R}$.

The smaller the common-mode voltage gain the better is the performance of the differential amplifier. The differential output common-mode gain is close to zero

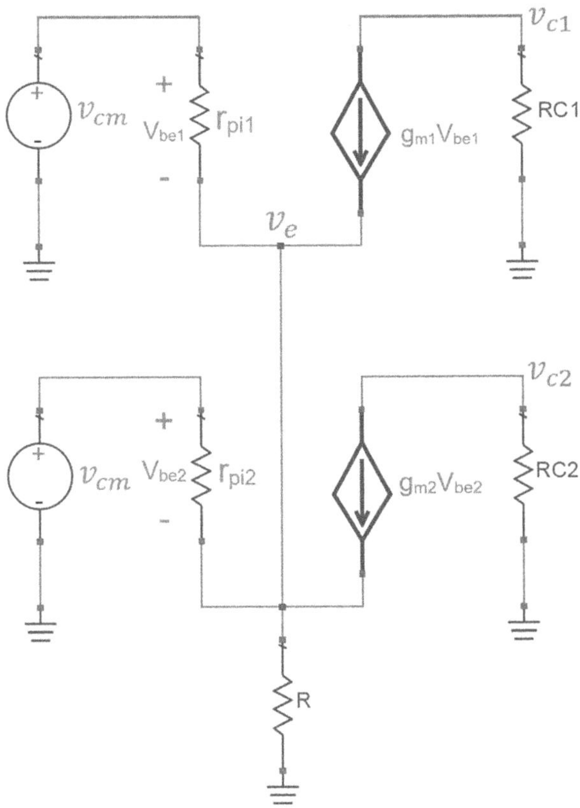

FIGURE 6.8 Small-signal diagram of the differential amplifier in common-mode analysis. As an approximation, the r_o resistors are not shown.

under perfect symmetry conditions. Slight asymmetric conditions may make the differential output common-mode gain nonzero, albeit small. The single-ended common mode gain strongly depends on R, the output resistance of the current source I. The larger R, the smaller the single-ended common-mode gain. For instance, changing the current source from a simple current mirror to a Wilson current mirror drastically reduces the single-ended common-mode voltage gain.

Let us look again at the cascade of differential amplifiers, shown in Figure 6.7. The main reason for creating such a two-stage differential amplifier is not so much to double the overall differential gain but rather to drastically lower the overall common-mode gain. The differential output implemented in stage 1 nominally sets the common-mode voltage gain of stage 1 to zero. The single-ended voltage gain of stage 2 is finite, but the overall common mode gain is still zero.

The performance of any differential amplifier is often characterized by a single measure called the **common mode rejection ratio (CMRR)**, which is the absolute value of the ratio between the differential-mode voltage gain and the common-mode differential gain. The larger the CMRR, the better the differential amplifier is.

For a differential amplifier that features a **differential output, CMRR → ∞**. For a differential amplifier that has a **single-ended output, CMRR ≈ g_mR**.

6.3 DESIGN EXAMPLE OF A SINGLE-STAGE DIFFERENTIAL AMPLIFIER

Example 6.1

Need to design a BJT differential amplifier (shown in Figure 6.9) to meet the following design specifications:

1. The differential input's amplitude is no larger than 10 mV.
2. The differential input resistance must be $R_{id} \geq 10$ kΩ.
3. The single-ended differential voltage gain must be $|A_{VdSE}| = 50$.
4. The single-ended common-mode voltage gain must be $|A_{VCMSE}| \leq 0.05$.
5. The common-mode input voltage is $v_{CM}(t) = 2 + \sin(2\pi \cdot 500t)$ [V].

It goes without saying that all BJTs must operate in active mode.

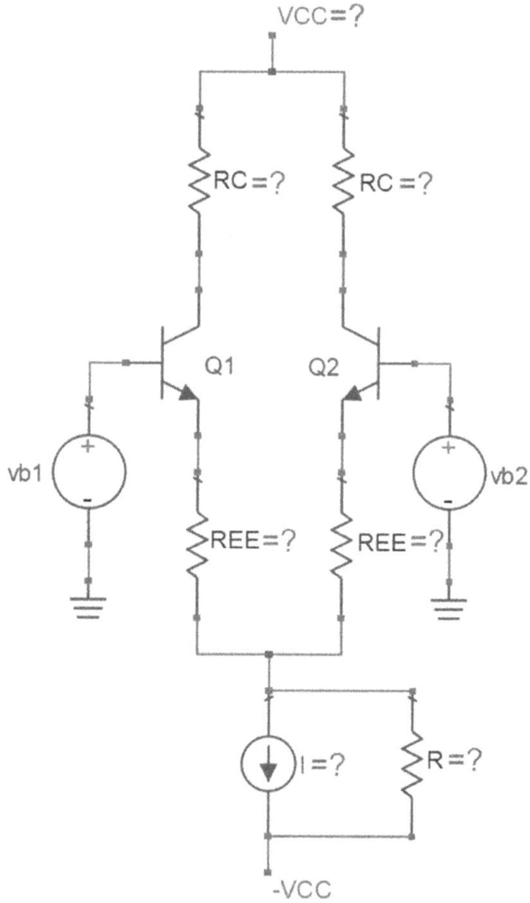

FIGURE 6.9 "?" indicates the unknown value of each design parameter.

Solution

Among the design parameters, R_C, R_{EE}, and V_{CC} are physical values that need to be determined as part of the design. The parameters I and R (the output resistance of the current source I) need to be determined, but thereafter, the current source is to be realized by means of either a simple or an advanced current mirror current source.

The first step in the design is to quantitatively interpret the specifications. As we do the interpretations, let us make some convenient approximations: $V_{BE}=0.7\,V$ if BJT is ON, $\beta=100$ if BJT is in active mode, and the Early voltage $V_A=100\,V$ if BJT is in active mode. We also assume $V_T=25\,mV$.

In designs involving BJTs, we sometimes assume $V_A \to \infty$. Not here though! As we need to determine R, knowing that R depends on the r_o small-signal parameter, we have to keep V_A finite.

Specification 1 (that indicates the maximum expected differential input amplitude) is related to the selection of a degeneration resistor R_{EE}. Here, of course, because the differential signal's amplitude is no larger than 10 mV, we can take $\mathbf{R_{EE}=0}$.

As in a CE amplifier design, an interpretation of an input resistance specification (Specification 2) leads to finding the needed parameters r_e and r_π and eventually to the needed bias current I. In the interpretation of the specification $R_{id}\geq 10$ kΩ, let us assume equality, trusting that the actual value of β (expected to be much larger than the assumed value of 100) will surely bring up the value of the differential input resistance:

$R_{id}=2r_\pi=10 \to r_\pi=5\text{ k}\Omega \to r_e\approx r_\pi/\beta\approx 50\ \Omega\approx V_T/(I/2) \to \mathbf{I\approx 1\ mA}$.

The value of I provides the other small-signal parameters: $g_m=(I/2)/V_T=20$ mA/V and $r_o=r_{o1}=100/0.5=200$ kΩ. We keep in mind that r_o of the BJTs that will constitute the current mirror current source is twice smaller (100 kΩ) because the mirror BJT's current is I (and not I/2).

The next step in the design is to interpret Specification 3 involving the desired single-ended differential gain. By now, after all the small-signal parameters have been found, the only unknown is R_C: $|A_{VdSE}|=50\approx g_m(R_C||r_{o1})/2\approx 10\cdot(R_C||200) \to R_C=(5^{-1}-200^{-1})^{-1}\approx 5.13$ kΩ. In the implementation, let us take $\mathbf{R_C=5.1\ k\Omega}$.

For Specification 4, regarding the single-ended common mode gain, the only remaining unknown is R, the output resistance of the current mirror current source: $|A_{VCMSE}|\leq 0.05 \to |A_{VCMSE}|\approx(R_C||r_{o1})/2R\approx 5/2R\leq 2.5/0.05=50$ kΩ. Noting that the output resistance of a simple current mirror current source, for a current of I = 1 mA, is $R=r_o=100$ kΩ, we conclude that a **simple current mirror current source will do**. Specification 4 is met by a wide margin. See Figure 6.10 (left).

To meet Specification 5, which calls for handling a known common-mode input signal, the supply value V_{CC} must be chosen properly. The given common-mode input signal has a DC level of 2 V and an AC component of amplitude 1 V and a frequency of 500 Hz. In this example, the dual supplies are symmetrical. In general, it does not have to be so. The positive supply V_{CC} must be large enough to assure that Q_1 and Q_2 operate in active mode for the largest expected common-mode input: $V_{CMmax}=2+1=3\,V \to V_{Emax}\approx 3-0.7=2.3\,V$. A choice of $V_{CE1}\approx I_{C1}R_{C1}=0.5\cdot 5.13=2.565\,V$ assures that the DC operating point of Q_1 and Q_2 can be placed sufficiently far away from the borderlines of saturation and cutoff modes. The calculation of V_{CC} concludes as follows: $V_{CC}\geq V_{Emax}+V_{CE1}+I_{C1}R_{C1}=7.43\,V$. We select $\mathbf{V_{CC}=7.5\ V}$.

The negative DC supply needs to ensure that the mirror's output transistor Q_3 stays in active mode for the smallest value of the common-mode input. Q_4 is in active

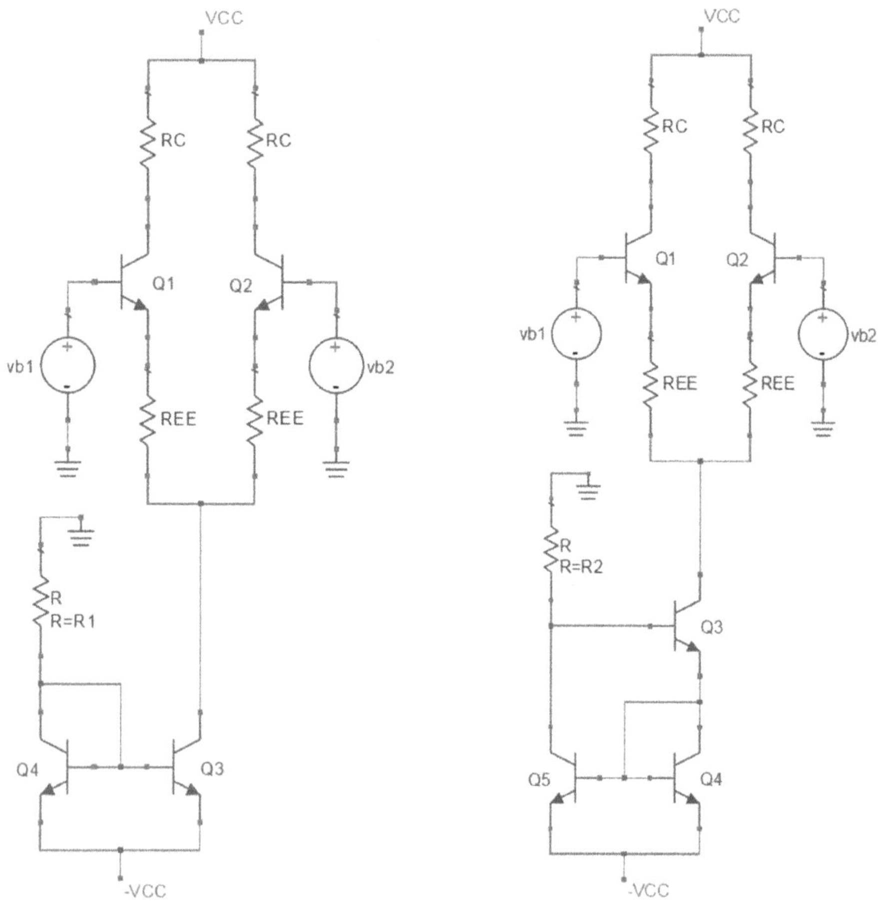

FIGURE 6.10 BJT differential amplifier with a bias current that is generated by a current mirror current source: (left) simple mirror, (right) Wilson mirror.

mode because the collector and base are short-circuited. $V_{CMmin}=2-1=1\text{ V} \rightarrow V_{Emin}\approx 1-0.7=0.3\text{ V}$. With $-V_{CC}=-7.5\text{ V}$, we check that $V_{B3}\approx -7.5+0.7=-6.8\text{ V} < V_{C3}=V_{Emin}=0.3\text{ V}$. Q_3 is in active mode.

Finally, for $I_{ref}=1\text{ mA}$, $\mathbf{R_1}=(0-(-V_{CC}+V_{BE4}))/I_{ref}=\mathbf{6.8\ k\Omega}$.

Example 6.1 follow-up: Let us revisit Specification 4 regarding the single-ended common mode gain. A specified gain of no more than 0.05 led to $R=50\text{ k}\Omega$, easily accomplished by a simple current mirror current source. Had the specification been a single-ended common-mode gain that is no larger than 0.01, R would have to be at least 250 kΩ, which cannot be met by a simple mirror. Recalling that in a Wilson current mirror current source (shown in Figure 6.10 (right)), $R_{wilson}\approx(\beta/2)R_{simple}=50\cdot 100=5\text{ M}\Omega$. It seems that the revised specification is now met by an unnecessarily huge margin, almost an "overkill". There is no way to gradually improve the gain – just a jump to the other extreme. We also note that $R_2=(0-(-V_{CC}+V_{BE4}+V_{BE3}))/I_{ref}=6.1\text{ k}\Omega$.

6.4 ADS AC AND TRANSIENT SIMULATION TESTING OF A DIFFERENTIAL AMPLIFIER DESIGN: DIFFERENTIAL MODE TESTING, COMMON-MODE TESTING, TRANSIENT SIMULATION TOTAL ANALYSIS; CURRENT MIRROR DC CURRENT SOURCE BIASING OF A DIFFERENTIAL AMPLIFIER

Example 6.1 (continued)

Our manual design of Section 6.3 involved multiple approximations. Let us test the design to see how well the specifications have been met. We start with AC simulation testing, shown in Figure 6.11.

As our design starts having more parts connections, it is sometimes difficult to keep all the wires without creating a "bird's nest" looking diagram. The solution is to use **wire labels**, the same node naming technique we have been using, but now we would name two or more nodes with the same name. **When nodes have the same name, they are electrically connected**, it is that simple.

Notice in Figure 6.11 that the wire label V_{cm} occurs twice. This means that those two nodes are connected via a wire label.

Note the righthand side "$-v_d/2$" (voltage source V_dR) using 180° phase shift.

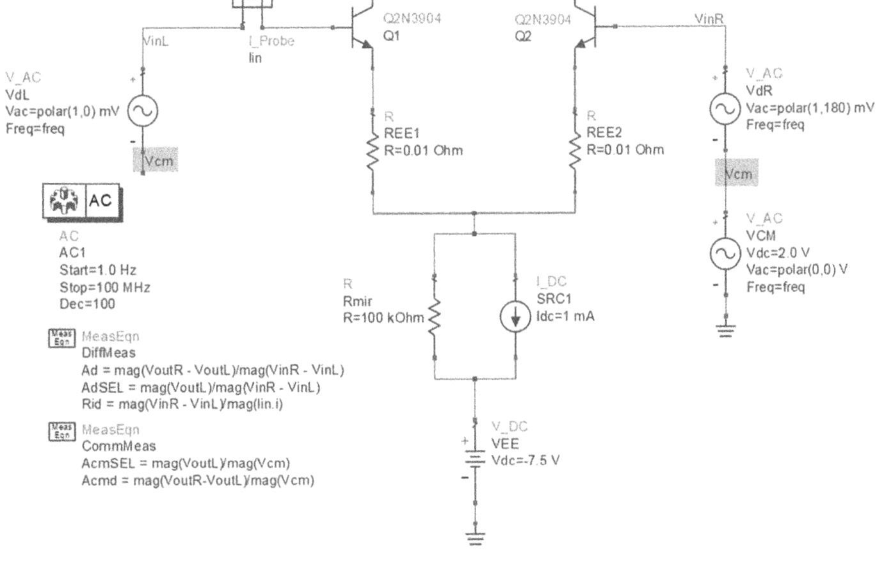

FIGURE 6.11 AC simulation schematic for the design of Example 6.1.

There are two groups of "MeasEqn": One valid for differential mode analysis and one valid for common mode analysis.

In the manual design, we saw that there was no need to use degeneration resistors. For possible future specifications changes, it is a good practice to hold on to the R_{EE} resistors, and if not needed, just make them very small (say 0.01 Ω). See Figure 6.12.

Figure 6.12 has everything that is needed for differential mode analysis and common mode analysis. For each type of analysis certain blocks will have to be disabled.

In order to start the testing with differential mode analysis, let us observe the modifications that need to be made in the schematic, shown in Figure 6.12. The changes are marked in Figure 6.13.

V_{cm} source has been short circuited and CommMeas deactivated.

Note that when we short circuit V_{cm} the DC value is "shorted" as well. Does this impact on the differential amplifier response? We shall see. The results for the differential gain A_d are shown in Figure 6.14 (left).

Figure 6.15 presents the alternate way of disabling the common mode input for the sake of conducting a differential mode analysis.

We do not deactivate the entire common mode signal; we just take zero amplitude of each common mode signal (here there is only one), except for DC.

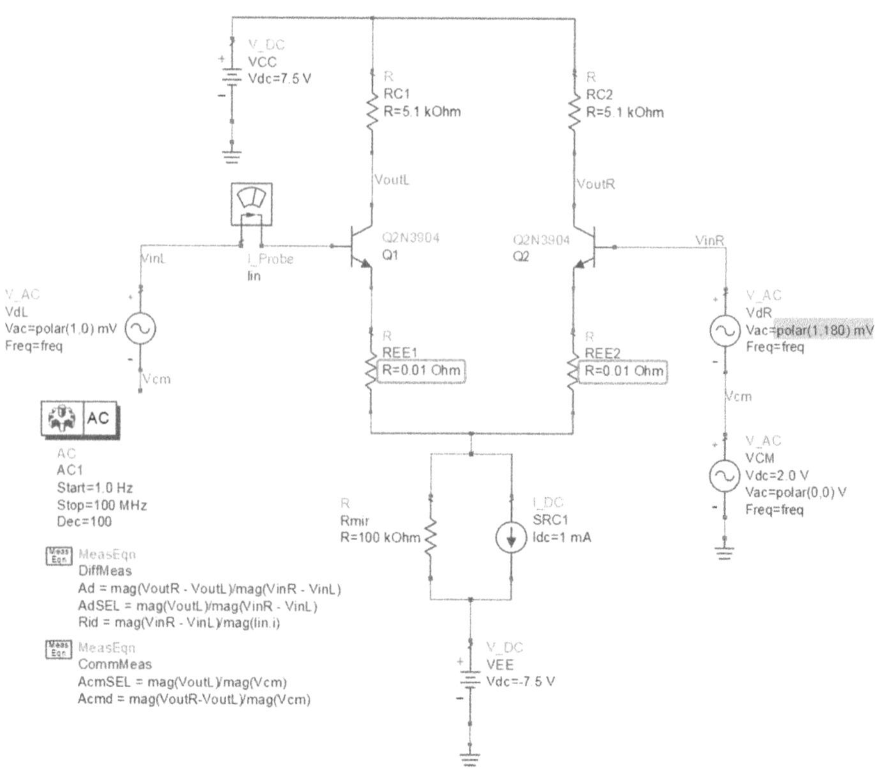

FIGURE 6.12 More preparations for an AC Simulation testing of the designed amplifier.

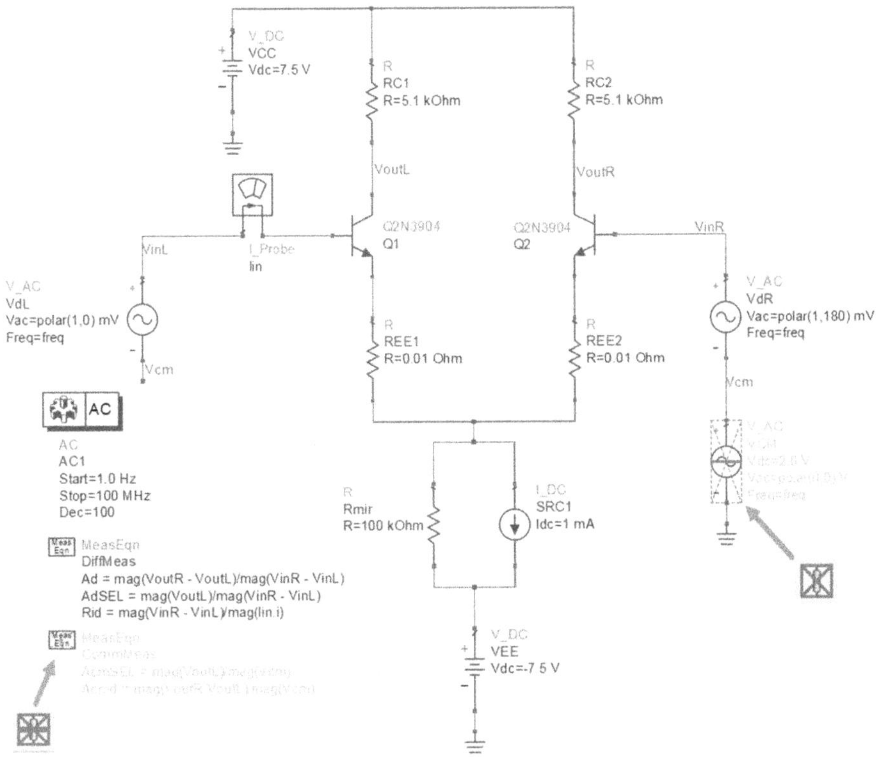

FIGURE 6.13 Differential mode analysis in AC simulation.

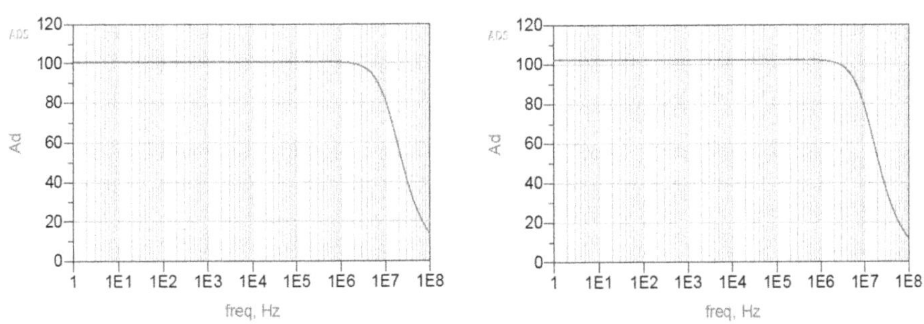

FIGURE 6.14 Two slightly different BJT differential mode differential output voltage gains: (left) short circuiting the input common mode voltage, including its DC level, (right) retaining the input common mode DC, while nulling the other common mode signals.

The results shown in Figure 6.14 indicate that the method of disabling the common mode does matter! The differential output Ad shown in Figure 6.14 (right) is the correct one.

Need to keep in mind that the DC part of the common mode input is part of the differential amplifier's biasing, along with V_{CC} and I. Short-circuiting the DC

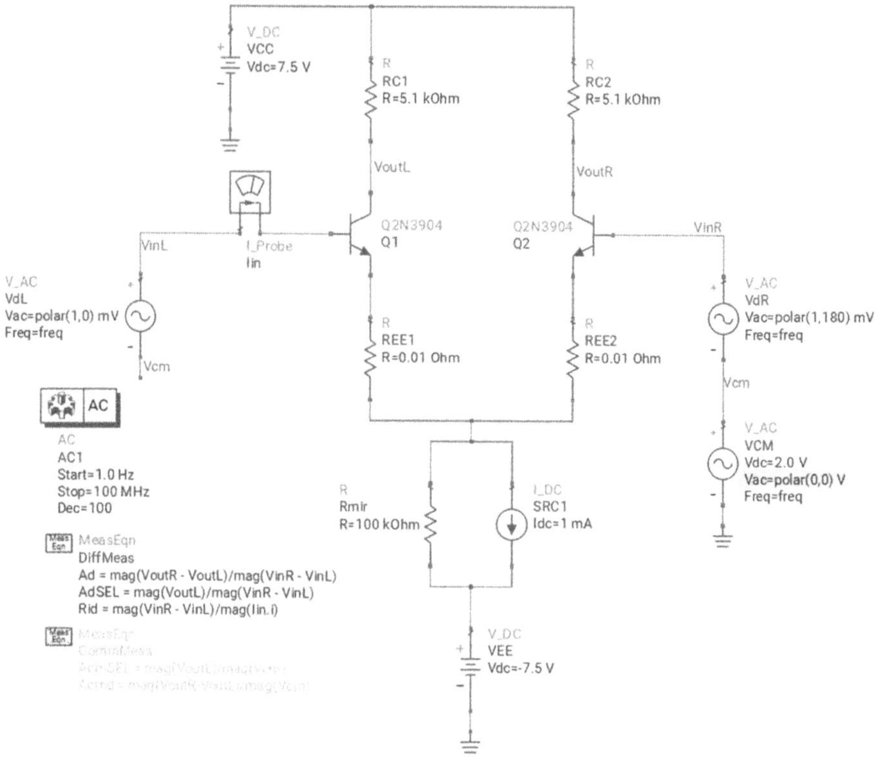

FIGURE 6.15 Disabling a common mode input by assignment of zero amplitude. The common mode DC is retained.

common mode signal, as happens when we disable by shorting, can create incorrect results. Here (based on Figure 6.14) the difference between the incorrect and correct results is very small.

One more comment about the shape of the frequency response of A_d:

No low frequency range (as we had in a CE amplifier due to the coupling and bypass capacitors), as there are no large capacitors. Differential amplifiers welcome differential DC inputs as well.

Two more differential mode analysis results (of the schematic shown in Figure 6.15) are shown in Figure 6.16.

Figure 6.17 presents the schematic for conducting common mode analysis. This time the differential input signals are deactivated with a short circuit, the amplitude of the common mode input is restored, and the MeasEqn that was valid for differential mode analysis gets deactivated, whereas the MeasEqn that is valid for common mode analysis gets activated.

As seen in Figure 6.17, we created measurement equations for both the common mode differential output voltage gain and for the common-mode single-ended voltage gain. In Figure 6.18 we present two versions of the common-mode differential output voltage gain.

We see that $A_{VCMdiff} = 0$. The results look noisy, what we call "numerical noise". These results (Figure 6.18 (left)) are for a perfectly balanced differential amplifier.

What if there were a slight imbalance, say $RC_1 \neq RC_2$? See Figure 6.19.

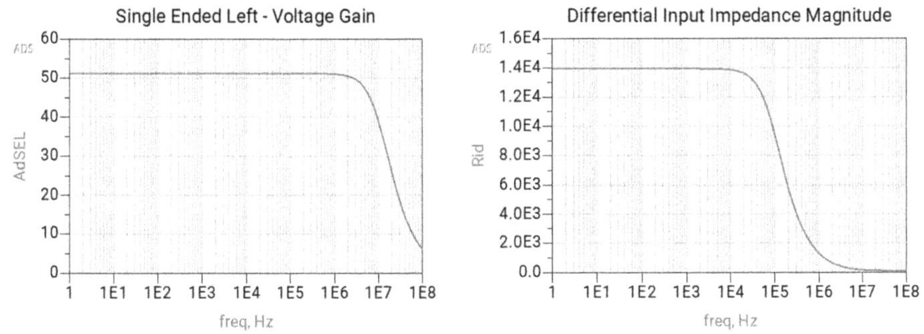

FIGURE 6.16 Differential mode single ended voltage gain (left). Differential mode differential input resistance (right).

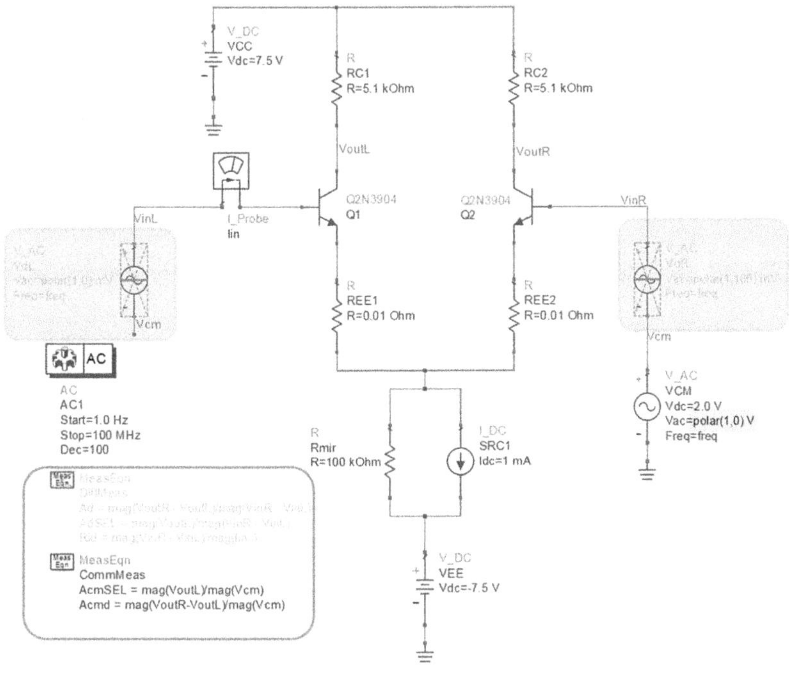

FIGURE 6.17 Schematic for common mode analysis.

Let us go back to differential mode analysis to inspect the impact of the slight imbalance in the RC resistors. We kept the amplifier unbalanced $RC_1 \neq RC_2$. The results are shown in Figure 6.20.

The results indicate that the differential gain and impedance have hardly changed, indicating the robustness of the differential amplifier.

ADS will always run a DC analysis for some simulations, so there is no need to explicitly instantiate or run a dedicated DC analysis to get DC results whenever using AC or HB, but do not try it with Transient. See Figure 6.21 for DC annotations, without bringing in a DC simulation controller.

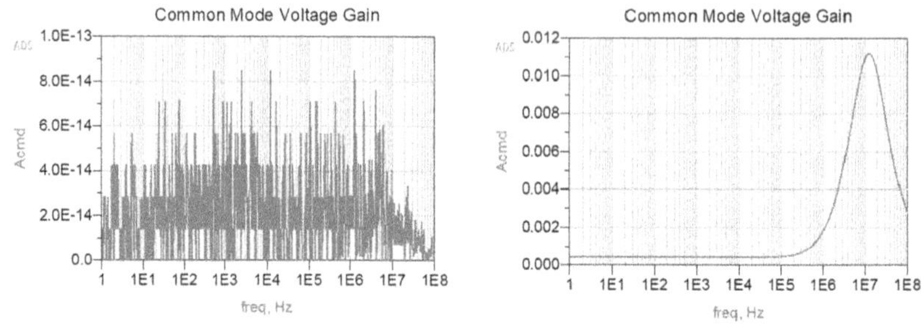

FIGURE 6.18 Common mode voltage gain: Differential output with perfectly balanced circuit (left). Differential output with slight imbalance in RC (right).

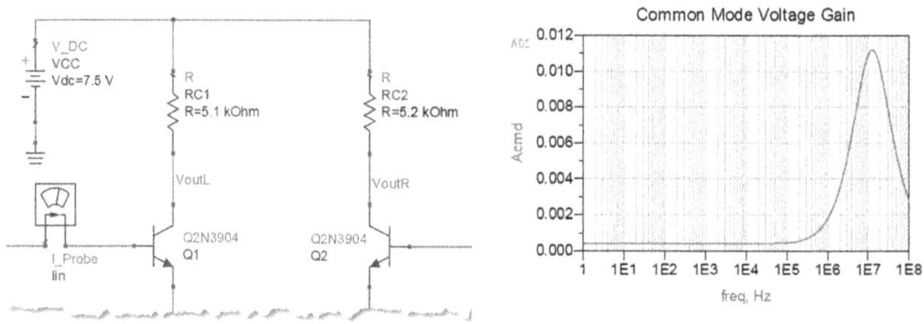

FIGURE 6.19 Modification of RC_2 by about 2%. $A_{VCMdiff}$ is small but greater than zero.

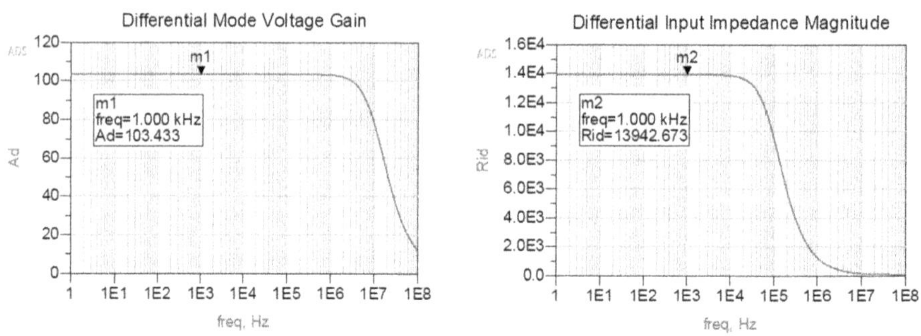

FIGURE 6.20 Results of the differential-mode analysis of an RC imbalanced amplifier.

Because of the current that flows into R_{mir}, each transistor receives slightly more than 0.5 mA.

Is this a realistic effect with a true current mirror circuit?

We need to test a real mirror, not an ideal current source I_{EE} with parallel resistance R_{mir}. We shall come back to it after testing the amplifier in transient simulation.

In preparation for transient simulation, we need to separate differential mode simulation from common mode simulation. In addition, formulas were created for the difference of voltages. Voltage gain formulas are calculated as ratios of peak-to-peak amplitudes.

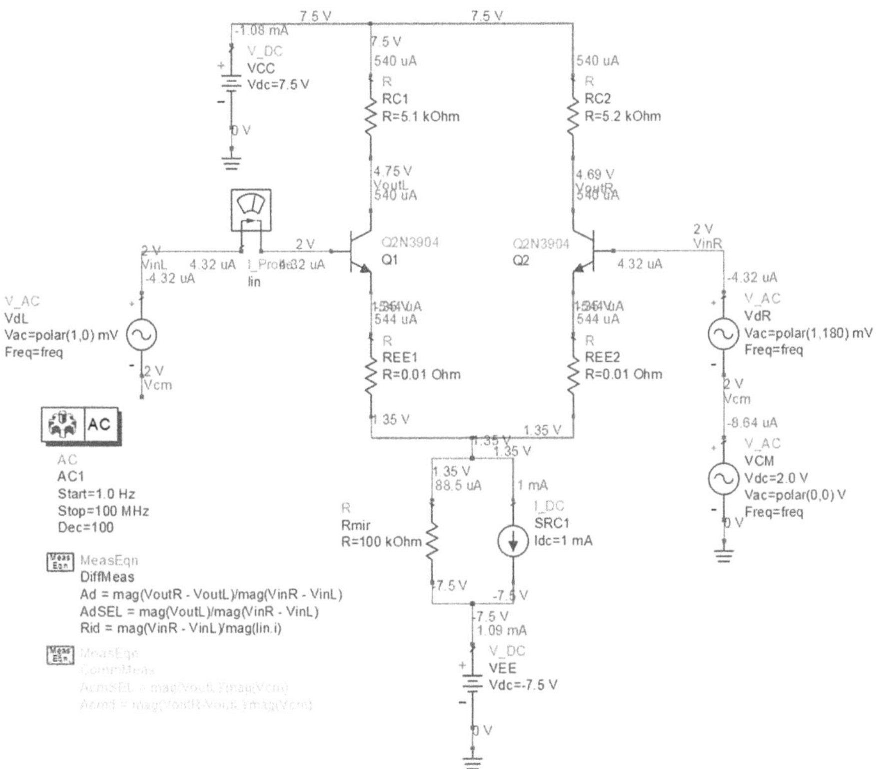

FIGURE 6.21 Annotated DC results obtained after the last AC simulation.

See Figure 6.22 for the schematic and Figure 6.23 for the results.

The results are close to expectations based on our knowledge from the AC simulations. The A_d gain is however lower; AC differential gain ≈ 102. What happened? Transient simulation is signal level sensitive; this is one way we assessed distortion earlier.

We leave it as an exercise to the reader to assess how much distortion there is and what happens with higher input signal amplitudes.

A transient simulation common-mode analysis of the perfectly balanced differential amplifier is studied next (Figure 6.24).

As seen in Figure 6.25, the amplifier produced a zero A_{cmd} (common-mode differential gain) and a somewhat small A_{cmSEL} (single-ended common-mode gain).

Transient simulation of the differential amplifier allows us a glimpse into total signal performance (differential mode and common mode mixed together). See Figure 6.26.

We see in the results, shown in Figure 6.27, a trace of V_{cm} in the differential V_{out}. Why?

$|V_d| \cdot |A_{dSEL}| = 0.02 \cdot 50 = 1\ V$

$|V_{cm}| \cdot |A_{cm}| = 1 \cdot 0.025 = 0.025\ V$

V_{out} is predominantly V_d, but the effect of V_{cm} is not so negligible.

This is a result of R_{mir} not being large enough. One can test this by making R_{mir} larger, re-running the analysis, and see consecutive peaks come closer in value.

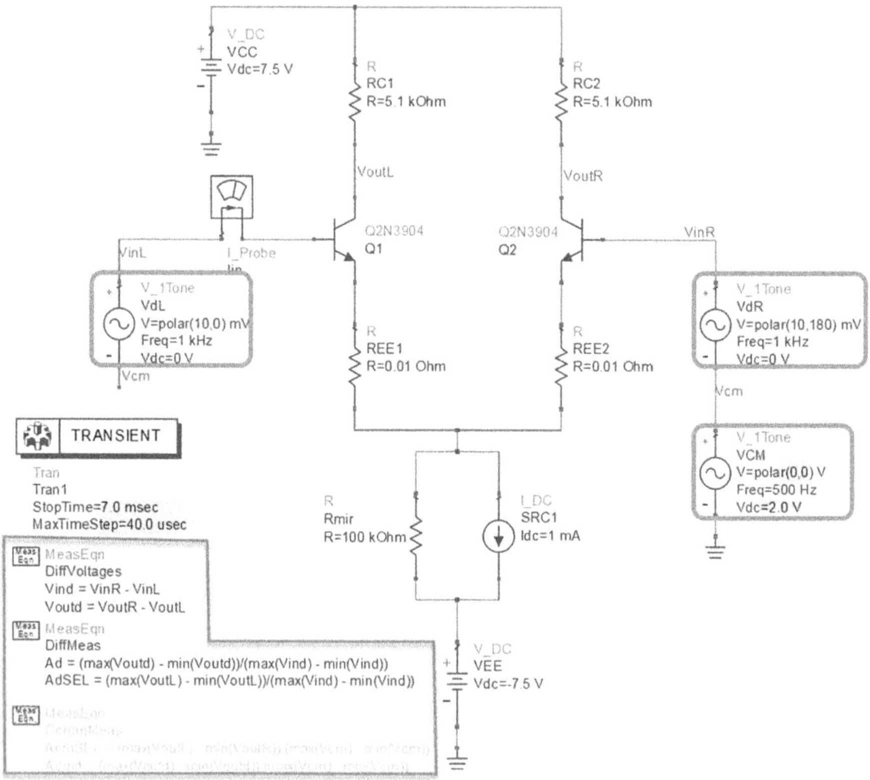

FIGURE 6.22 Transient simulation setup.

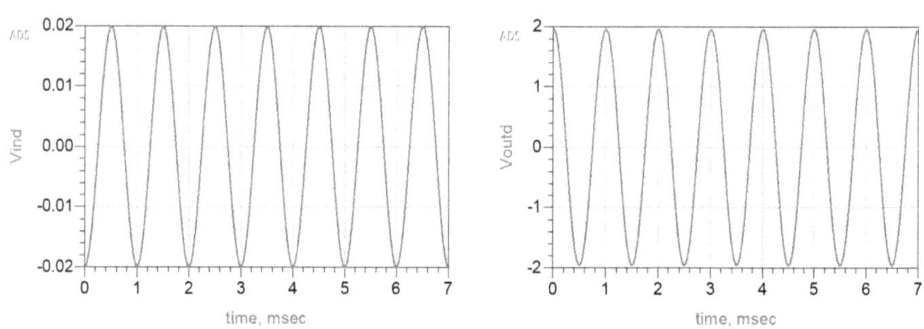

Ad	AdSEL
98.169	49.084

FIGURE 6.23 Transient differential mode results.

As we see from the markers m_1 and m_2 (in Figure 6.27), we note the dissimilar V_{outd} values. It is due to the V_{cm} impact.

In the schematic shown in Figure 6.28, we replace the generic current source I and its parallel resistance R by a simple current mirror current source.

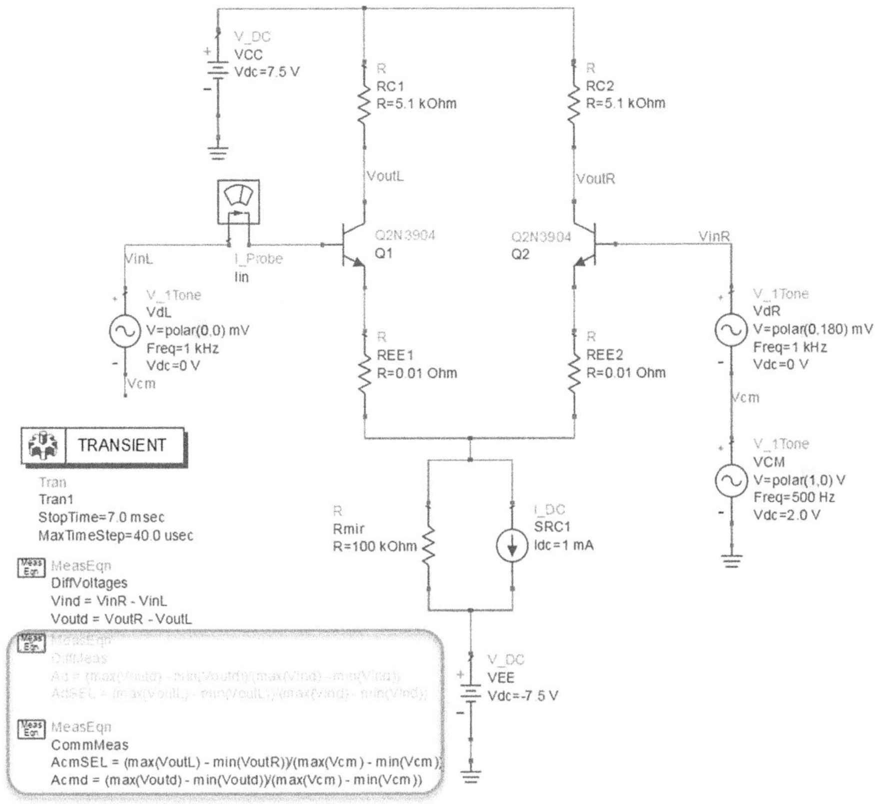

FIGURE 6.24 Transient simulation setup – common mode.

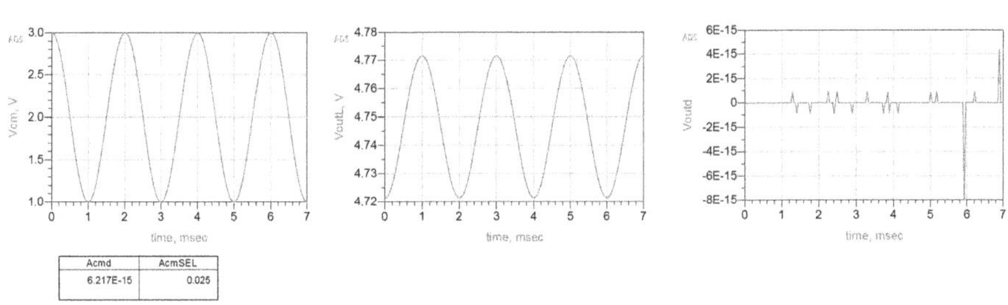

FIGURE 6.25 Transient simulation common mode results.

In Figure 6.28, notice the new wire labels V_{ee} to connect the negative DC supply.

DC analysis results of the differential amplifier biased by a current mirror current source are shown in Figure 6.29.

The current I, generated by the current mirror, is slightly off (1.1 mA, in place of 1 mA). Need to tweak the value of R_1, left as an exercise for the reader.

The results shown in Figure 6.31 look similar to what we did with the generic DC current source. We need, of course, to repeat the DM and CM analyses, which we leave to the reader.

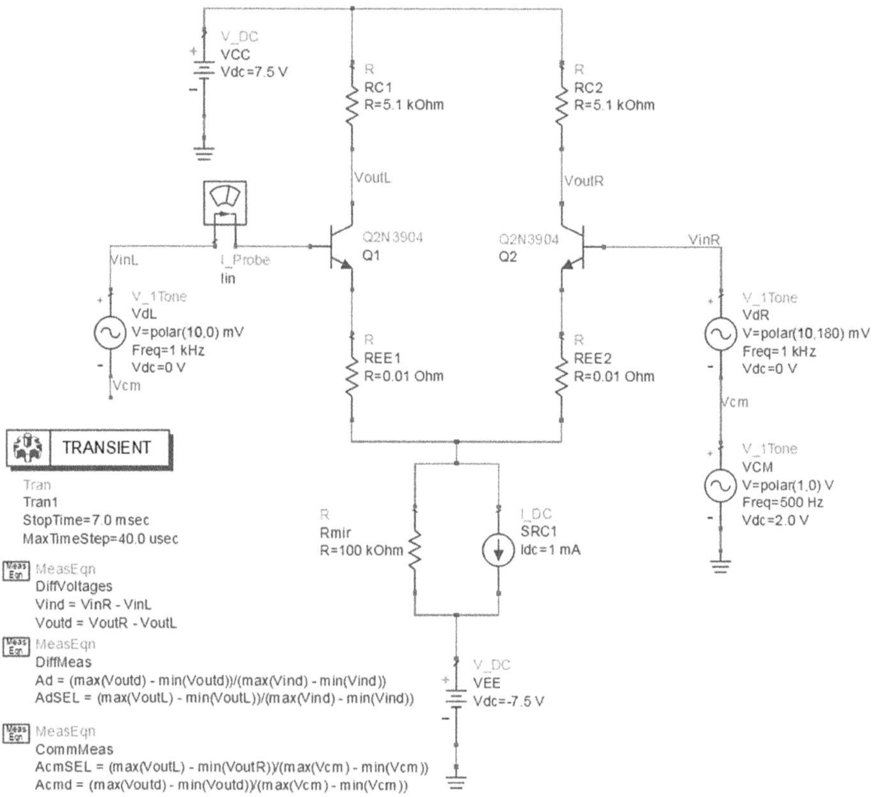

FIGURE 6.26 Total signals analysis of a differential amplifier.

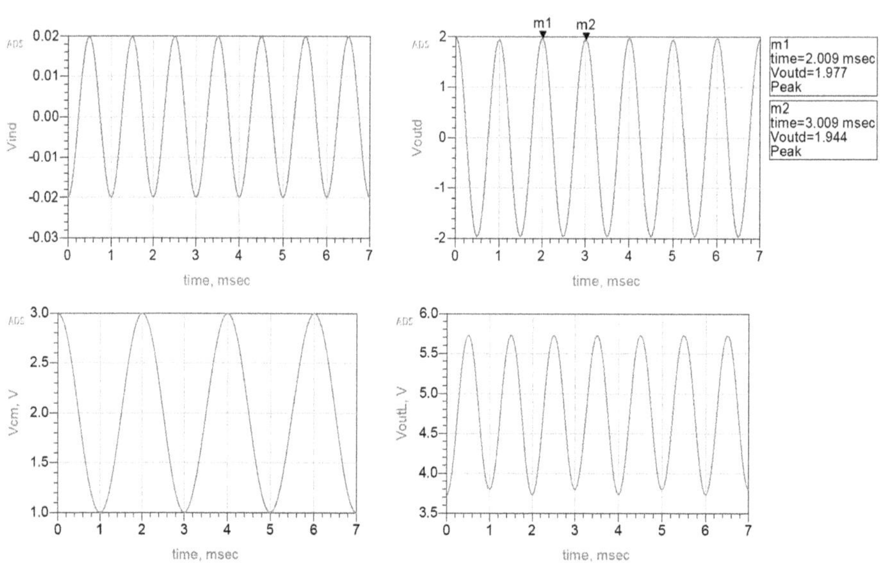

FIGURE 6.27 Total signal analysis results.

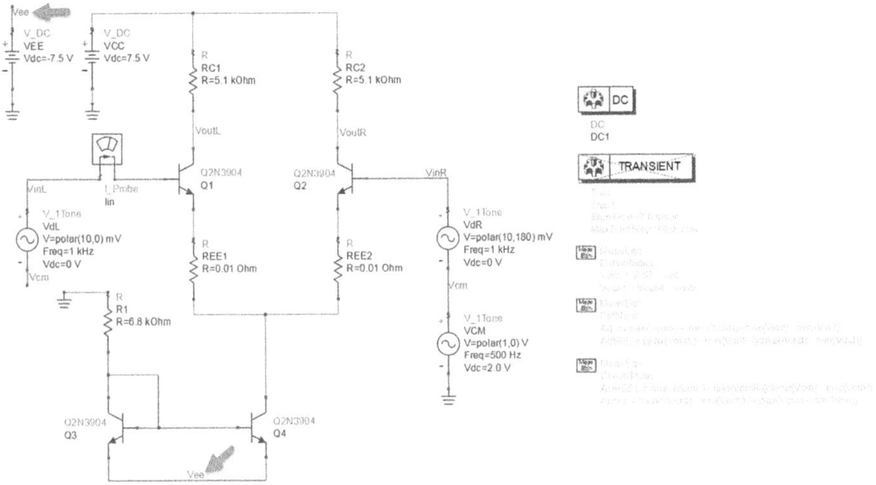

FIGURE 6.28 Replacing the generic current source by a simple current mirror.

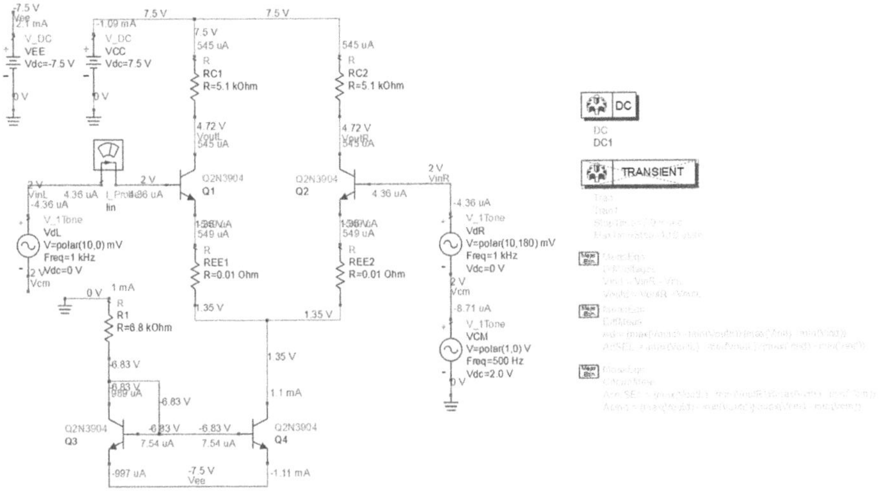

FIGURE 6.29 DC annotation of the differential amplifier with current mirror current source.

6.5 MULTISTAGE BJT DIFFERENTIAL AMPLIFIER

Example 6.2

The following example of a multistage differential amplifier, shown in Figure 6.32, is described in detail in Sedra and Smith's book [Sedra1]. The amplifier consists of four stages: The first two stages are a cascade of NPN BJT differential amplifiers intended to create a differential output in the first stage. We remember that a

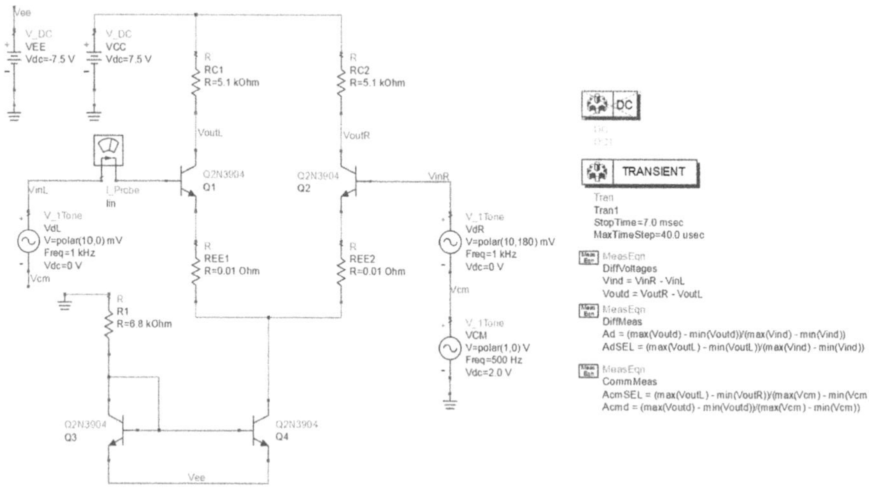

FIGURE 6.30 Transient simulation setup of the differential amplifier with current mirror I.

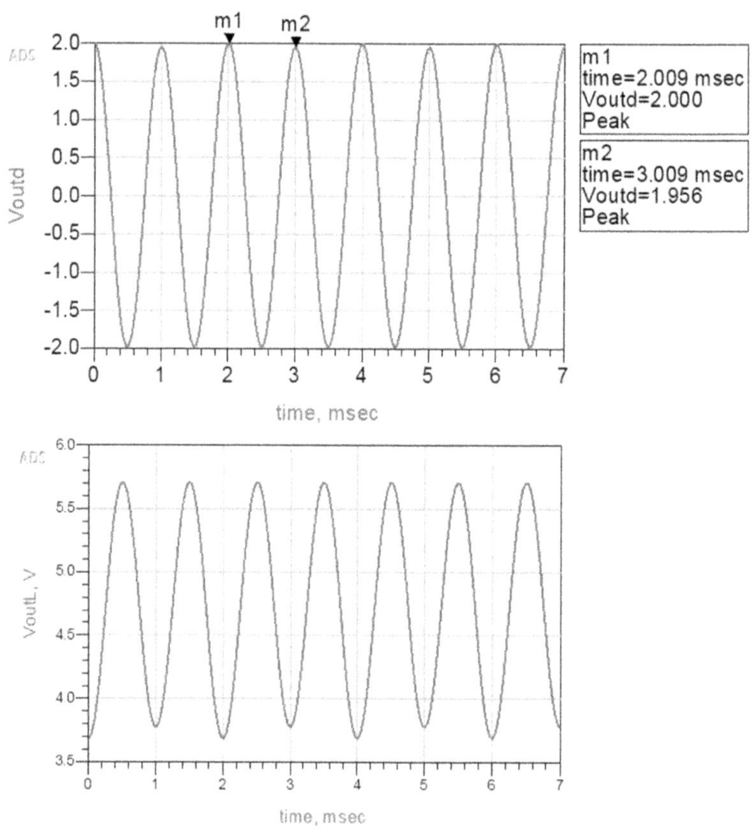

FIGURE 6.31 Total signals in a transient simulation of the differential amplifier with current mirror current source.

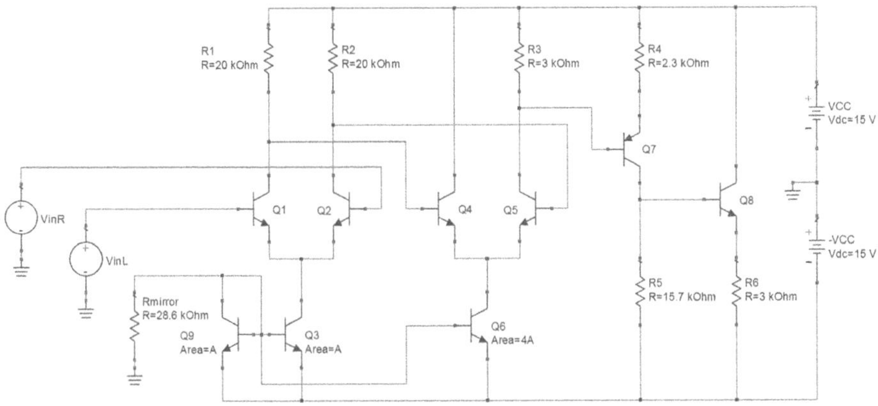

FIGURE 6.32 Sedra and Smith's example of a multistage differential amplifier.

differential amplifier with a differential output has nominally a zero common-mode gain. The third stage, centering around a PNP BJT, is a voltage level shifter, which is part of an output zero offset adjustment sub-circuit. The last stage is an emitter follower serving as a power amplifying output stage. The biasing of the first two stages is done by means of simple NPN current mirror current sources.

It can be shown that all BJTs of the above amplifier operate in active mode. In the circuit of Figure 6.32 we assume that the forward voltage of each of the BJT BE junctions is 0.7 V. It is also assumed that for each BJT, $\alpha = 1$ and $V_A = \infty$. Stage 1 is therefore biased by a DC current of 0.5 mA, whereas Stage 2 is biased by a DC current of 2 mA. Note that the current mirror's reference current of 0.5 mA is created by the negative supply $-V_{CC}$ and the resistor R_{mirror}.

Whatever is the BE junction area A of Q_3, the BE junction area of Q_6 needs to be 4A, or Q_6 may be replaced by four parallel output transistors, each with a BE junction area of A.

Assuming that the amplifier's inputs, V_{inR} and V_{inL}, are zero and that in all the BJTs $I_C = I_E$, the DC collector voltages of Stage 1 (Q_1 and Q_2) are 10 V. Note that the output of Stage 2, coming from the collector of Q_5, is single-ended, and the collector DC voltage of Q_5 is 12 V.

In any differential amplifier, if both input signals are zero, the amplifier's output must be zero. This is called "**zero offset**". I_C of Stage 3 is ideally 1 mA and therefore the resistance R_5 must be equal to 15.7 kΩ to create zero offset voltage at the emitter of Q_8.

Example 6.3

The amplifier of Example 6.2 can serve as a primitive operational amplifier. As such, this amplifier will accompany us in several subsequent ADS activities covered by the book: when we learn about ADS optimizers, about bandwidth, about the Miller effect and cascode amplifiers, and about feedback compensation.

All the above simulation needs require a slightly simplified version of the Figure 6.32 amplifier, shown in Figure 6.33.

Let us explain the motivation for the above circuit changes.

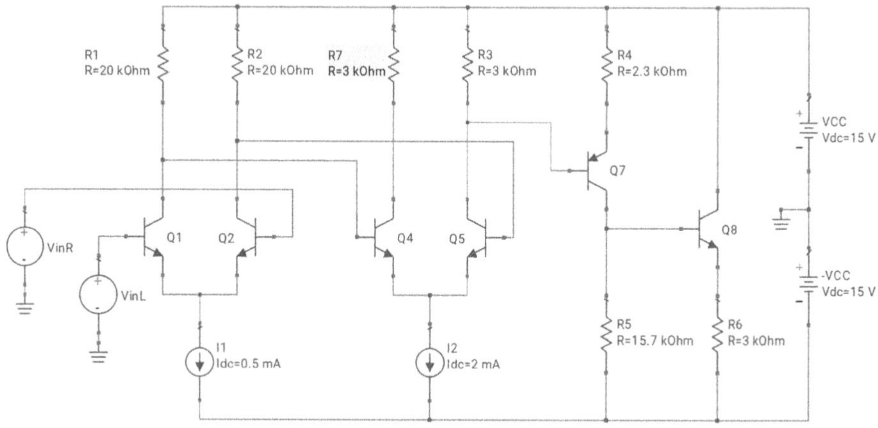

FIGURE 6.33 Simplified version of Sedra-Smith's multistage differential amplifier.

The first modification involves the replacement of the current mirrors with ideal current sources.

Recall that R_{out} of each current mirror (with output transistors Q_3 and Q_6) determines the single-ended common-mode voltage gain of each differential amplifier. It has no effect on the amplifier's differential voltage gain. Due to the differential output of Stage 1, the common mode gain of Stage 1 is nominally zero, and therefore, the overall multistage amplifier's common mode voltage gain is nominally zero.

The second modification is in Stage 2. In Figure 6.32 no collector resistor is taken for Q_4, arguing that such collector voltage does not matter, as we take a single-ended output from the right BJT Q_5. This is true in ideal calculations, neglecting the Early effect. In a simulation the lack of R_C symmetry in Stage 2 does matter because of the Early effect. We therefore restored symmetry by taking $R_7 = R_3$.

The third modification has to do with the zero offset adjustment done by the PNP collector resistor R_5. The value of R_5 in Figure 6.32 strongly depends on the $|V_{BE}| = 0.7\,V$, $V_A = \infty$ and $\alpha = 1$ assumptions. In reality another value of R_5 may need to be selected. We replaced $R_5 = 15.7\ k\Omega$ by a parameter $R_5 = RX$. See Figure 6.35.

Recall the two-stage voltage amplifier in which a CE amplifier stage was connected to a load R_L via an emitter follower stage. The overall voltage gain becomes the product of each stage's voltage gain, assuming that each stage's load is the input resistance of the next stage. This is approximately true in any multistage amplifier.

In a differential multistage amplifier, the above method of calculating the overall voltage gain applies to the two analyses – differential mode and common mode. In Sedra-Smith's example, we have already determined the overall common mode voltage gain to be zero.

What remains to be found is the overall differential voltage gain. Let the overall voltage gain A_d be $A_d = A_1 \cdot A_2 \cdot A_3 \cdot A_4$. A_1 is the differential output differential gain of Stage 1, loaded by the differential input resistance of Stage 2. A_2 is the single-ended differential voltage gain of Stage 2, loaded by the input resistance of Stage 3. A_3 is the voltage gain of Stage 3, loaded by the input resistance of Stage 4. A_4 is the voltage gain of Stage 4, assuming no load ($R_L = \infty$).

If we let R_5 of Figure 6.33 be equal to R_5 of Figure 6.32, and if we assume that all BJTs have the same $\beta = 100$ and that $V_T = 25\,mV$ and $V_A = \infty$ (meaning that $r_o = \infty$)

then the overall differential mode voltage gains come the same in both figures. Gains A_2, A_3, and A_4 are found based on known formulas. Only the gain calculation of A_1 is non-trivial [Sedra1]. Let us conveniently go backwards from the last stage towards the first stage.

Stage 4 is an emitter follower. Regardless of the value of R_L, we can approximate $\mathbf{A_4 \approx 1}$. The input resistance R_{in4} of Stage 4 (with no load) is approximately $\mathbf{R_{in4} \approx r_{\pi 8} + (\beta+1)R_6}$. Numerically, $R_{in4} \approx 303$ kΩ (neglecting $r_{\pi 8} \approx 500$ Ω resulting from $I_{C8} \approx 5$ mA).

Stage 3 is a PNP CE amplifier that has no bypass capacitor. Therefore, $\mathbf{A_3 \approx -(R_5 || R_{in4})/R_4}$. Numerically, $A_3 \approx -(15.7||303)/2.3 = -14.93/2.3 \approx -6.49$. The input resistance R_{in3} of Stage 3 is approximately $\mathbf{R_{in3} \approx r_{\pi 7} + (\beta+1)\ R_4}$. Numerically, $R_{in3} \approx 234.8$ kΩ (with the inclusion of $r_{\pi 7} \approx 2.5$ kΩ resulting from $I_{C7} \approx 1$ mA.)

As will be confirmed by simulation later on in Chapter 6, the base of Q_2, receiving the input voltage signal $v_{inR}(t)$, is the inverting (–) input of this op-amp. The base of Q_1, receiving the input voltage signal $v_{inL}(t)$, is the non-inverting (+) input of this op-amp.

The single ended differential voltage gain of Stage 2 is $\mathbf{A_2 \approx -(g_{m5}/2) \cdot (R_3 || R_{in3})}$. Numerically $A_2 \approx -(40/2) \cdot (3||234.8) = -59.2$. The differential input resistance of Stage 2 is $\mathbf{R_{id2} = 2r_{\pi 4}}$. Numerically, $R_{id2} \approx 5$ kΩ, as the collector current of Q_4 is approximately 1 mA.

The small signal diagram of the differential-mode analysis of Stage 1 is shown in Figure 6.34.

The differential input resistance R_{id2} of Stage 2 loads Stage 1 by being connected between the collectors of Q_1 and Q_2. That is, R_{id2} is a floating load, created by the way a differential output of stage 1 is implemented. Such a case has not yet

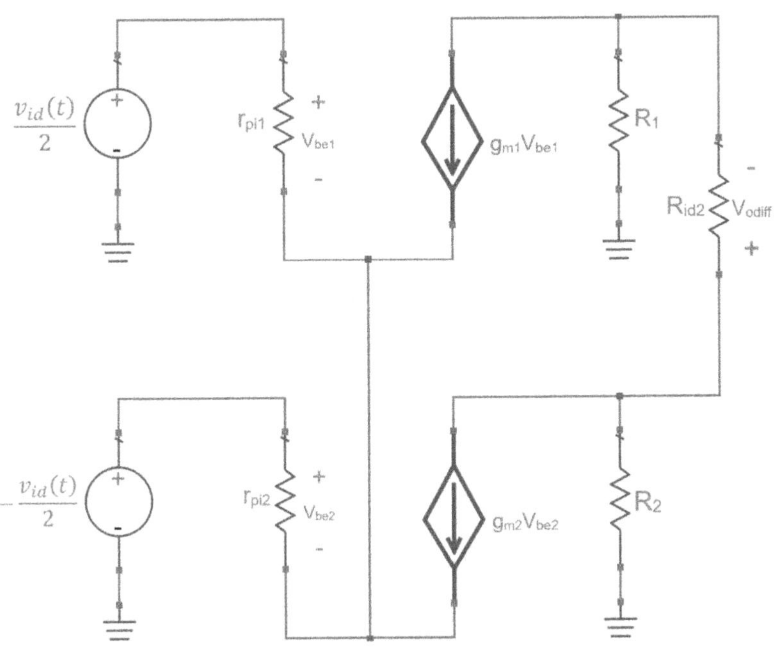

FIGURE 6.34 Small-signal diagram of Stage 1 differential-mode analysis.

been discussed as part of the suite of small-signal differential voltage gain formulas that have been developed or outlined, all assuming symmetric load resistors, each connected between collector and ground.

In Figure 6.34, R_1 and R_2 are the two symmetric R_C resistors. The small signal parameters of Q_1 and Q_2 are symmetric as well, including the r_o resistors that are neglected in the derivation. The polarity of the differential output voltage v_{odiff} is defined as shown in Figure 6.34, depending on the differential output implementation.

As we have seen, in all differential-mode analyses that we have done so far, the small-signal emitter voltage is $v_e=0$, a "virtual ground". Consequently $v_{be1}=v_{id}/2$ and $v_{be2}=-v_{id}/2$.

The differential output differential voltage gain is $A_1=v_{o,diff}/v_{id}$. It is found approximately (as we neglect the r_o resistors) from a KVL equation in the loop that is formed by R_1, R_{id2} and R_2:

$0=R_1(v_{o,diff}/R_{id2}-g_{m1}v_{id}/2)+v_{o,diff}-[-g_{m2}(-v_{id}/2)-v_{odiff}/R_{id2})]\cdot R_2$ that simplifies into $g_m(v_{id}/2)(R_1+R_2)=v_{o,diff}\cdot(1+\ R_1/R_{id2}+R_2/R_{id2})$. Finally, $\mathbf{A_1=[(R_1+R_2)||R_{id2}]/(r_{e1}+r_{e2})=(g_m/2)[(2R_C)||R_{id2}]}$. Numerically, $A_1\approx[(2\cdot 20)||5]/(2\cdot 0.1)=22.22$. Note that with no differential load the differential output differential gain of Stage 1 would reach 200 approximately.

The differential input resistance of Stage 1, $\mathbf{R_{in1}=2r_{\pi 1}}$ is the differential input resistance R_{id} of the whole amplifier. Numerically, $R_{id}=R_{in1}\approx 20\ k\Omega$.

The overall differential voltage gain A_d, based on all the approximations that were made, is $A_d\approx 22.22(-59.2)\cdot(-6.49)\cdot 1\approx 8537$.

6.6 ADS: DC SIMULATION ANALYSIS OF THE MULTISTAGE AMPLIFIER: NULLING OF OUTPUT OFFSET; AC SIMULATION OF THE MULTISTAGE AMPLIFIER: VOLTAGE GAIN AND BANDWIDTH OF EACH STAGE; INPUT AND OUTPUT RESISTANCE MEASUREMENTS OF THE AMPLIFIER; VOLTAGE GAIN TWEAKING BY MEANS OF ADS TUNING

Example 6.3 (continued)

Figure 6.35 presents the ADS schematic of the simplified multistage amplifier of Figure 6.33. The changes implemented with respect to Figure 6.32 are: No mirrors, just ideal DC current sources, symmetrical RC resistors in Stage 2, and RX is a swept variable.

Note the use of wire labels, shown highlighted in Figure 6.35. In Section 6.5 we provided numerical DC node voltages and component currents based on the simplifying assumptions: $V_{BE}=0.7\,V$, $V_A=\infty$ and $R_{CP}=15.7\ k\Omega$. We also assumed $\beta=100$, which affects the AC results.

The actual DC results are shown in Figure 6.36:

We observe the need to change the value of the offset adjustment resistor R_{CP}. The choice of $R_{CP}=15.7\ k\Omega$, based on ideal assumptions, is not good, as it leads to $V_{out}=660\,mV$. The reason that the manual calculation offset adjustment didn't work is that $I_{CP}=0.988\,mA$ and not 1 mA.

Let us do a parametric sweep for RX with a goal of bringing V_{out} to be within 10 mV of 0 V. See Figure 6.37 for the schematic and Figure 6.38 for the results.

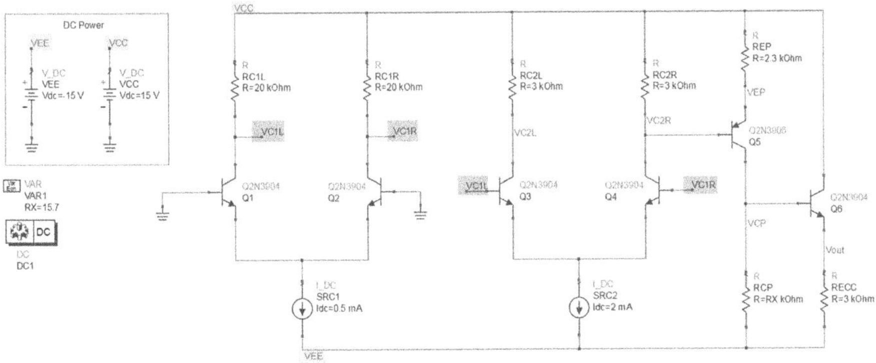

FIGURE 6.35 Schematic for DC simulation of the simplified multistage amplifier.

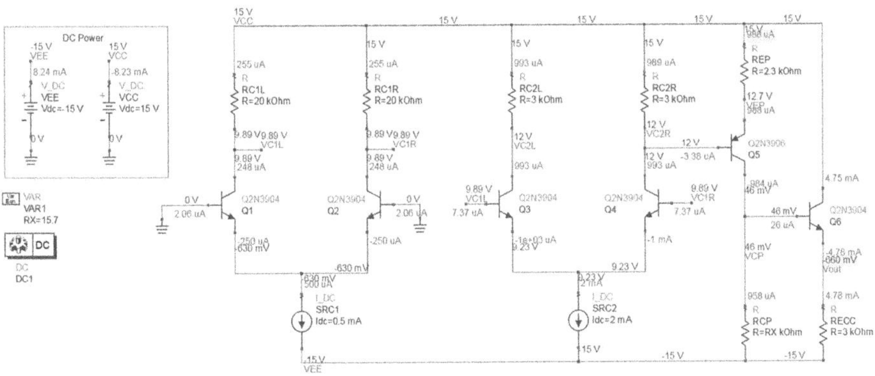

FIGURE 6.36 Actual DC results for the multistage amplifier where both its input voltage nodes are grounded.

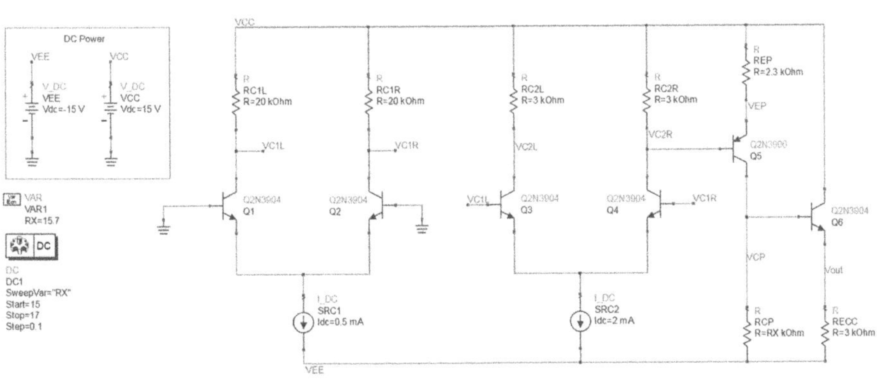

FIGURE 6.37 Schematic showing the DC sweep of RX for the sake of reducing the output offset error.

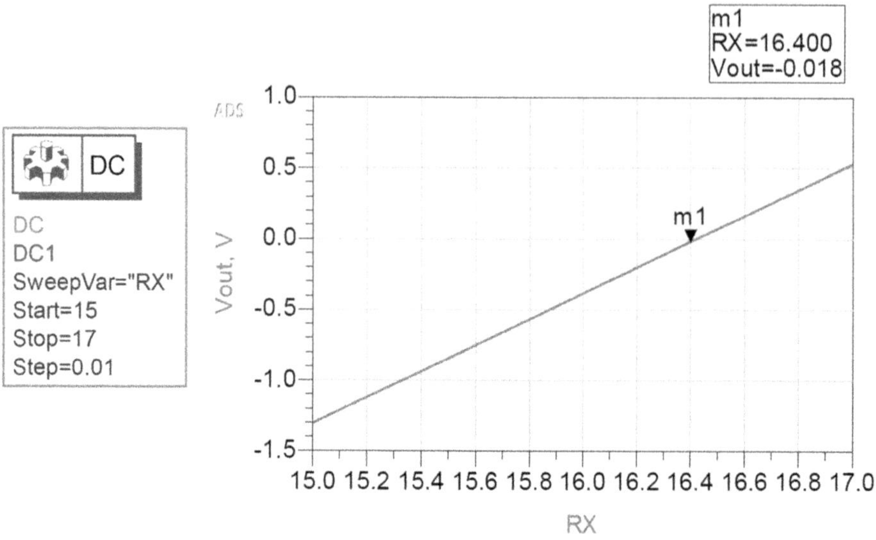

FIGURE 6.38 Simulation results showing a new value for R_{CP}, using RX = 16.4 kΩ, still having a large offset error of −18 mV.

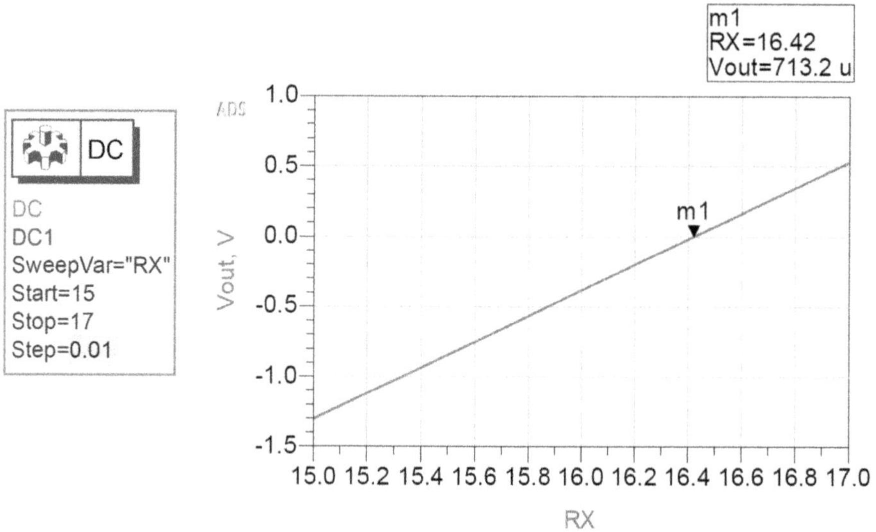

FIGURE 6.39 Sweeping RX with a resolution of 0.01 kΩ provides one more decimal point to R_{CP}, R_{CP} = 16.42 kΩ, bringing down the DC offset of V_{out} to 713 μV.

The sweep step of 0.1 kΩ seems to be too coarse.

We do not get a good enough resolution for the 0 ± 10 mV placing of marker m_1.

Finer step sweep results in increased resolution, as shown in Figure 6.39.

Here is a little DC simulation simplifying trick.

See Section 6.7 on how to use Tuning to accomplish the lowering of the offset DC voltage.

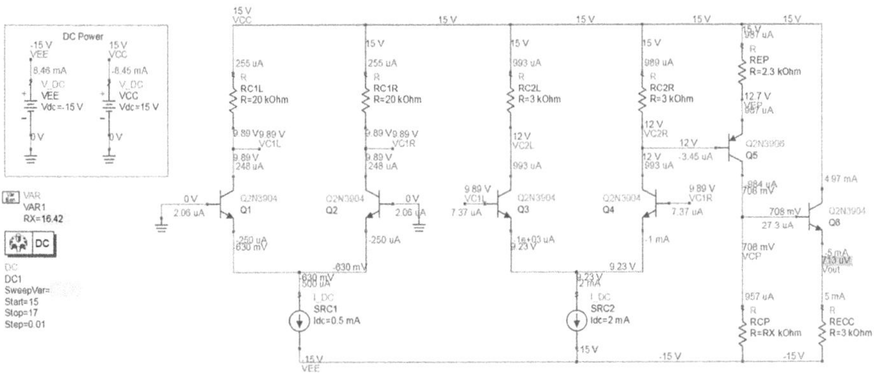

FIGURE 6.40 The DC internal sweep was bypassed by making SweepVar blank.

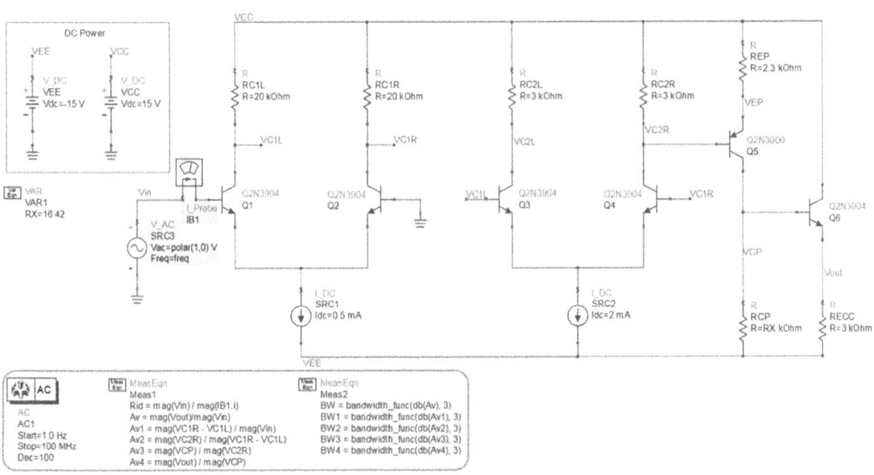

FIGURE 6.41 AC simulation of the multistage amplifier.

Figure 6.41 shows the multistage amplifier schematic, preparing for AC simulation by creating all the needed measurement equations. The simulation will show many voltage gain frequency response plots, the differential input resistance, and the bandwidth of each amplifier's stage.

This is a differential-mode analysis, where conveniently the entire $V_d(t)$ is injected into one side of the amplifier.

Because Stage 1 has differential output, the whole amplifier has a common-mode gain of zero. That is why there is no need to attach resistors in parallel to the DC current sources. It also eliminates the tiny common-mode signal created by the non-anti-symmetric input feeding.

Figure 6.42 brings all amplifier stages' voltage gain frequency response plots and the respective bandwidths. It also shows the overall amplifier voltage gain and bandwidth. We recall that in Section 6.5 we presented numerical values for the mid-band voltage gains, based on the assumptions $\beta = 100$ and $V_A = \infty$.

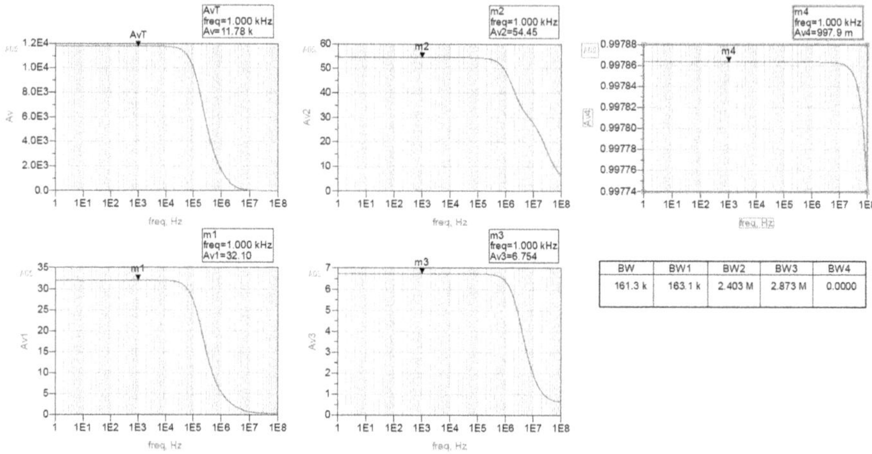

BW	BW1	BW2	BW3	BW4
161.3 k	163.1 k	2.403 M	2.873 M	0.0000

FIGURE 6.42 AC simulation results for the multistage amplifier.

Placing markers in the mid-band frequency range of each stage's gain frequency response gives us the following results: $A_v = 11{,}780$ (compared to 8537 obtained in the manual calculations), $A_{v1} = 32.1$ (compared to $A_{v1} = 22.22$ obtained manually), $A_{v2} = 54.4$ (compared to $A_{v2} = 59.2$ in absolute value), $A_{v3} = 6.75$ (compared to the manual $A_{v3} = 6.49$) and $A_{v4} = 0.998$ (compared to the manual $A_{v4} = 1$). Interestingly, $A_v = A_{v1}A_{v2}A_{v3}A_{v4} = 32.1 \cdot 54.4 \cdot 6.75 \cdot 0.998 = 11{,}764$. The stage gains product for a multistage amplifier is not 100% accurate. See Section 6.8 for further explanation.

We see that among the four stages, it is Stage 1 that has the largest deviation between the actual and calculated voltage gains. Why is it so?

Recall that in Section 6.5 we developed a formula for the small-signal differential-mode voltage gain of a differential amplifier that has differential output. As often happens, this voltage gain is strongly affected by R_{id2}, the differential input resistance of Stage 2, which strongly depends on β. Our assumption of $\beta = 100$ caused a large underestimation of R_{id2}.

Let us now comment on the measured bandwidths. The overall BW is dictated by the stage that has the lowest BW, which is Stage 1.

Why does Stage 1 have a relatively poor BW?

Why does Stage 2 have a better BW?

This will be better understood a little farther on in Chapter 6.

Stage 3 (with the PNP BJT) is a CE amplifier, but it has a large degeneration resistor and a good BW.

Stage 4 has phenomenal BW, but it cannot be computed because, as far as f goes, it is still lower than the 3 dB f_c of the emitter follower.

The input resistance of Stage 1 is also the differential input resistance R_{id} of the entire multistage amplifier. See Figure 6.43.

We again note the large underestimating of the calculated R_{id} (20 kΩ) compared to the actual $R_{id} = 29.5$ kΩ. Here again the underestimation of β is the cause of such a deviation.

The multistage amplifier can serve as a crude op-amp. In any op-amp, the larger R_{id} and the larger A_d the better the op-amp's performance.

In op-amps, it is desired that the output resistance be as small as possible. Figure 6.44 presents the schematic for AC simulation for R_{out} measurement of the

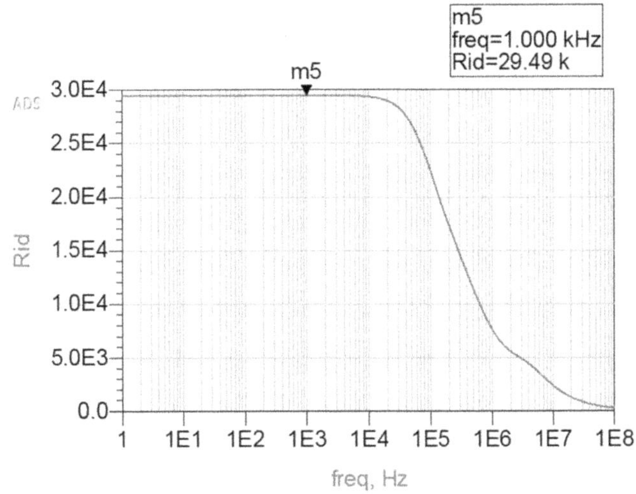

FIGURE 6.43 R_{id} frequency response.

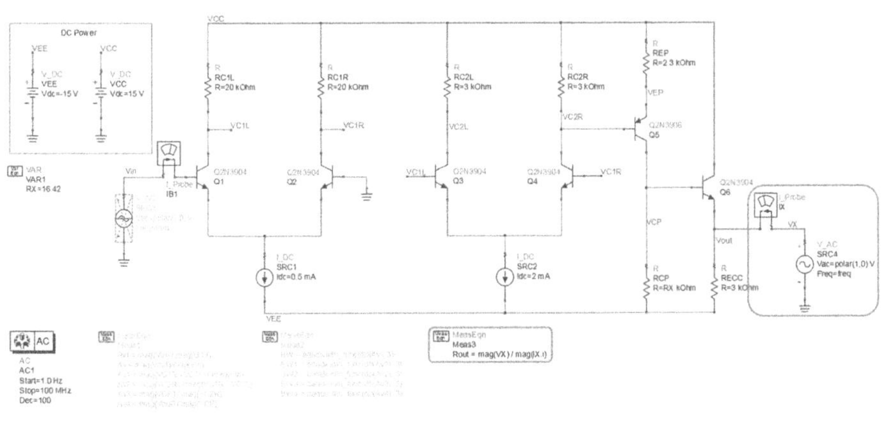

FIGURE 6.44 Output resistance measurement of the multistage amplifier.

amplifier. Figure 6.45 presents the results for R_{out}. The mid-band R_{out} is found to be around 85 Ω. We skipped the theoretical R_{out} derivation. We just note that the dominant influence on R_{out} comes from R_{CP} divided by β (and not from R_{ECC}).

6.7 ADS TIPS AND TRICKS 6A: PARAMETER OR VARIABLE TUNING

ADS has the capability to allow the "tuning" of parameters and variables. During a tuning session, the user can vary variables and parameters manually and observe the outputs. Depending on the simulation, the tuning updates of the output measurements are "real time", meaning as we modify the values, the data display updates at the same time. Let us illustrate how tuning works, solving the output DC offset voltage adjustment, done a little earlier.

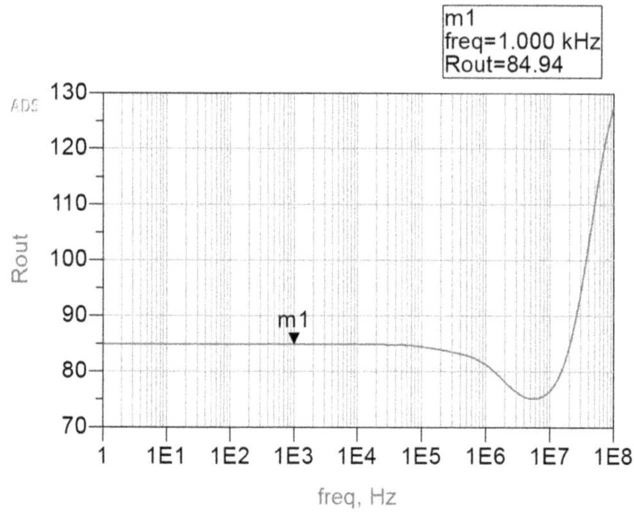

FIGURE 6.45 Multistage amplifier's output resistance results.

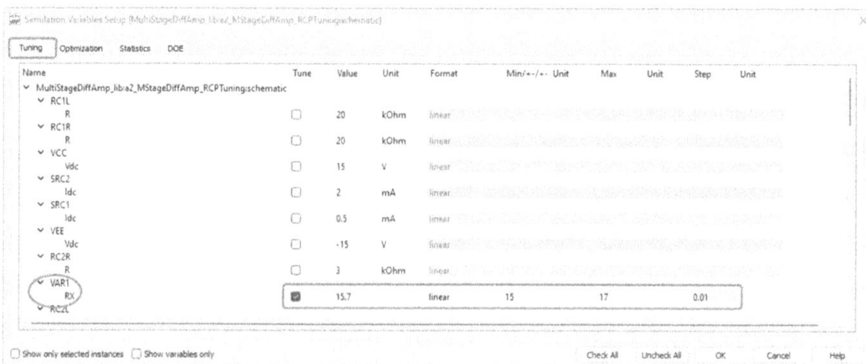

FIGURE 6.46 Simulation variables table (demonstrated earlier in conjunction with yield and sensitivity analyses).

Step 1: Set tuning variables.

In the simulation variables table, we select the tuning tab (Figure 6.46).

We find the variables or parameters we wish to tune and enable them.

We then set the format, min, max, and step.

Note, alternatively, we could set tuning under the properties of each parameter and variable (Figure 6.47).

The schematic will now show the {t} notation (Figure 6.48).

Step 2: Need to make sure that we have a simulation that runs, and that we can plot or list the desired measurement results, such as

freq	Vout
0.0000 Hz	-659.9 mV

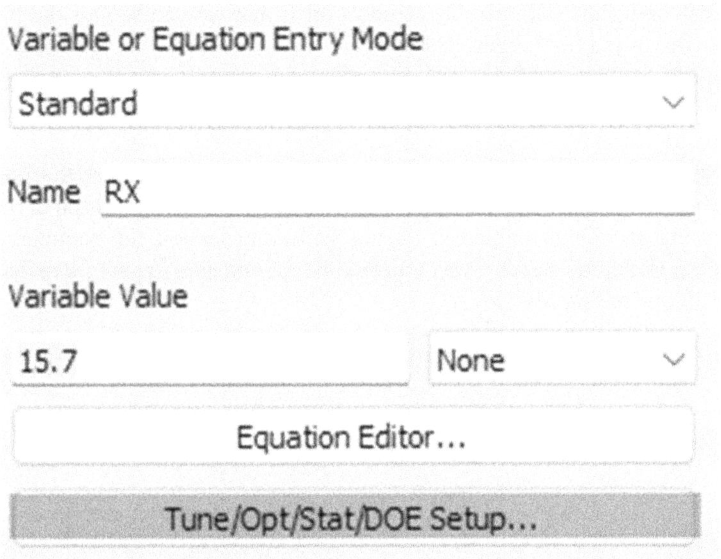

FIGURE 6.47 Alternate tuning designation, going parameter by parameter.

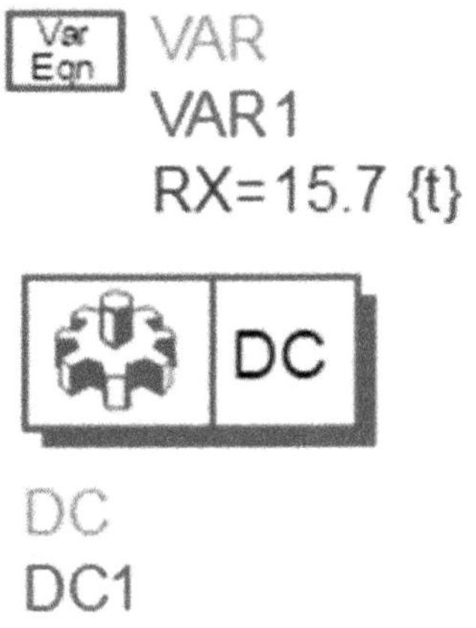

FIGURE 6.48 Tuning designation symbol {t} showing within VAR.

Step 3: On the schematic, from the menu, we select Simulation >>Tuning … or the tuning icon from the toolbar.

Step 4: The tuning session will commence. In addition to the tuning window, ADS will bring to the forefront the data display.

We suggest sizing the data display window such that we can see all the plots and tables that are of interest (Figure 6.49).

While our example includes just one table and one parameter, multiple parameters can be tuned during the same session and multiple measurements assessed (Figure 6.50).

During the tuning session, if we select the schematic, we will notice the crosshairs. This allows us to add more parameters or variables to the tuning session in real time.

Depending on what is happening during the tuning session, for example, we cannot meet a certain requirement, we can add other parameters to gain more degrees of freedom for our tuning session.

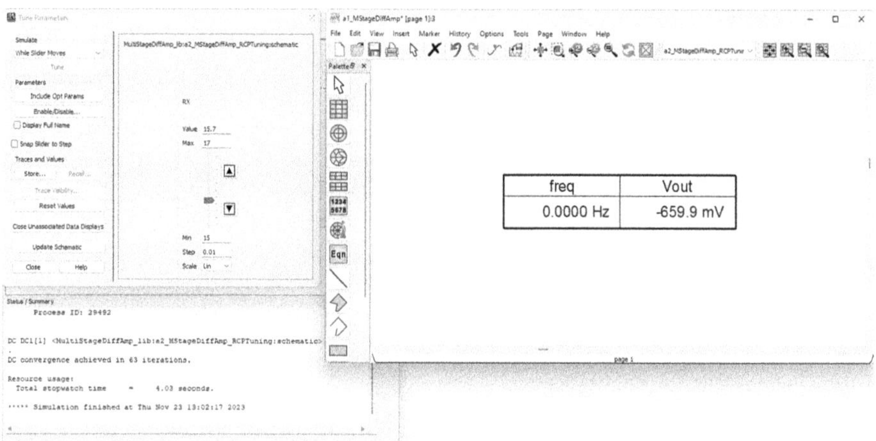

FIGURE 6.49 Tuning session begins.

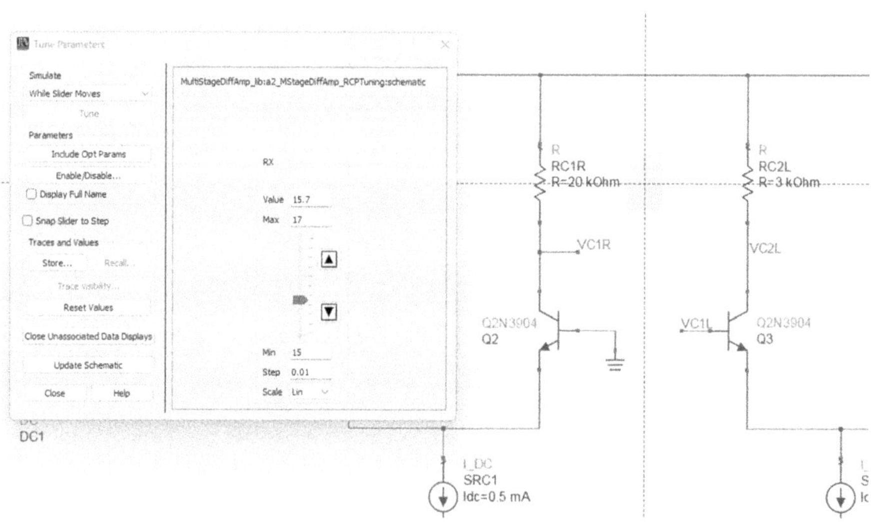

FIGURE 6.50 Tuning session continues. Optionally, one may add more variables and parameters in real time from the schematic.

When we have a slow simulation, we may not want the real-time update of the data display. We can select other options under the simulate pull-down in the Tuning window.

During our tuning session, we have access to change the max, min, step, and scale (aka Format).

We can modify the value by either typing in the new value in the Value field, clicking on the up/down arrows for the parameter, or clicking and sliding the blue arrow (Figure 6.51).

Once we are satisfied with the results obtained, we can ask the tuning window to "Update Schematic" and "Close" the session.

Further information on tuning is available in the ADS documentation [Keysight1].

In Figure 6.52, we show a few steps of our tuning to set the lowest offset within a 0.01 kΩ step size. The results are the same as those obtained earlier by DC parameter sweeping.

Tuning is used frequently when one, or several parameter(s), is needed. It is more efficient than sweeping when more than one parameter is needed to accomplish a requirement.

6.8 ADS TIPS AND TRICKS 6B: USING ADS MARKERS IN EQUATIONS

In the discussion about the multistatge gain (around Figure 6.42), a roundoff inaccuracy was noted. The resulting, manually calculated gain was $A_v = 11{,}764$, whereas we expected $A_v = 11{,}780$ from the total gain marker on the A_v plot. This 0.14% error

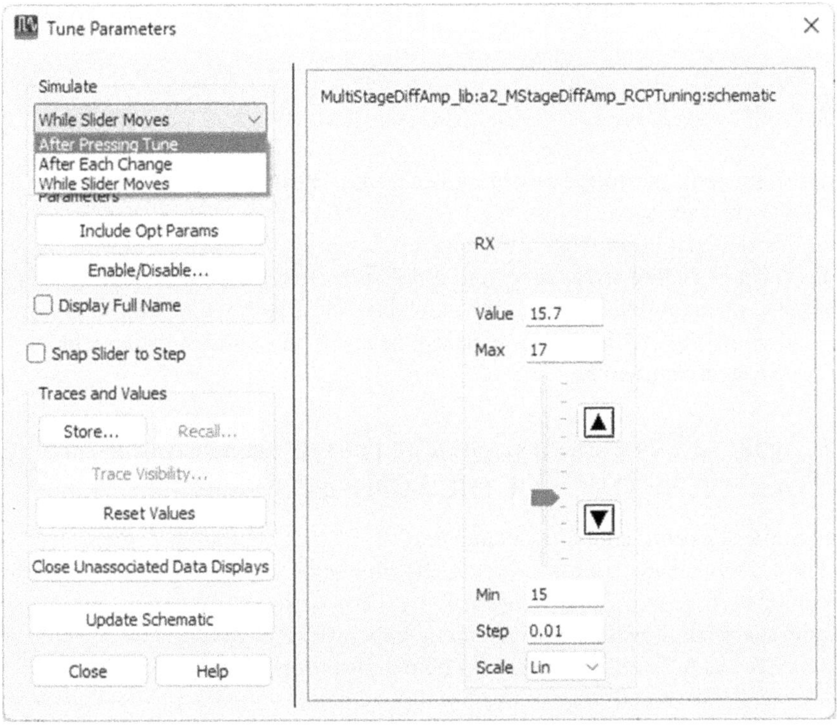

FIGURE 6.51 Modifying values of parameters.

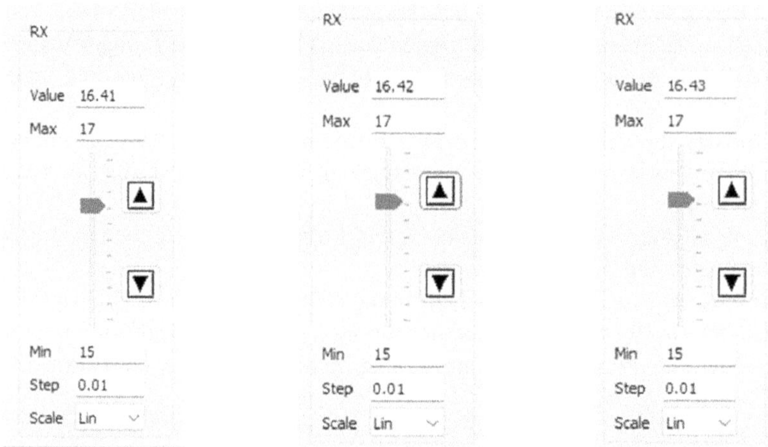

freq	Vout
0.0000 Hz	-8.443 mV

freq	Vout
0.0000 Hz	713.2 uV

freq	Vout
0.0000 Hz	9.868 mV

FIGURE 6.52 Demonstration of a few steps in the tuning of RX to obtain small V_{out} offset.

Eqn AvT_calc = m1*m2*m3*m4

freq	AvT_calc
1.000 kHz	11780.316

FIGURE 6.53 Demonstration of using markers for full ADS accuracy calculations.

is not significant, but to the new designer it may cause a pause and a feeling like the tool is not accurate. To use the full available accuracy, let us use the marker values in an equation in ADS. This ensures the use of the full accuracy of the tool and reduces roundoff errors. Creating a data display equation and calling it A_vT_calc made of the product of the four individual stage gain makers m_1 thru m_4, i.e. A_vT_calc = m_1*m_2*m_3*m_4. We then placed the result on a column listing table to see the results as in Figure 6.53.

6.9 ADS: TRANSIENT SIMULATION OF THE MULTISTAGE AMPLIFIER: STUDY OF THE AMPLIFIER SWING

Section 6.9 is a continuation of Section 6.6.

The reader may question why we do HB simulation and not transient simulation. HB simulation gives us the advantages that (i) it is steady state, (ii) time domain waveforms are still available using ts() function, (iii) HB produces DC results, and (iv) it allows us to assess quantitatively output distortions.

We recall that (based on assuming $\beta = 100$) we obtained (theoretically) that the single-ended differential gain of the entire amplifier is $A_V = 8520$. In ADS AC Simulation we measured an overall gain of $A_V = 11{,}780$. We argued that the underestimation of the β values caused a large deviation between the manual calculation results and the exact simulation results.

We further observed that the gain A_1 is influenced strongly by the floating load R_{id2} (which is strongly β dependent). The need to increase the offset adjustment resistor R_{CP} causes A_3 to slightly increase.

In HB and transient simulation, the size of the input amplitude matters.

Let us initially choose V_{in} with an amplitude of 1 mV. Is it small enough to avoid amplifier output voltage clipping distortion?

Theoretically, the amplitude of V_{out} is expected to be $(1\,mV)\cdot 8520 = 8.52\,V$. According to ADS, the expected V_{out} amplitude is 11.78 V. Either way, is there any expected clipping with supplies of ±15 V? Intuitively not.

Time domain simulation results of the schematic shown in Figure 6.54 are presented in Figure 6.55.

Recall the measured gains of each stage: $A_1 = 32.1$, $A_2 = 54.4$, $A_3 = 6.75$, $A_4 = 0.998$. Clipping depends on the output voltage amplitude at each stage, compared to the "swing" of the stage.

Note in Figure 6.55 that the output of Stage 2 is not clipped. The output of Stage 3 is.

Let us do some approximate calculations, keeping in mind that V_{CE} must be less than V_{CC} and larger than V_{BE}.

Stage 1: $V_E = -0.7\,V$, $V_{CEmax} = (10 + 0.7) + 0.0321$ → plenty of "swing".

Stage 2: amplitude $(V_{cout}) = 1\cdot 32.1\cdot 54.4 = 1.746\,V$, $V_E = 10 - 0.7 = 9.3\,V$, $V_{CEmax} = (12 - 9.3) + 0.746 = 4.446\,V$, $V_{CEmin} = (12 - 9.3) - 1.746 = 0.954\,V$. Still enough "swing" on the lower side, but not much.

Stage 3: amplitude $(V_{cout}) = 1\cdot 32.1\cdot 54.4\cdot 6.75 = 11.787\,V$, $V_E = 12.7\,V$, $V_C = 0.7\,V$, $V_{EC\,max} = (12.7 - 0.7) + 11.787 = 23.787\,V$, $V_{ECmin} = 12 - 11.787 = 0.213\,V$ → clipping on the upper part. Need to reduce the input's amplitude.

With V_{in} amplitude of 0.1 mV, Stage 3 $V_{ECmax} = 12 + 1.1787 = 13.1787\,V < 15\,V$.

The deeper reason why we chose HB Simulation over Transient Simulation will become clear as we begin to discuss design optimization.

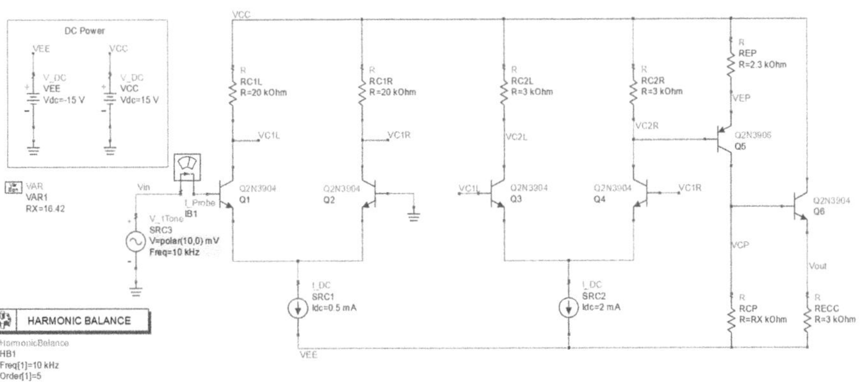

FIGURE 6.54 Schematic of the multistage amplifier in HB simulation.

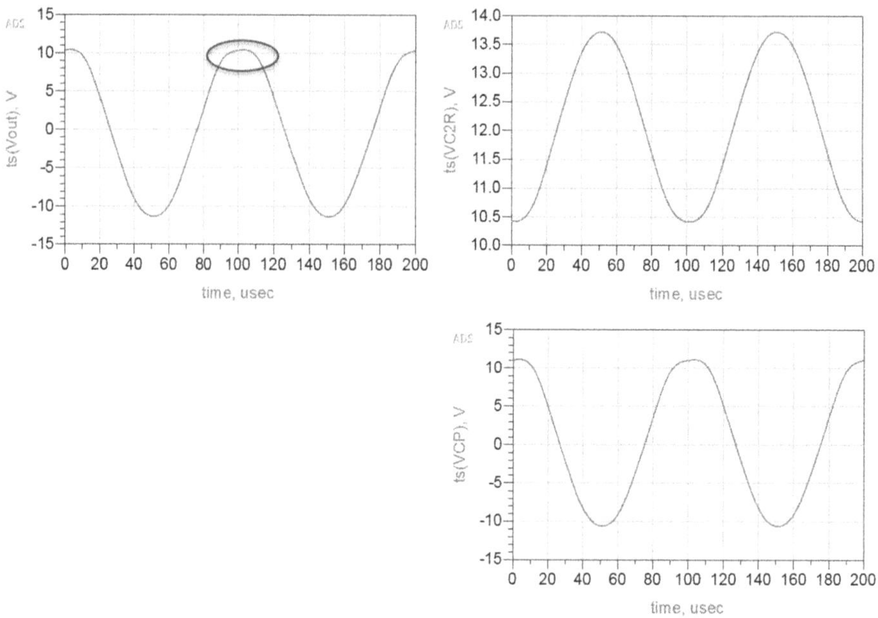

FIGURE 6.55 Output voltage of stages 2 and 3 and 4, with an AC input amplitude of 1 mV.

6.10 PREPARATION FOR REDESIGN OF THE MULTISTAGE AMPLIFIER EXAMPLE; GENERAL INTRODUCTION TO OPTIMIZATION; GENERAL INTRODUCTION TO THE ADS OPTIMIZER

In any design, we need to determine the design parameters and the design goals.

Example 6.3 (continued)

In the original multistage amplifier circuit (Figure 6.33), and with the ADS activities performed in Section 6.6, the two currents are set to $I_1 = 0.5$ mA and $I_2 = 2$ mA, and for these choices we had to adjust $R_{CP} = 16.42$ kΩ in order to obtain an acceptably small output voltage offset. We also obtained $A_V = 11{,}780$ and $R_{id} = 29.5$ kΩ, which were both better than our manual expectations and calculations. The larger, the merrier, as the multistage amplifier is essentially an op-amp.

Can we further "improve" the op-amp by using I_1, I_2 and R_{CP} as design parameters? Does the improvement, whatever it is, come for free, or are there some design tradeoffs? For instance, we may be able to increase both A_v and R_{id}, but maybe the overall bandwidth and swing may become smaller?!

Let us do an amplifier's performance improvement in "baby steps". Say that the desired new values for A_v and R_{id} are A_v to be around 12,500–13,000 and R_{id} to be above 30 kΩ. Of course, we continue to want all BJTs to be in active mode and the output DC offset to be within 10 mV of 0.

If the amplifier improvement design is done manually, assisted by ADS, some preparation may be needed to get a better insight into the set of design parameters and their effect on the design goals.

Here are some intuitive observations (that can be easily verified via AC simulation of the multistage amplifier):

- If I_1 becomes smaller, R_{id} becomes larger, but the gain A_1 becomes smaller, and vice versa: If I_1 is made larger, A_1 increases, but R_{id} decreases.
- For any changes in I_1 that keep the first-stage BJTs in active mode, there is no need to modify the zero offset resistor R_{CP}.
- Increasing I_1 too much, without lowering R_C (of Stage 1), may drive Stage 1 BJTs to saturation, which is totally unacceptable.
- If we change I_2, we always need to modify R_{CP}. If I_2 decreases, we have to increase R_{CP}, and if I_2 increases, we need to reduce R_{CP}.
- If I_2 decreases, A_1 increases because R_{id2} increases.
- If I_2 decreases, A_2 decreases.
- If I_2 decreases, the overall gain A_v can go either way. It often increases as both A_1 and A_3 increase, even though A_2 decreases.
- So, if we slightly lower I_1 and I_2 and increase R_{CP}, we tend to serve the goals of increasing both R_{id} and A_v.

Getting a better feeling, in what direction (up or down) each design parameter should go, may be helpful in manual design. It is less critical in a design that uses an optimizer.

Optimization is always present in any engineering problem. Whenever we design an amplifier to meet a certain gain and input resistance specifications, in a real sense we are trying to "optimize" the design.

Recall the sweeping of R_{CP} that we did earlier to reduce the DC offset. The process we followed was to find the value of R_{CP} that met our *goal* of an output DC offset voltage of $0 \pm 10\,\text{mV}$, or closer to 0.

There are many situations involving multiple goals, many design parameters, and multiple constraints. Such design optimization can be a daunting task to do manually.

Let us touch briefly on the mathematics of optimization. A complete treatment of such a subject is beyond the scope of this book. Many well-written textbooks on optimization were written, such as [Belegundu1], [Yang1].

An optimization problem can be generically cast as $\min[f_i(x)]$ where $i = 1, 2, \ldots, M$ and x is any real vector of length N. The minimization is subject to K equality constraints $\Phi_k(x) = 0$ for $j = 1, 2, \ldots, K$ and L inequality constraints $\Psi_l(x) \leq 0$ for $l = 1, 2, \ldots, L$. [Yang1].

The real-valued elements of vector x are called **design** (or decision) **variables**. Design variables could be continuous, discrete, or a mix of the two.

These are the optimizable values in the problem.

The functions $f_i(x)$ are called the **objective functions**, and in the case of $M = 1$, there is only a single objective. The objective function is sometimes called the **cost function** or **error function**. The equality constraint functions Φ_k and the inequality constraint functions Ψ_l are functions of the design vector x.

If we write the inequalities the other way (≥ 0), we can also formulate the objectives as a maximization problem. Minimization of $f_i(x)$ is the same as maximization of $-f_i(x)$.

Let us next discuss the implementation of optimization in ADS. Figure 6.56 lists what is needed to execute an optimization in ADS [Keysight1].

As the optimizer strives to minimize the error function, let us introduce how ADS defines the error function [Keysight1]. The error function is based on **residuals**

- **Simulation Controller** – make sure the simulation runs and produces reasonable results.
- **Enable components** for optimization – these are the design variables mentioned in the introduction.
- **Goal(s)** – can be MeasEqn, voltages, or currents.
- **Optim controller** – executes the optimization and specifies things needed for the optimization like optimizer type, the error function.

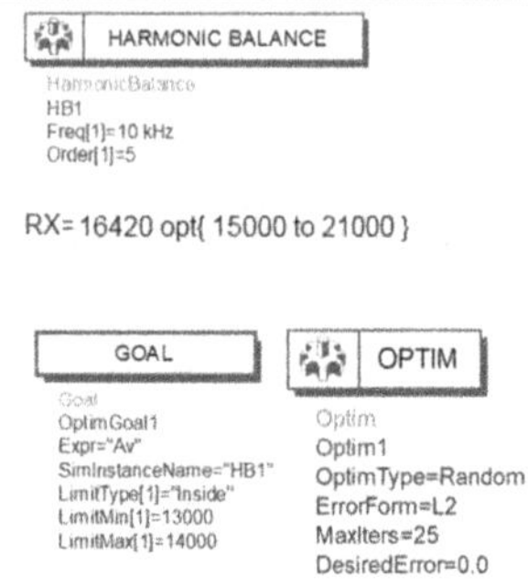

FIGURE 6.56 ADS optimization elements.

ADS has several error function (EF) formulations

Least Squares (L2): Each residual is squared, and all terms are then summed.

Negated Least-Squares: drives values to their extreme – effectively maximizes the error function. The goal is to find a worst typical response for a given set of parameters.

Minimax: attempts to minimize the largest of the residuals. This tends to result in equal ripple responses.

Minimax L1: is similar to L1 but cannot be less than zero, so it accounts for the most severely violated cases.

Least Pth: The Least Pth EF formulation is similar to L2, except that instead of squaring the residuals, it raises them to the Pth power with P=2 , 4 ,6 etc.

NOTE: default is to equally weight and normalize all goals.
Optimizer will continue simulations until the maximum iterations (MaxIters) is reached or the error function (DesiredError) reaches the value specified (typically zero).

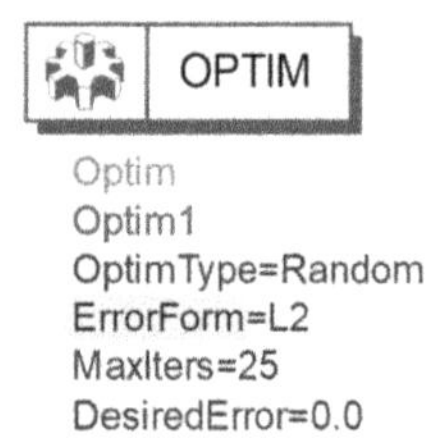

FIGURE 6.57 Error function choices for the OPTIM block.

$r_i = W_i |s_i - g_i|$ where r_i is the ith residual, s_i is the ith simulated response (e.g., simulated $A_v = 95$), g_i is the desired response for the ith goal (e.g. goal i $A_v = 100$) and W_i is the ith weighting factor. The larger W_i the greater the contribution of the ith goal to the error function value. The weight factor prioritizes goals in a multiple-goal optimization. See Figure 6.57 for several tentative error functions and Table 6.1 for a list of optimizer types [Keysight1].

As certain optimizers search for the global goal, miss or getting-stuck events may happen. Let us illustrate it by looking at minimizing a 1D function (Figure 6.58).

An optimizer approaching from the left may believe local_min1 is the minimum since to the right of it the values begin to increase.

In Figure 6.58, the same can be said if approaching from the right for local_min3. An optimizer approaching from the left may get over the first hump past local_min1, only to get stuck at local_min2, since the values to the left of it rise so fast. A best-in-class optimizer will find the global_min, which is, after all, the purpose of the optimization i.e. **find the global minimum**.

To provide some insight into search methods, let us briefly discuss random, gradient and hybrid search methods (Figure 6.59).

TABLE 6.1
Optimizer Types

Optimizer	Search Method	Error Function
Random	random	least-squares L2
Gradient	gradient	least-squares L2
Random minimax	random	minimax LI
Gradient minimax	gradient	minimax LI
Quasi-Newton	quasi-Newton	least-squares L2
Least Pth	quasi-Newton	least P^{th}
Minimax	mini-max	mini-max
Random max	random	negative L2
Hybrid	random/gradient	least-squares L2
Discrete	Discrete	least-squares L2
Genetic	n/a	least-squares L2
Simulated annealing	n/a	least-squares L2

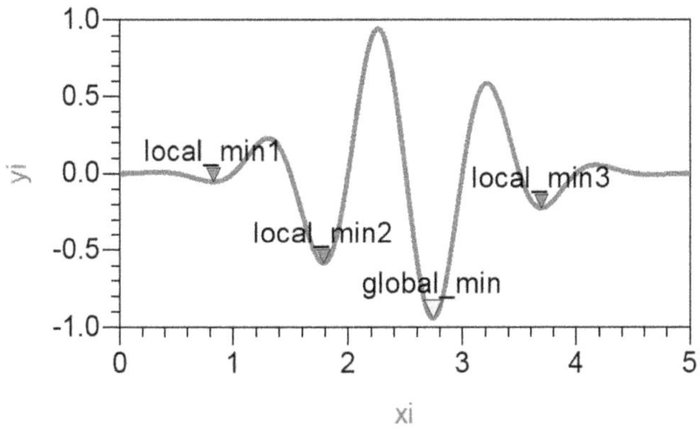

FIGURE 6.58 Illustration of 1D global and local minima.

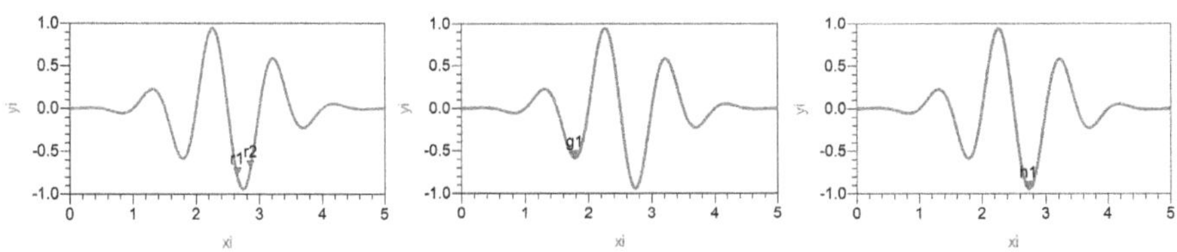

FIGURE 6.59 1D illustration of random (left), gradient (middle), and hybrid (right) search methods.

Random search, illustrated in Figure 6.59 (left), often gets us close to the global minimum, but due to its random nature, it does not quite get to the global minimum.

Note that random is not totally random. It uses some adaptation that helps it move closer to the goal.

Gradient search, illustrated in Figure 6.59 (middle), continues to move until the numerical derivative is zero, or very close to zero. Gradient can often get stuck at a local minimum and may never find the global minimum.

Hybrid search, illustrated in Figure 6.59 (right), is a combination of random and gradient searches. This method will start out with random to try to get close to the global minimum and then switch to gradient to reach the global minimum. It may interactively switch between the two search methods, as needed. For instance, after gradient gets the search to be stuck in a local minimum, a random jump releases the search from that local minimum.

Recommendation to the user to try search methods in the given order:

1. Always start with Random Optimizer.
2. If random struggles, or fails, try gradient.
3. If gradient is slow, or fails, try hybrid.
4. If needed, try other optimizers. See Table 6.1.

Using optimization in engineering problems is not always an exact science and is not guaranteed to succeed every time. A certain effort is always required.

As in many situations, there is no substitute for practice and experience. The more you try a certain type of optimization problem, with the same type of technology, the better you become at optimizing its performance.

The optimizer is not "magical" and is bound by the limitations of the problem. For example, asking an optimizer to find a gain, of a common emitter amplifier, greater than $A_v = g_m r_o$, will never succeed, since this is the maximum theoretical gain for a CE amplifier (a.k.a. **intrinsic gain**). In other words, the optimizer still requires users to understand the design theory and underlying design limitations. The preparation for the "improved design", done earlier, is not just for manual design.

Another common reason that the optimizer may fail to find any solution is that one or more design variables may have a mistaken search limit (or limits).

Optimization also requires an uncertain level of luck sometimes.

You may discover that, whenever doing experimentation (trial and error), that a certain combination of steps works best. Write it down before it gets lost, or you forget.

6.11 ADS: THE ADS OPTIMIZER: SIMULATION CONTROLLER, COMPONENTS ENABLING, SOLUTION GOALS, THE OPTIM CONTROLLER, ERROR FUNCTIONS, OPTIMIZATION VARIABLES, OPTIMIZATION TYPES AND SEARCH METHODS, GOALS MEASUREMENT EQUATIONS USING HARMONIC BALANCE SIMULATIONS, SETTING OF GOALS CONSTRAINTS AND WEIGHTS, THE OPTIMIZATION COCKPIT AND SOLUTION APPROACH DEMONSTRATIONS

Example 6.3 (continued)

The simplified version of Sedra and Smith's multistage differential amplifier has been studied in ADS DC, AC and HB simulations. The results for A_v, R_{id}, and the offset adjustment can be considered the "**benchmark**" (or "baseline") results with respect to which we wish to improve:

We wish to improve the differential gain from $A_v = 11{,}780$ to 12,500–13,000, and the differential input resistance R_{id} from 29.5 kΩ to anywhere above 30 kΩ. Baby steps indeed. We continue to adjust the zero-input DC part of V_{out} to be within ±10 mV of zero, and we want all BJTs to stay in active mode.

We keep in mind that ADS optimizer also works well with large deviations from the benchmark results and whenever there is no benchmark at all.

A_v and R_{id} goals can be monitored in either AC or transient simulations. The zero offset and active mode goals are monitored in DC simulation. Shall we run the optimizer while, say, both AC simulation engine and DC simulation engine are simultaneously active? How is the error function computed in such a mixed-mode simulation?

Indeed, the optimizer engine does not mind having AC simulation and DC simulation being both enabled. The search, though, may be slower and more computationally intensive. It would be better, though, to have only one simulation engine running along the optimization engine.

This is how we come to meet the interesting ability of **harmonic balance simulation** to **provide a unified platform to achieve goals and constraints of multiple domains.**

AC and DC, in our case. See the measurement equations of Figure 6.60 that will soon be explained. Another amazing feature of harmonic balance simulation is its ability to drastically reduce the need for elaborate DC constraints measurements. Keeping all the transistors in active mode normally implies inequality DC constraints for each transistor, checking its base-collector and base-emitter junction voltages. We know, however, that even if one BJT glitches to cutoff or saturation mode, $V_{out}(t)$ of the entire amplifier surely becomes drastically distorted. Therefore, **monitoring the largest side harmonics (typically harmonics 2 and 3) with respect to the main harmonic in HB simulation is indirectly equivalent to monitoring, in DC Simulation, all BJTs for active mode** (or saturation mode in amplifiers built with MOSFET transistors)**.** This too is shown in Figure 6.60.

Note that in Figure 6.60 an input voltage magnitude of 0.1 mV is taken, as per our earlier discussion on design swing limits.

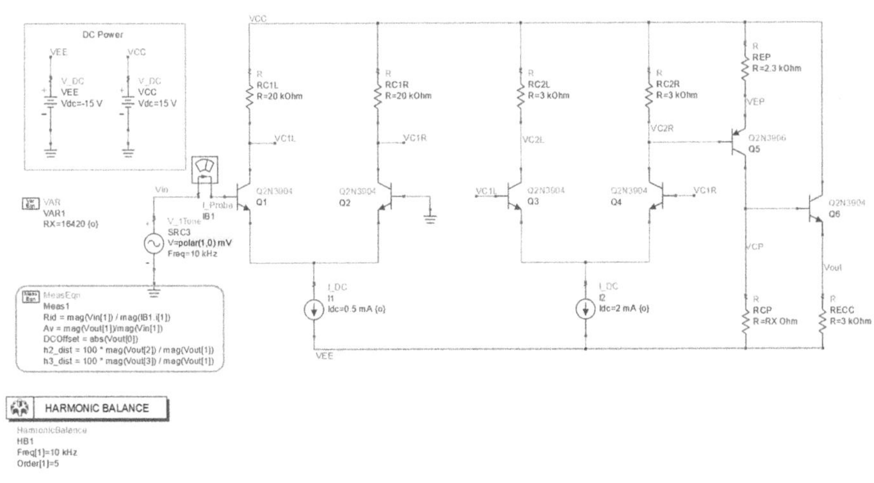

FIGURE 6.60 Preparation for running HB simulation together with the optimizer.

Figure 6.61 is a magnified view of the MeasEqn employed in the schematic of Figure 6.60.

$\mathbf{R_{id}}$ is the "magnitude of the input impedance", which we often call "input resistance", since at these low frequencies the imaginary part of the input impedance is usually negligible. Note the use of the [1] index, i.e., we take the fundamental frequency values of voltage (and current) harmonic balance results for this calculation.

$\mathbf{A_v}$ is the amplifier's total voltage gain. Again, this needs harmonic1 information. Hence [1] index.

DCOffset is the DC offset voltage. Notice the use of index zero. We use the abs() function. The reason is that the optimizer can only deal with real value numbers and $V_{out}[0]$, from HB results, is complex by definition, even though it can only be positive or negative.

h2_dist and h3_dist are the second and third harmonic distortions in percentage (%). This is why we use [2] and [3] indices. Asking for small side harmonics percentage (say 1%) is equivalent to asking the circuit to be linear.

Remember that **goals in ADS are the constraints of the optimizer.** If we do not track these constraints, the optimizer may end up with a solution that exceeds the constraints. In other words, the optimizer must keep track of all the goals for a successful optimization.

Table 6.2 shows the values of all the goals before the optimization process starts. We see that the DC offset (being less than 1 mV) and distortion results (being less than 2%, even also 0.1%) meet the requirements. A_v and R_{id} do not, as will be seen when the goals are formulated.

In Figure 6.62, we show the Simulation Variables table, in which we have to enable the optimizable variables. The enabled variables are marked by the symbol {o}, as in Figure 6.60.

Bringing in the Simulation Variables table, we then select the Optimization tab (Figure 6.62). Then we enable the variables that will be optimized and configure their limits. See Figure 6.63 for the I_1 variable.

The five goals of our optimization example are presented in Figure 6.64.

We shall discuss the details of how to set up each of the goals.

```
Meas Eqn  MeasEqn
          Meas1
          Rid = mag(Vin[1]) / mag(IB1.i[1])
          Av = mag(Vout[1])/mag(Vin[1])
          DCOffset = abs(Vout[0])
          h2_dist = 100 * mag(Vout[2]) / mag(Vout[1])
          h3_dist = 100 * mag(Vout[3]) / mag(Vout[1])
```

FIGURE 6.61 Measurement equations needed for the optimization.

TABLE 6.2
Goal Values Based on the Baseline Results

DCOffset (V)	A_v	R_{id} (kΩ)
336.2 u	11.76k	29.35k
h2_dist	h3_dist	
0.037	0.008	

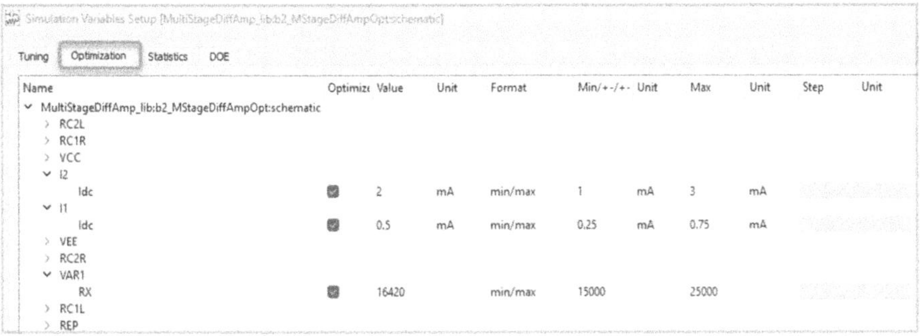

FIGURE 6.62 All simulation variables, including the three optimizable variables.

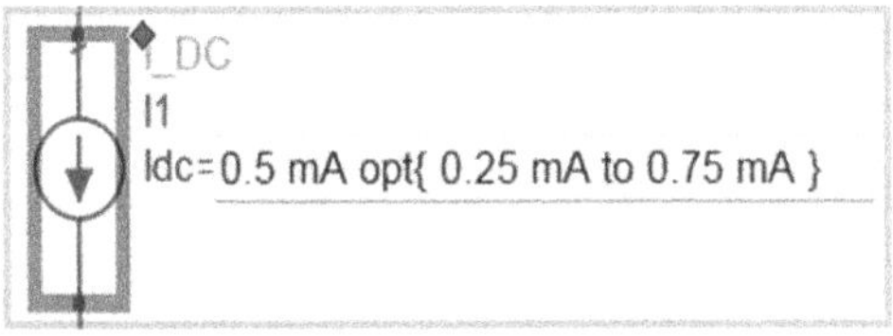

FIGURE 6.63 The resulting optimizable I_1, with its limits.

FIGURE 6.64 The five goals (constraints) of the multistage amplifier example, planned to be solved with the optimizer.

Goals and the optimization controller are found in the Optim/Stat/DOE palette. See Figure 6.65 (left). Double-clicking on the GOAL block brings the dialog box (Figure 6.65 (right)) for setting all the needed GOAL information.

Goals require:

1. Expression – a valid ADS expression or MeasEqn variable
2. Analysis – the name of the analysis that will evaluate the expression
3. Weight – weighting factor
4. Sweep variables – swept values used in the evaluation of the expression For use when we have swept analyses like AC or transient analyses. Can be used with parameter sweeps.
5. Limits – the requirement limits (optimization constraints)

The limit types are:

1. Greater than: Expression > Min
2. Less than: Expression < Max

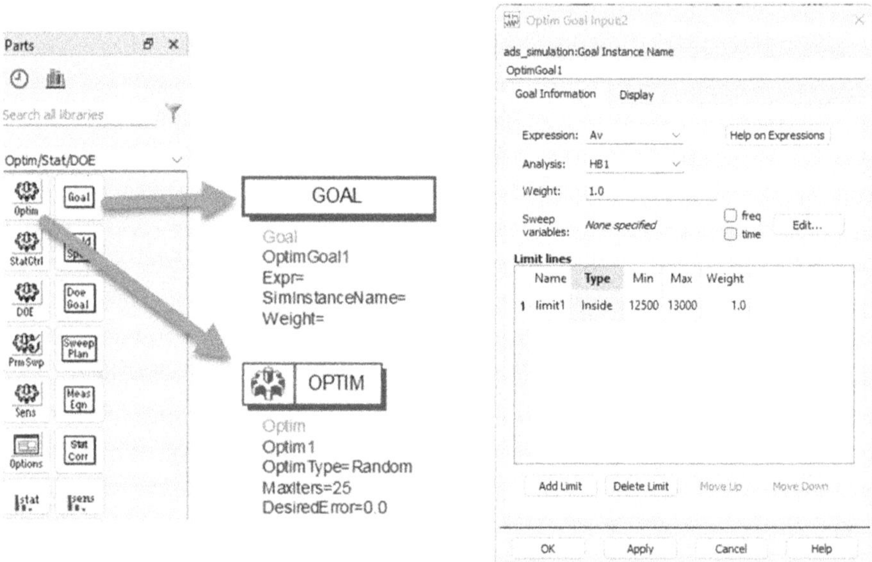

FIGURE 6.65 (Left) Palette location for GOAL and OPTIM. (Right) Dialog box for creating GOAL details.

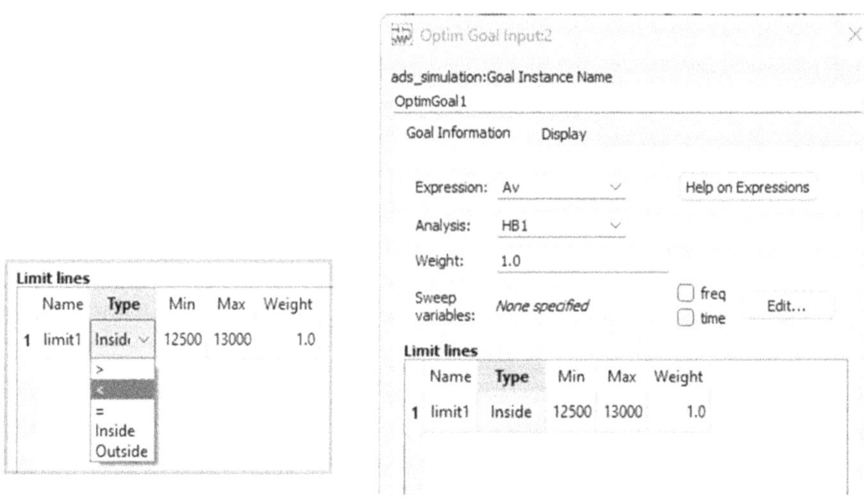

FIGURE 6.66 Assignment of Goal1 limits.

3. Equal: Expression = Min = Max
4. Inside: Min < Expression < Max
5. Outside: Expression < Min OR Expression > Max

Figure 6.66 shows the limits assignment for Goal 1.

Figure 6.67 is a summary of all the inequality conditions that are implemented in our problem's goals.

The Goals we have implemented are:

- OptimGoal1: 12500 < Av < 13000
- OptimGoal2: Rid > 30000
- OptimGoal3: DCOffset < 10e-3 (10 mV)
- OptimGoal4: h2_dist < 2 (%)
- OptimGoal5: h3_dist < 2 (%)

❖Remember the Goals are the requirements that we intend to have our design meet.

FIGURE 6.67 Summary of all goals' inequality conditions.

It is a common practice to designate goal values, and values of optimizable variables, in the {V, A, Ω} system, as done for R_{id} and DCOffset.

{V, A, Ω} is not, however, a strict rule. See for instance (in Figures 6.62 and 6.63) the units of I_1 and I_2. Both are in mA.

Finally, as a last preparation step, let us discuss the OPTIM controller. In Figure 6.68, we see the OPTIM block used in the upcoming optimization solution and its dialog box, obtained by double-clicking on the OPTIM block.

Under the Setup tab, one can select the Optimization Type from the pulldown menu.

Under the Optimization Goal and Variable Setup section, the user may decide not to use all goals or all optimizable variables.

The above ADS feature allows for maximum flexibility. For some problems, it is convenient to break the optimization by variables or goals. The use of this feature is considered fairly advanced, and we will not discuss it for now.

Stopping Criteria:

The optimization will stop under two conditions:

1. The number of iterations has been met
2. The desired error has been met

Note: some optimizers will stop if they are not making progress in minimizing the error.

Under the Parameters (see Figure 6.69), we recommend keeping the Output Data section selection as shown here. We only deactivate portions of it if too much data is being saved.

Further options for data control can be found in the Output Data Control section.

Normalize goals automatically: Provides a built-in method to make the contributions from all associated goals equal to the error function (more on it very soon).

Set best values for parent optimization: Used for the advanced optimization features, including swept optimization (studied in chapter 7) and programmable optimization.

Enable optimization cockpit: Used to decide the optimization running mode: **optimization cockpit** mode (real-time communication mode), or batch mode.

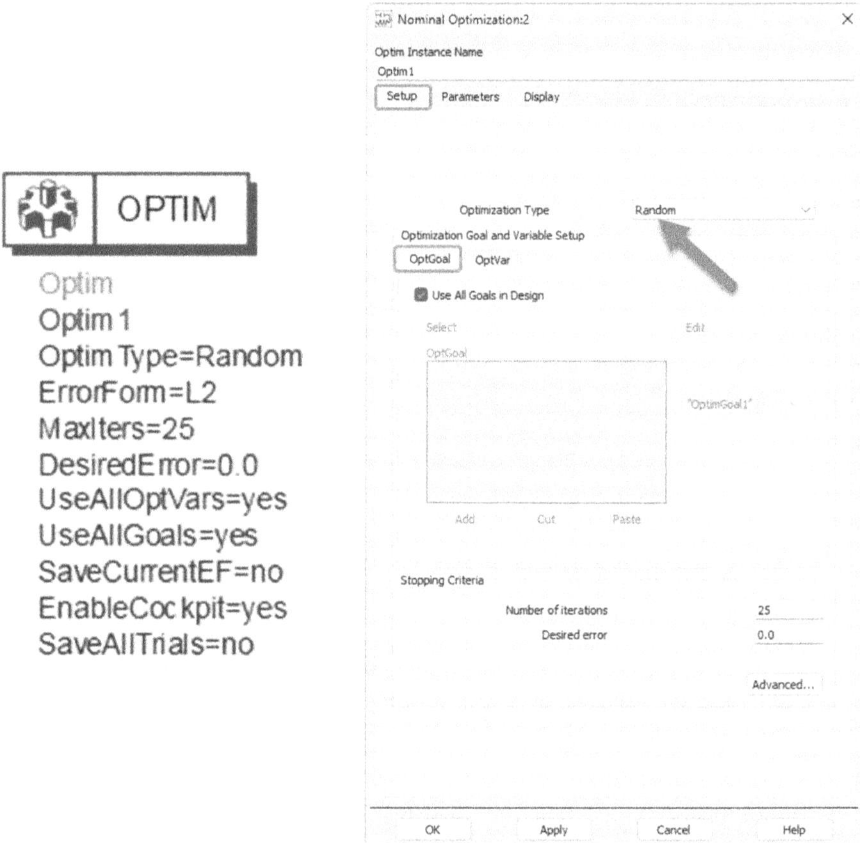

FIGURE 6.68 The OPTIM block and its properties dialog box.

By default, Enable Optimization Cockpit is selected, and the cockpit will pop up when one clicks the Optimize button. We can use the optimization cockpit to monitor and control the optimization process (more will be shown later).

Special notes on *Normalize goals automatically*, shown in Figure 6.70, (notes are taken from the ADS documentation [Keysight1]).

Useful when there is more than one Goal component associated with the optimization controller. The desired performances can be of different orders of magnitude, which can result in the error function being biased to the responses with larger values. To prevent any one goal from dominating the error function, use either of the following methods:

1. Manually set up the weight factor correctly for each Goal component according to the performance values.
2. Let the weight factor for each Goal component be set automatically to the appropriate value for the performance level.

It is strongly suggested that one selects the "Normalize goals automatically" check-box, whenever the optimization goals use the default weighting factor. The effect

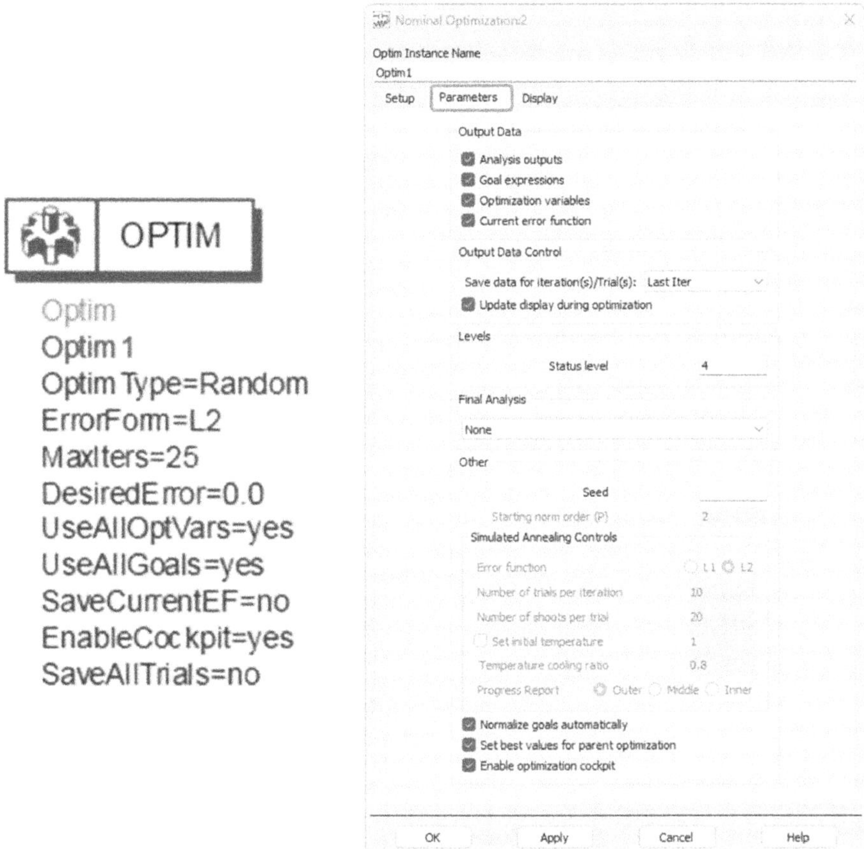

FIGURE 6.69 Selection of optimization data parameters.

of this option is to produce an internal weighting factor for each goal. This prevents any single goal from dominating the error function.

So, here we are in Figure 6.71, ready to run the optimization, by pressing the "run optimization button"  from the toolbar.

This will bring up the **Optimization Cockpit** used to visualize and control the optimization. This is a real-time cockpit, and the widgets seen, all update dynamically as the optimization executes.

Figure 6.72 is a view of the cockpit as the optimization begins.

Elapsed time is zero, the design variables are all in their initial values, Error is maximal for every one of the goals, and the initial Error history is displayed.

The zoomed-in view of the design variables zone (Figure 6.73) reveals that some controls are available in real time to the user. As seen in Figure 6.73 (left), one can pause the optimization to perform changes, and subsequently continue.

Let us elaborate on the various parts of the optimization Cockpit.

Once the OPTIM stops, one can run a simulation to inspect interim results. "Update Design ..." will update the schematic with latest values for the optimizable parameters. One can store, recall and revert to different states of the optimizer. "Options" gives us some more control on how to visualize things in the cockpit.

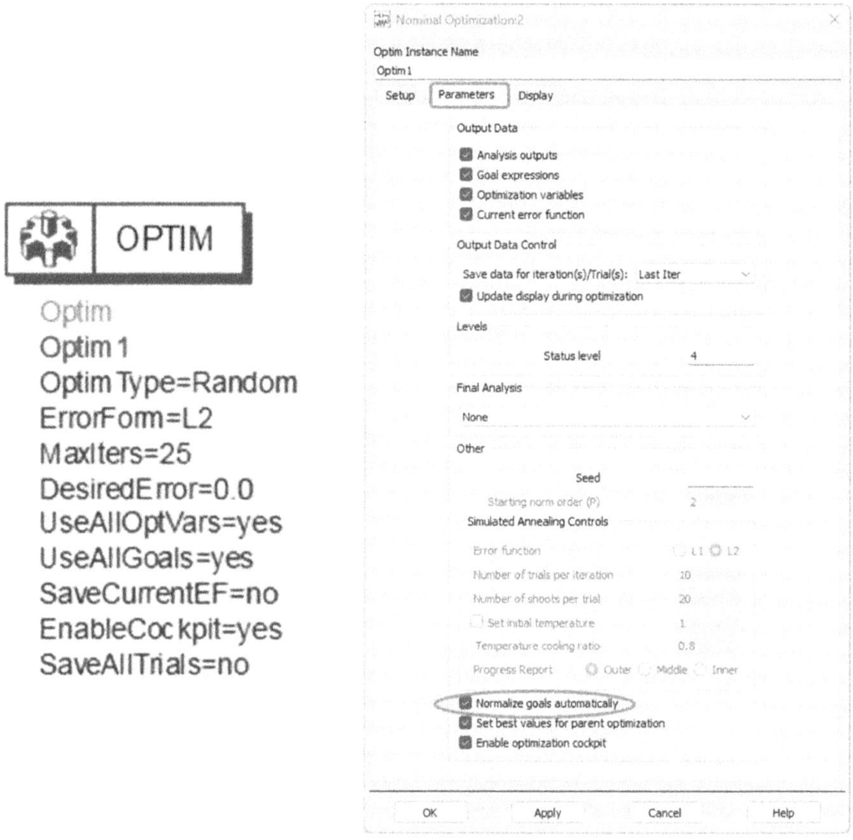

FIGURE 6.70 Optim parameters and "Normalize goals automatically".

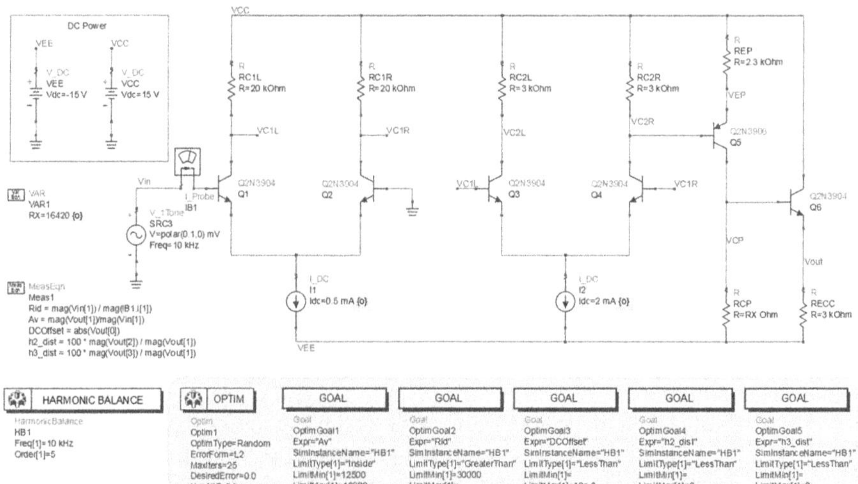

FIGURE 6.71 The schematic is ready to run the optimization.

FIGURE 6.72 View of the cockpit as the optimization begins.

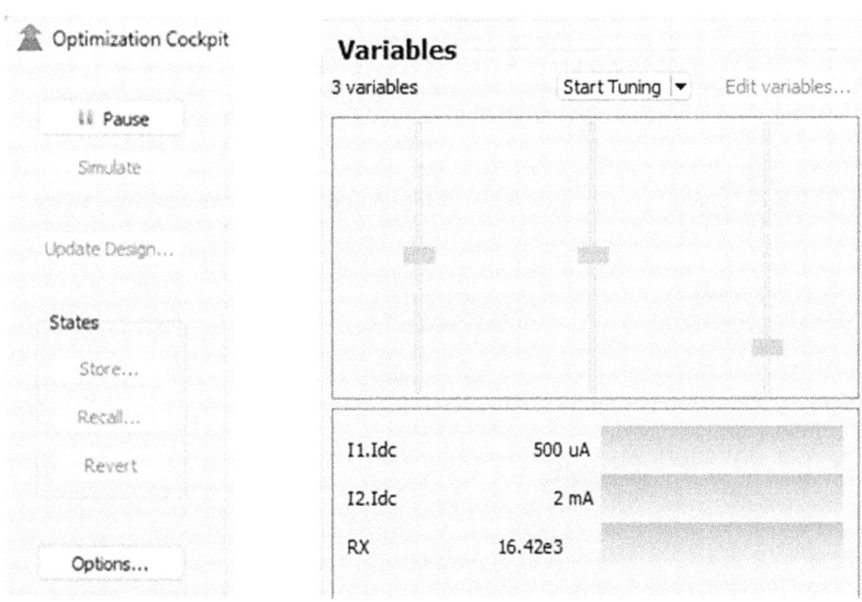

FIGURE 6.73 Some controls of the optimization (left). Optimized variables (right).

The Variables section (Figure 6.73 (right)) is a view of the status of the optimized variables as the optimization executes. In pause state one can manually tune the variables to assist a slow converging optimization. We can also edit an optimized variable, for example, change its limits and then continue the optimization.

Let us follow the events that unfold in the optimization Cockpit.

Figure 6.74 shows what happens as the optimizer just starts.

Then it stopped. Figure 6.74 captures this as well.

Note the name of the optimizer instance "Optim1", also showing that we use a "Random" optimizer type. Stopping reason is shown, along with the number of iterations, and elapsed time. The cockpit awaits us to either accept the results or make changes and continue the optimization process. Note that this has been a fairly fast optimization, taking only 5 s.

Simultaneously with the information shown in Figure 6.74, the contents of Figure 6.75, which is about goals, are shown on another part of the Cockpit's views.

In Figure 6.75, the left side is goals status information, as the optimization starts. On the right side, the information is shown as the optimization stopped.

The cockpit keeps a "Error History", of the progression of the error, in plot form. Next to the error history, it shows bar charts of the contribution of each goal to the error value.

Also shown are the Goals and where is the response relative to the goal limits. See blue asterisks and next to these is the Error history for each goal. This is all useful information that helps us understand why optimization fails or succeeds.

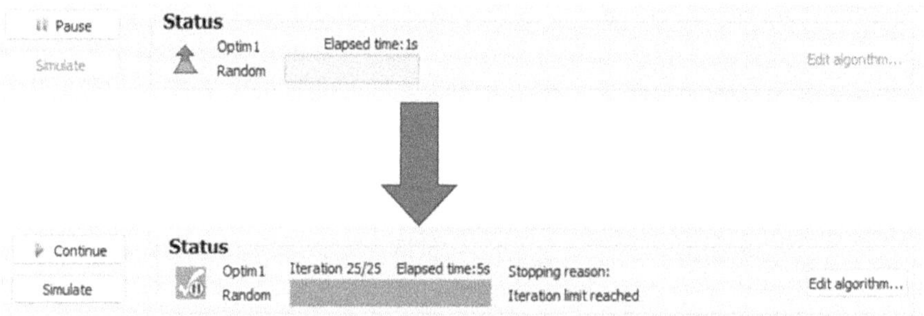

FIGURE 6.74 View of part of the Cockpit as the optimization just starts and as it stops.

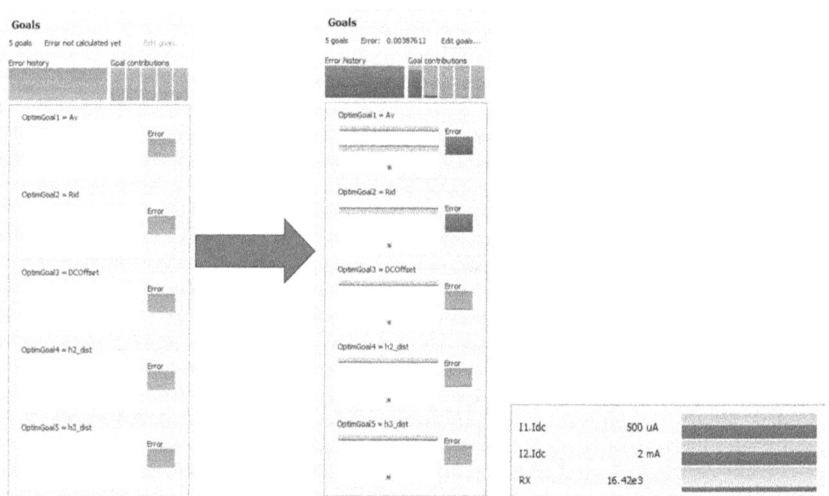

FIGURE 6.75 Cockpit view of "Goals attaining" information.

Notice that in the variables section the Cockpit also keeps track of the optimizable parameters' history (Figure 6.75 on the right side).

So, as seen in Figure 6.76, in the first optimization round, in which the optimization type was Random, the optimizer stopped after 25 iterations, as the stopping criteria were met.

Based on the error history information, the error never made any progress, which means that the Random optimizer struggled. We could increase the number of iterations and give Random another chance by selecting the edit algorithm button.

Notice in Figure 6.76 that the optimizable parameter history shows that these values were not perturbed, which further indicates that Random is not performing well for our problem. We choose to close the Cockpit and try the Gradient optimizer. See Figure 6.77.

When closing the optimizer, we are given the choice (Figure 6.77 (left)) to either Update the Design or not. If we choose to update, the optimizable parameters will update their value to the value determined by the optimizer. We chose not to update since the optimizer made no progress towards meeting the goals.

On the schematic we update the optimizer type to Gradient (Figure 6.77 (right)) and re-run the optimization.

Figure 6.78 captures the events occurring during the second run of the optimizer.

The optimizer stopped because the Gradient optimizer is no longer making improvements after 19 iterations. We suspect that a local minimum is to be

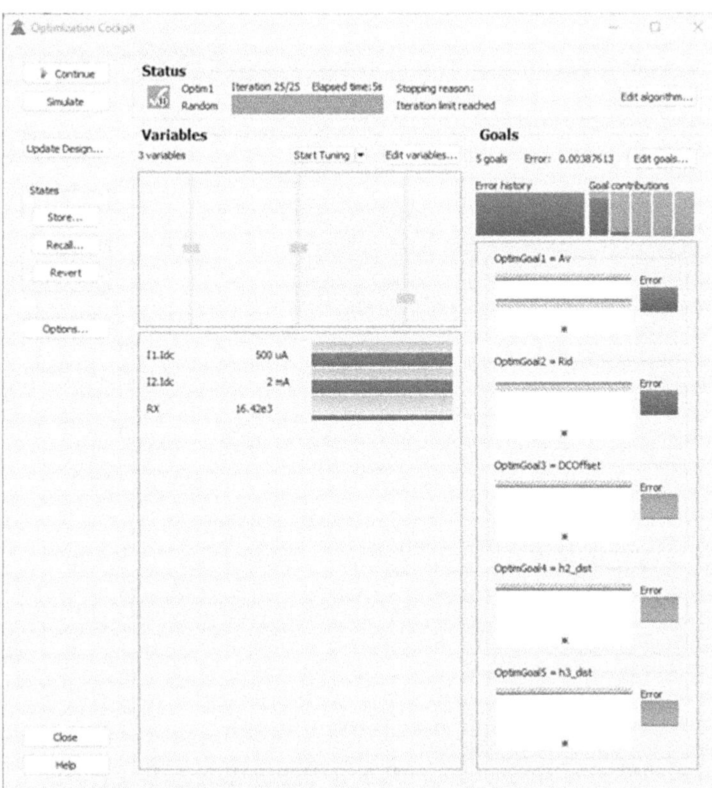

FIGURE 6.76 Full cockpit view after the optimizer stopped for the first time.

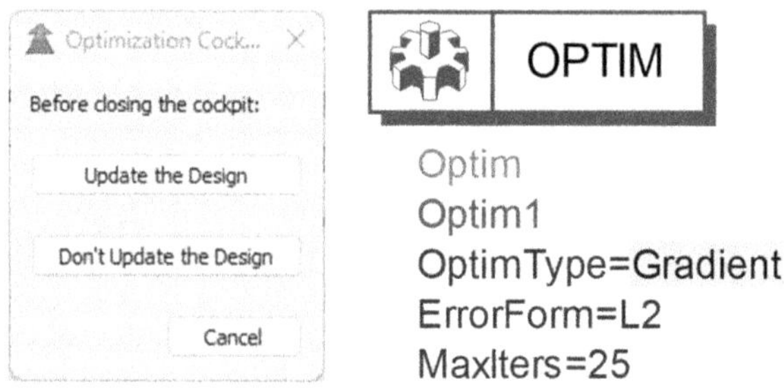

FIGURE 6.77 The next steps following the first optimizer's stopping.

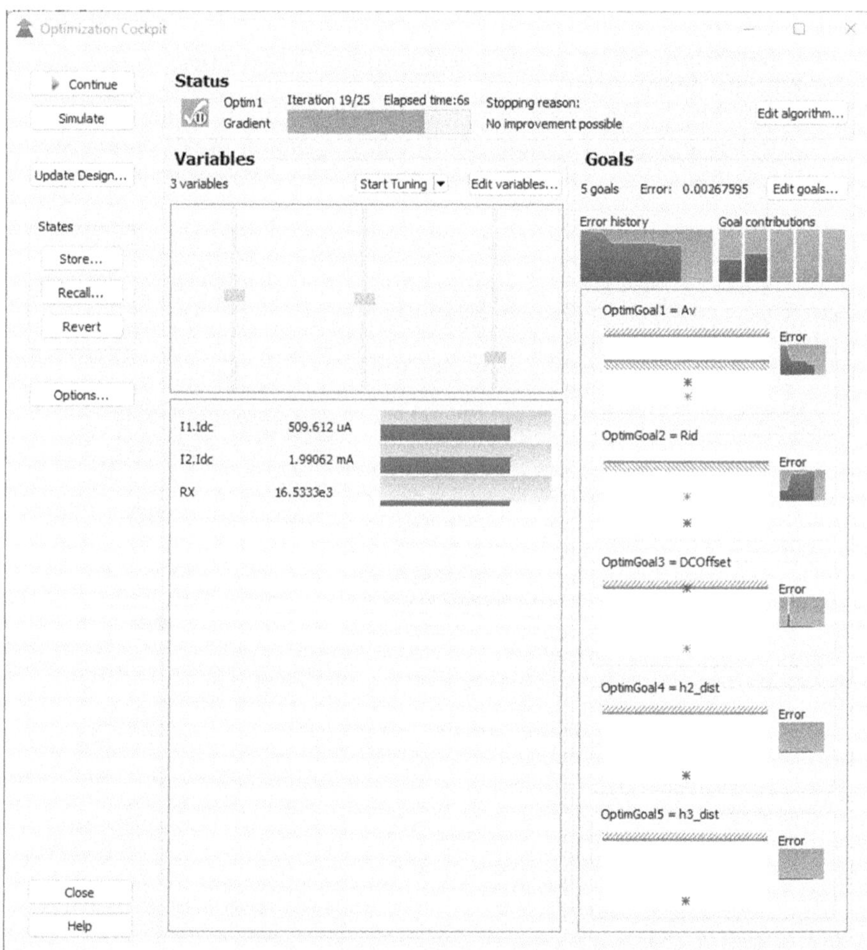

FIGURE 6.78 Cockpit View after the second stop.

blamed. Notice that this time the optimized variables did change slightly, and the error did reduce some. The A_v and R_{id} goals now both contribute almost equally to the total error.

While there was some progress, we chose to not update the design and move on to select the Hybrid optimizer (Figure 6.79).

Success!!! A eureka moment. The Hybrid optimizer has reduced the Error to zero and surprisingly it did so in a single iteration, where the two prior optimizers had failed.

The optimization went so well that we are tempted to be somewhat greedy and increased the input voltage amplitude to say 0.5 mV from 0.1 mV, a 5× increase (Figures 6.80 and 6.81).

We have not updated the optimizable variables just yet.

Again, as in Figure 6.81, for the 0.5 mV input amplitude the optimization succeeds quickly!

How about we try to achieve 1 mV input amplitude? See Figures 6.82 and 6.83.

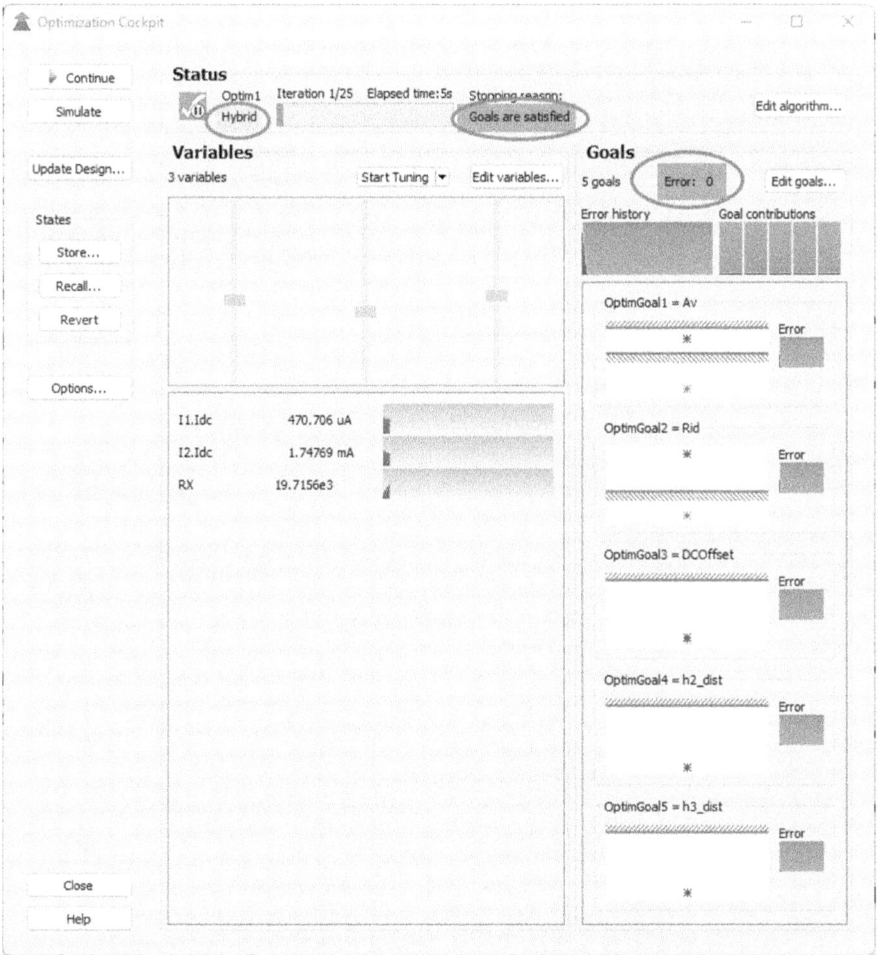

FIGURE 6.79 Cockpit view of the third round of optimization.

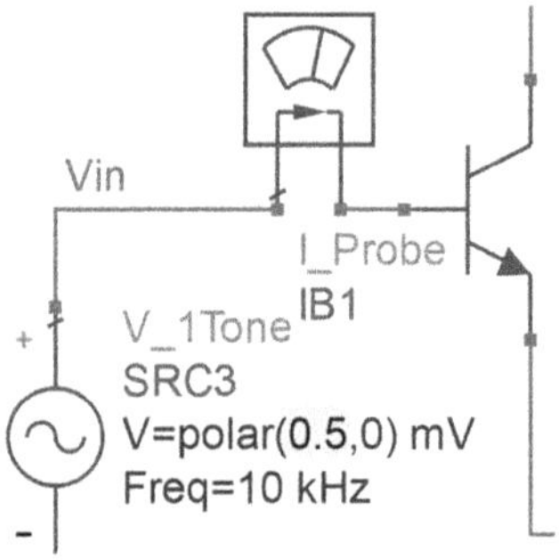

FIGURE 6.80 Increasing the input amplitude from 0.1 to 0.5 mV.

FIGURE 6.81 Cockpit view after increasing the input amplitude from 0.1 to 0.5 mV.

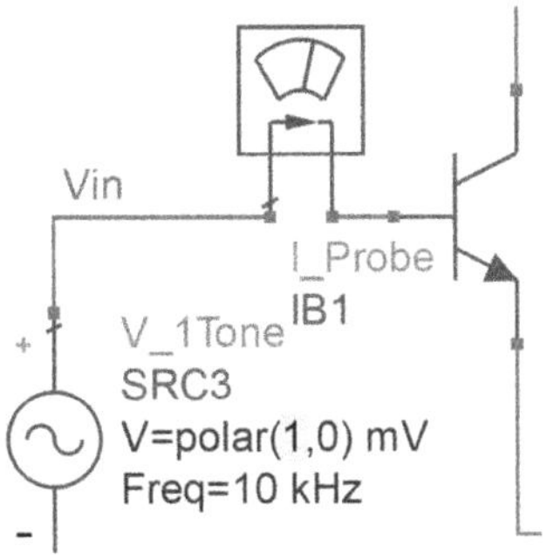

FIGURE 6.82 Increasing input amplitude from 0.5 to 1 mV.

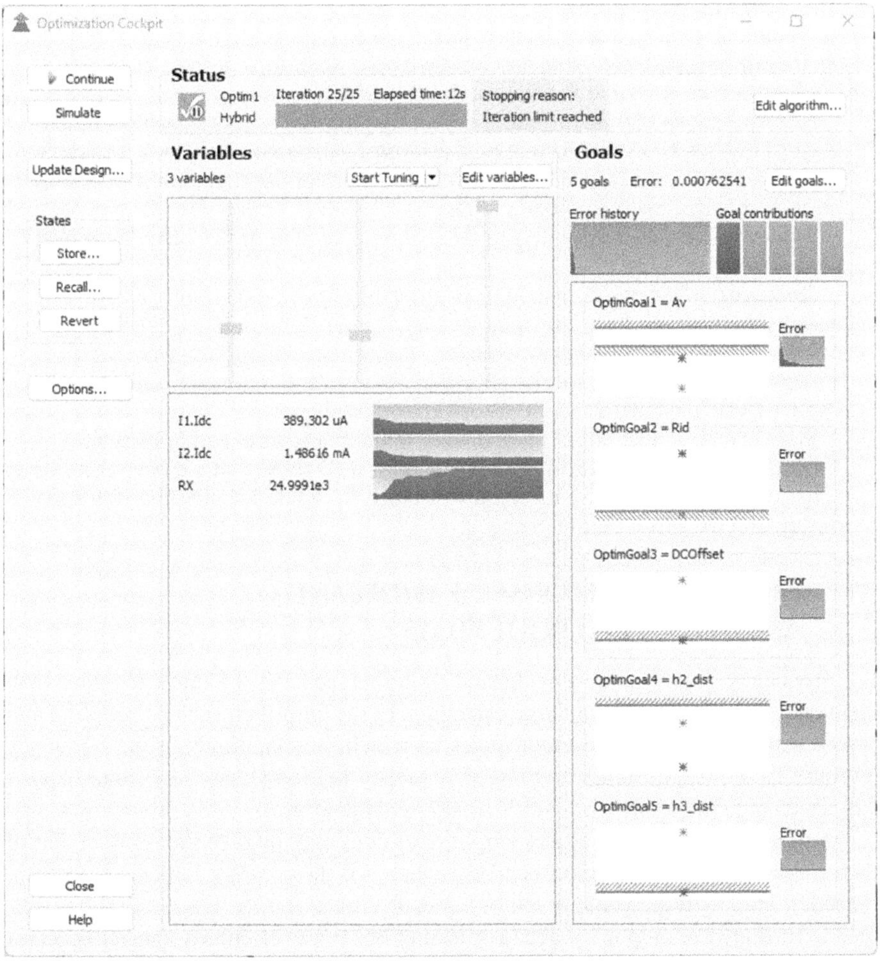

FIGURE 6.83 Cockpit view of the history of the first run after increasing the input amplitude to 1 mV, using the optimized variable values obtained for an input amplitude of 0.1 mV.

optIter	DCOffset	Av	Rid
1	974.3 u	12.67 k	31.03 k

optIter	h2_dist	h3_dist
1	0.198	0.257

I1.Idc	I2.Idc	RX
470.2 u	1.748 m	19.72 k

FIGURE 6.84 Results after implementing the optimization of the resulting variables.

The optimization did not succeed, and we know why. An input amplitude of 1 mV causes output clipping.

We are tempted to make changes, to see if we could get to the goals. After all, we are just missing the A_v goal and not by much. Wait, look at the history of RX. The RX value has railed to the highest designated search value.

From our earlier discussion, about output clipping that occurs at an input amplitude of 1 mV, we know that increasing this resistor value may increase A_3 and drive Stage 3 into saturation, an undesired outcome. The clipping effect is also visible in the goals that deal with harmonics ratios.

We revert back to 0.5 mV (Figure 6.82) and update the design after re-running the optimization.

Once we updated the design, ADS saved all the optimization results. Figure 6.84 presents the final results of the optimization. OptIter is the number of iterations it took the optimizer to arrive at the solution. Notice that all our goals were satisfied. The optimizable variables are also shown in our results. Other optimization results are saved, and we encourage the reader to explore these.

More examples that utilize an optimizer solution are presented in Chapter 7.

In conclusion the results in Figures 6.85 and 6.86 show that we have met our increased gain A_v and increase input resistance R_{id} while keeping the distortion low and the DC offset output in check.

6.12 FREQUENCY RESPONSE OF WIDEBAND AMPLIFIERS AND BODE PLOTS

The frequency response of a wideband amplifier consists of three regions: Low Frequency Region (created by large capacitors of the amplifier, such as coupling and bypass capacitors), Midband Frequencies (at which the magnitude of the frequency response is almost constant), and High Frequency Region (created by small capacitors which are part of the transistor's small-signal diagram).

An impedance analysis, of any amplifier circuit, may involve the replacement of each capacitance C by its impedance 1/sC (where s is the complex Laplace Transform variable). The analysis would result in a transfer function that relates the input voltage (or current) and output voltage (or current). For linear circuits a transfer function is a ratio of s polynomials (recall Section 1.14). The roots of the transfer function's numerator are called **zeros**, and the roots of the transfer function's denominator are called **poles**. See Example 6.4 and [Gable1].

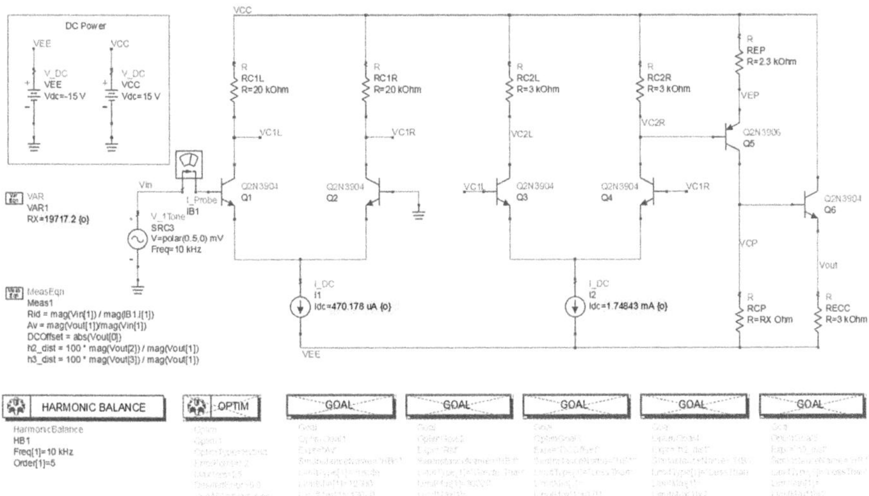

FIGURE 6.85 "Improved" multistage amplifier final design.

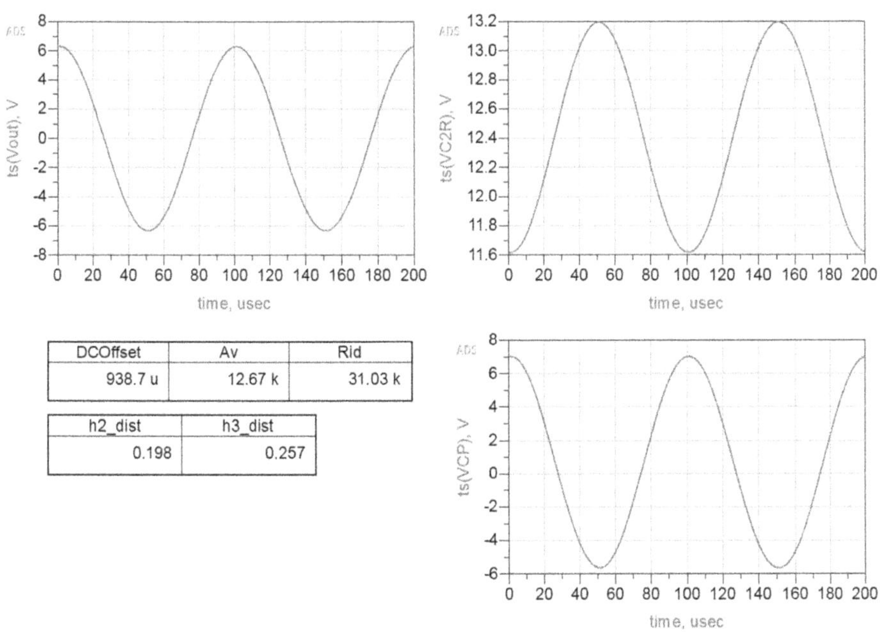

DCOffset	Av	Rid
938.7 u	12.67 k	31.03 k

h2_dist	h3_dist
0.198	0.257

FIGURE 6.86 Performance of the redesigned multistage amplifier.

Example 6.4

The amplifier circuit shown in Figure 6.87 has similarities to some of the circuits that will be discussed in detail further on in this section. Need to find the transfer function $V_O(s)/V_S(s)$ and its poles and zeros. The transconductance gain is $g_m = q > 0$.

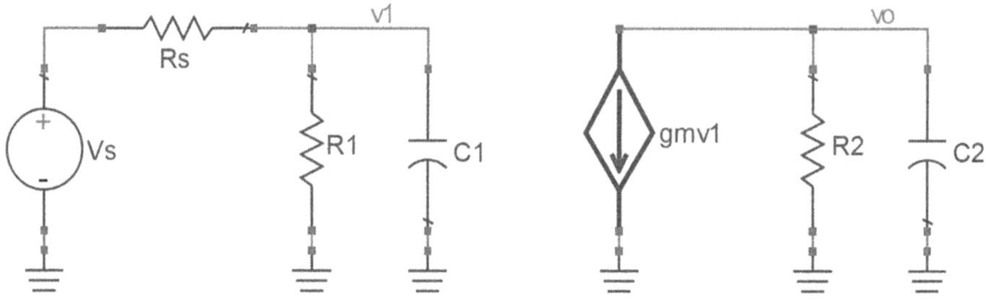

FIGURE 6.87 A fictional voltage amplifier featuring two high-frequency capacitors.

The circuit analysis consists of two parts: An "**input circuit**" analysis, for which we can find the transfer function $V_1(s)/V_S(s)$, and an "**output circuit**" analysis, for which we shall find the transfer function $V_O(s)/V_1(s)$. The overall transfer function Vo(s)/Vs(s) can be found by taking a product of the input and output circuits' transfer functions: $Vo(s)/Vs(s)=[V_1(s)/Vs(s)]\cdot[Vo(s)/V_1(s)]$.

We first observe that the impedance of a parallel connection between R_1 and C_1 is $R_1||(1/sC_1)=R_1/(1+sC_1R_1)$. Likewise, for R_2 and C_2. $V_1(s)$ relates to Vs(s) by a voltage division between R_S and $(R_1||1/sC_1)$, resulting in a first order transfer function: $\mathbf{V_1(s)/Vs(s)=[(R_1/(R_S+R_1)]\cdot\{1/[1+sC_1(R_1||R_S)]\}}$. This transfer function has one pole and no zeros. The input circuit's time constant τ_{input} consists of the input circuit's capacitor C_1 multiplied by the equivalent resistance "seen" by that capacitor, when the input $v_S(t)$ is short-circuited. The pole location $s=-p_1=-1/\tau_{input}$.

The transfer function $V_O(s)/V_1(s)$ is obtained via Ohm's Law involving the current $-qV_1(s)$ and the impedance $R_2||(1/sC_2)$ and it too is first order: $\mathbf{V_O(s)/V_1(s)=-qR_2\cdot 1/(1+sC_2R_2)}$. The output circuit's time constant $\tau_{output}=C_2R_2$. Again, R_2 is the resistance "seen" by C_2 when $v_S(t)=0$, which nulls $v_1(t)$ and consequently it nulls the dependent current source.

The overall transfer function $V_O(s)/Vs(s)$ is second order, having two poles $s=-1/\tau_{input}$ and $s=-1/\tau_{output}$ and no zeros. Example 6.4 will be revisited soon when we discuss the **bandwidth** of CS and CE amplifiers.

Substituting $s=j\omega$ into any transfer function results in obtaining its frequency response. The unit of the radian frequency ω is [rad/s], and the unit of the frequency f is [Hz], where $\omega=2\pi f$. A frequency response is a complex function of ω (or f, if we scale the f frequency axis by a factor of $1/2\pi$). A frequency response plot has two parts, the magnitude plot and the phase plot. These plots provide, for each AC input frequency ω, the ratio of output and input amplitudes, and the output to input phase shift. The frequency axis is usually taken on a logarithmic scale. A magnitude frequency response is a smooth curve. A magnitude **Bode plot** involves a **piecewise linear approximation of the actual magnitude frequency response plot**. Some books may discuss piecewise linear approximation of phase curves [Bower1], a topic which is not needed in this book.

Example 6.5

The example presents illustrations of Bode plots of two typical amplifier magnitude high frequency response plots.

Both amplifiers have no large capacitors, hence no low frequency region. Both have a midband magnitude frequency response voltage gain of 100. We define dB_Gain as **dB_Gain = 20log$_{10}$(Gain)**. Then both amplifiers have a dB_Gain of 40 dB. At the high frequency region, both amplifiers have two poles and one zero.

The first amplifier $G_1(s)$ has widely disjoint poles and zeros.

Its poles are at $\omega = 100$ krad/s and at $\omega = 20$ Mrad/s, and its zero is at $\omega = 200$ Mrad/s. The second amplifier $G_2(s)$ has poles and zeros that are closer to one another. Poles are at $\omega = 1$ Mrad/s and $\omega = 4$ Mrad/s, and a zero is at $\omega = 15$ Mrad/s.

MATLAB™, used to plot the magnitude frequency response of the two amplifiers (Figure 6.88 (left)), calls the plots "Bode magnitude", but these are actually the exact smooth frequency response plots. The piecewise linear approximations of the true frequency response plots are sketched in Figure 6.88 (right).

A **Bode plot consists of line segments that have slopes that are integer multiples of 20 dB/dec.**

The intersection of consecutive lines creates "**corner frequencies**".

The corner frequencies are **the locations of the poles and zeros of the transfer function.**

Each pole and zero contributes to the frequency response plot. Because of the logarithmic frequency scale, the contributions are added up. The log scale creates "**decades**": Frequencies with a ratio of 10:1 have the same length on the frequency axis of the plot.

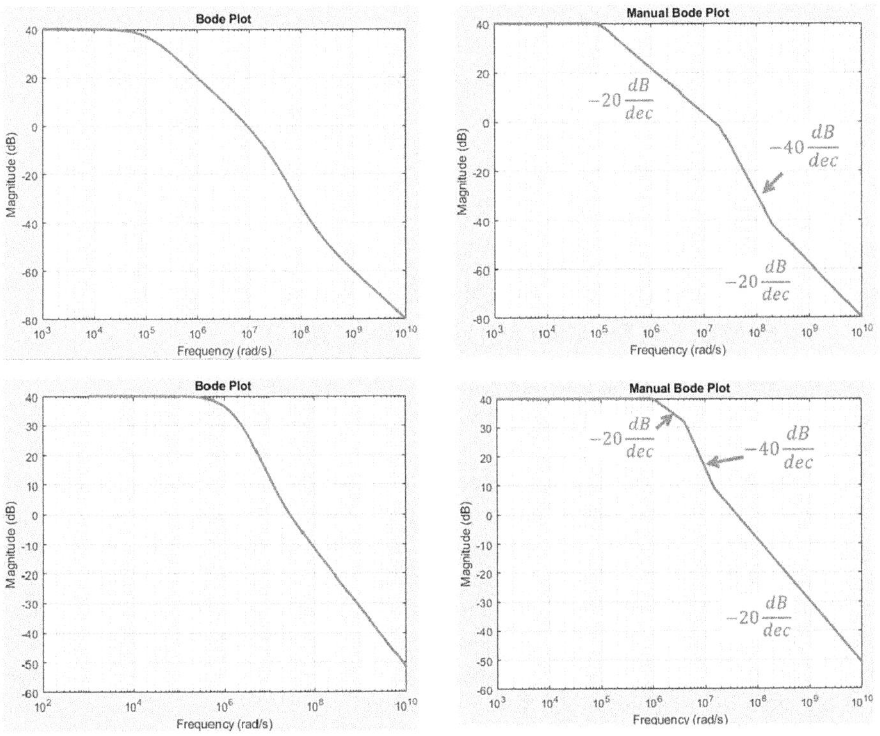

FIGURE 6.88 Amplifier with widely disjoint poles/zeros (up), and with poles/zeros closer to one another (down). Exact magnitude frequency response plots (left) vs. piecewise linear Bode plot approximations (right).

Observing the upper magnitude frequency responses of $G_1(s)$ in Figure 6.88, and because the poles are zeros are very sparse, we see that the pole at the corner frequency of $\omega = 0.1$ Mrad/s contributes two lines, one that has a slope of 0 dB/dec (to the left of the corner frequency) and another that has a slope of −20 dB/dec (to the right of the corner frequency), and due to the log scale, the actual magnitude frequency response plot converges fast to these straight line approximations. The largest error between the exact frequency response and the Bode plot happens at the corner frequency itself and it is equal to −3 dB. We see that the Bode plot is a good approximation of the exact plot at this frequency range.

As we continue to scan the frequency axis (of Figure 6.88 (up)) from left to right, the next corner frequency comes from the pole at $\omega = 20$ Mrad/s. This second pole contributes two lines. The one that is to the left of the second corner frequency is horizontal and the one that is created to the right of the corner frequency has a slope of −20 dB/dec. The two contributions of the first two poles add up and that is how the slope of the line that runs between $\omega = 20$ Mrad/s to the next corner, created by the zero, becomes −40 dB/dec. At frequencies that are larger than the zero's corner frequency of $\omega = 200$ Mrad, the slope changes from −40 to −20 dB/dec (because zero contributes a line that goes at a slope of +20 dB/dec to the right of its corner frequency).

In any case at which the transfer function poles and zeros are sufficiently distant from one another the straight-line portions of the true frequency response are visible, the Bode plot approximates well the true magnitude frequency response and one may easily find the transfer function from line fitting to measured frequency response data.

As seen in Figure 6.88 (down) regarding $G_2(s)$, whenever the poles and zeros are closer one to another, there is no obvious easy experimental way, like the one described for $G_1(s)$, to recover the Bode plot from the smooth frequency response. Bode plots allow approximate computations along the magnitude frequency response, that help us conceptualize tasks such as stability compensation of op amps (to be studied in Section 6.21).

6.13 TRANSISTOR HIGH FREQUENCY SMALL SIGNAL CAPACITANCES AND MILLER'S THEOREM

The basic BJT hybrid-pi small-signal model is shown in Section 3.9.

The model can be extended, as shown in Figure 6.89, to include high-frequency capacitance effects.

In active mode, capacitor C_μ is the depletion capacitor of the BJT reverse-biased BC junction. It is bias voltage dependent.

In active mode, the BE junction is forward biased. A diffusion capacitor C_π forms in parallel to the r_π resistor. What is the physical meaning of C_π? We recall that in active mode, the majority carriers (electrons for NPN or holes for PNP) diffuse from the emitter to the base and most of these get swept into the collector. The small difference between I_C and I_E, which constitute the base current I_B, is due to recombination of electrons and holes in the base region, and due to base majority carriers, that diffuse from the base to the emitter. The base current does not form instantly – it depends on the recombination time and on the number of carriers diffused from emitter to base and vice versa. There is some time constant "RC" to account for the finite rise time of the base current. The capacitance part of the time constant is C_π.

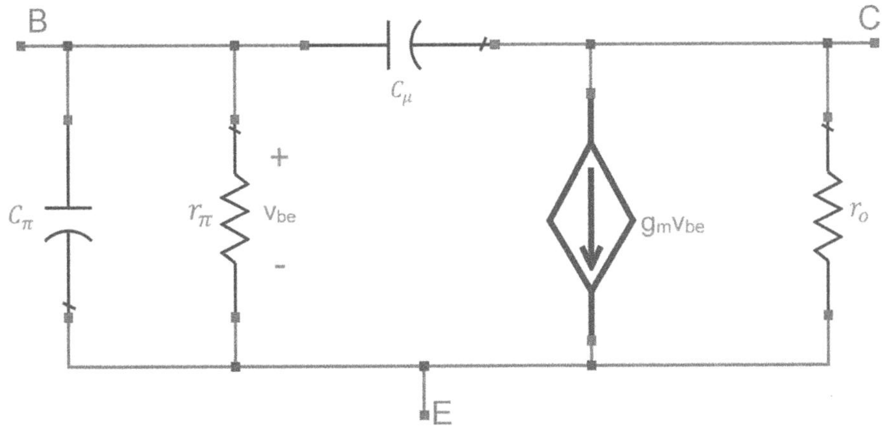

FIGURE 6.89 The BJT hybrid-pi model with two high-frequency capacitors.

It is bias-voltage dependent because the number of carriers that diffuse from one BJT region to another is voltage dependent. The resistance part of "RC" of the BE junction is of course r_π. See [Sedra1].

Section 4.7 deals with MOSFET small-signal parameters. Figure 4.34 displays the detailed MOSFET Level 1 small-signal model. It includes the five high frequency capacitors C_{DB}, C_{GB}, C_{GD}, C_{GS}, and C_{SB}. C_{GB} consists of a series connection of the gate oxide capacitance and a depletion capacitance that is created between the channel and the substrate. C_{GB} is negligible in terms of its effect on the high frequency performance of MOSFET amplifiers [Razavi1]. In Section 4.8, we observed, by ADS simulation, how each of the three capacitors C_{DB}, C_{GD} and C_{GS} contribute to the bandwidth of a CS small signal voltage amplifier, in which the source and the substrate are short circuited.

Let us focus first on the bandwidth of the CS and CE voltage amplifiers, which are sometimes affected by what is called the **Miller effect**. Both types of amplifiers, and consequently MOSFET and BJT differential amplifiers, which suffer from that effect, under certain circumstances (to be explained soon), have small-signal capacitors that bridge between the amplifier's input terminal (Gate or Base) and the amplifier's output terminal (Drain or Collector). In the case of a CS amplifier, it is the C_{GD} capacitor. In a CE amplifier it is the C_μ capacitor.

Miller's theorem is a general result, taken from Linear Electrical Network Theory, that helps us get a better insight into the effect created by components that bridge the input and output, of any network. The setup for the theorem is shown in Figure 6.90. See [Sedra1].

In Figure 6.90, Y, Y_1, and Y_2 are admittances, each may represent some passive linear circuit (i.e. some RLC network). Admittance is the inverse of an impedance, mentioned earlier in Section 1.14. Y is floating, connected between terminals 1 and 2. Y_1 is connected between terminal 1 and ground, and Y_2 is connected between terminal 2 and ground. Any pair of two-port networks are equivalent, if the networks have the same set of $\{I_1, V_1, I_2, V_2\}$ values.

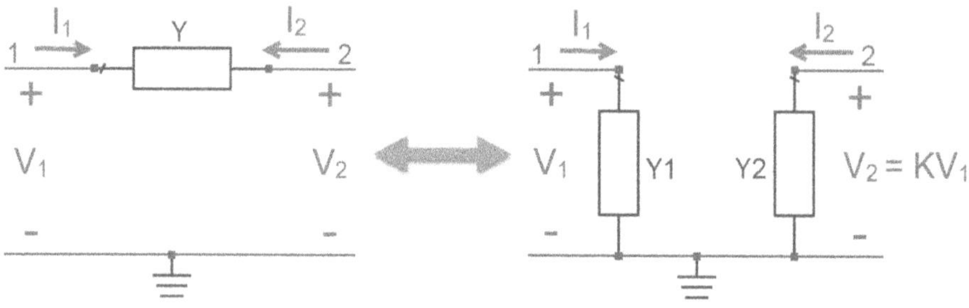

FIGURE 6.90 The setup of Miller's theorem: For a given voltage gain K, between terminal 1 and 2, under what conditions are the left and right two-port networks equivalent?

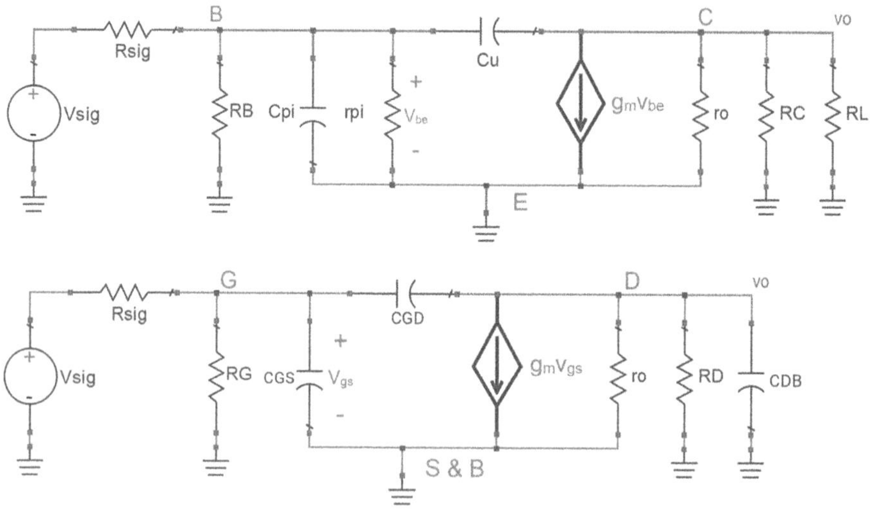

FIGURE 6.91 Two high-frequency amplifier small signal diagrams: (up) NPN CE amplifier with single supply and no degeneration resistor, (down) NMOS CS amplifier with R_D load.

The voltage gain, from terminal 1 to terminal 2, is $K = V_2/V_1$ and it **is considered as given. What should Y_1 and Y_2 (expressed in terms of Y and K) be to create equivalency of the two networks of** Figure 6.90?

The answer that is provided by the theorem is simple and straightforward to prove (simply by equating the respective currents): $\mathbf{Y_1 = Y\cdot(1 - K)}$ **and** $\mathbf{Y_2 = Y\cdot(1 - K^{-1})}$.

For example, if Y is a capacitor, $Y = j\omega C$, then Y_1 and Y_2 are capacitors as well: $Y_1 = j\omega C(1 - K)$ and $Y_2 = j\omega C(1 - K^{-1})$. The capacitances of Y_1 and Y_2 both exist physically, if the given value of K is negative.

So, how do we apply Miller's theorem to the analysis of electronic amplifiers? What is Y and what is K?

Figure 6.91 (down) shows the small signal diagrams of a (NMOS or PMOS) CS amplifier with RD load, with the high frequency capacitors included. Figure 6.91 (up) shows the small signal diagram of a (NPN or PNP) CE amplifier with the high frequency capacitors included.

Note that many other CS or CE amplifier architectures lead to the same small-signal diagram – single or dual supplies, DC current source biasing, to name some.

The CE amplifier transfer function $V_o(s)/V_{sig}(s)$ can be shown, by impedance analysis, to be second order (two poles and a zero). The transfer function for the CS amplifier is third order. A simpler high frequency analysis, of such wideband amplifiers, rather than finding the transfer function, involves determining the amplifier's bandwidth only. In other words, and very often, **only the lowest high frequency pole is of interest**, being near the upper edge of the amplifier's midband frequency range. Using Bode plots concepts; the **bandwidth of an amplifier is defined as the corner frequency of the lowest pole of the high frequency transfer function**.

As in Example 6.4, in both amplifiers (shown in Figure 6.91), there is an "input circuit" that describes how the input signal source creates the voltage that controls the dependent current source (v_{be} or v_{gs}), and there is an "output circuit" relating the dependent current source to the output voltage.

Unlike Example 6.4, though, in both amplifiers, the input and output circuits are bridged by a capacitor (C_μ or C_{GD} respectively). We may use Miller's theorem to separate the bridging capacitor into an input circuit grounded capacitor, and an output circuit grounded capacitor, considering **K to be the midband voltage gain from the base (or gate) to the collector (or drain)**. The resulting application of Miller's theorem is shown in Figure 6.92.

In Figure 6.92, for each amplifier, after applying Miller's theorem, the input and output circuits are no longer bridged. The respective input capacitors are: $\mathbf{C_{input} = C_{GS} + C_{GD}(1 - K)}$ for the CS amplifier, and $\mathbf{C_{input} = C_\pi + C_\mu(1 - K)}$ for the CE amplifier. The respective output capacitors are: $\mathbf{C_{output} = C_{DB} + C_{GD}(1 - K^{-1})}$ for the CS amplifier and $\mathbf{C_{output} = C_\mu(1 - K^{-1})}$ for the CE amplifier.

The midband frequency respective values of K (obtained by considering each of the high frequency capacitors to be open circuit) are $\mathbf{K = -g_m(R_D \| r_o)}$ for the CS amplifier, and $\mathbf{K = -g_m(R_C \| R_L \| r_o)}$ for the CE amplifier.

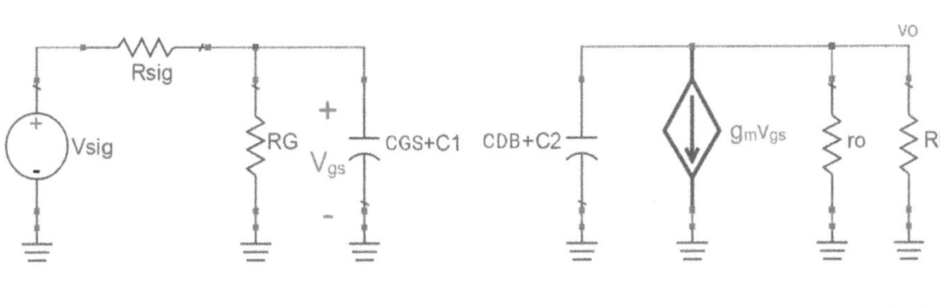

FIGURE 6.92 The CS and CE amplifiers' small-signal diagrams, following an application of Miller's theorem to the circuits of Figure 6.91.

We note that whenever K is sizeable, the input capacitance may become significantly enlarged, whereas the output capacitance stays more or less unchanged. This is called the **Miller effect**.

For the same DC current, g_m of a CE amplifier may be by one to two orders of magnitude larger than g_m of a CS amplifier with an R_D load, which causes K of a CE amplifier to be much larger than K of a $CS + R_D$ amplifier.

Following the application of Miller's theorem, to separate the bridging capacitor into input and output capacitors, the diagrams shown in Figure 6.92 look similar to the diagram of Example 6.4 shown in Figure 6.87. We may use the results of Example 6.4 to find the input circuit's time constant and the output circuit's time constant.

At this point it should be stressed that Miller's theorem assumes that the voltage gain K is known *exactly*. Taking midband frequencies gain K, in order to find the first high frequency pole, immediately renders such K, and the entire use of Miller's theorem to solve for the amplifier's bandwidth, as a mere approximation. The true frequency response magnitude of $V_O(s)/V_1(s)$ (where subscript 1 represents b or g, depending on if we refer to a CE or CS amplifier) at the first high frequency corner is 3dB below the dB value of K, if the first pole of the transfer function is sufficiently far away from all the other high frequency poles and zeros. If there are two nearby lowest high-frequency poles, the K error may become as large as 6dB.

Needless to say, that all other poles and zeros, that may be obtained by taking the product of the input circuit's transfer function and the output circuit's transfer function (as done in Example 6.4) are not valid, as the gain at these corner frequencies is typically vastly below K. Note that in Example 6.4, there is no room for Miller's theorem, as there is no bridging capacitor. Therefore, the poles are both valid.

Miller's theorem, that helps us remove the bridging capacitor, which complicates an exact impedance analysis, greatly simplifies the bandwidth finding analysis, by creating first order input and first order output circuits for the amplifier (Figure 6.92). So, as we find the input circuit's time constant, and the output circuit's time constant, the only valid next step is to **adopt the larger among both time constants and discard the smaller time constant**. The resistance part of each time constant may be found as in Example 6.4.

In the CE amplifier $R_{input} = r_\pi || R_B || R_{sig}$ and $R_{output} = R_C || R_L || r_o$.

In the CS amplifier with R_D load $R_{input} = R_G || R_{sig}$ and $R_{output} = R_D || r_o$.

We notice that, in either amplifier, whenever the input signal voltage source is ideal (that is $R_{sig} = 0$) the resistance part of the input time constant becomes 0, rendering the input time constant "undefined". A solution that goes around such a problem, is to consider a small contact resistance r_x for the base (or the gate, respectively) [Sedra1]. See Figure 6.93. In the SPICE code of BJT transistors such r_x is denoted R_B, and its typical value is in the order of magnitude of a few tens of Ω.

A revised version of the input and output circuits resistance part of the time constant is:

In the CE amplifier $\mathbf{R_{input} = r_\pi \,||\, [r_x + (R_B || R_{sig})]}$ **and** $\mathbf{R_{output} = R_C || R_L || r_o}$.

In the CS amplifier with R_D load $\mathbf{R_{input} = r_x + (R_G || R_{sig})}$ and $\mathbf{R_{output} = R_D || r_o}$.

So, as determined approximately by an analysis assisted by Miller's theorem, **the amplifier's bandwidth equals either the input circuit's pole's corner frequency, or the output circuit's pole's corner frequency, whichever is smaller.** Let us

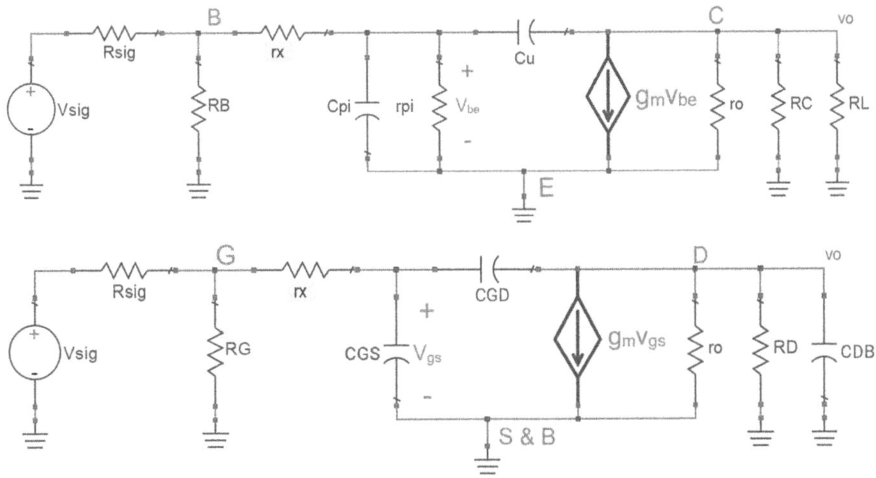

FIGURE 6.93 Adding a small base or gate terminal contact resistance r_x to the CE and $CS+R_D$ amplifiers.

compare the time constants of the input and output circuits and choose the larger of the two.

CE amplifier's input circuit time constant is: $\boldsymbol{\tau}_{\mathbf{input,CE}} \approx \mathbf{R}_{\mathbf{input,CE}}\cdot\mathbf{C}_{\mathbf{input,CE}} = \{\mathbf{r}_\pi||[\mathbf{r}_x + (\mathbf{R}_B||\mathbf{R}_{sig})]\}\cdot[\mathbf{C}_\pi + \mathbf{C}_\mu(1-\mathbf{K})]$ where $\mathbf{K} = -\mathbf{g}_m(\mathbf{R}_C||\mathbf{R}_L||\mathbf{r}_o)$.

CE amplifier's output circuit time constant is: $\boldsymbol{\tau}_{\mathbf{output}} \approx \mathbf{R}_{\mathbf{output,CE}}\cdot\mathbf{C}_{\mathbf{output,CE}} = [\mathbf{R}_C||\mathbf{R}_L||\mathbf{r}_o]\cdot[\mathbf{C}_\mu(1-\mathbf{K}^{-1})]$.

So, which side of the amplifier's circuit "wins" in determining the amplifier's bandwidth? That is, has the larger time constant.

We see that the Miller effect causes C_μ to become "amplified" by two orders of magnitude typically because $|K|$ is typically large. The input circuit's capacitance is by two orders of magnitude larger than that of the output circuit's capacitance, if K is large.

Comparison of the input and output circuit's resistance is less conclusive, due to the role played by the input signal source's internal resistance R_{sig}.

Whenever R_{sig} is not small (say, hundreds of Ω to a few kΩ) and whenever R_C is a few kΩ large, the input and output resistances have more or less the same order of magnitude. In this case the input time constant is very much larger than the output time constant. The resistor r_x is negligible.

Whenever R_{sig} is zero, or just small, the small resistor r_x dominates the resistance of the input circuit. The time constant of the output circuit becomes larger than that of the input circuit: Input circuit's capacitance is still by two orders of magnitude larger than the output circuit's capacitance. However, the output circuit's resistance may be two to three orders of magnitude larger than the input circuit's resistance.

Similar calculations and conclusions may be made for the CS amplifier with R_D load.

We note that because the g_m parameter of a CE amplifier is by one to two orders of magnitude larger than that of a CS amplifier, so does the Miller gain K. There will be more cases in which the CS output circuit time constant "wins" over the CS input circuit time constant, meaning to say that a CS amplifier may have a Miller effect, but the effect is not large enough to cause the input circuit to determine the amplifier's bandwidth.

The question is not just about which time constant, input circuit's or output circuit's, is tied to the amplifier's bandwidth. It is also about the bandwidth value itself: Does the CE amplifier (or CS amplifier) have a "good bandwidth" (relative to other types of amplifiers) or an "inferior bandwidth"? What we have found is that due to the Miller effect the bandwidth of a CE (and CS) amplifiers may be relatively small, whenever the input signal source is far from being ideal. Whenever a CE amplifier is driven by a good voltage signal source, the Miller effect (which is still significant) does not prevent the amplifier's bandwidth from being large.

6.14 ADS: MILLER EFFECT IN CE AMPLIFIERS; TUNING FOR CIRCUIT TESTING

Example 6.6

Let us consider an arbitrary CE amplifier (Figure 6.94).

Let us watch A_v, R_{in} and BW (defined in the MeasEqn of the schematic shown in Figure 6.94). We shall take several other R_{source} values, as well, and watch the same measures. Based on the theory of Miller effect in CE amplifiers, let us see if we can understand the actual BW results.

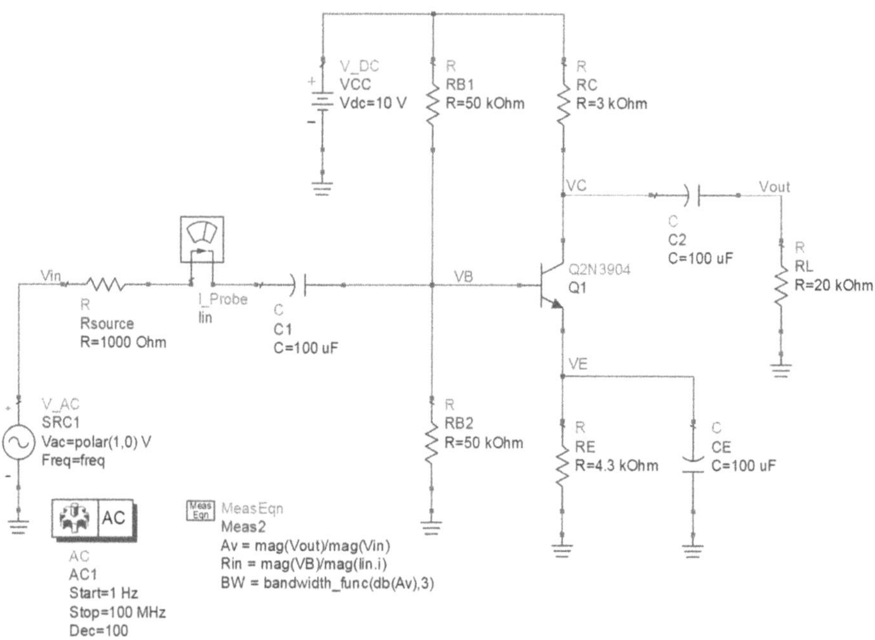

FIGURE 6.94 CE amplifier in which we shall take several input source resistor values.

As shown in Figure 6.95, with $R_{sig}=1\ k\Omega$, we have midband absolute voltage gain $A_v=72.5$, midband input resistance $R_{in}=3.6\ k\Omega$, and BW=819 kHz. The theory predicts that in such a CE amplifier case, capacitor C_μ is being "amplified" by a factor of $1+g_m(R_C||R_L||r_o)=1+40\cdot2.54=102.6$. The "input circuit's time constant" is way larger than the "output circuit's time constant". How do we know that the input circuit's time constant is the one determining the BW?

Let us slightly change R_{source} (Figure 6.96). If the BW changes, it is a sign that the input circuit plays a major role in determining the bandwidth. A change of R_{source} has no effect on the output circuit's time constant.

We will be simulating the same circuit with various values of R_{source}.

A good way to do this is to use *Tuning* (introduced in Chapter 5 and Section 6.6), because there will be no need to constantly rerun the simulation. Once we find a R_{source} value that we like, we can update the schematic with such a value. Of course, in this example we don't look for one R_{source} value.

Notice in Figure 6.96 how we added a variable Rsrc and assigned it to the value of the resistance R_{source}. When setting Rsrc for tuning, the step value of 10 is somewhat arbitrary since we will type in the specific values of interest.

R_{source} has no effect on R_{in}, so we discontinue displaying the R_{in} plot. Once one launches the tuning session the idea is to view the data display and manually change the tuned value and press the enter key. We do not use the arrows nor the blue slider (in Figure 6.99).

For **Rsrc = 2 kΩ** (shown on the left side of Figure 6.97), let us keep in mind to compare to Figure 6.95 for the case $R_{source}=1\ k\Omega$. The voltage gain is affected. We recall the CE voltage gain formula in which the effect of R_{source} comes in through a voltage division between R_{source} and R_{in}. The larger R_{source} the smaller the voltage gain. We also see that BW became smaller. So, it is the input circuit's time constant that affects the overall BW. The larger R_{source} the larger τ_{in} and the lower BW. Note that R_{source} has no effect on the Miller's gain K. So even though the overall voltage gain became smaller the small signal capacitances were not affected.

For **Rsrc = 2 Ω** (shown on the right side of Figure 6.97) we note that the voltage gain became larger. This is because with a tiny R_{source} there is almost no voltage division between R_{source} and R_{in}. We see that BW became relatively outstanding (BW=18.3 MHz), yet we still don't know which side of the amplifier circuit (input or output) determined that BW. Let us change Rsrc slightly. If the BW is affected, then the input circuit still dominates.

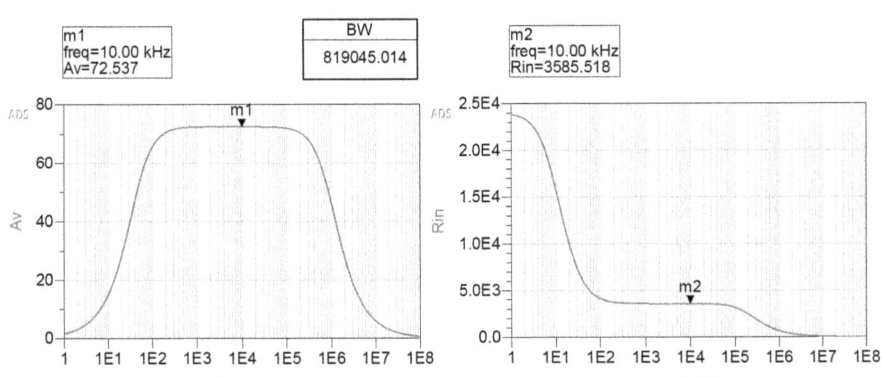

FIGURE 6.95 CE amplifier driven by a non-ideal voltage source with $R_{sig}=1\ k\Omega$.

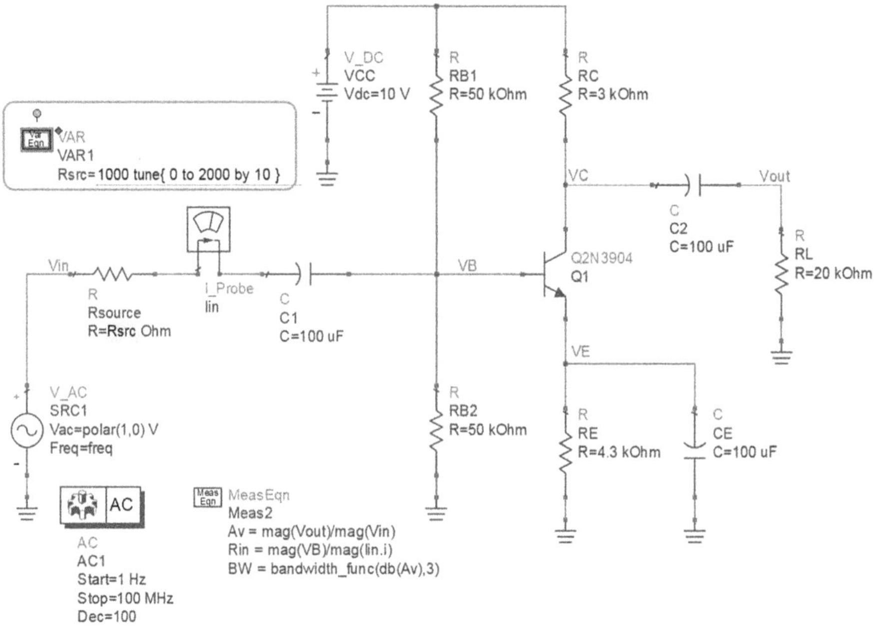

FIGURE 6.96 Tuning used for testing a circuit.

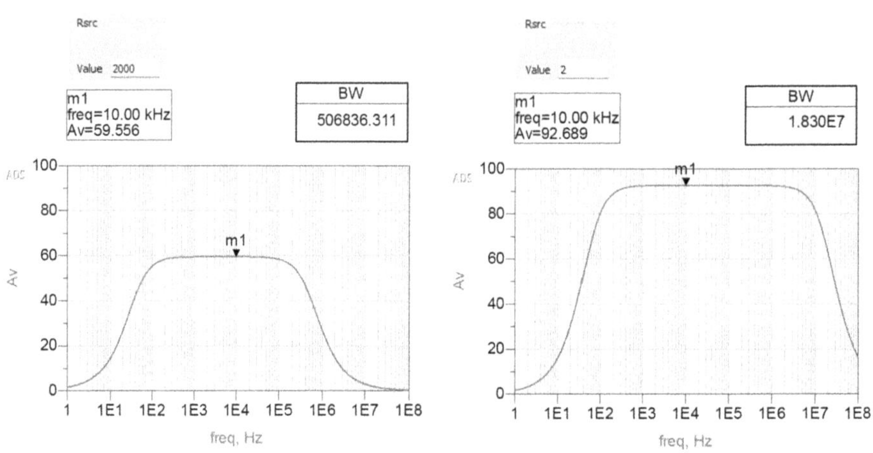

FIGURE 6.97 Results for two tuning steps: (left) Rsrc = 2000 Ω, (right) Rsrc = 2 Ω.

For **Rsrc = 1 Ω** (shown on the left side of Figure 6.98) we note that the BW became slightly larger. BW = 18.8 MHz. So, it is the input circuit's time constant that still dominates. So, yes, there is a Miller "amplification" of C_μ but because the input circuit's resistance is small the time constant is small. It is larger than the output circuit's time constant.

For **Rsrc = 100 Ω** (shown on the right of Figure 6.98) – a source resistor that is not so large and not so small. The BW reduces to BW = 5.1 MHz. It is still much higher compared to the case where R_{source} was in the kΩ range. Miller effect

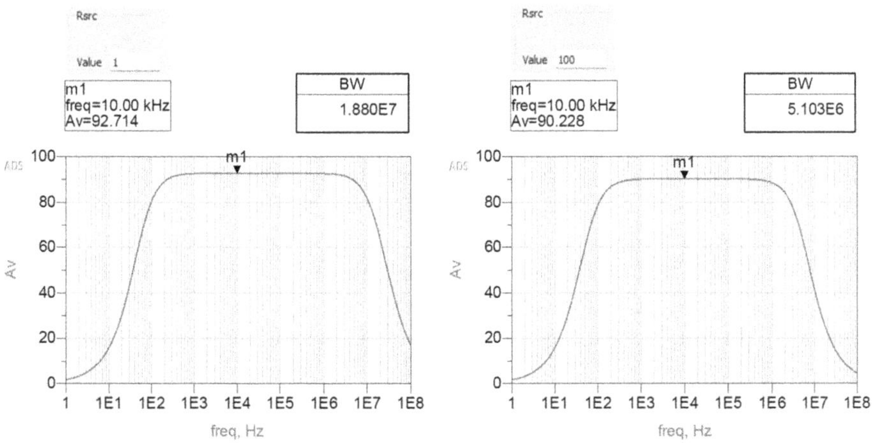

FIGURE 6.98 Two more tuning cases: (left) Rsrc = 1 Ω, (right) Rsrc = 100 Ω.

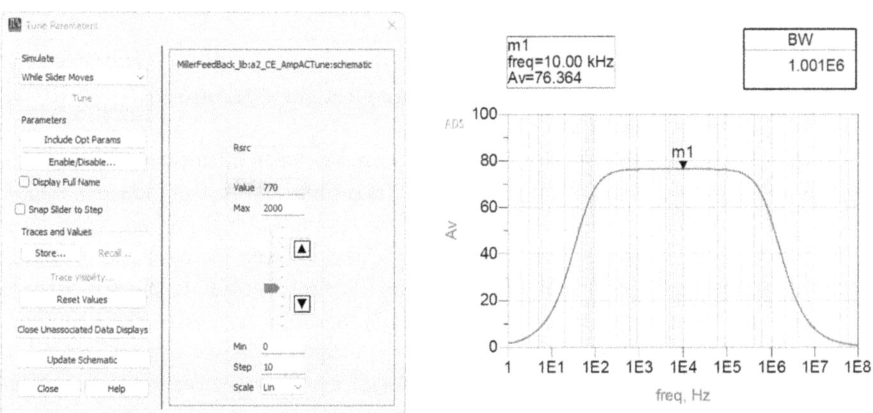

FIGURE 6.99 From tuning that is used for testing a circuit to tuning used in design.

drastically reduces the BW whenever the input circuit's equivalent resistance is large, combined with large voltage gain.

Before we leave the tuning session, say that we wish to know the value of R_{source} that gives a BW of 1 MHz. Use the blue slider (shown in Figure 6.99) to get close and then the arrows to get to about a BW of 1 MHz.

We can now confidently say that for $R_{source} < 770\ \Omega$ the CE amplifier will have a BW > 1 MHz and $A_v > 75$.

6.15 EMITTER FOLLOWER (CC) AND COMMON BASE (CB) AMPLIFIERS; BJT CASCODE AMPLIFIERS; BANDWIDTH ENHANCEMENT

By now, after studying Sections 6.13 and 6.14, we know that CE and CS amplifiers sometimes suffer from the Miller Effect. That is, the small-signal capacitor that bridges between the transistor's small signal input circuit to the transistor's output circuit, is effectively equivalent to a (sometimes) much larger input circuit capacitance. This effect is sometimes responsible for a relatively low bandwidth that such amplifiers experience, and that happens whenever the amplifier's input voltage signal source's resistance is non-small.

Remembering the half-circuit analysis of differential amplifiers, operating in differential mode (Section 6.2), a half-circuit, which is either CE or CS amplifier, tells us that these amplifiers too may sometimes perform poorly in terms of their differential mode bandwidth. In Sections 6.5 and 6.6, we met the multistage differential amplifier, and we recall that its simulation revealed that Stage 1 (a differential amplifier with differential output) had a bandwidth that was much lower than the bandwidth of each one of the other stages, resulting in a poor differential-mode overall bandwidth for the amplifier.

Are there more basic voltage amplifiers that experience a Miller effect? Since the theory of this subject, for BJT amplifiers, is very similar to that of MOSFET amplifiers, let us narrow the coming discussion to BJT amplifiers.

Figure 6.100 shows one of the architectures of an emitter follower amplifier (discussed earlier in Section 5.6) and its high frequency small signal diagram.

We recall that in an emitter follower, as in a CE amplifier, the input voltage signal enters the base terminal, and the collector is at ground for small signals. Unlike CE amplifiers, where the output voltage is taken from the collector, in an emitter follower amplifier the output is taken from the emitter.

Let us reemphasize that in Miller's theorem the **bridging admittance Y must be truly floating between the input and output terminals**. We see in the small signal diagram of the emitter follower that the collector side of C_μ is at ground. Therefore, C_μ is not a bridging component. It is C_π that bridges between the base and the emitter. So, yes, one may apply Miller's theorem to C_π, however due to the low voltage gain of the amplifier (being near 1, as we recall) there is very small, close to none, Miller effect. In other words, neither of the two emitter follower capacitances create an effect of a large capacitance. **Emitter followers therefore enjoy a very high bandwidth.**

Another basic single stage voltage amplifier, shown in Figure 6.101, is the **Common Base amplifier** (which has similar properties to a MOSFET Common Gate amplifier). On the left is one of the CB amplifiers architectures, and on the right is its small signal diagram that includes the high frequency capacitances.

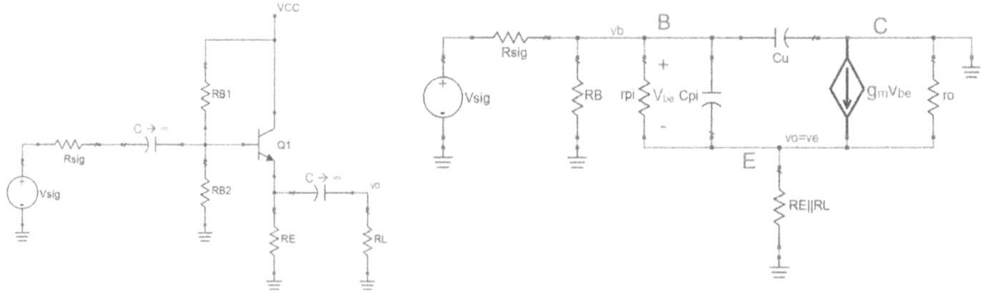

FIGURE 6.100 (Left) NPN emitter follower with a single supply. (Right) High-frequency small-signal diagram of the emitter follower.

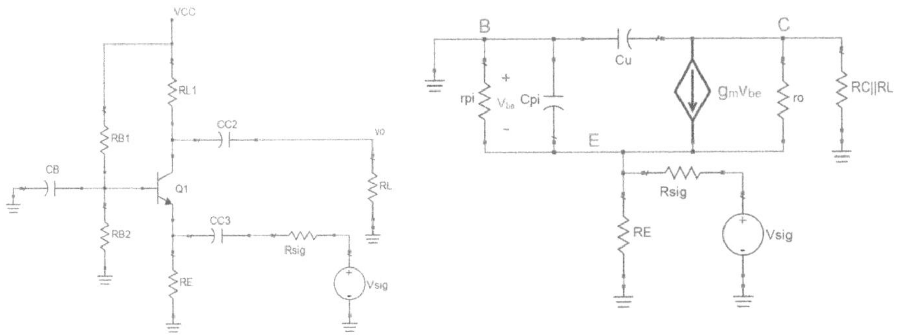

FIGURE 6.101 (Left) NPN single supply CB amplifier. (Right) High frequency small signal diagram of the CB amplifier.

The common base amplifier is called so because of the large capacitor C_B that grounds the base for small signals. We note that the input voltage is injected into the emitter, whereas the output voltage is taken from the collector.

We further note that for each of the BJT capacitors one of their small signal diagram terminals is grounded.

There is no Miller effect at all in a CB amplifier (and in a CG amplifier). Like an emitter follower, a **CB amplifier has a relatively an outstanding bandwidth**.

Let us briefly summarize some of the properties of a CB amplifier: (i) It has a small input resistance $\mathbf{R_{in} \approx r_e}$, (ii) It is a **non-inverting voltage amplifier** with a voltage gain of $\mathbf{A_v \approx (R_C||R_L)/(r_e + R_{sig})}$, (iii) It has a **current gain of 1** and a **sizeable output resistance**. With its small input resistance, it can therefore serve as a **current buffer**, (iv) Its input-circuit's time constant is approximately $\boldsymbol{\tau_{in} \approx C_\pi(r_e||R_{sig})}$, which is very small and therefore it is the output circuit that most often determines the CB bandwidth, (v) Indeed, the output circuit's time constant is $\boldsymbol{\tau_{out} = C_\mu(R_C||R_L)}$.

Finally, let us discuss the **cascode amplifier**. It is, either in BJT or MOSFET formats, a two-stage amplifier. For BJT implementation, the cascode is a CE amplifier followed by a CB amplifier. Likewise, for MOSFET implementation, a cascode is a CS amplifier feeding into a CG amplifier. A cascode amplifier may replace any CE amplifier that suffers from the Miller effect and consequently has a relatively low bandwidth. It retains the same R_{in} as that of the CE amplifier and, more or less, the same small-signal voltage gain. The key benefit of such a replacement is a dramatic enhancement of the amplifier's bandwidth.

Let us explain it more, referring to Figure 6.102 for one of the architectures of a BJT cascode amplifier. Note that Q_1 is part of a CE amplifier and Q_2 is part of a CB amplifier. Resistors R_1, R_2 and R_3 create bias voltages for the two bases. There are two large coupling capacitors, a bypass capacitor C_E for R_E of the CE amplifier and a base grounding capacitor C_B for the CB amplifier.

The two BJTs (in Figure 6.102) have more or less the same DC I_C current, and thus the small signal parameters of Q_1 and Q_2 are, respectively, the same, including the high frequency capacitors.

So, from time to time, we can conveniently omit the numerical subscripts shown in Figure 6.103.

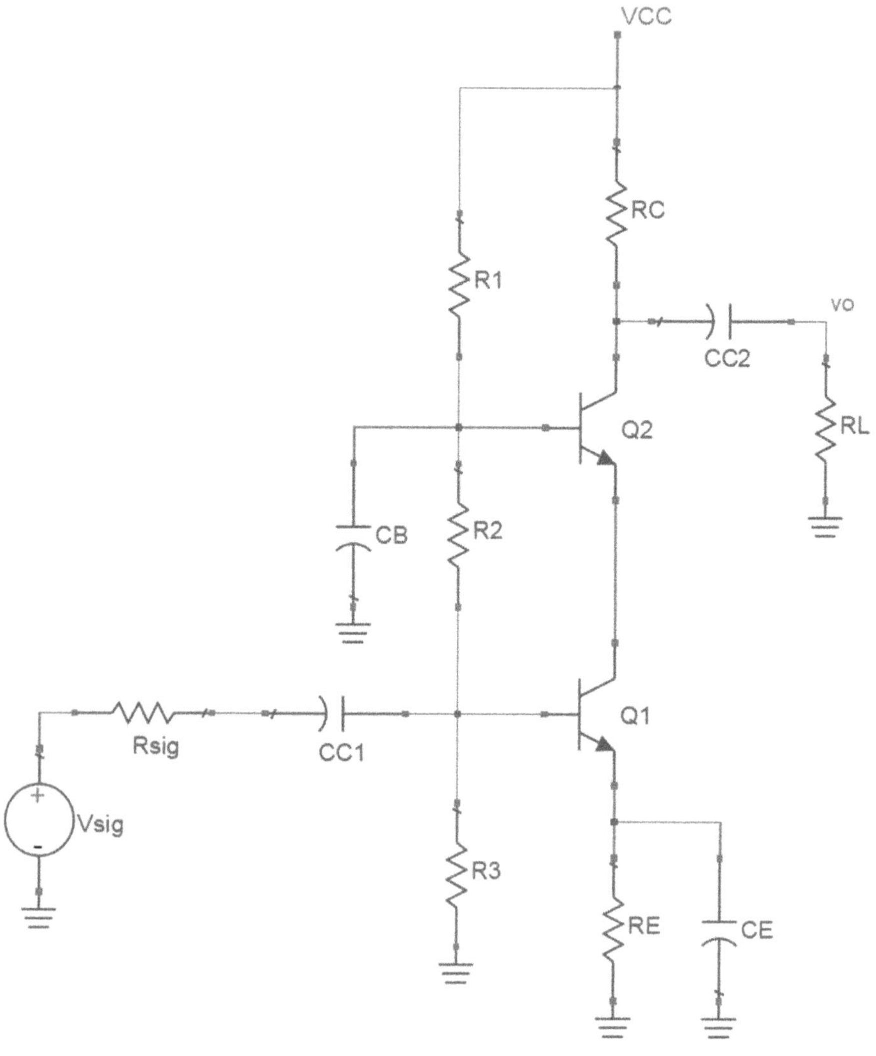

FIGURE 6.102 Single supply NPN implementation of a cascode amplifier.

So, what is the key idea that enables bandwidth enhancement by a cascode amplifier? The CE stage "sees" a load resistance which is the small input resistance $R_{in} \approx r_e$ of the CB stage. Therefore, the voltage gain of the CE stage is small (near −1). The small signal collector current of Q_1 is more or less the same as the small signal collector current of Q_2. This current "sees" the same overall amplifier's resistive load as that seen by the original CE amplifier (that is $R_C \| R_L \| r_o$). That is how the cascode and the CE amplifiers manage to have the same small-signal voltage gain.

The original CE amplifier (without the CB stage) sometimes suffers from the Miller effect (that is, large K → large input circuit capacitance → small bandwidth if R_{sig} is not small).

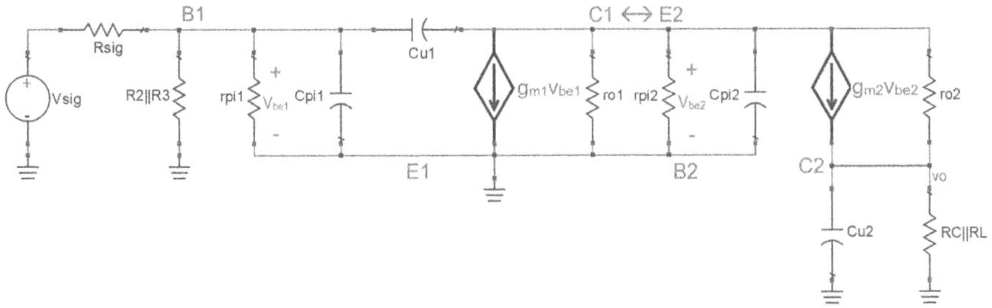

FIGURE 6.103 High frequency small signal diagram of a BJT cascode amplifier.

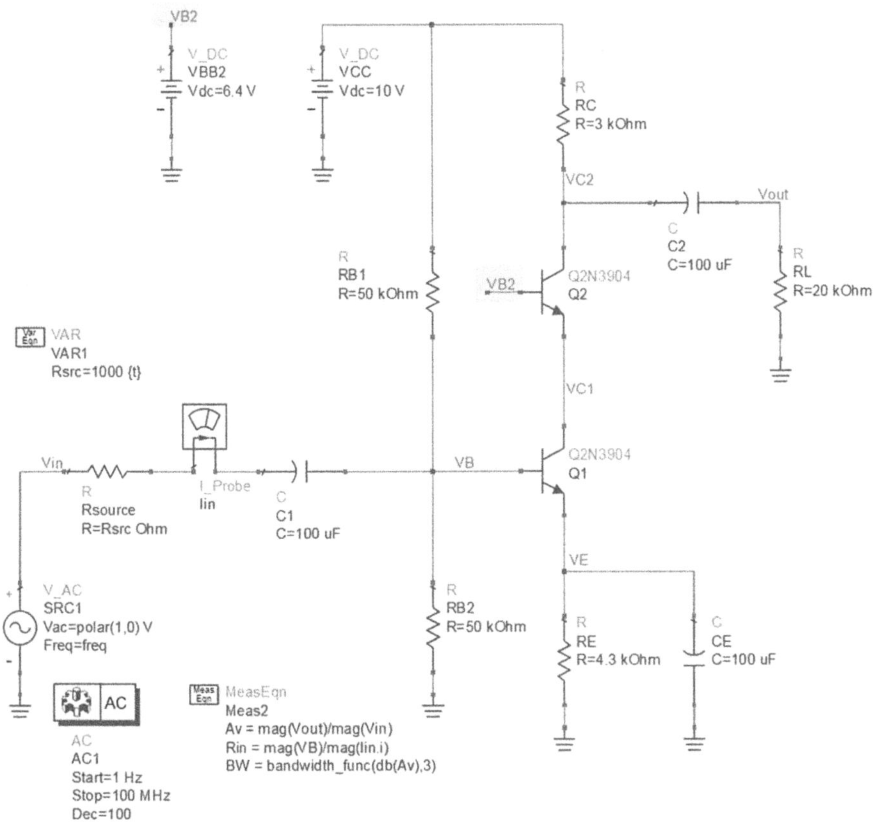

FIGURE 6.104 A single supply BJT cascode amplifier.

The cascode amplifier, on the other hand, escapes the "wrath" of the Miller effect. The CE stage does have a Miller effect but $K=-1$, and thus $C_{\mu 1}$ is not "amplified" significantly, only by a factor of 2. The second stage is a CB amplifier which has no Miller effect. All the high frequency capacitances, shown in Figure 6.103, stay effectively small.

Let us list the three cascode amplifier's high frequency time constants. We recall that in a CE amplifier's input circuit, as shown in Figure 6.93, we add the base contact resistance r_x.

As we recall, r_x may dominate the input circuit's time constant, only whenever R_{sig} is very small. The same happens for the CE first stage of a cascode amplifier. The resistor r_x is not shown in Figure 6.103, but it is there, as in Figure 6.93.

The invoking of Miller's theorem to Figure 6.103 (with K=−1) allows us to separate the bridging capacitor $C_{\mu 1}$ into an input circuit capacitance $2C_\mu$ and into a "mid-circuit" capacitance $2C_\mu$. We indeed see that the cascode amplifier, due to Miller's theorem, applied to the CE stage, has three time-constants – input circuit's, mid-circuit's and output circuit's time-constants. Which time-constant dominates the other two, to determine the cascode amplifier's bandwidth?

The cascode amplifier's input circuit's time-constant is: $\boldsymbol{\tau_{input,cascode} \approx R_{input,cascode} \cdot C_{input,cascode} = \{r_\pi \| \{r_x + [(R_2 \| R_3) \| R_{sig})]\} \cdot [C_\pi + 2C_\mu]}$.

Using Kirchhoff's Current Law for the impedances and dependent current sources, connected to the CE collector node, which is also the CB emitter node, yields the cascode amplifier's mid-circuit's time-constant: $\boldsymbol{\tau_{mid,cascode} \approx R_{mid,cascode} \cdot C_{mid,cascode} \approx r_{e2} \cdot [C_\pi + 2C_\mu]}$.

The cascode's amplifier's output circuit's time-constant is: $\boldsymbol{\tau_{output,cascode} \approx R_{output,cascode} \cdot C_{output,cascode} = [R_C \| R_L \| r_o] \cdot C_\mu}$.

By considering typical orders of magnitude for the various resistances and capacitances, we observe that the mid-circuit's time constant is almost always the smallest among the three, and therefore it has nothing to do with the cascode amplifier's bandwidth. We further observe that if the cascode amplifier is driven by a good voltage signal source (that is, R_{sig} is small), then the output circuit's time constant is likely to dominate the input circuit's time constant. If R_{sig} is modest to large, then both time-constants, input circuit's and output circuit's, are of the same order of magnitude.

To conclude this piece of theory with a smile, let us make up a new English word: "**cascodize**". It means **a replacement of a CE amplifier by a cascode amplifier with similar properties**. Likewise, we may cascodize a CS amplifier by creating a MOSFET cascode amplifier. So, a CE amplifier, that happens to really be suffering from the Miller effect, can be cascodized in order to obtain a much-improved bandwidth.

Recall the multistage amplifier of Sections 6.5 and 6.6: Stage 1 (a differential amplifier with differential output) had a low differential-mode bandwidth, Stage 2 (a single ended differential amplifier) had a much higher bandwidth, Stage 3 (a PNP CE amplifier) had a very good bandwidth, and Stage 4 (an emitter follower) had a phenomenal bandwidth. So, if we wish to enhance the multistage amplifier's overall bandwidth, do we simply cascodize Stage 1?

It's not so simple, noting (in the simulation examples of Sections 6.5 and 6.6) that $R_{sig}=0$, and it raises a question how come Stage 1 had a low bandwidth to begin with?

Should we cascodize Stage 2 instead? We recall the floating load of stage 1. If stage 2 suffers from Miller effect, the large input capacitance of stage 2 will strongly influence the output time- constant of stage 1. So, here is how a cascodizing of stage 2 causes a bandwidth enhancement of stage 1.

Under what circumstances (involving various R_{sig} scenarios) is it sometimes helpful to cascodize both Stage 1 as well as Stage 2?

The above questions guide our ADS activities of Section 6.16.

6.16 ADS: TUNING OF VAR PARAMETERS; "CASCODIZING" OF CE AMPLIFIERS BY USING WIRE LABELS; STUDY OF THE BANDWIDTH OF A CE AMPLIFIER AND ITS DEPENDENCE ON THE INPUT VOLTAGE SIGNAL SOURCE RESISTANCE; DESIGN OF BJT CASCODE AMPLIFIER; "CASCODIZING" STRATEGIES OF A MULTISTAGE DIFFERENTIAL AMPLIFIER

Example 6.7

Let us make a **comparison between a CE and cascode amplifiers** using the schematic shown in Figure 6.104.

The circuit of Figure 6.104 shows a cascode amplifier. The BJT of the CE stage (Q_1) is biased exactly the same as biasing a standalone CE amplifier.

The CB stage (Q_2) is biased in part by a battery V_{BB2}. Note the use of a wire label V_{B2}. By the choice of R_{B1}, R_{B2} and R_E we expect the current I_C to be around 1 mA.

So, for Q_2 to be in active mode, we need V_{C2} to be $\geq$ than V_{B2}. Therefore $V_{B2} \leq 7\,V$. For Q_1 to be in active mode: $V_{B2} - V_{BE} \geq V_{B1} = 5\,V$. So, $V_{B2} \geq 5.7\,V$. Let us choose V_{B2} to be the midpoint between the above two limits (5.7–7). Hence $V_{B2} = 6.4\,V$.

Let us first check the DC conditions. The results are shown in Figure 6.105. We see that the DC voltages and currents came out to be more or less as planned.

Next, let us check the AC simulation performance of the cascode amplifier. Results are shown in Figure 6.106.

Bandwidth comparisons between a CE amplifier and a matching cascode amplifier can be done most conveniently with the help of wire labels. In the cascode amplifier of Figure 6.104, if we wish to quickly jump to a CE amplifier, we may change two labels: V_{B2} (at the base of Q_2) → V_{C1}, and V_{C2} → V_{C1}. Now Q_2 is totally short circuited and disconnected from the V_{BB2} battery. So, the schematic of Figure 6.104 becomes that of a CE amplifier. Then if we change R_{source}, and wish to jump back to a cascode, we change Q_2's base and collector labels to V_{B2} and V_{C2}, respectively. And so back and forth.

The cascode results of Figure 6.106 are compared to the CE amplifier results of Figure 6.95 (or the quickly altered Figure 6.104).

There is almost the same voltage gain and R_{in}. The cascode gain is slightly larger though. BW of the cascode is significantly larger than that of the CE amplifier.

Here $R_{source} = 1\ k\Omega$, so there is a significant BW-lowering Miller effect in the CE amplifier.

How will the two amplifiers compare if we change R_{source} to 1 Ω?

What we see in Figure 6.107 is almost the same performance. Cascode is slightly superior though. Slightly larger A_v and Larger BW: Cascode BW = 20.6 MHz, C_E BW = 18.8 MHz.

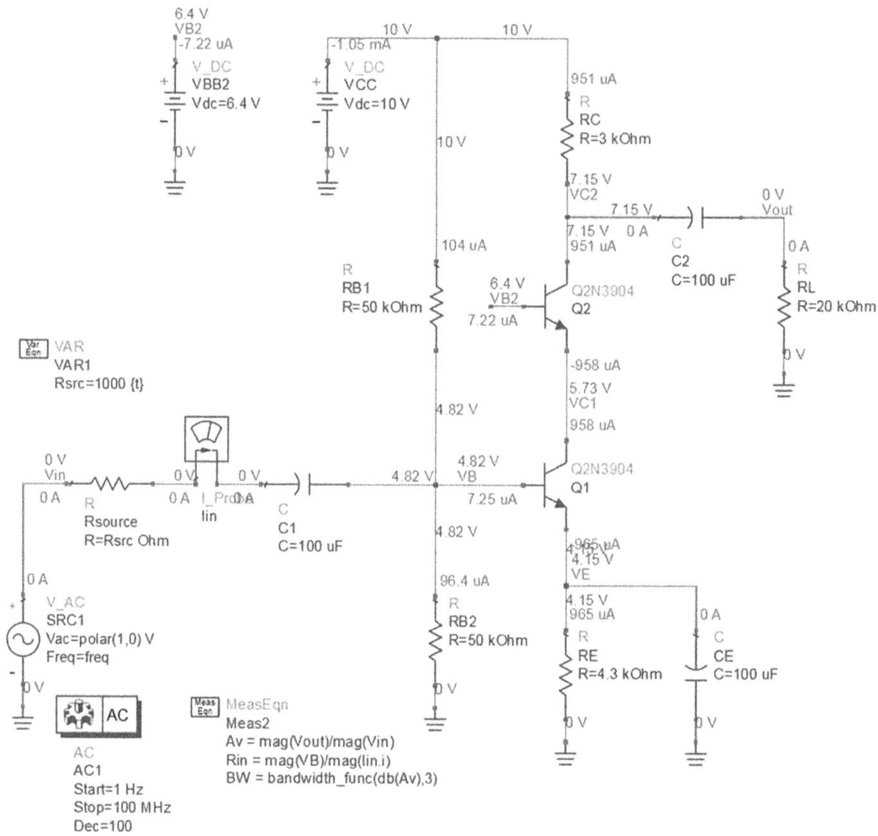

FIGURE 6.105 DC conditions of the BJT cascode amplifier.

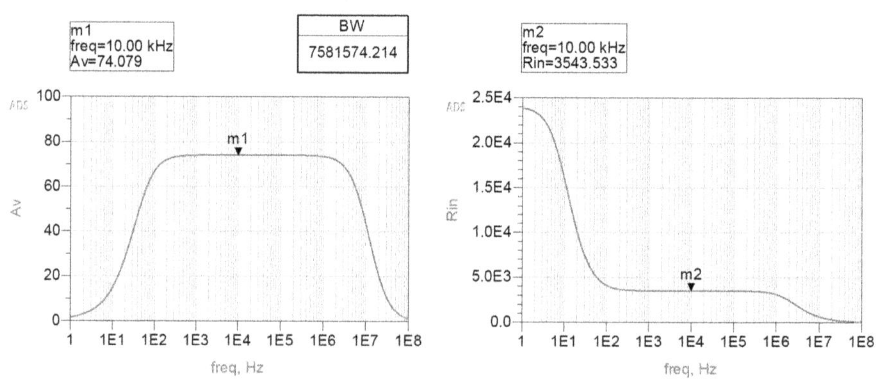

FIGURE 6.106 A_v and R_{in} of the cascode amplifier.

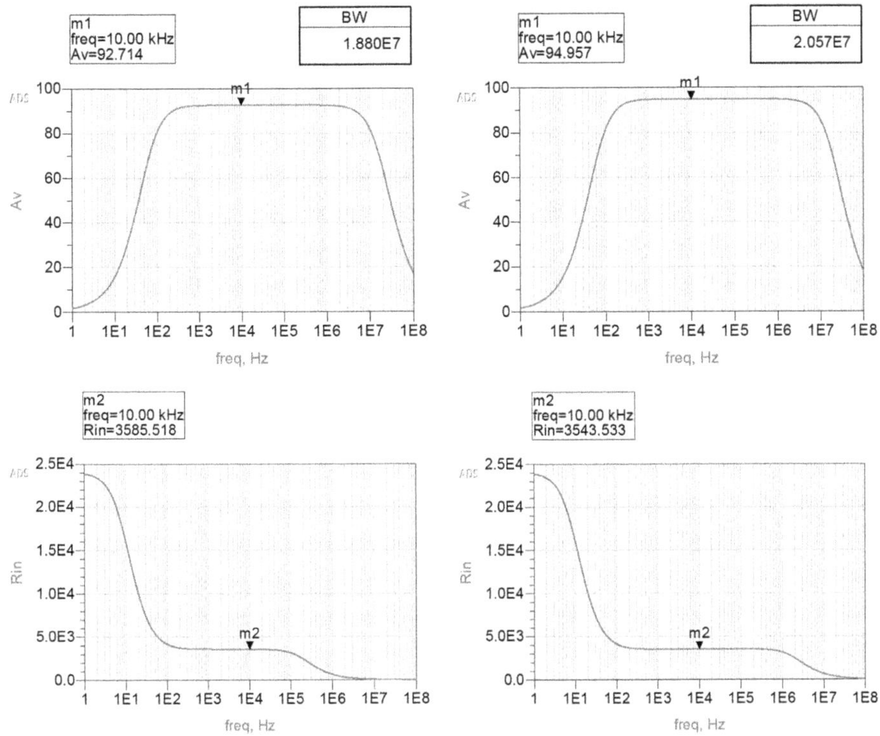

FIGURE 6.107 Comparison between CE (left) and cascode (right) when $R_{sig} = 1\ \Omega$.

Example 6.8

Finally, as promised, let us discuss the **cascodizing of the multistage amplifier**, shown in Figure 6.108. The design implements the cascodizing of both stage 1 and stage 2.

Depending on R_{source} (added symmetrically) what of the three options is the best in terms of BW (Cascodizing stage 1, cascodizing stage 2, cascodizing both)?

There are two batteries, one for stage 1 and one for stage 2, tuned each to the mid-value between two saturation borderline limits. Check why we chose values of $V_{BB1} = 6\,V$ and $V_{BB2} = 11\,V$.

Notice that we use wire labels to connect these two batteries as well, and the CB amplifiers' node labels to quickly and selectively switch stage 1 and/or stage 2 between cascode and regular differential amplifiers.

We can even cascodize stage 3, if we want to go crazy.

If we cascodize any of stages 1 or 2, do we have to readjust the offset resistor?

Probably not, if we only cascodize stage 1. Most likely yes whenever we cascodize stage 2.

The DC results for cascodizing the first two stages of the multistage amplifier are shown in Figure 6.109. We see that the DC offset came out to be –152 mV. An adjustment of RX is needed. The AC results of Figure 6.110 should be compared to those of Figure 6.42.

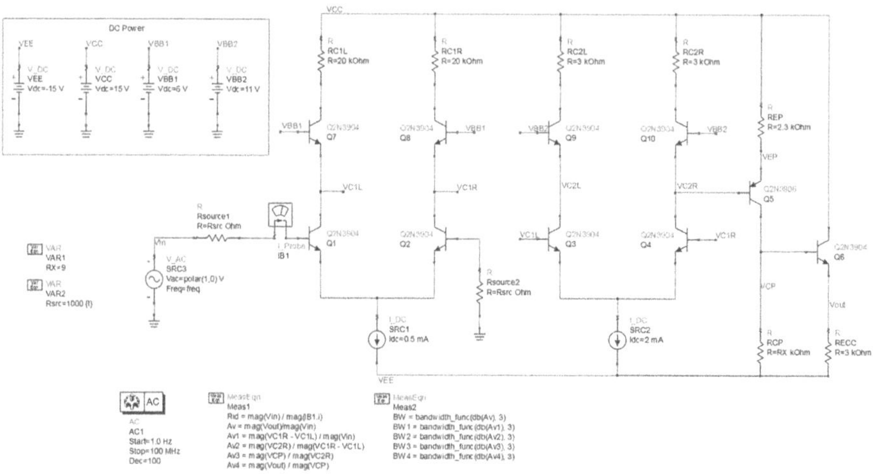

FIGURE 6.108 Cascodizing of the first two stages of the multistage amplifier.

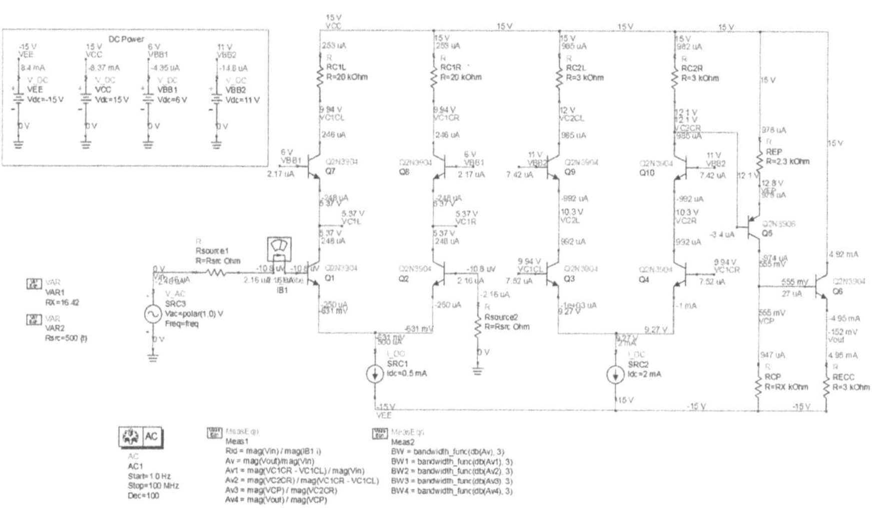

FIGURE 6.109 DC results of the multistage amplifier in which we cascodize the first two differential amplifier stages.

Compared to the non-cascode version, the BW of the cascode version is higher: 1.067 MHz vs. 161.3 kHz for non-cascode.

Let us conclude this section with a few more comments/observations.

Wire labeling of BJT nodes allows for an ability to quickly go back and forth between cascode and CE amplifiers. For example, observe the node names of the left cascode of the first stage of the multistage amplifier (Figure 6.108).

Q_7 is the BJT of the CB amplifier, and its nodes are labeled V_{BB1} (base), V_{C1L} (emitter) and V_{C1CL} (collector). Changing both labels V_{BB1} and V_{C1CL} to V_{C1L} totally short circuits Q_7. The resulting amplifier is a CE amplifier. Restoring the original names of the base and collector of Q_7 takes us back to a cascode configuration.

BW	BW1	BW2	BW3	BW4
1.067 M	1.805 M	2.328 M	2.889 M	0.0000

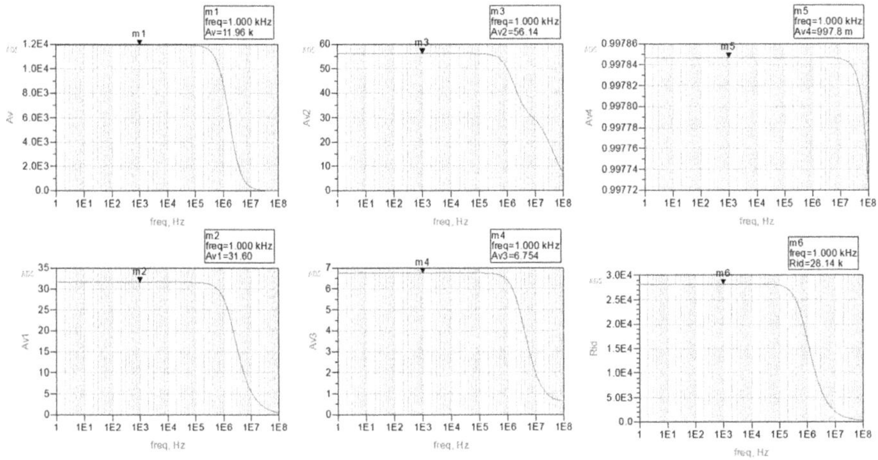

FIGURE 6.110 AC results of the multistage amplifier in which we cascodize the first two stages.

For the multistage amplifier, if the symmetric $R_{sig}=0$, we may be temporarily amazed to discover that a cascodizing of the two sides of Stage 1 is useless: BW actually becomes slightly smaller. Then we realize that with $R_{sig}=0$ the input circuits of the Stage 1 differential amplifiers have very little time constant. The poor bandwidth of Stage 1 is due to its output Stage 1 circuit's time constants.

Indeed, the differential load of Stage 1 (which determines the output circuit's time constants of Stage 1) is strongly affected by the Miller effect of Stage 2! The cascodizing of only Stage 2 causes a nice increase of the multistage amplifier's overall bandwidth.

If the adding of symmetric R_{sig} is modest to large, then both stages, 1 and 2, suffer from Miller effect. Cascodizing both Stage 1 and Stage 2 yields a better bandwidth than the one obtained if we cascodize only stage 2.

6.17 BASICS OF FEEDBACK ANALYSIS OF ELECTRONIC AMPLIFIERS

A linear circuit (such as RLC) may have multiple external inputs (that is, voltage and current signal sources connected to the circuit), and every node voltage or component current of the circuit may be chosen to be an output. Using Laplace transform [Gable1], for every pair of an input X(s) and a chosen output Y(s), there is a transfer function T(s) relating multiplicatively X(s) to Y(s): **Y(s) = T(s)·X(s)**.

Let us consider two linear circuits having transfer functions of $T_1(s)$ and $T_2(s)$, respectively. One way of connecting the two circuits is in **cascade**. That is, the output signal of $T_1(s)$ becomes the input signal of the second circuit. The overall transfer function of a **cascade connection** is $\mathbf{T_1(s) \cdot T_2(s)}$. Another way of connecting $T_1(s)$ and $T_2(s)$ is in **parallel**, meaning that both $T_1(s)$ and $T_2(s)$ are fed by the same input signal, and that their outputs are combined. The overall transfer function for a **parallel connection** is $\mathbf{T_1(s) + T_2(s)}$. See [Dorf1].

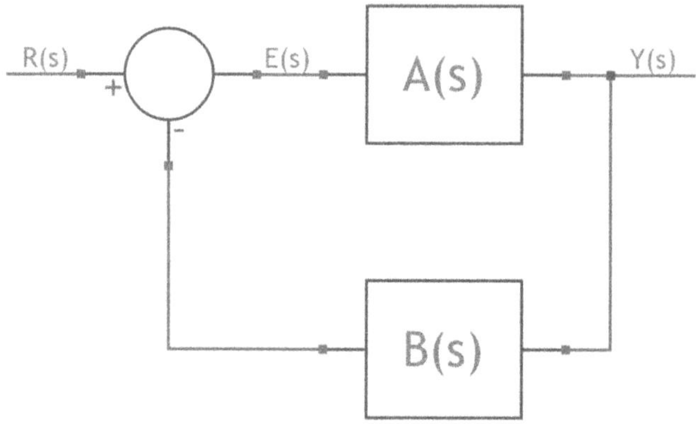

FIGURE 6.111 Negative feedback configuration of the two transfer functions A(s) and B(s).

A third basic connection between two transfer functions is the **feedback connection**, shown in Figure 6.111. The block diagram of Figure 6.111 can be analyzed using the above basic properties of transfer functions (of cascade and parallel connections): $Y(s) = A(s) \cdot E(s) = A(s) \cdot [R(s) - B(s) \cdot Y(s)]$. Solving for Y(s) yields equation (6.1).

The resulting transfer function between the input R(s) and the output Y(s) in a negative feedback connection is:

$$\frac{Y(s)}{R(s)} = \frac{A(s)}{1 + A(s)B(s)} \tag{6.1}$$

As seen in Figure 6.111, the transfer function A(s) represents the connection between the signal E(s) and the output signal Y(s). We call **A(s)** the **forward amplifier** (or sometimes we call it the **open-loop amplifier**). The transfer function B(s) has Y(s) as an input, and it creates a feedback signal that gets subtracted from the input R(s), to create the signal E(s). We call **B(s)** the **feedback network** (or sometimes we use the longer name **feedback sensor's transfer function**).

In order to implement a feedback connection there must be some kind of a "signals summing device". If the feedback signal, that is the output of B(s), is subtracted from R(s) we have **negative feedback**. If the summing device performs an addition of R(s) and the feedback signal, we call the connection a **positive feedback** connection. We shall be interested in negative feedback circuits. Positive feedback is often destabilizing. It may sometimes be used to create an electronic oscillator or a Schmitt Trigger (the latter has been studied in Section 1.22), but it is to be avoided whenever implementing a feedback amplifier.

The transfer function B(s) is often implemented by a suitable resistive network, and in such a case $B(s) = B$, a constant gain. If we are interested in the amplifier's performance within the mid-band frequency range then $A(s) = A$, a constant gain. By equation (6.1), a negative feedback amplifier with resistive feedback network, operating deep inside the mid-band frequency range, has a gain of **$A/(1 + A \cdot B)$**. **If $AB \gg 1$ then the feedback amplifier's gain equals approximately 1/B.** We have seen it, for instance, in the case of an op-amp voltage amplifier, where the overall gain was determined by the feedback network, due to the large differential gain of the op-amp (Chapter 1). **In order to implement a negative feedback amplifier configuration, in the presence of a subtraction mechanism, of the feedback signal from the input signal, the gains A and B must be either both positive or both negative.**

More often than not, we will be interested in the bandwidth of the feedback amplifier. A special case is whenever A(s) (of Figure 6.111) is a first order transfer function $\mathbf{A(s) = A_0/(1 + s/\omega_h)}$ where A_0 is the mid-band gain of A(s) and ω_h is the corner frequency of the first pole in the high frequency range. Assuming that $\mathbf{B(s) = B}$ = const. then by (6.1) the closed loop transfer function is $\mathbf{Y(s)/R(s) = [A_0/(1 + A_0B)] \cdot [1/(1 + s/(\omega_h(1 + A_0B))]}$. That is, due to the feedback, the open-loop gain A_0 reduces, in closed-loop operation, by a factor of $(1 + A_0B)$ while the open loop bandwidth ω_h increases in closed-loop operation by the same factor $(1 + A_0B)$. The so-called **gain-bandwidth product** equals $A_0\omega_h$, for any value of B, and it is a property of A(s) only. So, the larger the closed-loop gain 1/B, the smaller the closed-loop bandwidth. See [Razavi1]. Very often the "gain-bandwidth product property" is obtained by design, assuring that the closed-loop bandwidth $\omega_h(1 + A_0B)$ is sufficiently lower than all the corner frequencies of the high-frequency poles and zeros of A(s). We shall come back to this feedback compensation topic later on in this chapter.

There are four types of electronic amplifiers: **voltage amplifiers**, **current amplifiers**, **transconductance amplifiers,** and **transimpedance amplifiers**. In a transconductance amplifier, the input signal is a voltage, and the output signal is a current. In a transimpedance amplifier, the input signal is a current, and the output signal is a voltage. We shall be interested in feedback implementation of such amplifiers. See [Sedra1].

Any amplifier is essentially an interface between an input signal source r(t) and an output signal y(t) associated with a load R_L. The variable y(t) may be the voltage drop of R_L or the current flowing through R_L. The reader surely remembers encountering the subject of ideal vs. non-ideal voltage and current sources in an earlier Circuits course. An ideal voltage source has an internal resistance of zero. An ideal current source has an internal resistance of infinity [Nielson1].

Amplifiers that receive a voltage input (and these are voltage or transconductance amplifiers) are designed to have a large input resistance R_{in}, in order to make the gain less sensitive to finite internal resistance values of the voltage signal source. Likewise, amplifiers that receive a current input (and these are current or transimpedance amplifiers) are designed to have a small input resistance R_{in}, to make their gain less sensitive to the internal resistance of the current signal source.

Amplifiers that deliver a voltage output (that is, voltage or transimpedance amplifiers) are designed to have a small output resistance R_{out}, to make the gain less sensitive to variations in the R_L values. Similarly, amplifiers that deliver current (that is, current or transconductance amplifiers) are designed to have a large R_{out} to reduce their gain sensitivity to R_L.

Crucial aspects of such designs are the selection of resistive networks that constitute the feedback gain B, and the selection of how such networks, respectively, are to be connected to the open-loop amplifier A(s).

The feedback network "measures" the amplifier's output signal y(t).

It then converts the output signal to the type of signal, current or voltage, that needs to be subtracted from the input r(t).

To explain the output measuring idea, let us refer the reader again to an earlier Electrical Engineering course, this time around, a basic Circuits Lab. Recall the operation of a multimeter [Nilsson1]. To measure a certain node's voltage, the voltmeter's + probe must be attached to that node. The − probe connects to ground to provide a zero reference for the measurement. To measure a current, the ammeter must allow the measured current to flow through it, from the ammeter's incoming terminal to its outgoing terminal. That is, the path through which the current, that needs to be measured, flows, is opened up in order to insert the ammeter device. The same is done by the feedback resistive network B. If y(t) is a voltage signal, then the input node of B is connected to the node of y(t). Such a connection is called a **shunt connection**. If y(t) is a current, then y(t) must flow through one (or more) of the resistors of B. Such resistor(s) will constitute a **series connection** to the load through which y(t) flows.

The feedback network B converts the measured output signal y(t) into a feedback signal, proportional to y(t), that has the same type as that of the input r(t). That feedback signal needs to be subtracted from r(t), but how is the "summing device" (of Figure 6.111) implemented?

By KVL, voltages are added up, one to another, by being stacked in series (a series connection). For one voltage to be subtracted from the other, that voltage must have an opposite polarity from the other voltage. Alternatively, a subtraction of voltages is attained by inserting both as inputs to a differential amplifier, as a differential amplifier is sensitive to the differential input.

By KCL, on the other hand, currents are added up by flowing all into the same node (a shunt connection). In addition to becoming a subtraction, the feedback current signal must arrive at the addition location, having a polarity that is opposite to that of the input current signal r(t). In the following Figures 6.112–6.115, let us explore the simplest possible B network structure needed for each amplifier type and observe some properties of each amplifier type.

The example, shown in Figure 6.112, illustrates essential parts of a typical **feedback voltage amplifier**, also called a **series-shunt amplifier**, as the feedback network is connected in series at the input and in shunt at the output.

The simplest feedback network B consists of two resistors arranged in an L shape.

This L-shaped feedback network is how it is done in a non-inverting op-amp amplifier (studied in Chapter 1). R_{f1} connects in shunt to v_o. The feedback voltage signal v_f, created by the voltage divider of R_{f1} and R_{f2}, becomes one of two input signals of a differential amplifier (be it BJT or MOSFET), whereas v_{in} enters the opposite input terminal of the differential amplifier. A differential amplifier configuration is an attractive way of implementing the subtraction of voltages.

The forward voltage amplifier A(s) consists of the differential voltage gain of the first amplifier stage (noting that a non-inverting single-ended output is often chosen) and the gains of other stages, needed to create a large mid-band open-loop gain (that is $A{\cdot}B \gg 1$) as well as keeping $A > 0$. The voltage amplifier, which is part of the forward transfer function A(s), may be a single-input voltage amplifier (as shown in Figure 6.112) or a multistage differential amplifier. The input signal to the feedback

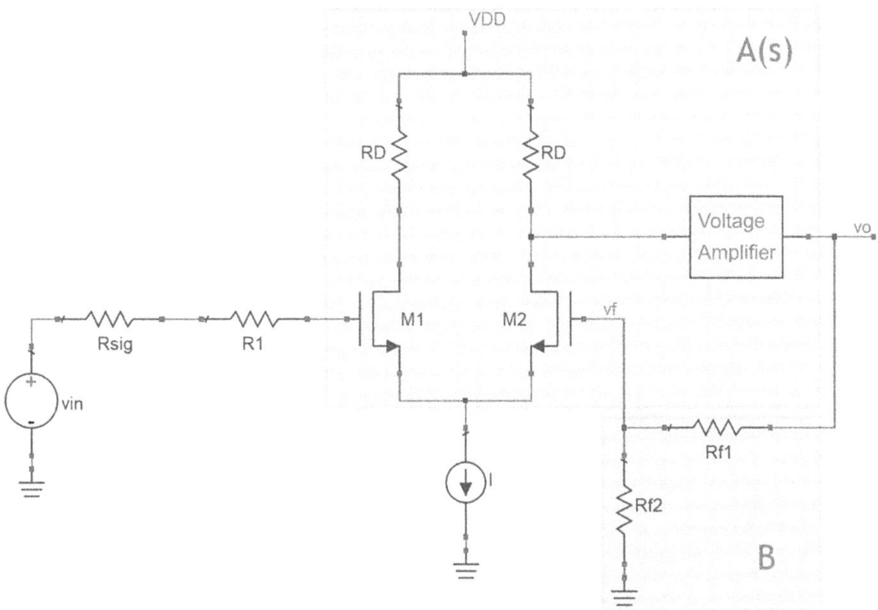

FIGURE 6.112 An illustration of a feedback voltage amplifier.

network B is the output voltage v_o, and the feedback voltage gain is $\mathbf{B = v_f/v_o = R_{f2}/(R_{f1} + R_{f2})}$. We note that $B > 0$. The loop gain is negative: $-A{\cdot}B$, due to the subtraction performed at the first differential stage. The feedback is indeed negative.

The feedback network B loads the right-hand input of the differential amplifier. Need to add to R_{sig} of v_{in} (the input signal that connects to the left-hand side of the differential amplifier) a specific series resistor R_1, to assure that the input ports of the differential amplifier maintain amplifier symmetry. We shall come back to this point.

Figure 6.113 illustrates the essential parts of a **feedback current amplifier**, also called a **shunt-series amplifier**.

In a current amplifier, the load R_L, carrying the current i_o, is usually floating (that is, not connected to ground). The resistive feedback network B is again a kind of L-shaped, consisting of the resistors r_{f1} and r_{f2}. The current division created between r_{f1} and r_{f2} samples linearly the load current i_o to create the feedback current i_f. To cause i_f to be subtracted from i_{in} (as shown in Figure 6.113, where i_f flows away from i_{in}), the feedback network gain B must be negative, as i_f flows inwards into B. Therefore, the current amplifier A(s) must be inverting to create a negative feedback. Here, both A and B must be negative.

The input signal to the feedback network B is i_o and hence $\mathbf{B = i_f/i_o = -r_{f1}/(r_{f1} + r_{f2})}$. Again, we note that $A{\cdot}B > 0$. We often need the loop gain to be large, obtained sometimes by having A(s) be a multistage amplifier.

We further note from Figures 6.112 and 6.113 that A must have the same types of input and output signals, as those of the overall feedback amplifier: In a feedback voltage amplifier, the open-loop transfer function A(s) has a voltage input and a voltage output, and in a feedback current amplifier, the open-loop A(s) has current input and current output. Likewise, and consistently, the types of input and output signals of the B networks.

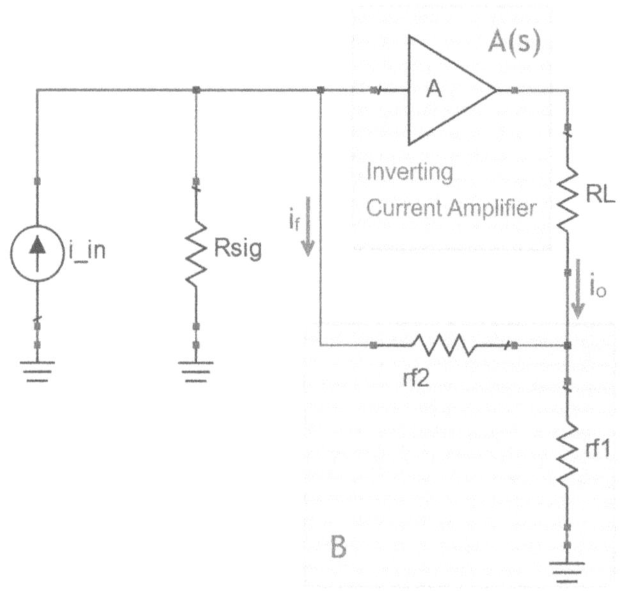

FIGURE 6.113 An illustration of a current feedback amplifier.

Moving on to **feedback transconductance amplifiers**, also called **series-series amplifiers**, the simplest possible passive B feedback network needs to have no less than three resistors r_{f1}, r_{f2}, and R_f, and it is п-shaped, as shown in Figure 6.114. The feedback network B samples the load current i_o, as we have just seen for the feedback current amplifier (Figure 6.113). It then needs a third resistor to create voltage feedback that is part of the differential input of the amplifier, as we have seen in a feedback voltage amplifier (Figure 6.112).

The gain B of the п-shaped feedback network is a transimpedance gain (from current i_o to voltage v_f). It is obtained by first considering the current division between r_{f1} and $r_{f2}+R_f$ that i_o, the input to the feedback network, encounters. Second, the feedback voltage v_f is created by the current flowing through resistor R_f. The resulting transimpedance feedback gain is $\mathbf{B=v_f/i_o=R_f \cdot r_{f1}/(r_{f1}+r_{f2}+R_f)}$.

Again (as in Figure 6.112), resistor R_1 is selected so that the two inputs of the differential amplifier are symmetric. The forward transconductance gain A must be positive (and it is usually large), and, like in a voltage amplifier, the negative feedback is obtained from the signal subtraction performed by the differential amplifier input stage.

The selected B network architecture determines the type of the feedback amplifier.

The fourth feedback amplifier type (illustrated in Figure 6.115) is a **transimpedance amplifier**, also called a **shunt-shunt amplifier**.

Among the four simple resistive networks B studied here, the one used to create a transimpedance amplifier is the simplest, consisting of a single bridging resistor R_f shown in Figure 6.115. As in the current amplifier, shown in Figure 6.113, in order to create feedback current i_f that opposes the polarity of the input current i_{in}, the feedback transconductance gain (from a voltage v_o to a current i_f) is negative.

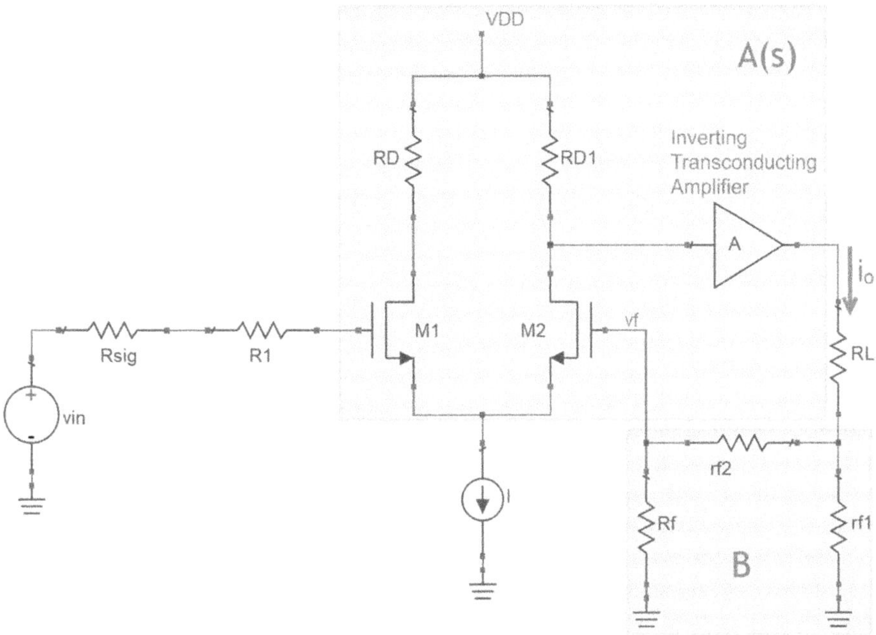

FIGURE 6.114 An illustration of a feedback transconductance amplifier.

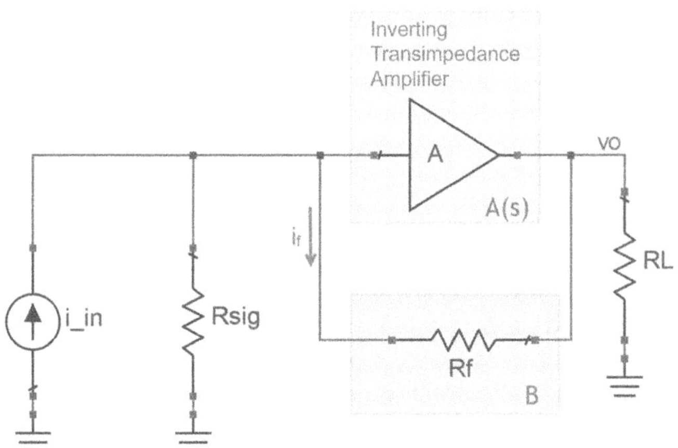

FIGURE 6.115 An illustration of a feedback transimpedance amplifier.

Therefore, the forward transimpedance amplifier A(s) must be an inverting amplifier. The mid-band gain of A(s) is taken to be large, as done in all four types of amplifiers. That is, $A < 0$ (i.e., large negative gain) and $|A \cdot B| \gg 1$, where the transconductance feedback gain is $\mathbf{B = i_f/v_o = -R_f}$.

Resistive B networks are suitable for any breadboard or PCB, discrete implementation of a feedback amplifier. For MOSFET ("system on a chip") implementation of feedback amplifiers, the feedback B network is often implemented with capacitors, using capacitor network architectures that remind one of the resistive networks discussed in Figures 6.112–6.115. See [Razavi1] and [Gray1].

The menu of B networks is indeed short. Once a specific B network is selected, combined with a negative feedback implementation and a large loop gain, the resulting effect on the closed-loop input resistance R_{inf} and closed-loop output resistance R_{outf} is dramatic. Each of the resistances increases or decreases by a factor of $1+AB$, going in the correct direction: For example, in a voltage amplifier, R_{inf} increases compared to R_{in} (of the open loop amplifier), and R_{outf} decreases compared to R_{out} of the open loop amplifier. As another example, in a transimpedance amplifier, R_{inf} and R_{outf} both decrease from the open-loop values, each by a factor of $1+AB$.

Table 6.3 summarizes the closed-loop input and output resistances of the four feedback amplifiers.

It remains to clarify though regarding Table 6.3, what exactly is meant by A·B, R_{in}, and R_{out}. Are these open-loop entities? Can these be inter-dependent?

Let us start with the "ideal" cases, illustrated in Figures 6.116–6.119. By "ideal" we mean that (i) the open-loop forward gain A and the feedback gain B are independent, (ii) the effect of the input internal resistance R_{sig} is negligible with respect to the effect of the open-loop input resistance R_{in0}, and (iii) the effect of the forward amplifier output resistance R_{oo} is negligible with respect to the load R_L whenever we measure R_{inf}.

In an ideal feedback voltage amplifier, shown in Figure 6.116, it is assumed that $R_{sig} \ll R_{in0} = v_e/i_e$ and $R_{oo} \ll R_L$.

The closed-loop input resistance, of Figure 6.116 (left), is derived by neglecting the voltage division between R_{in0} and R_{sig} and by neglecting the voltage division between R_{oo} and R_L: $R_{inf} = v_{in}/i_e = (v_f + v_e)/i_e = (B_v v_o + v_e)/i_e = (B_v A_v v_e + v_e)/i_e = (1 + A_v B_v) \cdot R_{in0}$.

The closed-loop output resistance R_{outf} is measured as shown in Figure 6.116 (right).

The measurement involves a test output current signal i_x, nulling of v_{in}, and neglecting of R_{sig}, which makes $v_e = -v_f$. Then $R_{outf} = v_x/i_x = v_x/[(v_x - A_v v_e)/R_{oo}] = v_x/[(v_x + A_v B_v v_x)/R_{oo}] = R_{oo}/(1 + A_v B_v)$.

The ideal feedback current amplifier, shown in Figure 6.117, has an open-loop inverting current model that includes a current-dependent-current-source $A_i i_e$. "Inverting", as the directions of $A_i i_e$ and i_o are opposite, is shown in Figure 6.117 (left). Note the direction of the feedback current i_f away from the current subtraction node, which, as explained earlier, necessitates that $B_i < 0$, and this in turn necessitates that the forward gain is $A_i < 0$.

It is assumed that $R_{sig} \gg R_{ino}$ and that $R_{oo} \gg R_L$.

TABLE 6.3
Closed Loop R_{in} and R_{out} of the Four Amplifier Types

	R_{inf}	R_{outf}
Voltage	$R_{in}(1+AB)$	$R_{out}/(1+AB)$
Current	$R_{in}/(1+AB)$	$R_{out}(1+AB)$
Transconductance	$R_{in}(1+AB)$	$R_{out}(1+AB)$
Transimpedance	$R_{in}/(1+AB)$	$R_{out}/(1+AB)$

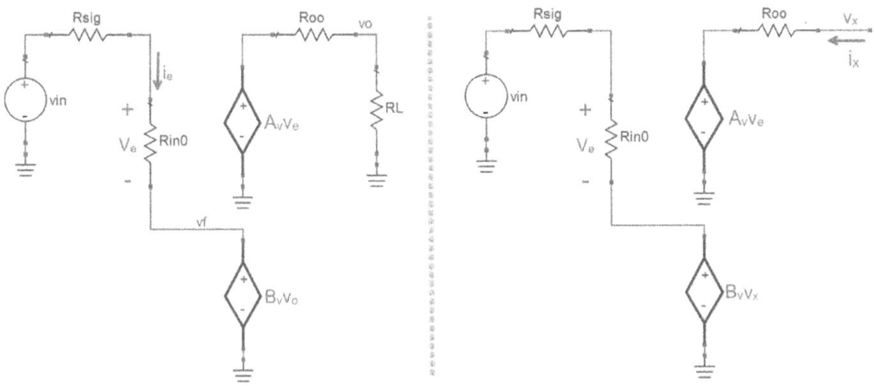

FIGURE 6.116 Ideal feedback voltage amplifier (left) and R_{outf} measurement (right).

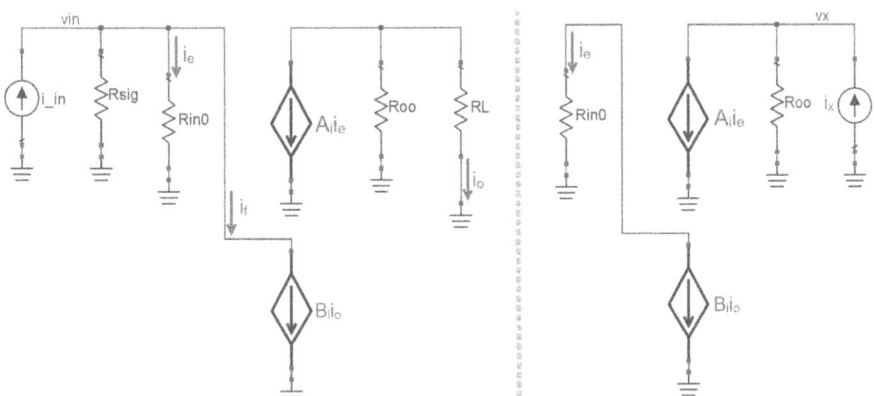

FIGURE 6.117 Ideal feedback current amplifier (left) and R_{outf} measurement (right).

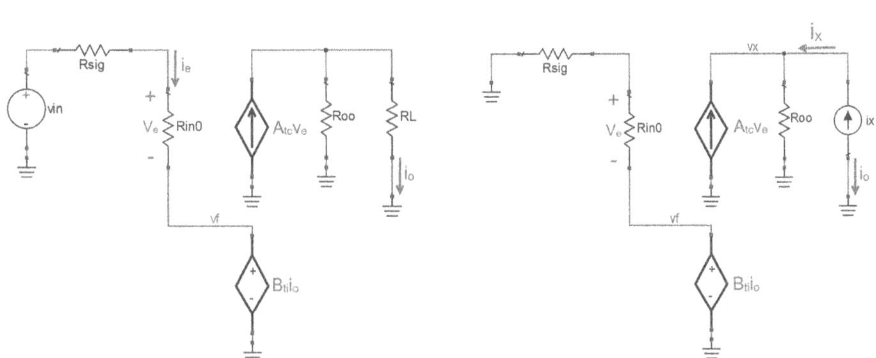

FIGURE 6.118 Ideal feedback transconductance amplifier (left) and R_{outf} measurement (right)

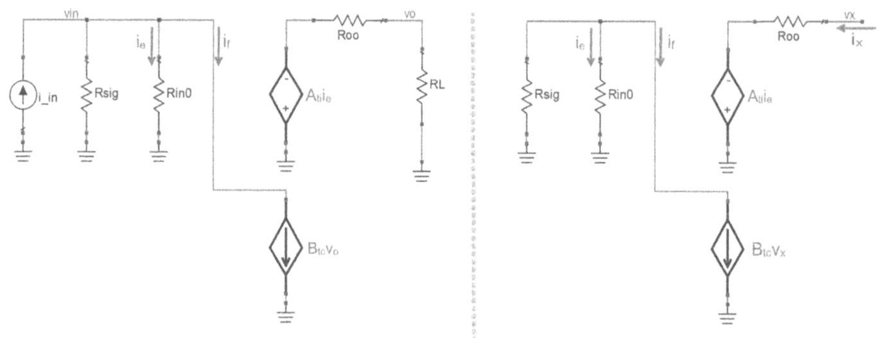

FIGURE 6.119 Ideal feedback transimpedance amplifier (left) and R_{outf} measurement (right).

The derivation of the closed-loop input resistance R_{inf} follows: $R_{inf} = v_{in}/i_{in} = i_e \cdot R_{in0}/(i_e + i_f) = i_e \cdot R_{in0}/(i_e + B_i i_o) = i_e \cdot R_{in0}/(i_e + B_i A_i i_e) = R_{in0}/(1 + A_i B_i)$.

For the derivation of the closed-loop output resistance R_{outf}, we note that $i_e = -i_f$, $i_f = B_i i_o$ and $i_o = -i_x$. Then $R_{outf} = v_x/i_x = (i_x - A_i i_e) \cdot R_{oo}/i_x = [i_x - A_i(-i_f)] \cdot R_{oo}/i_x = [i_x - A_i(-B_i i_o)] \cdot R_{oo}/i_x = [i_x - A_i(B_i i_x)] \cdot R_{oo} \rightarrow R_{outf} = R_{oo}(1 - A_i B_i)$ where $B_i < 0$ and in the inverting current-controlled-current-source (as seen by its current direction) the current gain coefficient is $A_i > 0$. R_{outf} increases by a factor of $1 + |A_i B_i|$.

In Figure 6.118, we note the open-loop transconductance gain A_{tc} and the feedback transimpedance gain B_{ti}. The loop gain $A_{tc}B_{ti}$ is unitless. The open-loop transconductance amplifier is non-inverting, shown in the dependent source current direction. It is assumed that $R_{in0} \gg R_{sig}$ and that $R_{oo} \gg R_L$. Then:

$R_{inf} = v_{in}/i_e = (v_f + i_e R_{in0})/i_e = (B_{ti} i_o + i_e R_{in0})/i_e = (B_{ti} A_{tc} v_e + v_e)/i_e = (1 + A_{ti} B_{tc}) R_{in0}$.

$R_{outf} = v_x/i_x = [(i_x + A_{tc} v_e) \cdot R_{oo}]/i_x = [i_x + A_{tc}(-v_f)] \cdot R_{oo}/i_x = [i_x + A_{tc}(-B_{ti} i_o)] R_{oo}/i_x = (i_x + A_{tc} B_{ti} \cdot i_x) R_{oo}/i_x = R_{oo}(1 + A_{tc} B_{ti})$.

Finally, for the transimpedance amplifier (Figure 6.119), it is assumed that $R_{sig} >> R_{in0}$ and $R_{oo} << R_L$.

We further note that B_{tc} should be negative and the forward transimpedance amplifier is made inverting (See Figure 6.119).

$R_{inf} = v_{in}/i_{in} = R_{in0} i_e/(i_e + i_f) = R_{in0} i_e/(i_e + B_{tc} v_o) = R_{in0} \cdot i_e/[i_e + B_{tc}(-A_{ti} i_e)] = R_{in0}/(1 - A_{ti} B_{tc})$. B_{tc} is negative, and the coefficient A_{ti} is positive. R_{inf} is by $(1 + |B_{tc} A_{ti}|)$ smaller than R_{in0}.

We neglect R_{sig} in the output resistance calculation.

$R_{outf} = v_x/i_x = (-A_{ti} i_e + R_{oo} i_x)/i_x = (-A_{ti}(-i_f) + R_{oo} i_x)/i_x = (-A_{ti}(-B_{tc} v_x) + R_{oo} i_x)/i_x = A_{ti} B_{tc} R_{outf} + R_{oo}$ which leads to $R_{outf} = R_{oo}/(1 - A_{ti} B_{tc})$. R_{outf} is smaller than R_{oo} by a factor of $1 + |A_{ti} B_{tc}|$.

With the 1 plus loop gain factor, Table 6.3 can be easily memorized, noting that whenever the feedback B network is shunt connected, say at the amplifier's input, it causes the input resistance to decrease. Why? Any network that is connected in parallel to any resistance decreases the overall resistance. The same happens in a shunt connection to the amplifier's output. Likewise, a serial connection increases resistance. So, for instance, in a shunt-shunt amplifier, the B network decreases both input and output resistance. In a series-shunt amplifier, the B network increases the input resistance and decreases the output resistance, and so on.

In each of the above ideal amplifiers, the feedback B network is represented by a dependent source. The B network's gain does not affect, and is unaffected by, the A(s) amplifier. In practice, though, as one replaces the ideal dependent source feedback network with a resistive network, there are mutual loading effects: A(s) is affected by the B network. The B network itself is affected by the forward amplifier. For most practical amplifiers, though, the effects of the forward amplifier on the operation of the feedback network are very small.

We refer to [Sedral] for a rigorous mathematical derivation, via appropriate two-port networks, of the ways we test how the B networks load the A(s) amplifier. We shall instead look at the bottom-line results of the above derivations, and explain their intuitive meaning.

Result 1: The loading by the B-network of the amplifier's input is modelled by a resistor R_{fi} that is connected to the input in the same way that the B-network is connected to the amplifier's input. For instance, if the B-network connects in series to the amplifier's input, the resistor R_{fi} is also connected in series to the input side of the amplifier.

Likewise:

Result 2: The loading by the B-network of the amplifier's output is modelled by a resistor R_{fo} that is connected to the output in the same way that the B-network is connected to the amplifier's output.

Result 3: R_{fi} is the equivalent resistance "seen" at the amplifier's B-network, looking from the amplifier's input side, as the feedback is eliminated at the B-network's amplifier's output side. If the B-network is shunt-connected to the amplifier's output, this connection node is grounded for the computation of R_{fi}. If the B-network is series-connected to the output, the series connection is disconnected for the R_{fi} calculation.

Likewise:

Result 4: R_{fo} is the equivalent resistance "seen" at the amplifier's B-network, looking from the amplifier's output side, as the feedback action is eliminated at the B-network's amplifier's input side. If the B-network is shunt-connected to the amplifier's input, this connection node is grounded for the calculation of R_{fo}. If the B-network is series-connected to the amplifier's input, the series connection is disconnected for the calculation of R_{fo}.

Figures 6.120–6.123 summarize, for each amplifier type, how the B-network is connected to the amplifier, how R_{fi} and R_{fo} are computed, and the circuit used for calculating the loaded forward gain A_{loaded}. B and A are negative whenever a current is the feedback signal.

The closed-loop input resistance and output resistance again increase, or decrease, by a factor of $1+A{\cdot}B$, compared to the respective open-loop input and output resistance, following Table 6.3. However, unlike the ideal A value, the correct practical A value is $A=A_{loaded}$.

The open-loop R_{in} depends on R_{in0} and R_{fi}, but it is not dependent on the "outside the amplifier" R_{sig}. The open-loop R_{out} depends on R_{oo} and $R_{fo,}$ and it should not depend on the "outside the amplifier" R_L. A_{loaded} though does depend on both R_{sig} and R_L. We refer to [Sedral] for the post-processing mathematical manipulations that remove R_{sig} from R_{inf} and remove R_L from R_{outf}. In most applications, though, R_{sig} is negligible compared to R_{in0}, and R_L affects the calculation of the closed-loop output resistance negligibly. In all the examples that follow, we played it simple, leaving R_{sig}

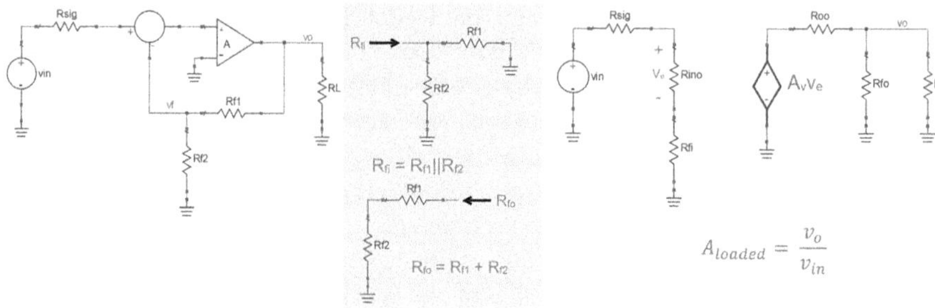

FIGURE 6.120 Voltage amplifier B-network loading effect on A.

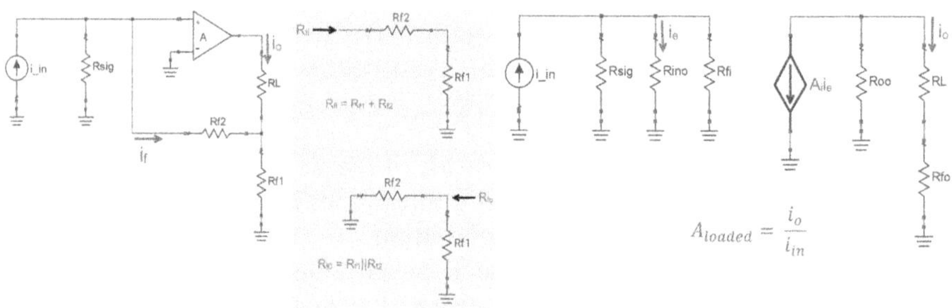

FIGURE 6.121 Current amplifier B-network loading effect on A.

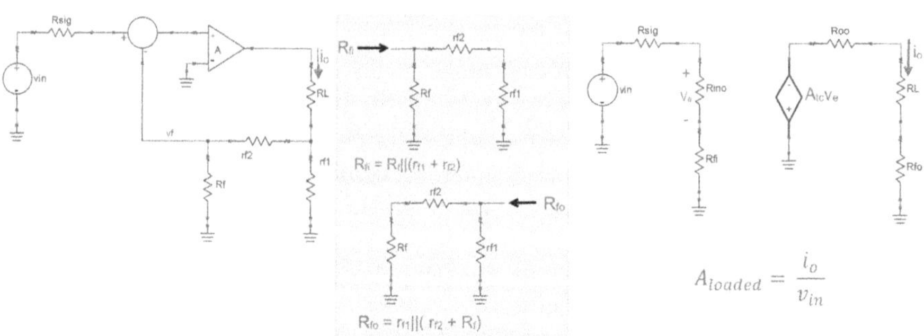

FIGURE 6.122 Transconductance amplifier B-network loading effect on A.

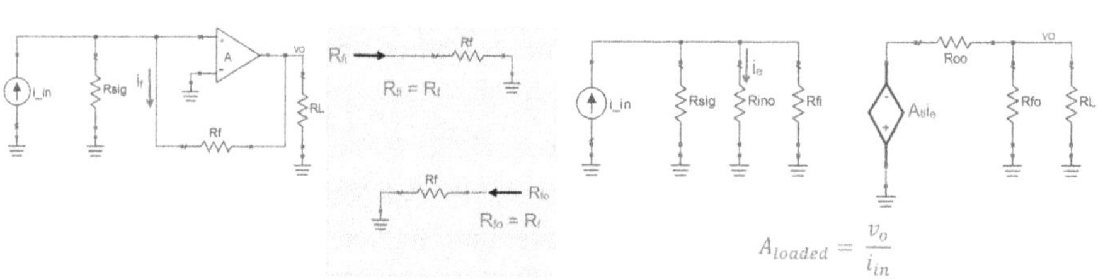

FIGURE 6.123 Transimpedance amplifier B-network loading effect on A.

and R_L in the closed-loop input and output resistance calculations, as an approximation. See all the feedback design examples that are coming up.

Finally, let us mention the open-loop and closed-loop bandwidth calculations. In Section 6.21, we shall discuss the stabilizing of the closed-loop amplifier by a "dominant pole" control strategy involving a "–20 dB/dec crossing over of the frequency axis at the 0 dB value of A(ω)B". As a result of the stabilization design strategy, the relationship between the open-loop bandwidth (which is often very low – for instance, the dominant pole of the 741 op-amp lies near $BW_{OL} \approx 10\,Hz$) and closed-loop bandwidth, is according to the Gain·Bandwidth product, discussed earlier in this section.

Example 6.9

The Non-Inverting Op-Amp Amplifier Revisited

In Section 1.7, we showed that if the op-amp is ideal (that is, has infinite differential voltage gain, infinite differential input resistance, and zero output resistance), then the non-inverting amplifier closed-loop voltage gain is equal to $1 + R_2/R_1$. We repeat Figure 1.41, as a reminder.

Figure 1.41 (repeated) Non-inverting op-amp voltage amplifier.

Say that the required voltage gain is 15. Then $R_2/R_1 = 14$. Let us take $R_1 = 1$ kΩ and $R_2 = 14$ kΩ. In Example 6.9, we study the voltage amplifier that uses a non-ideal op-amp, using the tools of feedback analysis.

We first observe (in Figure 1.41) the L-shaped B-network connected to the op-amp in a series-shunt configuration. This is indeed a feedback voltage amplifier. The B-network voltage gain is $B = v_f/v_o = R_1/(R_1 + R_2) > 0$. The op-amp is a differential amplifier, providing the subtraction between the input signal and the feedback signal.

Let us find the open-loop and closed-loop voltage gains A, considering the loading by the B-network of the forward amplifier. See Figure 6.124.

To include the non-ideal parameters of the op-amp, we use a three-parameter amplifier model with assumed parameters $\mu = 10^6$, $R_{in0} = 1\,M\Omega$ and $R_{oo} = 100\ \Omega$. We note by the polarity of v_e and the voltage-depended-voltage-source output that the op-amp is arranged in a non-inverting configuration $A > 0$. The feedback is

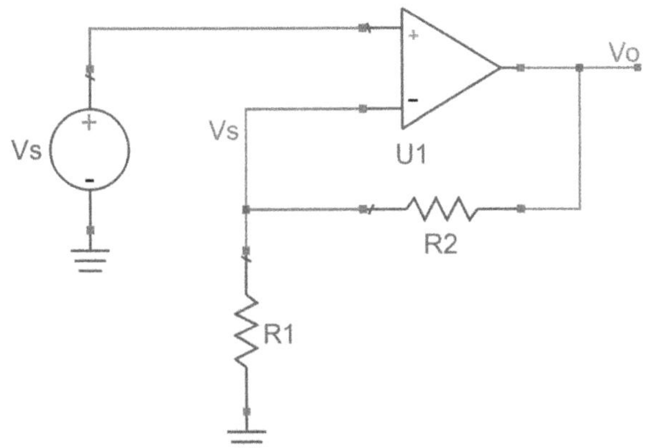

FIGURE 1.41 Non-inverting op-amp amplifier.

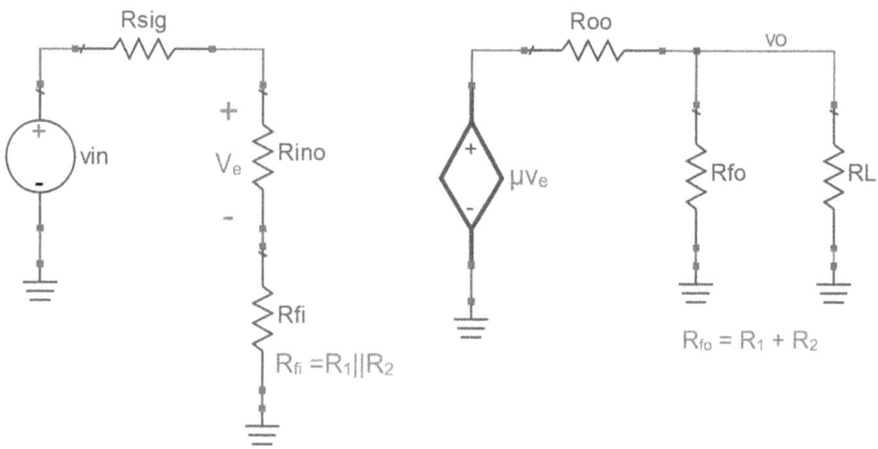

FIGURE 6.124 The loaded A diagram for an op-amp feedback voltage amplifier.

indeed negative. We also refer to Figure 6.120 regarding the ways that the loading resistances R_{fi} and R_{fo} are found.

The open-loop A calculation proceeds as follows: $A_{loaded} = v_o/v_{in} = [R_{ino}/(R_{sig} + R_{ino} + R_{fi})] \cdot \mu \cdot [(R_L||R_{fo})/(R_{oo} + (R_L||R_{fo})] \approx \mu > 0$, if R_{fi} and R_{sig} are $\ll R_{ino}$ and if $R_{oo} \ll R_L$ and R_{fo}. The approximations are justified for the B-network resistances chosen to be in the few kΩ range, for loads R_L that are a few kΩ large at least, and for R_{sig} that may be as large as a few tens of kΩ.

The values of A_{loaded} and B are now substituted in the closed-loop feedback gain equation to find the closed-loop voltage gain: $A_{CL} = A_{loaded}/(1 + A_{loaded}B) \approx \mu/[1 + \mu(R_1/(R_1 + R_2)] \approx 1 + R_2/R_1 = 15$.

We shall conclude the analysis of the non-inverting op-amp voltage amplifier by finding its closed-loop input resistance, output resistance, and bandwidth, for the numerical parameters used in this example. Let's assume that $R_{sig} = 1$ kΩ, $R_L = 10$ kΩ and $BW_{OL} = 10$ Hz.

The loop gain of the amplifier is approximately $A_{loaded}B \approx \mu \cdot [R_1/(R_1 + R_2)] = 66667$.

$R_{in} \approx R_{sig} + R_{ino} + R_{fi} \approx R_{ino} = 1\ \text{M}\Omega \rightarrow R_{inf} \approx R_{ino} \cdot (1 + A_{loaded}B) \approx 67\ \text{G}\Omega$.

$R_{out} \approx R_{oo}||R_{fo}||R_L \approx R_{oo} = 100\ \Omega \rightarrow R_{outf} \approx R_{oo}/(1 + A_{loaded}) \approx 1.5\ \text{m}\Omega$.

$BW_{CL} \approx BW_{OL}(1 + A_{loaded}B) \approx 10 \cdot 66{,}667 = 667$ kHz.

Example 6.10

The Inverting Op-Amp Amplifier Revisited.

In Section 1.1, we showed that, for an inverting amplifier, if the op-amp is ideal (that is, has infinite differential voltage gain, infinite differential input resistance, and zero output resistance) then the closed-loop voltage gain is equal to $-R_f/R_s$. Say that the required voltage gain is −15. Then $R_f/R_s = 15$. Let us, for this example, take $R_s = 1$ kΩ and $R_f = 15$ kΩ. Another choice that will be considered is: $R_s = 6$ kΩ and $R_f = 90$ kΩ. In this example, we study the amplifier that uses a non-ideal op-amp, using the tools of feedback analysis. Figure 1.1 is repeated as a reminder.

Figure 1.1 (repeated) An inverting op-amp amplifier.

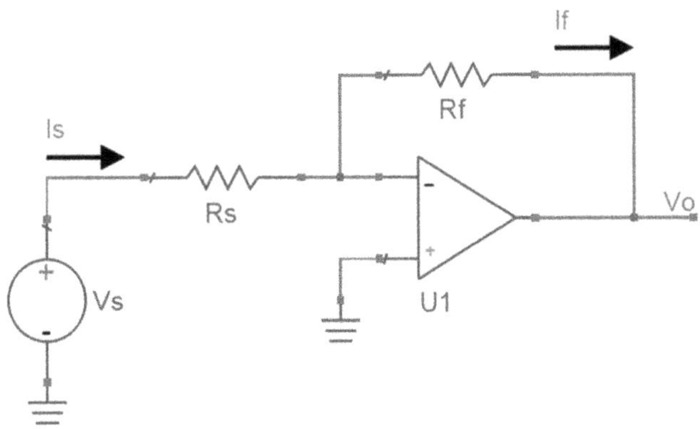

FIGURE 1.1 Inverting op-amp amplifier.

We right away run into several perceived difficulties as we observe the B-network. Isn't R_f connected in a shunt-shunt configuration? What is the role of R_s? Isn't the existence of a voltage signal source $v_s(t)$ a manifestation that we are dealing with an amplifier that has a voltage input?

Yes, these are shunt-shunt B-network connections.

The entire B-network consists of R_f only. It is truly a transimpedance amplifier.

There is no role for R_s within the B-network. It can be appended to the input signal's internal resistance. Furthermore, we recall that it is the B-network, and the way it is connected to the forward amplifier, that dictates the type of amplifier. In a transimpedance amplifier, the input voltage signal source can, and actually should, be transformed into an equivalent current signal source. For signal source transformation, see [Nilsson1].

A persistent reader may still argue that it is clearly an inverting voltage amplifier. We even simulated it once and observed its proper operation as a voltage amplifier. The answer is that a transimpedance amplifier can sometimes provide good voltage amplification. However, for a systematic feedback analysis, we have to treat the amplifier as what it truly is: transimpedance. Let us defer the voltage amplification capability of this transimpedance amplifier to later in the analysis of this example.

In Example 6.10, let us assume that the non-ideal op-amp has the same parameters as in the previous Example 6.9. Also, same R_{sig} and R_L as in Example 6.9.

Then (see Figure 6.123): $B = i_f/v_o = -1/R_f$ and $R_{fi} = R_{fo} = R_f$. Let us choose $R_f = 90$ kΩ to be consistent with one of the previous choices of R_s and R_f for the, soon to be discussed again, inverting voltage amplifier. Because of $R_{sig} = 1$ kΩ, let us adjust $R_s = 5$ kΩ, seeing that the inverting voltage amplifier's gain is $-R_f/(R_{sig} + R_s)$.

The calculation of the loaded transimpedance gain proceeds as follows: $A_{loaded} = v_o/i_{in} = [R_{in0}||R_{fi}||(R_{sig} + R_s)]\cdot(-\mu)\cdot[(R_L||R_{fo})/(R_{oo} + (R_L||R_{fo})] \approx (R_{sig} + R_s)\cdot(-\mu)$, as R_{fi} and R_{in0} are much larger than $R_{sig} + R_s$, and as $R_{oo} \ll R_L$ and R_{fo}. The approximations are justified for the B-network resistances that have $R_f = 90$ kΩ (but not so much for a choice of $R_f = 15$ kΩ as R_{fi} in such a case cannot be neglected).

The values of A_{loaded} and B are now substituted in the closed-loop feedback gain equation to find the closed-loop voltage gain: $A_{CL} = A_{loaded}/(1 + A_{loaded}B) \approx -\mu(R_{sig} + R_s)/[1 - \mu(R_{sig} + R_s)(-1/R_f) \approx -R_f = -90$ [V/mA] due to the large value of μ.

The voltage gain, provided by the transimpedance amplifier, can be found directly from the closed-loop transimpedance gain. See the transformed source in Figure 6.125. $A_{vCL} = v_o/v_s = v_o/[i_{in}(R_{sig} + R_s)] = A_{CL}/(R_{sig} + R_s) = -R_f/(R_{sig} + R_s) = -15$ as expected.

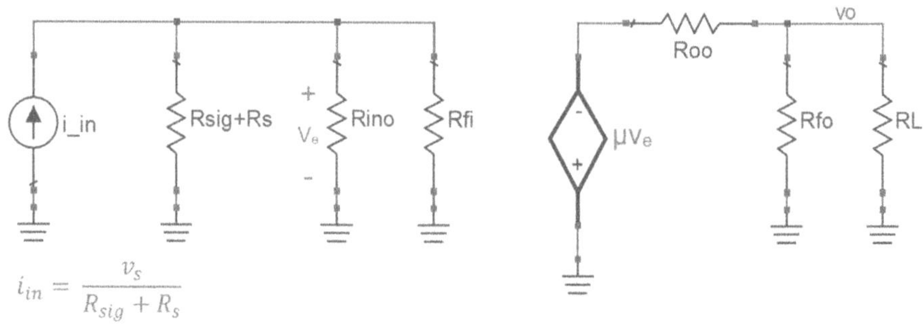

FIGURE 6.125 The loaded A diagram for an op-amp feedback transimpedance amplifier.

We shall conclude the analysis of the op-amp inverting transimpedance amplifier, also serving as an inverting voltage amplifier, by finding its closed-loop input resistance, output resistance, and bandwidth, for the numerical parameters used in this example.

The loop gain of the amplifier is approximately $A_{loaded}{\cdot}B \approx \mu{\cdot}(R_{sig}+R_s){\cdot}(1/R_f) = 66{,}667$.

$R_{in} \approx (R_{sig}+R_s)||R_{ino}||R_{fi} \approx R_{sig}+R_s = 6\ k\Omega \rightarrow R_{inf} \approx (R_{sig}+R_s)/(1+A_{loaded}B) \approx 90\ m\Omega$.

$R_{out} \approx R_{oo}||R_{fo}||R_L \approx R_{oo} = 100\ \Omega \rightarrow R_{outf} \approx R_{oo}/(1+A_{loaded}) \approx 1.5\ m\Omega$.

$BW_{CL} \approx BW_{OL}{\cdot}(1+A_{loaded}B) \approx 10{\cdot}66{,}667 = 667\ kHz$.

One last comment about R_{inf}, based on Figure 1.1. Recalling from Section 1.1 the "virtual ground", and how the input signal $v_s(t)$ and the resistor R_s give rise to a current $v_s(t)/R_s$ that forces itself on the resistor R_f, thus creating the voltage amplification of the amplifier: The closed loop input resistance of the voltage amplifier is clearly $R_{inf} = R_s$. The R_{inf} calculation done for the transimpedance amplifier is not valid for a feedback voltage amplifier.

6.18 DESIGN OF FEEDBACK AMPLIFIERS

The typical design specifications of any feedback amplifier are:

i. The type of amplifier: voltage, current, transconductance, or transimpedance.
ii. The minimum (or maximum) closed-loop input resistance, as needed by the amplifier's type. For instance, in a voltage amplifier, we set a minimum value for the needed large R_{inf}, whereas in a current amplifier, we set a maximum value for the needed small R_{inf}.
iii. The minimum (or maximum) closed-loop output resistance, as needed by the amplifier's type. For instance, to create a small enough R_{outf} of a transimpedance amplifier, or a large enough R_{outf} of a transconductance amplifier.
iv. The exact closed-loop gain. It is an equality condition. For example, the exact transconductance closed-loop gain that is needed.
v. The minimum needed closed-loop bandwidth. All four amplifiers are typically wideband amplifiers.

A typical feedback amplifier design approach is to create a large enough loop gain $A_{loaded}{\cdot}B \gg 1$, and by that to make the closed loop gain approximately equal to $A_{CL} \approx 1/B$. As we have seen, even for the most basic op-amp voltage amplifiers, a high loop gain design strategy yields a closed loop gain that depends primarily on the feedback resistors (that may be taken as precision resistors), and to a much lesser extent, on large tolerance entities, such as the op-amp total open loop differential gain, that may fluctuate badly.

A discrete feedback amplifier's high-gain forward amplifier A(s) may be dictated (say, by a requirement to use a specific op-amp, such as the μA741 op-amp), and in that case, the designer just needs to select the feedback B network and its component values. Sounds easy. If some of the specs are not met, there is a need to increase A some more, possibly by cascading more than one of the required op-amps.

In a "system on a chip" feedback amplifier design, the transistor circuits that constitute the A(s) open-loop amplifier may be taken from a library of ready-made VLSI designs of op amps, or OTAs (operational transconductance amplifiers). Such circuits may or may not be modifiable.

Some op-amps, such as the μA741, contain a stability compensation component that allows such A(s) to be connected to a large array of feedback B networks, while maintaining closed-loop stability. Other op-amps may require the designer to add specific compensation to be able to handle a specific B network. More about it later in this chapter.

Example 6.11

Design of a feedback voltage amplifier with a given Op-Amp and fluctuating input signal resistance and resistive Load.

The internally stabilized op-amp has a simplified model shown in Figure 6.126. The op-amp's open-loop input resistance is $R_i = 1\ M\Omega$. The open loop output resistance is $R_o = 100\ \Omega$, the open loop differential voltage gain is $\mu = 10^6$, and the open loop bandwidth is $f_p = 100\,Hz$.

Using the above op-amp, there is a need to construct a feedback voltage amplifier to interface between an input signal source $v_s(t)$, the internal resistance of which may vary in the range $0 \leq R_s \leq 1\ k\Omega$, and a resistive load R_L that varies $1\ k\Omega \leq R_L \leq 5\ k\Omega$, as shown in Figure 6.127.

The op-amp connects to a voltage amplifier B network to create a negative feedback amplifier.

The quantitative specifications of this feedback voltage amplifier include one equality requirement and three inequality specs:

i. Closed-loop input resistance $R_{if} \geq 4\,M\Omega$
ii. Closed-loop output resistance $R_{of} \leq 10\ \Omega$
iii. Closed-loop voltage gain $A_{CL} = 20$ [V/V]
iv. Closed-loop bandwidth $BW_{CL} \geq 1\,MHz$

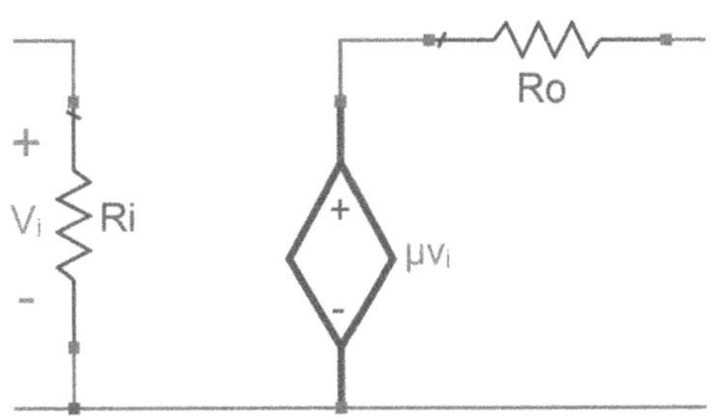

FIGURE 6.126 A simplified model of an internally stabilized op-amp.

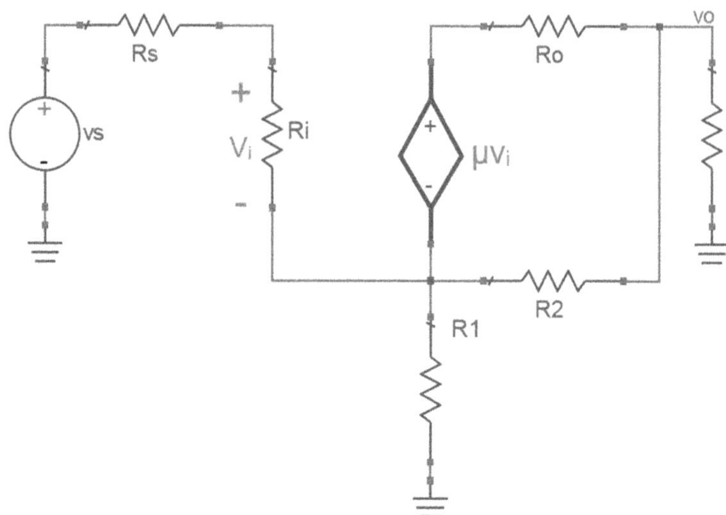

FIGURE 6.127 Feedback voltage amplifier noting the L-shaped B-network.

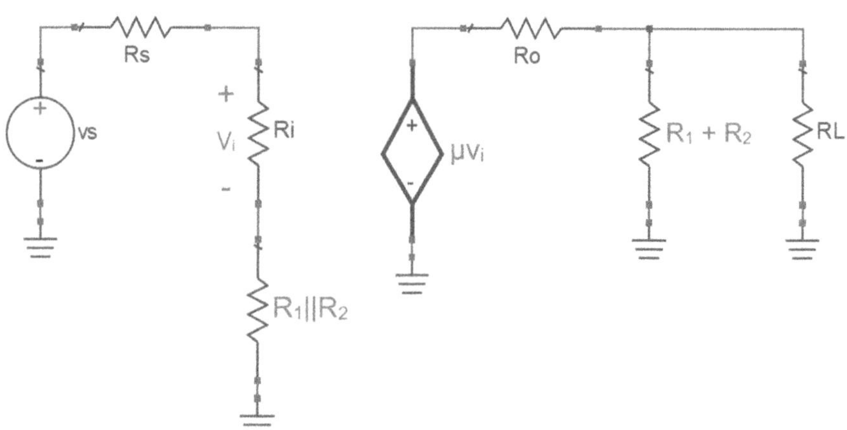

FIGURE 6.128 A_{loaded} computation diagram.

The first step in the design is to select the correct feedback network architecture and to connect the B network correctly: Done, according to Figure 6.127.

The second step in the design, following a high loop gain strategy, is to interpret the equality closed loop gain requirement (iii): $A_{CL} \approx 1/B = 20 \rightarrow B = 1/20 = R_1/(R_1 + R_2) \rightarrow R_2/R_1 = 19$. This second step is not over yet: Shall we choose $R_1 = 1$ kΩ and R_2 =19 kΩ? What will be the effect of such a choice on the feedback network loading of the A amplifier? Let us choose the resistors' values as done above, but we shall revisit this choice if needed.

Now, at any order, let us interpret the three inequality conditions.

Step 3: R_{if} interpretation.

We shall add R_s to the calculation of the open-loop input resistance (for simplicity) and the R_{fi} feedback network loading of $R_1||R_2$, connected in series to R_i, as shown in Figure 6.128.

$R_{in,OL} \approx R_s + R_i + (R_1||R_2) \approx R_i = 1\ M\Omega$.

We see that the fluctuations of R_s and the B-network loading at the input of the A amplifier are both insignificant, thus negligible.

Now (as we need to do for the two other inequality specs), we can estimate the minimum needed value of the A gain:

$4 \leq R_{if} \approx (1 + A_{loaded}B) \cdot R_{in,OL} \approx (1 + A_{loaded}B) \cdot 1\ M\Omega \rightarrow 1 + A_{loaded}B \geq 4$. As $B = 1/20$, we finally find that $A_{loaded} \geq 60$.

The op-amp trivially meets this requirement, but let us complete the interpretation of all the specs.

Step 4: R_{of} interpretation.

We shall add R_L to the open-loop output resistance calculation (again for simplicity) and the R_{fo} feedback network loading of $R_1 + R_2$ connected in shunt to v_o.

$R_{in,OL} \approx R_o||R_L||(R_1 + R_2) \approx R_o = 100\ \Omega$.

We see that the fluctuations of R_L are totally insignificant compared to R_o. It is also evident that the choices of values of R_1 and R_2, done earlier, create no significant loading of A(s).

$10\ \Omega \geq R_{of} \approx R_{in,OL}/(1 + A_{loaded}B) = R_o/(1 + A_{loaded}B) \rightarrow 1 + A_{loaded}B \geq 10 \rightarrow A_{loaded} \geq 180$.

We see that between the two specifications (closed-loop input and output resistance), it is R_{of} that requires a larger value of A_{loaded}.

Step 5: BW_{CL} interpretation.

$10^6\ Hz \leq BW_{CL} = BW_{OL}\ (1 + A_{loaded}B) = 100 \cdot (1 + A_{loaded}B) \rightarrow 1 + A_{loaded}B \geq 10^4 \rightarrow A_{loaded} \geq 2 \times 10^5$.

Clearly, summarizing steps 3–5, the forward gain needed to meet the closed-loop bandwidth requirement dominates the input and output resistance specs.

This is because of the typically low open-loop bandwidth of stabilized op-amps.

Step 6: A_{loaded} computation – is it large enough to meet all the specs?

$A_{loaded} = v_o/v_s = \{R_i/[R_s + R_i + (R_1||R_2)]\} \cdot \mu \cdot \{(R_L||(R_1 + R_2)/[R_o + (R_L||(R_1 + R_2)] \approx \mu = 10^6$.

The loading by the B network of A(s) is very insignificant. The op-amp A gain is much larger than the gain needed to meet the three inequality specifications. It means that the feedback amplifier would meet all the specs with large margins. We leave the calculations to the reader.

Example 6.12

Feedback Current Amplifier.

We shall demonstrate analytically, and then in ADS simulation, how a feedback current amplifier operates, which utilizes a stability compensated voltage op-amp as the A(s) amplifier. Such a choice of A(s) open-loop amplifier is far from being ideal for implementing a current amplifier. It would be easier to have an A(s) that has a small R_{in} and a large R_{out}. Instead, the voltage amplifier provides the opposite: large R_{in} and a small R_{out}. Let us assume that the op-amp's parameters are $R_i = 1\ M\Omega$, $R_o = 10\ \Omega$, $\mu = 10^6$, and $BW_{OL} = 10\ Hz$. See Figure 6.129.

Let us further assume that $R_s = 10\ k\Omega$ and $R_L = 10\ k\Omega$. Note the L-shaped B network that is characteristic of feedback current amplifiers.

The quantitative design specifications are

i. Closed-loop input resistance $R_{if} \leq 100\ \Omega$
ii. Closed-loop output resistance $R_{of} \geq 100\ k\Omega$
iii. Closed-loop current gain $A_{CL} = -100$ [mA/mA]
iv. Closed-loop bandwidth $BW_{CL} \geq 1\ MHz$

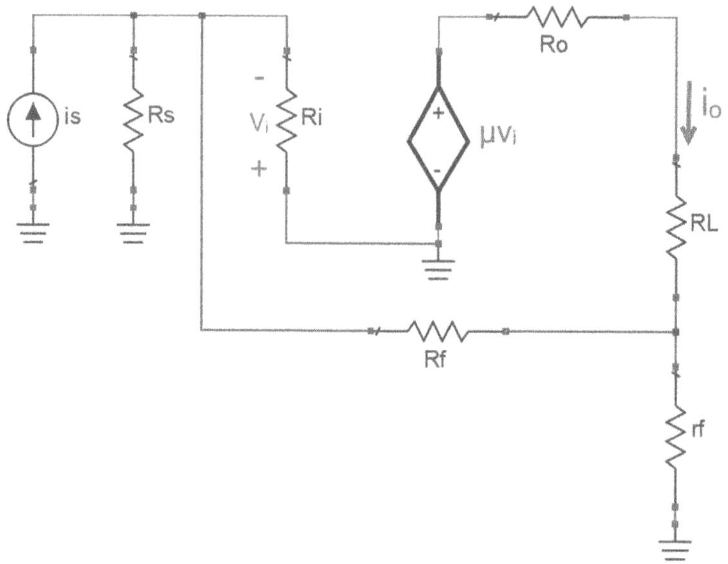

FIGURE 6.129 Feedback current amplifier built with a voltage amplifier op-amp.

The design commences by selecting the correct B network for the current amplifier and by connecting it correctly to the op-amp. Note that the B network samples the current output i_o from the floating load R_L and that its gain is negative, because of assuming that the positive direction of the feedback current i_f is inwards into the feedback network. Consequently, the op-amp must be connected in its inverting configuration, as noted by the polarity of v_i and μv_i. The resulting closed-loop current amplifier will be inverting as well. Thus, the feedback is negative.

The second step is to determine the value of the feedback gain B and the values of the B network resistances.

$A_{CL} = -100 = 1/B \rightarrow B = -1/100 = -r_f/(R_f + r_f) \rightarrow R_f/r_f = 99$. Let us choose $R_f = 990\ \Omega$ and $r_f = 10\ \Omega$, keeping the option of revisiting this choice.

Design steps 3–5 are similar to those of Example 6.11:

Step 3: Rinf spec.

$R_{in,OL} \approx R_s||(R_f + r_f)||R_i = 10||1||\cdot 1000\ k\Omega \approx 908\ \Omega$.

It is unusual to see the feedback network's loading of A(s) dominating the open-loop input resistance. The voltage op-amp, used for constructing a current amplifier, is to be blamed.

$100\ \Omega \geq R_{in,CL} \approx R_{in,OL}/(1 + A_{loaded}B) \rightarrow 1 + A_{loaded}B \geq 9.08 \rightarrow |A_{loaded}| \geq 808$ [mA/mA].

Step 4: Routf spec.

$R_{out,OL} \approx R_o||[(R_f||r_f) + R_L] = 0.01||[0.0099 + 10]\ k\Omega \approx 10\ \Omega$.

$R_{out,CL} \approx R_{out,OL}(1 + A_{loaded}B) \geq 100\ k\Omega \rightarrow 1 + A_{loaded}B \geq 10^4 \rightarrow |A_{loaded}| \geq 10^6$ [mA/mA].

Step 5: Closed-loop bandwidth spec.

$BW_{OL} = f_p = 10\,Hz$.

$BW_{CL} \approx BW_{OL}(1 + A_{loaded}B) \geq 1\ MHz \rightarrow 1 + A_{loaded}B \geq 10^5 \rightarrow |A_{loaded}| \geq 10^7$ [mA/mA].

Step 6: A_{loaded} calculation (Figure 6.130).

Let us compute A_{loaded}, using units of $k\Omega$ for the resistances:

$A_{loaded} = i_o/i_s = [-(R_s||(R_f + r_f)||R_i]\cdot(-\mu)\cdot[1/(R_o + R_L + (R_f||r_f)] = -(10||1||1000)\cdot(-10^6)\cdot[1/(0.01 + 10 + 0.0099)] \approx 0.9\cdot 10^6$.

Again, the bandwidth spec dominates.

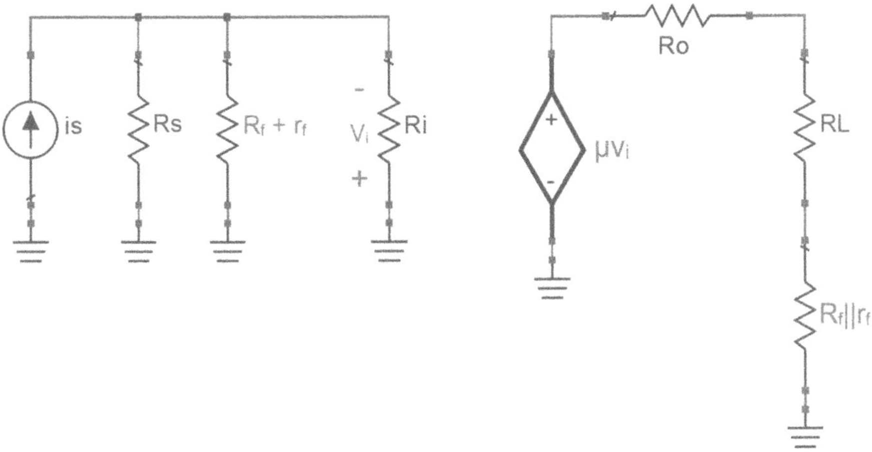

FIGURE 6.130 A_{loaded} calculation of a feedback current amplifier, built with a voltage op-amp.

This time, though, the open-loop voltage gain of the op-amp is insufficient. So, what do we do?

We may look for a better op-amp, one with a larger μ value. We need $\mu = 10^7$.

Another option is to modify (relax) some of the design specifications: For instance, is a closed loop bandwidth of 100 kHz acceptable?

One more solution is to combine two op-amps in cascade. We may let the first one be inverting and the second one be non-inverting, to ensure that the feedback remains negative.

6.19 DEPENDENT SOURCES: VCVS, VCCS, CCVS, CCCS; THE GAIN-BANDWIDTH PRODUCT; ADS DESIGN AND TESTING OF A FEEDBACK CURRENT AMPLIFIER USING VCVS DEPENDENT SOURCE

ADS offers four dependent sources (Figure 6.131):

- VCVS (Voltage-Controlled Voltage Source)
- VCCS (Voltage-Controlled Current Source)
- CCVS (Current-Controlled Voltage Source)
- CCCS (Current-Controlled Current Source)

These components are useful for some circuits discussed in the first Circuits course [Nilsson1]. VCVS can be used to emulate an ideal op-amp and a non-ideal one. Figure 6.132 shows the VCVS internal model and Edit Instance Parameters box.

Default values are shown in the component properties box. Only the dependent source voltage gain G = 1 is displayed by default. Usually, we like to display other parameters such as R_1, R_2, and F.

The default VCVS would become an ideal op-amp if we change G to say $1e^6$ and change F (the first cutoff frequency) to say F = 10 Hz. Gain G is complex. It has magnitude and phase. Figure 6.133 illustrates a non-ideal op-amp that is like the μA741 op-amp.

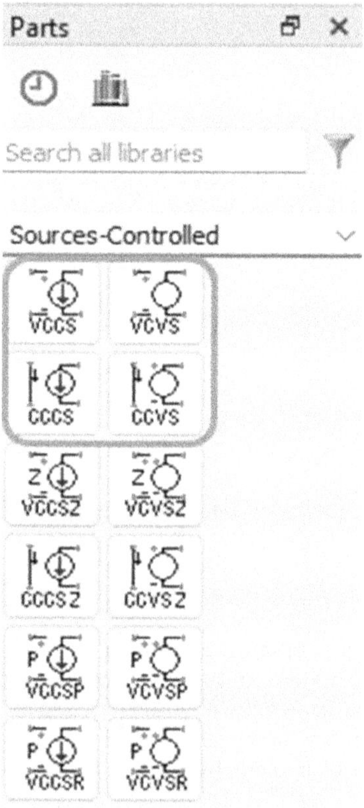

FIGURE 6.131 Source-controlled components palette.

Let us observe (with the schematic of Figure 6.134) the open-loop frequency response of a µA741-like op-amp. The results are shown in Figure 6.135.

Figure 6.135 shows the open loop frequency response in two forms: (i) Linear A_v vs. log of the frequency, (ii) Plotting of A_v in dB [20·log(A_v)] in the Y axis and log scale in X provides the plot where we can see the corner frequency and the –20 dB/dec drop. The corner –3 dB gain drop at F_c is also noticeable. View (i) is less informative here.

We are now ready to simulate the setup of Example 6.12 – see Figure 6.136. The schematic is based on the design done in Example 6.12.

To remind the reader, the design specs were:

$A_{ICL} = -100$ [mA/mA]
$R_{if} \leq 100\ \Omega$
$R_{of} \geq 100\ k\Omega$
$BW_{CL} \geq 1\ MHz$

See the inverted L-shaped feedback network, with a small R_{ff} that samples the output current I_L.

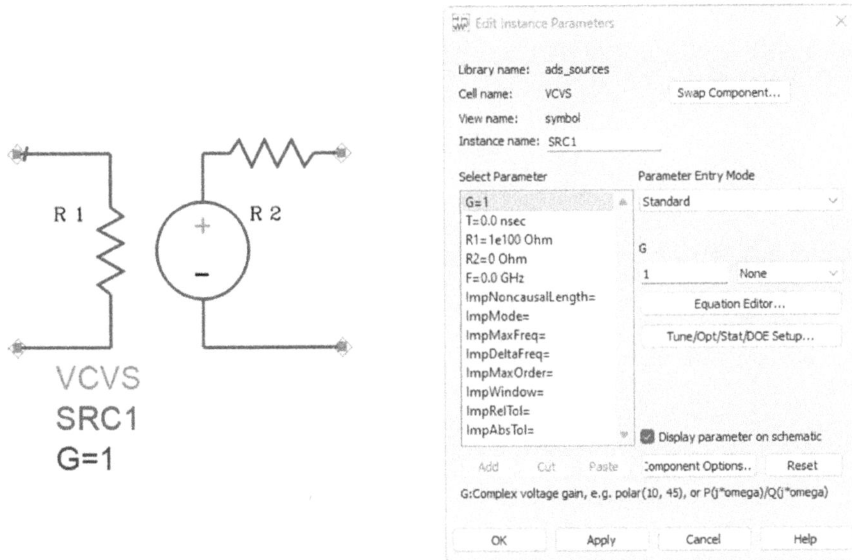

FIGURE 6.132 VCVS model and instance parameters.

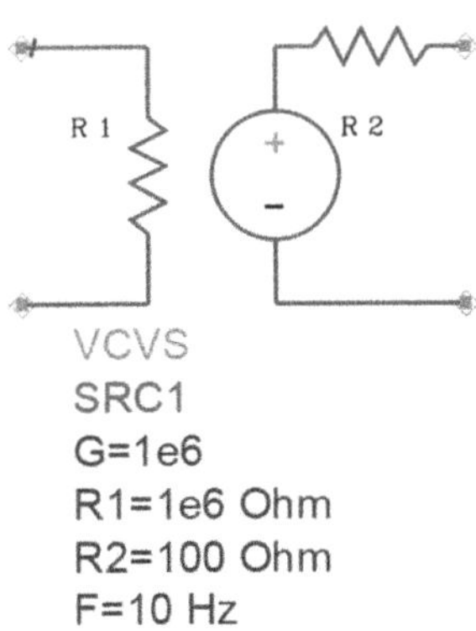

FIGURE 6.133 VCVS that emulates a non-ideal op-amp.

Notice that the load R_L is floating. Also, notice how, via G, we implement the use of the (–) input of the op-amp.

Figure 6.137 summarizes the performance of the current amplifier. The A_i and R_{if} performance is not good, and the closed-loop BW is by two orders of magnitude lower than what was designed. It's all due to the too-low op-amp's open-loop gain. Let's check R_{of} (output resistance) first (Figures 6.138). Then explore increasing the gain or lowering the specs.

We see in Figure 6.138 (right) that at low frequencies, Rout exceeds the spec by orders of magnitude, but the BW is catastrophically low.

The current gain of 100 is large, and therefore, the BW drops due to the Gain·BW product created by the compensated op-amp.

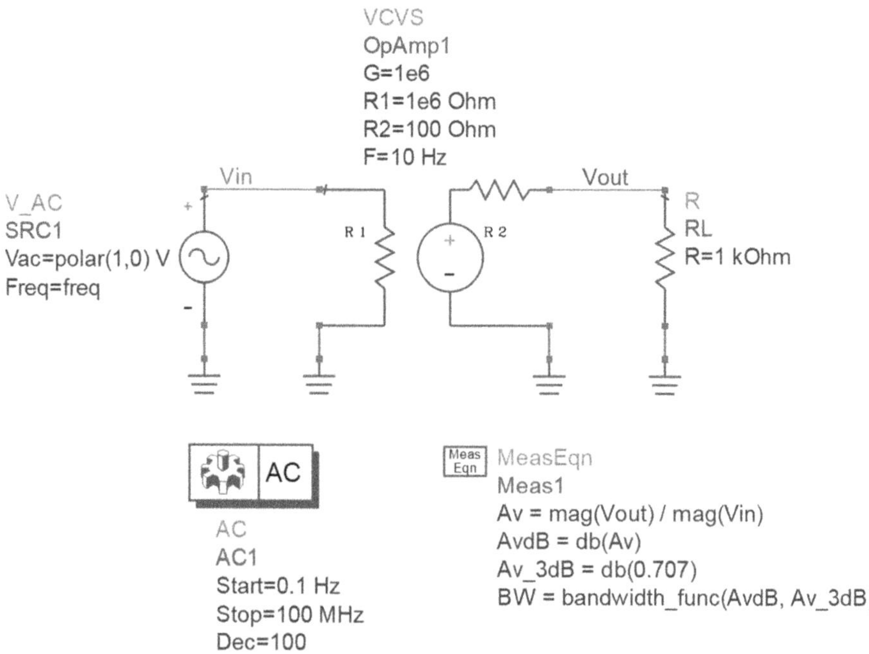

FIGURE 6.134 AC simulation of an open loop 741-like op-amp.

Let us try an open-loop gain of 10^7 of the op-amp – Figure 6.139.

We see that the closed-loop BW improves by one order of magnitude. In order for the BW to approach 1 MHz, we need an op-amp with a gain of more than 10^8. It practically means two cascaded op-amps, one with a positive voltage gain and one with a negative voltage gain.

In Figure 6.140, we show another test of the current amplifier, involving a lowering of the specified closed-loop current gain. We see that it is not the gain·BW product to be blamed. The poor BW performance is due to the insufficient open-loop gain.

6.20 FEEDBACK COMPENSATION FOR ELECTRONIC AMPLIFIERS AND NYQUIST'S STABILITY THEOREM

One of the most celebrated results in classical control systems theory is **Nyquist's Stability Theorem** [Bode1], [Ogata1], [Bower1], [Dorf1]. Let us refer to Figure 6.111 that as we recall, shows a single loop negative feedback system with a forward transfer function A(s) and a feedback transfer function B(s). Let us further recall equation (6.1) (that presents the closed loop transfer function of such a system as $Y(s)/R(s) = A(s)/[1 + A(s)B(s)]$. What Harry Nyquist set out to explore was the relationship between the open loop frequency response $A(j\omega)B(j\omega)$ and the number of unstable closed loop poles, which are roots of $1 + A(s)B(s) = 0$ that lie in the right half of the complex s-plane. In other words, the question is: **With the negative feedback connection being open, can we predict, based on the open-loop frequency response of the loop transfer function A(s)B(s), whether or not the closed-loop system is**

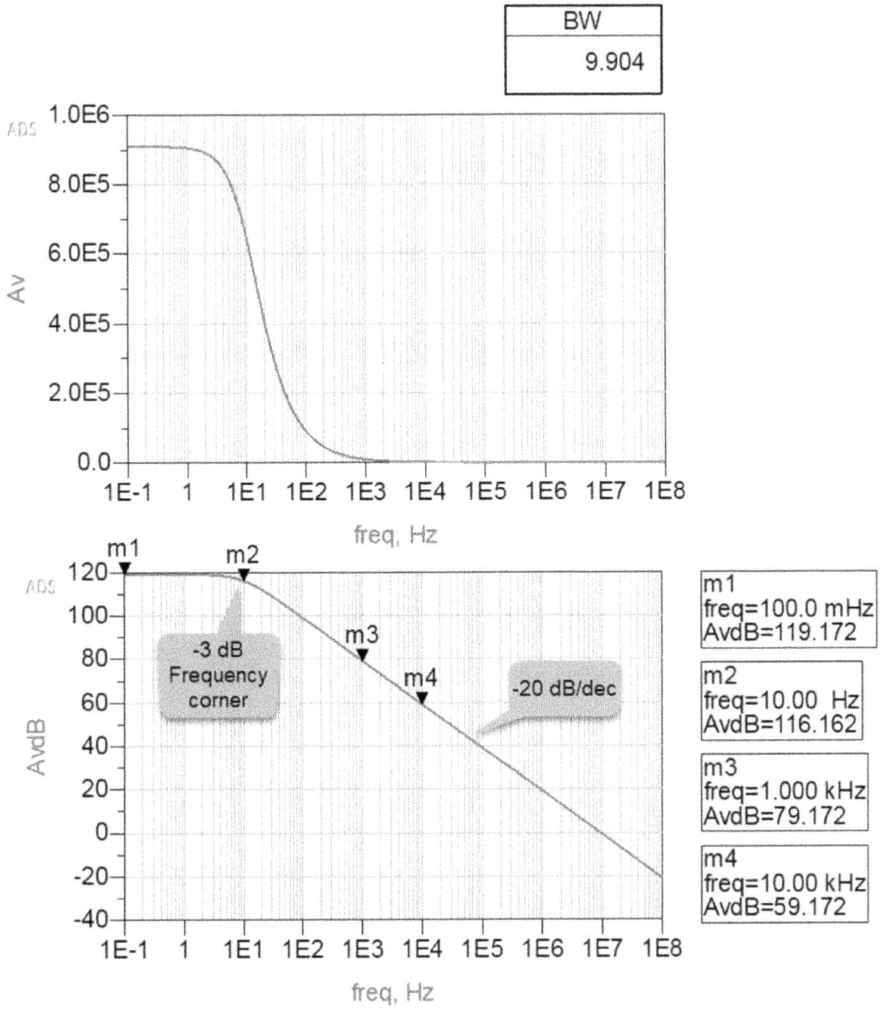

FIGURE 6.135 Open-loop frequency response of the 741-like op-amp.

stable? The answer is **Yes.** The number of unstable closed-loop poles depends on the number of encirclements of the point $-1+j0$ by the Nyquist plot of the loop transfer function A(s)B(s). Let us explain the key ideas of Nyquist's theorem.

Nyquist's theorem is derived from basic concepts of Complex Functions Theory, namely Contour Mapping and Cauchy's Principle of the Argument theorem. A **contour** is a continuous path in the complex s-plane. Nyquist's contour is a closed path that surrounds the entire right half of the s-plane. See Figure 6.141 for illustrations of typical Nyquist contours.

The left contour (of Figure 6.141) encounters no poles or zeros of A(s)B(s). The contour shown on the right of Figure 6.141 illustrates the by-passing of imaginary poles and zeros of A(s)B(s) at an infinitesimally small radius.

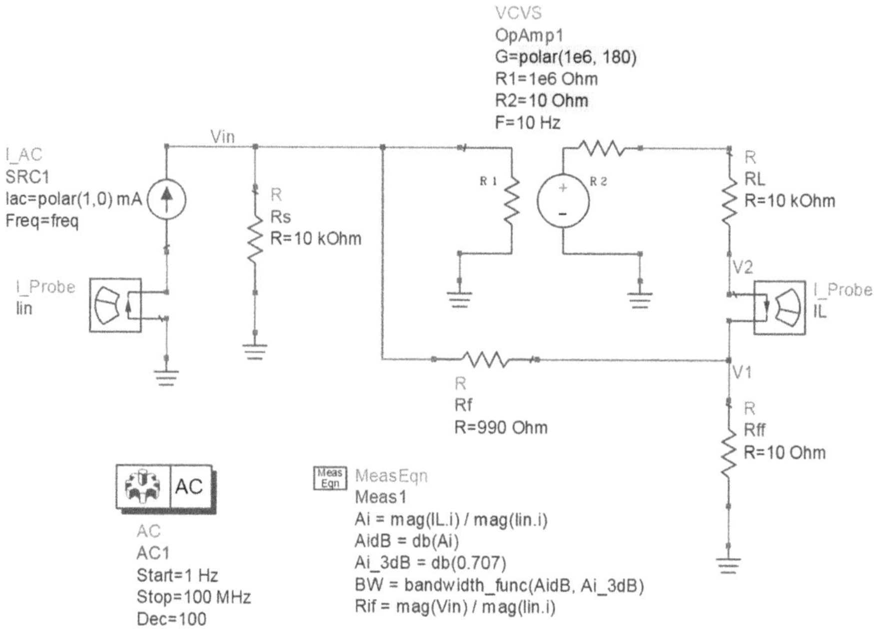

FIGURE 6.136 Feedback current amplifier built with a voltage op-amp.

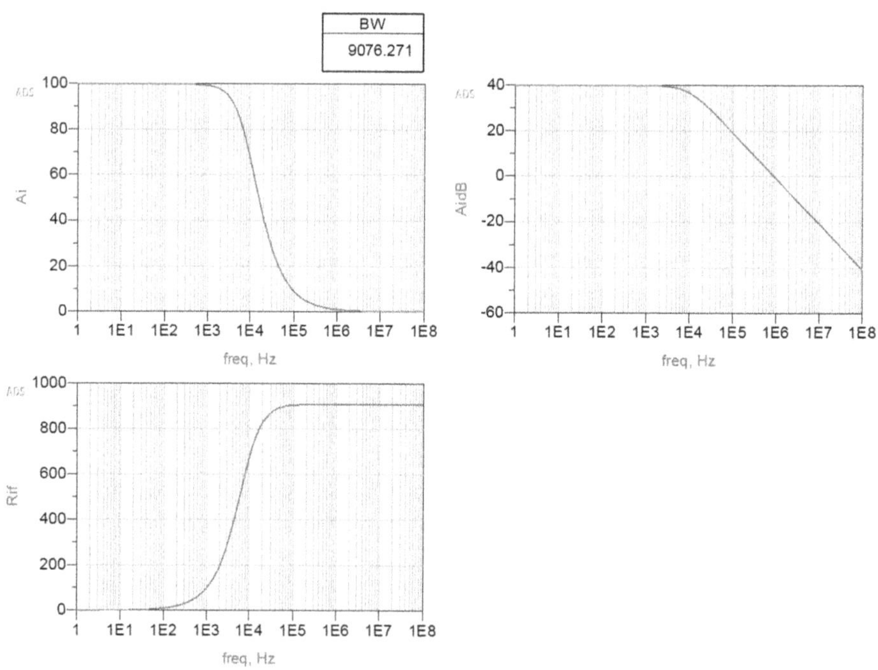

FIGURE 6.137 Current gain (two views) and R_{if}.

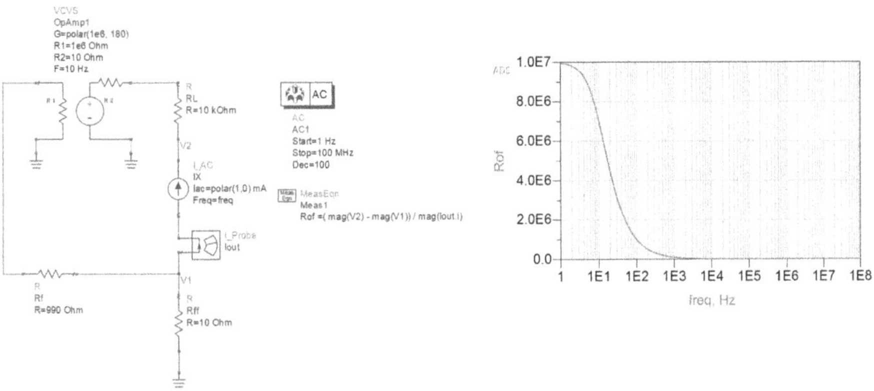

FIGURE 6.138 Closed loop output resistance measurement of the feedback current amplifier.

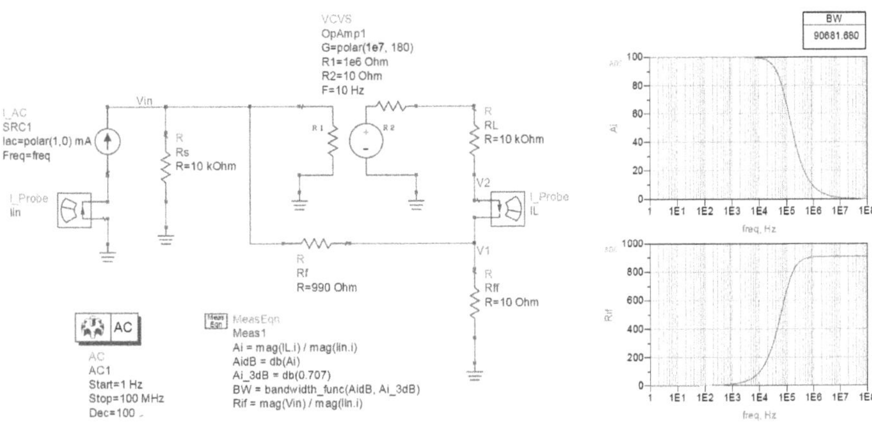

FIGURE 6.139 Current gain and input resistance performance of the current amplifier after an increase of the op-amp's open loop gain.

A **contour mapping** is a contour that results from substituting a complex function F(s) into every point s along a contour in the s-plane. See Figure 6.142.

The mapping A(s)B(s) of a Nyquist contour is called the Nyquist plot. It is obtained by substituting the loop transfer function A(s)B(s) into the points s along the Nyquist contour. That is F(s) = A(s)B(s). The only restriction in creating a Nyquist contour is that the contour must not cross any pole or zero of A(s)B(s), as shown on the right of Figure 6.141, where such poles and zeros have to be bypassed at an infinitesimally small radius.

Cauchy's Principle of the Argument theorem stems from the observation that a contour that encircles some of the poles and zeros of a mapping F(s) causes the mapped contour to encircle the origin (0, 0) of the F(s) plane a certain number of times, in the clockwise (or sometimes in the counterclockwise) direction, depending

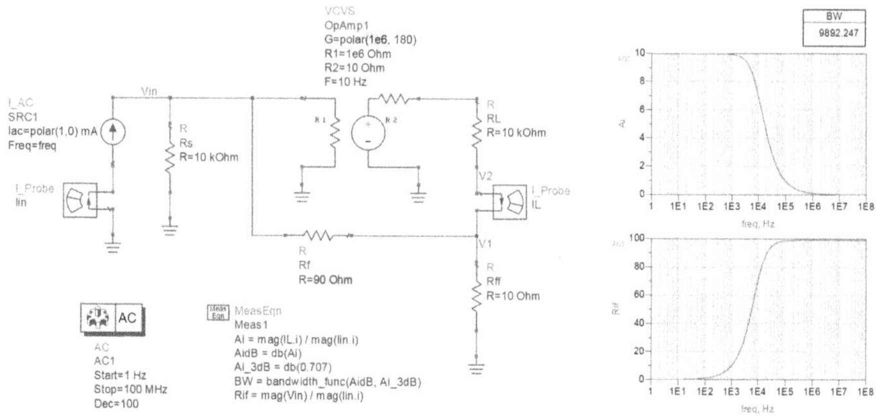

FIGURE 6.140 Rather than increasing the open loop gain we try to lower the specified current gain.

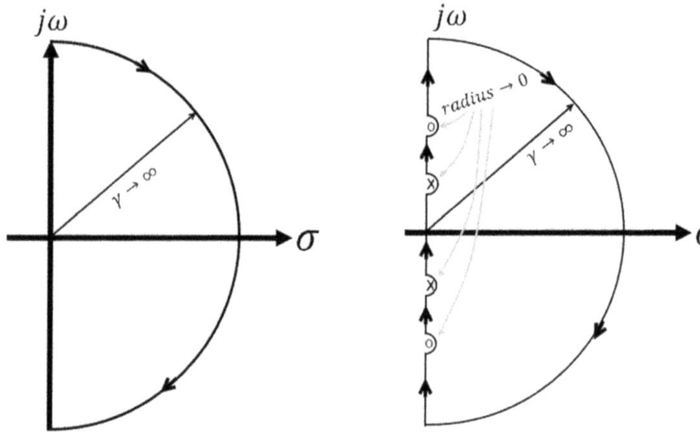

FIGURE 6.141 Nyquist's contours encircling the right half of the s plane at an infinite radius.

on the difference between the number of encircled zeros of F(s) and the number of encircled poles of F(s). See illustrations shown in Figures 6.143 and 6.144. This is so because a contour that encircles a pole (or a zero) causes the mapped contour to accumulate phase, which translates into complete encirclements of the origin. Poles and zeros that are not encircled do not cause the mapped contour to accumulate phase. Let us elaborate. Note that in any complex plane, such as the F(s) plane, every mapped contour point has a magnitude and a phase. The magnitude is the distance of the point from the origin. The phase is the angle created between that radius vector and the positive real line. As a point moves along an F(s) contour, all the poles and zeros of F(s) contribute to the total phase at each mapped point. Poles and zeros that are not encircled by the contour cause their phase contributions to the mapped

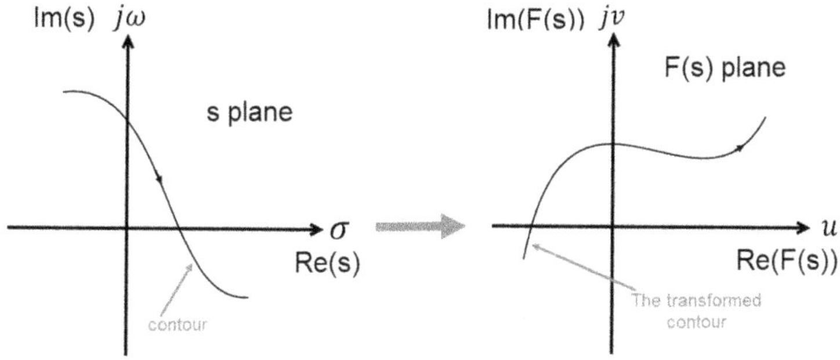

FIGURE 6.142 A contour on the s-plane (left) and a mapped contour in the F(s) plane (right).

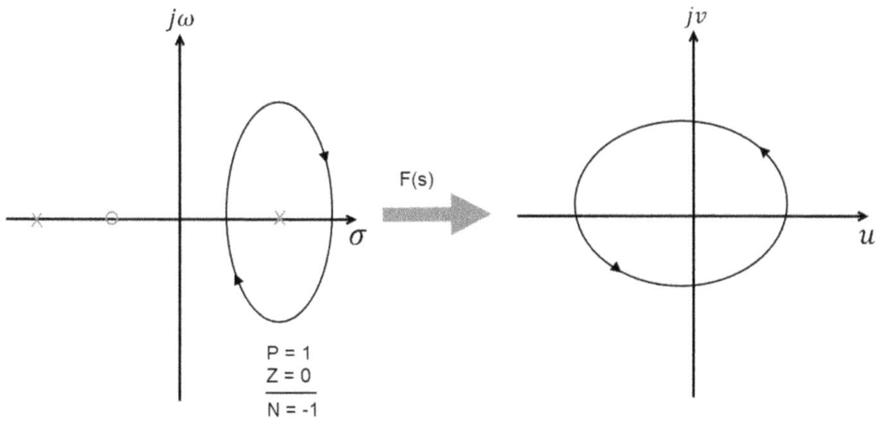

FIGURE 6.143 In the s-plane one pole is encircled in the clockwise direction (P = 1) and no zeros are encircled (Z = 0). As a result, the F(s) mapped contour encircles the origin once in the counterclockwise direction (N = −1).

contour to change up or down, but with no accumulation. If the contour encircles one of the poles or zeros of F(s), the phase accumulates. In Figure 6.143, as the s-plane contour encircles one of the poles, it causes phase accumulation in the F(s) contour. The other poles and zeros that are not encircled do not cause phase accumulation in the F(s) plane. A more complicated example is shown in Figure 6.144.

A portion of the Nyquist plot (such as the ones related to the Nyquist's contours shown in Figure 6.141) is the entire frequency response A(jω)B(jω) of the loop transfer function, represented in polar form Re{A(jω)B(jω)} vs. Im{A(jω)B(jω)}. It is obtained by mapping all the points s = jω for all positive and negative values of ω. See an example of such a polar plot in Figure 6.145 for the loop transfer function of $A(s)B(s) = 1/(s+1)^3$. The mapped contour curve that starts at ω = 0 and ends at ω = ∞ is the polar frequency response for positive frequencies. Note that the magnitude of the frequency response becomes smaller and smaller, falling from |A(j0)B(j0)| = 1 to |A(j∞)B(j∞)| = 0. The symmetric curve (about the real axis) is the mapped contour for negative frequencies.

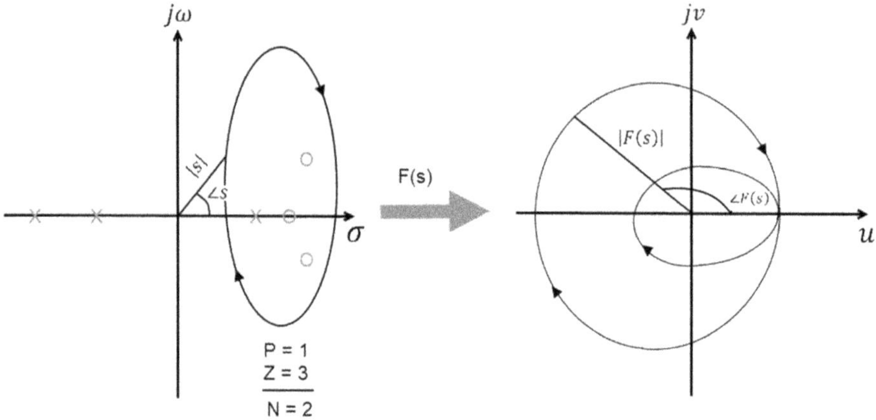

FIGURE 6.144 Encircling of three zeros and one pole results in two encirclements of the origin in the clockwise direction.

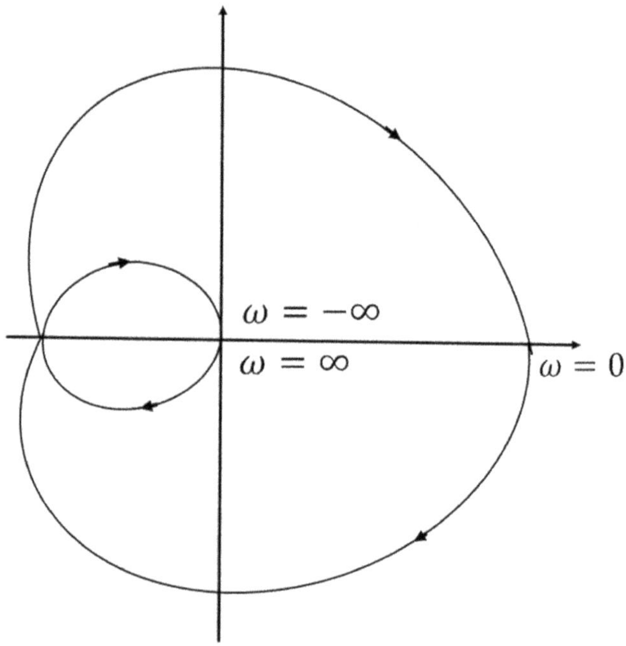

FIGURE 6.145 Nyquist plot for the loop transfer function $A(s)B(s) = 1/(s+1)^3$.

The phase starts at 0^0 and falls monotonically to -270^0, as seen in the mapped contour of A(s)B(s) for positive frequencies. The frequency response for negative frequencies is the mirror image of that for positive frequencies, where the real line is the axis of symmetry.

One last subtlety, before we can state Nyquist's Stability Theorem, is the relationship between the Nyquist plot of A(s)B(s) and that of 1 + A(s)B(s). For studying closed-loop stability, we are interested in finding how many unstable roots (zeros) $1+A(s)B(s)=0$ has. We are given the number P of unstable poles of A(s)B(s), and presumably we are able to sketch (or draw precisely) the Nyquist plot of A(s)B(s). Observation of $1+A(s)B(s)$ reveals that it has the same set of poles as A(s)B(s). That is, the same P. We further observe that the Nyquist plot of $1+A(s)B(s)$ has the same shape as that of A(s)B(s), except that the origin (0, 0) in the $1+A(s)B(s)$ plane is shifted to (–1, 0) in the A(s)B(s) plane. That is, we draw Nyquist's plot for A(s)B(s), as before, but we count encirclements of (–1, 0) in place of encirclements of the origin in the $1+A(s)B(s)$ plane.

Let N be the number of encirclements in the clockwise direction of the A(s)B(s) plane point $-1+j0$ by the A(s)B(s) mapped contour, P be the number of right half s-plane poles of A(s)B(s), and Z be the number of right half plane roots of $1+A(s)B(s)=0$. Z is the unknown that we look for, as it shows the number of unstable closed-loop poles. **Nyquist's Stability Theorem** says that $\mathbf{Z-P=N}$. In other words, for closed loop stability ($Z=0$), and for some given positive integer P, the Nyquist plot must encircle the point (–1, 0) P times in the counterclockwise direction.

The theorem applies to all loop transfer functions A(s)B(s), including ones that have free integrators (poles at $s=0$), unstable poles, right half plane zeros, and even a pure time delay.

In the example shown in Figure 6.145, we have $P=0$ (as A(s)B(s) is stable). The Nyquist plot encircles (0, 0) once in the clockwise direction, but (0, 0) is the wrong point to look at. We need to see if the Nyquist plot encircles the point (–1, 0). We note that the intersection point of the Nyquist plot with the real line is (–1/8, 0). That is $N=0$, and therefore $Z=0$. The closed-loop system is stable.

In feedback electronic amplifiers, though, $B(s)=B$, a constant feedback network's gain, and A(s), the forward amplifier, is generally an all-pole transfer function, and it is always designed to be open-loop stable ($P=0$, using the notation of Nyquist's theorem). Let us elaborate a little about these properties of A(s).

A single-stage voltage amplifier, such as a CS amplifier, has typically two high-frequency poles and a single right half-plane high-frequency zero [Sedra1], [Razavi1]. That zero and one of the two poles are at corner frequencies that are much larger than the bandwidth of the amplifier, and therefore tend to be insignificant in terms of their influence on the amplifier's bandwidth. A multistage amplifier A(s) consists typically of multiple stages that each have their own bandwidth (typically in the form of a dominant pole). These dominant pole locations may be of similar orders of magnitude. A high mid-band gain of A(s), and multiple real poles, each coming from a different forward amplifier's stage, cause the magnitude frequency response mag[A(jω)] to decrease monotonically. First very slowly, at mid-band frequencies, and then faster at higher frequencies, and so happens to Phase[A(jω)] as well.

Example 6.13

See Figure 6.146 of two amplifiers' frequency response illustrations, by means of Nyquist's plots, done using MATLAB®. Both systems have $B=1$. Both amplifiers have a DC gain of 40 dB, and both are all-pole of order 3. Amplifier $A_3(s)$ has its three poles at ω frequencies of $0.1e^6$, $20e^6$, and $200e^6$ Mrad/s. Amplifier

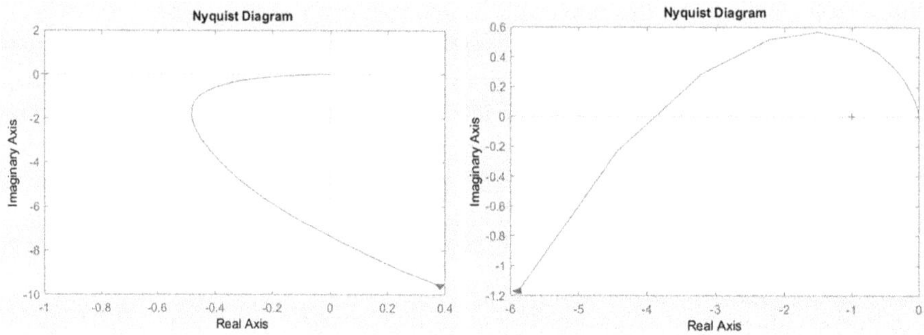

FIGURE 6.146 (For Example 6.13) Zoomed-in Nyquist plots for (left) stable closed-loop system with $A_3(j\omega)$ a forward transfer function and (right) unstable closed-loop system with forward transfer function $A_4(j\omega)$.

$A_4(s)$ has its three poles at ω frequencies of 1e[6], 4e[6], and 15e[6] Mrad/s. Both have monotonically decreasing magnitude and phase. For the same examples, see also Figure 6.147 that uses Bode plots.

Such all-pole and stable A(s) transfer functions, in particular, and any stable A(s) in general, require only a **"baby version" of Nyquist's theorem**. First, as $P=0$, then closed-loop stability ($Z=0$) happens if $N=0$. That is, the point $\{Re[A(j\omega)B(j\omega)]=-1, Im[A(j\omega)B(j\omega)]=0\}$ is not encircled by the Nyquist's plot. This is mathematically equivalent to requiring that **the polar plot of A(jω)B(jω), for non-negative frequencies, leaves the complex point −1 + j0 (above) to its left side, for closed loop stability.**

The zoomed-out Nyquist plots look a lot like the positive frequency part of Figure 6.145. The left plot leaves (−1, 0) on its left, whereas the right plot leaves (−1, 0) on its right, as ω varies from 0 to ∞.

In each of the polar frequency plots, at each frequency ω, the length of the radius vector from the frequency plot to the origin (0, 0) is a point on the magnitude frequency response plot $|A(j\omega)B|$ vs. ω. At every frequency ω along the polar frequency response plot, the angle between the radius vector and the positive real line is a point on the phase frequency response Phase$[A(j\omega)B]$ vs. ω. There we are. Back to Bode plots (discussed earlier).

Finally, explaining Nyquist's theorem's "baby version", for a stable A(s)B(s), is its representation in terms of magnitude and phase frequency response plots using the concepts of **gain margin** and **phase margin**. The frequency axis for ω in [rad/s] or for f in [Hz] is logarithmically scaled.

Figure 6.147 shows the magnitude and phase plots of $A_3(j\omega)$ and $A_4(j\omega)$, with special frequencies ω_c and ω_π marked along with the gain and phase margins. The figure shows the monotonically decreasing magnitude and phase frequency response functions of the loop transfer function A(s)B.

There are two frequencies of particular interest: The first is the **crossover frequency** $\boldsymbol{\omega_c}$ [rad/s] at which the magnitude frequency response of the loop gain equals 1: $\mathbf{|A(j\omega_c)B| = 1 = 0\,dB}$. The other frequency, denoted as ω_π [rad/s], is a frequency at which the loop transfer function's frequency response phase function equals −180°: $\mathbf{Phase[A(j\omega_\pi)B] = -180°}$.

See Figure 6.147, which shows the two special frequencies (above) along with the gain and phase margins, to be explained next.

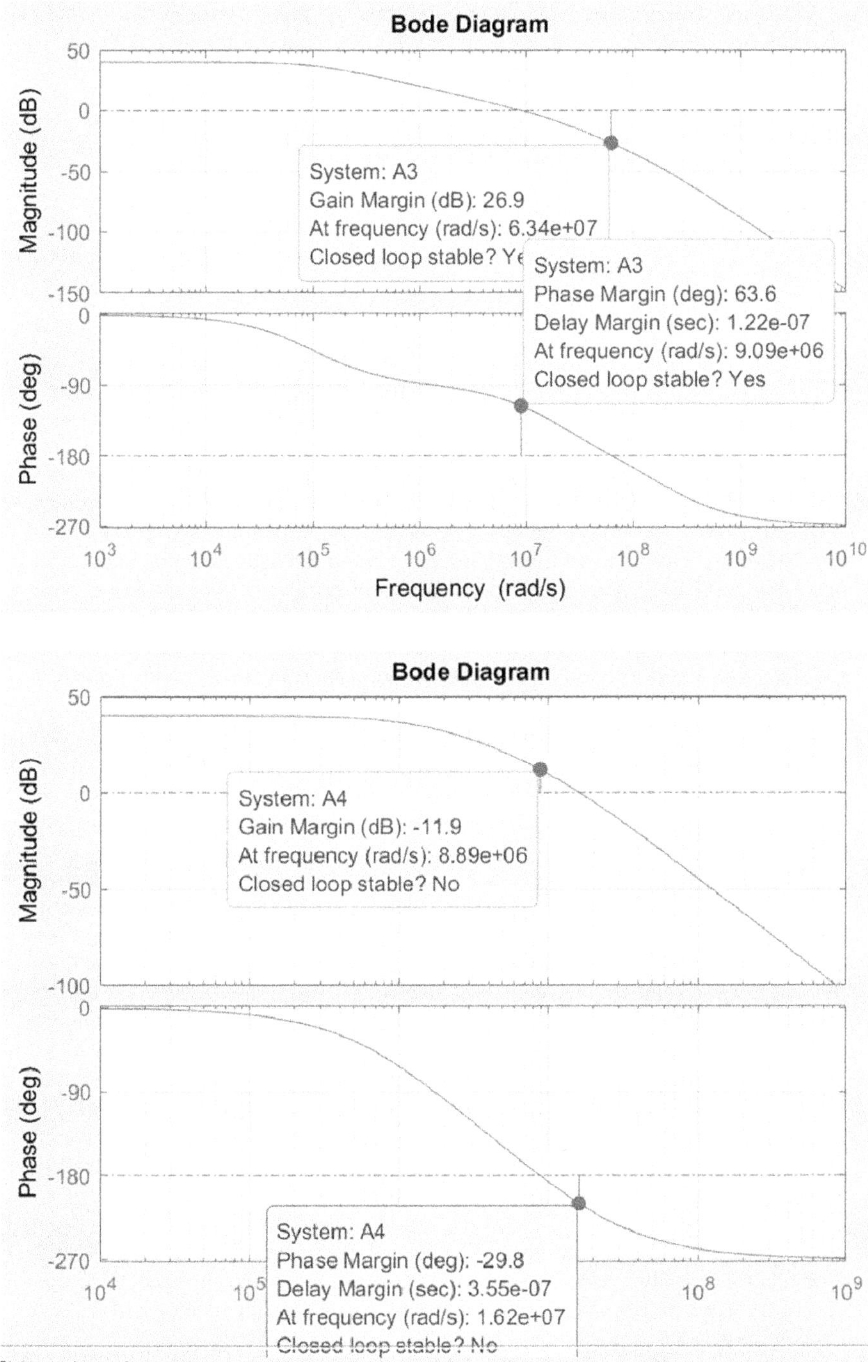

FIGURE 6.147 Frequency response of two all-pole forward amplifiers A(s) shown in Figure 6.146. The measured gain and phase margins are shown as well.

Of course, any integer multiple of 360° may be added to the phase without changing its value. The feedback gain B, whenever using a resistive network, is real. B may be positive or negative, depending on the type of feedback network. We noted earlier that in order to create negative feedback, either B and A (the midband forward gain) are both positive, or both are negative.

For gain-margin calculations, it is safe to assume that the phase curve of A(jω) B depends on A(jω) only.

Definition of GM, the Gain Margin of A(jω)B: $GM = -20\log_{10}(|A(j\omega_\pi)B|)$ dB $= 1/|A(j\omega_\pi)B|$.

Definition of PM, the Phase Margin of A(jω)B: PM = 180° + Phase[$A(j\omega_c)$].

Let us explain both definitions in more detail.

At the frequency $\omega = \omega_\pi$, at which the loop frequency response's phase crosses the −180° axis, and the loop frequency response's magnitude is less than 1 (or 0 dB), the gain margin is greater than 1, or in dB it is positive. Sometimes the phase of A(jω) does not reach −180° (say, whenever A(s) has no more than two poles), then the gain margin becomes undefined.

At the frequency $\omega = \omega_c$, at which the magnitude frequency response crosses over the 0 dB (gain 1) axis, if the phase of A(jω)B is larger (that is, more positive) than −180°, then the phase margin is positive. Phase margin can become undefined if the magnitude curve |A(jω)B| does not reach the 0 dB axis. Practically, this never happens. See Figure 6.147 for examples of positive and negative gain and phase margins.

Nyquist's theorem for the case of a stable A(s)B(s) implies that **closed loop stability** goes hand in hand with a **positive gain margin (in dB)** and a **positive phase margin**.

In a closed-loop stable system, both gain and phase margins are positive. The special frequencies, discussed above, satisfy $\omega_c < \omega_\pi$. The phase margin is measured at a frequency that is lower than the frequency at which the gain margin is measured. Vice versa for a closed-loop unstable system, at which both margins are negative. See Figure 6.147.

Gain and phase margins provide, in addition, measures of **relative stability**, showing how far the system is from the borderline of instability. The gain margin shows by how many dB we can increase the loop gain A·B to reach the borderline of instability. The phase margin shows by how many phase degrees of A(jω)B(jω) the system is distant from instability. Equivalently, the phase margin shows the maximum pure time delay that may be added (multiplicatively) to A(s)B(s) without losing closed-loop stability.

Any of the above, relative measures of stability, convey, indirectly and qualitatively, what to expect in terms of a closed-loop transient response. For instance, positive gain margin and phase margin, which are "too small" are associated with undesired closed-loop step (or pulse) transient features (such as a large overshoot and a long oscillatory convergence to a steady state).

In fact, both gain margin and phase margin measures seem to convey the same kind of relative stability information. Can any of them be redundant? Yes, indeed. One relative stability measure will do. "Will do" for what? For the analytic design of, say, stability compensation of a feedback electronic amplifier. Which of the two relative stability measures is easier to use? Answer: Phase Margin.

Let us elaborate on the tools that are available for a stability compensation design. To use gain margin in an analytic design, one has to plot (or sketch) the loop transfer function's phase plot. Specifically, one has to find (or estimate) the frequency ω_π. Then substitute it into the loop transfer function's magnitude frequency response plot. On the other hand, in order to use phase margin, one has to sketch the magnitude plot (an easier task than sketching of the phase plot) and find (or estimate) the crossover frequency ω_c (again, easier than finding ω_π). The magnitude Bode plot allows convenient computations along the plot, as will be explained next. The connection between phase margin and transient response features implies that a good design, to create a positive phase margin, does not attempt to maximize the phase margin. A "too-large" phase margin may be associated with an unacceptably slow and sluggish closed-loop step response. A good design rather aims at obtaining a phase margin that can practically be between 55° and 65°, more or less.

6.21 BODE PLOT DOMINANT POLE DESIGN FOR FEEDBACK AMPLIFIERS STABILITY COMPENSATION

We often refer to the piecewise linear approximation of the magnitude frequency response, with a logarithmic scale of the frequency, as a Bode plot. A Bode plot consists of all the corner frequencies of the respective transfer function, interconnected by straight lines that have slopes of −N·20 [dB/dec], N being an integer. That is, only slopes that are integer multiples of 20 dB/dec are physical. Bode magnitude plots allow a very simple computational approach. We refer to [Bowerl] for equation (6.2) below. Figure 6.148 illustrates a single segment of a Bode plot.

Let us consider two points (A_1, ω_1) and (A_2, ω_2) on a −N·20 dB/dec slope Bode plot segment, for some specific integer N. For example, N = 1 is a line of slope −20 dB/dec, N = 3 is a slope of −60 dB/dec and N = −1 is a slope of 20 dB/dec. The gains A_1 and A_2 are not in dB. Writing the above straight-line equation in the form of "y = ax + b": $20\log A_2 = 20\log A_1 - N\cdot 20\cdot[\log\omega_2 - \log\omega_1]$ which, after a few cancellations, yields the helpful formula:

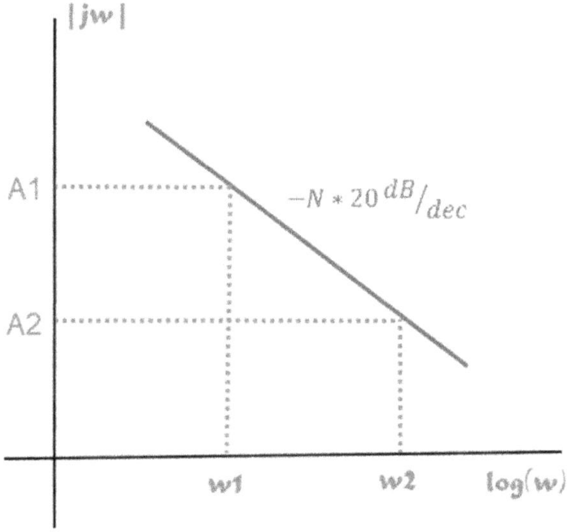

FIGURE 6.148 A segment of a Bode plot. It is a straight line in logarithmic scale of ω.

$$\frac{A_2}{A_1} = \left(\frac{\omega_1}{\omega_2}\right)^N \tag{6.2}$$

Example 6.14

An amplifier has a mid-band gain of 33 dB and a bandwidth of 730 kHz. The amplifier's other poles and zeros are at much higher frequencies (say hundreds of MHz). What is the amplifier's crossover frequency f_c, and what is its phase margin?

Solution

It is okay to scale the X axis in Hz, rather than rad/s. Let us assume that the amplifier's Bode plot is as shown in Figure 6.149.

Let us convert the 33 dB gain to numbers: $20\log_{10}A = 33 \rightarrow A = 10^{33/20} = 44.69$.

Now we apply equation (6.2) with $N=1$ to the −20 dB/dec segment. The points are: ($A_2 = 44.69$, $f_2 = 730$ kHz), ($A_1 = 1$, $f_1 = f_c$). Then $44.69/1 = f_c/730 \rightarrow f_c = 32.624$ MHz. Finally: $PM = 180 + \arg(A) = 180 - \tan^{-1}(f_c/730) = 180 - \tan^{-1}(32{,}624/730) = 180 - 88.72 = 91.28°$.

Example 6.15

An amplifier's Bode plot (magnitude and phase) is given in Figure 6.150. Find the transfer function G(s), the crossover frequency ω_c in rad/s, and the phase margin.
Solution

We find G(s) from the corner frequencies (scanned from low to high) and the changes of slope. We have a double pole at $\omega = 5$ rad/s. Then the slope changes from −40 to −20 dB/dec. Therefore, we have a zero at $\omega = 20$ rad/s. The DC gain of the amplifier is 12 dB = 3.98. One last subtlety before we write down G(s): $1 + s/a$ and $1 - s/a$ have the same Bode magnitude plot. We distinguish the two based on

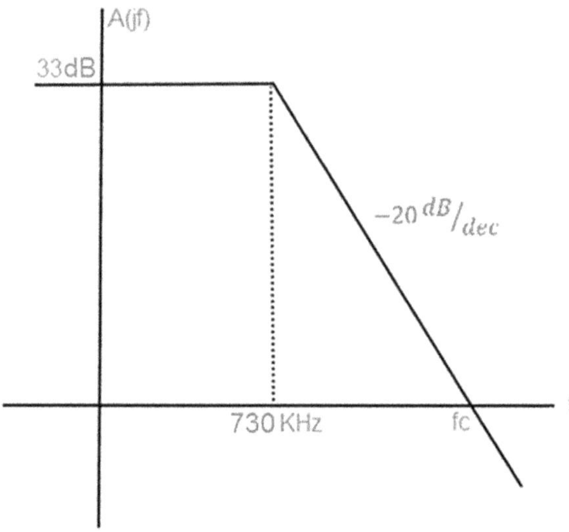

FIGURE 6.149 Amplifier's Bode plot.

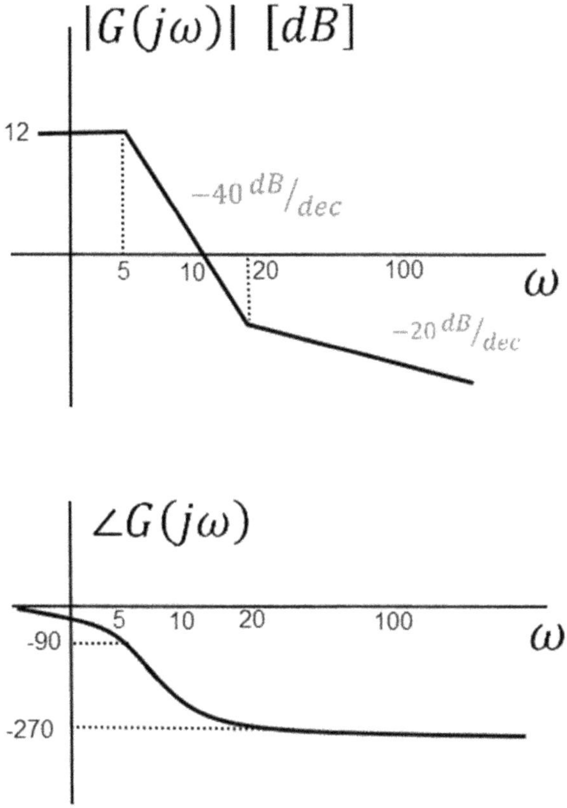

FIGURE 6.150 Amplifier's Bode plot – magnitude and phase.

the given phase plot. A left half plane zero at $\omega = 20$ rad/s contributes a positive phase. Had the zero been LHP, the phase plot would turn up and eventually converge to −90°. Instead, the phase plot went further down and converged to −270°. The zero at $\omega = 20$ rad/s is therefore a right half-plane zero. Consequently, the G(s) transfer function is $G(s) = 3.98\ (1 - s/20)/(1 + s/5)^2$.

Now, along the −40 dB/dec segment (N=2), we calculate $3.98/1 = (\omega_c/5)^2 \rightarrow \omega_c = 9.97$ rad/s. Finally: $PM = 180 - 2\tan^{-1}(\omega_c/5) - \tan^{-1}(\omega_c/20) = 180 - 2\tan^{-1}(9.97/5) - \tan^{-1}(9.97/20) = 180 - 126.73 - 26.5 = 26.77°$. RHP zeros in a closed-loop control system tend to be destabilizing, as these lower the open-loop phase margin.

Example 6.16

A "minimum phase" amplifier has a magnitude Bode plot shown in Figure 6.151. Find the amplifier's transfer function T(s) and its mid-band gain in dB.

Solution

"Minimum Phase" is just a name, given to transfer functions which have all their poles and zeros located in the left half s-plane. In a minimum phase transfer function, there is no ambiguity, as we have seen in the previous example, and

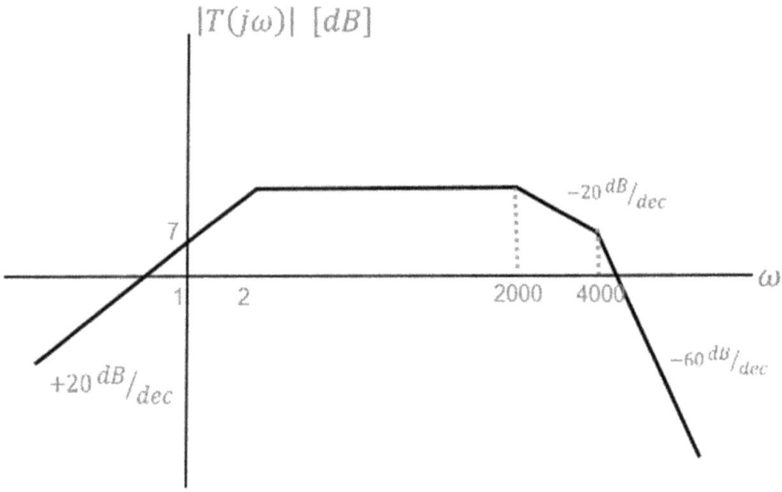

FIGURE 6.151 An amplifier's Bode plot showing the low, mid-band and high frequency ranges.

therefore, the magnitude frequency response is sufficient for finding the transfer function.

Our earlier discussion of Bode plots included only simple poles, simple zeros, and DC gain as building blocks of the overall frequency response. There are more types of building blocks. For instance, a pole at s=0 (representing a free integrator) has a –20 dB/dec line as its magnitude frequency response. The line passes through the point {ω=1 rad/s, magnitude=0 dB}. The phase associated with 1/s equals −90° at all frequencies ω. Similarly, a zero at s=0 (representing a differentiator) has a +20 dB line magnitude frequency response, going through the point {ω=1 rad/s, magnitude=0 dB}. The phase of s equals 90° at all frequencies.

Scanning the corner frequencies and slopes from left to right, we inspect that T(s) has a zero at s=0 and some DC gain K. As the slope changes at ω=2 rad/s from +20 to 0 dB/dec, we see that T(s) has a pole at s=–2. At ω=2000 rad/s the slope changes again: From 0 to –20 dB/dec, implying a second pole at s=−2000. Finally, we observe that at ω=4000 rad/s the slope changes from –20 to −60 dB/dec, due to two more poles of T(s), located at s=−4000. A double pole at s=−4000. So far $T(s)=K{\cdot}s/(1+s/2){\cdot}(1+s/2000){\cdot}(1+s/4000)^2$. What is the value of K?

We recall the Bode plot of a simple pole, consisting of two-line segments: One with a slope of 0 dB/dec to the left of the corner frequency, and another with a slope of –20 dB/dec to the right of the corner frequency. A Bode line segment of a magnitude of 0 dB and a slope of 0 dB/dec contributes 0 dB to the overall Bode plot at all frequencies that are lower than the corner frequency. We keep in mind that in log scale all the basic contributions add up. A 0 dB horizontal line contributes a multiplication by 1 to the overall transfer function, at all frequencies that are lower than the corner frequency.

Coming back to T(s): At all frequencies that are lower than ω=2 rad/s, the four poles at s=–2, s=–2000, and s=–4000 contribute 0 dB to the magnitude frequency response mag[T(jω)]. At these frequencies, the only contributions to the magnitude plot come from zero s=0 and the gain K. We recall that the +20 dB/dec magnitude response line passes through {ω=1 rad/s, magnitude=0 dB}. A gain of $K>1$ elevates the $s=0$ line, whereas a gain of $0<K<1$ lowers the $s=0$ line. We notice a dB gain K of 7 dB at $\omega=1$ rad/s. For inclusion of K into T(s), we convert 7 dB to $K=2.238$.

In Figure 6.151, we see three frequency response ranges for the amplifier. The low frequency range, determined by the zero $s=0$ and the pole at $s=-2$, has a +20 dB/dec magnitude frequency slope. The mid-band frequency response, characterized by a constant gain, spans between $\omega=2$ rad/s and $\omega=2000$ rad/s. The high frequency range starts at $\omega=2000$ rad/s. The slope changes from −20 to −60 dB/dec at $\omega=4000$ rad/s.

In order to find the mid-band gain A_0 (in numbers, not in dB), we may calculate based on two points: ($\omega=1$ rad/s, $A=2.238$) and ($\omega=2$ rad/s, magnitude $=A_0$) and a slope multiplier $N=-1$: $A_0/2.238=2/1 \rightarrow A_0=4.476=13.02$ dB.

Example 6.13 (revisited)

Let us look again at Figure 6.147, showing the frequency response of two forward amplifiers, $A_3(s)$ and $A_4(s)$. The closed loop circuits have several similarities: the same DC gain of the A(s) amplifier, the same number of poles of an all-pole A(s), and the same feedback gain B. So, why is it that the closed loop circuit with $A_3(s)$ is stable and the closed loop circuit with $A_4(s)$ is unstable?

$A_3(s)$ has sparse poles having the corner frequencies 0.1, 20, and 200 (all Mrad/s). In $A_4(s)$, the poles at 1, 4, and 15 (Mrad/s) are arranged more densely, forming a cluster of poles. We note from the magnitude frequency response of $A_3(s)$ (the top plot of Figure 6.147) that past $\omega=0.1$ Mrad/s the Bode plot has a −20 dB/dec slope, and that relatively slowly decreasing segment is long, spreading over more than two decades. Then, near $\omega=20$ Mrad/s, the slope changes to −40 dB/dec and a more negative phase is added. In fact, we note that at the crossover frequency ω_c the slope of the Bode plot is still −20 dB/dec. The resulting phase margin is significantly positive. The magnitude frequency response of $A_4(s)$ decreases more rapidly (due to the cluster of $A_4(s)$ poles). All the cluster poles simultaneously contribute a negative phase. As mag[$A_4(j\omega)$] crosses over the 0 dB axis, at a slope of −60 dB/dec, the phase margin is negative, hence instability.

A feedback amplifier, consisting of a multistage forward amplifier A(s), may become closed-loop unstable. The poles, contributed by each stage of A(s), very often form a cluster similar to the one observed in Example 6.13. Such feedback amplifiers must be compensated to become stable or to remain stable. A control design carried out by Bode plots is often called **Loop Shaping**. In a general loop shaping process, the open loop transfer function A(s)B is manipulated to meet certain design specifications. Such a manipulation may involve the adding, deleting, and shifting of open loop poles and zeros, and it is conveniently done with the aid of Bode plots.

In a stability compensation loop shaping control design, the **two main specifications** are:

a. **Determining the new crossover frequency f_c.**
b. **Creating a "good" (recall – not so large and not so small) positive phase margin.**

The crossover frequency specification relates to the fastness of the closed-loop circuit – what is the closed loop bandwidth? How small are the closed-loop time constants?

The phase margin specification deals with other closed-loop time domain features – how damped or underdamped is the closed-loop circuit? Does the closed-loop pulse response have an overshoot? How large or small is that overshoot? Is the pulse response's settling time too long, or in other words, is the transient time long and consists of slowly decaying oscillations? Does the closed-loop frequency response have a peak resonance? How large is the resonance?

In a typical compensation design, the A(s) poles of the original stages cannot, in general, be deleted to be replaced by others. What can be done, though, is to add other poles that may dominate the original poles.

The most important Loop Shaping guideline is: **Always cross over at a slope of −20 dB/dec**. Other slopes (which are, as we recall, integer multiples of 20 dB/dec) either cause instability (like the −60 dB/dec slope shown for $A_4(s)$ in Example 6.13) or may stabilize but without creating a large enough phase margin (as may happen with a −40 dB/dec crossing over slope). The −20 dB/dec slope must extend a sufficient length before $f=f_c$ and after $f=f_c$.

The **Dominant Pole Control Design** strategy aims at assuring a −20 dB/dec crossing over slope of A(s)B, by adding an open loop pole to A(s) at a frequency ω_p that is much smaller than the corner frequencies of all the poles of A(s). That low-frequency pole is created at one of the stages of A(s), replacing the original bandwidth of that stage by ω_p. The frequency ω_p becomes the new overall open-loop ($A_{compensated}(s)B$) bandwidth. At frequencies that are lower than $\omega=\omega_p$, the Bode plots of A(s)B and $A_{compensated}(s)B$ merge. That is, the DC gain of $A_{compensated}(s)B$ is equal to the DC gain of the uncompensated A(s)B. The open loop compensated new bandwidth ω_p must be sufficiently small, so that the long −20 dB/dec line, that starts at $\omega=\omega_p$ at a high DC gain of A(s)B, and traverses several frequency decades till it crosses over at $\omega=\omega_{c,new}$ at a gain of 1, is not disrupted by any of the original poles of A(s).

For example, in the μA741 op-amp, the open loop DC gain is approximately 10^6, and the compensated open loop bandwidth is at approximately $\omega_p \approx 2\pi f_p$, where $f_p \approx 10\,\text{Hz}$. If $B=0.1$ in order to create a closed loop gain of 10, the −20 dB/dec line of $A_{compensated}(s)B$ goes from 10^6 at $f_p=10\,\text{Hz}$ to a gain of 1 at $f_c=1\,\text{MHz}$, and further beyond.

Example 6.17

Let us demonstrate loop shaping design. See Figure 6.152 for a block diagram of a multistage feedback amplifier. A description of an electronic amplifier by means of a block diagram is highly idealized. We have seen in Sections 6.5 and 6.6 how interconnected stages load one another, and we have seen in Section 6.17 how the feedback network B loads the forward amplifier A(s). In any block diagram description, each transfer function block retains its gain and poles despite the interconnections.

Solution

Each A(s) stage is characterized by its DC gain and by its bandwidth. As the DC gains multiply, the overall DC gain of A(s) becomes 3000, and we observe that $A{\cdot}B=3000{\cdot}0.1=300 \gg 1$, suitable for a high loop gain design strategy that makes the closed loop gain equal $1/B=10$. The feedback amplifier, if stable, is intended to create a closed-loop voltage gain of 10.

Each amplifier stage contributes a single pole. We note that the three poles are close to one another, creating a cluster.

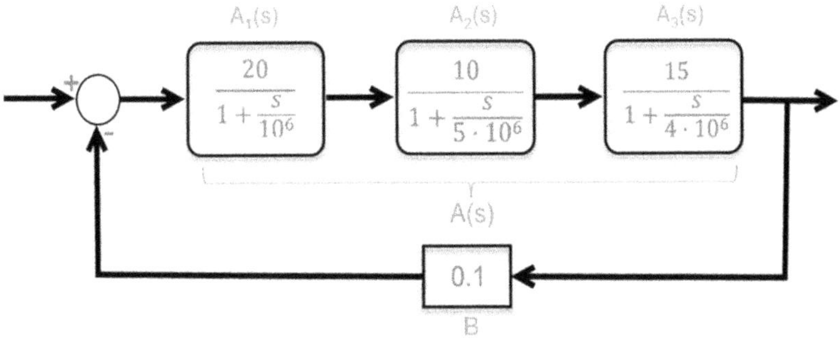

FIGURE 6.152 Block diagram of a three-stage feedback amplifier.

In Figure 6.153, we draw the (yet uncompensated) Bode plot of the open-loop transfer function A(s)B. Let us calculate the crossover frequency ω_c and then the phase margin of A(s)B.

The piecewise linear Bode plot approximates the true magnitude frequency response. Therefore, the sequence of calculations (to follow), using equation (6.2), is all approximations. As we sketch $|A(j\omega)B|$, we cannot tell off-hand if the plot crosses over the 0 dB axis at a slope of −40 or −60 dB/dec. Once we are able to estimate the gains x and y (see Figure 6.153) it will become evident.

Along the −20 dB/dec line segment N = 1: we calculate $300/x \approx 4/1 \rightarrow x \approx 75$. Note that the gains 300, x and y are not in dB.

Along the −40 dB/dec line segment N = 2: we calculate $x/y \approx (5/4)^2 \rightarrow y \approx 48$. We now see that the crossing over slope is −60 dB/dec. Had y been smaller than 1, the crossing over slope would be −40 dB/dec.

Finally, along the −60 dB/dec line N = 3: we calculate $y/1 \approx (\omega_c/5)^3 \rightarrow \omega_c \approx 18.17$ Mrad/s.

Each pole of A(s)B contributes a negative phase at the crossover frequency. The contribution of each pole is $\tan^{-1}$(crossover_frequency/corner_frequency_of_the_contributing_pole). Let us calculate the uncompensated phase margin:

$PM = 180 + \arg(A(j\omega_c)B) = 180 - \tan^{-1}(18.17/1) - \tan^{-1}(18.17/4) - \tan^{-1}(18.17/5) = 180° - 86.84° - 77.58° - 74.61° = 180° - 239.03° = -59.03°$.

As we see, for this −60 dB/dec crossing over Bode plot, as the magnitude of $|A(j\omega_c)B|$ becomes 1, the phase of $A(j\omega_c)B$ drops below −180°, thus creating a negative phase margin. The closed-loop circuit would be unstable.

The dominant pole compensation idea is demonstrated in Figure 6.154.

By equation (6.2) for N = 1, for any choice of the dominant pole at ω_p, the new crossover frequency $\omega_{c,new}$ maintains the same ratio: $\omega_{c,new}/\omega_p = 300/1$. In fact, it is more convenient to first choose a location for $\omega_{c,new}$, and then find ω_p. The phase contribution of the dominant pole at ω_p is $-\tan^{-1}(\omega_{c,new}/\omega_p) = -\tan^{-1}(300) = 89.81°$. If $\omega_{c,new}$ is chosen to lie much to the left of the amplifier stages poles, then the amplifier's poles negative phase contributions would be very small and the resulting phase margin would be around 80°–90°, which is considered too large. So, $\omega_{c,new}$ rather be to the left of the smallest uncompensated amplifier's pole, but not by a large distance (as will be seen soon). That allows the amplifier's poles to contribute negative phase more significantly, and by that bring the phase margin to an acceptable value.

We observe in Figure 6.154 that the first stage's pole (at $\omega_1 = 1$ Mrad/s), which happens to be smaller than the other two stages' poles, causes the compensated

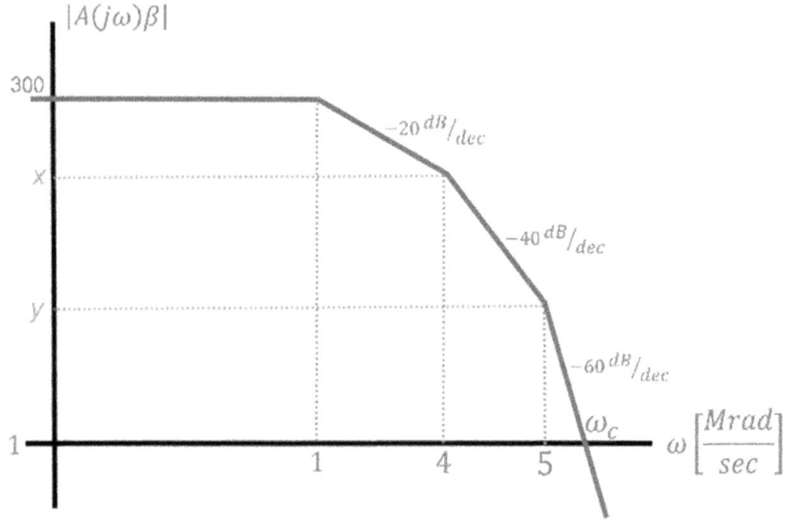

FIGURE 6.153 Magnitude Bode plot of A(s)B.

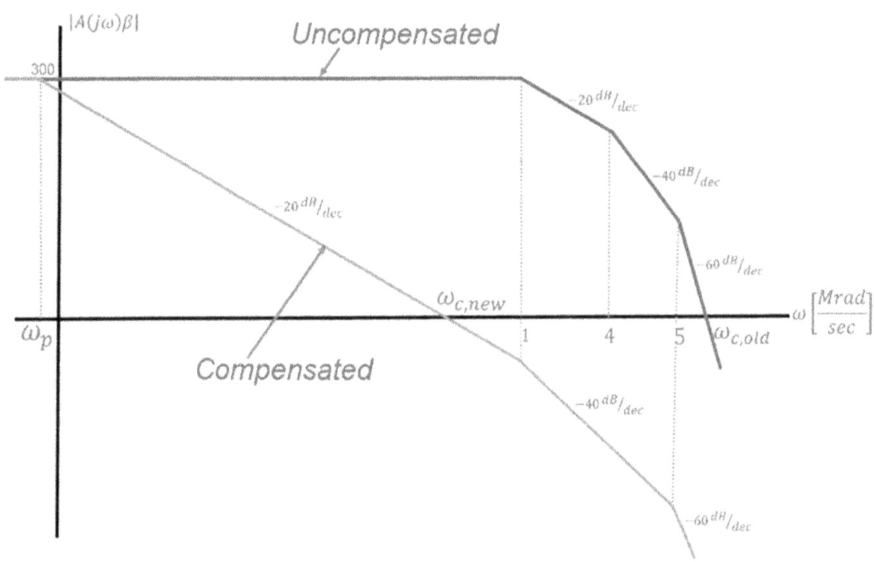

FIGURE 6.154 The uncompensated and compensated Bode plots for the feedback amplifier. Both Bode plots are open-loop magnitude frequency plots.

Bode plot to bend from a slope of −20 to −40 dB/dec at $\omega = 1$ Mrad/s, which is made to be not too close to the new crossover frequency (as will be calculated soon).

We then observe that the third stage's pole (at $\omega_3 = 4$ Mrad/s) is ignored. What does it mean?

It simply means that the compensation design opts to implement the dominant pole ω_p at stage 3 of the multistage amplifier A(s). This dominant pole replaces the original stage's pole.

Finally, the pole of stage 2 ($\omega_2 = 5$ Mrad/s) further bent the Bode plot from −40 to a −60 dB/dec slope. In which stage to create ω_p is a matter of convenience, and very often several possible choices may be equally convenient. Let us now explore where to place the new crossover frequency to obtain a favorable phase margin.

First Try: Let $\omega_{c,new} = 0.1\omega_1 = 100$ krad/s → $300/1 = \omega_{c,new}/\omega_p$ → $\omega_p = 333$ rad/s. Also, PM = 180 − tan^{-1}(300) − tan^{-1}(0.1/1) − tan^{-1}(0.1/5) = 180 − 89.8 − 5.7 − 1.1 = 83. 4°. It is probably too large. We can make the closed-loop amplifier faster by choosing a larger value of $\omega_{c,new}$.

Second Try: Let $\omega_{c,new} = 0.2\omega_1 = 200$ krad/s → $300/1 = \omega_{c,new}/\omega_p$ → $\omega_p = 667$ rad/s. Also, PM = 180 − tan^{-1}(300) − tan^{-1}(0.2/1) − tan^{-1}(0.2/5) = 180 − 89.8 − 11.3 − 2.29 = 76.6°. Still seems a little too large.

Third Try: Let $\omega_{c,new} = 0.5\omega_1 = 500$ krad/s → $300/1 = \omega_{c,new}/_{\omega p}$ → $\omega_p = 1.67$ krad/s. Also, PM = 180 − tan^{-1}(300) − tan^{-1}(0.5/1) − tan^{-1}(0.5/5) = 180 − 89.8 − 26.6 − 5.57 = 57.9°. We can stop here. It is good enough.

Finally, let us comment that **the new crossover frequency of the compensated open-loop amplifier is a good approximation of the closed-loop bandwidth**. Let us explain why. Let the closed loop magnitude frequency response be mag[M(jω)] = mag(A(jω)B)/mag[1 + A(jω)B]. As mag[A(jω)B] ≫ 1 if $\omega \ll \omega_{c,}$ then mag(M(jω)) ≈ 1 at these frequencies. As mag[A(jω)B] ≪ 1 if $\omega \gg \omega_{c,}$ then mag[M(jω)] ≈ mag[A(jω)B] at these frequencies. Hence, the bandwidth of M(s) is near the crossover frequency of A(s)B.

6.22 ATTACHING FEEDBACK NETWORK TO THE MULTISTAGE DIFFERENTIAL AMPLIFIER; ADS TRANSIENT SIMULATION DETECTION OF CLOSED-LOOP INSTABILITY; AC SIMULATION OF THE OPEN-LOOP AMPLIFIER LOADED BY THE FEEDBACK NETWORK: UNWRAPPING OF THE FREQUENCY RESPONSE PHASE PLOT, MEASUREMENT OF THE CROSSOVER FREQUENCY AND THE PHASE MARGIN; COMPENSATION PLANNING: ALTERNATING AC SIMULATION COMPENSATION OPEN LOOP DESIGN AND TRANSIENT SIMULATION CLOSED LOOP TESTING OF THE PULSE RESPONSE

Example 6.18

We plan to use the familiar multistage differential amplifier as the forward block of a feedback voltage amplifier with a voltage gain of 10. Let us plan on using the op-amp non-inverting configuration. See Figure 6.155.

Let us assume that the inverting input of the "op-amp" is the base of BJT Q2. This assumption can be easily justified by a direct open-loop AC simulation measurement. Note the two resistors R_{f1} and R_{f2}, inserted to obtain a closed-loop voltage gain of 10. For input symmetry, we add $R_{f1}||R_{f2}$ to the base of Q_1.

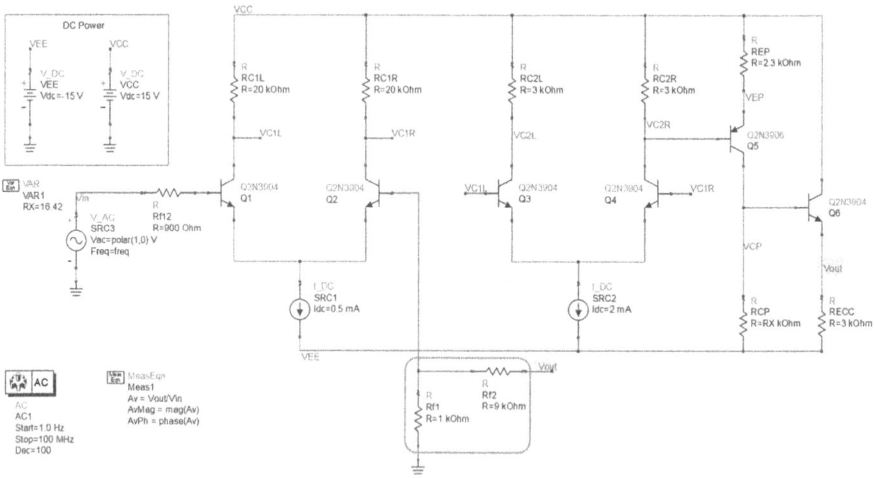

FIGURE 6.155 Feedback voltage amplifier that uses a multistage differential amplifier as the forward amplifier and a two-resistor voltage amplifier B feedback network.

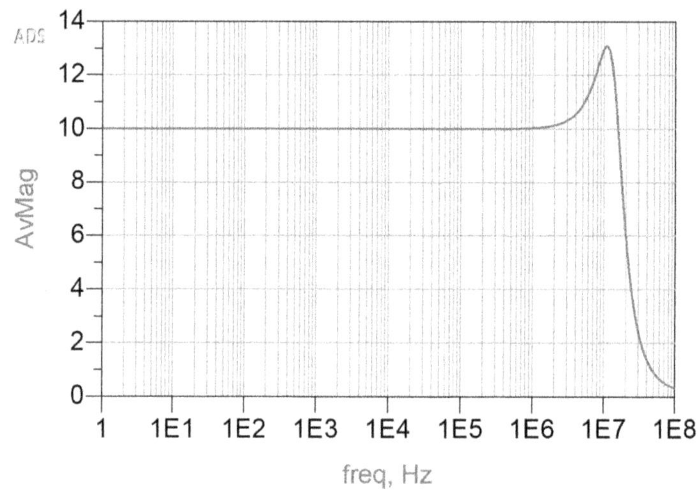

FIGURE 6.156 The uncompensated magnitude frequency response of the closed loop voltage amplifier.

Let us start with an AC simulation of the uncompensated closed-loop amplifier (as shown in Figure 6.155), yielding the closed-loop voltage gain frequency response shown in Figure 6.156. It appears that the closed-loop amplifier (that is based on the multistage amplifier) performs nicely. It has a mid-band frequency voltage gain of 10, and a bandwidth of nearly 10.5 MHz (or 18.5 MHz), depending on which side of the resonance peak we measure the 3 dB frequency. The peak is about 30% above 10, and f = 10.5 MHz near the peak.

Why compensate? Let us not rush to quick conclusions. We also need to check the closed-loop performance in Transient Simulation, as shown in the schematics of Figure 6.157. Is transient simulation capable of detecting potential amplifier instability?

Sometimes it does, and sometimes it doesn't. It all depends on the correct choice of the Transient simulation parameters.

We apply a small step input voltage.

We know that for an AC input amplitude of 0.1 mV, the output of the amplifier does not clip. We check to see if there is any high-frequency and large-amplitude oscillation of the amplifier. The non-obvious Transient Simulation timing parameters, shown in Figure 6.157, are for the time being assumed to have been obtained by some trial-and-error. We shall revisit it at the end.

Later, we shall establish the connection between these parameters and the open-loop frequency response crossover frequency. The Transient simulation result is shown in Figure 6.158. Yes, the uncompensated closed-loop voltage amplifier is unstable!

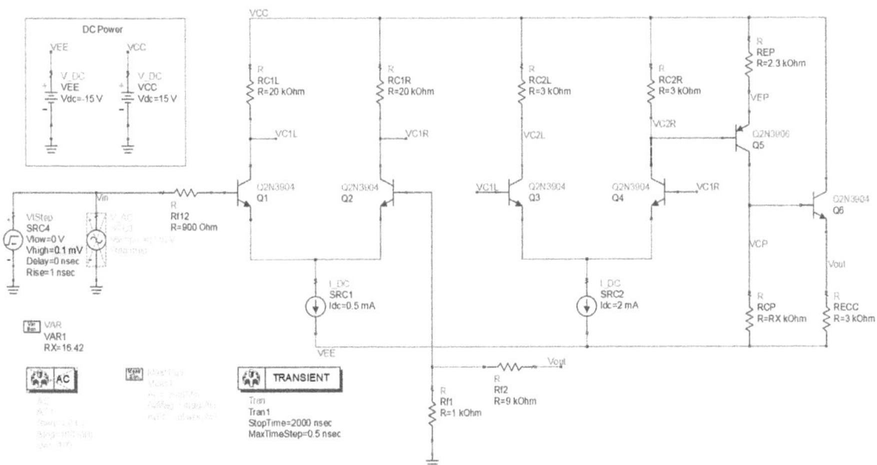

FIGURE 6.157 Transient simulation of the uncompensated feedback voltage amplifier.

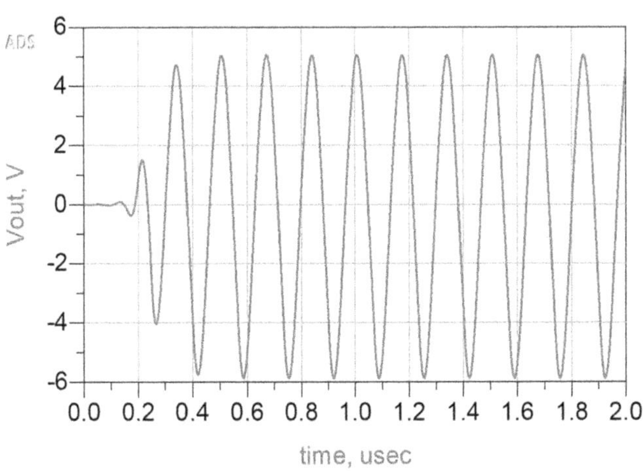

FIGURE 6.158 The unstable feedback amplifier oscillates at a high frequency and a large amplitude.

The amplifier oscillates with a peak-to-peak amplitude of nearly 11 V and a period of nearly 0.15 μs. The oscillation frequency is about 6.67 MHz.

Let us observe again the uncompensated frequency response of the loop transfer function A(s)B, this time around with the B feedback network loading of A(s) (See Figure 6.159), and measure what the crossover frequency and phase margin are.

Note that with the earlier feedback resistors of $R_{f1} = 9$ kΩ and $R_{f2} = 1$ kΩ, the loading of the amplifier is:

$R_{f1} + R_{f2} = 10$ kΩ connected in shunt to the output.

$R_{f1} || R_{f2} = 900$ Ω connected in series to the inverting input of the multistage amplifier (base of Q_2).

We symmetrically add 900 Ω to the base of Q_1.

Let's denote the gain of the loaded open-loop forward amplifier as A(f). The magnitude and Phase of A(f), shown in Figure 6.160, are not particularly informative in the frequency response present form. What is hard to do with these measurements, in their current form?

1. Measuring f_c at which $|A(f_c)| = 1$.
2. Measuring the phase at $f = f_c$.
3. Seeing what the slope is of |A(f)| near crossover frequency.

Let us pause to discuss how the measurement of the crossover frequency ought to be done.

We measure |A(f)|, but we really need to measure |A(f)|B, where B is the feedback network gain (which is 0.1, if we build a closed-loop amplifier with a gain of 1/B = 10). So, $f = f_c$ at which |A(f)|B = 1, is really the frequency f at which |A(f)| = 1/B = 10.

In other words, both the crossover frequency and consequently the phase margin are to be measured not with respect to the 0 dB axis of |A(f)| but with respect to the |A(f)| = 1/B axis. In our example, our new X-axis will be at A = 10 (or 20 dB).

We alter the MeasEqn block of the AC simulation of Figure 6.159, as shown in Figure 6.161:

Let us draw |A(f)| with a logarithmic Y-axis, for easier collection of measurements.

Let us also work with a continuous phase plot so we need to unwrap the phase.

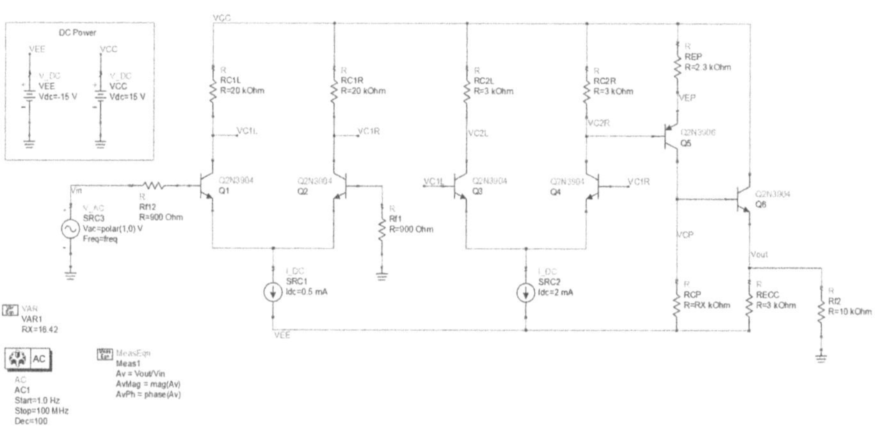

FIGURE 6.159 Open loop, still uncompensated, frequency response of the forward amplifier, with feedback network loading of A(s) included.

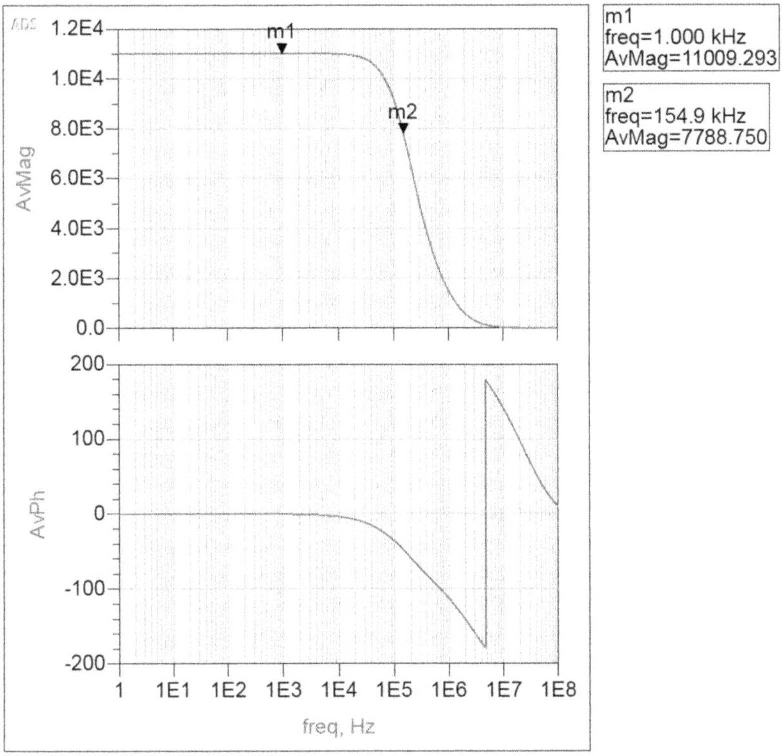

FIGURE 6.160 Frequency response of the uncompensated and loaded A(f).

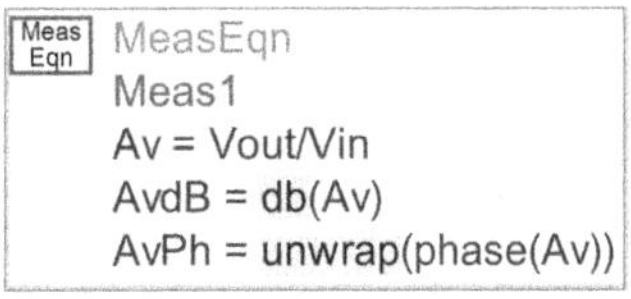

FIGURE 6.161 A new MeasEqn to the schematic of Figure 6.159.

After the modification of the MeasEqn, we re-simulate and obtain the frequency response results shown in Figure 6.162.

We see that the mid-band gain of A(f) is 80.825 dB = 11000. At $f_c = 16.98\,\text{MHz}$ we have $|A(f_c)| = 20$ dB = 10. This is the "crossover frequency" for the sake of creating a closed-loop voltage gain of 10. The phase measured at $f = f_c$ is –251.296°. The phase margin is $PM = 180 + \text{phase}(f_c) = -71.296°$. That's why the uncompensated feedback amplifier is unstable.

We can see that |A(f)| falls initially at a slope of –20 dB/dec. At around f = 1 MHz, the slope changes to −40 dB/dec. This is the slope at the crossover frequency of $f_c = 16.98\,\text{MHz}$.

Note that in Section 6.6 the amplifier's gain was 11,800, and now it is 11,000. This is because of the resistors at the bases of stage 1: The feedback network loading and its symmetrical copy.

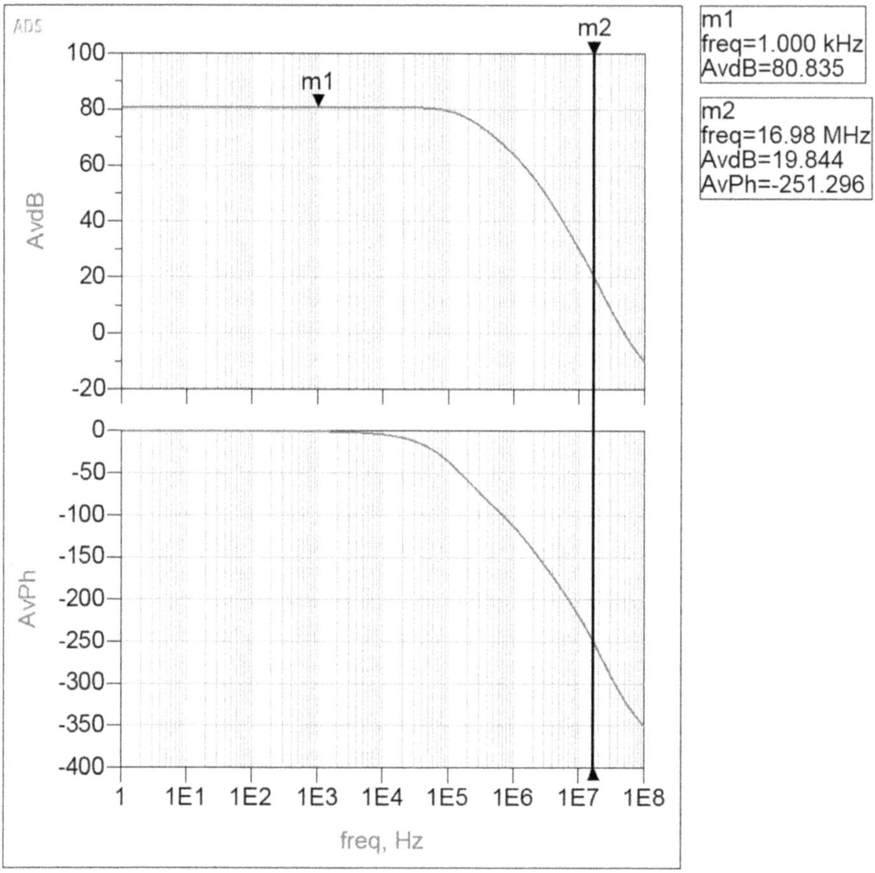

FIGURE 6.162 A second view of the results of the AC simulation of Figures 6.160 and 6.161.

Let us now plan how to implement stability compensation.

With a mid-band gain of 11,000 and an elevated "new X-axis" at 10, the ratio between the "new crossover frequency f_{cnew}" and the "compensation pole corner frequency f_p" (because of the –20 dB/dec drop to be created after we compensate) is $f_{cnew}/f_p = 11{,}000/10 = 1100$.

Need to place $f = f_{cnew}$ at a lower value than the pole corner frequencies of A(f). Recall from Section 6.6 that the lowest pole of A(f), $f = f_h$, was around 160 kHz (influenced by the BW of Stage 1).

Let's try to place f_{cnew} at $\mathbf{f_{cnew}} = 0.5 \cdot f_h = \mathbf{80\,kHz}$. Later as we test the performance of the compensated amplifier, we may explore other ratios: 0.3, 0.7, etc.

The compensation is done by means of a single capacitor C_f, connected between a stage input (or output) node and ground.

So, we start with an estimate that $\mathbf{f_p} = f_{cnew}/1100 = 80{,}000/1100 = \mathbf{72.7\,Hz}$. In which amplifier's stage shall we place the compensation capacitor C_f? Not in stage 1 or stage 2, as it may disrupt the differential amplifying actions. So, it has to be stage 3 (the PNP voltage level shifter) or stage 4 (the emitter follower output stage).

Refer to Figure 6.159 for components and node names.

- Possible location A: Between node "VC2R" and ground. (It is the input of Stage 3).
- Possible location B: Between node "VCP" and ground. (It is the input of Stage 4).

We recall (from Sections 6.5–6.6) the large input resistance values of stages 3 and 4.

Capacitor C_f in location A "sees" an equivalent resistance of $R_{eqA} = R_{C2R} || R_{in3} \approx R_{C2R} = 3\ k\Omega$.

Capacitor C_f in location B "sees" an equivalent resistance of $R_{eqB} = R_{CP} || R_{in4} \approx R_{CP} = 16.4\ k\Omega$.

Let's arbitrarily choose location A. The compensation can work equally well with location B.

So, we started with an estimate that $f_p = f_{cnew}/1100 = 80{,}000/1100 = 72.7\,\text{Hz}$. The compensation pole is $f_p = 1/(2\pi \cdot R_{eqA} \cdot C_f)$. Then $\mathbf{C_f} = (2\pi \cdot 3000 \cdot 72.7)^{-1} = 0.729e^{-6}$ F = **729 nF**.

If later on we find that we must, for example, decrease f_{cnew} from $0.5f_h$ to $0.25f_h$, then we simply double C_f (computed above). If we need to increase f_{cnew} to say $0.6f_h$ then we reduce C_f (above) by a factor of 6/5.

We shall now conduct an open-loop AC simulation of the compensated amplifier. See Figure 6.163. The frequency response produced reasonably close results, shown in Figure 6.164, validating the design calculations done earlier.

Recalling the design values: $f_{cnew} = 80\,\text{kHz}$, $f_p = 72.7\,\text{Hz}$, to be compared to the readings of markers m_2 and m_3 of Figure 6.164.

Why are the slightly off results? Maybe R_{eqA} was not exactly 3 kΩ.

The phase of the compensated A(f) at the new crossover frequency $f = f_{cnew} = 75.86\,\text{Hz}$ is Phase = −107.696°. The phase margin is PM = 180 + Phase $(f_{cnew}) = 180 - 107.696 = 72.304°$.

Now we need to check the closed-loop performance in transient simulation and watch the shape of the pulse response. The fine adjustments of C_f are done by switching back and forth between open-loop AC simulation (compensation design) and closed-loop transient simulation (compensation validation). We shall contrast the closed-loop overshoot and settling-time time-domain features against the open-loop frequency-domain phase margin.

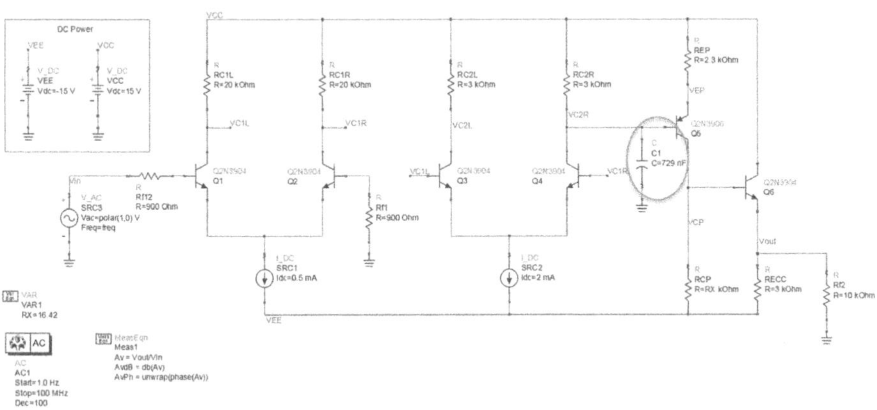

FIGURE 6.163 The compensated loaded forward amplifier – an AC simulation.

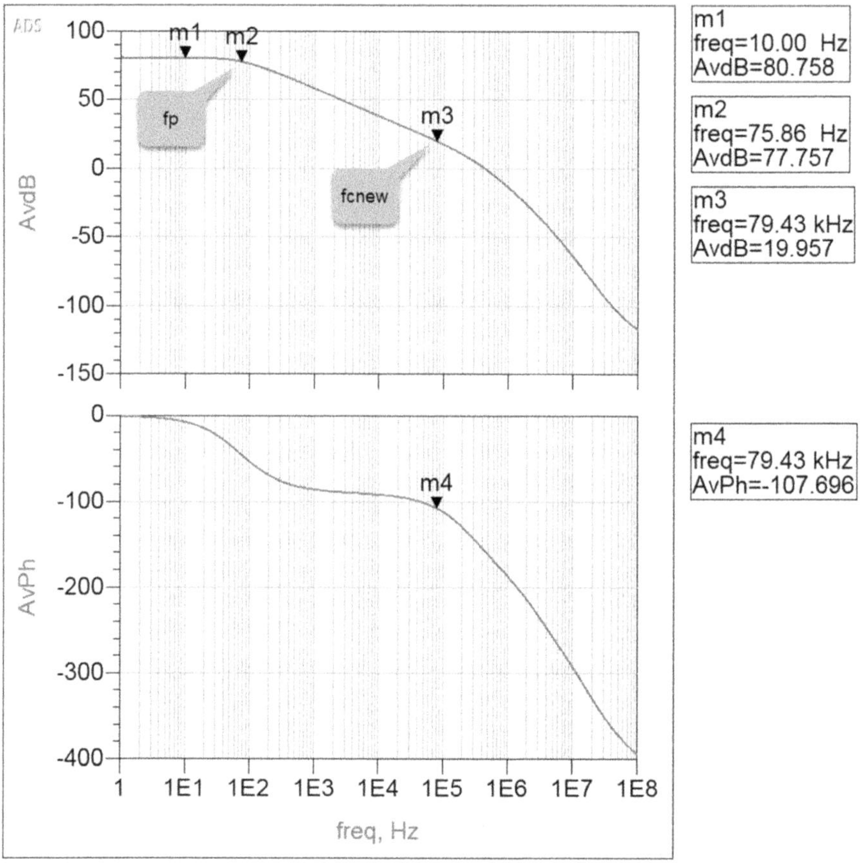

FIGURE 6.164 Compensated amplifier's open-loop frequency response results.

Let us plan how to test the compensated closed-loop amplifier in transient simulation.

- It is known that the open-loop crossover frequency f_c is very close to the closed-loop bandwidth BW_{CL}. We shall demonstrate it later on.
- Therefore, $1/f_c$ is an indicator of the closed-loop time constant.
- It takes four time constants for a step response to reach steady-state [Dorf1].
- The pulses that we input into the closed loop amplifier will, of course, be amplified by a factor of $1/B = 10$. We will be interested in assessing the shape of the amplified output pulses: Do they have overshoots and decaying oscillations? Slow rise time?
- The closed-loop pulse response shape features (such as overshoot and settling time) depend on the obtained phase margin.

It is time to do quantitative planning for the Transient simulation.

- We have found (Figure 6.164) that $f_{cnew} = 75.86$ kHz $\approx BW_{CL}$
- The closed loop time constant is estimated to be $\tau \approx 1/(2\pi BW_{CL}) = 2.1$ μs

- The going-up pulse response transient time (sometimes called "settling time") is around four time-constants. Likewise, the transient for the going-down pulse. Total estimated transient time (per one input pulse) $\approx 8 \cdot 2.1 = 16.8$ μs.
- Therefore, let each input pulse have a width of 20 μs and an amplitude of 1 mV. The choice of input pulse amplitude will be discussed soon.
- With 50% duty cycle, and taking three pulses, the simulation stop time will be $3 \cdot 2 \cdot 20 = 120$ μs.
- The transient simulation maximum step size is taken as $\tau/100 = 21$ ns.

See Transient simulation parameters in Figure 6.165. The pulse response result, for $C_f = 729$ nF is shown in Figure 6.166. We recall from the previous open-loop AC simulation (Figure 6.164) that PM = 73°. The pulse response seems a little slow. Let us half C_f and watch Figure 6.167.

The pulse response for twice the f_{cnew} appears not to be an improvement, due to the nearly 10% overshoot. Figure 6.168 presents three more pulse response plots, for smaller values of C_f, all are clearly worse.

We see that as we lower C_f (and obtain larger f_{cnew} values), the pulse response overshoot increases, and the length of decayed oscillations transient time increases as well.

For the same five C_f values, Figures 6.169 and 6.170 contrast phase margin values against pulse response plots.

In the summary table (of Figure 6.169), Column 1 shows C_f in [nF].

Column 2 shows the open-loop dB gain at which phase margin values were collected.

Column 3 lists the crossover frequencies, and Column 4 shows the phase margin values.

The best two capacitor values appear to be $C_f = 729$ nF, corresponding to a phase margin of 73.07°, and $C_f = 365$ nF, corresponding to a phase margin of 59.3°. The former has no overshoot, but it has a slower rise time, whereas the latter has an overshoot of nearly 8%, but a much faster rise time. The decision on which compensation capacitor to choose may be subjective. An in-between choice (say a phase margin of 65°) may be the best.

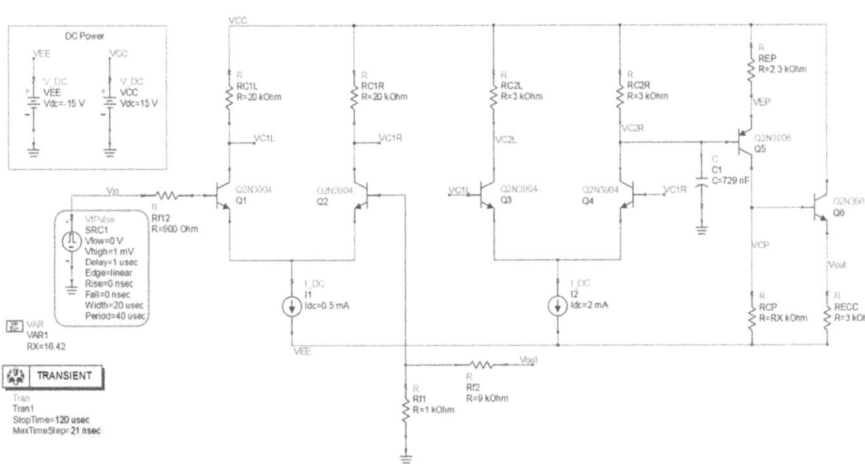

FIGURE 6.165 Schematic for transient simulation of the compensated closed-loop amplifier.

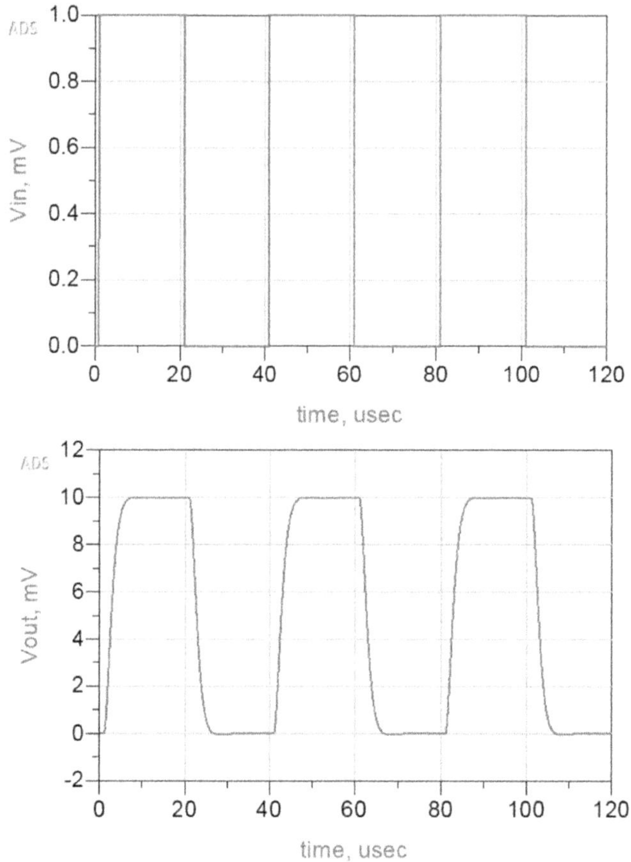

FIGURE 6.166 Closed loop pulse response for $C_f = 729$ nF and a phase margin of 73°.

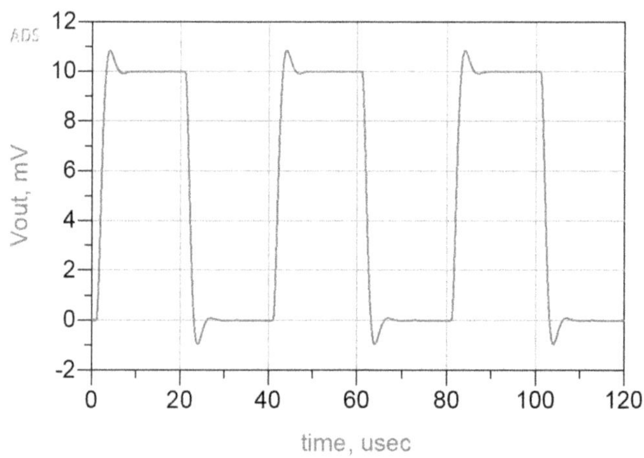

FIGURE 6.167 Closed-loop pulse response for $C_f = 365$ nF.

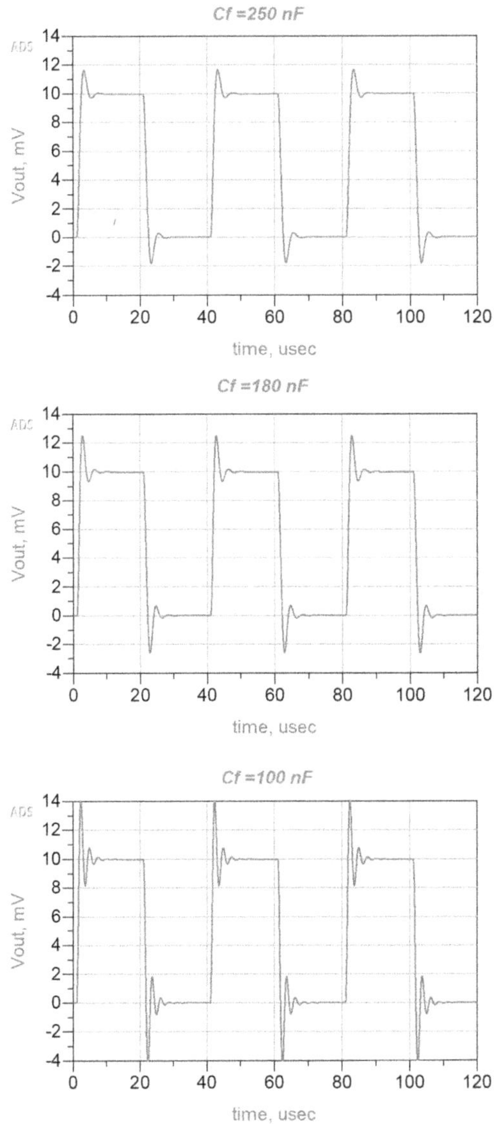

FIGURE 6.168 Pulse response for three more capacitor values.

An interesting exploration activity for the interested reader can be to explore how the various pulse response features depend on the input pulse amplitude. For the 1 mV amplitude, used above, the pulse response is almost linear in the sense that its time features follow the phase margin values according to Linear Control Systems theory [Dorf1]. Raising the input amplitude to, say, 10 mV may be close to violating the amplifier's small-signal assumption. For a 10 mV input amplitude, $C_f = 365$ nF may perform a lot better than $C_f = 729$ nF.

Finally, as promised, let us revisit the instability detection of the closed-loop uncompensated amplifier.

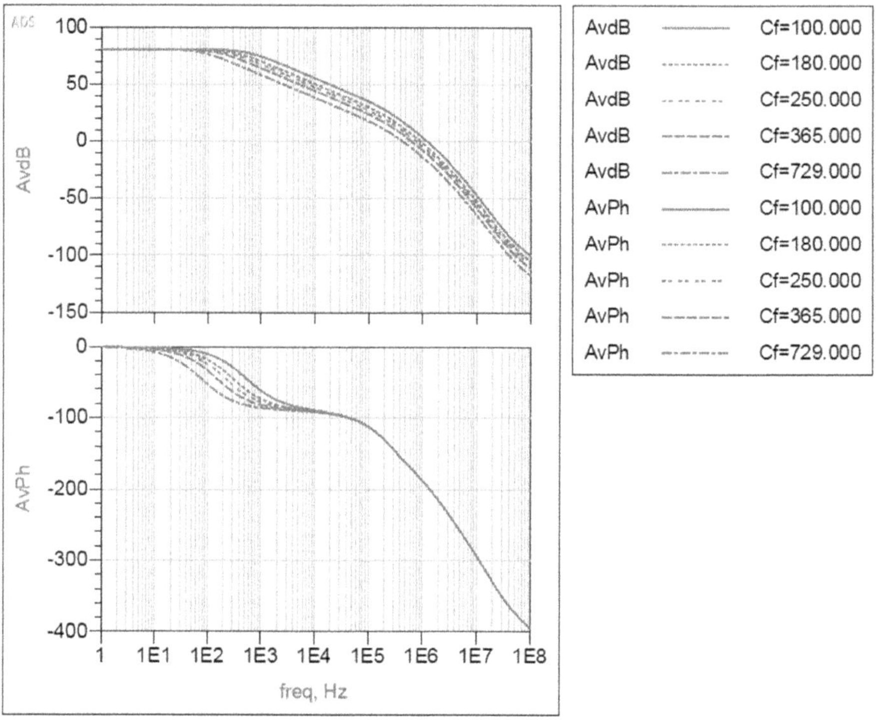

Summary

CfVals	AvMagVals	FpVals	PhMagVals
729.0	20.39	75.86 k	73.07
365.0	20.06	144.5 k	59.30
250.0	20.04	195.0 k	50.59
180.0	20.12	245.5 k	43.05
100.0	19.98	363.1 k	29.28

FIGURE 6.169 Five compensated frequency response plots.

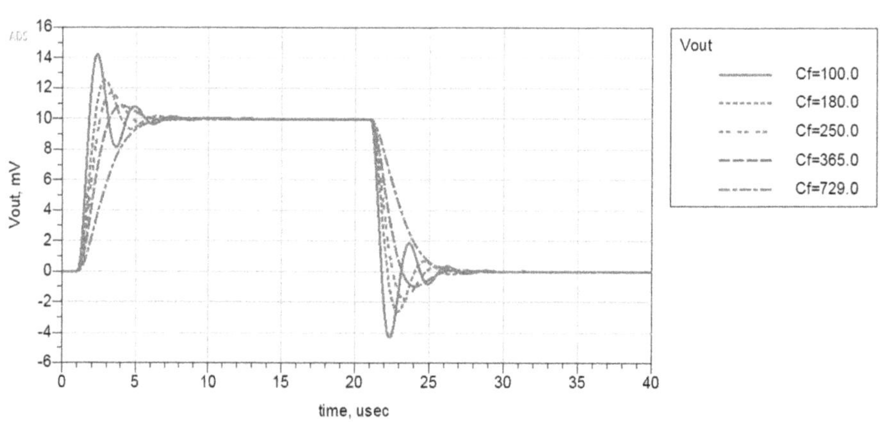

FIGURE 6.170 Five pulse response plots for the same C_f values of Figure 6.169.

Let us discuss how to choose the transient simulation timing parameters as we try to detect the instability phenomenon that occurs for the uncompensated closed-loop amplifier.

Figures 6.156 and 6.158 revisited: The uncompensated closed-loop time response and frequency response.

We see from the closed-loop frequency response that the BW is around 10 MHz. The closed-loop time constant is therefore around 0.1 μs. The oscillation period may be around 0.1 μs or some small multiple of it. The simulation stop time can be taken around 1 or 2 μs. For the MaxTimeStep we take 1/100 of the time constant: 0.1 μs/100 = 1 ns. So, it was not really a trial-and-error search.

From Figure 6.158: Transient simulation parameters that enabled the detection of amplifier instability.

REFERENCES

[Belegundu1] Ashok D. Belegundu and Tirupathi R. Chandrupatla, *Optimization Concepts and Applications in Engineering*, 2nd Edition, Cambridge University Press, 2011.

[Bode1] Hendrik W. Bode, *Network Analysis and Feedback Amplifier Design*, Van Nostrand, 1945.

[Bower1] J.L. Bower and P.M. Schultheiss, *Introduction to the Design of Servomechanisms*, Wiley, New York, 1958.

[Dorf1] Richard C. Dorf and Robert H. Bishop, *Modern Control Systems*, 13th Edition, Pearson, 2016.

[Keysight1] *Keysight Advanced System Design (ADS) Release 2025*, Keysight Technologies, Inc. https://www.keysight.com/us/en/products/software/pathwave-design-software/pathwave-advanced-design-system.html

[Ogata1] Katsuhiko Ogata, *Modern Control Engineering*, 5th Edition, Pearson, 2010.

[Sedra1] Adel S. Sedra and Kenneth C. Smith, *Microelectronic Circuits*, 7th Edition, Oxford University Press, 2015.

[Yang1] Xin-She Yang, *Optimization Techniques and Applications with Examples*, Wiley 2018.

CHAPTER 6 PROBLEMS – SINGLE STAGE AND MULTISTAGE DIFFERENTIAL AMPLIFIERS, OPTIMIZER DESIGN, MILLER EFFECT, AND FEEDBACK

PROBLEM 6.1: DESIGN OF A SINGLE STAGE DIFFERENTIAL AMPLIFIER

Design a BJT differential amplifier, using BJT 2N3904 transistors. There is no load ($R_L = \infty$), and the signal source's output resistance is $R_S = 0$. Determine the needed bias current I, the minimum needed DC current source's output resistance R_{EE}, the R_C resistors and the supply V_{CC}. There is also a negative supply $V_{EE} = -V_{CC}$ to which the current source I is flowing. Decide which DC current source to use in the design (whether a simple BJT current mirror or a Wilson current source) and add the implementation of I to the design. For the mirror use a simple V_{CC} and R arrangement to create I_{ref}.

The design specifications are:

a. Single-ended differential voltage gain of 27 V/V.
b. Differential input resistance ≥ 6.5 kΩ.
c. Single-ended common-mode voltage gain ≤ 0.003 V/V.
d. Differential input signal's amplitude may reach 38 mV (which means that degeneration resistors are needed). Use a 700 Hz signal.
e. The Common-mode input signal is a combination of DC and AC. The DC part can be as large as 2.1 V and the AC part has a frequency of 2.4 kHz and an amplitude of up to 2.2 V.

1a: Demonstrate in ADS (Transient simulation) that your design is correct, and that the small-signal differential voltage gain and common-mode voltage gain of the amplifier indeed meet the specifications. Also measure the small-signal differential input resistance R_{id}. Be sure to check (in Harmonic Balance simulation) what the maximum differential input amplitude is, for which no side harmonic is larger than 10% of the main harmonic.

1b. Find the above voltage gains and the differential input resistance using AC simulation. It is okay to put the full differential signal into the left amplifier transistor.

Problem 6.2: Almost Similar to Problem 6.1

Design a BJT differential amplifier, using BJT 2N3904 transistors. There is no load ($R_L = \infty$), and the signal source's output resistance is $R_S = 0$. In Determine the needed bias current I, the minimum needed DC current source's output resistance R_{EE}, the R_C resistors and the supply V_{CC}. There is also a negative supply $V_{EE} = -V_{CC}$ to which the current source I is flowing. Decide which DC current source to use in the design (whether a simple BJT current mirror or a Wilson current source) and add the implementation of I to the design. For the mirror use a simple V_{CC} and R arrangement to create I_{ref}.

The design specifications are:

a. Single-ended differential voltage gain of 31 V/V.
b. Differential input resistance ≥ 5.1 kΩ.
c. Single-ended common-mode gain ≤ 0.002 V/V.
d. Differential input signal's amplitude may reach 11 mV (which means that degeneration resistors are probably not needed). Use a 750 Hz signal.
e. The Common-mode input signal is a combination of DC and AC. The DC part can be as large as 2.4 V and the AC part has a frequency of 3 kHz and an amplitude of up to 1.8 V.

Demonstrate in ADS (both Transient simulation and AC simulation) that your design is correct, and that the small-signal differential voltage gain and common-mode voltage gain of the amplifier indeed meet the specifications. Also find the small-signal differential input resistance R_{id}. Be sure to check what is the maximum differential input amplitude for which no side harmonic is larger than 10% of the main harmonic.

Problem 6.3: Multistage Amplifiers – Gain, Input Resistance and DC Conditions

Refer to Figure 6.33.

Preparation Activity 1: Simulate the schematics in DC simulation, with both inputs grounded. Is the amplifier output 0 V? If not, check the annotated values of the appropriate DC currents and voltages, and adjust the 15.7 kΩ resistor to obtain an output voltage that is as small as possible (a few mV will do). The adjustment may be done using parametric sweep. Let us consider an output voltage that is within ±10 mV from 0 as "close enough".

Preparation Activity 2: Connect a VtSine source to one of the amplifier's inputs. Keep the other input grounded. Use a 1 kHz signal. What is the largest amplitude of the input signal before any clipping distortions appear? It should be around 0.9 mV. *Comment*: Later on, as you try to implement larger and larger voltage gains, you will always have to reduce the input amplitude to a sufficiently small amplitude to avoid any clipping effects. A safe value for the amplitude can be for instance 0.1 mV, but it all depends on what the gain is [input amplitude x gain should not exceed $|V_{CE}|$ of any of the BJTs].

Preparation Activity 3: Technically speaking, we need to connect $v_d/2$ and $-v_d/2$ to the inputs to study the correct differential gain and input resistance. However, if the only input is v_d connected to one of the bases, the other input is grounded, and there is no additional common-mode signal, the common-mode signal is very small, and we can neglect its effect on the output, not to mention that the first stage, with its differential output configuration attenuates common-mode signals very well. Assume therefore that the actual circuit configuration is a good approximation to a pure differential gain testing. What is the differential voltage gain of the amplifier?

Preparation Activity 4: Using AC simulation find the voltage gain and bandwidth of each of the four amplifier stages. Which of the stages determines the overall bandwidth of the amplifier?

Preparation Activity 5: Find the differential input resistance.

Preparation Activity 6: Find the amplifier's output resistance.

General Instructions:

In the assignments below you may not change the DC supplies ±15 V. *Don't use cascode amplifiers – presumably you don't know yet the theory of these circuits.* Make sure, after each circuit re-design adjustment, that you re-adjust the output DC offset voltage to be close enough to zero. Whenever you decide to change a component value you have to explain what design idea prompts the change. Always verify (in DC simulation) that all the BJTs stay in the active mode.

3a: Decrease I_2 from $I_2 = 2\,mA$ to $I_2 = 1.7\,mA$ and adjust the collector resistor of Stage 3 to achieve zero output offset. Observe: What happens to the overall $R_{id,}$ the R_{id} of Stage 2, the gain of Stages 1, 2 and 3 and the overall amplifier gain? Draw conclusions, as you will need to refer to it in the design assignment 6.3c.

3b: Increase I_1 from $I_1 = 0.5\,\text{mA}$ to $I_1 = 0.85\,\text{mA}$. Observe: Do you need to adjust anything in Stage 3 to keep a zero-output offset? What happens to the overall $R_{id,}$ the R_{id} of Stage 2, the gain of Stages 1, 2 and 3 and the overall amplifier gain? Again, draw conclusions for later reference.

3c: The ground rules for this assignment are: (i) All BJTs must stay in active mode, (ii) You should maintain zero output offset voltage.

The allowed design parameters are I_1, I_2, R_{C1}, R_{C3}, $R_{E,pnp}$, $R_{C,pnp}$. [*You will primarily use I_1 and I_2* based on what you learned from Problem 6.3a and 6.3b, and adjust $R_{C,pnp}$ as necessary. Normally you will not have to touch the other parameters.]

Consider the input resistance and voltage gain obtained in the **Preparation Activities (for $I_1 = 0.5\,\text{mA}$ and $I_2 = 2\,\text{mA}$)** as benchmark values R_{id0} and A_{v0}. Need to modify the amplifier to be "better than the benchmark amplifier". That is, both $R_{id} > R_{id0}$ *and* $A_v > A_{v0}$ must be achieved. Let us set a Reward Function:

$$C = \frac{A_v}{A_{v0}} + \frac{R_{id}}{R_{id0}} \quad \text{where} \quad \frac{A_v}{A_{v0}} > 1.2 \quad \text{and} \quad \frac{R_{id}}{R_{id0}} > 1.2$$

to help you assess quantitatively how much better your design is compared to the benchmark. Initially, $C = 2$, and as a minimum performance your design must reach a value of $C = 1.2 + 1.2 = 2.4$. Notice that both A_v and R_{id} must increase simultaneously.

Based on your observations in Problem 6.3a and 6.3b: Can your design reach a reward value of $C = 3.5$ or better? Explain how (preferably step by step – don't just jump to the final design) and demonstrate in ADS that it works.

3d: Study both designs – the one done in the preparation activities against your design that provided better voltage gain and better differential input resistance. As you know, most things in life do not come for free. What is the price that your circuit pays compared to the original design? These are the three issues that you have to study: (i) Is your amplifier overall bandwidth lower or higher? (ii) Is your amplifier dissipating more power at quiescence than the original amplifier or not? (iii) Does your amplifier have less "swing"? (Let's explain the swing consideration: As you know, depending on the voltage gain A_v and the DC supplies $\pm V$, there is a maximum allowed input signal amplitude M, so that $A_v \cdot M \leq V$ to avoid output signal clipping (and be aware that often the clipping phenomenon starts at Stage 3 or 2). Obviously, if you are able to increase A_v from A_{v1} to A_{v2}, M is expected to reduce by a factor of A_{v2}/A_{v1} and this is normal. If M reduces by a larger factor, it means that you lost some swing – that is, in your circuit there is now less room for the amplified signal due to some changed node DC voltages). In this part, do not just answer verbally what the prices are, but actually show the measurements that demonstrate the prices that you pay for obtaining a larger Reward Function C.

3e: Repeat the design of Problem 6.3c using the ADS Optimizer in Harmonic Balance simulation.

Problem 6.4: Bandwidth of CE Amplifiers

Your starting point may be any CE amplifier circuit that is available to you. For instance, the one in Section 5.3, or the one in Problem 5.1, or the ones in Chapter 3 and its problems. In AC simulation find the voltage gain and bandwidth of the CE amplifier, for three values of the input signal source internal resistance R_{sig}: Large (such as $R_{sig} = 1.5\ k\Omega$), medium (such as $R_{sig} = 120\ \Omega$) and small (such as $R_{sig} = 3\ \Omega$).

4a: Explain the bandwidth results from a Miller Effect viewpoint.

4b: For each of the above R_{sig} values can you improve the bandwidth, by cascoding the amplifier, without decreasing the voltage gain, in each case? When constructing the cascode amplifier it is okay not to change the base resistors of the CE. Just add a CB amplifier on top of the CE amplifier collector and below R_C and it is okay to use VDC at the base of the CB stage.

4c: Are the three experiments of Problem 6.4a providing enough data to estimate the BJT high frequency capacitances C_π and C_μ? If so, find the two capacitances.

Problem 6.5: Bandwidth Enhancement of the Multistage Differential Amplifier

Consider the multistage amplifier shown in Figure 6.33.

Now add symmetrically two resistors $R_{sig} = 500\ \Omega$ to each base of the first stage. That is, R_{sig} is added to the input signal (assumed to be connected to left first base). R_{sig} that you connect to the right first base is connected to ground. The adding of these R_{sig} resistors slightly lowers the overall gain (compared to the benchmark value found in Problem 6.3).

There is a need to improve the amplifier's bandwidth, without further lowering the total voltage gain. Explain (in terms of Miller effect) and study by simulations which of the three strategies works best:

Strategy A: Cascoding the first differential amplifier only.

Strategy B: Cascoding the second differential amplifier only. [You may need to readjust the offset resistor]

Strategy C: Cascoding both differential amplifiers. [Again, you may need to readjust the offset resistor].

Now change R_{sig} to $2\ \Omega$ and look again at the three cascading strategies.

Problem 6.6: Design of a Feedback Current Amplifier

Construct a non-ideal op-amp using a VCVS block with a voltage gain of $A_v = 3{\cdot}10^5$. Add to it a differential input resistance $R_{in} = 150\ k\Omega$, a differential input capacitance $C_{in} = 6\ pF$ (or some bandwidth F) and a series output resistance $R_{out} = 60\ \Omega$.

Use the non-ideal op-amp to implement the forward gain A of a closed-loop **current amplifier** of gain $A_{CL} = -10$ mA/mA. For feedback network B use an appropriate resistors network. The other design specifications are: $R_{in,f} \leq 90\ \Omega$, $R_{out,f} \geq 50\ k\Omega$ and closed-loop bandwidth ≥ 2 MHz.

Assume that the input signal is an AC current source with source resistance of $R_{sig} = 100\ k\Omega$. The load resistance is $R_L = 500\ \Omega$.

Adopt a high loop gain design strategy, analyze the four design specifications and decide on the value of the feedback network B and how large A must be. If the op-amp does not provide a large enough A recommend how many identical op-amps must be taken in cascade in order to meet all the specs.

In this problem, assuming that you can get by with a single op-amp, internal feedback stabilization is not an issue.

Problem 6.7: Design of a Feedback Trans-Impedance Amplifier

Construct a non-ideal op-amp using a VCVS (voltage-controlled voltage source) with a voltage gain of $A_v = 2.5 \cdot 10^5$. Add to the two-port network a differential $R_{in} = 300$ kΩ, a differential $C_{in} = 4$ pF and $R_{out} = 70$ Ω.

Use the above non-ideal op-amp to implement the forward gain A of a closed-loop **transimpedance amplifier** of gain $A_{CL} = -25$ V/mA. For feedback network B use an appropriate resistors network. The other design specifications are: $R_{in,f} \leq 90$ Ω, $R_{out,f} \leq 80$ Ω and closed-loop bandwidth ≥3.5 MHz.

Assume that the input signal is an AC current source with source resistance of $R_{sig} = 150$ kΩ. The load resistance is $R_L = 6$ kΩ.

Adopt a high loop gain design strategy, analyze the four design specifications and decide on the feedback network B and how large A must be. If the op-amp does not provide a large enough A recommend how many identical op-amps must be taken in cascade in order to meet all the specs.

In this problem, assuming that you can get by with a single op-amp, thus the op-amp is a first order system, internal feedback stabilization is not an issue.

Problem 6.8: Design of a Voltage Feedback Amplifier combined with Stabilization

In this problem we use the circuit shown in Figure 6.33 to implement the forward amplifier A. Don't use any of the modifications done in earlier this chapter problems. Assume that the input signal is a voltage source with $R_{sig} = 100$ Ω. Let $R_L = \infty$. Section 6.6 includes a full demonstration of the design of such an amplifier, just for a different closed loop voltage gain.

Need to design a **non-inverting** feedback voltage amplifier with gain of $A_v = 26$.

The main task in this problem is to internally stabilize the amplifier by means of a capacitor inserted between the base of Stage 3 (the pnp level shifter) and ground. You have to measure correctly |A(f)B|, including the loading of A by the B network. Of course, you plot |A(f)| and replace the 0 dB axis with a parallel one that goes at 1/B. Find the crossover frequency and phase margin. Demonstrate the instability resulting if you don't stabilize. Test your design both in the frequency domain and in the time domain. For the latter use a voltage pulses signal (with carefully selected timing parameters based on the resulting f_c frequency).

Problem 6.9: Repetition of Problem 6.8 with Slightly Different Parameters

As in Problem 6.8, except for $R_{sig} = 50$ Ω and the required closed loop voltage gain of $A_v = 23$.

7 Introductory Concepts in ADS Simulation-Based Submicron Design of MOSFET Amplifiers

7.1 MOSFET SHORT-CHANNEL DC CIRCUITS

We start by comparing the DC performance of long-channel and short-channel CS amplifiers with an R_D load. The drain-to-source current (I_{DS}) versus the gate-to-source voltage (V_{GS}) is studied first (see schematic in Figure 7.1).

The intent is to locate the value of the zero-bias threshold voltage V_{TO} on the plot for NMOS (Figure 7.2) and PMOS (Figure 7.4) devices on the I_{DS} vs. V_{GS} plot. In the Level 1 model (i.e., a long-channel MOSFET), a square law current for $V_{GS} > V_{TO}$ (Figures 7.2 and 7.4) is expected.

Then small channel modeling with the BSIM3 model (Figure 7.5) will be introduced. We shall discuss how to compare data taken from multiple datasets. Finally, let us introduce a powerful ADS design tool: Simulator calculation of DC operating point and small-signal model parameters.

In Figure 7.1, note the Short1 "component", enabling us to place a label of V_S, despite the source being grounded.

We saw before that for the Level 1 NMOS MOSFET model, with $V_{TO} = 0.7\,V$, as V_{GS} is swept, it results in the I_{DS} vs. V_{GS} curve in Figure 7.2 (left) and zoomed-in in Figure 7.2 (right).

Zooming in to the X and Y axes of the graph to observe the occurrence of the zero-bias threshold V_{TO} is done by means of lowering the sweep stop limit from 3 to 1 V.

We observe that once $V_{GS} > V_{TO}$, the current increases following the square law $I_{DS} \propto (V_{GS} - V_{TH})^2$. The same is true for the Level 1 curve of a PMOS I_{DS} vs. V_{GS} (see Figures 7.3 and 7.4). Level 1 models from [Razavi1] were used.

Now, see Figure 7.5 for a BSIM3 model card for a 0.5 μm NMOS device. We gratefully acknowledge the list of BSIM3 parameters obtained from MOSIS™.

One possible way of creating the BSIM3 populated model card, shown in Figure 7.5, is to bring in the empty shell BSIM3 model card from the ADS components palette and manually write the parameters into it.

To study the small-geometry effects in CMOS technology, we will use this NMOS (and, later in this chapter, its PMOS companion) BSIM3.1 model.

Recall that the Level 1 model is only valid for channel lengths that are greater than 2 μm. The BSIM3.1 models we will be using are valid to 0.5 μm.

DOI: 10.1201/9781003606772-7

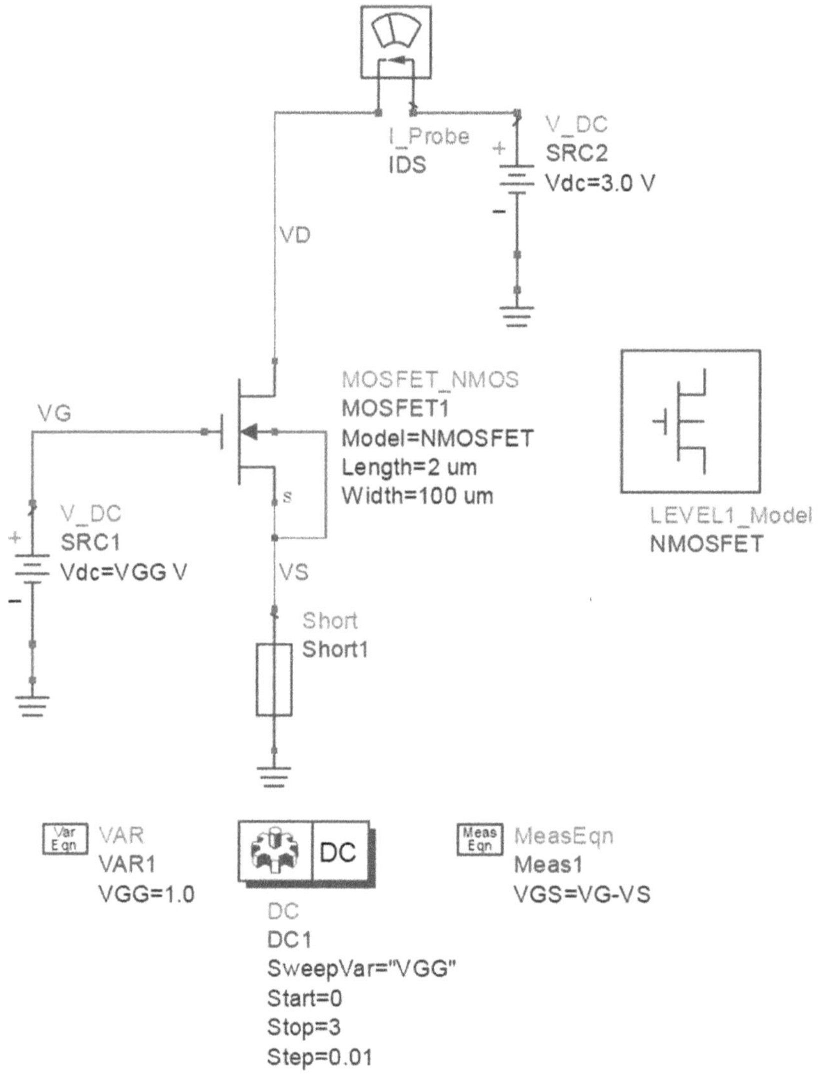

FIGURE 7.1 Schematic for finding NMOS I_{DS} vs. V_{GS} curve using Level 1 model.

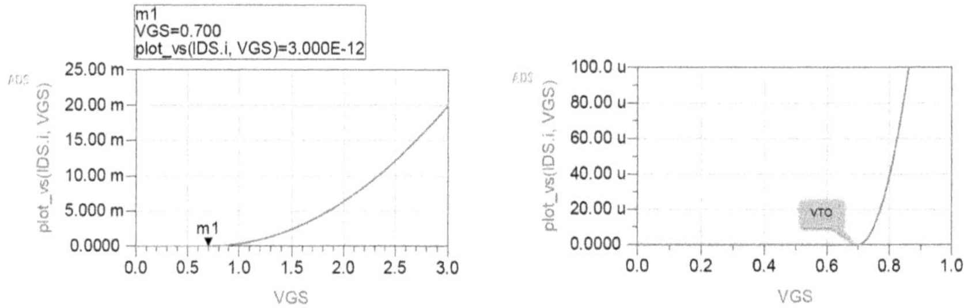

FIGURE 7.2 I_{DS} vs. V_{GS} according to NMOS Level 1 model with $V_{TO} = 0.7\,V$.

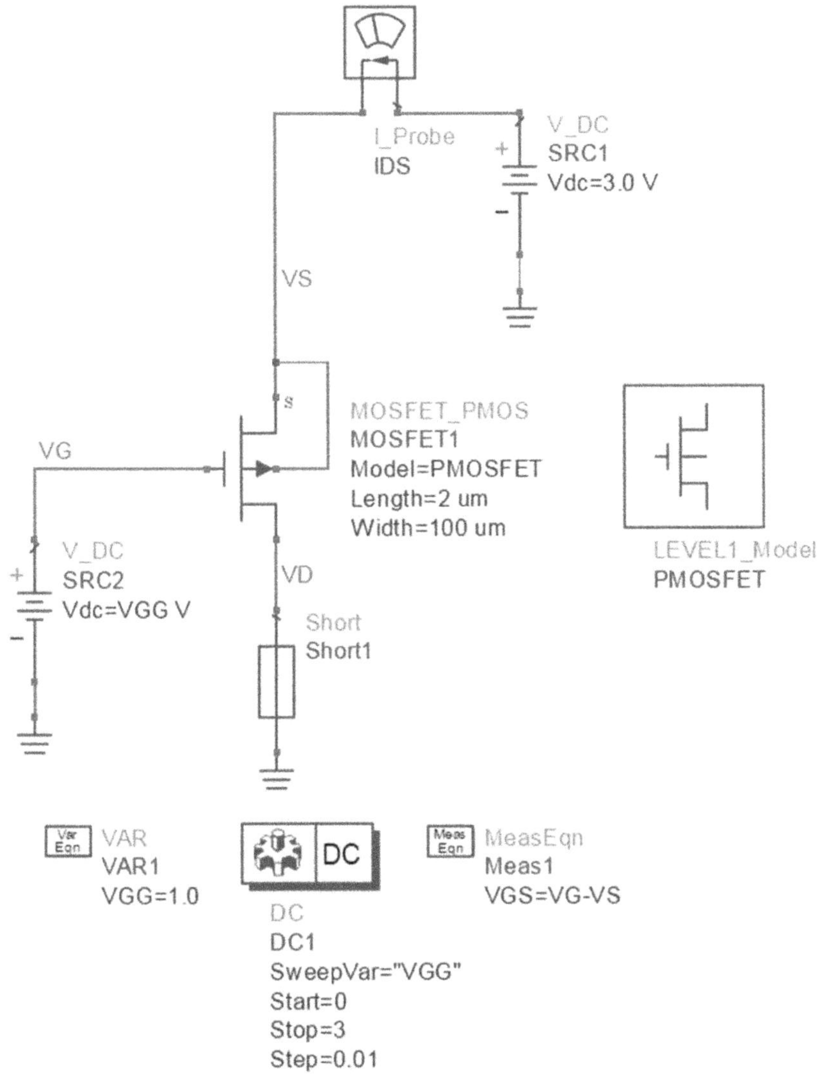

FIGURE 7.3 Schematic for Level 1 PMOS I_{DS} vs. V_{GS} curve.

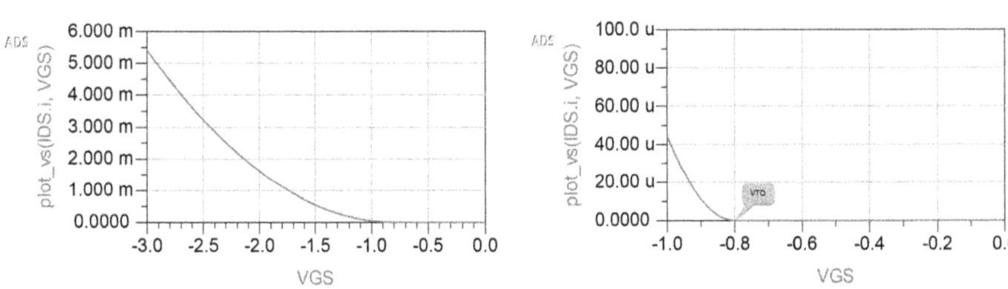

FIGURE 7.4 PMOS Level 1 I_{DS} vs. V_{GS}, with zooming in on V_{TO} (right side).

```
VERSION = 3.1           TNOM     = 27             TOX      = 1.42E-8
XJ      = 1.5E-7        NCH      = 1.7E17         VTH0     = 0.6183921
K1      = 0.9255775     K2       = -0.104436      K3       = 18.5584262
K3B     = -8.3197932    W0       = 1E-8           NLX      = 1E-9
DVT0W   = 0             DVT1W    = 0              DVT2W    = 0
DVT0    = 2.7304721     DVT1     = 0.4274991      DVT2     = -0.1404076
U0      = 449.8409373   UA       = 1E-13          UB       = 1.625666E-18
UC      = 9.189696E-12  VSAT     = 1.593774E5     A0       = 0.6050318
AGS     = 0.1251831     B0       = 2.59246E-6     B1       = 5E-6
KETA    = -2.939472E-3  A1       = 0              A2       = 0.3736451
RDSW    = 1.387179E3    PRWG     = 0.0286265      PRWB     = 0.0186862
WR      = 1             WINT     = 2.75951E-7     LINT     = 1.873294E-8
XL      = 0             XW       = 0              DWG      = -1.792465E-8
DWB     = 3.430273E-8   VOFF     = -3.127712E-3   NFACTOR  = 0.8595359
CIT     = 0             CDSC     = 2.4E-4         CDSCD    = 0
CDSCB   = 0             ETA0     = 4.033049E-3    ETAB     = -4.984872E-4
DSUB    = 0.1256715     PCLM     = 2.4898822      PDIBLC1  = 0.1417403
PDIBLC2 = 2.327707E-3   PDIBLCB  = -0.0323405     DROUT    = 0.6905503
PSCBE1  = 6.253885E8    PSCBE2   = 1.842509E-4    PVAG     = 0
DELTA   = 0.01          RSH      = 81             MOBMOD   = 1
PRT     = 0             UTE      = -1.5           KT1      = -0.11
KT1L    = 0             KT2      = 0.022          UA1      = 4.31E-9
UB1     = -7.61E-18     UC1      = -5.6E-11       AT       = 3.3E4
WL      = 0             WLN      = 1              WW       = 0
WWN     = 1             WWL      = 0              LL       = 0
LLN     = 1             LW       = 0              LWN      = 1
LWL     = 0             CAPMOD   = 2              XPART    = 0.5
CGDO    = 2.33E-10      CGSO     = 2.33E-10       CGBO     = 1E-9
CJ      = 4.241708E-4   PB       = 0.97673        MJ       = 0.4457941
CJSW    = 3.269038E-10  PBSW     = 0.1            MJSW     = 0.1111321
CJSWG   = 1.64E-10      PBSWG    = 0.1            MJSWG    = 0.1111321
CF      = 0             PVTH0    = 0.0844136      PRDSW    = 6.9092819
PK2     = -0.03041      WKETA    = -0.0154037     LKETA    = 1.63827E-3
NOIMOD  =1              AF       =1               KF       = 1.00E-31
```

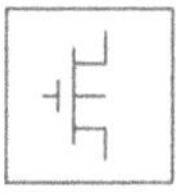

FIGURE 7.5 BSIM3 model card for 0.5 μm NMOS device.

We modified the NMOS device to use the BSIM model. The channel length was set to 0.6 μm, instead of 0.5 μm, because the effect of lateral diffusion is more significant at this smaller geometry; we do so in order to maintain a $L_{eff} = 0.5\,\mu m$.

The schematic for obtaining the I_{DS} vs. V_{GS} curve of a 0.5 μm NMOS, and the resulting curve, are presented in Figure 7.6.

Recall $L_{eff} = L - 2L_D$ for Level 1 modeling. Note that the I_{DS} vs. V_{GS} curve indicates a $V_{TO} \approx 0.6\,V$.

In Figure 7.7, we compare, on the same scale, the I_{DS} vs. V_{GS} curves obtained with the Level 1 (long channel) and BSIM3 (short channel) models.

Notice that on the left zoomed-out plot, the Level 1 current clearly follows a square law, while the BSIM3 does not so much.

This is typical of smaller geometry devices. Looking at the plot on the right, zoomed in, something seems out of place for the BSIM3-modeled NMOS. We were expecting to see the current increase after about 0.6 V; however, the onset is much lower now.

The comparison continues in Figure 7.8.

In Figure 7.8, setting the Y-axis on a log scale, we notice that the BSIM3 current is not flat even before a complete channel inversion occurs at $V_{GS} = V_{TO}$. This is, in fact, the correct behavior that was never captured using the Level 1 model. The onset of the channel is not abrupt, as the Level 1 model assumes. All models other than Level 1, including BSIM models, account for “sub-threshold conduction”. In other words, the BSIM3 model corrected certain deficiencies of the Level 1 model while also increasing the model fidelity. After all, the BSIM3 model has many more parameters.

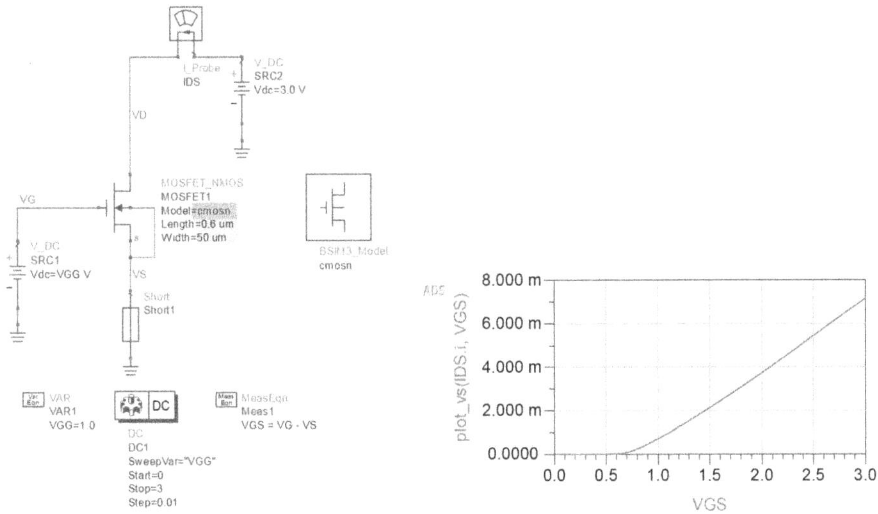

FIGURE 7.6 Measuring I_{DS} vs. V_{GS} using a BSIM3 NMOS model card.

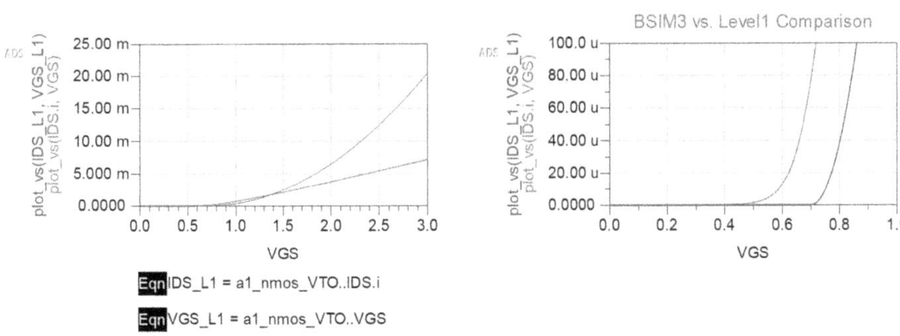

FIGURE 7.7 I_{DS} vs. V_{GS} NMOS curves, of Level 1 and BSIM3 models, compared on the same scale.

Note that the sharp turn of the Level 1 model current can sometimes create simulation convergence issues. For the BSIM3 model that we use, the actual $V_{TO}=0.618$, as seen in Figure 7.5, third column and second row.

For a full description of how to include two or more separate datasets on the same plot, see Section 7.2.

We will continue to use the Level 1 model for its academic value, but will at times use the BSIM3 model to demonstrate small-geometry effects.

A powerful analysis and design tool, available in ADS, is the **Simulator DC Device Operating Point**.

Frequently, during our study of CMOS circuits, certain parameters are needed that are based on the DC operating conditions of the device.

If a DC analysis is performed, device operating point data can be saved for certain components. For CMOS devices, typical values that are saved that will be useful in

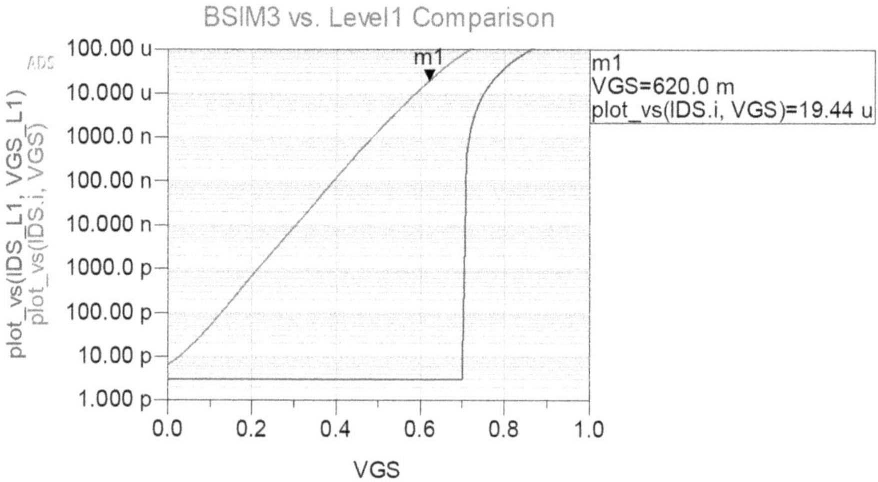

FIGURE 7.8 Further zooming in of the compared I_{DS} vs. V_{GS} using Y-axis log scale.

TABLE 7.1
Small-Signal Parameters That Can Be Measured After Enabling the Device Operating Point [Keysight1]

Capbd	Small-signal bulk drain capacitance	Farads
Capbs	Small-signal bulk source capacitance	Farads
CgdM	Small-signal gate drain capacitance	Farads
CgdM	Small-signal gate bulk capacitance	Farads
CgsM	Small-signal gate source capacitance	Farads
Gds	Small-signal drain source conductance	Siemens
Gm	Small-signal V_{DS} to I_{DS} transconductance	Siemens
Gmb	Small-signal V_{BS} to I_{DS} transconductance	Siemens
V_{Dsat}	Saturation voltage	Volts
V_{TH}	Threshold voltage	Volts

our studies are indicated in Table 7.1. Note that there are more parameters that are saved for these devices. For more information, see the ADS documentation on the MOSFET models [Keysight1]. As device geometries get smaller and model complexity increases, having these small-signal parameter values, provided by the simulator, is very convenient.

Let us learn how to activate the DC Device Operating Point. See Figure 7.9 for the dialog boxes needed for the activation of the tool and the selection of the needed parameters. The DC Device Operating Point is activated by selecting such an option in the DC simulation controller, used in the schematic. Likewise, we activate the AC small-signal parameters. Notice that the detailed operating point is selected in order to get more than just the DC voltage, current, and power values. By selecting "Detailed" (as shown in Figure 7.9), we get the device parameters. The selection of

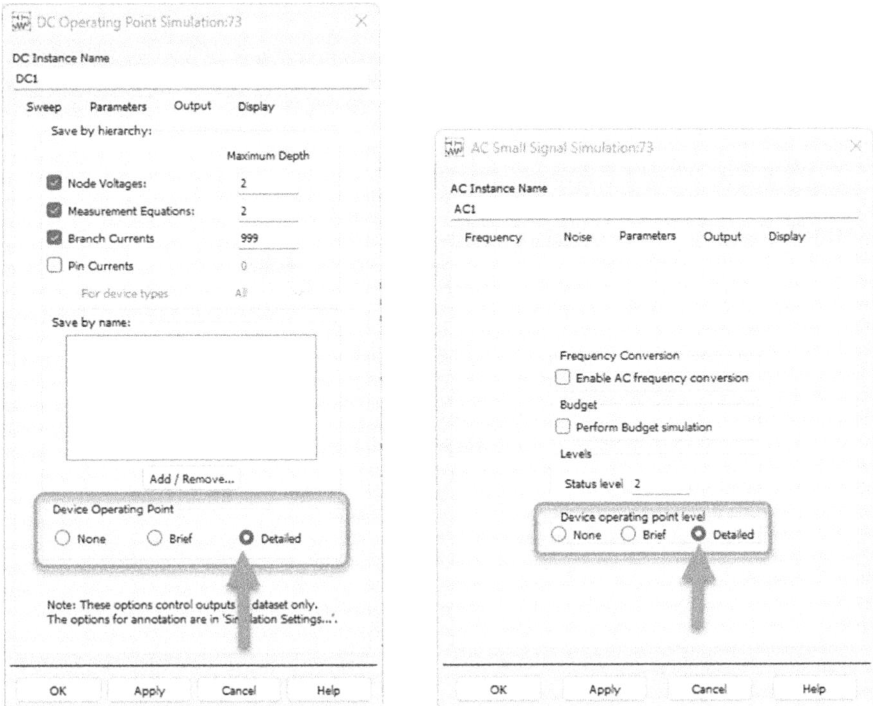

FIGURE 7.9 Activation of DC device operating point and AC small-signal parameters.

the detailed device operating point DC simulation option is displayed in the schematic, as seen in Figure 7.10, for Level 1 modeling.

We wish to conduct a study of the Device Operating Point for the NMOS device shown in Figure 7.10 with an aspect ratio W/L = 100. Note that the *detailed* operating point has been activated.

In Figure 7.11, we look at the selection of V_{Dsat}, g_m, and r_o by means of post-processing equations. In addition, I_{DS} is available from the current probe.

Notice that the simulation includes a sweep of V_{DD} from 0 to 5 V for a fixed $V_{GG} = 1$ V.

Since this is the first time that we use the device operating point, let us inspect how ADS saves the data, using the "what()" functions. See Figure 7.11 and recall our earlier explanation of the "what()" function in Section 5.5.

IDS.i is placed here for comparison. IDS.i only depends on the swept value, in this case V_{DD}.

As observed in Figure 7.11, the device operating point values VDsat, Gm, and GDS depend not only on VDD but also on the index. The value of the index was assigned by ADS and is just the way the tool has "decided" to save the data. However, as we learned in Section 5.5 (in conjunction with yield analysis), we often need to reduce the data dimensionality to plot it properly. Notice, in the data display equations, the use of index [0] in the definitions of VDSsat, gm, and ro. Let us first inspect, in

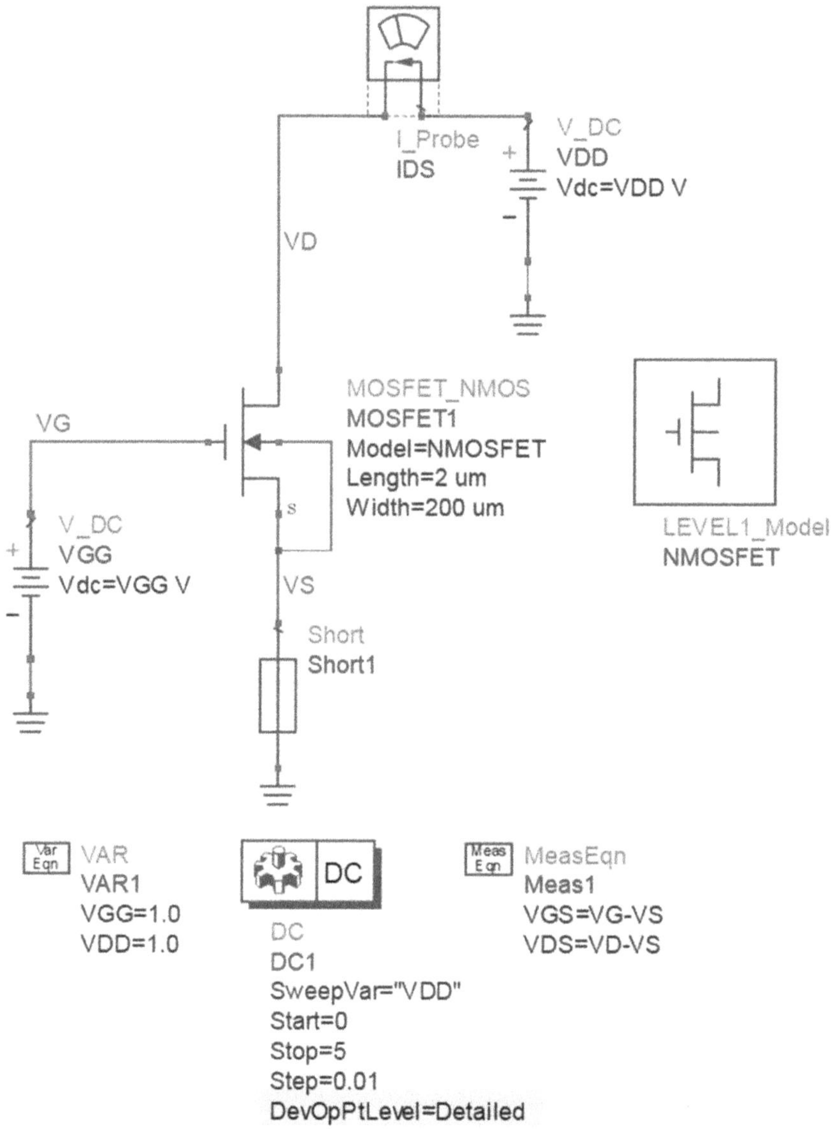

FIGURE 7.10 Selection of device operating point, with a detailed level, in a V_{DD} sweep study.

Figure 7.12, the operating point and small-signal parameter curves for a CMOS Level 1 device. The plotting results (shown in Figure 7.12) are typical of Level 1 CMOS modeled parts. We note that V_{DSsat} does not change with V_{DD}. Indeed, the borderline V_{DS} value between saturation and triode modes is constant. The I_{DS} current, gm, and r_o have a visible discontinuity for $V_{DD}=V_{DSsat}$, as predicted by Level 1 theory, and is applicable for long-channel devices.

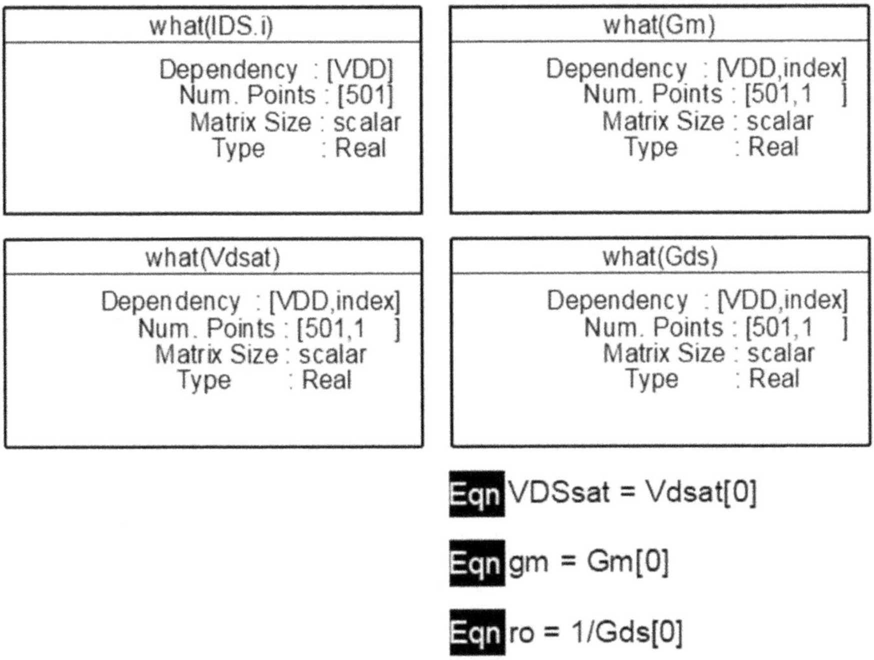

FIGURE 7.11 Look at how ADS saves the operating point and small signals parameters data.

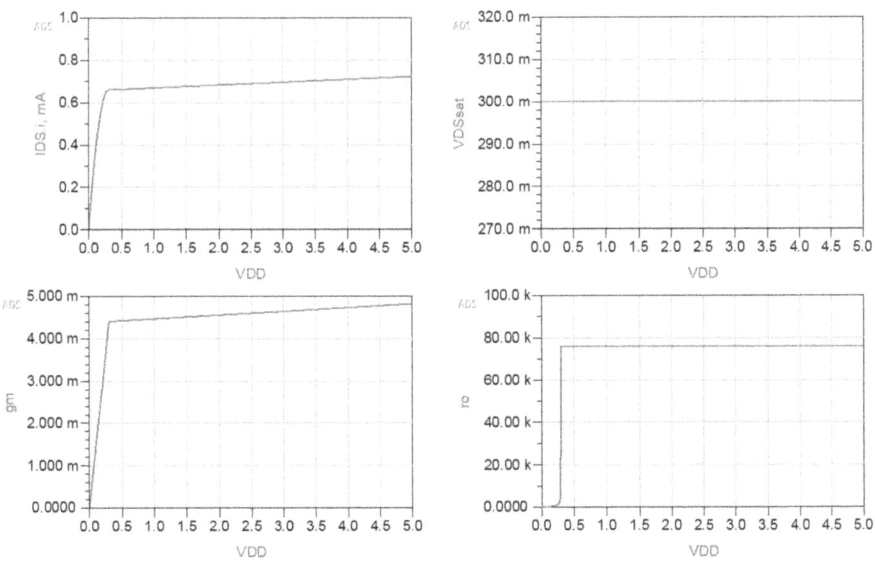

FIGURE 7.12 Curves of I_{DS}, V_{DSsat}, g_{m}, and r_o with device operating point Level 1 NMOS.

Now, we wish to compare the above plots to a small-geometry device modeled next. In Figure 7.13, we enable a detailed device operating point analysis of a 0.5 μm NMOS, using its BSIM3 model.

We want to conduct a study of the Device Operating Point for the smaller channel length NMOS device shown with a W/L ≈ 100. Again, we use the BSIM3.1 model and turn on the detailed operating point. The resulting curves are displayed in Figure 7.14 and are ready to be compared to the Level 1 curves presented in Figure 7.12.

The first notable result in Figure 7.14 is that there is no discontinuity of V_{DSsat}. In fact, V_{DSsat} itself is no longer a constant. At about $V_{DD}=0.4$ V, it becomes "somewhat constant" at a value of 278 mV, more or less.

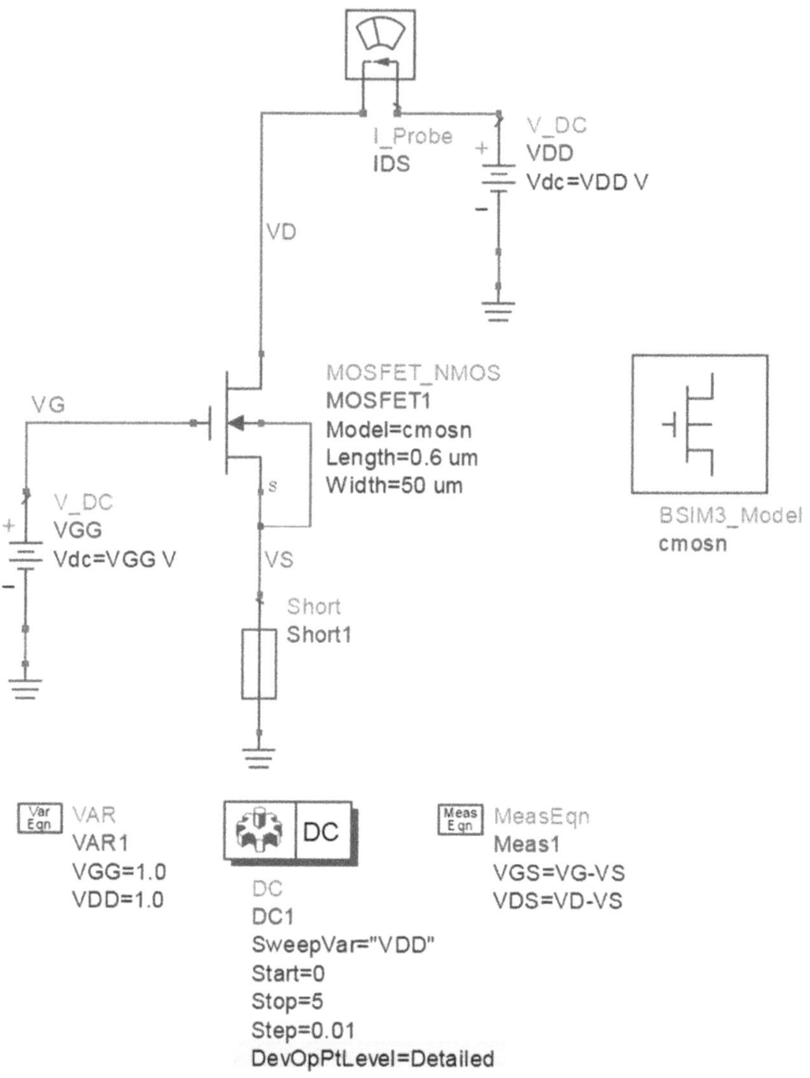

FIGURE 7.13 Detailed device operating point with BSIM3 model card.

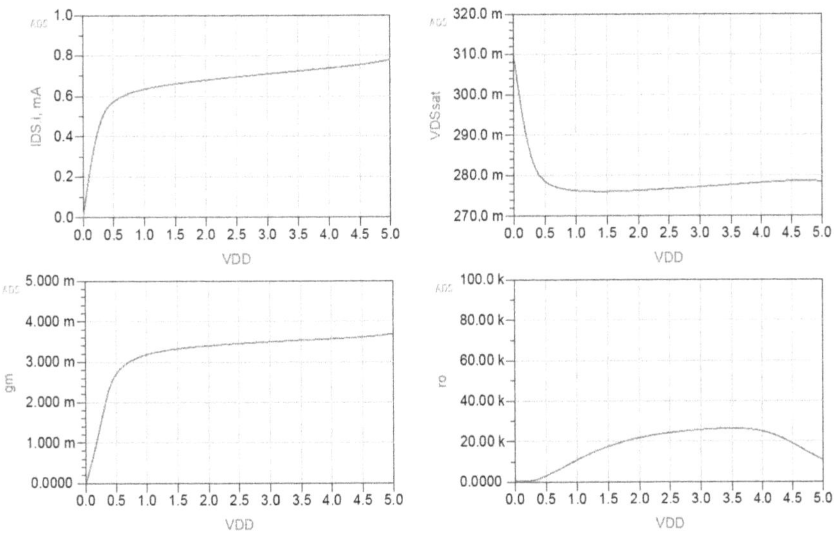

FIGURE 7.14 Curves of I_{DS}, V_{DSsat}, g_m, and r_o when using device operating point for a BSIM3 NMOS.

Notice that, compared to the larger device, the transconductance (g_m) is somewhat lower. The output resistance (r_o), however, is significantly lower. Smaller gm and smaller r_o values are some of the significant challenges as the size of the CMOS devices becomes smaller.

Let us now study the small-geometry impact on the performance of a common source (CS) amplifier with an R_D load. Our starting point will be the design done in Section 4.8 to obtain a voltage gain of $A_v = -30$. See Figure 7.15.

How do we repeat the above design using the 0.5 μm BSIM3.1 modeled process, and get the same voltage gain, $A_v = -30$?

Initially, we naively start with the same design, just replacing the transistor model and resizing the NMOS transistor, keeping the same aspect ratio W/L = 819.6/2 = 409.8. See Figure 7.16 for the new schematic and the respective annotated DC results. The bias conditions seem reasonable. The gain is lower, though $A_v \approx 11.3$, but the bandwidth is significantly higher.

The higher bandwidth makes sense, as a smaller device results in lower high-frequency capacitances. How to tweak the design (with a BSIM3 model) to meet the A_v requirement?

Let us attempt to redesign the CS amplifier, using optimization, to meet the gain requirement.

First, a few assumptions about the optimization search ranges:

1. The MOS process allows $1\,\mu m \leq W \leq 800\,\mu m$.
2. Accurate and stable bias available is $0.65\text{ V} \leq V_{bias} \leq 3.0\text{ V}$ (where V_{bias} is the gate DC voltage).
3. R_D resistor values are $1.0\text{ k}\Omega \leq R_1 \leq 20.0\text{ k}\Omega$.

Using these assumptions, we set up the optimizable variables as shown in Figure 7.17.

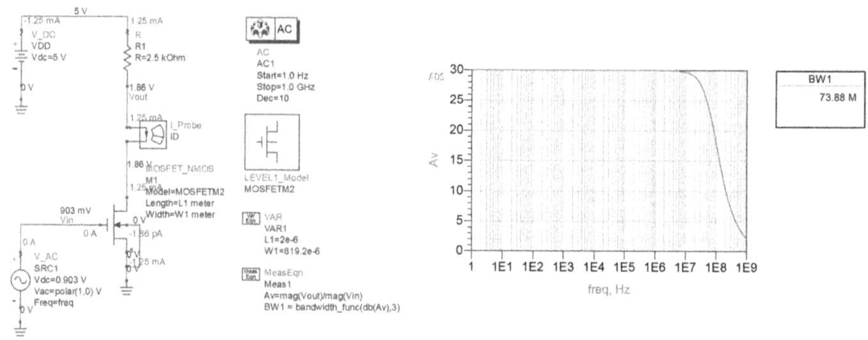

FIGURE 7.15 The CS amplifier with R_D load designed in Section 4.3.

Notice that in the optimization schematic of Figure 7.18, both AC and DC simulations are enabled simultaneously. The MeasEqn for A_v implicitly calls the AC values for V_{in} and V_{out} by preceding them with "AC". We do this so that Av is not calculated during DC evaluation, as Av makes no sense for DC simulation.

We use the hybrid optimizer, as after a few attempts, we found that for this circuit it works well.

There are two goals: the A_v goal and a DC V_{out} goal. We restrained V_{out} to be between 2.0 and 3.0 V to make sure that we have enough voltage swing.

Next, we run the optimizer. The cockpit view is presented in Figure 7.19.

The optimization cockpit shows a successful optimization in which all goals were met (Figure 7.19).

Let us update the design with these values and see the results of the actual implementation of the design in Figure 7.20. Indeed, successful optimization leads to a successful implementation, yielding $A_v = -30$.

We notice a reduction in BW, most likely caused by the amplifier output time constant increase. The bias current reduced dramatically from 1.11 mA to 165 μA, a consequence of the smaller geometry. So, what is next?

The CS with R_D resistor design provides many good insights into the operation of the CS amplifier. The reader may be wondering about the fairly high value of R_D (17.3 kΩ) in our example. A CMOS process would not use such high values resistor internally on the chip, maybe externally.

As we continue our journey through basic submicron CMOS amplifier design, we investigate other forms of bias and provide loads to the amplifiers using CMOS devices. As mentioned, and whenever possible, we would like to start with a Level 1 model design for the formulation of the main concepts. We then move to the small geometry to gain an understanding of the smaller technology impacts beyond Level 1 modeling. This will be explored in Section 7.3.

7.2 ADS TIPS AND TRICKS 7A: PLOT FROM MULTIPLE DATASETS

There are situations in which, after running several simulations, one would want to compare the results.

ADS data display allows users to create plots and tables with data coming from different simulation results (datasets).

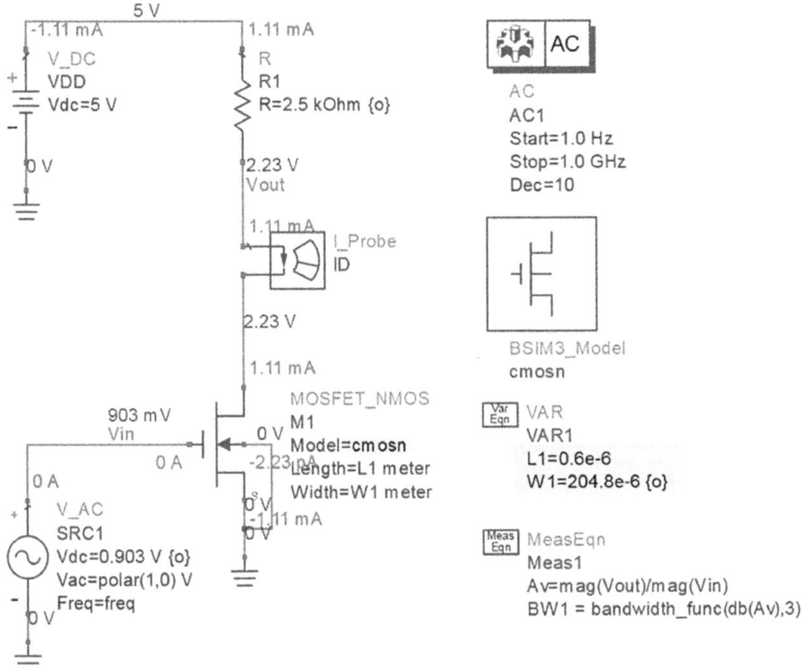

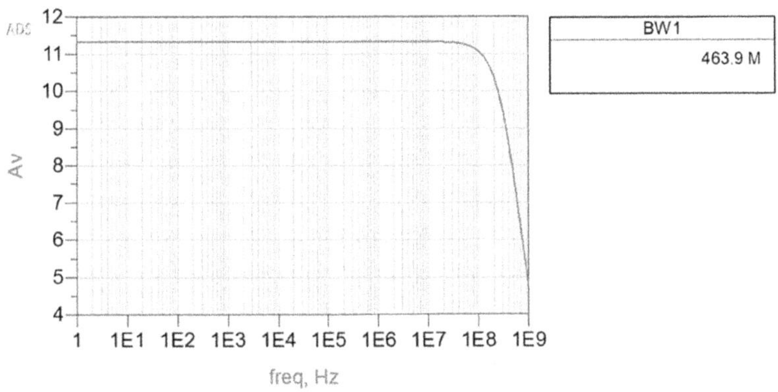

FIGURE 7.16 A design that worked for Level 1 model fails with a BSIM3 model.

Let us say we wish to compare IDS from schematic "a1_nmos_VTO" (Figure 7.1) and "a3_nmos_BSIM3_VTO" (Figure 7.6).

Default Dataset is shown in Figure 7.21.

In the ADS data display, the default dataset will be used for all plots unless the user indicates otherwise.

In the shown plot (Figure 7.21), both $I_{DS.i}$ and V_{GS} come from the a3_nmos_BSIM3_VTO dataset. See the dialog box in Figure 7.22.

To add a plot (or plots) taken from another dataset(s), the user must *implicitly* indicate the name of that other dataset(s).

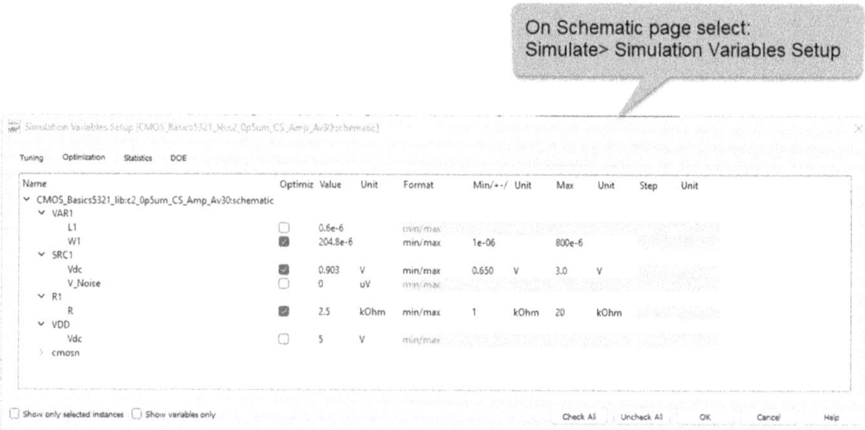

FIGURE 7.17 CS amplifier with R_D load, with BSIM3 model, optimization design parameters are W_1, $V_{DC,}$ and R (VAR_1 is W_1, SRC_1 is V_{DC} at the gate, and R_1 is R which is R_D).

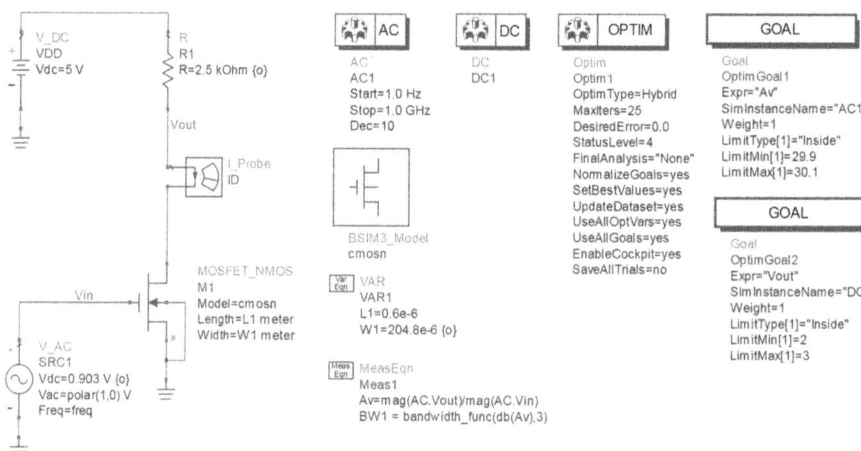

FIGURE 7.18 Small-geometry CS amplifier with R_D load optimization design.

Let us observe how all this is done.

The *first step* is to open the plot properties window (Figure 7.22).

Under the Datasets and Equations pull-down menu, we select the new dataset. In our example we are selecting *al_nmos_VTO*. Again, see Figure 7.22.

In the *second step* we select the variable we wish to plot in the normal fashion (Figure 7.23). We selected $I_{DS.i}$. Then after ">>Add Vs>>", we put ">> VGS".

In Figure 7.23, note that the data now shows the explicit dataset name in a *dataset_name..variable* format.

This format can produce very long variable names, and the plot may not be visually pleasing (see Figure 7.24).

Let us see how to work around this. See Figure 7.25 first.

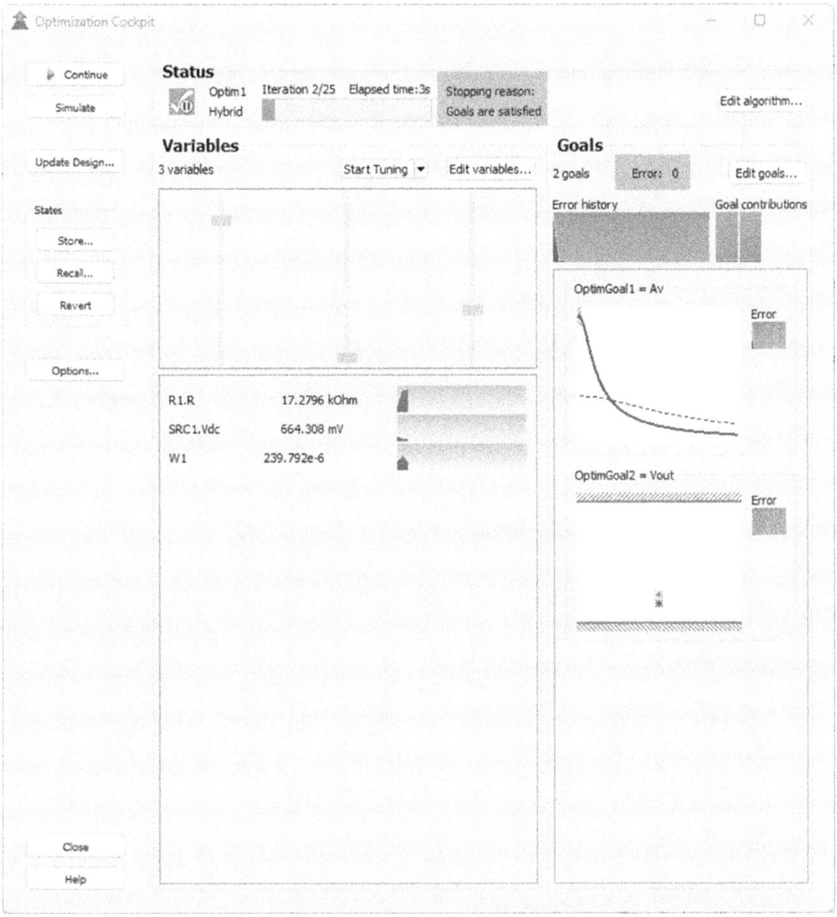

FIGURE 7.19 Cockpit view of the CS amplifier design optimization.

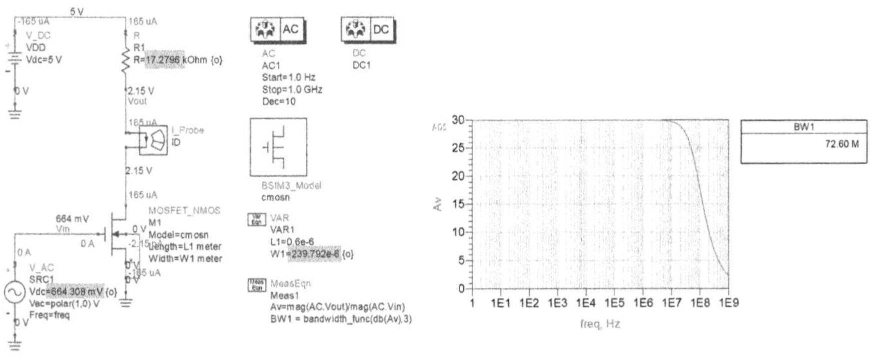

FIGURE 7.20 Optimization design of the CS amplifier with R_D load, with BSIM3 model, was successful.

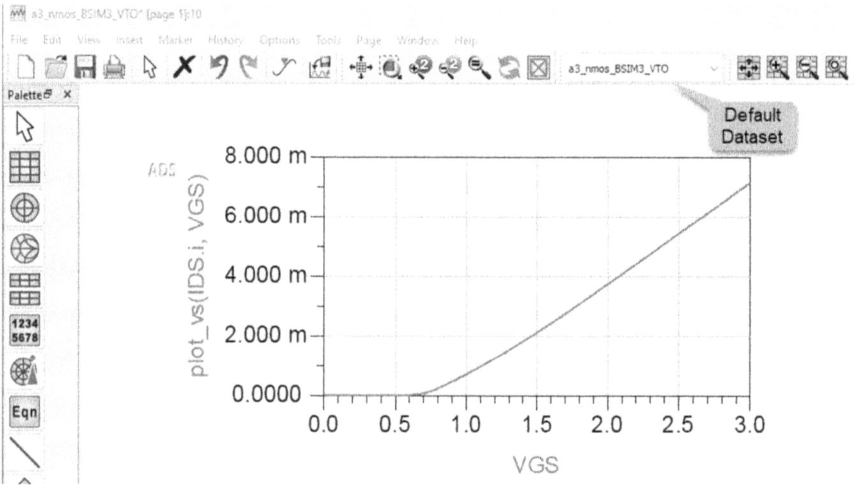

FIGURE 7.21 Arbitrarily chosen "Default Dataset" (Figure 7.6).

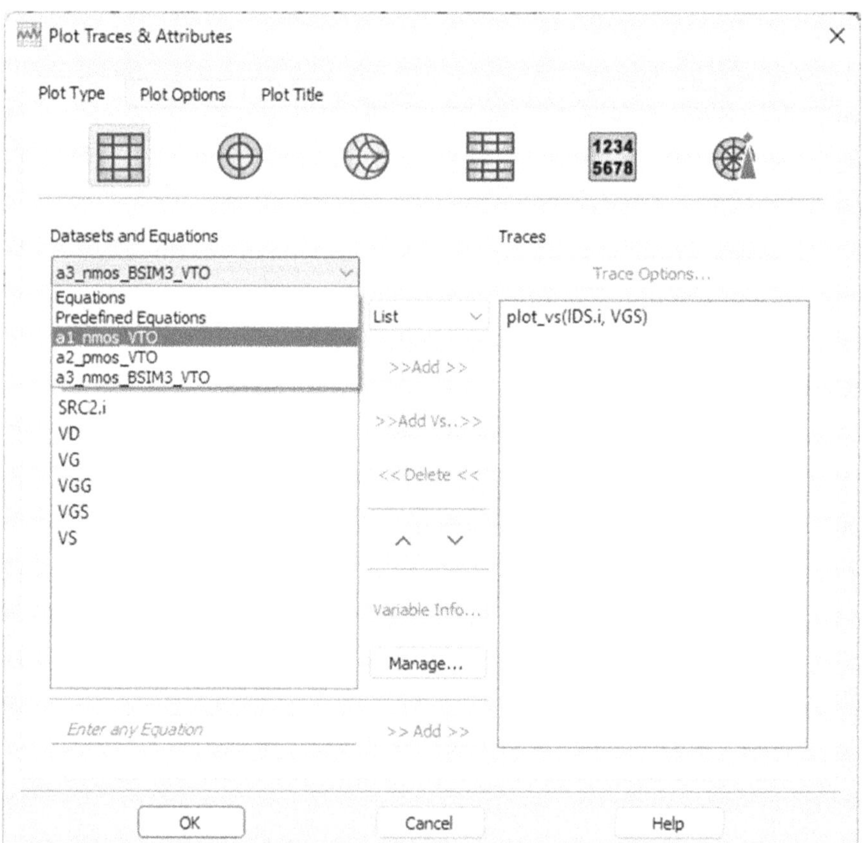

FIGURE 7.22 Selection of a second dataset to join the default dataset.

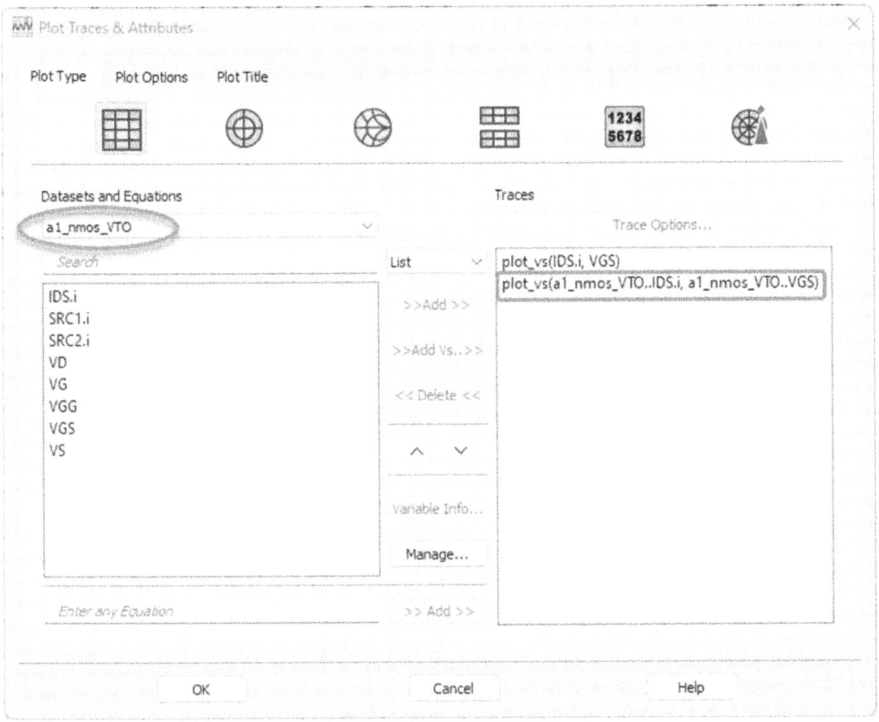

FIGURE 7.23 Selection of the plotted variables.

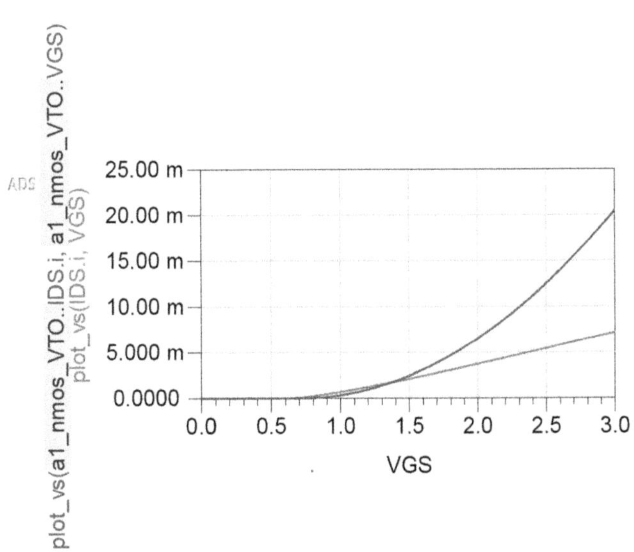

FIGURE 7.24 Resulting joint Level 1 and BSIM3 plots – Attempt 1.

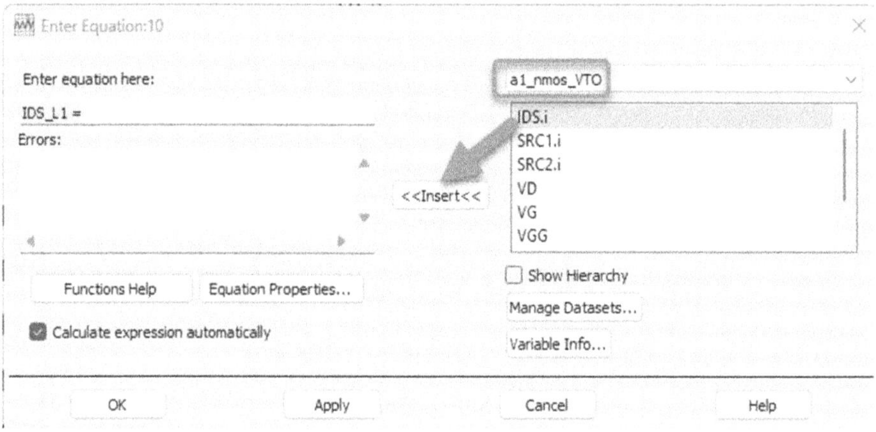

FIGURE 7.25 Working with equation (EQN) in the data display to shorten the Y-axis title.

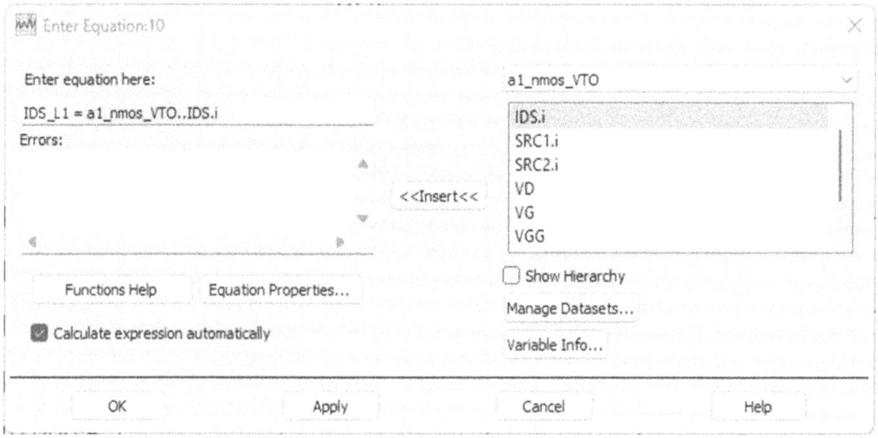

FIGURE 7.26 Selection of a variable name while referring to an implicit dataset value.

Whenever working with an equation (EQN) in the data display, one can also refer to an implicit dataset.

We start by naming the new variable and placing the equal (=) sign, as seen in Figure 7.25.

From the dataset pull-down menu, we select the dataset related to this new variable.

Then we select the variable. Here we select $I_{DS.i}$ and click <<Insert<< (as seen in Figure 7.26).

The new variable will refer to the implicit dataset value.

We repeat the aforementioned steps for the VGS_L1 equation. In the plot properties, we then plotted from the equations IDS_L1..Add Vs.. >> VGS_L1

The more visually pleasing second plotting attempt is displayed in Figure 7.27.

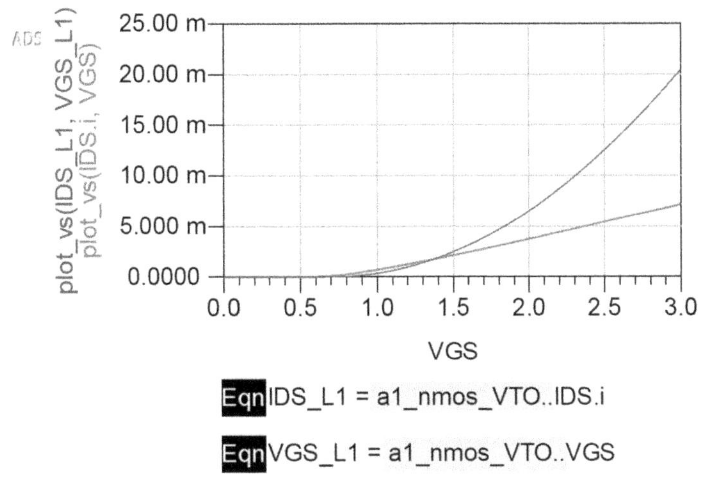

FIGURE 7.27 The joint Level 1 and BSIM3 plot – Attempt 2.

7.3 SIMULATION-BASED DESIGN OF A CS AMPLIFIER WITH R_D LOAD

Example 7.1

We revisit the design of a CS amplifier with R_D load done in Section 7.1.

New design specifications for a CS amplifier with R_D load:

$A_v = -20$, $V_{DD} = 5\,V$, $I_{DS} = 1\,mA$, Length $= 2\,\mu m$

We first use the Level 1 NMOS Model, as shown in Figure 7.28.

The design variables are R_D, W_1, and V_{GG}.

The first step in this Level 1 analytical design is to select R_D:

The small-signal voltage gain for this topology is

$$A_v = -g_m\left(\frac{r_o R_D}{r_o + R_D}\right)$$

where

$$r_o = \frac{1}{\lambda I_{DS}} = \frac{1}{(0.01)\left(1\times10^{-3}\right)} = 100\ k\Omega$$

To determine R_D, we need to set V_{DS}. Let us select $V_{DS} = V_{DD}/2 = 2.5\,V$, which seems to provide a reasonable voltage swing.

We know that

$$I_{DS} = \frac{V_{DD} - V_{DS}}{R_D}$$

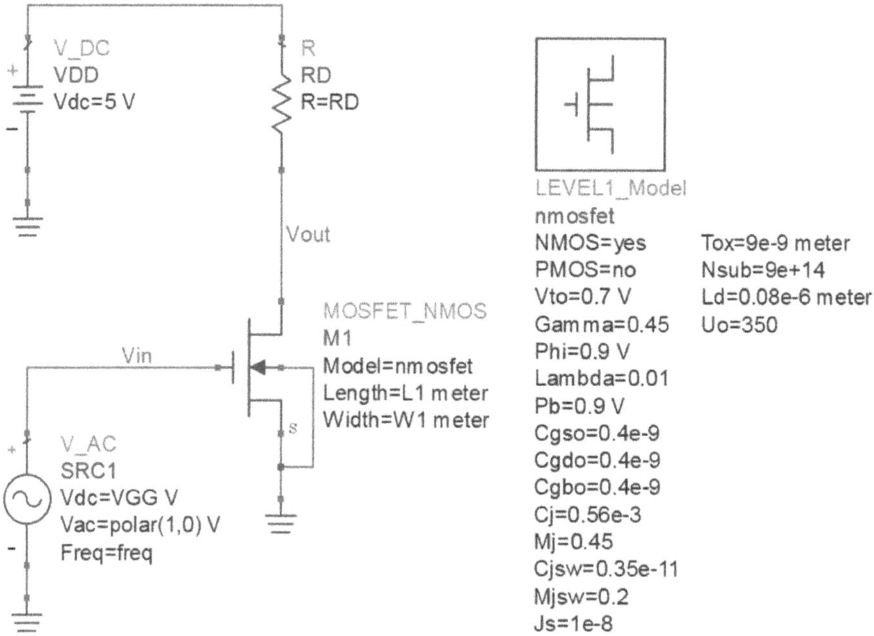

FIGURE 7.28 Level 1 design of a CS amplifier with R_D load with new specifications.

Therefore,

$$R_D = \frac{V_{DD} - V_{DS}}{I_{DS}} = \frac{5 - 2.5}{1 \times 10^{-3}} = 2.5 \text{ k}\Omega$$

The second step in the design is to determine W:

From the gain equation, we can determine g_m

$$g_m = -A_v \left(\frac{r_o + R_D}{r_o R_D} \right) = 20 \left[\frac{(100 \times 10^3) + (2.5 \times 10^3)}{(100 \times 10^3)(2.5 \times 10^3)} \right] = 8.2 \text{ mA/V}$$

We can use g_m to determine W/L

$$g_m = \sqrt{\frac{2\mu C_{ox} \frac{W}{L} I_{DS}}{1 + \lambda V_{DS}}}$$

rearranging for W/L

$$\frac{W}{L} = \frac{g_m^2 (1 + \lambda V_{DS})}{2\mu C_{ox} I_{DS}} = \frac{(8.2 \times 10^{-3})^2 (1 + (0.01)(2.5))}{2(350)\left(\frac{3.45 \times 10^{-13}}{9 \times 10^{-7}} \right)(1 \times 10^{-3})}$$

$$\frac{W}{L} = 256.8$$

Let L = 2 μm, then W = 513.6 μm.

The third step in the analytic design is to determine V_{GS}.

From the current equation below, we can determine V_{GS}.

$$I_{DS} = \frac{\mu C_{ox}}{2}\left(\frac{W}{L}\right)(V_{GS} - V_{TH})^2(1 + \lambda V_{DS})$$

rearranging for $(V_{GS} - V_{TH})^2$

$$(V_{GS} - V_{TH})^2 = \frac{2I_{DS}}{\mu C_{ox}\left(\frac{W}{L}\right)(1+\lambda V_{DS})} = \frac{2(1\times 10^{-3})}{(350)\left(\frac{3.45\times 10^{-13}}{9\times 10^{-7}}\right)(256.8)(1+(0.01)(2.5))}$$

$$(V_{GS} - V_{TH})^2 = 56.63 \times 10^{-3}$$

$$(V_{GS} - V_{TH}) = 0.238$$

$$V_{GS} = 0.238 + 0.7 = 0.938 \text{ V}$$

Now, let us verify the analytic design results in a Level 1 amplifier simulation (Figure 7.29) using the model from [Razavi1].

The results for the voltage gain (shown in Figure 7.30) came somewhat close, an aproximate *10% error.*

We saved the DC operating point (not shown in Figure 7.29, but results are shown in Figure 7.30). These results indicate why the design is off.

Note that Id in the table comes from the DC operating point results, that is, there is no need to add a current probe on the schematic to get I_{DS}.

The current is higher than 1 mA by 8.6%, a large error. The results for g_m and r_o also show are large error. What happened?

In our design calculations, we passively assumed that L was indeed L_{eff}.

For Level 1, $L_{eff} = L_{drawn} - 2L_D$.

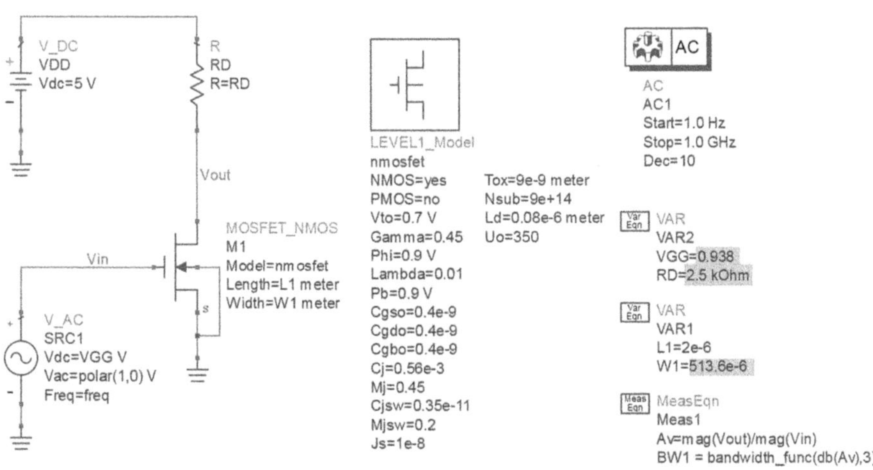

FIGURE 7.29 Checking by simulation the analytic design of a CS amplifier with R_D load.

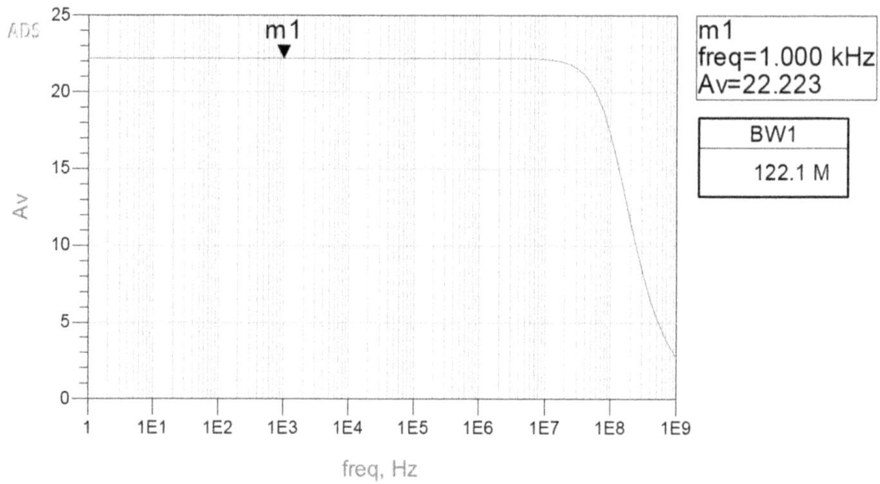

Eqn ro = 1/Gds

Id	Gm	ro	Vdsat	Vth
1.086 m	9.125 m	94.20 k	238.0 m	700.0 m

FIGURE 7.30 CS with R_D amplifier verification results, not just for the voltage gain but also for the operating point variables and parameters.

Var Eqn VAR
VAR1
L1=2.16e-6
W1=513.6e-6

FIGURE 7.31 Revised L value considering the effect of L_D.

Therefore, on the schematic, we should compensate for L_D (lateral diffusion) by adding $2L_D$ to the value of length. Let's try it – See Figure 7.31, and the new results are shown in Figure 7.32.

We change the length value to compensate for $2L_D$; length now becomes 2.16 µm as shown in Figure 7.31.

After making this one change, we re-ran the simulation, and now the results match the analytic design very well. The minor differences are most likely related to round-off values during the design calculations.

Example 7.2

Let us now repeat the design of a CS amplifier with an RD load, this time around with a BSIM3 model. See Figure 7.33.

Can the design procedure, used with Level 1 in Example 7.1, be repeated for the devices with smaller geometry, using the BSIM3 models?

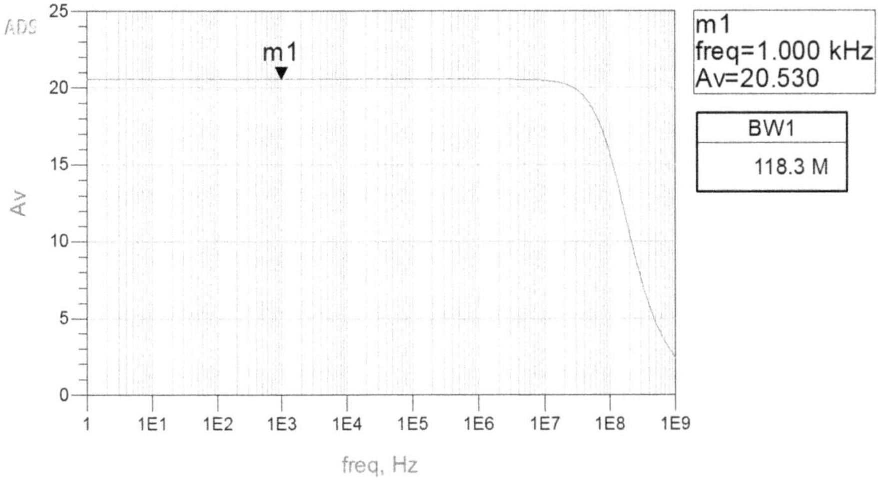

Eqn ro = 1/Gds

Id	Gm	ro	Vdsat	Vth
1.001 m	8.412 m	102.4 k	238.0 m	700.0 m

FIGURE 7.32 Revised verification results following the revision of L.

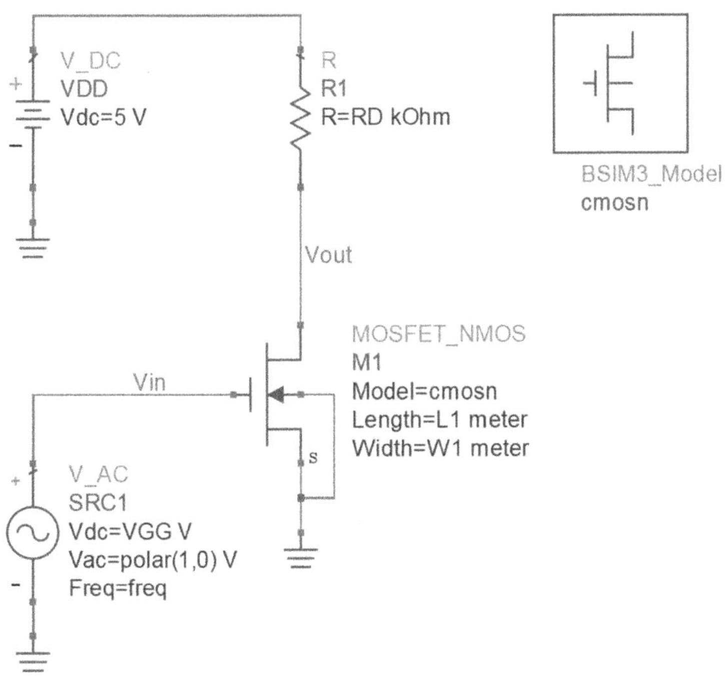

FIGURE 7.33 CS amplifier with R_D load and BSIM3 model – to be designed.

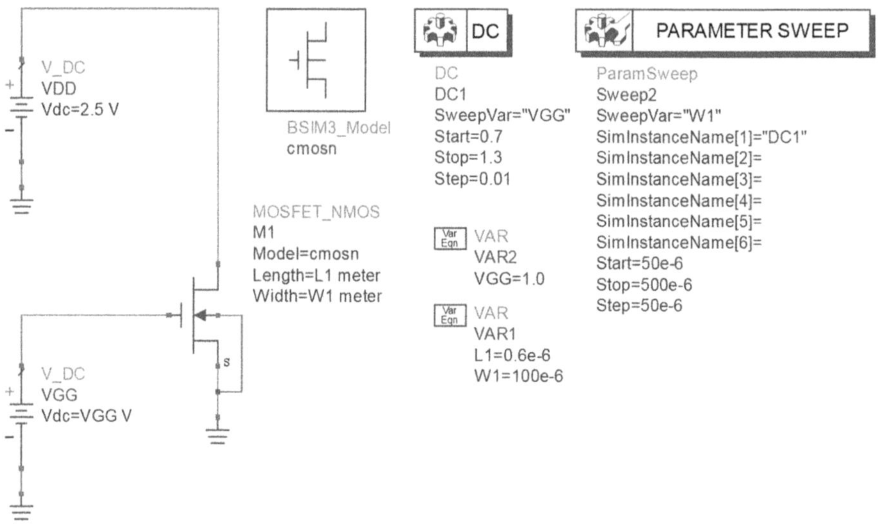

FIGURE 7.34 Schematic for conducting a MOSFET small-signal characterization.

The model and foundry PDK complexity are really not conducive to the same analytical approach as that followed for large devices using Level 1 modeling.

The reality is that foundries also use model binning, for example, as the effective lengths and widths of the devices needed change, some of the model parameters also change. A more comprehensive modeling discussion is beyond our scope.

What we will do next is to use the DC operating point results from the simulator.

We also use the understanding that comes from Level 1 intuition to show the boundaries of the devices in terms of gain. This is a structured procedure for the design of the CS with R_D load.

The first step is to perform a **characterization of the NMOS** using parameter sweeps of the design variables, as shown in Figure 7.34. In this example, we sweep V_{GS} and transistor width (W_1) as the main design variables. We fix $V_{DS}=2.5$ V. We also keep $V_{DD}=5$ V.

Hence, the values of the V_{DD} source and the V_{DS} voltage, used in the equations of Figure 7.35, in the data display, enable the subsequent characterization experiment.

For now, we run the sweeps and perform some calculations in the data display.

The reason behind the design parameter limits, and steps of the sweeps, will be discussed soon.

The ADS data display is a very useful data post-processing tool. We use it here to design the CS with the R_D load amplifier. Notice the similar procedure followed, as done with the Level 1 equations. Let us explain Figure 7.35 in more detail.

We start by declaring V_{DD} and V_{DS}. There is no need to specify the value of I_{DS} with a current probe, as we use the I_D taken from the DC operating point swept analysis. Next, we calculate the R_D value needed for all the swept iterations. Transconductance g_m and output resistance r_o are determined from the DC operating point.

Note that we have all the values needed to calculate the gain at each swept value, and we call it A_{vo}.

```
Eqn VDD = 5
Eqn VDS = 2.5
Eqn IDS = Id[0]
Eqn RD = (VDD - VDS) / IDS
Eqn gm = Gm[0]
Eqn ro = 1/Gds[0]
Eqn Avo = gm * (RD*ro) / (RD + ro)
Eqn Avi = gm * ro
```

FIGURE 7.35 Pre-characterization data post-processing.

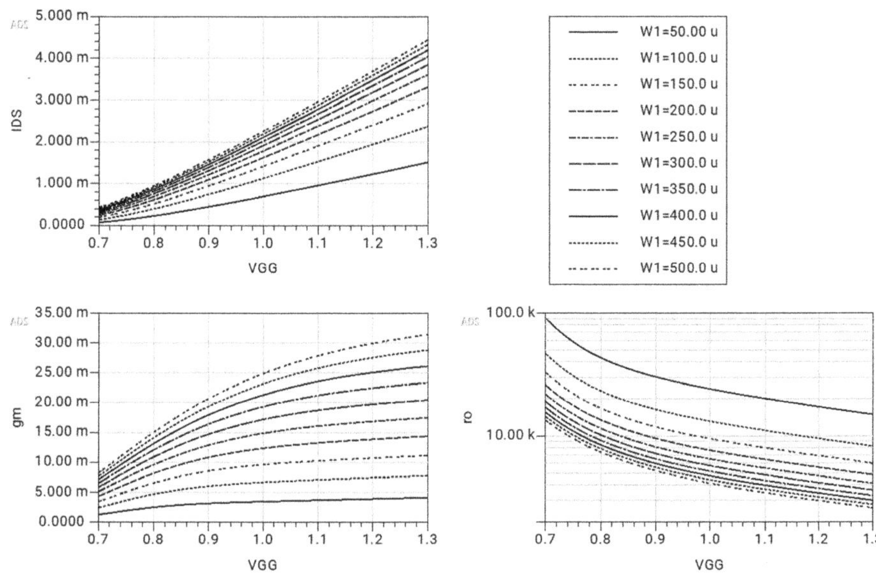

FIGURE 7.36 Partial characterization data of the short-channel NMOS: I_{DS}, g_m, and r_o.

The equation for the **voltage gain A_{vo}** follows a **universal and model-independent formula** [Razavi1]: **A_{vo} = (short circuit transconductance)·(output resistance).**

The value of A_{vi} is the **intrinsic gain** of the MOSFET transistor. It is the maximum gain available for this transistor at all swept values. This is the gain if R_D is replaced by an ideal current source.

We now run a double sweep, shown in the schematic of Figure 7.34. V_{GG} is the primary swept variable, and W_1 is the secondary swept variable.

We can get double-sweep plots for I_{DS}, g_m, and r_o, as shown in Figure 7.36. This is called the **Characterization Data of the NMOS.**

Figure 7.36 shows us the reason for the limits in the swept analysis. Obviously, some trial-and-error precedes the characterization in its final form.

Here is what we observe:

If we reduce V_{GG} any further, we get closer to V_{TH}, and the channel begins to go into **"moderate inversion"** (a V_{GS} region that extends from a little below threshold to a little above threshold), and in such a region, the value of gm becomes low, reducing the achievable gain.

Before continuing with the design solution, aided by the characterization of the transistor, and since we mentioned moderate inversion, let us add some more terminology [Binkley1], [Enz1], [Jespers1]:

The I_{DS} vs. V_{GS} curve runs continuously in three regions: **weak inversion** (in which much of I_{DS} consists of diffusion currents, and the region extends below V_{TH} minus 100 mV, more or less), **moderate inversion** (where I_{DS} is a mix of diffusion and drift currents, and it extends from 100 mV below threshold to 100 mV above threshold, more or less), and **strong inversion** (where I_{DS} consists mostly of drift currents, the region runs above 100 mV above threshold, and the I_{DS} obeys the square law dependence on V_{GS}).

As seen in Figure 7.36, increasing the channel width reduces r_o, which in turn reduces the available gain.

Increases in V_{GG} and W_1 increase the current, and hence they increase the amplifier's power dissipation.

In Figure 7.37, we see one more part of the device characterization, the intrinsic gain A_{vi}.

The Avi results, which show the attainable voltage gain value for each pair of design parameters, look very promising for designing an amplifier with $A_v = -20$.

Can the CS amplifier with R_D load meet the gain requirement, and how do you actually use this characterization data to design it?

See Figure 7.38.

We plot the double sweeping results for the expected actual gain A_{vo} and R_D (Figure 7.38).

Recalling that the R_D calculation will guarantee a $V_{DS} = 2.5$ V, this value of R_D is used to calculate A_{vo}.

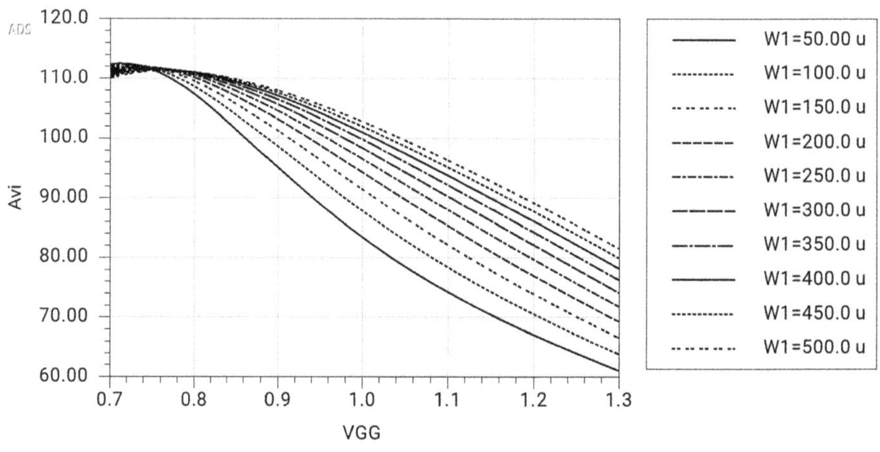

FIGURE 7.37 Another part of the device characterization – the intrinsic gain A_{vi}.

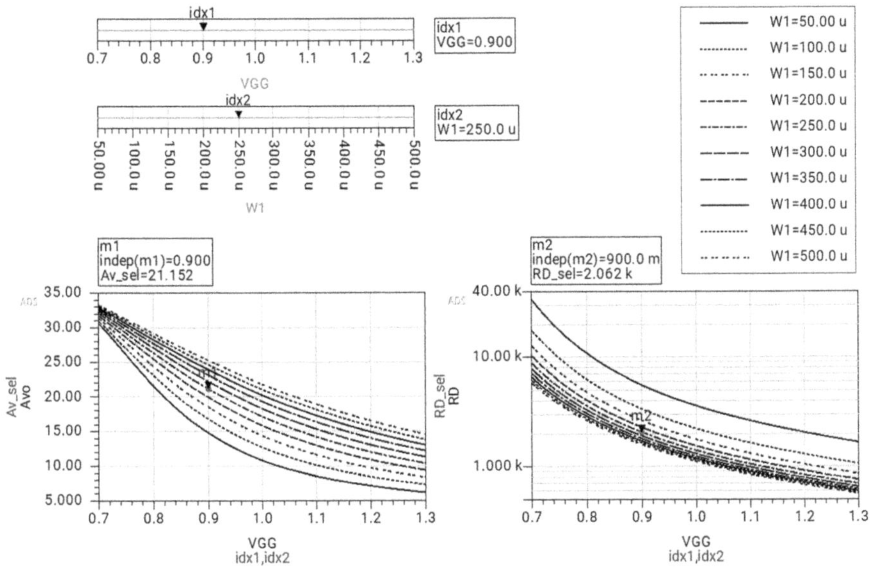

FIGURE 7.38 Simulation-based design using sliders and characterization data.

Inspection of the A_{vo} plot in Figure 7.38 shows several opportunities (i.e., sets of possible design parameters to meet the requirement of $A_v = -20$).

On the data display, we have created two **sliders** $idx_1 \Rightarrow V_{GG}$ and $idx_2 \Rightarrow W_1$ for the design parameters. Note that as the sliders move, so do the blue dots, with the marker updates on the A_{vo} and R_D plots. The marker readouts provide the gain and R_D values for the V_{GG} and W_1 selected by the sliders. We thus have our three design parameters that are needed, V_{GG}, W_1, and R_D, for the gain A_{vo} shown by marker m_1.

See Section 7.4 on how to create and use sliders.

So far, the design (shown in Figure 7.38) shows the following concluded results:

The sliders indicate that for:

$V_{GG} = 0.9\,V$, $W_1 = 250\,\mu m \rightarrow R_D = 2.062\ k\Omega$

We also expect a gain of $A_v = -21.15$. A design based on characterization of the device is merely a "wishful thinking". Let's validate how good the design results are with an AC simulation, see Figure 7.39.

The bias point is right on, in line with our expectation of $V_{DS} = 2.5\,V$. The gain, however, is about half of what we expected, $A_v = -10.2$. What happened?

This time around (unlike the debugging done in the Level 1 design of the CS amplifier with R_D, in which the culprit parameter turned out to be L_D), the act of debugging for the short-channel amplifier is way more involved.

To understand why our design did not meet the requirements, we took a look at the MOSFET transistor small-signal model (Figure 7.40) used in the ADS simulation [Keysight1].

At low frequencies, used to measure gain, the capacitors of Figure 7.40 have no impact.

Notice that in addition to r_o, there are two resistors that impact the small-signal voltage output, namely *resistors* r_d *and* r_s.

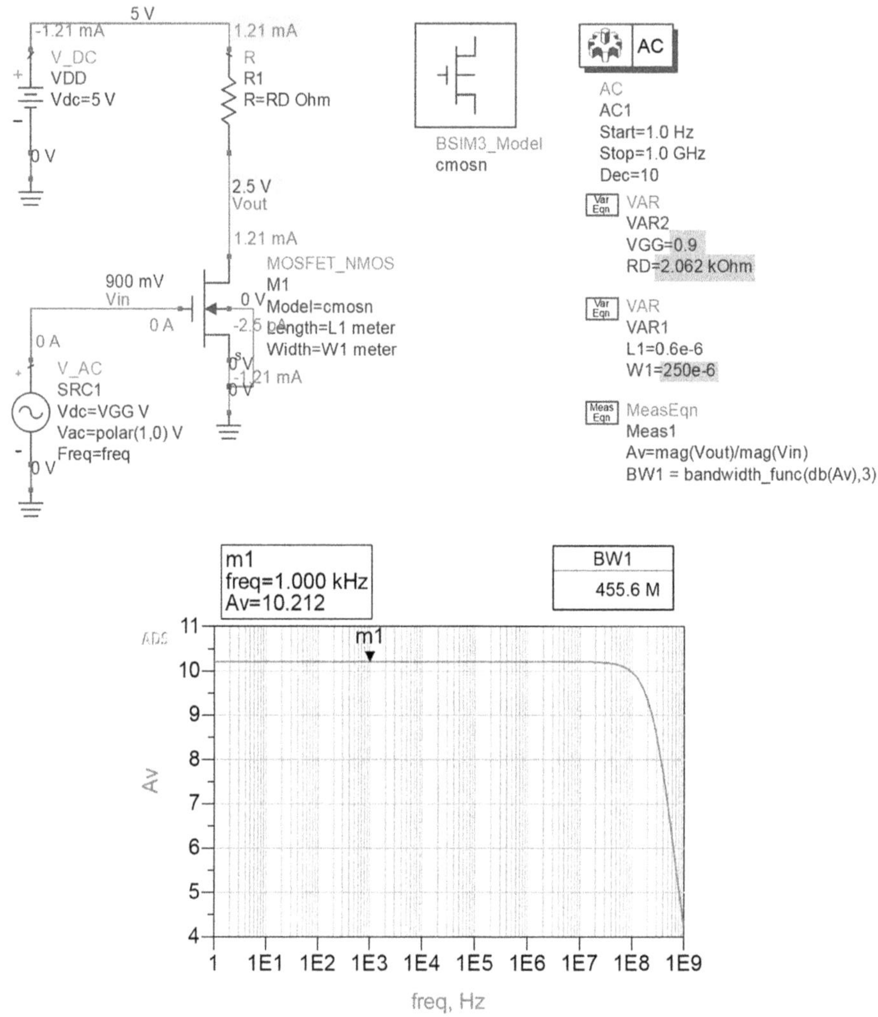

FIGURE 7.39 (Top) Design validation simulation – (bottom) the design failed.

Resistors **r_d and r_s** account for the **intrinsic contact resistance of MOSFET's drain and source,** respectively. It includes such effects as metal contact and semiconductor resistivity. All things kept equal, the smaller the contact area, the higher these resistance values will be, and therefore their reduction impact on the amplifier voltage gain increases as well.

One of the challenges for smaller geometry technologies is the increase in contact resistance values. How to account for these resistances during the design?

With the inclusion of the rd and rs contact resistance, the amplifier's output resistance changes; hence, the voltage gain equation has to change.

$$A_v = -g_m\left(\frac{r_o R_D}{r_o + R_D + r_{con}}\right) \tag{7.1}$$

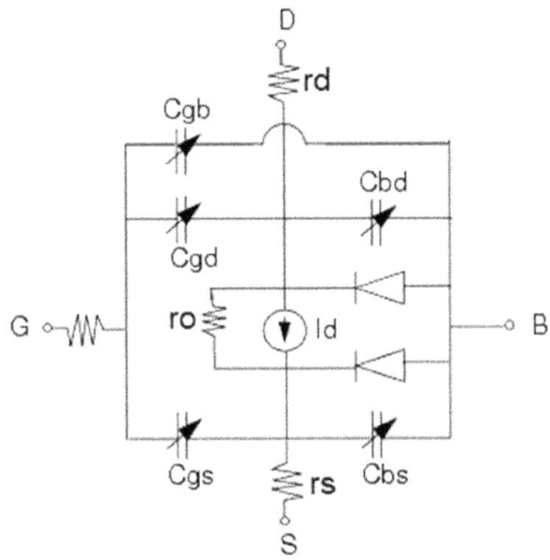

FIGURE 7.40 MOSFET transistor small-signal model used in the ADS simulation.

where

$$r_{con} = r_d + r_s$$

For proof of the modified A_v formula, see Section 7.5.

The determination of the value of r_{con} needs to be added to the small-signal characterization of the short-channel MOSFET. Let us see how this is done by modifying the data display equations and consequently, the characterization plots.

Our starting point will be Figure 7.34, slightly modified. The modifications of Figure 7.41, with respect to Figure 7.34, are highlighted in Figure 7.42.

A single-frequency AC analysis, done at a very low frequency of 1 Hz, was added (Figures 7.41 and 7.42). This analysis will allow us to calculate the total resistance as follows:

$$r_{total} = \frac{v_{ds}}{i_{ds}} = r_o + r_{con}$$

Once we measure r_{total}, we can easily calculate r_{con}, since we already know r_o from the DC operating point.

Note that we took the opportunity to update the W_1 sweep with a **sweep plan**. See Section 7.6 for a discussion of sweep plans. Suffice to say that here we have a dual step size sweep of W_1, first from 10 to 90 μm in steps of 10 μm and then from 100–500 μm in steps of 50 μm.

The source connected to the drain was updated from a DC battery to a V_AC voltage source. We now also apply the needed DC voltage of 2.5 V through this source.

The main idea is to calculate the value of r_{total} as

$$r_{total} = \frac{v_{ds}}{i_{ds}} = \frac{1}{(-SRC_{1.i})}$$

ADS will save the AC current from the source, so there is no need to add a current probe.

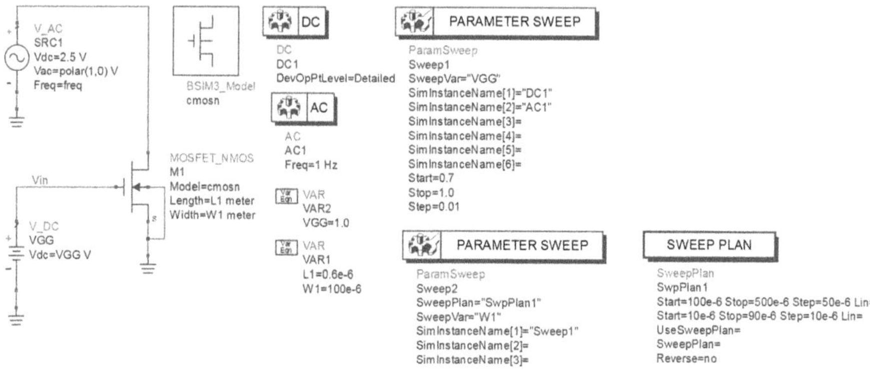

FIGURE 7.41 Pre-characterization parametric sweep, with a slight modification of the W_1 sweep plan and with a modification of the AC simulation parameters.

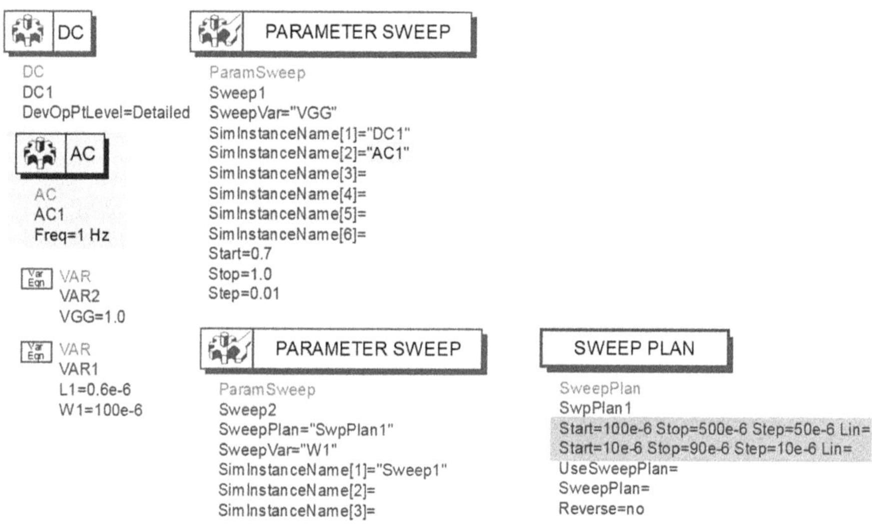

FIGURE 7.42 The modifications of Figure 7.41 with respect to Figure 7.34 highlighted.

Keep in mind that the ADS current convention is for a current leaving a source (as in the AC source node) to be considered having a negative polarity. Hence, the negative current sign when calculating r_{total} (i.e., $-SRC_{1.i}$). Also note, the numerator of r_{total} is 1 V, assuming that the AC voltage applied in the ADS AC simulation is 1 V magnitude, as shown in Figure 7.43, that is, SRC_1 has V_{ac}=polar(1, 0) V.

In Figure 7.44, we now update the data processing equations that precede the device characterization to include the calculation of r_{total}, based on the AC simulation analysis.

In Figure 7.44, the contact resistance r_{con} is calculated as the difference between r_{total} and r_o.

The gain A_{vo} is updated as well to include the impact of r_{con}, as shown in Figure 7.44.

In Figure 7.45, we observe the resulting double-sweep r_{con} characterization.

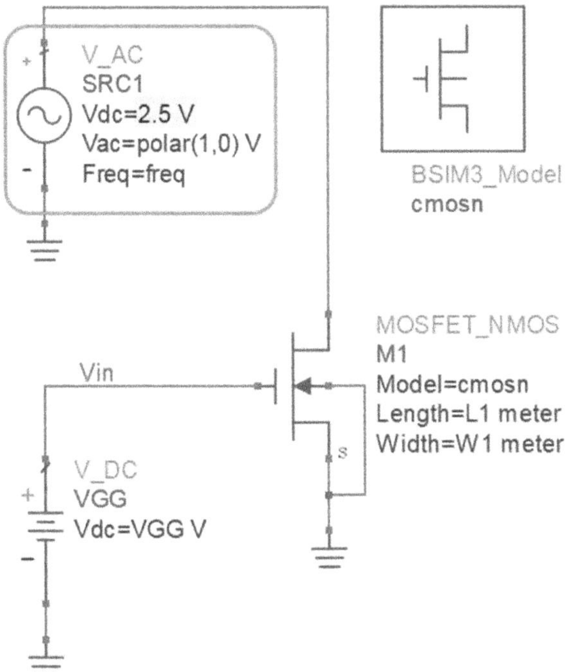

FIGURE 7.43 A schematic for measuring total contact resistance, as part of the short-channel MOSFET new characterization.

```
Eqn VDD = 5
Eqn VDS = 2.5
Eqn IDS = Id[0,0]
Eqn RD = (VDD - VDS) / IDS
Eqn gm = Gm[0,0]
Eqn ro = 1/Gds[0,0]
Eqn rtotal=real(-1/AC.SRC1.i[0])
Eqn rcon = rtotal- ro
Eqn Avo = gm * (RD*ro) / (RD + ro + rcon)
Eqn Avi = gm * ro
```

FIGURE 7.44 Revised pre-characterization data processing equations.

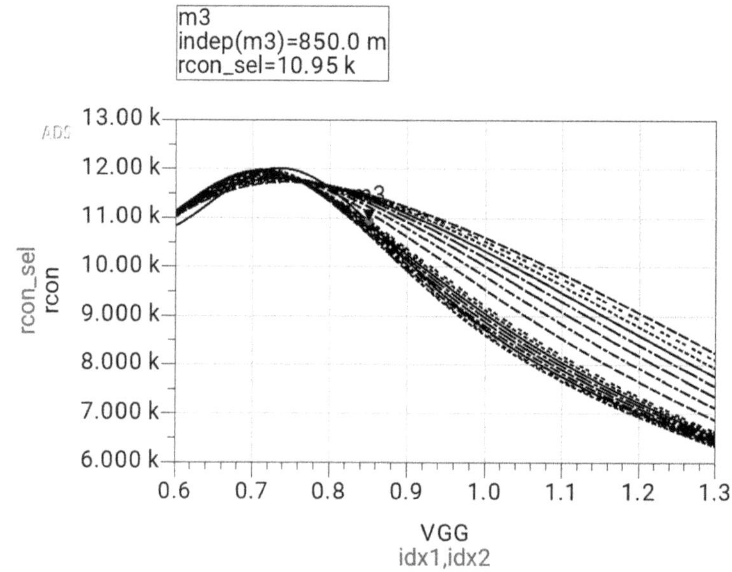

FIGURE 7.45 Characterization plot of the contact resistance r_{con}.

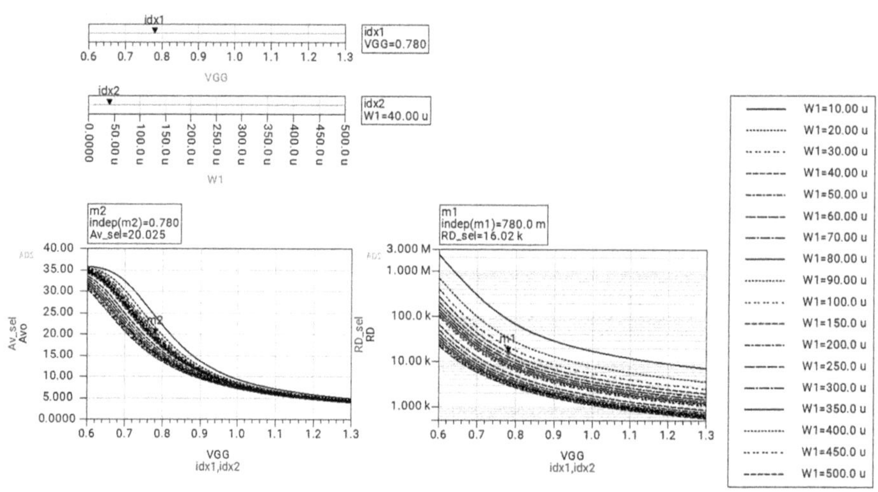

FIGURE 7.46 Revised design procedure based on the new MOSFET characterization.

Inspecting the results of r_{con} for the design (pursued earlier), we see by the blue marker dot $r_{con} = 10.95$ kΩ. This value is of a similar order of magnitude as R_D and r_o. No wonder it impacted the voltage gain so much.

It is time to revise the design of the short-channel CS amplifier with R_D load. The revised characterization and new design are shown in Figure 7.46.

We finally select $V_{GG} = 0.78$ V, $W_1 = 40\,\mu$m → $R_D = 16.02$ kΩ, and the obtained voltage gain $A_v = -20.025$.

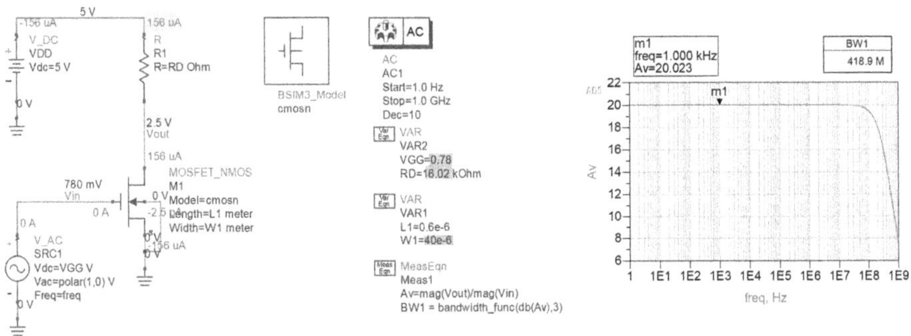

FIGURE 7.47 Validation of the revised design of the short-channel CS amplifier with R_D load. Left – Updated Schematic and Right - Updated Results

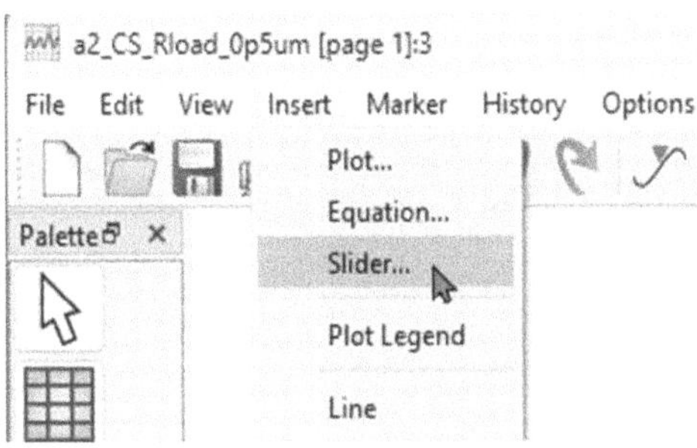

FIGURE 7.48 Data display menu for the Insert >, allowing a choice of a slider.

Again, let us validate this with an AC simulation. See schematics in Figure 7.47 (left).

The bias point is again in line with our expectation of $V_{DS}=2.5$ V.

From Figure 7.47 (right), the gain meets the requirement. It matches our prediction from our design methodology, using DC operating point and a simple AC low-frequency analysis characterization.

7.4 ADS TIPS AND TRICKS 7B: DATA DISPLAY SLIDERS

Sliders are a convenient way to parse through swept data.

The idea behind a slider is to index into data, say we have V_{out}[index]. See Section 5.5.

This could be, for instance, a frequency sweep for frequencies 1–10 kHz in 1 kHz steps. In this simple case, if one wants V_{out} at 5 kHz, we need to write $myV_{out} = V_{out}[5]$ because index = 5 corresponds to 5 kHz, simple enough.

What if we did a more complicated sweep, such as a log of frequencies? Now, the index is not so easy to determine manually, and who wants to type in a number every time we want a different frequency?

The slider will take care of all this.

In the data display, we access sliders from the menu Insert > Slider ... (Figure 7.48).

Once we place the slider on the data display, we will get the Slider Plot Trace & Attributes window (Figure 7.49).

We need to select a variable to use for indexing, for example, DC.V_{GG} (Figure 7.50).

Clicking >> Add >> moves the indexing variable to the Trace Options column (Figure 7.51). Note (Figure 7.50) that DC.V_{GG} has two dependencies, W_1 and V_{GG}. We select accordingly in the data to sweep with the slider (e.g., V_{GG}) (Figure 7.51).

In Figure 7.51, once we click OK, the slider will show up like in Figure 7.52.

One can now select m_1 with the mouse and "slide" it to a desired value.

To rename m_1, we double left-click on it and enter the new name.

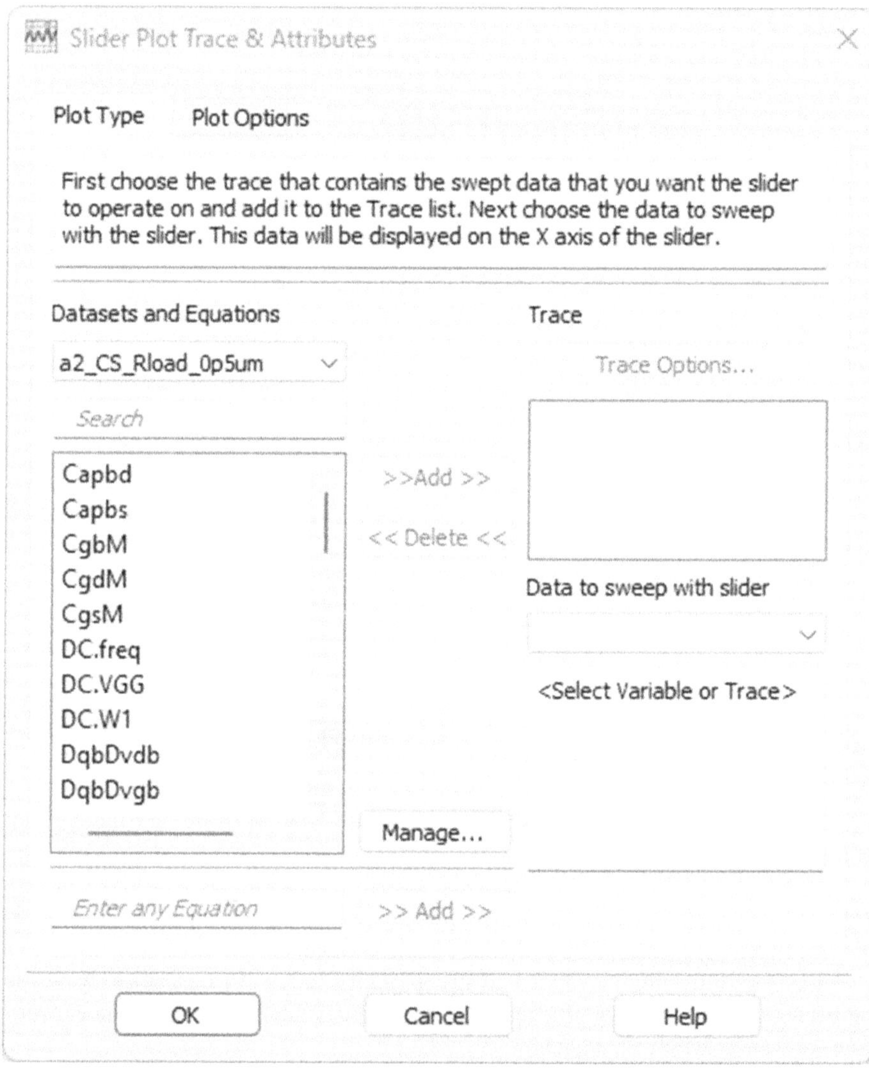

FIGURE 7.49 The slider plot trace and attributes window.

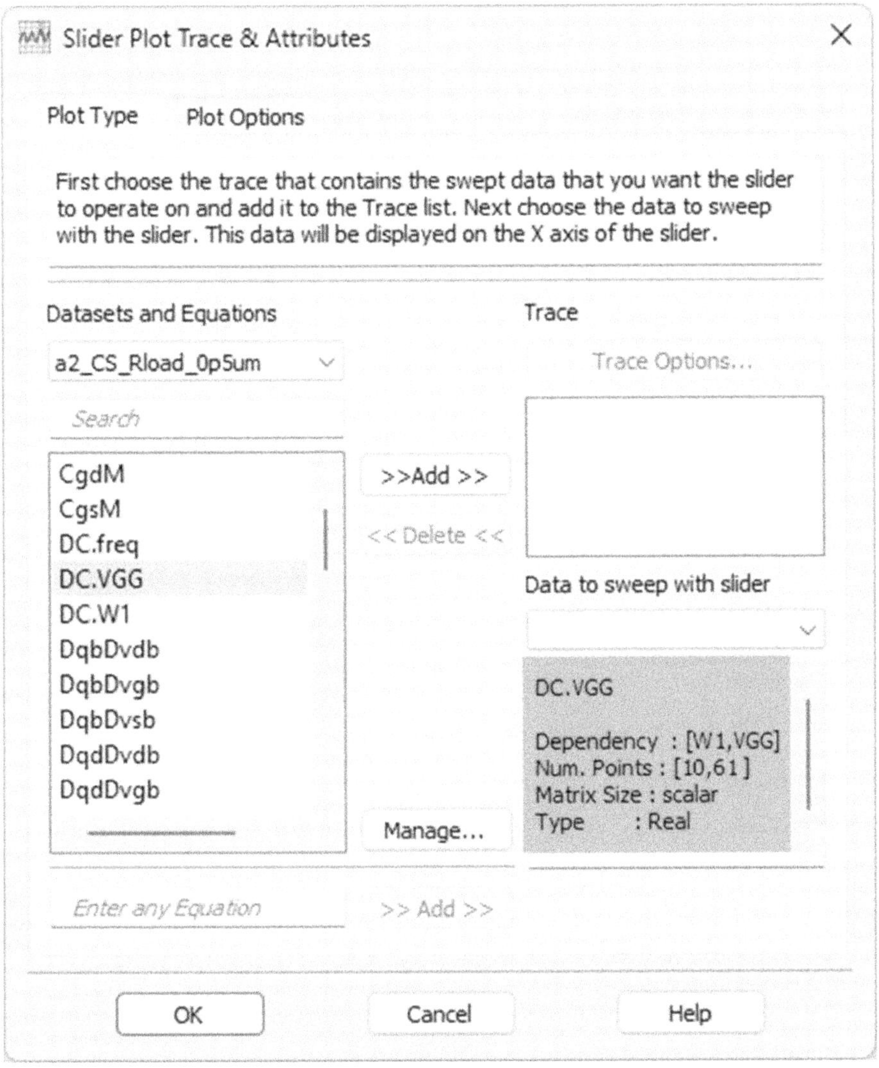

FIGURE 7.50 Selection of an indexing variable.

For a demonstration on how to use sliders, we use the variable I_{DS} (Figures 7.53 and 7.54).

I_{DS} has two dependencies, one of which is V_{GG}, which is being pointed at with the m_1 slider (Figure 7.52).

Let us create a variable myI_{DS} that selects the values of I_{DS} for the V_{GG} values selected on slider m_1. For that, we insert an Eqn for myI_{DS}, as shown in Figure 7.54, and the resulting equation is displayed in Figure 7.54.

Note that the slider created variables: m_1 (slider value) and $m_1_V_{GG}_$index (index for the selected slider value).

We use the $m_1_V_{GG}_$index to select all I_{DS} values for the V_{GG} selected by the slider.

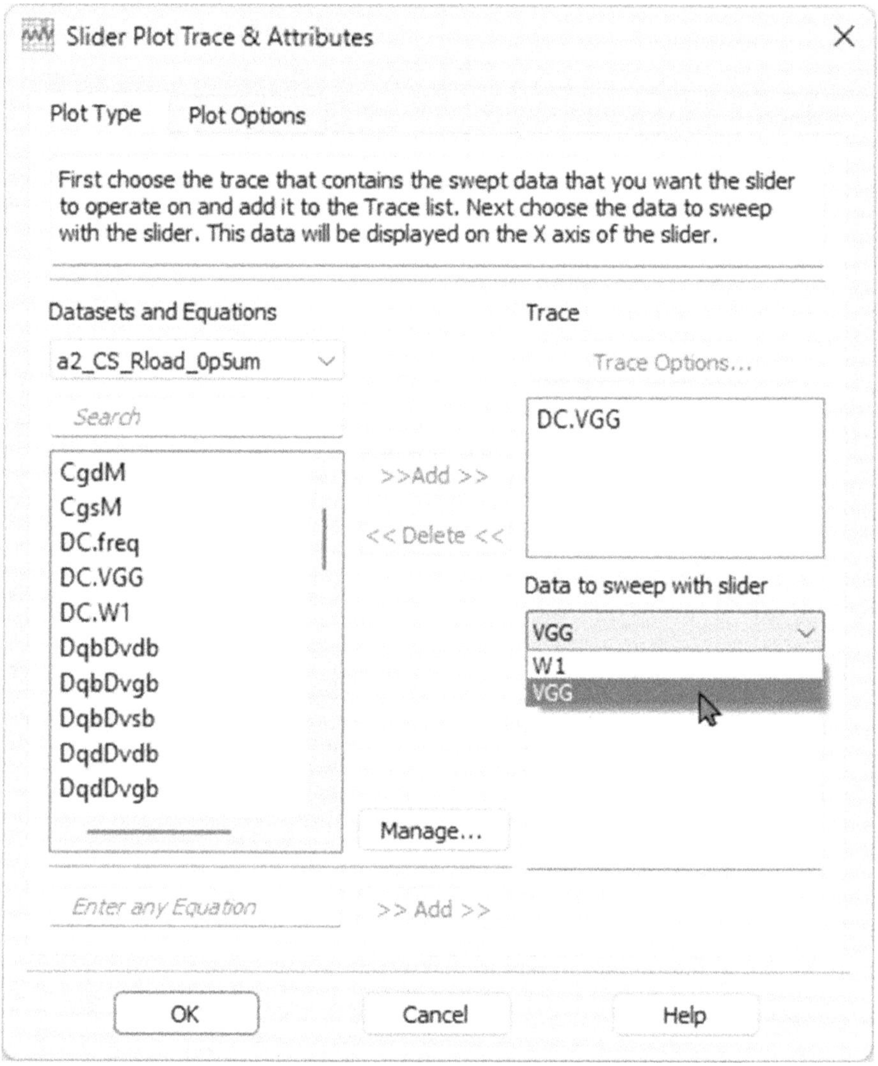

FIGURE 7.51 Trace options … window and data to sweep with slider menu.

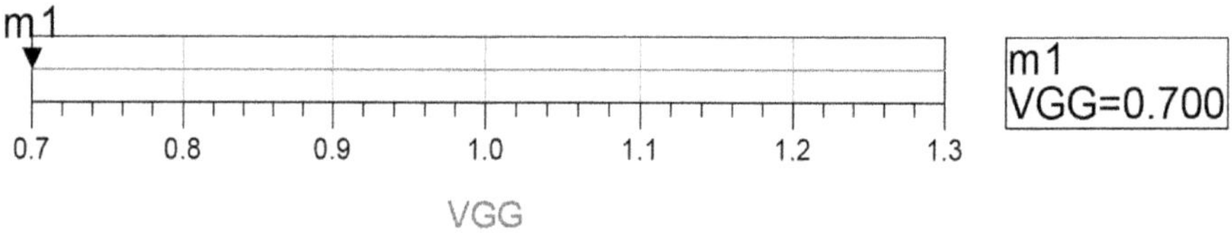

FIGURE 7.52 The slider for V_{GG}.

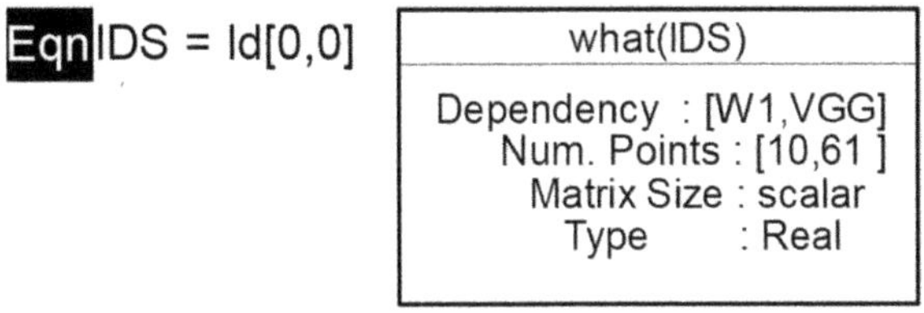

FIGURE 7.53 Demonstration of how a slider is used.

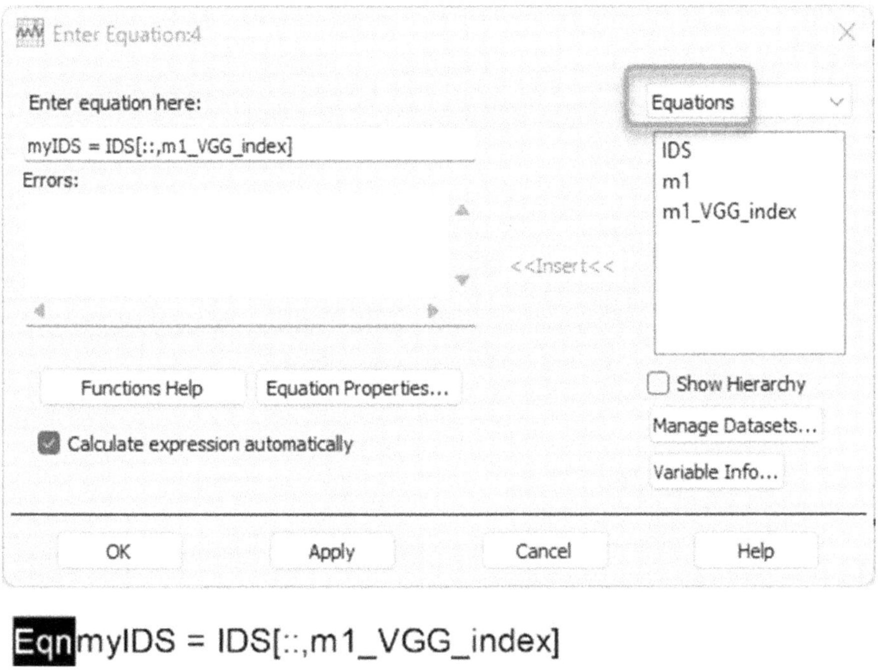

FIGURE 7.54 Insertion of an Eqn to create a double indexed new variable myI_{DS}.

Since I_{DS} depends on both W_1 and V_{GG}, as seen in the what() table in Figure 7.53, the symbol "::" in this expression (shown in Figure 7.54) selects "all the W_1 values".

We now list (or plot) the myI_{DS} data. In Figure 7.55, we demonstrate the slider's functionality.

Since I_{DS} has two dependencies, we can create a slider for W_1 and dual index into I_{DS} to get a single value for W_1 and V_{GG}. This is shown in Figure 7.56.

7.5 ADS TIPS AND TRICKS 7C: DERIVATION OF THE VOLTAGE GAIN EXPRESSION THAT INCLUDES CONTACTS RESISTANCE

We define the low-frequency contact resistance based on the MOSFET small-signal equivalent circuit repeated in Figure 7.57. Figure 7.57 also shows the CS amplifier with R_D load.

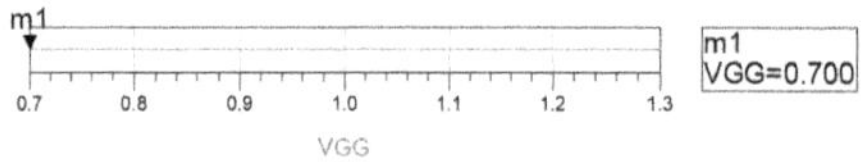

W1	myIDS
50.00 u	73.28 u
100.0 u	140.4 u
150.0 u	195.6 u
200.0 u	242.2 u
250.0 u	282.7 u
300.0 u	318.5 u
350.0 u	350.5 u
400.0 u	379.5 u
450.0 u	405.9 u
500.0 u	430.3 u

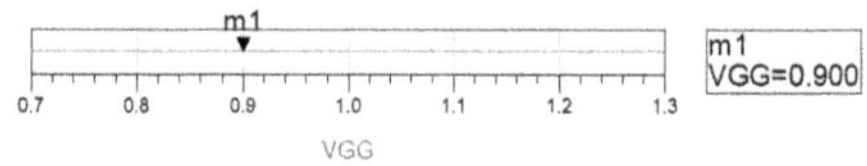

W1	myIDS
50.00 u	449.4 u
100.0 u	743.0 u
150.0 u	944.5 u
200.0 u	1.095 m
250.0 u	1.212 m
300.0 u	1.309 m
350.0 u	1.389 m
400.0 u	1.459 m
450.0 u	1.519 m
500.0 u	1.573 m

FIGURE 7.55 Tables of myI_{DS} values for 2 V_{GG} values (selected by the V_{GG} slider) and 10 W_1 values.

Eqn myIDS_W=IDS[m2_W1_index,m1_VGG_index]

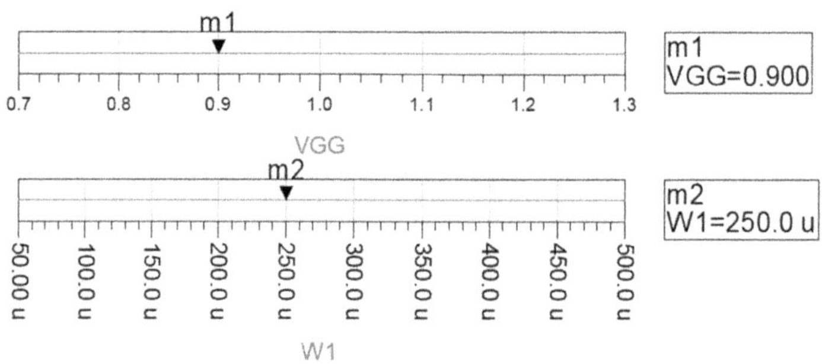

myIDS_W
1.212 m

FIGURE 7.56 Creation of a second slider for a dual index variable.

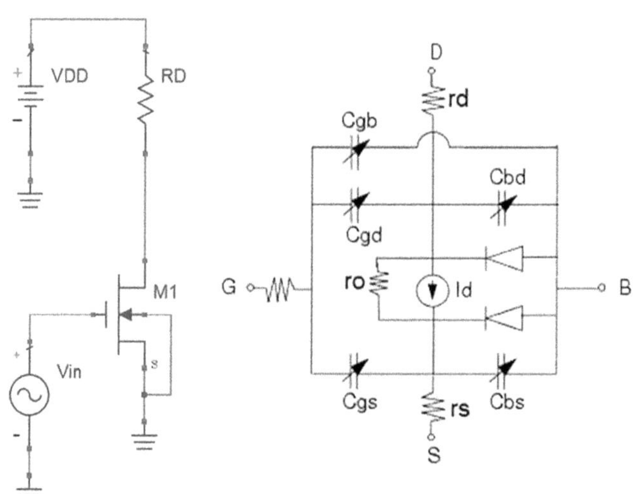

FIGURE 7.57 Circuit diagrams needed for the proof of the gain formula.

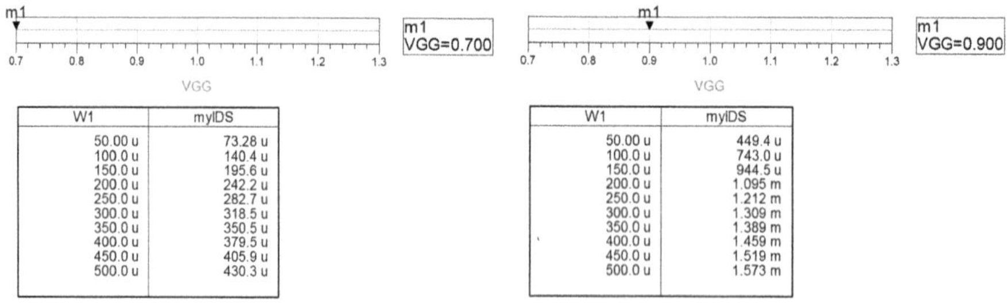

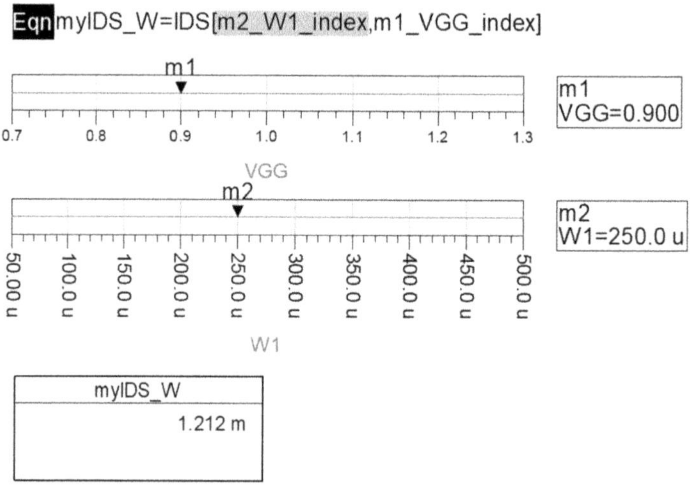

The contact resistance is $r_{con} = r_d + r_s$

By the universal voltage gain formula, the gain for the CS with R_D load is

$$A_v = -g_m R_{LOAD}$$

Since R_D and r_{con} are in series, R_{LOAD} can be written as

$$R_{LOAD} = \frac{r_o (R_D + r_{con})}{r_o + R_D + r_{con}}$$

The voltage across R_{LOAD} is

$$V_{LOAD} = V_{R_D} + V_{con}$$

The actual load is R_D, which forms a voltage divider with r_{con}; hence, we write the output voltage as

$$V_{out} = A_v \left(\frac{R_D}{R_D + r_{con}} \right) V_{in}$$

Substituting for the definition of A_v

$$V_{out} = \left[-g_m \frac{r_o (R_D + r_{con})}{r_o + R_D + r_{con}} \right] \left(\frac{R_D}{R_D + r_{con}} \right) V_{in}$$

The $(R_D + r_{con})$ terms are canceled, so one can replace R_D for any CS load, and the same gain expression (7.1) still applies.

7.6 ADS TIPS AND TRICKS 7D: SIMULATION SWEEP PLAN

ADS provides various mechanisms to do parameter sweeps. One can use the sweep built into the DC simulation controller or an external parameter sweep component.

These sweeps allow for single-point, linear, or logarithmic sweeps. In certain situations, these sweeps may be limited in terms of flexibility.

For example, what if we want three points that are not linearly or logarithmically related (e.g., 1, 5, 8). For such a case, would we have to run more than one simulation and collate the data?

Another case is when, in the entire variable space, one wants to use different resolutions for the sweep. For example, we want to study how a MOSFET current varies with channel width that runs from 1 to 500 μm. The sweep would have to be in 1 μm steps and have 500 points. Not difficult for the simulator, but what if we could run the first 10 points from 1 to 10 μm in steps of 1 μm and from 20 to 500 μm in steps of 10 μm for a total of 58 points? Of course, 58 points have less resolution, but if our interest is the trends, the lower resolution is not a problem.

There are other circumstances that one may encounter that. It is worth understanding how to conduct such more flexible sweeps.

Sweep plans are found in the palettes for the different simulation controllers. See Figure 7.58.

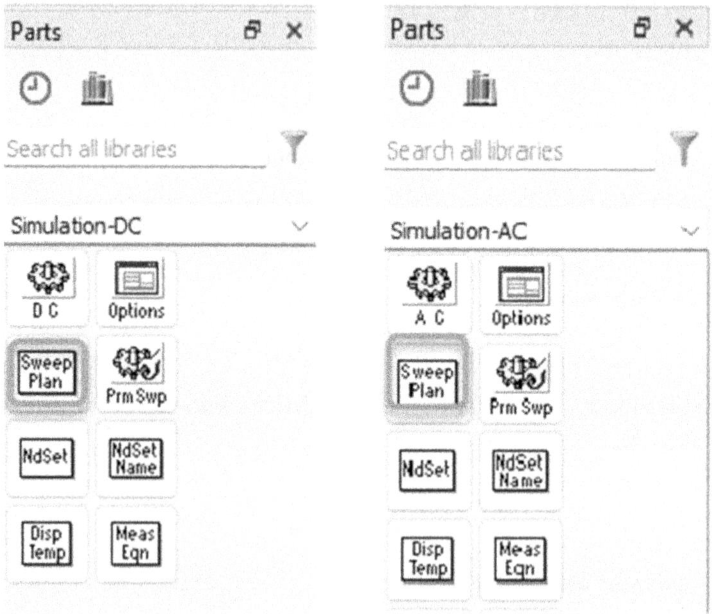

FIGURE 7.58 Sweep plan available in the component's palettes.

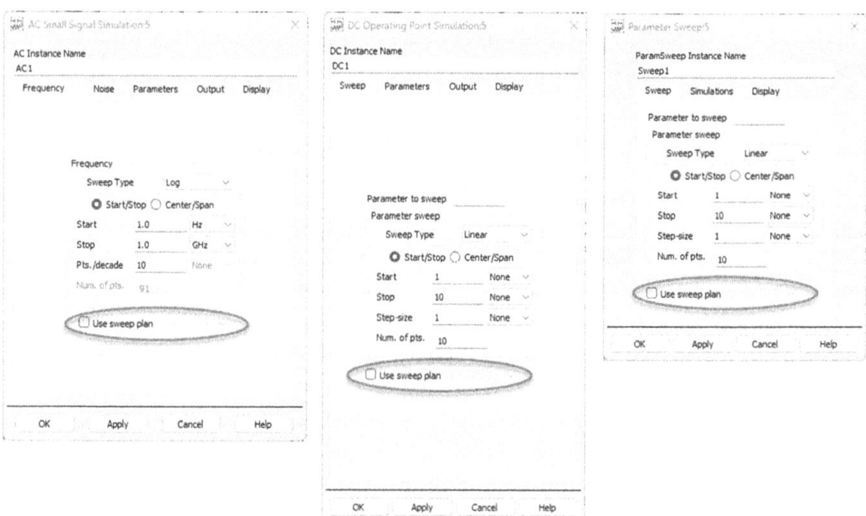

FIGURE 7.59 Examples of simulation controllers that allow a sweep plan.

Any controller that has a sweep capability has a provision to refer to a sweep plan. Just to show a few, see Figure 7.59.

We instantiate a sweep plan on the schematic, as shown in Figure 7.60.

Note that a sweep plan has different sweep types, and we can add as many different sweep intervals as needed.

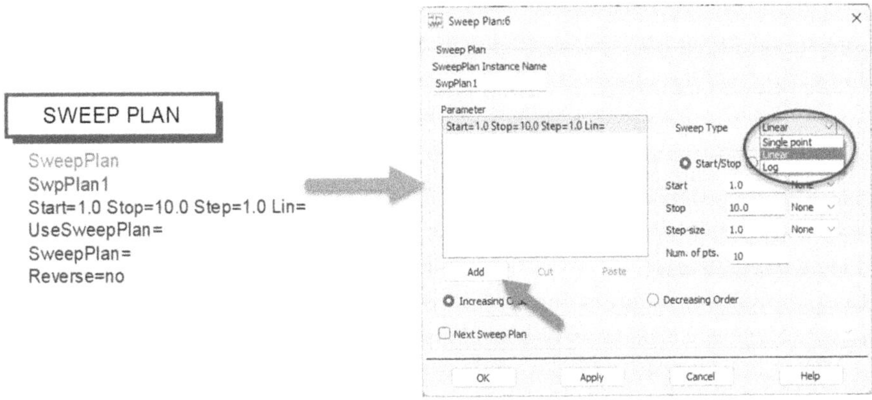

FIGURE 7.60 Sweep plan block and its dialog box.

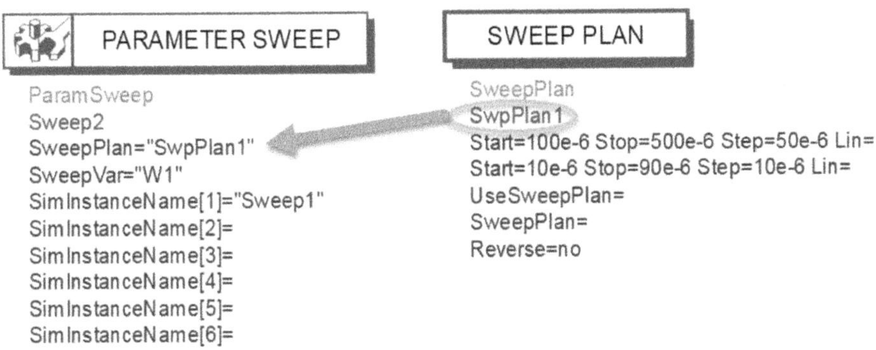

FIGURE 7.61 The sweep plan that allowed non-uniform W_1 sweep.

The sweep plan that we used in the modified characterization study, Example 7.3, is presented in Figure 7.61.

The variable being swept is W_1.

The first interval has nine points 10–90 μm in steps of 10 μm.

The second interval is from 100 to 500 μm in steps of 50 μm.

7.7 LEVEL 1 ANALYTIC DESIGN OF A CS AMPLIFIER WITH DIODE-CONNECTED LOAD

This section and the next one focus on the computer-aided design of a CS amplifier with MOSFET load, using the 0.5 μm technology BSIM3 model and the ADS design platform. It is of interest, though, to compare the designs to the analytic design, as well as the simulation-based design, that can both be carried out in Level 1 modeling. For the analytic analysis and design for such amplifier configurations, we refer to [Razavi1].

If one electrically shorts the base and the collector nodes of an NPN BJT transistor, the BJT becomes a diode. The shorted base and collector become the anode of the diode, and the emitter becomes the cathode. Similarly, if we short the base and collector of a PNP BJT, this shorted node becomes the cathode of a diode, and the emitter becomes the anode of the diode. In a MOSFET transistor, if we short the gate and the drain, we just give it the name, "**diode-connected MOSFET**", even though it is not a diode at all. What is special about such a shorting action is that the MOSFET, if ON, automatically enters saturation mode (for NMOS that has a positive threshold voltage, and for PMOS that has a negative threshold voltage). It never enters a triode mode. The MOSFET load of a CS amplifier, which we discuss in this section, is a diode-connected configuration. We met this configuration in Section 4.9, as we discussed the reference side of a current mirror current source. Such CS amplifiers, with a diode-connected load, come in several configurations. In a CS amplifier employing an NMOS input transistor M_1, the diode-connected load M_2 may be either an NMOS or a PMOS. See Figure 7.62. There may be CS amplifier configurations in which the input transistor is a PMOS, having its source connected to V_{DD}.

Three comments about Figure 7.62: (i) V_{out} consists of the amplified v_g voltage riding on top of a DC level, which is the M_1 drain DC voltage V_D; (ii) With an NMOS diode-connected load (left), we note that the substrate of M_2 (which is not shown)

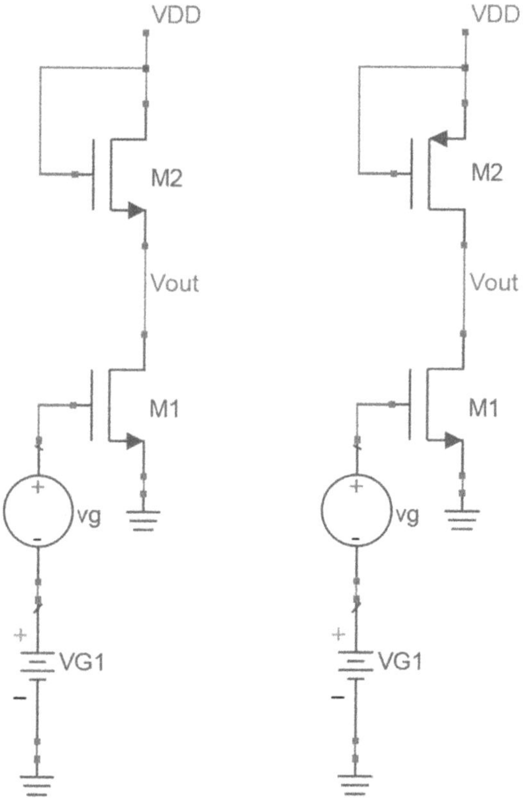

FIGURE 7.62 CS amplifier with NMOS input and a diode-connected MOSFET load: (left) diode-connected NMOS, (right) diode-connected PMOS.

must be connected to ground, whereas the source of M_2 is connected to V_{out}. Thus, the amplifier's load exhibits a body effect; (iii) with a diode-connected PMOS load (right), both source and substrate are connected to V_{DD}. There is no body effect. We choose to demonstrate both analytic and ADS-based designs on this PMOS load configuration.

The small-signal diagram shown in Figure 7.63 serves both amplifier circuits of Figure 7.62. In both, $v_{gs1}=v_{in}$ and $v_{gs2}=-v_{out}$. For the NMOS load, $v_{bs2}=-v_{out}$, whereas for the PMOS load, $v_{bs2}=0$.

An equivalent simpler view of the diagram in Figure 7.63 is shown in Figure 7.64.

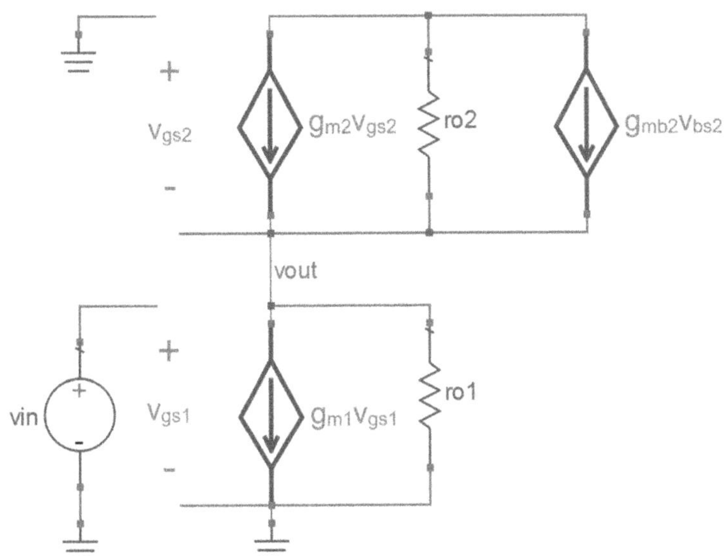

FIGURE 7.63 Small-signal diagram of any of the configurations of a CS amplifier with diode-connected load: NMOS load features a body effect, thus having a nonzero g_{mb2} parameter. For PMOS load, there is no body effect. We set $g_{mb2}=0$.

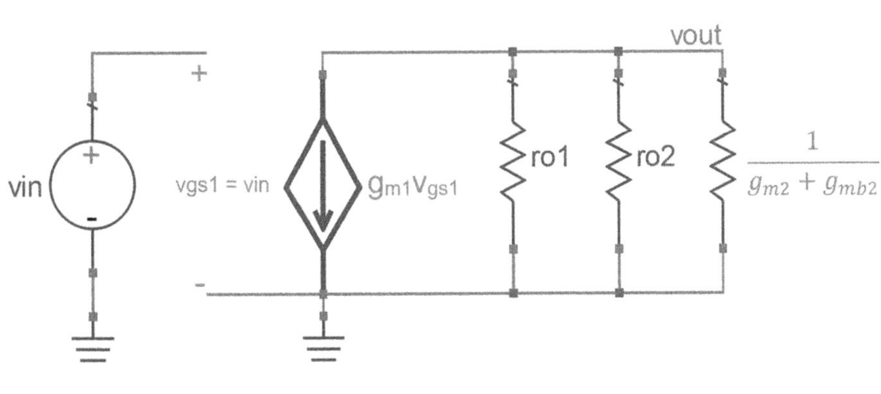

FIGURE 7.64 An equivalent small-signal diagram to the one in Figure 7.63.

The small-signal voltage gain, for both load configurations, is then $A_v = v_{out}/v_{in} = -g_{m1}[(g_{m2}+g_{mb2})^{-1}||r_{o1}||r_{o2})]$ for NMOS load and $\mathbf{A_v = v_{out}/v_{in} = -g_{m1}[(g_{m2}^{-1})||r_{o1}||r_{o2}]}$ **for PMOS load.**

For a long-channel PMOS load, $1/g_{m2}$ is typically much smaller than any of the r_o resistors. For a short-channel PMOS load, the r_o resistors are not so large and therefore need to be considered in the voltage gain.

For diode-connected PMOS load cases, in which r_{o1} and r_{o2} play a secondary role, a good approximation of the small-signal voltage gain is $\mathbf{A_v \approx g_{m1}/g_{m2}}$, which simplifies into

$$A_v = -\sqrt{\frac{\mu_n (W/L)_1}{\mu_p (W/L)_2}} \tag{7.2a}$$

The mobility ratio is approximately 3.5. This square root voltage gain formula implies that even a modest required gain value necessitates a large input transistor and a much smaller load transistor. That is, a large ratio of aspect ratios of the input transistor to the load transistor.

The input transistor M_1 must be kept in saturation mode by a proper choice of its drain voltage V_{D1} (i.e., the DC level of V_{out}). As commented earlier, the load transistor M_2, being diode-connected, is always in saturation mode. The same current I_D flows through M_1 and M_2.

$$\mu_n\left(\frac{W}{L}\right)_1 (V_{GS1} - V_{TH1})^2 \approx \mu_p\left(\frac{W}{L}\right)_2 (V_{GS2} - V_{TH2})^2$$

Therefore, an alternative version of the approximated voltage gain formula (7.2a) is

$$A_v = -\sqrt{\frac{\mu_n (W/L)_1}{\mu_p (W/L)_2}} \approx \frac{|V_{GS2} - V_{TH2}|}{V_{GS1} - V_{TH1}} \tag{7.2b}$$

We learn from equation (7.2b) that for a large voltage gain, which may be specified, a large ratio of overdrive voltages is required: A large overdrive of the load M_2 and a small overdrive of the input transistor M_1. Formula (7.2b) presents a substantial voltage "swing" constraint (meaning the ability to keep the amplified voltage v_{out}, riding on top of V_{D1}, from causing the amplifier to dynamically enter either triode or cutoff modes). Let us demonstrate numerically the impact of equation (7.2b).

Example 7.3

Say that $V_{DD} = 3\,V$, $V_{TH2} = -0.7\,V$, and $V_{GS1} - V_{TH1} = 200\,mV$. Then for a voltage gain of 10, by equation (7.2b), we need $V_{GS2} < -2.7\,V$. By the diode connection of M_2, $V_{DS2} < -2.7\,V$, which leaves only 0.3 V for V_{DS1}: $V_{DS1} < 0.3\,V$. To keep M_1 in saturation mode, $V_{DS1} > V_{GS1} - V_{TH1} = 0.2\,V$. Therefore, the room available for the amplified voltage is between 0.2 and 0.3 V, which is rather small. For certain (larger than 10) desired voltage gains, the design may quickly lead to a "No Solution" situation.

7.8 ADS: SWEPT OPTIMIZATION FOR LONG-CHANNEL AND SHORT-CHANNEL CS AMPLIFIERS WITH DIODE-CONNECTED LOAD

Example 7.4

Following up on the CS amplifier with R_D load Example 7.2, which we solved in Section 7.3, let us hold on to the design specifications and naively assume that the diode-connected MOSFET load replaces R_D. We shall start with a Level 1 design of the CS amplifier with a diode-connected load (shown in Figure 7.65).

Again, we require a voltage gain of $A_v = -20$. Let us try to obtain $I_{DS} = 1$ mA and hold on to the $V_{DD} = 5$ V supply. We also keep all transistor channel lengths at 2 μm whenever using the Level 1 model. The design variables are W_1, W_2, and V_{GG}.

Initially, we shall keep V_{GG} and W_1 from the CS with R_D load design (see Figure 7.66.)

The Level 1 MOSFET model cards are shown in Figure 7.67. We adopted Level 1 parameters provided in [Razavi1].

With the adoption of V_{GG} and W_1 (from the design of the CS amplifier with R_D load) and by using the approximation equation (7.1a) of A_v, the only unknown is W_2:

$$(W/L)_2 \approx \mu_n (W/L)_1 / \mu_p A_v^2 = 350 \cdot 256.3 / 100 \cdot (20)^2 = 2.2.$$

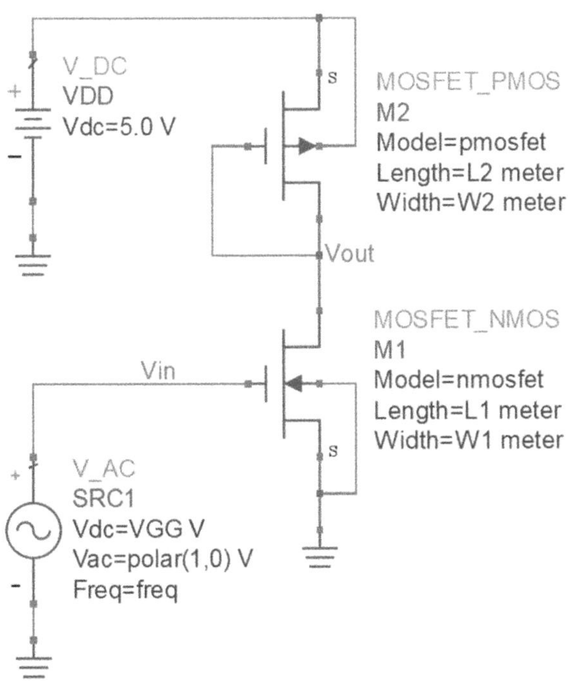

FIGURE 7.65 CS amplifier with PMOS diode-connected load.

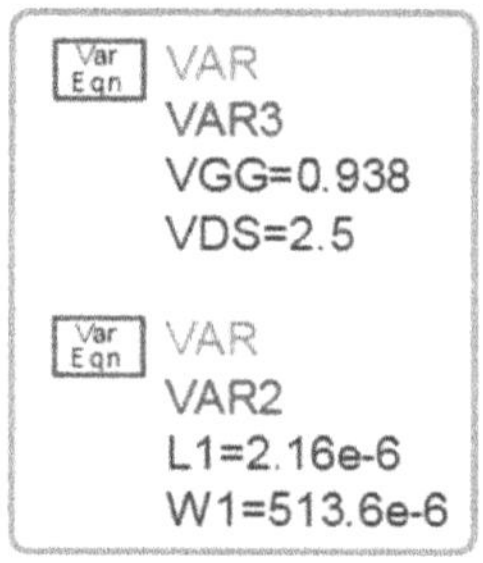

FIGURE 7.66 VAR variables taken from the final Level 1 design of the CS amplifier with R_D load.

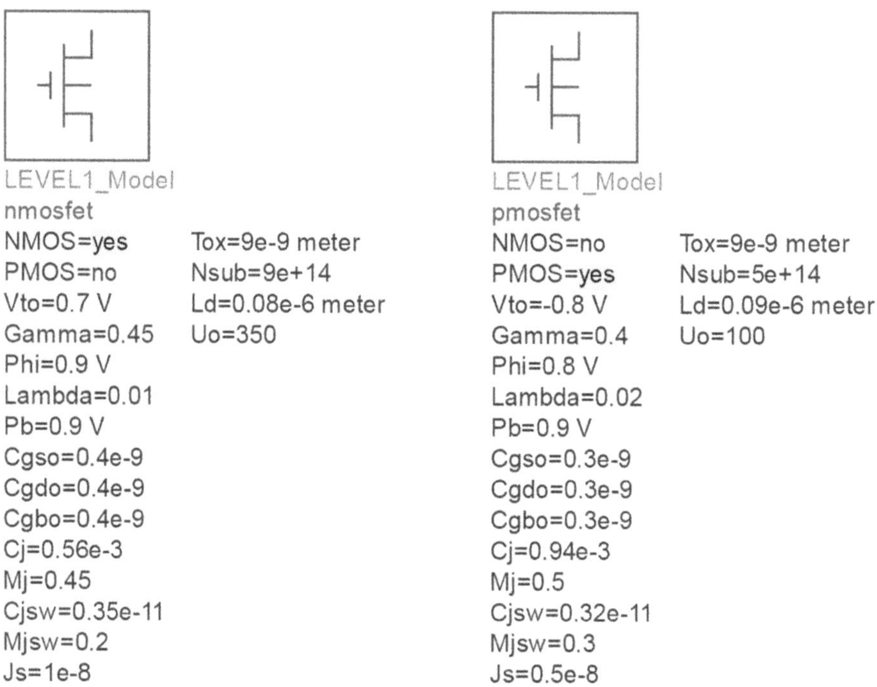

FIGURE 7.67 Model cards for Level 1 NMOS and PMOS adopting parameters recommended in [Razavi].

We let $L_{eff2}=2\,\mu m$. Then $W_2=4.4\,\mu m$.

Next, we will simulate the design to see how well we did versus the requirements (Figure 7.68).

Let us observe the DC solution annotation in an attempt to better understand what had gone wrong (Figure 7.69).

Annotating the DC solution on the schematic, we see that the DC bias point has changed dramatically from our expectations of $V_{DS}=2.5\,V$ and $I_{DS}=1.0\,mA$. This bias change has obviously affected the expected performance of the amplifier,

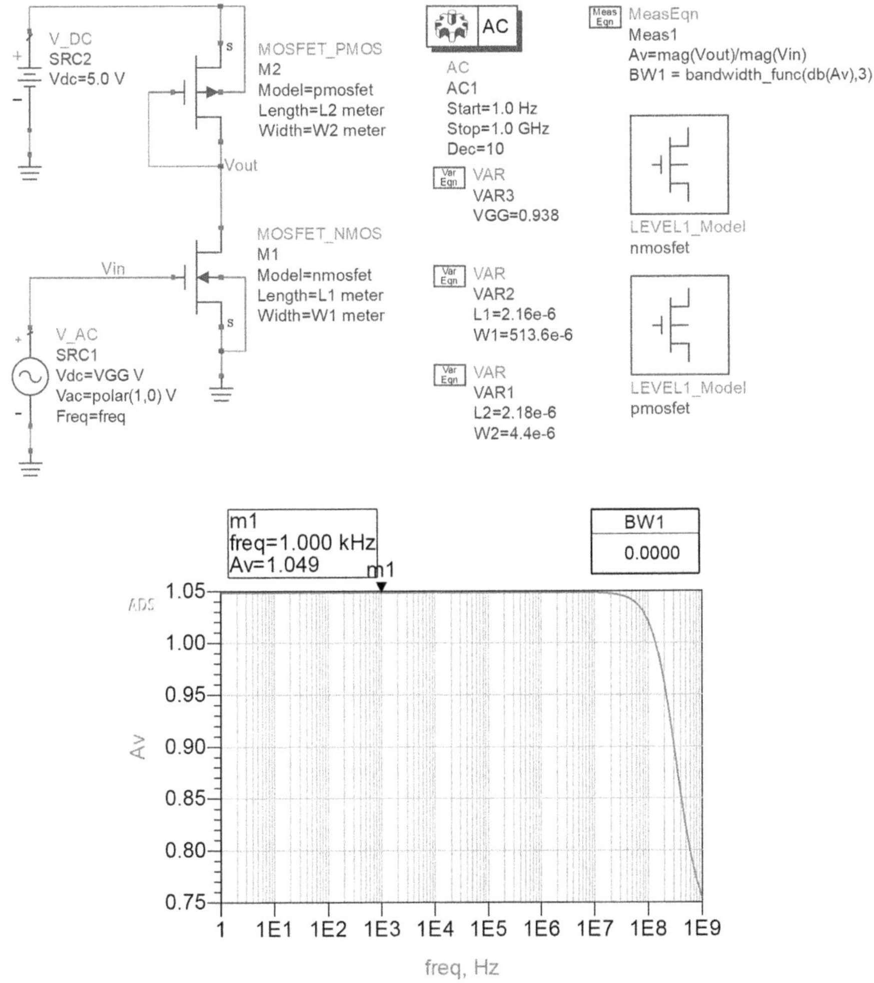

FIGURE 7.68 Schematics of the first design attempt of Level 1 CS amplifier with PMOS diode-connected load: (up) the schematics with design parameters inspired by the design of a CS amplifier with R_D load, (down) The Disastrous results – there is no voltage gain.

as M_1 is driven to triode mode. Recall that an NMOS transistor will go into triode mode if $V_{DS} \leq (V_{GS} - V_{TH})$. In our case here, $V_{GS} = 938\,mV$. The NMOS M_1 transistor is using a Level 1 model, so $V_{TH} = V_{TO}$, and from the model card, $V_{TO} = 0.7\,V$. Therefore, $V_{GS} - V_{TH} = 238\,mV$, which is greater than $V_{DS} = 128\,mV$; hence, the M_1 transistor is in triode mode.

We must redo the biasing. Let us try to modify V_{GG} by calculating a new value for V_{GS1}, using the gain equations (7.1a) and (7.1b) for this amplifier, and letting $V_{GS2} = V_{DS} - V_{DD} = -2.5\,V$. Rearranging equation (7.1b) for an expression for V_{GS1}:

$$V_{GS1} \approx \frac{|V_{GS2} - V_{TH2}|}{A_v} + V_{TH1} \rightarrow V_{GS1} \approx \frac{|-2.5 - (-0.8)|}{20} + 0.7 = 0.785\ V.$$

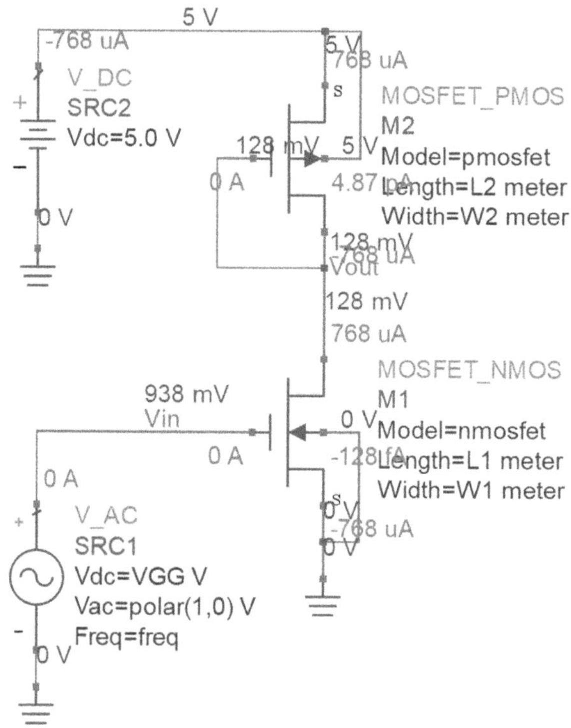

FIGURE 7.69 More results of the first attempt design – the resulting DC conditions.

As a second attempt, we updated this Level 1 design with this new value of V_{GS1}, obtained from the Level 1 gain formulas for diode-connected load, and rerun the simulation. See Figure 7.70. A definite improvement to the design. Note that we no longer require $I_{DS}=1$ mA, that is, it is now a free design variable.

Note $V_{DS}=2.5$ V even when assuming $\lambda=0$ in the calculations. We decided to retry using *Optimization* instead of further "tweaking" the design. See schematic in Figure 7.71.

We set the optimization design variables and their search limits as shown in Figure 7.72.

The NMOS width search range is set to $2\,\mu m \leq W_1 \leq 800\,\mu m$, and for $V_{GG,}$ it is set to $0.75\ V \leq V_{GG} \leq 4.0$ V.

The AC analysis is set to a single frequency of 1 Hz, as we are just interested in the low-frequency gain for OptimGoal 2. See Figure 7.71 for the optimization goals definitions.

The DC analysis goal was added to assure that $V_{GS2}=-2.5\pm0.01$ V.

Let's run the optimizer and see the results in Figure 7.73.

A quick note on the dataset values saved by the optimizer. The optimizer, as instructed in the simulation controller, automatically saves the error function value in FinalEF and the number of iterations executed in NumIters. See Figure 7.73.

The optimized variables are saved in OPTSOLNVALS.*variable name*. Again, see Figure 7.73.

The goals are saved in OptimGoalName, for example, the gain is saved as *OptimGoal2* (see schematic in Figure 7.71 and results in Figure 7.73).

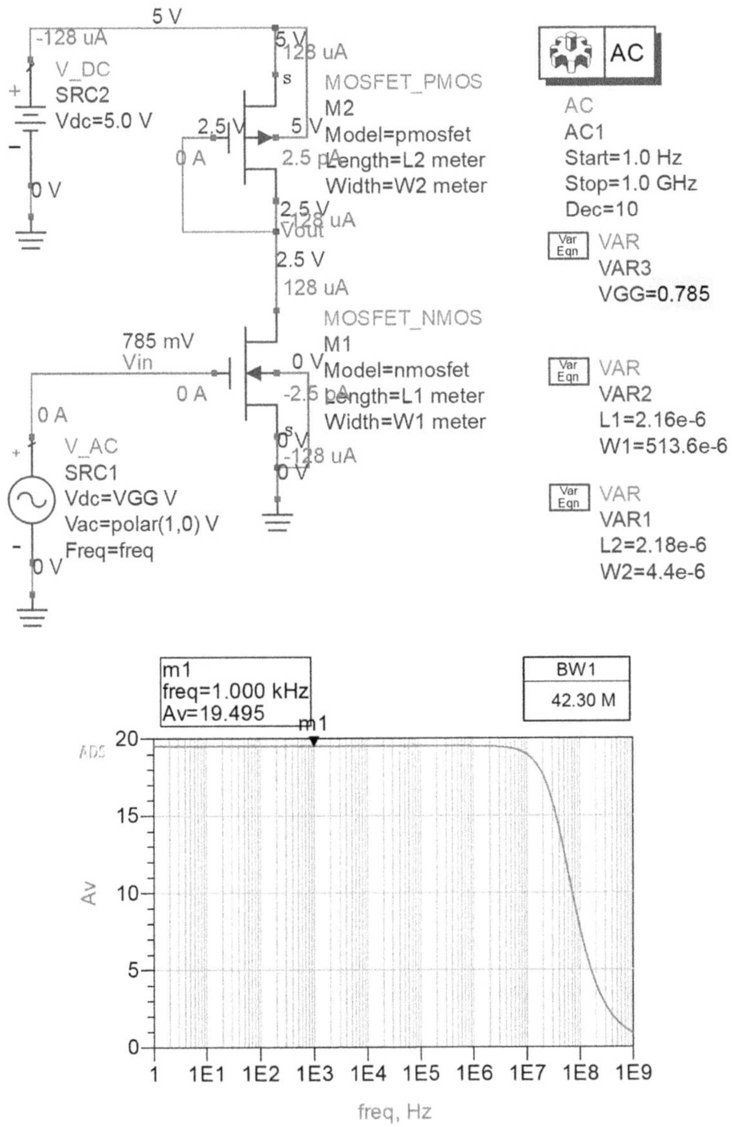

FIGURE 7.70 Second attempt of the Level design of the CS amplifier with a PMOS diode-connected load: (up) schematic with revised V_{GG} and W_2 values, (down) gain results.

As seen in Figure 7.73, the results show an error = 0 after six iterations. V_{DS} and A_v have met specifications. Note that, as in our calculation, we let I_{DS} be a free variable, getting $I_{DS} = 127.3\ \mu A$.

Next, we update the design with $V_{GG} = 0.7827\,V$ and $W_1 = 540.7\,\mu m$, which are close to our calculated values. Note that W_1 is increased to meet the new increased current value. Plots of the results are presented in Figure 7.74.

The change in current is negligible, just 1 μA. Bandwidth is somewhat less due to the increased width of M_1.

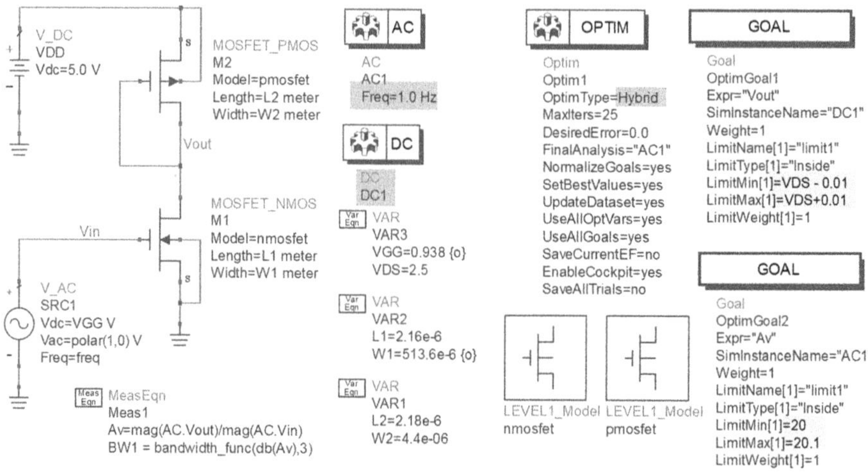

FIGURE 7.71 Third attempt: optimization schematic for tweaking the common source diode-connected load second attempt design.

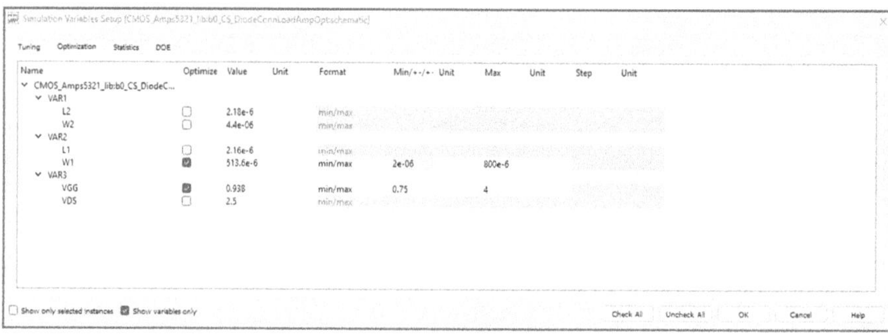

FIGURE 7.72 Optimization variables setup.

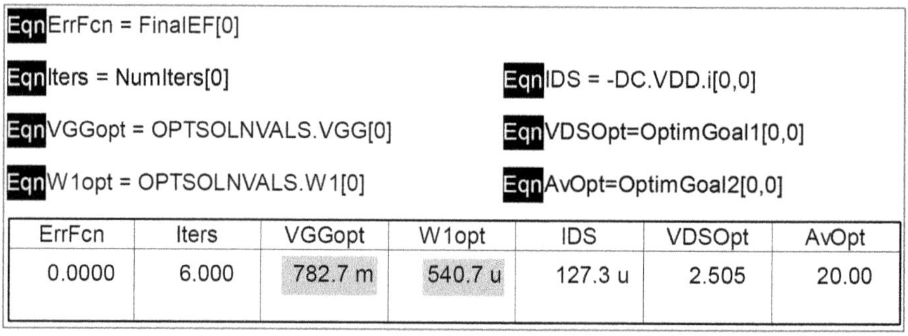

FIGURE 7.73 Post-processing of the optimization session results.

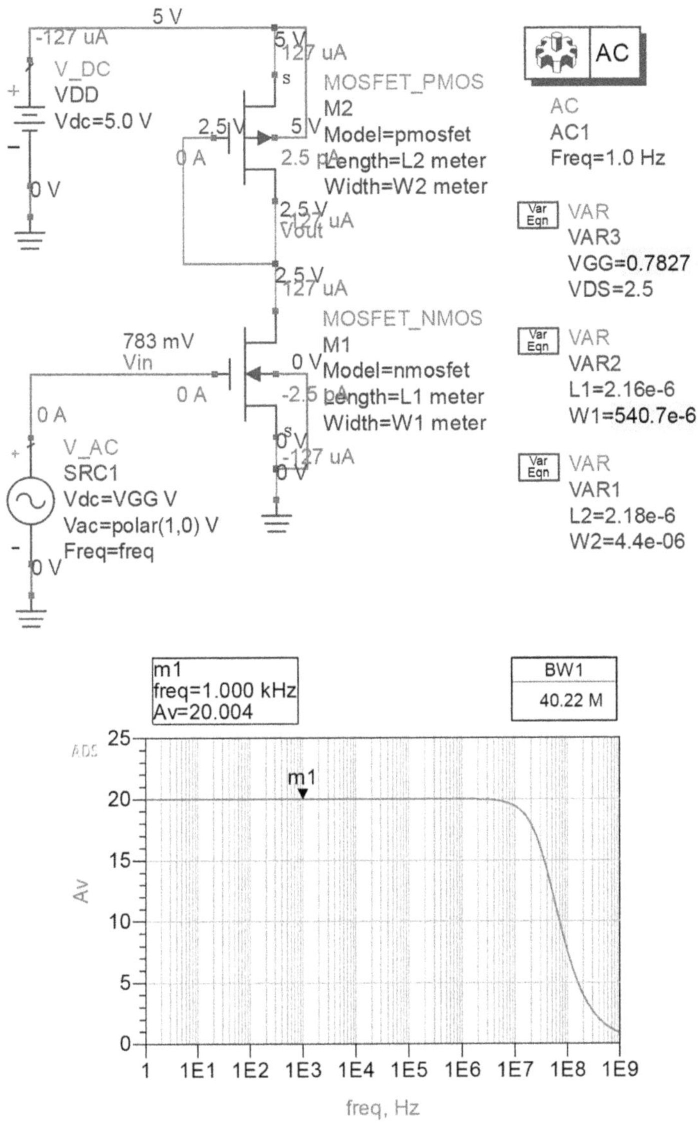

FIGURE 7.74 Revised results after substitution of the optimization optimal design parameters.

Example 7.5

What would happen if a value other than V_{DS} = 2.5 V was needed in Example 7.4? Will the design fail? We are still conducting Level 1 simulations.

These questions can be answered with a unique combined usage of the optimizer and parameter sweeps. The idea is to sweep V_{DS} but at each step, observe what the changes are that are needed for the design variables, in our case V_{GG} and W_1.

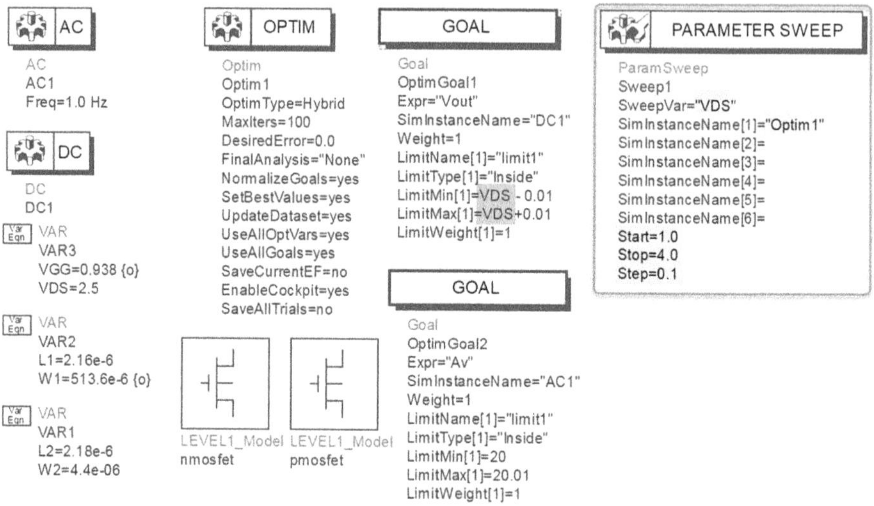

FIGURE 7.75 Level 1 design of CS amplifier with diode-connected PMOS load, combining optimization and parameter sweep.

This type of analysis is called a "***Swept Optimization***" and it is quite an advanced and powerful analysis tool.

Let us see how this is done for the CS amplifier with a diode-connected load amplifier. Refer to the schematic shown in Figure 7.75.

The swept optimization starts with adding a parameter sweep controller. As seen in Figure 7.75, we sweep variable V_{DS} from 1.0 to 4.0 V in 0.1 V steps.

Notice that OptimGoal1 limits depend on V_{DS}, so the optimization limits will change with the parameter sweep.

We keep the same optimizable limits for W_1 and V_{GG}. Let us run the optimization (by activating the cockpit) and observe the tabular results (Table 7.2).

The results show that the design is capable of working for $1.0\ V \le V_{DS} \le 3.1\ V$. How do we know? For these V_{DS} values, the optimizations converge, and the resulting error function is 0. See Table 7.2.

Starting at $V_{DS} = 3.2$ V and up, in our sweep, the error is no longer equal to zero.

Would increasing the number of optimization iterations help? No. In each run, for the larger V_{DS} values, the value of V_{GG} has reached the minimum limit of 0.75 V. Note also that the number of iterations went all the way up to the specified maximum of 100. The results indicate that V_{GG} needs to go lower, but we keep the limit to give the design some margin before V_{GG} becomes too close to V_{TH} of M_1. Recall that the threshold voltage may have very large tolerances.

A yield analysis of such a Level 1 design, in which V_{GG} is very close to V_{TH}, may feature large percentage of catastrophic failures in the form of transistor cutoff.

Repeating the optimization for many swept values of a parameter does not necessitate a convergence to a global optimum in each sweeping step. The nice thing about such a design is that there are no catastrophic failures. The gain drops whenever the optimization error is nonzero, but it does so gracefully and monotonically in a predictable way. Notice the changes in V_{GG} and W_1 as we swept V_{DS}. Since $V_{GS2} = V_{DS} - V_{DD}$, it means that the current must change, as clearly shown in

TABLE 7.2
Optimization Results for the Swept VDS Values

V_{OS}	ErrFcn	Iters	V_{GGopt}	W_{1opt}	I_{DS}	V_{OSOpt}	A_{vOpt}
1.000	0.0000	2.000	852.7m	588.1 u	465.2 u	1.005	20.01
1.100	0.0000	1.000	848.8m	584.8 u	439.2 u	1.093	20.00
1.200	0.0000	2.000	843.9m	581.8 u	409.0 u	1.199	20.01
1.300	0.0000	5.000	839.3m	578.2 u	381.7 u	1.298	20.00
1.400	0.0000	1.000	834.6m	575.4 u	355.0 u	1.399	20.01
1.500	0.0000	2.000	830.4m	572.2 u	331.5 u	1.491	20.00
1.600	0.0000	2.000	825.0m	568.8 u	303.3 u	1.606	20.01
1.700	0.0000	4.000	820.5m	565.4 u	280.2 u	1.704	20.00
1.800	0.0000	2.000	816.2m	563.0 u	259.8 u	1.795	20.01
1.900	0.0000	2.000	811.3m	559.5 u	237.0 u	1.901	20.01
2.000	0.0000	4.000	806.2m	555.9 u	214.6 u	2.010	20.00
2.100	0.0000	1.000	802.1m	553.5 u	198.0 u	2.095	20.01
2.200	0.0000	2.000	796.8m	549.8 u	176.7 u	2.208	20.00
2.300	0.0000	3.000	792.8m	547.2 u	161.7 u	2.294	20.00
2.400	0.0000	1.000	787.8m	543.9 u	144.1 u	2.399	20.00
2.500	0.0000	1.000	783.4m	541.3 u	129.5 u	2.491	20.01
2.600	0.0000	1.000	778.6m	538.3 u	114.6 u	2.591	20.01
2.700	0.0000	3.000	773.5m	534.7 u	99.60 u	2.698	20.00
2.800	0.0000	6.000	768.5m	531.6 u	86.01 u	2.803	20.00
2.900	0.0000	1.000	763.4m	528.5 u	73.31 u	2.909	20.00
3.000	0.0000	1.000	758.9m	525.9 u	63.06 u	3.001	20.01
3.100	0.0000	3.000	753.7m	522.6 u	52.14 u	3.109	20.01
3.200	1.962 u	100.0	750.0m	519.5 u	44.99 u	3.186	19.99
3.300	917.5 u	100.0	750.0m	510.4 u	44.20 u	3.195	19.82
3.400	3.321 m	100.0	750.0m	502.3 u	43.51 u	3.202	19.67
3.500	6.976m	100.0	750.0m	495.1 u	42.89 u	3.210	19.53
3.,600	11.68m	100.0	750.0m	488.7 u	42.34 u	3.216	19.41
3.700	17.27m	100.0	750.0m	483.0 u	41.85 u	3.222	19.30
3.800	23.59m	100.0	750.0m	477.9 u	41.41 u	3.227	19.20
3.900	30.53m	100.0	750.0m	473.4 u	41.02 u	3.231	19.11
4.000	37.96m	100.0	750.0m	469.4 u	40.67 u	3.235	19.03

the table. The values of V_{GG} and W_1 have changed to stay consistent with the current needed for M_2.

What would happen if instead of sweeping V_{DS}, one wished to sweep W_2 (width of the diode-connected PMOS transistor)? See the problem set at the end of the chapter.

Figure 7.76 provides a view of the optimization cockpit during the solution. The optimization cockpit works with swept optimization. In fact, the cockpit allows the user to examine individually any of the swept optimization iterations. In some cases, having this information helps guide decisions about changes needed for a successful analysis.

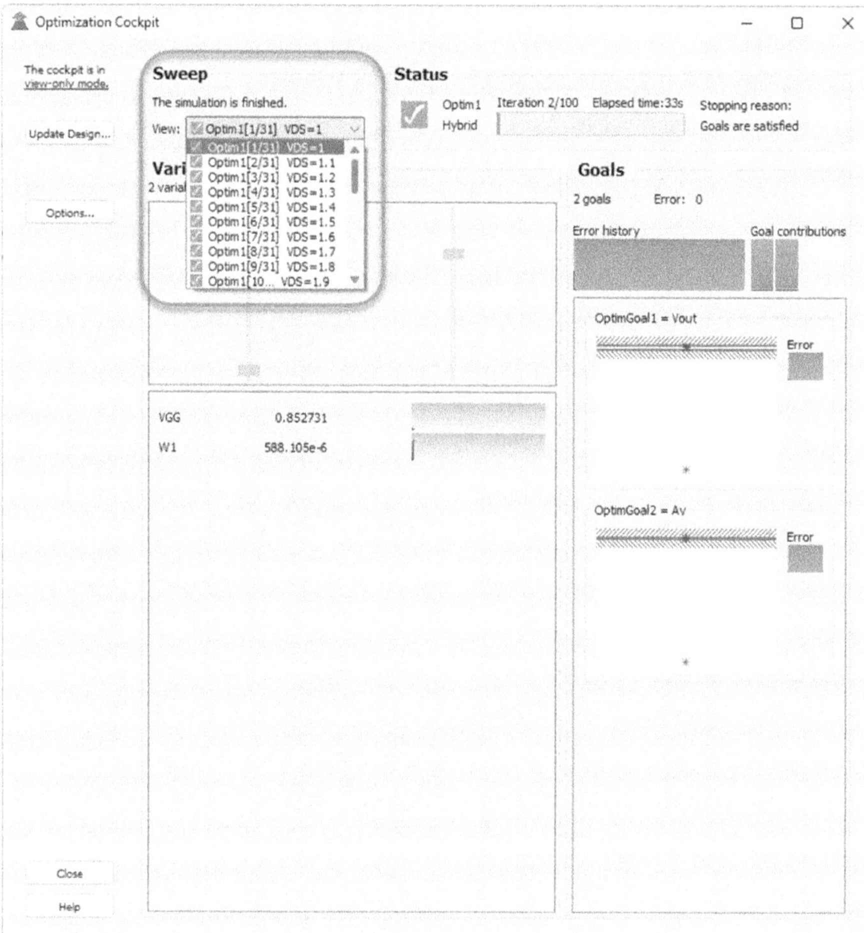

FIGURE 7.76 Swept optimization cockpit, with access to each sweep iteration.

Example 7.6

For the design of a short-channel (0.5 μm) amplifier, we recall Example 7.3 that featured the design of a CS amplifier with R_D load, in which the drain and source contact resistances played an important role in determining the voltage gain. The diode-connected load resistance acts as the R_D value in the CS amplifier with the R_D load case. In order to reduce the impact of the CS amplifier's contact resistance, the value of the diode-connected resistance (denoted as r_{diode}) should be made as large as possible.

For the small-geometry implementation, the understanding of the impact of the PMOS aspect ratio and bias voltage on the current and r_{diode} values, will assist us when the time comes to design the CS amplifier with a diode-connected load amplifier.

We start with a characterization sweep study of the diode-connected PMOS load's width, for a fixed V_{DS}, and study the r_{diode} and I_{DS} results. See schematics of Figure 7.77.

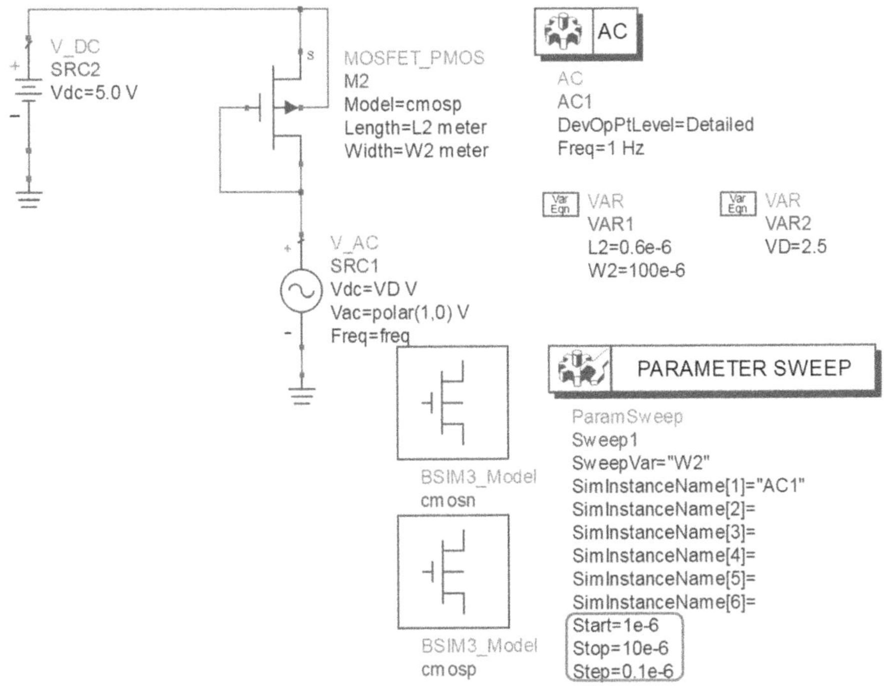

FIGURE 7.77 Preparation for a design of a short-channel CS amplifier with diode-lconnected load: A parameter sweep characterization study of the diode-connected PMOS load.

The V_AC source, with an amplitude of 1 V, in Figure 7.77, is intended for measurement of r_{diode}.

The W_2 sweep results are presented in Figure 7.78. Note the [0, 0] index of $I_{DS_}$diode$=I_D[0, 0]$; this current comes from the device operating point and depends on the device and the frequency. Since there is only one device and one frequency (0 Hz, DC), the indexes are [0, 0]. The results show that the r_{diode} impedance drops with an increase of the width W_2, while the current increases with W_2. These results are as we intuitively expected.

Since we wish to use a high value of r_{diode}, we somewhat arbitrarily choose $r_{diode}=5$ kΩ as the smallest value of interest.

Therefore, for $V_{DS}=2.5$ V, shown in Figure 7.77, the highest W_2 we use is 4 μm, as in Figure 7.78 (left), which results in a maximum I_{DS} value of $I_{DS}=193.8$ μA. If we design the CS amplifier's M1 transistor for a higher current, we no longer meet the 5 kΩ minimum load value.

Let us now study what happens for other values of V_{DS}. To perform this study, we added the parameter sweep of V_D, as found in Figure 7.79. Notice that the PMOS V_{DS} value is determined by VDS_PMOS=VD−VDD. Later on (see Figure 7.82), we will be sweeping V_{DS} of the NMOS amplifier, which is similar to the sweeping of V_D in Figure 7.77.

The double-sweep results of the diode-connected PMOS device have W_2 as the primary swept variable and V_D as the secondary swept parameter. The plots are presented in Figure 7.80.

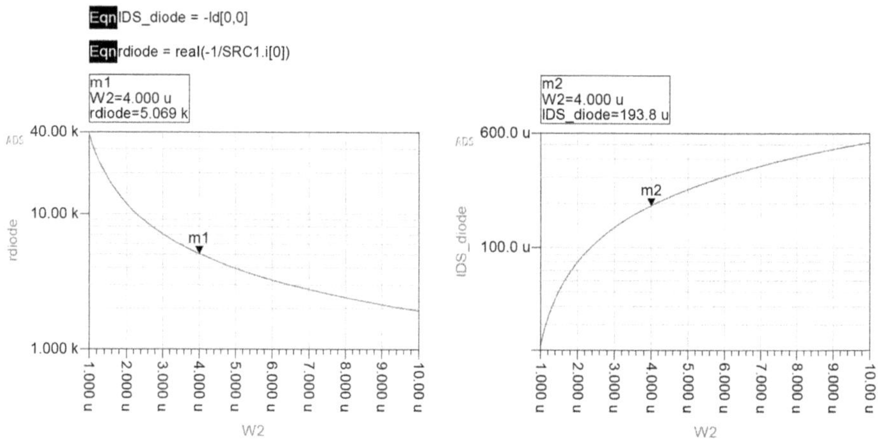

FIGURE 7.78 Short-channel I_{DS} and r_{diode} vs. W_2 sweeping results.

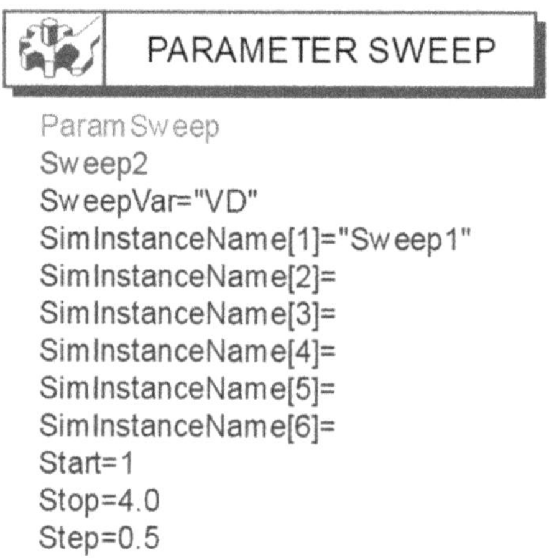

FIGURE 7.79 A V_D parameter sweep is added to the diode-connected PMOS sweep studies schematic.

Note that for $V_{DS}=4$ V, the PMOS $V_{GS2}=4-5=-1$ V. The current is very small, maximally around 3 μA.

The reason is that this V_{DS} value results in a PMOS V_{GS2} that is very close to V_{TH}, and the PMOS transistor is in moderate inversion mode. This is not good because to have a sizable CS amplifier voltage gain, M_1 will need to have higher current. We shall sweep V_D to make the PMOS stay in the strong inversion region.

From this point on, we will not use $V_{DS}=4$ V. We re-ran the diode-connected PMOS dual sweep, limiting V_D to 3.5 V. See Figure 7.81. Examining the results, keeping in mind that we want to keep $r_{diode} \geq 5$ kΩ: The highlighted area in the top plot indicates the values for r_{diode} that meet our requirement. This indicates that 1 μm$\leq W_2 \leq$7 μm. The highlighted area in the bottom plot shows the current values where our diode-connected device resistance requirement is met. Therefore, $I_{DS} \leq 0.5$ mA.

Note: The highlighted areas in Figure 7.81 are not a feature in ADS. The highlights were added outside of ADS while editing the figure.

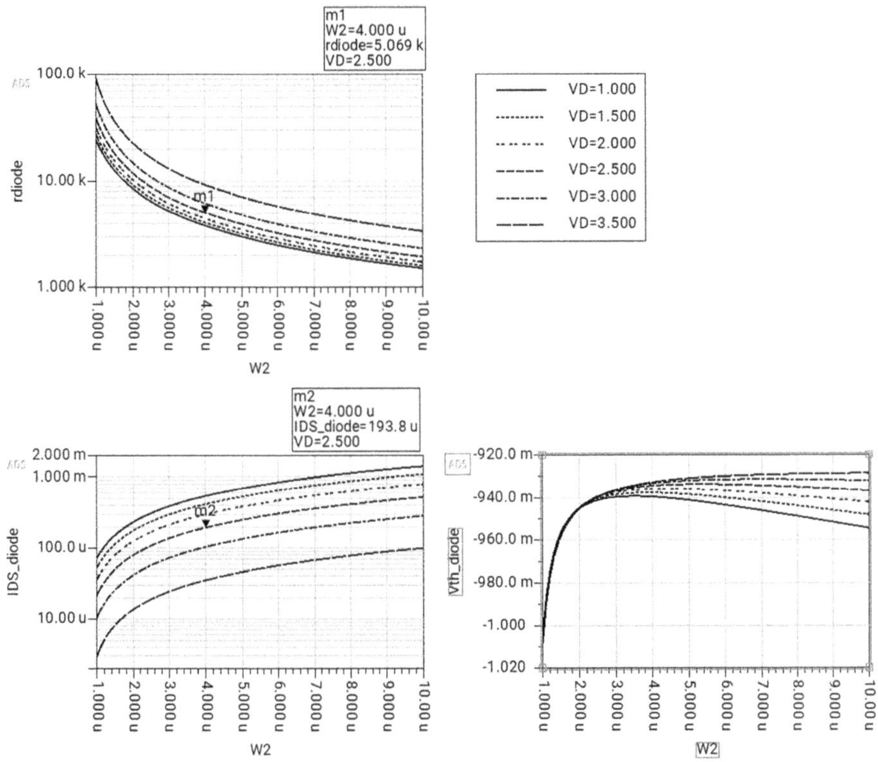

FIGURE 7.80 Sweeps of r_{diode}, I_{DS} and V_{TH} in terms of W_2 and V_D.

The optimization set up is very similar to the one done for the Level 1 CS amplifier with a diode-connected PMOS load. See Figure 7.82. Based on the diode-connected PMOS swept study (Figure 7.81), we keep $W_2 = 4\,\mu m$, close to the middle of the W_2 range.

The swept optimization results for the short-channel CS amplifier with diode-connected PMOS load are presented in Figure 7.83.

We see in Figure 7.83 that we missed the requirements. What happened?

Looking at the results for W_1opt, it becomes clear that the optimizer was trying to increase W_1 and "railed" at the largest possible value.

A "rookie" mistake was made when $r_{diode} \geq 5\ k\Omega$ was selected earlier.

The issue is related to the voltage division caused by the contact resistance (r_{con}) of the CS M_1 transistor, as we saw with the R_D load amplifier. Note (in Figures 7.84 and 7.85) that $r_{con} > 5\ k\Omega > r_{diode}$ when V_{DS} of M_1 is as small as 1 V. We need to change the limit set for r_{diode}. We require $r_{diode} \geq 10\ k\Omega$. Figure 7.84 shows the schematic for measuring r_{con} and zoomed-in r_{con} characterization plots for two values of V_{DS}. To clarify, Figure 7.77 is the setup to measure the PMOS resistance, whereas Figure 7.84 is for the NMOS.

With the new requirement of $r_{diode} \geq 10\ k\Omega$, the ranges of W_2 and V_D have reduced dramatically. See Figure 7.85. This indicates that we should use $1\,\mu m \leq W_2 \leq 2.7\,\mu m$, $I_{DS} \leq 350\,\mu A$. We will use $W_2 = 1\,\mu m$ to give us the best chance of success. We repeat the swept optimization. See results in Table 7.3.

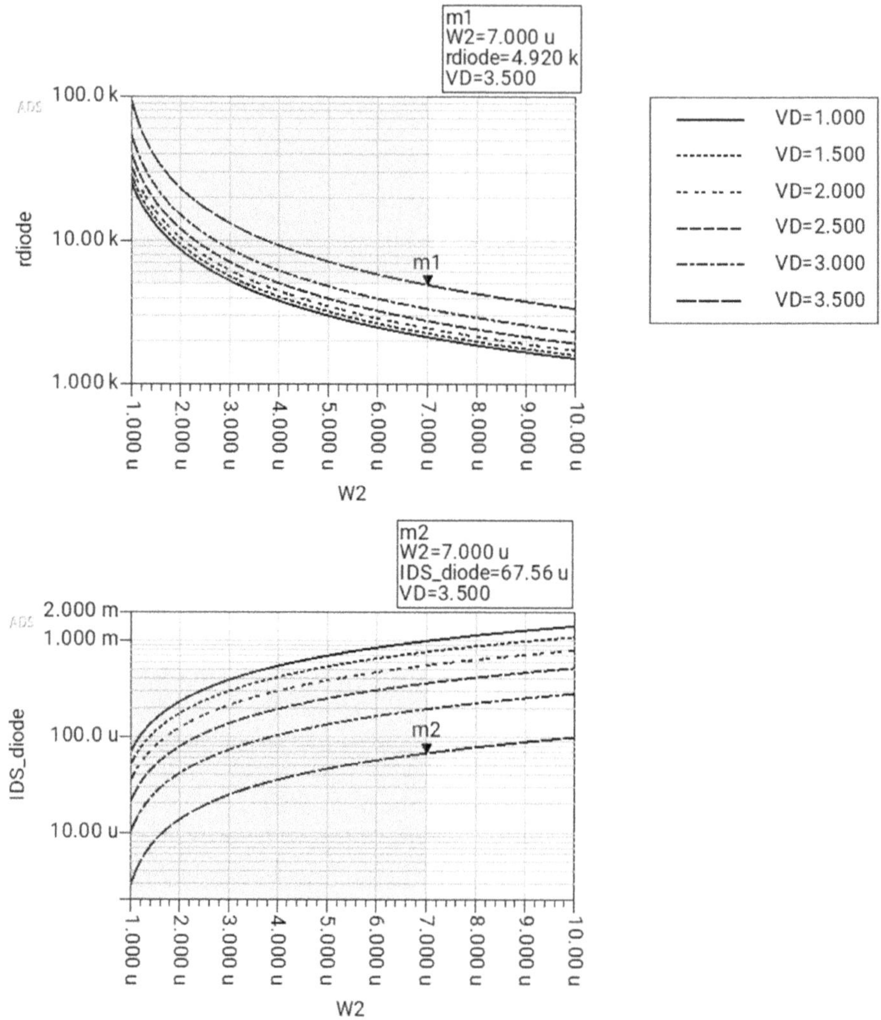

FIGURE 7.81 Sweep study of the diode-connected PMOS device, limiting V_{DS} to values no larger than 3.5 V.

The new results with $W_2 = 1\,\mu m$ show that for $1.0\,V \leq V_{DS} \leq 1.8\,V$, the CS amplifier with diode-connected load can meet the requirement of $A_v = -20$.

Notice that a gain of 20 is near the peak of the gain available for CS with a diode-connected load amplifier in this 0.5 μm process.

7.9 SIMULATION-BASED DESIGN OF A SMALL-GEOMETRY CS AMPLIFIER WITH CURRENT SOURCE LOAD

One of the configurations of a CS amplifier with a current source load (shown in Figure 7.86) involves an NMOS input transistor M_1 and a PMOS load transistor M_2. Any MOSFET that works in saturation mode functions as a current source. This is

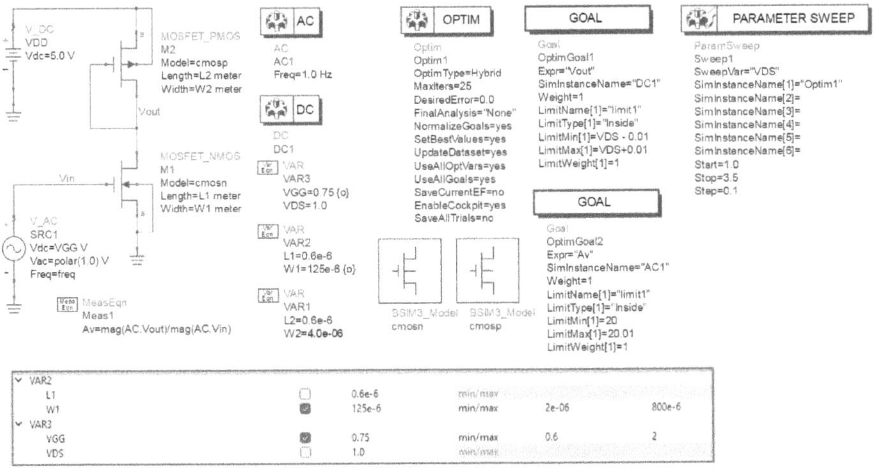

FIGURE 7.82 (Up) Schematic for the swept optimization study of a short-channel CS amplifier with diode-connected PMOS load. (Down) The design parameters W_1 and V_{GG}.

Eqn ErrFcn = FinalEF[0]

Eqn Iters = NumIters[0]

Eqn VGGopt = OPTSOLNVALS.VGG[0]

Eqn W1opt = OPTSOLNVALS.W1[0]

Eqn IDS = -DC.VDD.i[0,0]

Eqn VDSOpt=OptimGoal1[0,0]

Eqn AvOpt=OptimGoal2[0,0]

VDS	ErrFcn	Iters	VGGopt	W1opt	IDS	VDSOpt	AvOpt
1.000	45.93 m	25.00	713.4 m	800.0 u	539.2 u	1.015	15.71
1.100	45.48 m	25.00	707.6 m	800.0 u	515.6 u	1.106	15.73
1.200	45.78 m	25.00	702.6 m	800.0 u	495.3 u	1.185	15.72
1.300	46.83 m	25.00	696.7 m	800.0 u	471.6 u	1.278	15.67
1.400	48.53 m	25.00	690.8 m	800.0 u	448.6 u	1.370	15.60
1.500	50.83 m	25.00	685.0 m	800.0 u	426.2 u	1.460	15.51
1.600	53.70 m	25.00	679.3 m	800.0 u	404.7 u	1.549	15.39
1.700	57.13 m	25.00	673.7 m	800.0 u	384.0 u	1.635	15.26
1.800	61.11 m	25.00	668.2 m	800.0 u	364.2 u	1.718	15.12
1.900	65.66 m	25.00	662.9 m	800.0 u	345.4 u	1.798	14.97
2.000	70.79 m	25.00	657.7 m	800.0 u	327.6 u	1.875	14.80
2.100	76.50 m	25.00	652.7 m	800.0 u	311.0 u	1.948	14.63
2.200	82.79 m	25.00	648.0 m	800.0 u	295.5 u	2.017	14.46
2.300	89.66 m	25.00	643.5 m	800.0 u	281.2 u	2.082	14.29
2.400	97.10 m	25.00	639.2 m	800.0 u	268.0 u	2.142	14.12
2.500	105.1 m	25.00	635.3 m	800.0 u	256.0 u	2.198	13.95
2.600	113.6 m	25.00	631.6 m	800.0 u	245.0 u	2.249	13.79
2.700	122.6 m	25.00	628.1 m	800.0 u	235.1 u	2.296	13.63
2.800	132.0 m	25.00	625.0 m	800.0 u	226.2 u	2.339	13.48
2.900	141.9 m	25.00	622.1 m	800.0 u	218.2 u	2.378	13.34
3.000	152.1 m	25.00	619.4 m	800.0 u	211.0 u	2.414	13.21
3.100	162.5 m	25.00	617.0 m	800.0 u	204.6 u	2.445	13.09
3.200	173.3 m	25.00	614.8 m	800.0 u	198.9 u	2.474	12.98
3.300	184.2 m	25.00	612.8 m	800.0 u	193.9 u	2.500	12.87
3.400	195.4 m	25.00	611.0 m	800.0 u	189.4 u	2.522	12.78
3.500	206.6 m	25.00	609.4 m	800.0 u	185.4 u	2.543	12.69

FIGURE 7.83 Swept optimization results for the CS amplifier with diode-connected PMOS load.

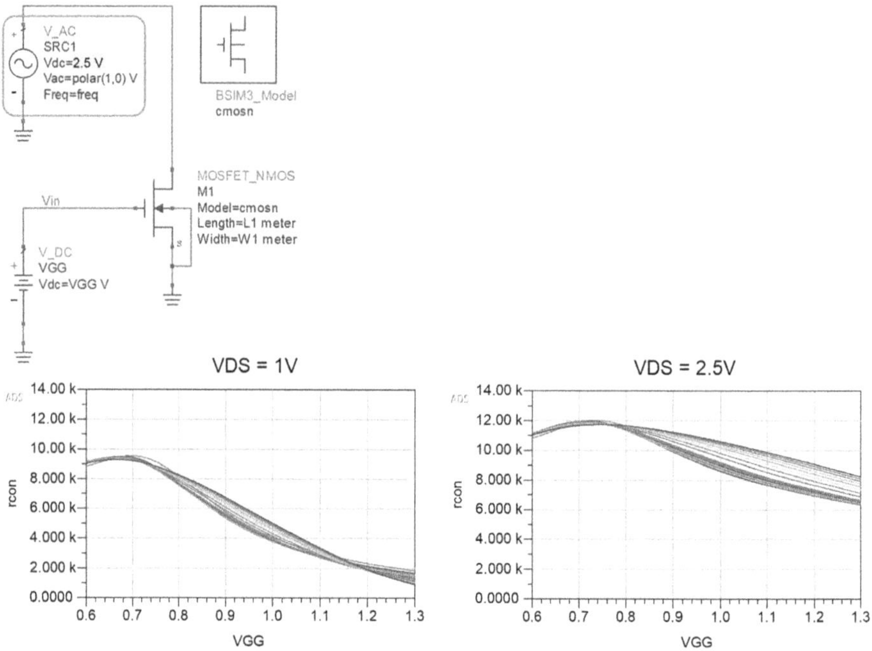

FIGURE 7.84 (Up) Schematic for measuring r_{con}. (Down left) for $V_{DS}=1$ V r_{con} is always smaller than 10 kΩ. (Down right) for $V_{DS}=2.5$ V there is a range of V_{GG} values for which r_{con} is larger than 10 kΩ.

done by selecting carefully the DC voltage V_{G2} at the gate of M_2. The drains of M_1 and M_2 are connected at a node that serves as V_{out} of the amplifier. The signal V_{out} consists of a DC voltage V_D and a small-signal v_{out}, which is the amplified v_{in}. V_D and V_{G1} are selected to keep M_1 in saturation mode and for the amplifier to have a good voltage swing.

As seen from the small-signal diagram (Figure 7.87)., the small-signal voltage gain of the amplifier is

$$A_v = v_{out}/v_{in} = -g_{m1}\cdot\left(r_{o1} \parallel r_{o2}\right) \tag{7.3a}$$

In Level 1 MOSFET modeling, $r_o = 1/\lambda I_D$. Substitution of the two r_o parameters into equation (7.3a), and recalling that in the Level 1 model $g_{m1} = 2I_D/(V_{G1} - V_{TH1})$, reveals a useful alternative formula for the small-signal voltage gain of the amplifier:

$$A_v = v_{out}/v_{in} = 2\Big/\left[\left(V_{G1} - V_{TH1}\right)\cdot\left(\lambda_1 + \lambda_2\right)\right] \tag{7.3b}$$

From equation (7.3b), it is seen that A_v is independent of I_D. The voltage gain depends only on the overdrive voltage of M_1 and on the channel length modulation parameters of the two transistors.

As seen from Figure 7.88, the I_D vs. V_{DS} curves of M_1 and the load curves of M_2 intersect at the Q points that are available for the amplifier. We see the crucial

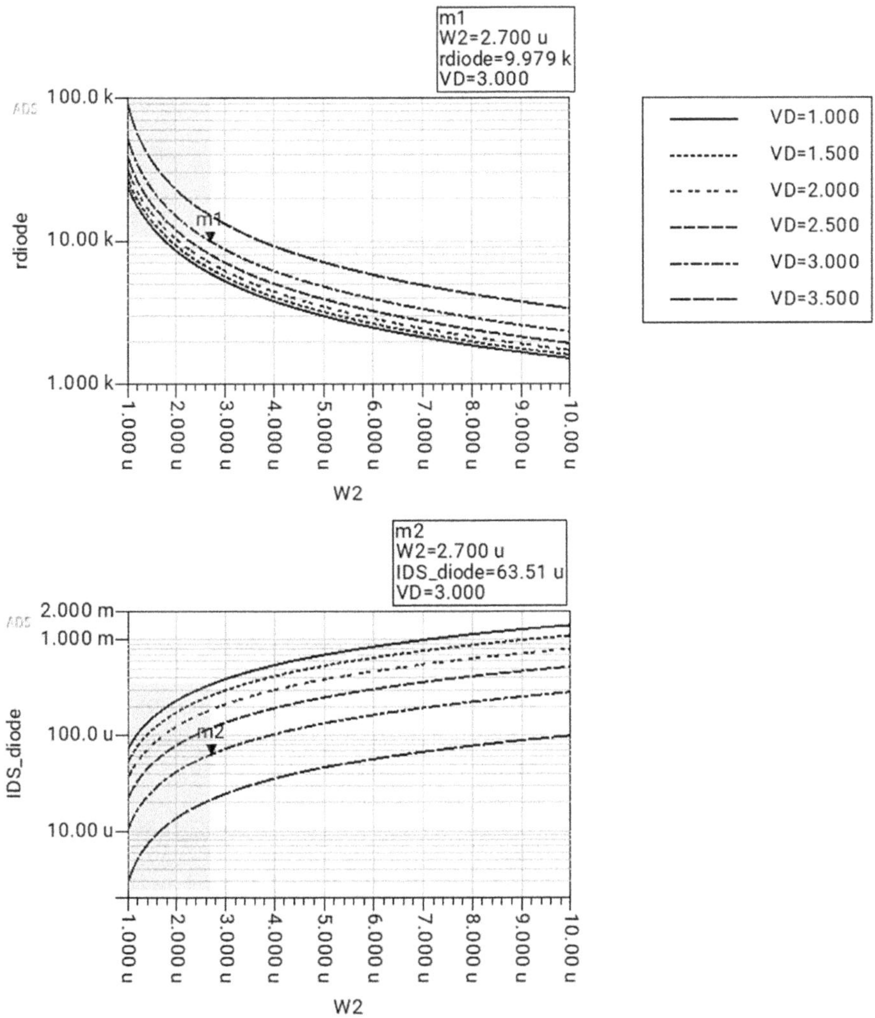

FIGURE 7.85 I_{DS} and r_{diode} characterization results.

role played by the channel length modulation effect in determining the DC conditions of the amplifier. Without the nonzero saturation mode slopes of these lines, the Q points become undefined. For instance, V_D becomes undefined for zero λ parameters.

We recall CS amplifiers with diode-connected loads and how only small enough voltage gains can be implemented. In Level 1 modeling, the two types of amplifiers, those with diode-connected loads and those with current source loads, complement one another. One is good for implementing low gains, whereas the other is good for large gains. An observation of equation (7.3b) reveals that small λ values enable relatively large voltage gains (around 50–100). It is obtained by selection of a reasonably small M_1 overdrive voltage.

TABLE 7.3
Swept Optimization Results – Second Attempt

V_{OS}	ErrFcn	Iters	V_{GGopt}	W_{1opt}	I_{DS}	V_{DSOpt}	A_{vOpt}
1.000	0.0000	25.00	715.1 m	49.17 u	72.52 u	994.3 m	20.01
1.100	0.0000	5.000	709.8 m	48.88 u	67.87 u	1.109	20.00
1.200	0.0000	1.000	703.8 m	49.85 u	64.12 u	1.203	20.00
1.300	0.0000	7.000	697.1 m	51.67 u	60.75 u	1.290	20.00
1.400	0.0000	14.00	687.4 m	55.38 u	56.78 u	1.395	20.00
1.500	0.0000	10.00	675.4 m	61.76 u	52.89 u	1.501	20.00
1.600	0.0000	18.00	661.4 m	72.10 u	49.33 u	1.600	20.00
1.700	0.0000	8.000	642.4 m	93.09 u	45.83 u	1.700	20.01
1.800	10.05 n	25.00	606.8 m	16 8.6 u	42.11 u	1.810	20.01
1.900	334.7 u	25.00	600.0 m	182.2 u	40.09 u	1.871	19.69
2.000	2.121 m	25.00	600.0 m	16 9.7 u	37.94 u	1.938	19.24
2.100	5.388 m	25.00	600.0 m	158.9 u	36.03 u	1.998	18.82
2.200	10.04 m	25.00	600.0 m	149.6 u	34.33 u	2.053	18.43
2.300	15.96 m	25.00	600.0 m	141.5 u	32.83 u	2.102	18.07
2.400	23.03 m	25.00	600.0 m	134.6 u	31.50 u	2.147	17.74
2.500	31.10 m	25.00	600.0 m	128.5 u	30.32 u	2.187	17.44
2.600	40.04 m	25.00	600.0 m	123.3 u	29.28 u	2.222	17.17
2.700	49.73 m	25.00	600.0 m	118.7 u	28.37 u	2.254	16.92
2.800	60.05 m	25.00	600.0 m	114.7 u	27.56 u	2.283	16.70
2.900	70.89 m	25.00	600.0 m	111.3 u	26.85 u	2.308	16.50
3.000	82.15 m	25.00	600.0 m	108.3 u	26.22 u	2.331	16.32
3.100	93.73 m	25.00	600.0 m	105.7 u	25.68 u	2.351	16.16
3.200	105.6 m	25.00	600.0 m	103.4 u	25.19 u	2.368	16.02
3.300	117.6 m	25.00	600.0 m	101.4 u	24.78 u	2.384	15.89
3.400	129.7 m	25.00	600.0 m	99.64 u	24.40 u	2.398	15.78
3.500	141.9 m	25.00	600.0 m	98.14 u	24.07 u	2.410	15.68

Example 7.7

Say that the channel length modulation coefficients are $\lambda_1 = 0.01$ [1/V], $\lambda_2 = 0.02$ [1/V], the threshold voltages are $V_{TH1} = 0.7$ V, $V_{TH2} = -0.8$ V, the mobility coefficients are $\mu_n = 350$ [cm^2·s/V], $\mu_p = 100$ [cm^2·s/V], and the DC voltage supply is $V_{DD} = 5$ V.

Then by equation (7.3b), a desired $A_v = -80$ is obtained by taking $V_{G1} - V_{TH1} = 0.833$ V. A wide range of V_D values keeps M_1 and M_2 in saturation mode. For instance, $V_D = 2.5$ V, which is good for large "swing" operation, may be a good choice (as will be seen soon).

On the other hand, if we wish $A_v = -15$, then $V_{G1} - V_{TH1} = 4.444$ V, an impossible value for which M_2 is in cutoff.

The ability of CS amplifiers with current source load to create large voltage gains is true only in long-channel MOSFET transistors, for which Level 1 models are valid. We recall, from Section 7.1, the rather low values of the r_o parameters obtained by analyzing amplifiers with 0.5 µm technology with BSIM3 models.

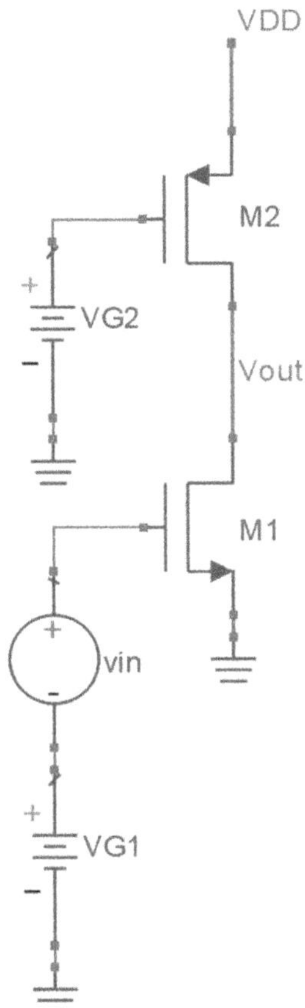

FIGURE 7.86 NMOS CS amplifier with a PMOS current source load.

A typical Level 1 design of a CS NMOS amplifier with a PMOS current source load, in which some specific (and possible) voltage gain A_v is desired, has the following design parameters: $(W/L)_1$, $(W/L)_2$, V_{G1}, V_{G2}, and V_D.

Step 1 of the design uses equation (7.3b) to determine V_{G1} for the required A_v value, as we have seen in **Example 7.7**.

Step 2 of the design involves the selection of V_{G2}, based on the selected value of V_{G1}. In general, and despite the same I_D that flows through M_1 and M_2, the transconductances g_{m1} and g_{m2} need not be the same. It is convenient to require $g_{m1} = g_{m2}$, and by that a specific value of V_{G1} dictates a specific value of V_{G2}: $g_{m1} = 2I_D/(V_{G1} - V_{TH1}) = g_{m2} = 2I_D/(V_{DD} - V_{G2} - |V_{TH2}|) \rightarrow \mathbf{V_{G1} - V_{TH1} = V_{DD} - V_{G2} - |V_{TH2}|}$.

Let us refer again to **Example 7.7**. We have seen that for a voltage gain of $A_v = -80$, we need $V_{G1} = 1.533\,V$. With the above transconductance equality strategy, we obtain $V_{G2} = 3.367\,V$.

In *Step 3* of the design, we select the desired DC level of V_D. We first set up the cutoff and saturation mode inequalities in order to determine the available range for V_D.

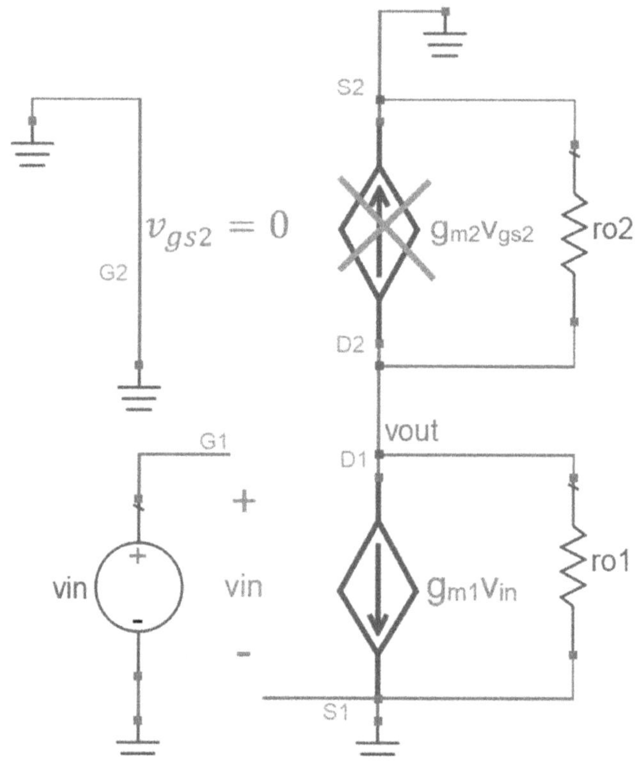

FIGURE 7.87 Small-signal diagram of a CS amplifier with current source load.

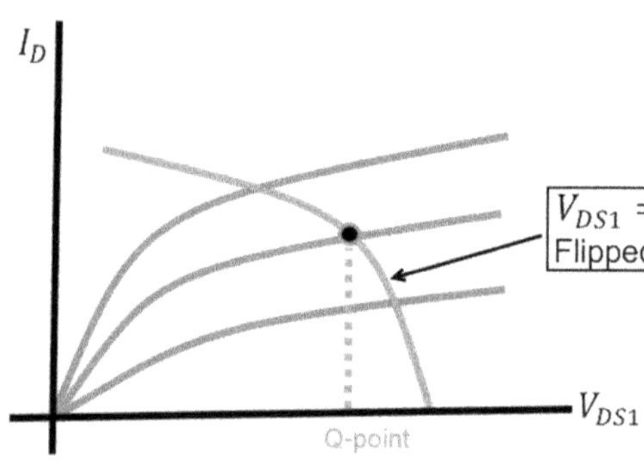

FIGURE 7.88 Curves of M_1 and a curve of M_2. Demonstration of the role played by the channel length modulation in determining the performance of the amplifier.

Continuing **Example 7.7**: For the PMOS to be ON, $\mathbf{V_{G2} - V_{DD} < V_{TH2}}$. That is $3.367 - 5 = -1.633\text{ V} < -0.8\text{V}$, which validates the correctness of the selection of V_{G2}. For the PMOS to be in saturation mode, $\mathbf{V_D - V_{DD} < V_{G2} - V_{DD} - V_{TH2}}$. That is, the upper bound of V_D is $V_D < 3.367 + 0.8 = 4.167\text{V}$. For the NMOS to be in saturation mode $\mathbf{V_D - V_S > V_{G1} - V_S - V_{TH1}}$. That is, the lower bound of V_D

is $V_D > 1.533 - 0.7 = 0.833$ V. Finally, for M_1 to be ON, $\mathbf{V_{G1} - V_S > V_{TH1}}$. Indeed, $1.533 > 0.7$. So, the available range for V_D is $0.833\text{ V} < V_D < 4.167$ V. For maximum swing, we may choose the midpoint $V_D = 2.5$ V.

Finally, in *Step 4* of the design, we determine the aspect ratios $(W/L)_1$ and $(W/L)_2$ needed to create an I_D current equality between M_1 and M_2.

Let us conclude **Example 7.7.** We combine current equality: $I_{D1} = 0.5\mu_n C_{ox} \cdot (W/L)_1 (V_{G1} - V_{TH1})^2 (1 + \lambda_1 V_D) = I_{D2} = 0.5\mu_p C_{ox} \cdot (W/L)_2 (V_{DD} - V_{G2} - |V_{TH2}|)^2 (1 + \lambda_2 (V_{DD} - V_D))$ with the transconductance equality $V_{G1} - V_{TH1} = V_{DD} - V_{G2} - |V_{TH}|$ to obtain $\mathbf{\mu_n (W/L)_1 (1 + \lambda_1 V_D) = \mu_p (W/L)_2 (1 + \lambda_2 (V_{DD} - V_D))}$. In conclusion, **Example 7.7**: Say that $L_1 = L_2 = 2$ µm. Let $W_1 = 2$ µm as well. The remaining unknown is W_2: $350 \cdot 1 \cdot (1 + 0.01 \cdot 2.5) = 100 \cdot (W_2/2) \cdot (1 + 0.02 \cdot 2.5)$, resulting in $W_2 = 6.833$ µm.

Unlike CS amplifiers with R_D load or diode-connected load, which often require unacceptably large MOSFET width values, we see that a CS amplifier with a current source load remains small. Its small dimensions also provide for a large bandwidth.

7.10 ADS: GAIN SWEPT OPTIMIZATION IN LEVEL 1 AND BSIM3 DESIGNS

The Level 1 model version of the CS amplifier with a current source load is studied first to get some general understanding of the design flow and function of this amplifier.

In this section, the characterization of the current source load is performed for the small-geometry technology using the 0.5 µm BSIM3 model. Specifically, we shall study the amplifier with different loads within the operating range.

Figure 7.89 presents the CS amplifier with a current source load in a configuration that has an input NMOS and a PMOS current source load.

Example 7.8

We shall start the simulations-based design with an example of a Level 1 design of the CS amplifier NMOS transistor M_1 and a PMOS current source load transistor M_2. It is similar to Example 7.7, except that the desired voltage gain is chosen to be relatively small, which makes the design more challenging. The design has the following specifications:

$A_v = -20$, $V_{DD} = 5$ V, $V_{DS} = 2.5$ V, $I_{DS} = 1$ mA, $L_{eff} = 2$ µm (of both transistors).

We deliberately chose a relatively small gain (−20) for such an amplifier to make it possible to compare it to our previously designed CS amplifier with a diode-connected load.

The design variables are the gate widths W_1 and W_2 of M_1 and M_2, respectively, and the gate DC voltages V_{GG1} and V_{GG2} of the two transistors, respectively.

Refer to formula (7.3a). Let's pretend we are unaware of equation (7.3b). Indeed, in simulation-based design, we tend to rely on the universal voltage gain formula.

$$A_v = -g_{m1}(r_{o1} \parallel r_{o2})$$

Calculate r_{o1} and r_{o2}

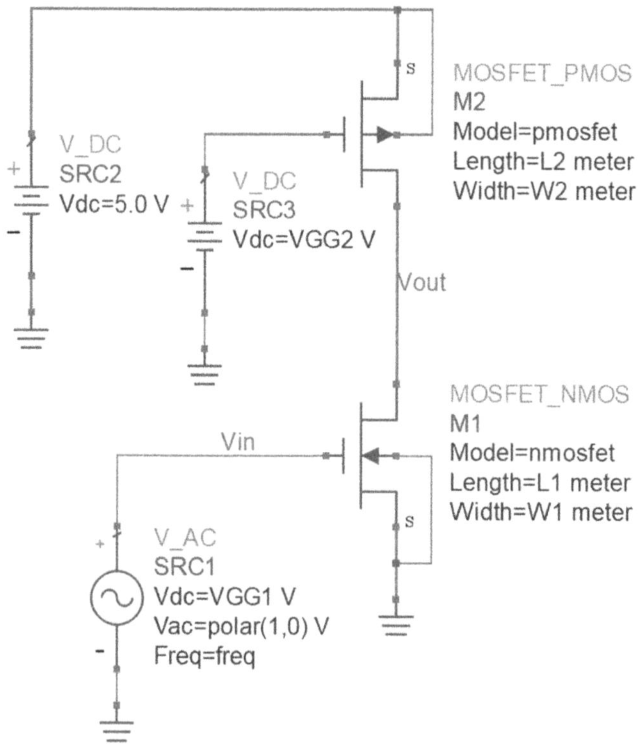

FIGURE 7.89 CS amplifier with NMOS input and PMOS current source load.

$$r_{o1} = \frac{1}{\lambda_n I_{DS}} = \frac{1}{(0.01)\left(1\times 10^{-3}\right)} = 100\text{ k}\Omega$$

$$r_{o2} = \frac{1}{\lambda_p I_{DS}} = \frac{1}{(0.02)\left(1\times 10^{-3}\right)} = 50\text{ k}\Omega.$$

Now, calculate the desired g_{m1}

$$g_{m1} = \frac{-A_v}{r_{o1}\|r_{o2}} = \frac{-(-20)\left(100\times 10^3 + 50\times 10^3\right)}{\left(100\times 10^3\right)\left(50\times 10^3\right)} = 0.6\text{ mA/V}.$$

For a MOSFET, we know that

$$g_{m1} = \sqrt{\frac{2\mu C_{ox}\dfrac{W}{L}I_{DS}}{1+\lambda V_{DS}}}.$$

Rearranging for W/L

$$\left(\frac{W}{L}\right)_1 = \frac{g_{m1}^2\left(1+\lambda V_{DS}\right)}{2\mu C_{ox} I_{DS}} = \frac{\left(0.6\times 10^{-3}\right)^2\left(1+(0.01)(2.5)\right)}{2(350)\left(\dfrac{3.45\times 10^{-13}}{9\times 10^{-7}}\right)\left(1\times 10^{-3}\right)}$$

$$\left(\frac{W}{L}\right)_1 = 1.4.$$

Let $L_{eff} = 2\,\mu m$, then $W_1 = 2.8\,\mu m$.

Calculating V_{GS1}

$$\left(V_{GS} - V_{TH}\right)^2 = \frac{2I_{DS}}{\mu C_{ox}\left(\dfrac{W}{L}\right)\left(1+\lambda V_{DS}\right)}$$

$$= \frac{2\left(1\times 10^{-3}\right)}{(350)\left(\dfrac{3.45\times 10^{-13}}{9\times 10^{-7}}\right)(1.5)\left(1+(0.01)(2.5)\right)}$$

$$\left(V_{GS} - V_{TH}\right)^2 = 9.7$$

$$\left(V_{GS} - V_{TH}\right) = 3.11$$

$$V_{GS1} = 3.11 + 0.7 = 3.95\ \text{V}$$

The V_{GS1} value is already problematic. It means that the saturation voltage for M_1 is $V_{DSat1} = V_{GS1} - V_{TH} = 3.95 - 0.7 = 3.25\,\text{V}$.

Note that the desired V_{DS} value is $V_{DS1} < V_{DSsat1}$; therefore, our original specification of $V_{DS} = 2.5\,\text{V}$ needs to be revised (increased) such that $V_{DS1} > V_{DSsat1}$. For voltage swing considerations, let $V_{DS} = 3.5\,\text{V}$.

Now that $V_{DS1} = 3.5\,\text{V}$, we could redo the calculations, but since V_{DS} contributes so little to g_m, it is not worth the effort.

For the PMOS, let $V_{GG2} = 3.0\,\text{V}$. Note that $V_{DS2} = V_{DS1} - V_{DD} = 3.5 - 5 = -1.5\,\text{V}$. $V_{DSsat2} = V_{GS2} - V_{TH} = 3 - 5 - (-0.8) = -1.2\,\text{V}$. Since $V_{DSsat2} > V_{DS2}$, we have voltage headroom for M_2 as well.

The verification of V_{DSsat} for both transistors, such that enough voltage headroom is allowed for voltage swing, is critical to the proper functioning of the CS with the current source load amplifier.

We calculate W/L for the PMOS M_2 transistor

$$\left(\frac{W}{L}\right)_2 = \frac{2I_{DS}}{\mu C_{ox}\left(V_{GS} - V_{TH}\right)^2\left(1+\lambda V_{DS}\right)}$$

$$= \frac{2\left(1\times 10^{-3}\right)}{(100)\left(\dfrac{3.45\times 10^{-13}}{9\times 10^{-7}}\right)(3 - 5 + 0.8)^2\left(1+(0.02)(-2.5)\right)}.$$

$$\left(\frac{W}{L}\right)_2 = 34.5$$

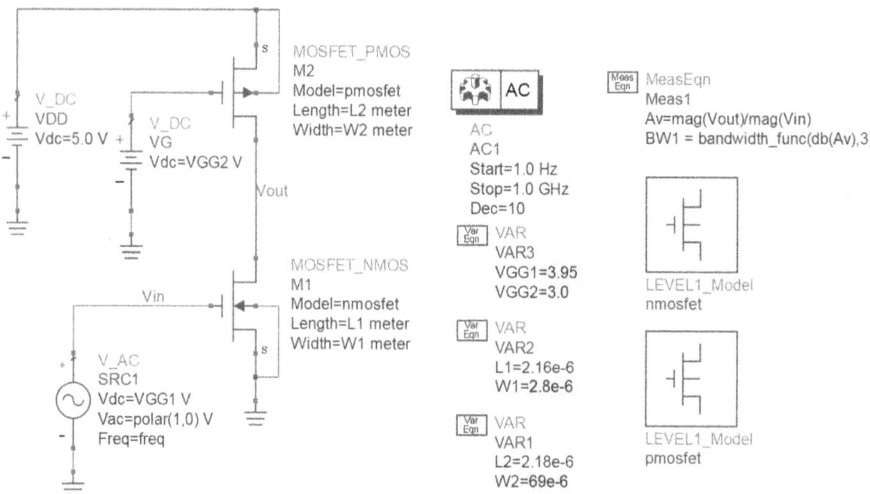

FIGURE 7.90 Design validation of Level 1 CS amplifier with current source load.

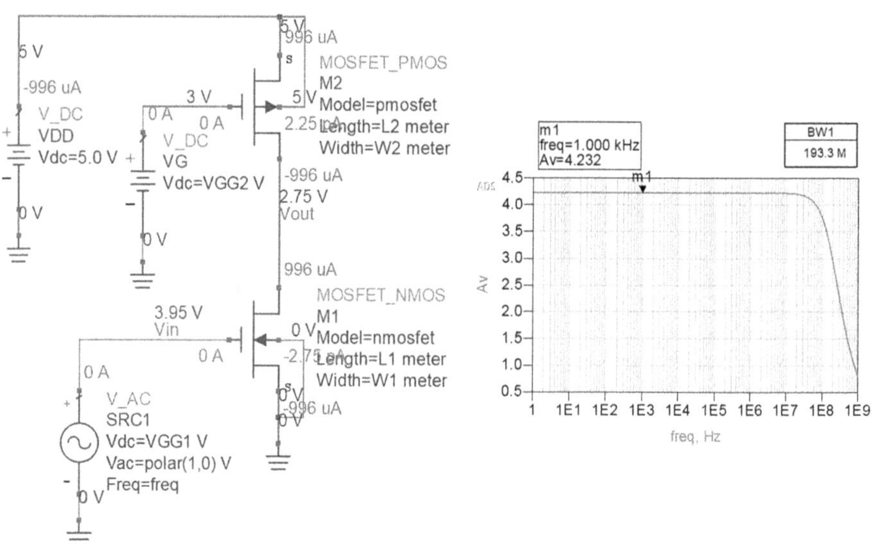

FIGURE 7.91 First round of validation results of the Level 1 CS amplifier with current source load manual design.

Let $L_{eff} = 2\,\mu m$; then $W_2 = 69.0\,\mu m$.

Let us validate the design in AC analysis (Figure 7.90). Note the actual L values, considering the L_D parameter.

The results shown in Figure 7.91 are disappointing. Both the gain and bias are not close to our calculations. We revisited the assumption that a change in V_{DS} from 2.5 to 3.5 V has a small impact by re-doing calculations with the new $V_{DS1} = 3.5$ V.

After recalculation, with the new V_{DS} value, we have the following small changes to the design variables (Table 7.1):

TABLE 7.4
Manual Design Updated Following the Selection of a New V_{DS} Voltage

V_{DS} (V)	2.5	3.5
V_{GS1}(V)	3.95	3.92
V_{GS2} (V)	3.0	3.0
W_1 (µm)	2.8	2.8
W_2 (µm)	69.0	70.3

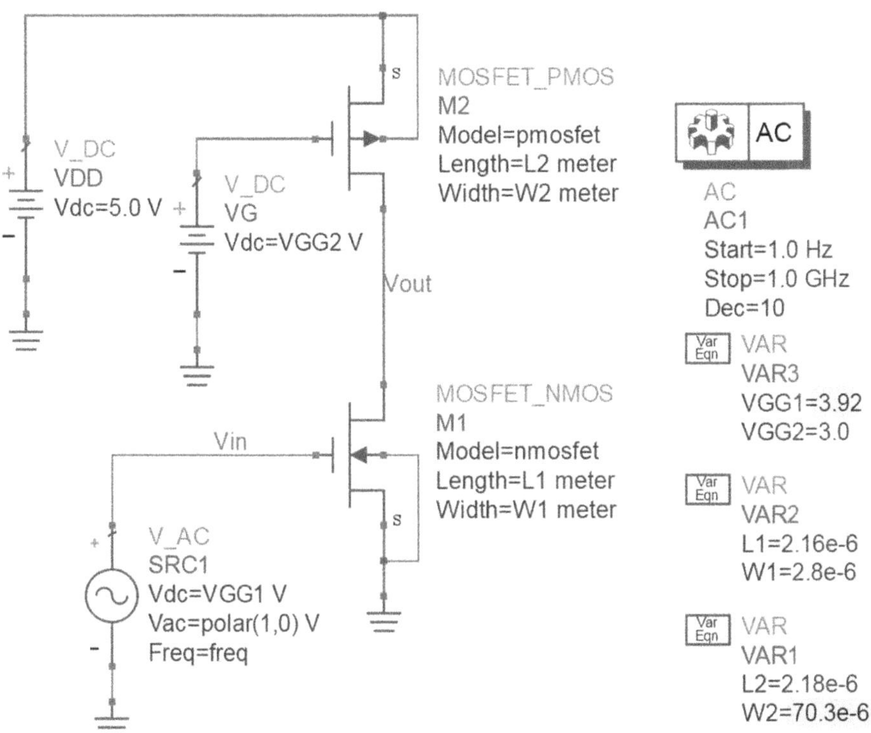

FIGURE 7.92 Round 2 of the design validation. Note the very tiny changes.

The schematic for Validation Round 2 is presented in Figure 7.92. Note the tiny changes in V_{GG1} and in W_2, and no changes in V_{GG2} and W_1. Could the design be so sensitive to small changes in the design parameters? The validation results in Figure 7.93 show that they are indeed so.

We see an improvement in the gain. However, $V_{DS1} = 3.21$ V, which is too close to $V_{DSsat1} = 3.22$ V. There is no voltage headroom for the output voltage to swing without driving M_1 into triode mode. *This is still a deficient design.*

Is what we experienced with this design indicating some flaw in our design decisions? The answer is yes. Moving away from the centered values of $V_{DS} = V_{DD}/2$

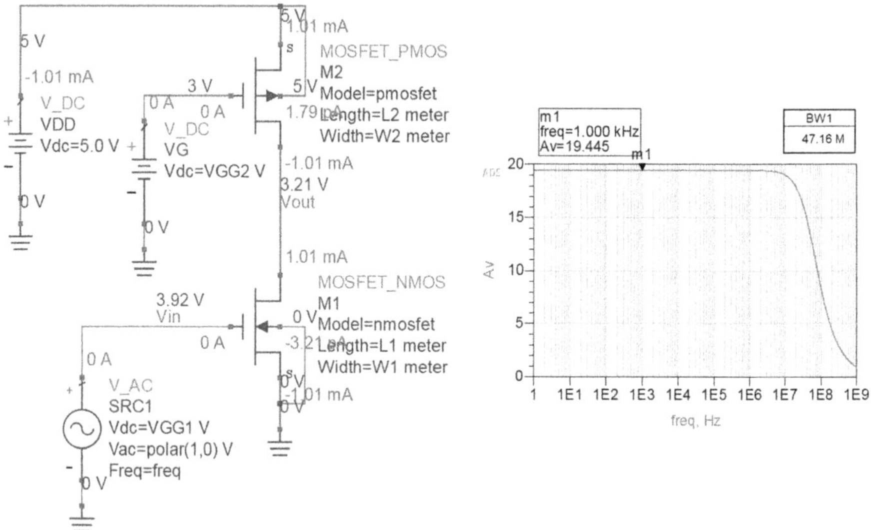

FIGURE 7.93 A seemingly successful validation results for the design of CS amplifier with current source load. Are we indeed done?

TABLE 7.5
Table to Show How the Design Parameters Change for Different Desired Voltage Gains

A_v	20	30	40	50
V_{GS1} (V)	3.95	2.87	2.33	2.0
V_{DSat1} (V)	3.25	2.17	1.63	1.3
V_{GS2} (V)	3.0	3.0	3.0	3.0
W_1 (μm)	2.8	6.2	11.0	17.2
W_2 (μm)	69.0	69.0	69.0	69.0

is where the design ran into issues. This was done in an attempt to satisfy the value of transconductance 0.6 mA/V required for the amplifier to meet the gain requirements at the desired I_{DS}. Let's say that, in order to lessen the voltage headroom issue, we are willing to double the I_{DS} DC current to 2 mA. Will this help? The simple answer is no. We have seen in Section 7.7 (Formula 7.3b) that the voltage gain A_v of the CS amplifier with current source load is independent of I_{DS}. It depends only on the overdrive voltage of M_1.

Table 7.5 shows a comparison of different values of A_v and the desired variables needed to meet the gain.

This data points to the fact that, for a successful CS amplifier with a current source load, in this 2 μm process, the lowest practical gain is about 30 for a reasonable output voltage swing headroom.

Example 7.9 (A Follow-up to Example 7.8)

Let us perform a study maintaining an $I_{DS}=1$ mA while sweeping the gain requirement. For this, we will use a **swept optimization of the variable gain** as shown in the schematic in Figure 7.94. We keep the same limit for the optimization design variables V_{GG1} and W_1 as used in the earlier example of the CS amplifier with diode-connected load. Notice that we are using OptimType = Simulated Annealing, as in this problem it resulted in the most efficient optimizer type.

As seen in Figure 7.95, the value of V_{GG} drops as we increase the gain, as we expected from equation (7.3b).

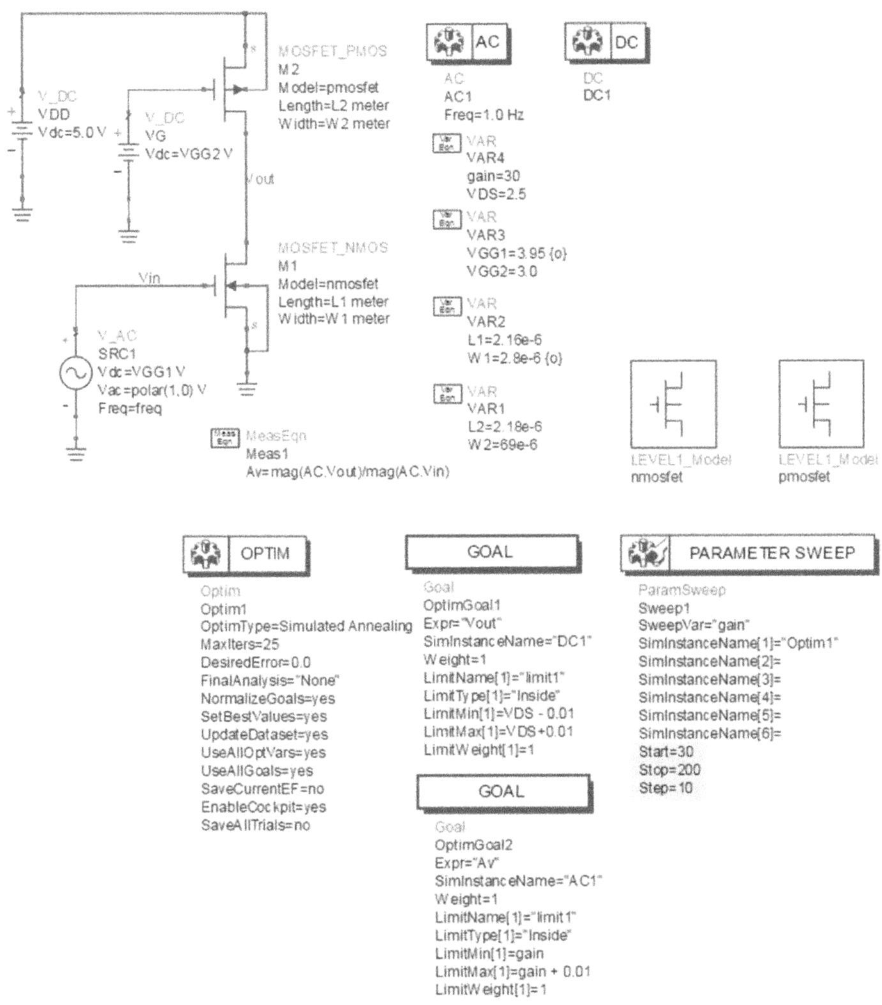

FIGURE 7.94 Schematic of a swept optimization design of a CS amplifier with current source load.

Eqn ErrFcn = FinalEF[0]

Eqn Iters = NumIters[0]

Eqn VGGopt = OPTSOLNVALS.VGG1[0]

Eqn W1opt = OPTSOLNVALS.W1[0]

Eqn IDS = -DC.VDD.i[0,0]

Eqn VDSOpt=OptimGoal1[0,0]

Eqn AvOpt=OptimGoal2[0,0]

gain	ErrFcn	Iters	VGGopt	W1opt	IDS	VDSOpt	AvOpt
30.00	0.0000	13.00	3.014	5.430 u	1.001 m	2.509	30.01
40.00	0.0000	5.000	2.436	9.652 u	1.001 m	2.495	40.00
50.00	0.0000	4.000	2.088	15.09 u	1.001 m	2.499	50.01
60.00	0.0000	4.000	1.857	21.72 u	1.001 m	2.498	60.01
70.00	0.0000	4.000	1.692	29.56 u	1.001 m	2.495	70.01
80.00	0.0000	5.000	1.568	38.61 u	1.001 m	2.493	80.01
90.00	0.0000	4.000	1.472	48.86 u	1.001 m	2.500	90.01
100.0	0.0000	5.000	1.394	60.32 u	1.001 m	2.494	100.0
110.0	0.0000	5.000	1.331	72.98 u	1.001 m	2.508	110.0
120.0	0.0000	4.000	1.279	86.86 u	1.001 m	2.508	120.0
130.0	0.0000	4.000	1.234	101.9 u	1.001 m	2.495	130.0
140.0	0.0000	4.000	1.196	118.2 u	1.001 m	2.493	140.0
150.0	0.0000	4.000	1.163	135.7 u	1.001 m	2.496	150.0
160.0	0.0000	4.000	1.134	154.4 u	1.001 m	2.507	160.0
170.0	0.0000	4.000	1.108	174.3 u	1.001 m	2.497	170.0
180.0	0.0000	4.000	1.086	195.4 u	1.001 m	2.501	180.0
190.0	0.0000	2.000	1.065	217.7 u	1.001 m	2.508	190.0
200.0	0.0000	3.000	1.047	241.3 u	1.001 m	2.506	200.0

FIGURE 7.95 Results of the CS amplifier design done by means of gain swept optimization.

Notice that the rate of change in V_{GG} slows down as the gain requirement increases.

We conclude that the CS amplifier with current source design involves setting the current with the PMOS load device, selecting the proper V_{GS} for the NMOS device, and sizing appropriately for the current. Of course, this works as long as the I_{DS} value results in reasonably high values of r_o for each device.

One more Level 1 study: What would happen if we reduced the required current to $I_{DS}=100\ \mu A$?

The straightforward way to do this is to reduce the $(W/L)_2$ by a factor of 10. We set $W_2=6.8\ \mu m$ and rerun the swept optimization. The results are shown in Table 7.6.

Table 7.6 shows that the optimizer "struggled" for gains between 30 and 50, as it could not get the error exactly to zero. Notice how W_1 is very close to the lower limits, and the gain is somewhat close to the intended value. Manual calculations using the design equations in Section 7.3 resulted in $W_1<2\ \mu m$ and explained why the optimizer struggled.

Example 7.10

Figure 7.96 shows the schematic for a small-geometry (0.5 μm) characterization of the PMOS current source load, using the BSIM3 model.

The first step is to study the current source PMOS in a similar way as it is done for the diode-connected load.

TABLE 7.6
Results of Gain Swept Optimization for a Redesign in Which the Current Is Lowered by a Factor of 10

Gain	ErrFcn	Lters	V_{GGopt}	W_{1opt}	I_{DS}	V_{OSOpt}	A_{vOpt}
30.00	276.3 m	25.00	1.761	2.486 u	97.51 u	3.824	30.29
40.00	271.6 m	25.00	1.761	2.486 u	97.56 u	3.813	40.35
50.00	22.91 m	25.00	1.906	2.000 u	100.1 u	2.481	57.58
60.00	0.0000	3.000	1.857	2.172 u	100.1 u	2.498	60.01
70.00	0.0000	5.000	1.692	2.956 u	100.1 u	2.505	70.01
80.00	0.0000	4.000	1.568	3.861 u	100.1 u	2.506	80.01
90.00	0.0000	5.000	1.471	4.886 u	100.1 u	2.496	90.00
100.0	0.0000	4.000	1.394	6.032 u	100.1 u	2.494	100.0
110.0	0.0000	4.000	1.331	7.298 u	100.1 u	2.505	110.0
120.0	0.0000	6.000	1.279	8.687 u	100.1 u	2.499	120.0
130.0	0.0000	5.000	1.234	10.19 u	100.1 u	2.503	130.0
140.0	0.0000	5.000	1.196	11.82 u	100.1 u	2.509	140.0
150.0	0.0000	5.000	1.163	13.57 u	100.1 u	2.506	150.0
160.0	0.0000	3.000	1.134	15.44 u	100.1 u	2.497	160.0
170.0	0.0000	4.000	1.108	17.43 u	100.1 u	2.502	170.0
180.0	0.0000	4.000	1.086	19.55 u	100.1 u	2.494	180.0
190.0	0.0000	4.000	1.065	21.78 u	100.1 u	2.505	190.0
200.0	0.0000	5.000	1.047	24.13 u	100.1 u	2.507	200.0

Based on a $V_{DD}=5\,V$, the V_{DS} for the transistor is set to 2.5 V.

For the characterization of the current source load, there are two design variables, namely the gate voltage V_{GG} and the width (W_2) of the PMOS transistor. Hence, we run a double-sweep simulation.

The aim is to study the PMOS current source resistance and current for the 0.5 μm PMOS transistor.

The double-sweep results are shown in Figure 7.97. For the CS amplifier with current source load, as we saw with the 2 μm process, the lower the I_{DS} current, the higher the values of r_o.

The transconductance of the NMOS transistor, however, does the opposite: the higher the current, the larger g_m becomes. This means that there is an area of operation for V_{GG} and width in which the CS amplifier current source load must operate to have a successful design.

Since the small-geometry equations based on the BSIM3 model are more complex than Level 1 modeling, we use the characterization data seen in Figure 7.97 to design our CS current load amplifier. Let's say that $r_o \geq 5$ kΩ and $I_{DS} \geq 100$ μA are part of the requirements for this design. The highlighted portion of Figure 7.97 indicates the area of operation.

Notice that W = 1 μm is outside the area of operation. The other widths studied only work for a portion of the VGG range.

Widths of 4 and 5 μm are closest to the center of the area of operation.

If $W_2 = 4$ μm is selected, then $2.0\,V \leq V_{GG} \leq 3.0\,V$.

If $W_2 = 5$ μm is selected, then $2.2\,V \leq V_{GG} \leq 3.2\,V$.

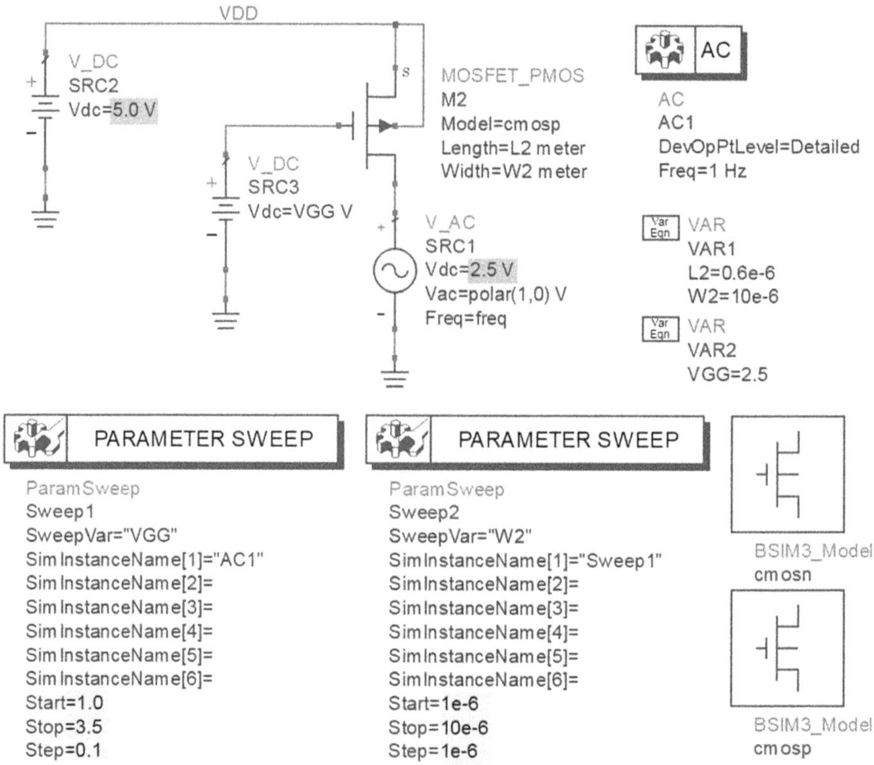

FIGURE 7.96 Small-geometry characterization study of the amplifier's PMOS current source load.

Next, we do a swept gain optimization. Figure 7.98 is the schematic for the small-geometry swept gain optimization design of the whole amplifier.

The gain is swept from 10 to 100 in steps of 10 with design variables W_1 and V_{GG1} as optimizable variables. The small-geometry swept gain optimization design result is presented in Figure 7.99. As seen in Figure 7.99, the PMOS current source load generates an $I_{DS} \approx 290.2\ \mu A$.

Further observations from Figure 7.99.

As seen in Figure 7.99, the optimizer was successful up to an $A_v = 60$, after which the optimizer did not reach the goals. V_{GG} has railed to the lower side of the values and can reduce no further, as per our setup, hence a nonzero error.

Question: Can we increase the current or make any other changes to the current source PMOS and meet all goals?

The results shown in Table 7.7 are taken from a study of the current source's operating range.

The experiment shows that gain is more or less restricted between 10 and 70.

Toward the lower gain ($A_v = -10$), W_1 is the limiting factor, and on the high side ($A_v = -70$), V_{GG1} is the limiting factor.

The following comments apply to the operating range selected for the CS amplifier with current source load:

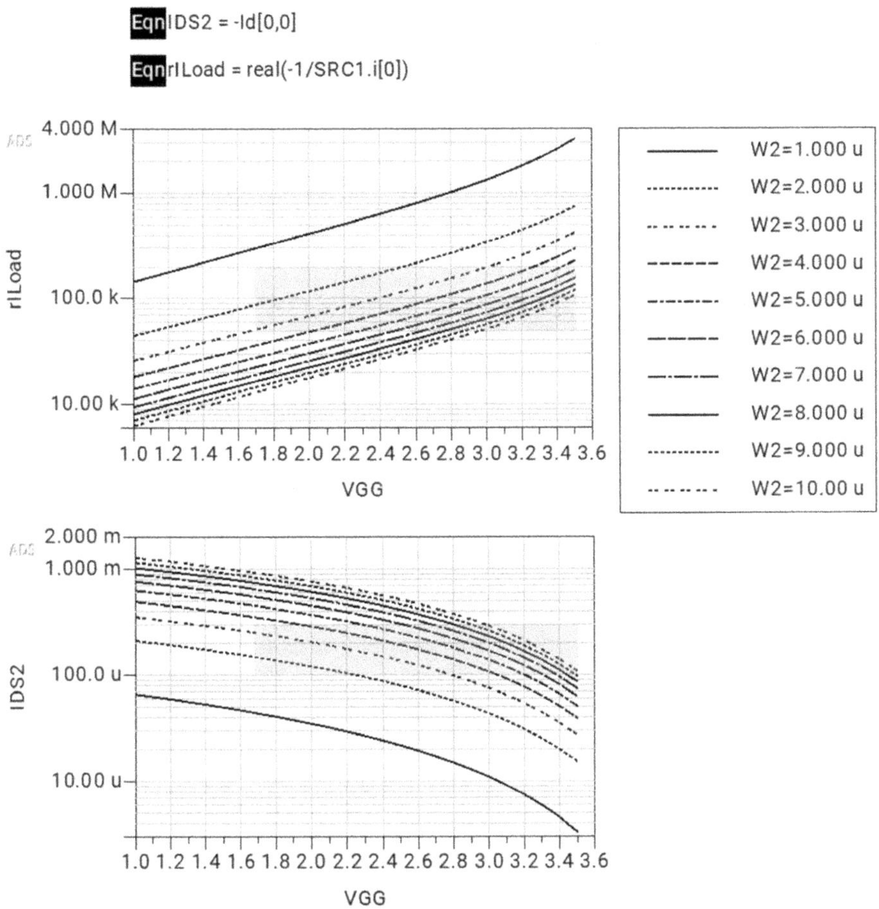

FIGURE 7.97 Results of the small-geometry characterization of the PMOS current source load.

- Increasing V_{GG2} reduces the I_{DS} current because $V_{GS2} = V_{GG2} - V_{DD}$, as we recall that $I_{DS} \propto (V_{GS} - V_{TH})^2$
- Reducing I_{DS} increases r_{o1} and r_{load}.
- r_{o1} reduces with V_{GG1}.
- Increasing I_{DS} increases g_{m1}.
- $W_1 \propto I_{DS}$, which in turn means that the contact resistance (r_{con1}) decreases with I_{DS}.

The accumulation of these effects is what limits the gain of the CS amplifier with a current source load.

For example, reducing the current increases r_{o1} and r_{load} that would tend to increase the gain; however, since r_{con} also increases and g_m decreases hence, the gain stays low.

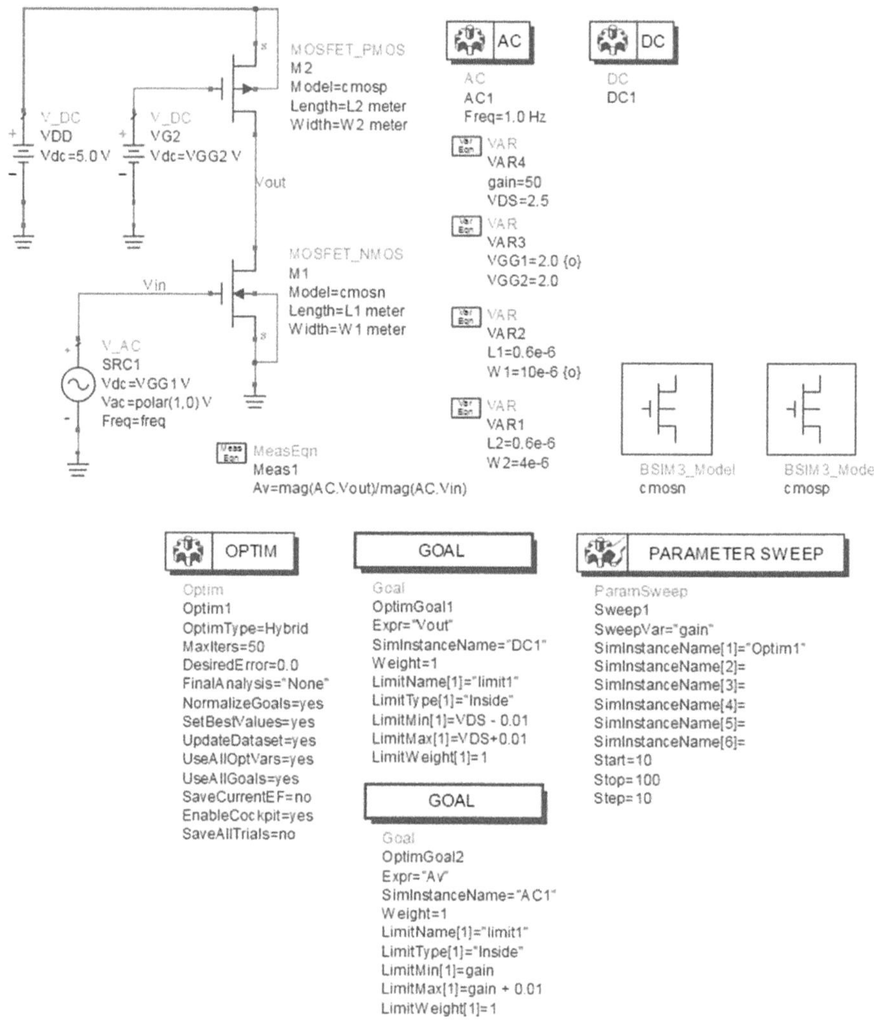

FIGURE 7.98 Small-geometry swept gain optimization design.

7.11 EPILOGUE

Submicron CMOS analog electronic circuit design has been the subject of intense research in the last two to three decades. As an epilogue to this chapter, let us outline some of the trends and some notable references.

As MOSFET transistors continue to miniaturize from one technology to the next, their models become more complex. The Compact Model Coalition (CMC – https://si2.org/compact-model-coalition/) consists of tens of semiconductor and EDA industry members. With each new technology, the CMC certifies certain MOSFET models as standard, such as the latest versions of BSIM, PSP, and HiSIM. Semiconductor foundries create MOSFET design kits that each support at least one standard MOSFET model. Such PDKs, which support the latest technology and earlier

Eqn ErrFcn = FinalEF[0]

Eqn Iters = NumIters[0]

Eqn VGGopt = OPTSOLNVALS.VGG1[0]

Eqn W1opt = OPTSOLNVALS.W1[0]

Eqn IDS = -DC.VDD.i[0,0]

Eqn VDSOpt=OptimGoal1[0,0]

Eqn AvOpt=OptimGoal2[0,0]

gain	ErrFcn	Iters	VGGopt	W1opt	IDS	VDSOpt	AvOpt
10.00	0.0000	6.000	2.081	3.391 u	290.2 u	2.507	10.01
20.00	0.0000	18.00	1.315	8.287 u	290.5 u	2.494	20.00
30.00	0.0000	5.000	1.040	16.30 u	290.2 u	2.506	30.01
40.00	0.0000	1.000	920.1 m	27.11 u	290.4 u	2.498	40.01
50.00	0.0000	11.00	841.7 m	45.20 u	290.4 u	2.495	50.00
60.00	0.0000	2.000	770.6 m	91.63 u	290.5 u	2.490	60.01
70.00	2.136 m	25.00	700.0 m	264.7 u	291.7 u	2.434	67.17
80.00	24.17 m	25.00	700.0 m	271.2 u	293.5 u	2.340	68.52
90.00	59.99 m	25.00	700.0 m	274.6 u	294.4 u	2.292	69.14
100.0	101.3 m	25.00	700.0 m	276.5 u	294.8 u	2.266	69.46

FIGURE 7.99 Small-geometry swept gain optimization design results.

technologies, serve as the interface between the analog CMOS designer and the semiconductor foundry. For example, this is how the designers get MOSFET models in simulation-based circuit design.

In [Rivas-Torres2], the MOSIS organization provided model cards supporting BSIM3 for a 0.5 μm technology, used in this chapter. In [Rivas-Torres1], a 180 nm PDK that supported BSIM3 and PSP was provided by the TowerJazz foundry.

The design variables for complex circuits are more of the same as what we have witnessed in CS amplifiers: transistor sizing, overdrive voltages, and V_{DS} voltages.

The universal voltage gain formula (short circuit transconductance times the output resistance [Razavi1]) works for single-stage amplifiers, as well as for any complicated multi-transistor amplifier circuits. Transconductance and output resistance of any circuit can be measured by simulations done with the inclusion of a PDK model for each transistor. The characterization of driving transistors and load circuits may constitute many sets of double-sweep plots [Binkley1] or pre-computed look-up tables [Jespers1].

The simulation-based characterizations are not always done with the help of a standard model PDK. The EKV model (named after Enz, Krummenacher, and Vittoz) is described in detail in [Enz1]. The model, which is not standard, has been popular, as it consists of relatively compact analytic relationships for long-channel models, as well as for short-channel effects, down to 65 nm technology, as reported. These analytic relationships supported comprehensive characterizations, assisted by user-interface sliders for the design parameters and a guiding "compass" that, based on the design goals, tells the designer in what direction the design parameters need to go [Binkley1].

Coming back to the universal voltage gain formula: Many researchers ([Binkley1], [Jespers1]) opt for gain expressions that involve transconductance efficiency (g_m/I_D) and normalized output conductance (g_{out}/I_D), which, with no parameter tolerances, is equivalent to the original voltage gain expression. One of the advantages of doing

TABLE 7.7
A Family of Design Results Obtained by Changing the PMOS Load V_{GG2} and W_2 Parameters

VGG2 = 2V, W2 = 4um

gain	ErrFcn	Iters	VGGopt	W1opt	IDS	VDSOpt	AvOpt
10.00	0.0000	6.000	2.081	3.391 u	290.2 u	2.507	10.01
20.00	0.0000	18.00	1.315	8.287 u	290.5 u	2.494	20.00
30.00	0.0000	5.000	1.040	16.30 u	290.2 u	2.506	30.01
40.00	0.0000	1.000	920.1 m	27.11 u	290.4 u	2.498	40.01
50.00	0.0000	11.00	841.7 m	45.20 u	290.4 u	2.495	50.00
60.00	0.0000	2.000	770.6 m	91.63 u	290.5 u	2.490	60.01
70.00	2.136 m	25.00	700.0 m	264.7 u	291.7 u	2.434	67.17
80.00	24.17 m	25.00	700.0 m	271.2 u	293.5 u	2.340	68.52
90.00	59.99 m	25.00	700.0 m	274.6 u	294.4 u	2.292	69.14
100.0	101.3 m	25.00	700.0 m	276.5 u	294.8 u	2.266	69.46

VGG2 = 2.5V, W2 = 4um

gain	ErrFcn	Iters	VGGopt	W1opt	IDS	VDSOpt	AvOpt
10.00	0.0000	1.000	2.268	2.263 u	193.7 u	2.501	10.00
20.00	0.0000	2.000	1.400	5.140 u	193.8 u	2.496	20.01
30.00	0.0000	5.000	1.086	9.798 u	193.6 u	2.509	30.01
40.00	0.0000	2.000	949.1 m	15.85 u	193.9 u	2.491	40.01
50.00	0.0000	1.000	864.3 m	24.91 u	193.7 u	2.509	50.00
60.00	0.0000	1.000	797.2 m	42.40 u	193.6 u	2.509	60.01
70.00	0.0000	1.000	720.4 m	107.8 u	193.8 u	2.494	70.00
80.00	10.22 m	25.00	700.0 m	151.0 u	194.8 u	2.419	72.24
90.00	39.36 m	25.00	700.0 m	152.5 u	195.3 u	2.375	72.63
100.0	76.89 m	25.00	700.0 m	153.4 u	195.6 u	2.351	72.83

VGG2 = 3V, W2 = 4um

gain	ErrFcn	Iters	VGGopt	W1opt	IDS	VDSOpt	AvOpt
10.00	107.0 m	50.00	1.878	2.000 u	102.8 u	3.096	12.29
20.00	0.0000	1.000	1.446	3.176 u	107.5 u	2.501	20.01
30.00	0.0000	3.000	1.109	5.787 u	107.6 u	2.499	30.00
40.00	0.0000	1.000	960.1 m	9.148 u	107.5 u	2.507	40.01
50.00	0.0000	2.000	872.3 m	13.87 u	107.5 u	2.509	50.00
60.00	0.0000	27.00	804.4 m	22.27 u	107.6 u	2.492	60.01
70.00	0.0000	14.00	733.1 m	46.66 u	107.6 u	2.498	70.01
80.00	6.952 m	50.00	700.0 m	74.86 u	107.9 u	2.456	73.42
90.00	33.97 m	50.00	700.0 m	75.31 u	108.1 u	2.427	73.57
100.0	70.45 m	50.00	700.0 m	75.56 u	108.2 u	2.411	73.64

VGG2 = 2.2V, W2 = 5um

gain	ErrFcn	Iters	VGGopt	W1opt	IDS	VDSOpt	AvOpt
10.00	0.0000	5.000	2.115	3.586 u	322.7 u	2.503	10.01
20.00	0.0000	3.000	1.331	8.822 u	322.6 u	2.508	20.00
30.00	0.0000	4.000	1.054	17.30 u	322.9 u	2.497	30.01
40.00	0.0000	2.000	928.6 m	29.01 u	323.0 u	2.493	40.01
50.00	0.0000	1.000	847.4 m	48.82 u	322.7 u	2.505	50.00
60.00	0.0000	6.000	774.6 m	100.1 u	322.6 u	2.508	60.01
70.00	1.909 m	50.00	700.0 m	311.2 u	324.1 u	2.443	67.24
80.00	24.13 m	50.00	700.0 m	318.4 u	325.8 u	2.356	68.33
90.00	60.39 m	50.00	700.0 m	322.0 u	326.7 u	2.313	68.82
100.0	102.1 m	50.00	700.0 m	324.0 u	327.2 u	2.289	69.08

VGG2 = 2.7V, W2 = 5um

gain	ErrFcn	Iters	VGGopt	W1opt	IDS	VDSOpt	AvOpt
10.00	0.0000	3.000	2.260	2.347 u	203.7 u	2.499	10.01
20.00	0.0000	1.000	1.399	5.356 u	203.7 u	2.497	20.00
30.00	0.0000	1.000	1.088	10.20 u	203.8 u	2.494	30.00
40.00	0.0000	2.000	948.2 m	16.65 u	203.7 u	2.499	40.01
50.00	0.0000	4.000	865.1 m	26.08 u	203.7 u	2.496	50.01
60.00	0.0000	13.00	797.7 m	44.60 u	203.7 u	2.501	60.01
70.00	0.0000	8.000	718.4 m	118.1 u	203.7 u	2.498	70.00
80.00	10.98 m	50.00	700.0 m	160.8 u	204.6 u	2.429	71.85
90.00	40.99 m	50.00	700.0 m	162.2 u	205.1 u	2.392	72.13
100.0	79.16 m	50.00	700.0 m	162.9 u	205.3 u	2.371	72.27

VGG2 = 3.2V, W2 = 5um

gain	ErrFcn	Iters	VGGopt	W1opt	IDS	VDSOpt	AvOpt
10.00	112.5 m	50.00	1.921	2.000 u	104.8 u	1.893	12.37
20.00	0.0000	2.000	1.411	3.188 u	100.6 u	2.505	20.01
30.00	0.0000	1.000	1.089	5.787 u	100.6 u	2.505	30.00
40.00	0.0000	3.000	948.5 m	9.106 u	100.6 u	2.494	40.01
50.00	0.0000	1.000	862.4 m	13.90 u	100.5 u	2.508	50.01
60.00	0.0000	5.000	794.1 m	22.86 u	100.5 u	2.506	60.00
70.00	0.0000	22.00	719.0 m	52.47 u	100.6 u	2.505	70.00
80.00	10.42 m	50.00	700.0 m	69.66 u	100.9 u	2.458	71.90
90.00	40.53 m	50.00	700.0 m	69.97 u	101.0 u	2.437	71.98
100.0	78.92 m	50.00	700.0 m	70.15 u	101.1 u	2.425	72.03

so is the very low sensitivity of the transconductance efficiency to the binning of the MOSFET model parameters. [Rivas-Torres1].

Finally, the smaller the W and L, the stronger the impact of node contact resistance and other short-channel effects.

REFERENCES

[Binkley1] David M. Binkley, *Tradeoffs and Optimization in Analog CMOS Design*, Wiley-Interscience, 2008.

[Enz1] Christian C. Enz and Eric A. Vittoz, *Charge-Based MOS Transistor Modeling: The EKV Model for Low-Power and RF IC Design*, John Wiley & Sons, 2006.

[Jespers1] Paul G. A. Jespers and Boris Murmann, *Systematic Design of Analog CMOS Circuits: Using Pre-Computed Lookup Tables*, Cambridge University Press, 2017.

[Razavi1] Behzad Razavi, *Design of Analog CMOS Integrated Circuits*, 2nd Edition, McGraw-Hill, 2017.

[Rivas-Torres1] Wilfredo Rivas-Torres, "An Empirical Methodology for Foundry Specific Submicron CMOS Analog Circuit Design", PhD Dissertation, Florida Atlantic University, December 2013.

[Rivas-Torres2] Wilfredo Rivas-Torres, "Submicron CAD Design and Analysis of MOS Current Mirrors", MS Thesis, Florida Atlantic University, May 2004.

CHAPTER 7 PROBLEMS – CMOS AMPLIFIERS

Note: This book edition provides full solutions only to all Chapter 7 problems solved with Level 1 model.

PROBLEM 7.1: DESIGN OF AN NMOS CS AMPLIFIER WITH PMOS CURRENT SOURCE LOAD

1a: Design an NMOS CS amplifier with PMOS current source load to achieve a small-signal voltage gain of $A_v = -64$. Assume that (i) the amplitude of the input voltage signal is sufficiently small (say less than 10 mV). (ii) There is no resistive load: $R_L = \infty$. (iii) The amplifier works with a single DC voltage supply of $V_{DD} = 5$ V. Demonstrate the amplifier's performance in simulations. Use an NMOS amplifier that has **Level 1 model** parameters described in Razavi's book (as shown below). Assume further that $L = 2\,\mu$m for both transistors.

In Table 7.8, ignore the given W values.

Helpful Comment: For your design calculations convenience, let us calculate K_P for each of the transistors so that you can add it to the model of each. From the given $T_{ox} = 9e^{-6}$ [cm], one can find that $C_{ox} = 4.13e^{-5}$ [mF/cm^2] for both transistors. We know that $K_P = U_0 C_{ox}$. Therefore, for NMOS, we have $K_P = 1.45e^{-2}$ [mA/V^2] and for PMOS we have $K_P = 4.13e^{-3}$ [mA/V^2]. In the SPICE code for each mode, because the units of K_P must be [A/V^2] simply add for NMOS $K_P = 1.45e^{-5}$ and for PMOS: $K_P = 4.13e^{-6}$.

Hints:

a. You may wish to try what may be called a "dummy battery" method: Fixating V_{DS}, and tuning W separately for each transistor in order to make I_D the same in both transistors. Once the I_D is made the same in both transistors,

TABLE 7.8
CMOS Model Parameters for Problem 7.1

PMOS	NMOS
+LEVEL = 1	+LEVEL = 1
+L = 2 u	+L = 2 u
+W = 20,000 u	+W = 200 u
+T_{OX} = 9e^{-9}	+T_{OX} = 9e^{-9}
+V_{TO} = −0.8	+V_{TO} = 0.7
+P_B = 9	+P_B = 9
+MJSW = 0.3	+MJSW = 0.2
+GAMMA = 0.4	+GAMMA = 0.45
+U_0 = 100	+U_0 = 330
+CJ = 0.94e^{-3}	+CJ = 0.56e^{-3}
+CGDO = 0.3e^{-9}	+CGDO = 0.4e^{-9}
+PHI = 0.88	+PHI = 0.9
+LAMBDA=0.02	+LAMBDA = 0.01
+CJSW = 0.32e^{-11}	+CJSW = 0.35e^{-11}
+NSUB = 5e^{+14}	+NSUB = 9e^{+14}

there is zero current in the battery, so it can be discarded. Check if without the V_{DS} battery, I_D and V_{DS} are as planned. Show the DC simulation tuning of the various W values.

b. Check your voltage gain in AC Sweep and tweak it to bring it exactly to −64. Do the tweaking by means of modifying V_{GS}. Hopefully, there is no need to bring back the dummy battery for re-tuning the W values, but more often than not, you may need to redo the tuning.

1b: Follow the ideas of Section 7.4 to design the CS amplifier (of Problem 7.1a) using a 0.5 µm BSIM3 model. The provided workspace schematics include the necessary model cards for both the NMOS and PMOS transistors.

Problem 7.2: Level 1 Analysis of the Bandwidth of a CS Amplifier with PMOS Current Source Load

Note the bandwidth analysis done in Chapter 4 for a CS amplifier with an R_D load.

Consider the amplifier designed in Problem 7.1. In the design, we assumed that the input voltage signal source's resistance is R_{sig} = 0 and we needed the gain to be A_v = −64. You may have chosen R_G = 1 MΩ, which can stay.

Change R_{sig} from R_{sig} = 0 to R_{sig} = 50 kΩ.

For the NMOS and PMOS, we used model cards based on Razavi's book recommended parameters. The models include some high-frequency capacitances, but they do not include all such effects. The parameters A_S (area of the source region), A_D (area of the drain region), P_S (perimeter of the source) and P_D (perimeter of the drain) all have default values of 0 in SPICE. Assuming rectangular layout of the transistors (once you know the gate's W and L dimensions), we can assume that the drain and the source have the same length L as the gate and the same width W as the gate – this allows you to compute the area and perimeter of each. Another missing parameter is CGSO that can be made equal to the parameter CGDO. CGSO should be added to the SPICE code of each model card.

Compare the amplifier's bandwidth in the following three cases and explain (in detail, referring to Miller's theorem) the differences:

a. As designed in Problem 7.1 (without adding A_D, A_S, P_D, P_S, and CGSO).
b. As designed in Problem 7.1 but add CGSO to each transistor (and leave it there for the remainder of this problem).
c. As designed in Problem 7.1 but need to add the drain and source regions dimensions.
d. As in Problem 7.2c but using 0.5 μm BSIM3 model.

EXTRA PROBLEMS (A LITTLE BEYOND THE SCOPE OF THIS BOOK)

Problem 7.3: CS Amplifier with Current Source Load Feeding into a Source Follower Amplifier

3a: In a CS amplifier with current source load, there is no way to add a resistive load R_L between the drains and ground. The amplifier will stop working if such a load is connected. You need a buffer: A source follower amplifier can be connected to the CS amplifier, and it can feed R_L (Figure 7.100).

A source follower (M_3 and M_4) has a voltage gain somewhere between 0 and 1 that is almost insensitive to R_L.

The source follower also acts as a DC voltage shifter between V_{D1} and V_{out} if there is no coupling capacitor between the CS amplifier and the source follower amplifier.

Let $R_L = 12$ kΩ. Design a source follower for the CS amplifier with current source load that you designed earlier. The overall voltage gain should be either −69 or −64. Keep in mind though that you will have to tweak the CS gain slightly upwards in order to compensate for the less-than −1 voltage gain of the next amplifier stage.

Comment: As much as possible try to avoid using any coupling capacitors. If you have no other choice (because, say, that R_L steals too much DC current from M_4) you may add a large capacitor "DCBlock" (from the "Lumped-Components" palette) between V_{out} and R_L, or wherever else a coupling capacitor is needed. Use **Level 1 model**.

3b: Repeat Problem 7.3a, but with 0.5 μm BSIM3 model.

Problem 7.4: CG (Common Gate) Amplifier Design

4a: A voltage signal source has an internal resistance of $R_{sig} = 50$ Ω. It connects to a CG amplifier with an R_D load via a coupling capacitor (DCBlock). Design the CG amplifier to meet two specifications: (i) $A_v = 25$, and (ii) $R_{in} = 50$ Ω. First do the design manually using approximations. Then use the ADS optimizer. For the

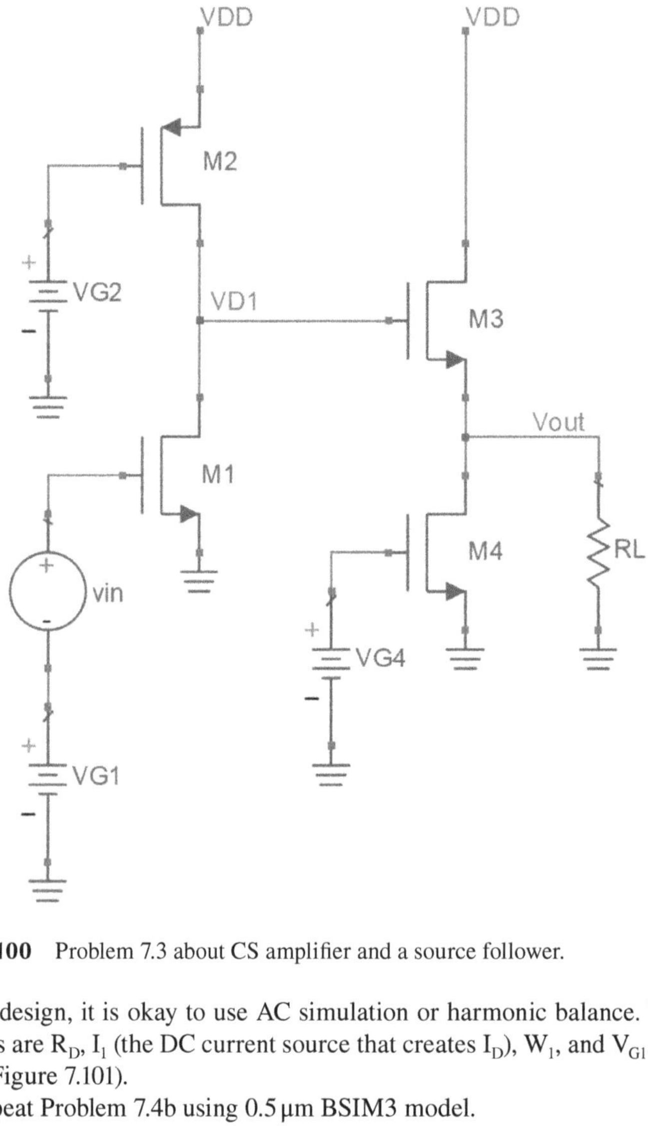

FIGURE 7.100 Problem 7.3 about CS amplifier and a source follower.

optimizer design, it is okay to use AC simulation or harmonic balance. The design parameters are R_D, I_1 (the DC current source that creates I_D), W_1, and V_{G1}. Use **Level 1** model (Figure 7.101).

4b: Repeat Problem 7.4b using 0.5 μm BSIM3 model.

Problem 7.5: Design of a Cascode Amplifier

5a: Consider a cascode amplifier consisting of two stacked NMOS transistors M_1 and M_2 (this is the driving stage) feeding into a single PMOS current source load M_3 (that replaces R_D shown in the diagram). V_{out} is taken from the drain that is common to M_2 and M_3. The specifications are (i) $A_v \geq 100$, (ii) each of the transistors must be in saturation mode. Use only the ADS optimizer in this design, and work with harmonic balance simulation. There are six design parameters: V_G of each MOSFET and W of each MOSFET. Use **Level 1 model** (Figure 7.102).

5b: Repeat Problem 7.5a using 0.5 μm BSIM3 model.

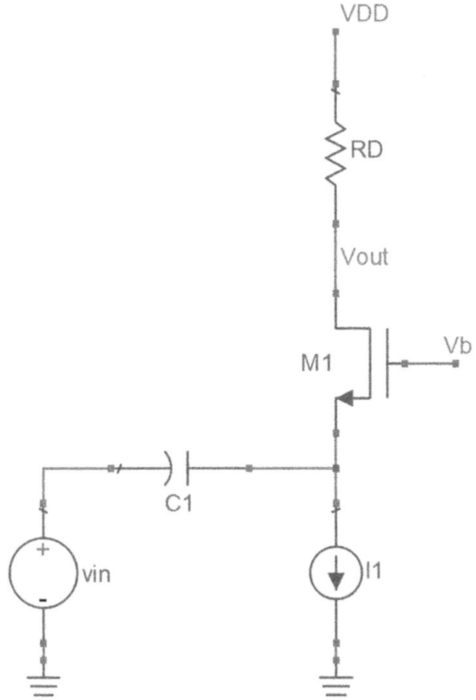

FIGURE 7.101 Problem 7.4 about common gate amplifier.

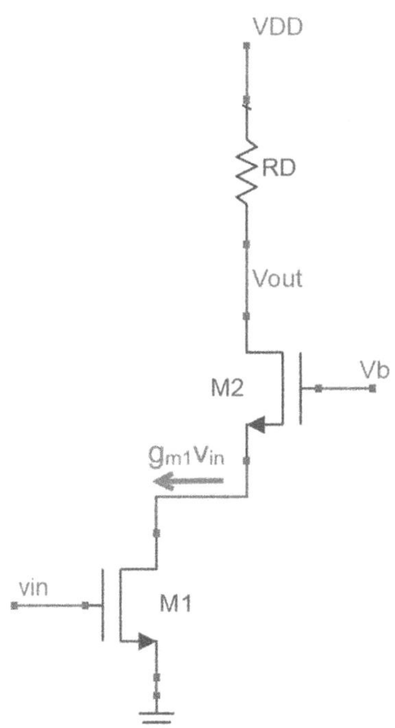

FIGURE 7.102 Problem 7.5 about CMOS cascode amplifier.

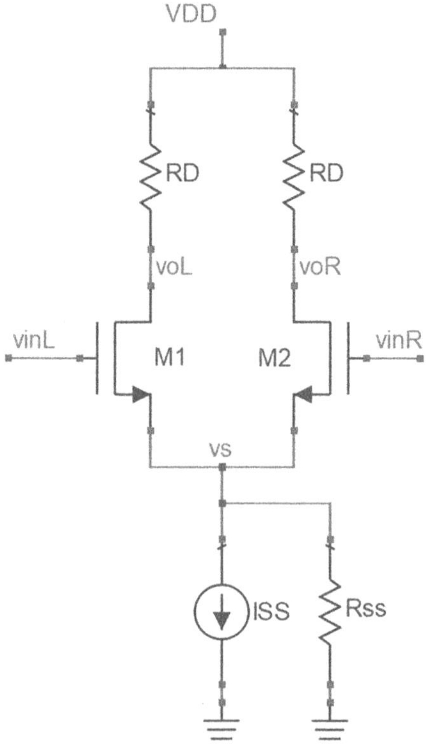

FIGURE 7.103 Problem 7.6 about CMOS differential amplifier.

Problem 7.6: Level 1 MOSFET Differential Amplifier and Yield Analysis

Solve manually using approximated formulas. Then use the ADS optimizer.

6a: Design a MOSFET differential amplifier with R_D loads. The single-ended differential gain should be around $A_{d,SE} = 20$, and the single-ended common-mode gain should be $A_{CM,SE} \leq 0.05$ (Figure 7.103).

6b: Replace the ideal I_{SS} DC current source (with parallel resistor R_{SS} by a simple NMOS current mirror driven by a reference current of $I_{ref} = 100\ \mu A$). Now R_{SS} will be equal to R_{out} of the current mirror, which is r_o of the tail transistor. Check again the differential mode and common-mode performance of the amplifier.

6c: Assume that R_D has a tolerance of ±10% (uniform distribution). Run yield analysis. "Success" means that the differential gain changes by no more than ±3% and the common-mode gain decreases (by any amount) or does not increase by more than 3%. How will the yield improve if the R_D tolerance becomes ±5% (uniform distribution) and ±1% (uniform distribution)?

Problem 7.7: Design and Testing of a Level 1 Gilbert Cell

Using ADS construct a Gilbert cell and demonstrate how it works as an analog multiplier. Choose input signals that produce a predictable and easy to verify multiplication

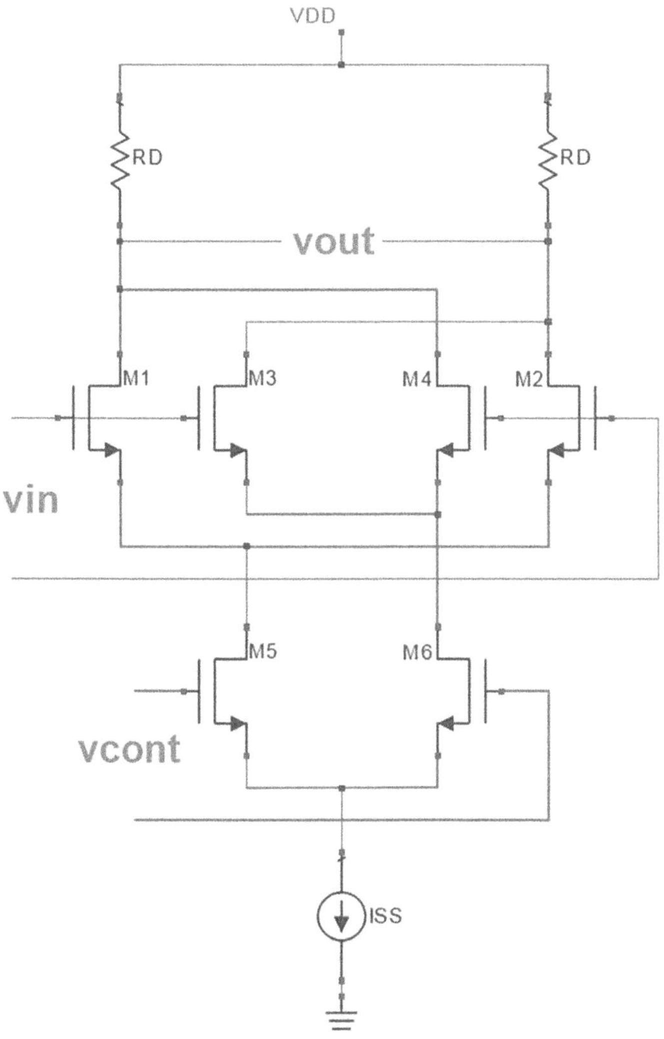

FIGURE 7.104 Problem 7.7 about Gilbert cell.

result. For instance, synchronized square wave and AC, or synchronized AC signals of the same frequency (Figure 7.104).

Problem 7.8: Simple Level 1 OTA (Operational Transconductance Amplifier) Applications

Use an abstract simplified ideal model of an OTA. In ADS, use a VCIS (voltage controlled current source) component. There is no need to try a practical CMOS differential amplifier implementation of an OTA (Figure 7.105).

Let $V_{DD} = 5\,V$ for all problems below.

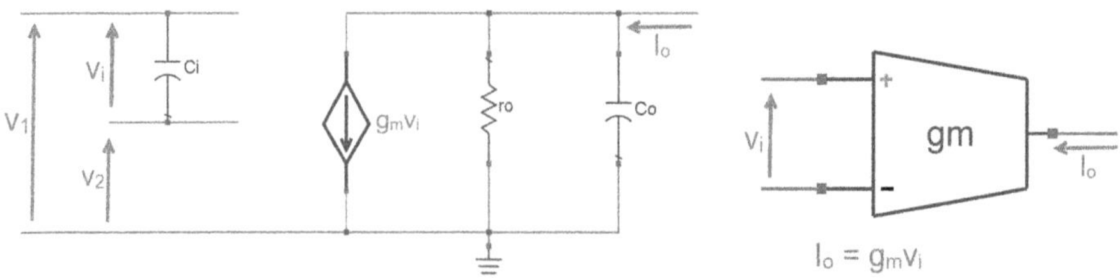

FIGURE 7.105 Problem 7.8 about idealized CMOS operational transconductance amplifier.

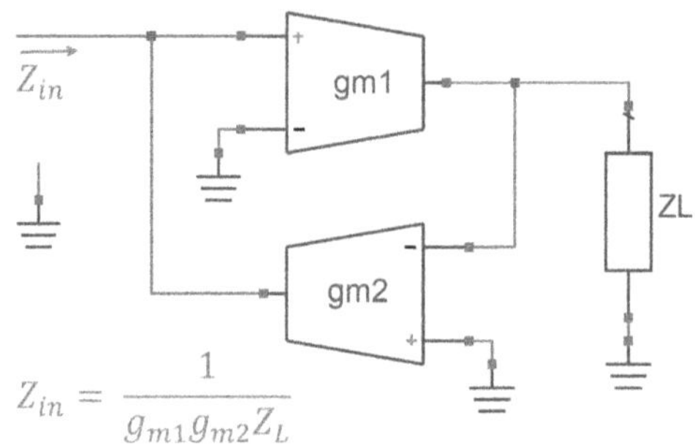

FIGURE 7.106 Creating an inductor using OTA loaded by a capacitor.

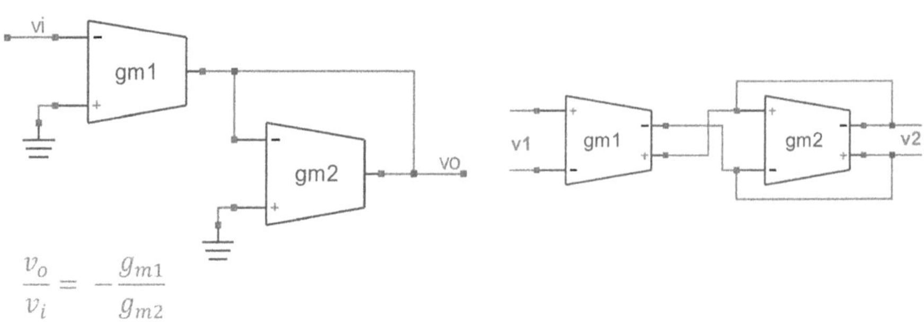

FIGURE 7.107 OTA implementation of inverting and non-inverting amplifiers.

8a: Using OTAs and a capacitor design, an inductor L = 100 μH. Run some experiment of your choice to convince that the device that you created indeed operates like an inductor. As you may know, C and L in parallel (with some initial energy) oscillate at a frequency that depends on C and L. If you add a resistor R in parallel to L and C, the oscillation dies out gradually. Show both an ideal oscillator and dissipative oscillator in ADS (Figure 7.106).

8b: Use OTAs only, and no other components, to implement (b_1) an inverting voltage amplifier with a gain of −4, (b_2) a non-inverting voltage amplifier with a gain of +4, (b_3) a linear combiner for two signals with a voltage addition ratio of 2:3 (Figures 7.107 and 7.108)

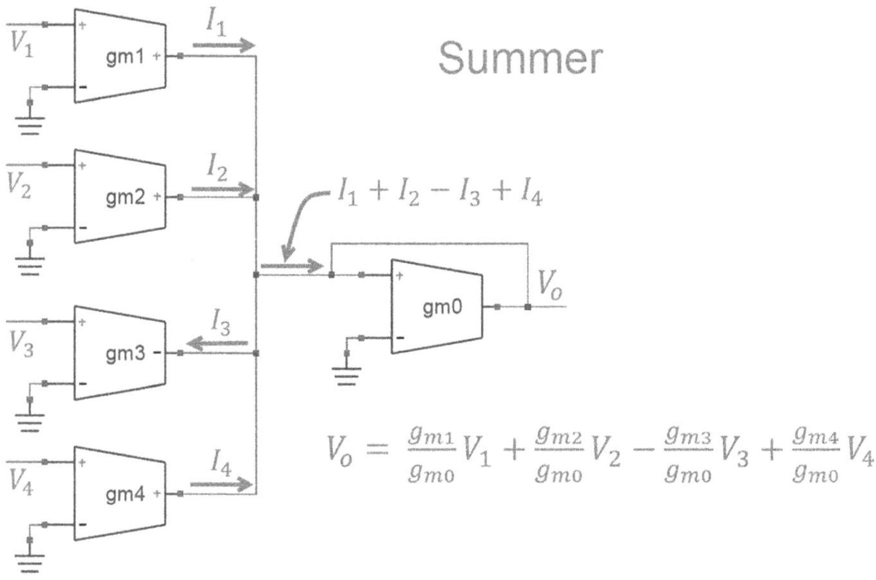

FIGURE 7.108 OTA implementation of a linear combiner (summer).

Index

Note: **Bold** page numbers refer to tables and *italic* page numbers refer to figures.

B

C

D

For Product Safety Concerns and Information please contact our EU
representative GPSR@taylorandfrancis.com
Taylor & Francis Verlag GmbH, Kaufingerstraße 24, 80331 München, Germany

www.ingramcontent.com/pod-product-compliance
Ingram Content Group UK Ltd.
Pitfield, Milton Keynes, MK11 3LW, UK
UKHW022211190926
475702UK00003B/313
9781032999326